D0094977

WITHDRAWN

Methods in Enzymology

Volume XXXI
BIOMEMBRANES
Part A

METHODS IN ENZYMOLOGY

EDITORS-IN-CHIEF

Sidney P. Colowick   Nathan O. Kaplan

Methods in Enzymology

Volume XXXI

Biomembranes

Part A

EDITED BY

Sidney Fleischer

DEPARTMENT OF MOLECULAR BIOLOGY
VANDERBILT UNIVERSITY, NASHVILLE, TENNESSEE

Lester Packer

DEPARTMENT OF PHYSIOLOGY-ANATOMY
UNIVERSITY OF CALIFORNIA, BERKELEY, CALIFORNIA

Advisory Board

Ronald W. Estabrook   Arnost Kleinzeller
H. Ronald Kaback   George Laties
Stephen C. Kinsky   George E. Palade

ACADEMIC PRESS New York San Francisco London 1974
A Subsidiary of Harcourt Brace Jovanovich, Publishers

COPYRIGHT © 1974, BY ACADEMIC PRESS, INC.
ALL RIGHTS RESERVED.
NO PART OF THIS PUBLICATION MAY BE REPRODUCED OR
TRANSMITTED IN ANY FORM OR BY ANY MEANS, ELECTRONIC
OR MECHANICAL, INCLUDING PHOTOCOPY, RECORDING, OR ANY
INFORMATION STORAGE AND RETRIEVAL SYSTEM, WITHOUT
PERMISSION IN WRITING FROM THE PUBLISHER.

ACADEMIC PRESS, INC.
111 Fifth Avenue, New York, New York 10003

United Kingdom Edition published by
ACADEMIC PRESS, INC. (LONDON) LTD.
24/28 Oval Road, London NW1

Library of Congress Cataloging in Publication Data
Main entry under title:

Biomembranes, part A-

    (Methods in enzymology, v. 31-
    Includes bibliographies.
    1.   Cell membranes.   2.   Cell fractionation.   3.   Cell
organelles.   I.   Fleischer, Sidney, ed.   II.   Packer,
Lester, ed.   III.   Estabrook, Ronald W., ed.
[DNLM: 1.   Cell membrane.   W1ME9615K v. 31 / QH601
B6193]
QP601.C733 vol. 31   [QH601]   574.1′925′08s   [574.8′75]
ISBN 0–12–181894–2   74-11352

PRINTED IN THE UNITED STATES OF AMERICA

WHEATON COLLEGE LIBRARY 188878
NORTON, MASS. 02766

Table of Contents

Section I. Multiple Fractions from a Single Tissue

Section II. Isolation of Purified Subcellular Fractions and Derived Membranes (from Mammalian Tissue Excluding Nerve)

A. Plasma Membrane

v

B. Golgi Complex

C. Rough and Smooth Microsomes

D. Nuclei

E. Mitochondria

F. Lysosomes

G. Peroxisomes

H. Secretory Granules

B. Specific Organelles and Derived Components

Section V. Preparations Derived from Unicellular Organisms

Section VI. General Methodology

Contributors to Volume XXXI

Article numbers are in parentheses following the names of contributors.
Affiliations listed are current.

M. R. ADELMAN (19), *Duke University School of Medicine, Durham, North Carolina*

PER-ÅKE ALBERTSSON (76), *Department of Biochemistry, University of Umea, Umea, Sweden*

VINCENT G. ALLFREY (23), *The Rockefeller University, New York, New York*

N. G. ANDERSON (78), *Molecular Anatomy Program, Oak Ridge National Laboratory, Oak Ridge, Tennessee*

NATHAN N. ARONSON, JR. (6), *Department of Biochemistry, Pennsylvania State University, University Park, Pennsylvania*

NANCY L. BAENZIGER (12), *Departments of Internal Medicine and Biochemistry, Washington University School of Medicine, St. Louis, Missouri*

MARCO BAGGIOLINI (34), *Research Institute, Wander Ltd., Bern, Switzerland*

S. F. BARTLETT (39), *Department of Biochemistry, University of Sheffield, Sheffield, England*

PIERRE BAUDHUIN (36), *Laboratoire de Chimie Physiologique, University of Louvain, Louvain, Belgium*

W. M. BECKER (51), *Department of Botany, University of Wisconsin, Madison, Wisconsin*

HARRY BEEVERS (58), *Division of Natural Sciences, University of California, Santa Cruz, California*

MYRA BERMAN-KURTZ (57), *Department of Biological Sciences, State University of New York at Albany, Albany, New York*

G. BLOBEL (19), *The Rockefeller University, New York, New York*

C. J. BOS (5), *Departments of Biochemistry and Electron Microscopy, Antoni van Leeuwenhoek-Huis, The Netherlands Cancer Institute, Amsterdam, The Netherlands*

WILLIAM E. BOWERS (35), *The Rockefeller University, New York, New York*

R. W. BREIDENBACH (58), *Department of Agronomy, University of California, Davis, California*

C. A. BURTIS (78), *Molecular Anatomy Program, Oak Ridge National Laboratory, Oak Ridge, Tennessee*

JOE H. CHERRY (61), *Agricultural Experiment Station, Department of Horticulture, Purdue University, Lafayette, Indiana*

EDWARD C. COCKING (60), *Department of Botany, University of Nottingham, University Park, Nottingham, England*

ZANVIL A. COHN (33), *The Rockefeller University, New York, New York*

CARL W. COTMAN (47), *Department of Psychobiology, University of California, Irvine, California*

ROBERT K. CRANE (9), *Department of Physiology, College of Medicine and Dentistry, Rutgers Medical School, Piscataway, New Jersey*

GUSTAV DALLNER (18), *Department of Pathology, Kungl. Universitetet i. Stockholm, Biokemiska Institution, Stockholm, Sweden*

ALEXANDER EICHHOLZ (9), *Department of Physiology, College of Medicine and Dentistry, Rutgers Medical School, Piscataway, New Jersey*

JANICE E. EKHOLM (15), *Department of Biochemistry, College of Medicine, University of Arizona, Tucson, Arizona*

P. EMMELOT (5), *Departments of Biochemistry and Electron Microscopy, Antoni van Leeuwenhoek-Huis, The Netherlands Cancer Institute, Amsterdam, The Netherlands*

JAMES B. FIELD (11), *Clinical Research Unit, University of Pittsburgh School of Medicine, Pittsburgh, Pennsylvania*

BECCA FLEISCHER (17), *Department of Molecular Biology, Vanderbilt University, Nashville, Tennessee*

SIDNEY FLEISCHER (1, 2, 27), *Department of Molecular Biology, Vanderbilt University, Nashville, Tennessee*

T. GALLIARD (53), *Agricultural Research Council Food Research Institute, Colney Lane, Norwich, England*

THEODORE E. GRAM (21), *Laboratory of Toxicology, National Institutes of Health, Bethesda, Maryland*

JOHN W. GREENAWALT (30, 66), *Department of Physiological Chemistry, The Johns Hopkins School of Medicine, Baltimore, Maryland*

F. CARVALHO GUERRA (28), *Centro de Estudos de Bioquimica do Instituto de Alta Cultura, Faculdade de Farmacia, Universidade do Porto, Porto, Portugal*

HERBERT F. HABERMAN (40), *Clinical Science Division, University of Toronto, Toronto, Canada*

DONALD J. HANAHAN (15), *Department of Biochemistry, College of Medicine, University of Arizona, Tucson, Arizona*

K. HANNIG (75), *Max-Planck-Institut für Eiweiss und Lederforschung, München, Germany*

W. G. HANSTEIN (77), *Scripps Clinic and Research Foundation, La Jolla, California*

Y. HATEFI (77), *Scripps Clinic and Research Foundation, La Jolla, California*

H.-G. HEIDRICH (75), *Max-Planck-Institut fur Eiweiss Und Lederforschung, München, Germany*

SHIGERU I. HONDA (55), *Department of Biological Sciences, Wright State University, Dayton, Ohio*

D. B. HOPE (42), *Department of Pharmacology, University of Oxford, Oxford, England*

J. D. JAMIESON (3), *Section of Cell Biology, Yale University School of Medicine, New Haven, Connecticut*

LEONARD JARETT (4), *Division of Laboratory Medicine, Washington University School of Medicine, and Barnes Hospital, St. Louis, Missouri*

H. R. KABACK (72), *Roche Institute of Molecular Biology, Nutley, New Jersey*

JEFFREY A. KANT (16), *Department of Biochemistry, The University of Chicago, Chicago, Illinois*

CHARLES B. KASPER (26), *McArdle Laboratory for Cancer Research, University of Wisconsin, Madison, Wisconsin*

WOLFGANG KEMMLER (38), *Städt. Krankenhaus München-Schwabing, München, Germany*

MARIJA KERVINA (1, 2), *Department of Molecular Biology, Vanderbilt University, Nashville, Tennessee*

A. M. KIDWAI (10), *Department of Pharmacology, Faculty of Medicine, University of Libya, Benghazi, Libya*

JOHN T. KNOWLER (25), *Department of Biochemistry, University of Glasgow, Glasgow, Scotland*

HAROLD KOENIG (49), *Neurology Service, Veterans Administration Research Hospital, and Department of Neurology, Northwestern University Medical School, Chicago, Illinois*

EDWARD D. KORN (71), *Section on Cellular Biochemistry and Ultrastructure, Laboratory of Biochemistry, National Heart and Lung Institute, National Institutes of Health, Bethesda, Maryland*

LADISLAV KOVÁČ (65), *Krajská psychiatrická liečebna, Pezinok okres Bratislava, Czechoslovakia*

G. KREIBICH (20), *Department of Cell Biology, New York University School of Medicine, New York, New York*

G. KRISHNA (7), *Laboratory of Chemistry and Pharmacology, National Heart and Lung Institute, National Institutes of Health, Bethesda, Maryland*

ROBERT R. KULIKOWSKI (57), *Depart-*

ment of Biological Sciences, State University of New York, Albany, New York

GEORGE G. LATIES (62), Department of Botanical Sciences and Molecular Biology Institute, University of California, Los Angeles, California

W. D. LOOMIS (54), Department of Biochemistry and Biophysics, Oregon State University, Corvallis, Oregon

CARL V. LUNDEEN (56), Division of Marine Science, University of North Carolina, Wilmington, North Carolina

W. H. MCSHAN (43), Department of Zoology, University of Wisconsin, Madison, Wisconsin

PHILIP W. MAJERUS (12), Departments of Internal Medicine and Biochemistry, Washington University School of Medicine, St. Louis, Missouri

JOSEPH P. MASCARENHAS (57), Department of Biological Sciences, State University of New York, Albany, New York

PHILIPPE MATILE (59), Institute for Allgemeine Botanik, ETH Universitätstrasse, Zurich, Switzerland

ARVID B. MAUNSBACH (32), Department of Cell Biology, Institute of Anatomy, University of Aarthus, Denmark

CHARLES W. MEHARD (29), Department of Marine Biology, University of California, San Diego, and Scripps Institute of Oceanography, La Jolla, California

GERHARD MEISSNER (22, 27), Department of Molecular Biology, Vanderbilt University, Nashville, Tennessee

I. ARADVINDAKSHAN MENON (40), University of Toronto, Clinical Science Division, Medical Sciences Building, Toronto, Canada

HAROLD L. MOSES (25), Department of Pathology, Mayo Clinic, Rochester, Minnesota

R. MUNSON (67), Department of Microbiology, University of Connecticut School of Medicine, Farmington, Connecticut

DAVID M. NEVILLE, JR. (8), Section of

Biophysical Chemistry, Laboratory of Neurochemistry, National Institute of Mental Health, Bethesda, Maryland

E. H. NEWCOMB (51), Department of Botany, University of Wisconsin, Madison, Wisconsin

PARK S. NOBEL (63), Department of Botanical Sciences, University of California, Los Angeles, California

WILLIAM T. NORTON (46), The Saul R. Korey Department of Neurology, Albert Einstein College of Medicine, New York, New York

DIETER OESTERHELT (69), Friedrich-Miescher-Laboratorium der Max-Planck-Gesellschaft, Tubingen, Germany

M. J. OSBORN (67), Department of Microbiology, University of Connecticut School of Medicine, Farmington, Connecticut

JAMES F. PERDUE (14), Lady Davis Institute for Medical Research, Jewish General Hospital, Montreal, Quebec, Canada

J. C. PICKUP (42), Department of Pharmacology, University of Oxford, Oxford, England

C. A. PRICE (52), Particle Separation Facility, Department of Biochemistry and Microbiology, (Cook College) Rutgers University, Piscataway, New Jersey

ERIC REID (73), Wolfson Bioanalytical Centre, University of Surrey, Guildford, Surrey, England

M. RODBELL (7), Laboratory of Nutrition and Endocrinology, National Institute of Arthritis and Metabolic Diseases, National Institutes of Health, Bethesda, Maryland

D. D. SABATINI (19, 20), Department of Cell Biology, New York University School of Medicine, New York, New York

MILTON R. J. SALTON (68), Department of Microbiology, New York University School of Medicine, New York, New York

GOTTFRIED SCHATZ (65), *Biocenter University of Basel, Basel, Switzerland*

A. N. SIAKOTOS (45, 48, 50), *Department of Pathology, Indiana University School of Medicine, Indianapolis, Indiana*

MURRAY SMIGEL (27), *Department of Molecular Biology, Vanderbilt University, Nashville, Tennessee*

A. D. SMITH (39), *Department of Pharmacology, University of Oxford, Oxford, England*

THOMAS C. SPELSBERG (25), *Department of Endocrine Research, Mayo Clinic, Rochester, Minnesota*

THEODORE L. STECK (16), *Department of Medicine, The University of Chicago, Chicago, Illinois*

DONALD F. STEINER (38), *Department of Biochemistry, The University of Chicago, Chicago, Illinois*

WALTHER STOECKENIUS (69), *Cardiovascular Research Institute, University of California, San Francisco, California*

BERNARD L. STREHLER (45), *Department of Biology, University of Southern California, Los Angeles, California*

A. M. TARTAKOFF (3), *The Rockefeller University, New York, New York*

J. R. TATA (24), *National Institute for Medical Research, Mill Hill, London, England*

T. O. TIFFANY (78), *Molecular Anatomy Program, Oak Ridge National Laboratory, Oak Ridge, Tennessee*

N. E. TOLBERT (74), *Department of Biochemistry, Michigan State University, East Lansing, Michigan*

OSCAR TOUSTER (6), *Department of Molecular Biology, Vanderbilt University, Nashville, Tennessee*

ANDRÉ TROUET (31), *Laboratoire de Chimie Physiologique, University de Louvain, Louvain, Belgium*

HIDEYUKI TSUKADA (37), *Department of Pathology, Cancer Research Institute, Sapporo Medical College, Sapporo, Japan*

BÖRJE UVNÄS (41), *Department of Pharmacology, Karolinska Institutet, Stockholm, Sweden*

W. J. VAN BLITTERSWIJK (5), *Departments of Biochemistry and Electron Microscopy, Antoni van Leeuwenhoek-Huis, The Netherlands Cancer Institute, Amsterdam, The Netherlands*

R. P. VAN HOEVEN (5), *Departments of Biochemistry and Electron Microscopy, Antoni van Leeuwenhoek-Huis, The Netherlands Cancer Institute, Amsterdam, The Netherlands*

A. E. WALSBY (70), *Marine Science Laboratories, Menia Bridge, Anglesey, Wales*

LEONARD WARREN (13), *Department of Therapeutic Research, University of Pennsylvania School of Medicine, Philadelphia, Pennsylvania*

ZENA WERB (33), *Strangeways Research Laboratory, Wort's Causeway, Cambridge, England*

ANDRES WIEMKEN (59), *Institute für Allgemeine Botanik, Zurich, Switzerland*

WILLIAM R. WILEY (64), *Pacific Northwest Laboratories, Battelle Memorial Institute, Richland, Washington*

CHARLES H. WILLIAMS (44), *Department of Medicine, University of Missouri, Columbia, Missouri*

ROBERT WILLIAMSON (73), *The Beatson Institute for Cancer Research, Glasgow, Scotland*

ROBERT WOOD (27), *Department of Molecular Biology, Vanderbilt University, Nashville, Tennessee*

KAMEJIRO YAMASHITA (11), *Department of Medicine, University of Tokyo School of Medicine, Tokyo, Japan*

Preface

A major portion of enzymology deals with enzymes that are components of biological membranes. The study of membrane-bound proteins requires specialized techniques and must be understood within the framework of the membrane and cellular organization. Biomembranes represent one of the most active, exciting, and important fields in contemporary biology. As in other fields, progress in the field of biological membranes depends on advances in technology. In this regard, there have been remarkable advances in the field of biomembranes, and it is quite fitting and timely that "Methods in Enzymology" devote several volumes to this subject.

Such advances include the growth and culture of cells, isolation of cell organelles (or derived purified subcellular fractions), the isolation and characterization of purified membranes and their components, the reconstitution of membranes from their phospholipid and protein components, as well as the use of model membrane systems. Several volumes on biomembranes have been planned. The first two (XXXI and XXXII) deal with techniques in the isolation and characterization of cells, organelles, membranes, and their components. A third volume will be devoted to electron transport and oxidative phosphorylation. It will update the very useful "Oxidation and Phosphorylation," Volume X which was edited by R. W. Estabrook and M. E. Pullman. Yet another volume will concentrate on methods for the study of properties of membranes, their architecture and function, including the study of biological transport and membrane receptors.

The original outline and organization of Volumes XXXI and XXXII benefited from the suggestions, advice, and help of a number of individuals. For this we are most grateful to our Editorial Advisory Board. Additional helpful comments came from a number of individuals including Drs. C. de Duve, J. O. Lampen, Oscar Touster, and Becca Fleischer. We are especially grateful to the contributors for making these volumes possible. The manuscripts from S. F.'s laboratory which are included in Volumes XXXI and XXXII benefited from the advice and helpful comments of Dr. Sidney Colowick. The excellent secretarial assistance of Mrs. Jean Talton of Vanderbilt and the friendly cooperation of the staff of Academic Press are gratefully acknowledged.

SIDNEY FLEISCHER
LESTER PACKER

METHODS IN ENZYMOLOGY

EDITED BY

Sidney P. Colowick and Nathan O. Kaplan

VANDERBILT UNIVERSITY
SCHOOL OF MEDICINE
NASHVILLE, TENNESSEE

DEPARTMENT OF CHEMISTRY
UNIVERSITY OF CALIFORNIA
AT SAN DIEGO
LA JOLLA, CALIFORNIA

METHODS IN ENZYMOLOGY

EDITORS-IN-CHIEF

Sidney P. Colowick   Nathan O. Kaplan

VOLUME VIII. Complex Carbohydrates
Edited by ELIZABETH F. NEUFELD AND VICTOR GINSBURG

VOLUME IX. Carbohydrate Metabolism
Edited by WILLIS A. WOOD

VOLUME X. Oxidation and Phosphorylation
Edited by RONALD W. ESTABROOK AND MAYNARD E. PULLMAN

VOLUME XI. Enzyme Structure
Edited by C. H. W. HIRS

VOLUME XII. Nucleic Acids (Parts A and B)
Edited by LAWRENCE GROSSMAN AND KIVIE MOLDAVE

VOLUME XIII. Citric Acid Cycle
Edited by J. M. LOWENSTEIN

VOLUME XIV. Lipids
Edited by J. M. LOWENSTEIN

VOLUME XV. Steroids and Terpenoids
Edited by RAYMOND B. CLAYTON

VOLUME XVI. Fast Reactions
Edited by KENNETH KUSTIN

VOLUME XVII. Metabolism of Amino Acids and Amines (Parts A and B)
Edited by HERBERT TABOR AND CELIA WHITE TABOR

VOLUME XVIII. Vitamins and Coenzymes (Parts A, B, and C)
Edited by DONALD B. MCCORMICK AND LEMUEL D. WRIGHT

VOLUME XIX. Proteolytic Enzymes
Edited by GERTRUDE E. PERLMANN AND LASZLO LORAND

Section I

Multiple Fractions from a Single Tissue

[1] Long-Term Preservation of Liver for Subcellular Fractionation[1]

By Sidney Fleischer and Marija Kervina

A procedure is described for the preservation of rat liver which makes possible the isolation of purified subcellular fractions after prolonged storage (months or longer).[2] The method consists of preparing a concentrated tissue homogenate, quick-freezing it using liquid nitrogen, and storage in a liquid nitrogen refrigerator. At the desired time, the sample is rapidly thawed and homogenized to further disintegrate the cells, and is then ready for separation into purified subcellular fractions. This method is designed to facilitate biochemical and biomedical studies at the subcellular level. It gives the investigator control in timing both for convenience as well as for collecting and storing samples from experiments or biopsy for later use. It also makes feasible the shipment of samples from one laboratory to another. The procedure has first been shown possible for rat liver. There is reason to believe that it will find general applicability with minor modification for other tissues as well.

The storage medium is fortified with dimethyl sulfoxide, a cryoprotective substance. Dimethyl sulfoxide has the advantage over other neutral solvents such as glycerol in that it is less viscous and penetrates rapidly. The prevention of ice crystal formation and local high concentrations of electrolytes or other components is believed to be important in the recovery of viable cells after freezing and thawing.

Rapid freezing and thawing minimizes the time during which degradative processes can occur. This approach was based in part on our previous experience in preserving delicate functions in chloroplast fragments[5] and mitochondria[6] for extended periods of time.

[1] These studies were supported in part by a Grant AM-14632 of the USPHS.
[2] The technology for storing mammalian cells in the frozen state has been achieved for many types of cells including sperm,[3] erythrocytes, normal and transformed tissue culture cells.[4] Such accomplishments have significant commercial value and are important to the researcher who is faced with the problem of preserving cell lines. These procedures make use of a programmed slow cooling rate and rapid thawing.
[3] C. Polge, A. U. Smith, and A. S. Parkes, *Nature (London)* **164**, 666 (1949).
[4] G. E. W. Wolstenholme and M. O'Conner, eds., "The Frozen Cell." Churchill, London, 1970.
[5] A. Wasserman and S. Fleischer, *Biochim. Biophys. Acta* **153**, 154 (1968).
[6] K. G. Walton, M. Kervina, D. S. Dow, and S. Fleischer, *J. Bioenerg.* **1**, 3 (1970).

Reagents and Solutions

> Amberlite MB-1 ion exchange resin, analytical grade. The resin is 5-fold washed with 3 volumes of deionized water prior to use.
> Dimethyl sulfoxide (DMSO), Certified A.C.S., Fisher, or comparable grade
> Mannitol and sucrose, reagent grade
> N-2-Hydroxyethylpiperazine-N'-2-ethanesulfonic acid (HEPES), A-grade, Calbiochem, La Jolla, California

The *storage medium* is 0.21 M Mannitol–0.07 M sucrose–20% (w/v) dimethyl sulfoxide (DMSO), pH 7.5. A solution of mannitol, sucrose, DMSO which will make 500 ml is treated with about 25 g of ion-exchange resin, Amberlite MB-1, to remove heavy metal ions. The mixture is swirled intermittently for several minutes and then filtered using Whatman No. 1 Filter Paper (W. and R. Balston Ltd., England) to remove the resin. The procedure is repeated once more. The pH is adjusted and the solution diluted to volume.

The *thawing medium* is 0.25 M sucrose–10 mM HEPES, pH 7.5. A sucrose solution in about four-fifths of final volume is treated with Amberlite MB-1 as described above for the storage medium. A concentrated solution of (0.20 M) HEPES neutralized with potassium hydroxide to pH 7.5 is then added. The pH and volume of the solution are then adjusted.

Equipment

> Potter-Elvehjem type homogenizer equipped with a piston-type Teflon pestle with stainless steel rod (Arthur H. Thomas and Co., Size C, Tissue Grinder, 55 ml volume). The glass grinding vessel has an inner diameter of 1.00 inch. The Teflon pestles are trimmed in a local machine shop to specified diameters, such as 0.974 and 0.988 inch. The pestle is driven using a motorized drive equipped with a tachometer. Since both hands are required in this operation, a foot pedal switch is used as the on/off control.
> Liquid nitrogen refrigerator. A type LR 35-9, which has nine canisters with a diameter of 2.5 inches each, produced by the Linde Division of Union Carbide Corp., is used in our laboratory. Liquid nitrogen level is kept above 4 inches from the bottom at all times.

Preservation Procedure

Homogenization. Two rats are decapitated using a guillotine (Harvard Apparatus Co., Millis, Massachusetts). The livers, excised immediately post-mortem (approximately 18–24 g), are weighed in a tared beaker and

1 volume of the storage medium is added, so that the final concentration of DMSO is 10%. The livers are diced with scissors and then homogenized in a Potter-Elvehjem glass-Teflon homogenizer, using a Teflon pestle of 0.974 inch diameter. Three passes, of about 5 seconds for each up-down stroke, are used with the pestle rotating at approximately 1000–1100 rpm. The homogenization and quick-freeze procedure are carried out in the cold room (0–4°).

Quick-Freeze Procedure. The homogenate is transferred into a polyethylene plastic bag (6 inches wide and 12 inches long, 0.002 inch thick polyethylene, Cole Parmer Instrument Co., Chicago, Illinois). The bag is placed on a flat surface to evenly distribute the contents in its lower portion over an area of about 7 inches by 6 inches. It is then rapidly slid horizontally into a liquid nitrogen bath while applying a twist so that the bag freezes in the shape of a cylinder of about 2 inches in diameter. The bag, with its contents, remains in the liquid nitrogen until the contents are well frozen (about 2 minutes) and is then rapidly transferred to a canister in a polyethylene cylindrical container. The canister, in turn, is placed into the liquid nitrogen refrigerator (approximate temperature of −196°). These operations should be carried out with dispatch; thus everything required should be available and in close proximity.

Quick-Thaw Procedure. When the bag is removed from liquid nitrogen, it is brittle and may crack if not handled with care. It soon warms up and regains its flexibility. The upper portion of the bag is cut away, and the homogenate in the form of flakes and powder is sprinkled into a beaker containing 4 volumes (based on original wet weight of liver) of 0.25 M sucrose–10 mM HEPES, pH 7.5, warmed to a temperature of 45°. Stirring is used to ensure rapid thawing so as to provide fresh surface for the flakes. If the sample had not been frozen uniformly, it may be desirable to break up the larger chunks of the frozen homogenate prior to thawing, using a hammer. When this is necessary a polyethylene sheet is used to shield the frozen chunks from direct contact with the hammer. Thawing is accomplished in 1 minute or less. After thawing, the mixture, at approximately 11°, is transferred to an Erlenmeyer flask and rapidly cooled to 1–2° using salt–ice water bath at −10° to −15°.

The thawed suspension is then rehomogenized with a Potter-Elvehjem homogenizer using Teflon pestles 0.974 and 0.988 inch in diameter, sequentially, for two and three passes, respectively, and rotating at 1000–1100 rpm. The homogenate is ready for filtration and separation into subcellular fractions as described this volume [2].[7]

Cell fractions containing purified nuclei, mitochondria, rough and

---

[7] S. Fleischer and M. Kervina, this volume [2].

smooth microsomes, Golgi complex, or plasma membranes can be prepared from the frozen homogenate. These cell fractions from livers preserved for nearly a year were comparable to those isolated from normal fresh livers as judged by enzymatic criteria, phosphorus to protein ratios, and polyacrylamide gel electrophoresis as well as by electron microscopy.[7]

[2] Subcellular Fractionation of Rat Liver[1,2]

By SIDNEY FLEISCHER and MARIJA KERVINA

A powerful approach to understanding the multiplicity of activities carried out by living cells is to fractionate the cell into its subcellular components. In this way, it is possible to localize specific cellular functions and to evaluate the role of each of the organelles in the operation of the cell.[3] Cell fractionation also facilitates the study of the biogenesis of cellular components, their intracellular transport,[4,5] membrane assembly, and hence, the factors that regulate or control such intracellular dynamics. In this article we describe the technology for isolating each of the known membrane-containing organelles as purified subcellular fractions (Fig. 1).[6]

[1] This work was supported in part by a grant from the National Institutes of Health AM 14632 [S. Fleischer and M. Kervina, 9th International Congress of Biochemistry, Stockholm, July 1–7, 1973 (Abstract p. 282)]. The skilled electron microscopy was performed by Mr. Akitsugu Saito. The helpful advice of Dr. Becca Fleischer is gratefully acknowledged.

[2] The term subcellular fractionation is used to mean separation of subcellular fractions that are referable to cell organelles. The term cell fractionation is avoided, since it may imply separation of different types of liver cells from one another.

[3] B. Fleischer and S. Fleischer, *in* "Biomembranes" (L. Manson, ed.), Vol. 2, p. 75. Plenum, New York, 1971.

[4] T. Peters, Jr., B. Fleischer, and S. Fleischer, *J. Biol. Chem.* **246**, 240 (1971).

[5] J. D. Jamieson and G. E. Palade, *J. Cell Biol.* **34**, 597 (1967).

[6] The methodology described is based on the accumulated experience in this field. Most notably, we modified and incorporated salient features from Stein *et al.,*[7] Cheveau *et al.,*[8] Fleischer and Fleischer,[9] Leighton *et al.,*[10] and Dallner.[11] The reader is referred to the essay by de Duve for a historical perspective of tissue (subcellular) cell fractionation.[12]

[7] Y. Stein, C. Widnell, and O. Stein, *J. Cell Biol.* **39**, 185 (1968).

[8] J. Cheveau, Y. Moulé, and C. Rouiller, *Exp. Cell Res.* **11**, 317 (1956).

[9] B. Fleischer and S. Fleischer, *Biochim. Biophys. Acta* **219**, 301 (1970).

[10] F. Leighton, B. Poole, H. Beaufay, P. Baudhuin, J. W. Coffey, S. Fowler, and C. de Duve, *J. Cell Biol.* **37**, 482 (1968).

[11] G. Dallner, *Acta Pathol. Microbiol. Scand.,* Suppl. 166, 1 (1963).

[12] C. de Duve, *J. Cell Biol.* **50**, 20 D (1971).

Rat liver is disrupted in a buffered sucrose solution, controlled shear is used, and the purified subcellular fractions are prepared sequentially using differential and gradient centrifugation. Either fresh liver or a liver homogenate which has been frozen and stored for many months according to Fleischer and Kervina[13] may be used. The fractionation procedure is conveniently carried out with two rat livers (18–24 g wet weight of liver). All operations are carried out either in a cold room or an ice bucket and using refrigerated centrifuges.

I. Preparation of Rat Liver Homogenates[14,16]

Animals. Male albino rats, Sprague-Dawley strain (Holtzman Company, Madison, Wisconsin) weighing 200–275 g were used. Unless otherwise specified, the rats were fed ad libitum. The livers from rats starved (about 16 hours, 5 PM to 9 AM) weigh about 35% less than those from

[13] S. Fleischer and M. Kervina, this volume [1].
[14] The disruption of tissues and cells to yield a homogenate is a key step. In principle, one would like to peel off the plasma membrane and tease apart the cell organelles. Since it would appear that rough and smooth endoplasmic reticulum, the nuclear membranes, Golgi, and plasma membranes are in part contiguous,[15] the connections between the different membranes should also be carefully severed. In practice, the mechanical shear which is applied to the liver tissue (rather than to single cells) is a balance between sufficient disruption for adequate recovery and excessive shear which would damage the cell organelles, thereby adversely affecting their fractionation properties. It appears remarkable enough that the shear can be controlled at least to the extent that all the membranous organelles can be isolated with reasonable integrity with one exception, the endoplasmic reticulum, which becomes disrupted and vesicles are formed. The use of the Potter-Elvehjem glass-Teflon type tissue grinder is, in our experience, the method of choice for liver. A size C tissue grinder (A. H. Thomas, Philadelphia, Pennsylvania) is used. The shear can be controlled by varying both the rate of rotation of the pestle and the clearance between the Teflon pestle and the precision tubing of the glass homogenizer. The Teflon pestle is trimmed to the desired clearance (0.026 and 0.012 inch) in a machine shop. The term clearance will be used to refer to the difference in diameter between the wall of the glass grinding vessel (1.000 inch) and the Teflon pestle. An untrimmed size C tissue grinder has a clearance of about 0.008 inch. During homogenization the pestle is rotated at near-constant speed using a high torque (series wound) motor equipped with speed control and tachometer while the glass vessel is moved up and down. A foot switch is used as the on/off control allowing both hands to control the homogenizer.
[15] G. E. Palade, *Proc. Nat. Acad. Sci. U.S.* **52**, 613 (1964).
[16] The formation of a homogenate is accompanied by release and activation of hydrolytic enzymes. It is reasonable to assume that the structural and functional integrity of the cell fractions can be optimized by working rapidly, using cold solutions (~0°), and not allowing the samples to warm up. The temperature control of the centrifuges are set close to zero (1.5°). The control should be calibrated periodically as a precaution against the sample being frozen during centrifugation. Centrifuge rotors are stored in the cold ready for use.

fed rats, mainly owing to utilization of the glycogen reserves while the body weight drops about 15%.

Reagents

Sucrose, 0.25 *M* neutralized to pH 7 with a solution of potassium hydroxide

Homogenization medium, 0.25 *M* sucrose–10 m*M* HEPES, pH 7.5

Fresh Liver. The rats are decapitated using a guillotine (Harvard Apparatus Co., Millis, Massachusetts). The livers are rapidly excised, immersed into ice-cold 0.25 *M* sucrose and cut into several large pieces. The pieces of liver are blotted on adsorbent paper, weighed, and placed into 5 volumes (w/v) of homogenization medium. The liver is further diced using a pair of scissors and is now ready for homogenization (see below).

Frozen Liver. The liver homogenate, previously quick-frozen in 1 volume of a storage medium, is thawed by the quick-thaw procedure using 4 volumes of warm homogenization medium[13] (based on original weight of liver) and is ready for homogenization.

Homogenization Procedure. The tissue in 5 volumes of homogenization medium is transferred to a 50-ml glass homogenizer (Potter-Elvehjem type tissue grinder, A. H. Thomas, Philadelphia, Pennsylvania).[14] The Teflon pestle is driven mechanically at approximately 1000–1100 rpm. Three passes each with a pestle of 0.026-inch clearance, are followed by three passes with a pestle of 0.012-inch clearance. The time of homogenization for each up-and-down stroke is about 5 seconds.

Filtration of Homogenate. The homogenate is gravity filtered through a 110-mesh nylon monofilament bolting cloth[17] (Tobler Ernst and Traber Inc., Elmsford, New York) aided by stirring with a glass rod. The residue on the nylon filter is washed with about 10 ml of homogenization medium.

II. Purification of Subcellular Fractions[18]

The fractionation procedure was designed to separate all the membranous cell organelles from the same homogenate (cf. flow diagram,

[17] The nylon mesh or cheesecloth, to be used as a filter, may be conveniently placed on top of a beaker and help in place with a rubber band at the upper edge. The filtrate is collected directly in the beaker.

[18] The concentrations of bivalent metal ions such as Mg^{2+} or Ca^{2+} can profoundly influence the subcellular fractionation. Their presence tends to keep membranes associated with one another, thereby adversely affecting separation. On the other hand, the presence of bivalent metal ions such as Mg^{2+} serves to stabilize the nuclei (3 m*M*)[19] and the binding of ribosomes with the rough endoplasmic reticulum (1 m*M*).[20] These opposing considerations must be balanced in the fractiona-

Fig. 1). The procedure requires two participants so that fractions are processed without delay. The amount of time required to isolate the purified fractions is given in Table III. The separation of cell fractions is accomplished using differential and sucrose gradient centrifugation.[21]

Step [A]. Fractionation of the Homogenate

The filtered homogenate is centrifuged at 3500 rpm in a JA-20 rotor (960 g at R_{av}, 7.0 cm; 1450 g at R_{max}, 10.8 cm) in the Beckman J-21 model centrifuge for 10 minutes. The supernatant ($Supt_1$) is decanted through several layers of cheesecloth. The pellet (residue 1, R_1) is used in [B] for isolation of plasma membrane, nuclei, and mitochondria (R_1). Supernatant 1 is used for the isolation of Golgi complex, mito-

---

tion procedure. The homogenate itself contains Ca^{2+} and Mg^{2+} and is not supplemented in this regard. Solutions used in the purification of nuclei are supplemented first with 1 mM $MgCl_2$ when working with the crude nuclear pellet containing plasma membrane, and then with 3 mM $MgCl_2$ with the more purified nuclear pellet (cf. B and B.3, respectively). On the other hand, EDTA is added to sequester Mg^{2+} and Ca^{2+}, so as to improve separation in the purification of plasma membrane and mitochondria.

[19] C. C. Widnell and J. R. Tata, *Biochim. Biophys. Acta* **87**, 531 (1964).

[20] T. Scott-Burden and A. O. Hawtrey, *Biochem. J.* **115**, 1063 (1969).

[21] The centrifugation time includes the time for acceleration but not deceleration. Some characteristics of Spinco rotors referred to in this article are compiled below.

Type	R_{min} (cm)	$R_{average}$ (cm)	R_{max} (cm)	No. of tubes per rotor	Volume per tube (ml)
SW 25.2	6.7	11.0	15.2	3	60
SW 41	6.6	10.9	15.2	6	13.2
SW 50	5.0	7.4	9.8	3	5.0
30	5.0	7.8	10.5	12	38.5
42.1	3.9	6.9	9.9	8	38.5
Titanium 50	3.8	5.9	8.1	12	13.5
Titanium 60	3.7	6.3	9.0	8	38.5
65	3.7	5.7	7.8	8	13.5

SW refers to swinging-bucket type rotors; the other rotors are of the fixed-angle type. The number of the rotor, to the left of the decimal point, denotes the maximum rated speed for the rotor in thousands. The number to the right of the decimal point further designates the type of rotor, if more than one is available. The centrifugal force (g) at any speed may readily be calculated using the equation

$$\text{Centrifugal force} = \frac{(0.1047 \times \text{rpm})^2}{980} \cdot R$$

where R = radius in cm.

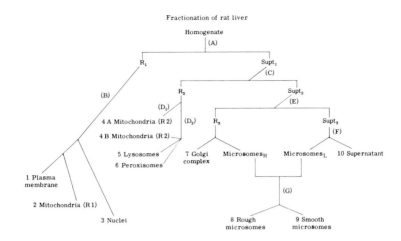

FIG. 1. Flow diagram for the subcellular fractionation of rat liver. The letters in parentheses refer to preparatory separation steps, and the numbers refer to the purification of the subcellular fractions. Both letters and numbers correspond to the outline in Section II of the text. Mitochondria (R_1) and mitochondria (R_2) refer to the mitochondrial fractions purified from residue 1 and residue 2, respectively.

chondria, peroxisomes and lysosomes, rough and smooth microsomes, and cell supernatant (cf. [C] Fractionation of Supernatant 1). Residue 1 and supernatant 1 are processed concurrently.

Step [B]. *Isolation of Plasma Membrane, Nuclei and Mitochondria* (R_1)

Reagents

0.25 M Sucrose, pH 7, neutralized to pH 7 with a solution of potassium hydroxide
0.25 M Sucrose–10 mM HEPES, pH 7.5
0.25 M Sucrose–10 mM HEPES–1 mM MgCl$_2$, pH 7.5
0.25 M Sucrose–10 mM HEPES–1 mM EDTA, pH 7.5
2.4 M Sucrose–10 mM HEPES–1 mM MgCl$_2$, pH 7.5
2.4 M Sucrose–10 mM HEPES–1 mM EDTA, pH 7.5

Procedure. Residue 1 (cf. [A]) is suspended in 0.25 M sucrose–10 mM HEPES–1 mM MgCl$_2$, pH 7.5 to a volume of 33.5 ml. A Dounce homogenizer[22] is used in this and all subsequent steps to resuspend fractions enriched in plasma membrane. High density sucrose, 54.5 ml (2.4 M

---

[22] Dounce homogenizers, type A, pestles (large clearance) (Kontes Glass Co.) are available in 7-, 15-, and 40-ml sizes.

sucrose–10 mM HEPES–1 mM MgCl$_2$, pH 7.5) is added to a final concentration of 1.60 M (45.7% w/w sucrose). The concentration of sucrose is adjusted to ±0.1% using a refractometer.[23] A two-layered step gradient is set up.[24] Forty-five milliliters of the suspended residue 1 in 1.60 M sucrose–10 mM HEPES–1 mM MgCl$_2$ are transferred into each of two SW 25.2 rotor tubes. The suspensions are overlayered with about 15 ml of 0.25 M sucrose–10 mM HEPES, pH 7.5, to fill the tube and are then centrifuged at 24,000 rpm (70,900 g at R$_{av}$) for 70 minutes in a SW 25.2 rotor. The band at the interface (0.25 M/1.6 M sucrose) is enriched in plasma membrane and mitochondria (Fig. 2E). Most of the 0.25 M sucrose layer is carefully removed by aspiration. The massive band, rimmed free from the edge of the tube with a spatula, can then be poured as a button, together with a small volume (ca. to 10 ml) of the liquid, into a Dounce homogenizer. The pellet of this first sucrose gradient is used for the preparation of nuclei (cf. [B] 3. Preparation of Nuclei).

The button containing both plasma membrane and mitochondria is resuspended gently in the Dounce homogenizer in approximately 2 volumes of cold water and enough 0.25 M sucrose–10 mM HEPES–1 mM EDTA, pH 7.5, to give a final volume of 70 ml. The suspension is centrifuged in two tubes at 4000 rpm (1200 g at R$_{av}$ 7.0 cm; 1900 g at

---

[23] An Abbé 3L refractometer, Bausch and Lomb, or equivalent, is used to adjust sucrose solutions where specified, to ±0.1% (w/w). The measurement is made on a drop of sample with the instrument thermostatted at 24° so that the density of the sucrose solution in the cold is somewhat greater.

[24] A gradient consisting of three layers is superior, in principle, since it affords an additional purification. In this case, R1 should be suspended in twice the volume (180 ml), and four tubes are required. The bottom layer is a 10-ml cushion consisting of 2.4 M sucrose–10 mM HEPES–1 mM MgCl$_2$, pH 7.5. Forty-five milliliters of the suspended pellet in 1.60 M sucrose–10 mM HEPES–1 mM MgCl$_2$ is layered next; this in turn is overlayered with enough 0.25 M sucrose–10 mM HEPES, pH 7.5, to fill the tube. The tubes are then centrifuged at 24,000 rpm for 70 minutes in SW 25.2 rotors. The plasma membranes, which form a band at the upper interface, are processed as described above except that they need to be combined from four tubes. The need for a larger volume, i.e., four tubes instead of two, when using the gradient consisting of three layers, is to prevent formation of a heavy plug at the lower interface which otherwise would trap the nuclei, preventing them from sedimenting into the bottom layer.

The bottom pellet is gently resuspended (cf. [B] 3) in 26 ml of 2.2 M sucrose–5 mM HEPES–3 mM MgCl$_2$, pH 7.5. The nuclei are centrifuged at 25,000 rpm (76,400 g at R$_{av}$) for 1 hour in a SW 41 rotor. The float at the top of the tube contains nonnuclear material and is discarded. The tubes are drained and wiped to remove any adhering scum. The nuclear pellet is carefully resuspended in 13 ml of the same medium and recentrifuged as before. The pellet containing purified nuclei is suspended in 2.0 ml of 2.2 M sucrose–1 mM MgCl$_2$ (pH ~7).

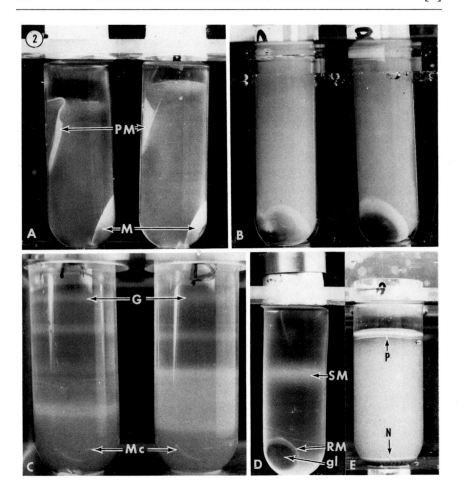

FIG. 2. Photographs illustrating the appearance of the tubes after centrifugation at different stages in the fractionation procedure. (A) The separation of plasma membrane from mitochondria by flotation of the plasma membrane (cf. step [B] 1). The plasma membrane (PM) is observed at the 0.25 M/1.45 M interface attached to the inner edge of the tube reflecting the use of a fixed angle rotor. The mitochondria (M) are in the pellet at the outer edge of the tube. Similar results were obtained for livers from fed (tube on left) and starved (tube on right) rats. (B) The sedimentation of the postnuclear supernatant (Supt$_1$) to obtain a mitochondrial pellet (R$_2$) (cf. step [D$_1$] 4a), purification of mitochondria (R$_2$). The tube on the right, prepared from livers of starved rats, shows two sharp layers in the pellet, an upper microsomal layer and a lower brown mitochondrial layer. In fed rats, the separation of layers is less distinct, (tube on the left) owing to the presence of glycogen. (C) The flotation of the Golgi fraction. The Golgi fraction (G) appears at the upper interface (29.0%/33.0%). The "heavy" microsomes (Mc) are prepared from the bottom layer in 43.7% sucrose (cf. step [E] 7, purification of Golgi complex). The experiment was carried out on livers from fed (tube on left) and starved (tube

R_{max}) for 10 minutes in a Beckman JA-20 rotor (first wash). For the second wash, the pellet is resuspended in about 70 ml of 0.25 M sucrose–10 mM HEPES–1 mM MgCl$_2$, pH 7.5 and centrifuged as before. The slightly turbid supernatant, containing some mitochondria, is discarded. The pellet, a mixture of both plasma membrane and mitochondria, is resuspended using 0.25 M sucrose–10 mM HEPES–1 mM EDTA, pH 7.5, to 22.1 ml, and 27.9 ml of 2.4 M sucrose–10 mM HEPES–1 mM EDTA, pH 7.5, are then added to bring the final concentration to 1.45 M (42.0% ± 0.1% sucrose).[23] A two-layer step gradient is prepared in 30-ml capacity tubes[25] consisting of 25 ml of the enriched plasma membrane suspension in 1.45 M sucrose and overlayered with enough 0.25 M sucrose–10 mM HEPES, pH 7.5 to fill the tubes. The tubes are centrifuged at 28,000 rpm (68,400 at R_{av} 7.8 cm) for 60 minutes in a Spinco type 30 rotor. The plasma membrane appears as a band at the 0.25 M/1.45 M sucrose interface, and mitochondria are in the bottom pellet (Fig. 2A). Both fractions require further purification as outlined below: Steps [B] 1 and [B] 2.

1. PURIFICATION OF PLASMA MEMBRANES

The material at the interface is collected from the inner edge of the tube (Fig. 2A), with a Pasteur pipette or a syringe equipped with a 90° bevel needle, resuspended in 0.25 M sucrose–10 mM HEPES–1 mM EDTA, pH 7.5, to approximately 35 ml (one tube) and centrifuged at 15,000 rpm (17,600 g at R_{av}) for 10 minutes in the Beckman JA-20 rotor. A third step gradient is used for the final purification. The pellet is resuspended in 9.8 ml of 0.25 M sucrose–10 mM HEPES–1 mM EDTA, pH 7.5, and mixed with 10.3 ml of 2.4 M sucrose–10 mM HEPES–1 mM EDTA, pH 7.5. The final concentration should be adjusted to 1.35 M (39.4% ± 0.1%) sucrose.[23] Ten milliliters are placed into each of two Spinco 65 rotor tubes (capacity of 13.5 ml each),

[25] Nalgene Oak Ridge-type polycarbonate centrifuge tubes with polypropylene screw closures are convenient for this purpose (331210, Beckman Instruments, Palo Alto, California).

on right) rats. Both conditions yield a good Golgi fraction. (D) The preparation of rough (RM) and smooth (SM) microsomes. The smooth microsomes appear at the 0.25 M/1.3 M sucrose interface. The rough microsomes are in the pellet sitting above a clear colorless layer of glycogen (gl) at the bottom of the tube. The appearance of the glycogen (gl) is clear and colorless and is somewhat misleading in the photograph due to the lighting (cf. [G], purification of rough and smooth microsomes). (E) Separation of R$_1$ into two fractions, an upper float at the 0.25 M/1.6 M interface containing plasma membrane (P) and mitochondria and a pellet containing nuclei (N) at the bottom of the tube (cf. step [B]).

overlayered with 0.25 M sucrose–10 mM HEPES, pH 7.5 to fill the tubes, and centrifuged for 30 minutes at 60,000 rpm (231,000 g at R_{av}) in a Spinco type 65 rotor. The plasma membrane is collected from the inter-face, diluted with approximately 2 volumes of cold water and additional 0.25 M sucrose to fill the tube, and sedimented for 10 minutes at 15,000 rpm (14,500 g at R_{av}). The whitish pellet of purified plasma membrane is resuspended in an appropriate volume ($\sim$5 ml) of 0.25 M sucrose (pH 7) to obtain approximately 1 mg of protein per milliliter.

2. Purification of Mitochondria [R₁]

Reagents

> Buffered sucrose–EDTA solution, 0.25 M sucrose–10 mM HEPES–
> 1 mM EDTA, pH 7.5
> 0.25 M Sucrose, pH 7

Procedure. The mitochondrial pellet obtained from the second sucrose gradient centrifugation (cf. [B]) requires further repurification. The pellet is resuspended in approximately 35 ml of buffered sucrose–EDTA solu-tion using a Potter-Elvehjem tissue grinder (0.012-inch clearance)[26] and centrifuged at 18,000 rpm in a JA-20 (25,300 g at R_{av}) rotor for 10 minutes. Two layers are clearly visible. The lower brown layer is the mitochondria. The upper light layer is swirled off with the aid of a fine-tipped glass rod and fresh buffered sucrose–EDTA solution and is dis-carded. The lower brown layer is resuspended in more buffered sucrose–EDTA solution, and the purification is repeated as described above to remove any remaining upper light layer. The mitochondria are finally resuspended in about 5 ml of 0.25 M sucrose, pH 7 to give a concentra-tion of approximately 14 mg of protein per milliliter.

3. Purification of Nuclei[8]

Reagents

> 2.2 M Sucrose–5 mM HEPES–3 mM MgCl$_2$
> 2.2 M Sucrose–1 mM MgCl$_2$

Procedure. Nuclei are purified from the pellet of the first sucrose gradient (cf. [B] and Fig. 2E). The pellet is resuspended in about 25 ml

---

[26] Potter-Elvehjem glass–Teflon type tissue grinders (A. H. Thomas or Kontes Glass Co., Vineland, New Jersey) or Duall Teflon-type tissue grinders (Kontes Glass Co.) are used, generally, to aid in resuspension of samples. These are available in a variety of assorted sizes. The Teflon pestle of the Potter-Elvehjem type tissue grinder can readily be trimmed to specified clearance to decrease the shear (cf. footnote 14).

of 2.2 M sucrose–5 mM HEPES–3 mM MgCl$_2$ with the aid of a loose-fitting glass–Teflon Potter-Elvehjem homogenizer (0.026-inch clearance). The pellet is tacky, and care must be taken to use minimal shear and yet to adequately resuspend the pellet. This is achieved by working the homogenizer up and down slowly, while hand-rotating the pestle about a quarter turn, clockwise and counterclockwise repeatedly, with the pestle anchored vertically with the chuck of the motor. The resuspended material is diluted to about 60 ml to fill one tube and is centrifuged in a SW 25.2 rotor (70,900 g at R_{av}) at 24,000 rpm for 1 hour. The float, containing nonnuclear material, is discarded. The tube is drained, and its walls are wiped to remove any adhering float. The nuclear pellet is resuspended as before in about 26 ml of the same solution and centrifuged in two tubes at 25,000 rpm for 1 hour in an SW 41 rotor (76,000 g at R_{av}) (two tubes). The white pellet of purified nuclei is gently resuspended in 2–4 ml of 2.2 M sucrose 1 mM MgCl$_2$, pH 7.

Step [C]. Fractionation of Supernatant 1

Supernatant 1 (cf. [A]) (~70 ml) is filtered through two double layers of cheesecloth to remove congealed lipid and is centrifuged at 18,000 rpm for 10 minutes in a Spinco-type 42.1 rotor (25,000 g at R_{av}). The sediment, residue 2 (R$_2$), can be used either for purification of mitochondria (R$_2$) (cf. [D$_1$] 4a) or for separation of mitochondria, lysosomes, and peroxisomes (cf. Section VIII, Addendum). Supernatant 2 is decanted and used in the purification of Golgi complex, microsomes, and cell supernatant (cf. [E]).

Step [D$_1$] 4a. Purification of Mitochondria-R$_2$

Reagents

Sucrose, 0.25 M, pH 7
(Buffered sucrose–EDTA solution) 0.25 M sucrose–10 mM HEPES–1 mM EDTA, pH 7.5

Residue 2 in two tubes, from [C] is combined and resuspended in approximately 25–30 ml of 0.25 M sucrose–10 mM HEPES–1 mM EDTA, pH 7.5 and recentrifuged, for 10 minutes at 18,000 rpm (Spinco type 42.1 rotor). A Potter-Elvehjem tissue grinder with a clearance of 0.012 inch is used to resuspend mitochondrial pellets.

When livers from starved rats are used, the pellet contains two distinct layers, an upper light portion and a lower brown portion, (Fig. 2B, tube on right side). The upper portion (mostly microsomes) is separated and discarded. This is readily achieved with the aid of a stirring rod and careful swirling with fresh buffer. The lower brown portion, enriched in mitochondria, is resuspended in 10 ml of the same buffer and resedimented. Any residual upper layer is again removed. The mitochondria are resuspended in approximately 5 ml of 0.25 M sucrose. On occasion, a small dark button is observed which is left behind by mechanically teasing away the mitochondria using a fine glass rod.

Livers from rats fed ad libitum contain large amounts of glycogen. The particulate glycogen sediments throughout the fractionation scheme complicating the separation somewhat. The difference is most noticeable in the purification of the mitochondria, where the light upper portion blends into the lower brown portion (Fig. 2B, tube on left). Glycogen—which appears as an opaque white, or colorless glassy, material—can be observed at the bottom of the pellet. Separation is achieved by resuspending and recentrifuging the pellet several times. This process permits the glycogen to preferentially associate with itself and the glycogen accumulates at the bottom of the tube. Concomitantly, the separation of the upper light portion from the lower brown mitochondrial portion of the pellet becomes sharper, and the upper fluff is removed as described above. The mitochondria are separated from the glycogen by somewhat more vigorous swirling than is required to tease away the upper light layer.

Step [D_2]. Use of R_2 for Purification of (4b) Mitochondria,
(5) Lysosomes, and (6) Peroxisomes

The separation of mitochondria, lysosomes, and peroxisomes[10] was not carried out as part of the continuous fractionation procedure. Instead, separate experiments, were carried out to show that it is feasible to use residue 2 for this purpose (cf. VIII, Addendum).

Step [E]. Separation of Fractions from Supernatant 2

7. PURIFICATION OF GOLGI COMPLEX[9]

Supernatant 2 (cf. [C]) is centrifuged for 30 minutes at 20,000 rpm (34,000 g at R_{av}) in two tubes using a Spinco type 30 rotor in order to sediment a "heavy" microsomal fraction which is enriched in the Golgi complex. The supernatant (Supt$_3$) is carefully decanted and is used to recover the remainder of the microsomes ("light" microsomes)[27]

---

[27] "Microsomes" are defined operationally as a membranous fraction which can be sedimented from a postmitochondrial supernatant at high speed (40,000 rpm for 1 hour). Although rough microsomes are largely referable to fragmented endoplasmic reticulum, smooth microsomes are composed not only of fragmented smooth endoplasmic reticulum, but also of membranes referable to other organelles including the Golgi complex, plasma membrane, fragmented mitochondria, lysosomes, and peroxisomes.

and cell supernatant (cf. [F] below). The pink sediment, residue 3 is significantly enriched in Golgi complex, although quantitatively derived mainly from endoplasmic reticulum. It is resuspended gently, using a loose-fitting Dounce homogenizer, in approximately 10 ml of 52% sucrose–0.1 M PO$_4$ buffer, pH 7.1, and the density is adjusted to 43.7% $\pm$ 0.1% sucrose.[23] This fraction, (10–14 ml) is placed in one SW 25.2 rotor centrifuge tube and is overlayered sequentially with 11–15 ml, 10 ml, 10 ml, and 12 ml of 38.7%, 36.0%, 33.0%, and 29.0% sucrose (each $\pm 0.1\%$) and centrifuged at 25,000 rpm for 53 minutes.[21] The Golgi fraction appears in the 29%/33% sucrose interface (cf. Fig. 2C). This layer is collected with the use of a hypodermic syringe equipped with a 90° bevel needle. The fraction in about 5 ml may be diluted with an equal volume of cold water and centrifuged at 40,000 rpm for 40 minutes in a type 65 rotor.[28] The pellet is resuspended in 1–2 ml of 0.25 M sucrose, pH 7, using a small, loose-fitting Dounce homogenizer. Heavy microsomes are recovered from the bottom layer of the gradient.

"Heavy" Microsomes. The upper four layers of the sucrose gradient (cf. [E] 7) are discarded using aspiration, and the bottom layer (43.7% sucrose) is diluted with 2 volumes of cold distilled water and centrifuged at 40,000 rpm for 1 hour (124,000 g at R_{av}) in a type 42.1 rotor. The pellet, "heavy" microsomes, is suspended in 6 ml of 0.25 M sucrose, pH 7 and will be used, in part, for the preparation of rough and smooth microsomes (cf. [G]).

Step [F]. Separation of "Light" Microsomes and Cell Supernatant from Supernatant 3

This fraction is obtained by centrifuging supernatant 3 (cf. Fig. 1) at 40,000 rpm for 1 hour (type 42.1 rotor). The pellet, the "light" microsomes, is resuspended in 6 ml of 0.25 M sucrose, pH 7, and will be used, in part, for the preparation of rough and smooth microsomes. The supernatant is saved (cf. 10, cell supernatant).

Step [G]. Purification of (8) Rough and (9) Smooth Microsomes[11]

Equal proportions of "heavy" and "light" microsomes (cf. [E] and [F] above), usually two-thirds of the total yield of each, are combined.

[28] The Golgi complex is sensitive to osmotic shock and mechanical handling, so that this dilution is not recommended if you are interested in preserving the structure (S. Fleischer and B. Fleischer, unpublished studies). To preserve morphology for electron microscopy, an aliquot of the undiluted sample is fixed directly with 0.1 volume of 25% glutaraldehyde in buffer (cf. Section V, Preparation of Samples for Electron Microscopy). Osmotic shock can be avoided with the use of an Amicon ultrafilter device (Amicon, Lexington, Massachusetts) to concentrate the sample.

The mixture is diluted to 28 ml with 0.25 M sucrose, and 1 M CsCl (0.42 ml) is added to bring its final concentration to 0.015 M. This suspension is layered into two tubes, each containing 20 ml of 1.3 M sucrose–0.015 M CsCl and then centrifuged at 58,000 rpm for 110 minutes in a Spinco 60 Ti rotor (237,000 at R_{av}). Special care must be taken to ensure that the tubes are completely filled or they may collapse.[29] The rough microsomes are in the pink sediment, and the smooth microsomes are at the interface (Fig. 2D). When the livers from fed rats are used, this pellet will consist of two distinct layers, a pink upper layer (rough microsomes) and a colorless clear bottom layer consisting of glycogen. The pink layer can be separated from the glycogen using vigorous swirling with the 5 ml of 0.25 M sucrose in which it is resuspended. The smooth microsomes at the interface (Fig. 2D) are recovered using a syringe equipped with a 90° bevel needle, diluted with an equal volume of 0.25 M sucrose, pH 7.0, and centrifuged 40,000 rpm (124,000 g at R_{av}) for 1 hour in a Spinco type 42.1 rotor. The pellet containing the smooth microsomes is suspended in about 3 ml of 0.25 M sucrose, pH 7, leaving behind the bottom clear glycogen layer, if present.

10. CELL SUPERNATANT

The supernatant which is obtained after sedimentation of the "light" microsomes (cf. [F]) is referable to the soluble phases of the cell.

III. Storage of Samples for Enzymatic and Chemical Analysis

The purified fractions are resuspended in a sucrose solution without HEPES buffer, which interferes with the estimation of protein by the Folin procedure.[30] It is best to carry out assays and analysis immediately. However, by this time fatigue has set in. If an enzyme is found not to be very stable, it should be assayed right away. We have found that many enzymatic activities can be preserved for prolonged periods by freezing the sample in liquid nitrogen and storing it in a liquid nitrogen refrigerator, or a low temperature ($-80°$) freezer. The liquid nitrogen refrigerator is preferred, but its capacity is limited. The low temperature freezer is usually adequate for most activities. "Serum tubes" made by NUNC of Denmark (distributed by Vanguard International, Box 312, Red Bank, New Jersey) are convenient for storage of samples. They are polypropylene tubes with a screw cap and silicone washer. The samples, divided

---

[29] The use of high speed polycarbonate tubes with 3-piece Noryl caps is recommended for this operation (Noryl cap and tube assembly, 335348, Beckman Instruments, Palo Alto, California).

[30] O. H. Lowry, N. J. Rosebrough, A. L. Farr, and R. J. Randall, *J. Biol. Chem.* **193**, 265 (1951).

into several aliquots, are quick-frozen by dipping them into liquid nitrogen. One tube of each sample can be thawed leaving several equivalent tubes available for later studies.

Quick-thawing is carried out with only several samples at a time, dipping them first into warm water (water bath ~45°), then into cooler water when they are nearly thawed and using continuous motion to ensure that no portion of the sample warms up appreciably above 0°. The assays should be carried out soon after thawing, especially if an enzymatic activity is of questionable stability. The remainder of thawed sample should be refrozen soon after aliquots are removed to limit further breakdown. It is marked "thawed 1X" and is reused mainly for chemical assays or for enzymatic assays known to be stable to the freezing and thawing process. Some functions that are unstable can sometimes be preserved by working out appropriate conditions for storage. For example, it is possible to preserve oxidative phosphorylation, respiratory control, and energized swelling of liver mitochondria, by adding dimethyl sulfoxide and bovine serum albumin, quick-freezing the mitochondria, and storing them in liquid nitrogen.[31]

IV. Analysis of Liver Cell Fractions

The purity of the cell fractions is characterized by a variety of techniques including chemical analysis, enzymatic assay, electron microscopy, and polyacrylamide gel electrophoresis. Cell fractions are routinely assayed for protein[30] using bovine serum albumin for a standard and total phosphorus,[32] which serves as an estimate of phospholipid, when the nucleic acid content is low. Thus it cannot be so used for nuclei and rough microsomes. A reliable phospholipid analysis requires lipid extraction and analysis (cf. Rouser and Fleischer[32]). The Clark oxygen electrode was used to measure oxygen uptake, phosphorylation efficiency (P/O ratios), and respiratory control ratios.[31,33] A number of diagnostic enzymes were assayed to estimate contamination. These are: galactosidase[34] and 5'-nucleotidase[35] for plasma membrane; succinate-cytochrome c reductase[36] for

[31] K. G. Walton, M. Kervina, S. Fleischer, and D. S. Dow, *Bioenergetics* 1, 3 (1970).
[32] P. S. Chen, T. Y. Toribara, and H. Warner, *Anal. Chem.* 28, 1756 (1956) as modified in G. Rouser and S. Fleischer, this series, Vol. 10, p. 385 (1967); the assay for phosphate is described on p. 404. The assay can be used to measure both organic phosphate or inorganic phosphate. For the latter, the perchloric acid digestion step is omitted.
[33] R. W. Estabrook, this series, Vol. 10, p. 41.
[34] B. Fleischer and S. Fleischer, *Biochim. Biophys. Acta* 183, 265 (1969).
[35] C. C. Widnell and J. C. Unkeless, *Proc. Nat. Acad. Sci. U.S.* 61, 1050 (1968).
[36] S. Fleischer and B. Fleischer, this series Vol. 10, p. 406 (cf. pp. 427–428 for cytochrome c reductase assays).

mitochondria; urate oxidase[37] and catalase[38] for peroxisomes, aryl sulfatase,[39] and acid phosphosphatase[40] for lysosomes; *N*-acetylglucosamine galactosyltransferase[41] for the Golgi complex; rotenone insensitive DPNH-cytochrome *c* reductase[36] and glucose-6-phosphatase[42] for endoplasmic reticulum. Mg^{2+} stimulated ATPase activity was measured as previously described.[36] Some of these enzymes are listed in Table I, which contains some examples of the more reliable diagnostic enzymes for cell organelles. Other enzymes may be used reflecting special interests of a laboratory, but which are localized in more than one organelle, such as galactosidase, which is also present in lysosomes. Mg^{2+}-stimulated ATPase is present in high specific activity in both mitochondria and plasma membranes.

Some departures from the assay procedures in the literature are as follows: The galactosidase assay is carried out at pH 8.6 as previously described,[34] except that *N,N*-bis(2-hydroxyethyl)glycine(Bicine)-HCl buffer is used at pH 8.6.[9] At this pH the plasma membrane enzyme gives optimal activity, and lysosomal activity is depressed. Rotenone insensitive DPNH-cytochrome *c* reductase is assayed as previously described for DPNH–cytochrome *c* reductase activity except that 5 μg of rotenone (5 μl of 1 mg/ml ethanol) are added to the sample in 1-ml assay mixture.[36] Glucose-6-phosphatase is carried out as previously described,[42] but the released inorganic phosphate is measured by the more sensitive phosphate analysis, ascorbic acid being used as the reducing agent.[32] In this way 100 μg of sample may be assayed and the enzymatic reaction is carried out for 5 and 10 minutes at 37°.

Polyacrylamide gel electrophoresis, under dissociating conditions, is useful for characterization of the cell fractions (cf. Fig. 13).[43,44]

V. Preparation of Samples for Electron Microscopy[45]

Reagents

Glutaraldehyde, 50% (Eastman Kodak)
Sodium cacodylate (Amend Drug and Chemical Company, Inc., New York, New York)

[37] F. Leighton, B. Poole, P. B. Lazarow, and C. deDuve, *J. Cell Biol.* **41**, 521 (1969).
[38] P. Baudhuin, this volume [36]. The separation of peroxisomes from lysosomes and mitochondria is described in detail.
[39] A. B. Roy, *Biochem. J.* **53**, 12 (1953).
[40] A. Trouet, this volume [31].
[41] B. Fleischer, this volume [17].
[42] M. A. Swanson, this series, Vol. 2, p. 541.
[43] W. L. Zahler, this series, Vol. 32 [7] in press.
[44] D. M. Neville, Jr., and H. Glossman, this series, Vol. 32 [9] in press.
[45] S. Fleischer, B. Fleischer, and W. Stoeckenius, *J. Cell Biol.* **32**, 193 (1967).

Glutaraldehyde, 25% (diluted from 50%) in 0.1 M cacodylate buffer, pH 7.2

Glutaraldehyde, 5% (diluted from 50%) in 0.2 M cacodylate buffer, pH 7.2

Buffered sucrose wash solution, 0.1 M cacodylate buffer in 0.25 M sucrose, pH 7

Michaelis buffer, 0.1 M Veronal acetate, pH 7.4, containing 2.4 mM CaCl$_2$ and 60 mM NaCl

Osmium tetroxide (Purified Anhydride Crystal, Fisher Scientific Co., St. Louis, Missouri)

Osmium tetroxide, 1% in Michaelis buffer, pH 7.4

Uranyl acetate, 0.5%, is diluted in Michaelis buffer, pH 6.0, and stored in the dark

Fixation of Subcellular Fractions using Glutaraldehyde. The sample, approximately 0.5 mg of protein suspended in approximately 0.15 ml of a sucrose solution in a small centrifuge tube,[46] is fixed by adding an equal volume of cold 5% glutaraldehyde in cacodylate buffer so that the final concentration is 2.5% glutaraldehyde in 0.1 M cacodylate buffer. It is left in the cold room overnight. It is preferable to fix the sample immediately after isolation to minimize the possibility of changes with time. The nuclei are in heavy (2.2 M) sucrose so that care must be taken to obtain good mixing. In order to preserve the morphology of the Golgi complex for

[46] The minimum amount of sample used is dictated by the need to obtain a small compact pellet which can readily be processed and handled. If too much sample is used, the pellet is too thick and the reagents may not readily penetrate. Small centrifuge tubes and adapters are available which are convenient for small volumes (Beckman Instruments, Palo Alto, California). For the fixed-angle rotors (types 65, 40, and Titanium 50), 2 ml tubes $\frac{5}{16}''$ × $1\frac{15}{16}''$ (303369) and adapters (303669) are available; for the swinging-bucket type SW 50 rotor, 0.8 ml tubes, $\frac{3}{16}$ × 1⅝ inches and adapters (305527) are available. If sample dilution is required prior to fixation, the sample should be diluted with the same concentration of sucrose solution to avoid changes due to osmotic shock. One of the advantages of using glutaraldehyde as the initial fixative, as compared with osmium tetroxide, is that samples can readily be fixed in the presence of high sucrose concentrations as a suspension which can be sedimented at lower force than the unfixed sample.

When the sample is more dilute than 0.5 mg/0.15 ml, a larger aliquot is required. If an angle rotor is used to sediment the fixed sample, the pellet becomes broadly distributed along the vertical wall of the tube. The sedimented material should then be resuspended in 0.15 ml of buffered sucrose wash and recentrifuged to form a small compact pellet. The problem is avoided if a swinging-bucket rotor is used.

The washes are carried out conveniently by adding and removing solutions with a Pasteur pipette. The volumes are not important as long as enough solution is used to wash down the walls and cover the sample (~0.5–1.0 ml).

electron microscopy an aliquot of the sample (cf. [E] 7) from the sucrose gradient in several milliliters is used directly, without dilution. In this case 25% glutaraldehyde in cacodylate buffer (0.1 volume per volume of sample) is added, and care is taken to ensure good mixing by gently inverting the tube, parafilm being used to cover the tube. The sample is then left in the cold room overnight.

Obtaining a Compact Pellet. The sample is then sedimented with sufficient force to obtain a compact pellet that will remain intact through numerous washes. The minimum amount of force consistent with a compact pellet is desired; avoid over compacting the sample. A good average for most samples is 10,000 rpm for 15 minutes. The fixed nuclei are first diluted with an equal volume of 0.1 M cacodylate solution prior to centrifuging the sample for 3000 rpm for 15 minutes.

The fixed Golgi sample is diluted with an equal volume of 0.1 M cacodylate buffer in 0.25 M sucrose (to decrease the density to facilitate pelleting) and is centrifuged for 10,000 rpm for 5 minutes. Part of the pellet becomes packed on the vertical wall of the tube. The supernatant is carefully removed. The sample is resuspended in 1–1.5 ml of the buffered sucrose wash, using a stirring rod, and is transferred to the smaller centrifuge tube and recentrifuged. Again the sample is spread along the wall. It is resuspended in about 0.2 ml of the wash buffer and recentrifuged. By now there is a compact pellet at the bottom of the tube.

Washing the Pellets to Remove Glutaraldehyde. The glutaraldehyde solution is removed with a Pasteur pipette. About 1 ml of buffered sucrose wash solution is added, care being taken to wash down the walls of the tube. After about 10 minutes, the solution is carefully removed with a Pasteur pipette and fresh wash solution is added. About five washes are used in a period of about 1 hour.

Fixation of Sample with Osmium Tetroxide. The last wash is removed and replaced with about 0.5 ml of 1% osmium tetroxide in Michaelis buffer. The pellet may be fixed overnight in the cold room or for 2 hours at room temperature.

Removal of Excess Osmium Tetroxide. The osmium tetroxide is removed and replaced with Michaelis buffer, care being taken to wash down the inner walls of the tube. The solution is removed and replaced with fresh solution and left for 10 minutes at room temperature.

"Block Staining" (Counterstaining). The Michaelis buffer is replaced with 0.5% uranyl acetate, pH 6.0, for 2 hours at room temperature. The samples are placed in a cabinet at room temperature, away from light. All succeeding steps are carried out at room temperature. The samples are dehydrated in a graded ethanol–water series, embedded in a hard plastic, such as Araldite or Epon, and polymerized at 60°. The pellet should be

embedded in known orientation so that top to bottom in cross section can be studied in thin sections to evaluate whether the sample is uniform or stratified (cf. Tartakoff and Jamieson[47]). Thin sections are cut with a diamond knife using an LKB Ultrotome or an equivalent ultramicrotome. The thin sections are double-stained, sequentially with uranyl magnesium acetate and lead citrate,[45] prior to electron microscopy.

VI. Characterization of Liver Cell Fractions

General Considerations

Cell organelles have specific roles in cellular function. Such specialization is reflected by their morphology, composition, and enzymatic activities (Table I), which are useful in characterizing the purified cell fractions. The nucleus which houses and replicates genetic information is rich both in DNA content and DNA polymerase. Electron transport and electron transfer components cytochrome $(a + a_3)$ can be used to characterize mitochondria. Lysosomes contain acid hydrolases. The Golgi complex, which is involved in secretion, catalyzes the enzymatic transfer of specific sugars,[48,49] such as galactose from UDP galactose to protein or free N-acetylglucosamine.[48] The rough endoplasmic reticulum which manufactures proteins is studded with ribosomes and is thereby rich in RNA. The endoplasmic reticulum carries out most of the lipid synthesis and manufactures practically all the triglyceride, phosphatidylcholine, phosphatidylethanolamine, and sphingomyelin of the cell and contains the enzymes specific to this role.[50,51] Glucose-6-phosphatase is localized in the endoplasmic reticulum. The plasma membrane and peroxisomes likewise have distinguishing enzymes. A number of processes occur predominantly in the nonparticulate "soluble" phase of the cell. These include the enzymes that catalyze de novo fatty acid biosynthesis, pentose phosphate pathway, and glycolysis.[52] The enzymes responsible for catalyzing these processes can therefore be assayed in the supernatant.

The purity of the isolated subcellular fractions can be characterized with regard to these distinguishing features. When a characteristic feature

[47] A. M. Tartakoff and J. D. Jamieson, this volume [3].

[48] B. Fleischer, S. Fleischer, and H. Ozawa, *J. Cell Biol.* 43, 59 (1969).

[49] H. Schachter, I. Jabbal, R. L. Hudgin, L. Pinteric, E. J. McGuire, and S. Roseman, *J. Biol. Chem.* 245, 1090 (1970).

[50] L. M. G. van Golde, B. Fleischer, and S. Fleischer, *Biochim. Biophys. Acta* 249, 318 (1971).

[51] L. M. G. van Golde, J. Raben, J. J. Batenburg, B. Fleischer, F. Zambrano, and S. Fleischer, *Biochim. Biophys. Acta,* in press.

[52] N. G. Anderson and J. G. Green, *in* "Enzyme Cytology" (D. B. Roodyn, ed.), p. 475. Academic Press, New York, 1967.

TABLE I

SELECTED DIAGNOSTIC CHARACTERISTICS OF CELL ORGANELLES[d,e]

Cell fraction	Characteristic	Reference
Plasma membrane	5′-Nucleotidase	i
	Alkaline phosphodiesterase I	j
Mitochondria[a]	Cytochrome oxidase	k
	Succinate-cytochrome c reductase	l
	Cytochrome $(a + a_3)$	m
Nuclei	High DNA content; DNA[b] polymerase	n
Lysosomes	Acid hydrolases (e.g., acid phosphatase)	k, o
Peroxisomes	Catalase	p
	Urate oxidase	q
Golgi complex	N-Acetylglucosamine galactosyltransferase	r, s
Endoplasmic reticulum (microsomes)	Rotenone insensitive DPNH-cytochrome c reductase[c]	h, l
	Glucose-6-phosphatase	t
	Choline phosphotransferase	u
Supernatant (soluble phase of cell)	Soluble enzymes, such as phosphoglucomutase	v, w
	Lactic dehydrogenase	x

[a] Mitochondria are composed of two membranes. The characteristics shown are for inner membranes. A characteristic enzyme for the outer membranes is monoamine oxidase.[f]

[b] Mitochondria also contain small amounts of DNA and DNA polymerase.[g]

[c] Although localized mainly in endoplasmic reticulum,[h] this activity is also present in the outer membrane of the mitochondrion.[f] Nuclear membranes are contiguous with, and also seem to contain, endoplasmic reticulum enzymes.

[d] B. Fleischer and S. Fleischer, in "Biomembranes" (L. Manson, ed.), Vol. 2, p. 75. Plenum, New York, 1971.

[e] D. B. Roodyn, ed., "Enzyme Cytology." Academic Press, New York, 1967.

[f] G. L. Sottocasa, B. Kuylenstierna, L. Ernster, and A. Bergstrand, this series Vol. 10, p. 448.

[g] "Biochemical Aspects of the Biogenesis of Mitochondria" (E. C. Slater, J. M. Tager, S. Papa, and E. Quagliarello, eds.). Adriatica Editrice, Bari, 1968.

[h] B. Fleischer and S. Fleischer, Biochim. Biophys. Acta 219, 301 (1970).

[i] C. C. Widnell and J. C. Unkeless, Proc. Nat. Acad. Sci. U.S. 61, 1050 (1968).

[j] N. N. Aronson and O. Touster, this volume [6].

[k] F. Appelmans, R. Wattiaux, and C. de Duve, Biochem. J. 59, 438 (1955).

[l] S. Fleischer and B. Fleischer, this series Vol. 10, p. 406 (cf. pp. 427–428 for cytochrome c reductase assays).

[m] S. Fleischer, B. Fleischer, A. Azzi, and Britton Chance, Biochim. Biophys. Acta 225, 194 (1971).

[n] J. R. Tata, this volume [24].

[o] A. Trouet, this volume [31].

[p] P. Baudhuin, H. B. Beaufay, Y. Rahman-Li, O. Z. Sellinger, R. Wattiaux, P. Jacques, and C. de Duve, Biochem. J. 92, 179 (1964).

[q] F. Leighton, B. Poole, P. B. Lazarow, and C. de Duve, J. Cell Biol. 41, 521 (1969).

[r] B. Fleischer, S. Fleischer, and H. Ozawa, J. Cell Biol. 43, 59 (1969).

[s] B. Fleischer, this volume [17].

[t] M. A. Swanson, this series, Vol. 2, p. 541.

is associated exclusively with one cell organelle, it can serve as a "marker" for that organelle. The term "marker enzyme" refers to an enzyme that is localized, ideally, in one cell organelle. The detection of this activity in another subcellular fraction is then considered to be due to contamination and the amount of contamination can be estimated. The percent contamination of a protein basis may be estimated by dividing the specific activity of the marker enzyme in the fraction where it is a contaminant by the specific activity of the marker enzyme of the purified fraction where it is localized. Marker enzymes are particularly valuable in cell fractionation to help guide purification and to estimate contamination of purified cell fractions. With the available technology for subcellular fractionation, the use of a reliable marker enzyme enables one to monitor many fractions usually with great sensitivity. Such a measurement is especially valuable for characterizing smooth vesicles where morphological identification becomes practically useless.

In general, membranes from smooth and rough microsomes are similar in composition and enzymatic activity. An enzymatic activity may be found in the smooth but not in the rough microsomes. This would indicate that the activity is referable not to endoplasmic reticulum, but to other smooth membranes, such as the Golgi complex or plasma membranes. When we first observed galactosyltransferase in smooth but not in rough microsomes, this was an indication that this enzyme could be localized in the plasma membrane or Golgi complex, but not in endoplasmic reticulum.[3,9] On the other hand, GDP mannosyltransferase is a rare instance of an enzyme which is specifically associated with the rough microsomes. This enzyme adds mannose to polypeptides and seems to be associated with the protein-synthesizing machinery of rough endoplasmic reticulum.[53,54]

Characterization by Electron Microscopy

Electron micrographs of the purified subcellular fractions are shown in Figs. 4–12. The isolated cell fractions unmistakably resemble their cor-

[53] J. Molnar, M. Tetas, and H. Chao, *Biochem. Biophys. Res. Commun.* 37, 684 (1969).
[54] C. M. Redman and M. G. Cherian, *J. Cell Biol.* 52, 231 (1972).
[55] C. R. Hackenbrock, *J. Cell Biol.* 30, 269 (1966).
[56] D. S. Dow, K. G. Walton, and S. Fleischer, *Bioenergetics* 1, 247 (1970).

[u] L. M. G. van Golde, B. Fleischer, and S. Fleischer, *Biochim. Biophys. Acta* 249, 318 (1971).
[v] H. G. Hers, J. Berthet, L. Berthet, and C. de Duve, *Bull. Soc. Chim. Biol.* 33, 21 (1951).
[w] N. G. Anderson and J. G. Green, *in* "Enzyme Cytology" (D. B. Roodyn, ed.), p. 475. Academic Press, New York, 1967.
[x] O. H. Lowry, this series, Vol. 4, p. 366.

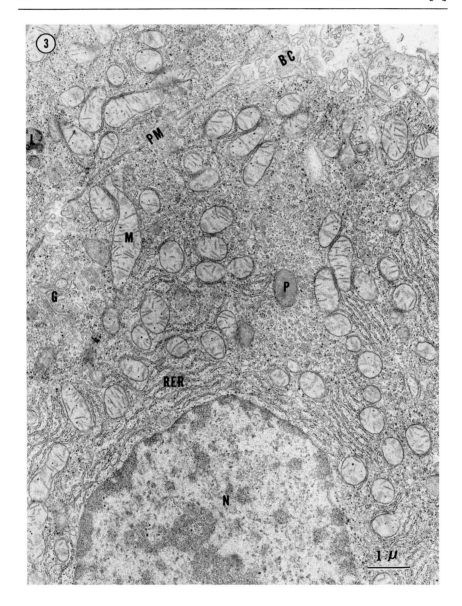

FIG. 3. Electron micrograph of a section of liver illustrating the appearance of cell organelles in the cell: nuclei, N; rough endoplasmic reticulum, RER; mitochondria, M; Golgi complex, G; plasma membrane, PM; peroxisomes, P; and lysosomes, L. The complex invaginations of the bile canaliculus (BC) portion of the plasma membrane can be observed. The length of the bar denotes 1 μm.

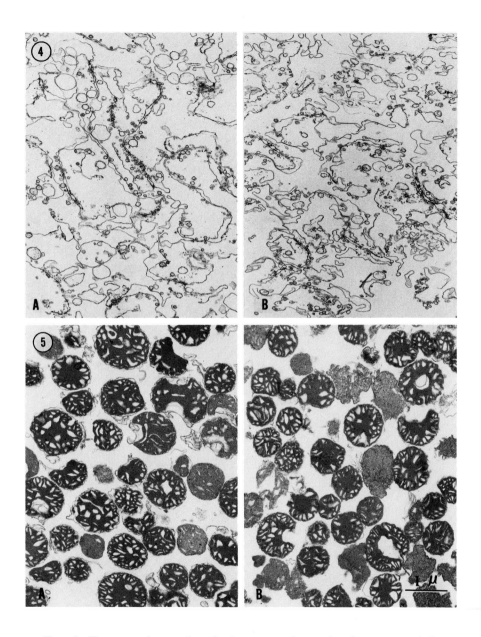

FIG. 4. Electron micrographs of plasma membrane fractions prepared from fresh (Fig. 4A) and frozen tissue homogenate of liver (Fig. 4B). The homogenates from both livers were processed in parallel and thereby treated in the cell fractionation procedure in nearly identical manner. The frozen liver was stored as a tissue homogenate which was quick-frozen and stored in liquid nitrogen for 293 days according to the procedure of Fleischer and Kervina (this volume [1]).

FIG. 5. Electron micrographs of mitochondria (R_1) prepared from fresh (Fig. 5A) and frozen livers (Fig. 5B) (cf. legend Fig. 4). Most of the mitochondria are in the condensed configuration. The length of the bar denotes 1 μm. Figures 4 and 5 are the same magnification.

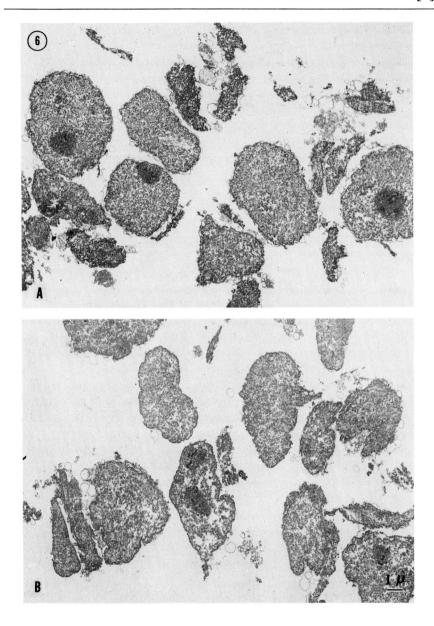

Fig. 6. Electron micrographs of nuclei prepared from fresh (Fig. 6A) and frozen (Fig. 6B) livers (cf. legend of Fig. 4). The length of the bar denotes 1 μm.

responding organelles. The isolated mitochondria appear mainly in the condensed configuration (Fig. 5). They can readily be converted back to the orthodox configuration,[55,56] which is more characteristic of mitochondria in the liver cell. The Golgi complex is sensitive to osmotic shock and handling (Fig. 7), but when properly prepared for electron microscopy (cf. Section V, Preparation of Samples for Electron Microscopy) shows remarkable preservation of structure (Fig. 8). Vesicles of rough microsomes correspond to the rough endoplasmic reticulum (Fig. 9). The smooth microsomes (Fig. 10), while largely referable to smooth endoplasmic reticulum, have a multiple subcellular origin and appear more heterogeneous than the rough microsomal fraction.

Protein Profiles

The protein profiles of the purified cell fractions, analyzed by polyacrylamide gel electrophoresis, are shown in Fig. 13. Each fraction has a distinctive pattern.

Lipid Analysis

The phospholipid content varies appreciably, ranging from 0.1 to 1.0 μmole per milligram of protein (Table II) and being lowest in nuclei and highest in the Golgi complex, in accordance with the membrane content of each organelle. The phospholipid to protein ratios for each cell organelle are provided in Table II and can be used to normalize specific activities (discussed below) on the basis of membrane phospholipid. This allows a more meaningful comparison for membrane-bound enzymes in organelles with large differences in membrane content, such as nuclei and rough microsomes.[51]

Most, but not all, of the phospholipid is referable to membrane. Soluble serum lipoproteins, which are secreted by the liver, have been shown to be present within the Golgi compartment.[57] The contents of the compartment can be removed[58] so that the Golgi membrane per se can be analyzed (cf. Table II).

The cholesterol content of membranes varies significantly in the different fractions. Both plasma membranes[59] and Golgi complex contain high amounts of cholesterol. The cholesterol content of mitochondria and rough microsomes is quite low, by comparison (Table II).

[57] R. W. Mahley, R. L. Hamilton, and V. S. LeQuire, J. Lipid Res. 10, 433 (1969).
[58] Subcellular fractions as isolated are frequently highly structured and contain nonmembranous components within their compartment. For analysis of membranes, such contents can be released from within the compartment, cf. G. Kreibich and D. D. Sabatini, this volume [20].
[59] E. L. Benedetti and P. Emelot, in "The Membranes" (A. J. Dalton and F. Haguenau, eds.), Vol. 4; p. 33. Academic Press, New York, 1968.

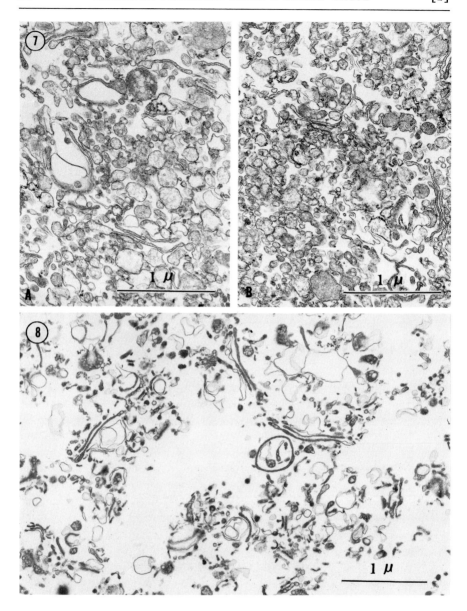

FIG. 7. Electron micrographs of Golgi complex fractions from fresh (Fig. 7A) and frozen (Fig. 7B) livers (cf. legend of Fig. 4) after osmotic shock. The samples after isolation from the gradient had good morphology (cf. Fig. 8) but were damaged by osmotic shock after dilution with distilled water and sedimentation, prior to fixation for electron microscopy (cf. Fig. 8). The length of the bar denotes 1 μm.

TABLE II
Lipid Content of Liver Cell Fractions[a]

Fraction	Phospholipid protein (μmoles/mg)	Cholesterol protein (μmoles/mg)	Cholesterol phospholipid (molar ratio)
Plasma membrane	0.825	0.285	0.345
Nuclei	0.097	—	—
Golgi complex	0.745	0.202	0.271
Golgi membranes[b]	1.003	0.267	0.266
Mitochondria	0.268	0.007	0.026
Rough microsomes[c]	0.481	0.025	0.052

[a] Studies of S. Fleischer, B. Fleischer, and F. Zambrano.

[b] Golgi membranes were prepared by disrupting the fraction to remove the soluble contents, using a Parr bomb as described by B. Fleischer (this volume [17]).

[c] One third of the protein of rough microsomes is ribosomal protein. The phospholipid to protein ratio for rough microsomal membrane (i.e., minus the ribosomes) would be 0.72 (0.481 × 1.5 = 0.722).

Enzymatic Activities

Some characteristic properties of the purified subcellular fractions from rat liver are summarized in Table III. Typical values of diagnostic enzymes are presented from which contamination of the purified fractions can be estimated. We will estimate the contamination of the Golgi complex with mitochondria, endoplasmic reticulum (microsomes), and plasma membrane to illustrate the use of marker enzymes for this purpose, as well as several pertinent considerations. The specific activity of galactosidase (diagnostic for plasma membrane)[60] for the Golgi complex fraction and for purified plasma membranes is 20 and 300 nmoles of galactose

[60] Galactosidase activity (measured at pH 8.6)[9] is used as a diagnostic enzyme for plasma membrane.[31] Some of this activity (with a more acid pH optimum) is also found in lysosomes. Contamination of the Golgi fraction with lysosomes was found to be quite small (<3%) using aryl sulfatase to estimate lysosomal contamination so that the galactosidase activity would then estimate plasma membrane contamination, if present. Thus, a diagnostic enzyme can still be useful even if not uniquely localized, providing the second organelle in which it is found is absent, or if the specific activity in the second organelle is very low comparatively.

Fig. 8. Electron microscopy of Golgi complex. The Golgi fraction from the gradient (cf. Fig. 2C) was prepared from fresh liver. The interface material was fixed directly with glutaraldehyde after removal from the upper interface (29.0%/33.0%), care being taken to minimize osmotic shock (cf. Section V, Fixation of Samples for Electron Microscopy). The length of the bar denotes 1 μm.

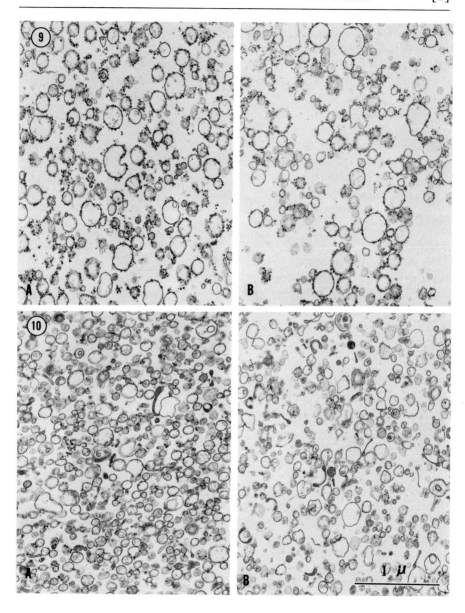

FIG. 9. Electron micrographs of rough microsomes from fresh (Fig. 9A) and frozen (Fig. 9B) livers (cf. legend of Fig. 4).

FIG. 10. Electron micrographs of smooth microsomes from fresh (Fig. 10A) and frozen (Fig. 10B) livers (cf. legend of Fig. 4). The smooth microsomes have a more heterogeneous appearance reflecting their multiple origin. The length of the bar denotes 1 μm. Figures 9 and 10 are the same magnification.

released per milligram per hour at 37°, respectively (data not given in Table II). The estimate of contamination of the Golgi complex fraction by plasma membrane is 7% ($20/300 \times 100 = 6.7\%$). In a similar manner, succinate–cytochrome c reductase ($0.005/0.443 \times 100 = 1.1\%$) and glucose-6-phosphatase ($0.025/0.244 \times 100 = 10.3\%$) can be used to estimate contamination of the Golgi fraction with mitochondria and endoplasmic reticulum, respectively. In the case of endoplasmic reticulum, there is a minor complication. Both rough and smooth microsomes are derived from this organelle; however, only the rough microsomes are generally obtained with high purity. The specific activity of glucose-6-phosphatase of rough microsomes is therefore used in this calculation. It is based on total protein which includes membrane proteins (two-thirds) and ribosomal proteins (one-third) so that the specific activity of glucose-6-phosphatase per milligram of membrane protein would be 0.366 ($1.50 \times 0.244 = 0.366$). The use of the higher specific activity value ($0.025/0.366 = 6.8\%$) assumes contamination with smooth (ER) membranes; this is reasonable since rough microsomes are rarely seen in the Golgi preparation.

Thus, the major contaminants of the Golgi fraction, mitochondria, endoplasmic reticulum, and plasma membrane, add up to less than 20% contamination. This is about as good a Golgi fraction as the current state of the art allows. The specific activity of galactosyltransferase activity of the Golgi fraction made by the procedure described is higher than has been reported previously.[9]

Mitochondria (R_1) which are prepared from the first low speed (960 g at R_{av} for 10 minutes) centrifugation of the homogenate are of high purity. They are practically free of lysosomes, peroxisomes, and microsomes, the latter estimated from the glucose-6-phosphatase activity (Table III). The rotenone-insensitive DPNH–cytochrome c reductase activity in these mitochondria is largely localized in the outer mitochondrial membrane.[61] This enzymatic activity can now be used to estimate microsomal contamination in mitochondria (R_2) by subtracting the rate which is referable to the outer membrane. By comparison, mitochondria (R_2) which are prepared from the supernatant of the first centrifugation are contaminated with endoplasmic reticulum as indicated by the higher rotenone-insensitive DPNH–cytochrome c reductase activity. This is also reflected by the greater phosphorus to protein ratio as compared to mitochondria (R_1) (12.9 vs. 10.4).

A comparison of specific activity of the marker enzymes in the purified fraction and the homogenate permits an estimation of the content of or-

[61] G. L. Sottocasa, B. Kuylenstierna, L. Ernster, and A. Bergstrand, this series, Vol. 10, p. 448.

TABLE III

Characterization of Purified Subcellular Fractions from the Fractionation of Rat Liver[a]

Fraction	Protein conc. (mg/ml)	Total protein	Bound P (µg/ mg) protein	Enzymatic activity						Preparation time (hours)
				RI DPNH-Cyt. c	Glucose-6-P'ase	Succinate Cyt. c	ATPase	Galact. transf.	5'AMP-ase	
Homogenate	25.9	2430	18.2	0.317	(0.096)	0.132	0.177	6	0.036	0.2
Mitochondria (R$_2$)	12.7	70	12.9	0.450	—	0.327	0.306	—	—	1.5
Golgi	0.45	0.97	26.8	0.119	(0.025)	0.005	—	764		6.5
Supernatant	9.8	578	14.9	0.004		0.002	—			2.0
Mitochondria (R$_1$)	13.5	93[b]	10.4	0.185[f]	(0.003)	0.433	0.420			6.0
Plasma membrane	1.00	5.3	23.1	0.139	(0.040)	0.020	1.18	(5)	0.99	7.0
Microsomes (heavy)	9.1	64.4	36.3	1.39		0.008				6.0
Microsomes (light)	9.1	56.2	32.4	0.79		0.004				2.0
Rough microsomes	4.2	37.5[c]	37.6	1.04	(0.244)[e]	0.006		(3)	0.050	10
Smooth microsomes	2.5	10.1[c]	29.8	1.64	—	0.010		(52)		11
Nuclei[d]	8.4	32.5	31.1	0.075	(0.036)	0.002		(2)		6.0

a Most of the values in the table are averages of data from 4–6 experiments, with subcellular fractions prepared from both fresh and frozen livers (21 g) processed in parallel and thereby treated in nearly identical manner. The enzymatic activities are expressed as micromoles per minute per milligram of protein for all assays except galactosyltransferase (galact. trans) and galactosidase, where nanomoles per hour per milligram of protein is used. Rotenone insensitive (RI) DPNH–cytochrome c reductase, succinate–cytochrome c reductase and ATPase were assayed at 30°C; galactosyltransferase, galactosidase, urate oxidase, glucose-6-phosphatase, 5′-nucleotidase (5′AMPase) and aryl sulfatase were assayed at 37°.

b The recovery of mitochondria (R1) from frozen liver is significantly lower than from fresh liver. The value in the table is an average. The yield of mitochondria (R1) as milligrams of protein is 117 ± 25 for fresh tissue and 60 ± 19 for frozen tissue, respectively (3 samples each).

c The yield is corrected to the total homogenate; i.e., if ⅔ of heavy and light microsomes are used in [G] to prepare the rough and smooth microsomes, the yield has been multiplied by 3/2.

d It is our general impression that nuclei prepared by the three-step gradient (the modified procedure) exhibits better morphology and is recommended when the emphasis is on the study of nuclei. The values in the table for the nuclei (except those enclosed in parentheses) were prepared using a two-step gradient.

e The specific activity of glucose-6-phosphatase for rough microsomes is expressed per milligram of protein, one-third of which is ribosomal protein. The specific activity corrected to membrane protein (minus ribosomes) is 0.366.

f The rotenone-insensitive DPNH–cytochrome c reductase of this highly purified preparation of mitochondria is largely localized in the mitochondrial outer membrane (G. L. Sottocasa, B. Kuylenstierna, L. Ernster, and A. Bergstrand, this series, Vol. 10 [72c], p. 448).

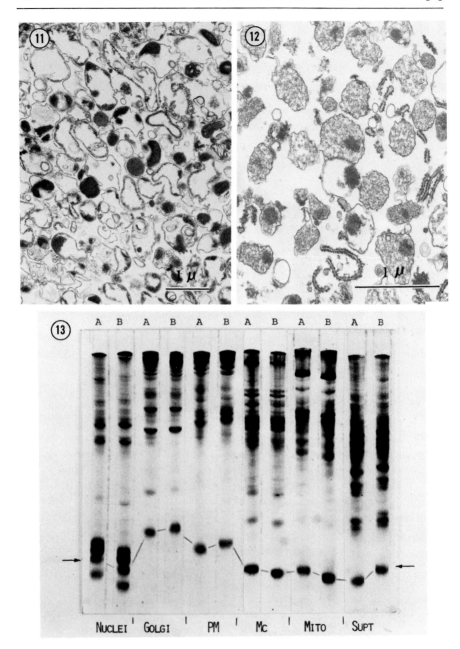

FIG. 11. Electron micrographs of lysosomes prepared by injecting rats with Triton. The Triton is selectively taken up by the lysosomes, decreasing their density

ganelles in the cell. Quantitative morphometry can be used to confirm such results.[62,63] For the Golgi complex this comes to about 1% ($6/764 \times 100 = 0.9\%$); that is to say, if the specific activity in the purified cell fraction can be enriched more than 100-fold over the homogenate, it accounts for less than one-hundredth of the cell's protein content.

The specific activity of glucose-6-phosphatase in nuclei (0.036) and rough microsomes (0.244) can be compared based on the amount of membrane phospholipid present in these two (Table II). The phospholipid/protein ratios for rough microsomal membranes (minus ribosomes) and nuclei are 0.722 and 0.097, respectively. Normalized to the same amount of phospholipid, the specific activity of glucose-6-phosphatase of the nuclei would be 0.27 ($0.722/0.097 \times 0.036 = 0.269$), i.e., about half the specific activity of smooth (endoplasmic reticulum) membranes.

Because of the small membrane content of nuclei, these calculations must be interpreted with some caution. A small amount of microsomal contamination might easily account for the observed activity. An independent method, such as the use of EM cytochemistry, is in order and, in fact, confirms that glucose-6-phosphatase is associated with liver nuclei.[64]

A number of assumptions are implicit in the calculations just illustrated. The key assumption is that the marker enzyme is localized exclusively in one organelle. This is rarely the case. Recent EM cytochemical studies have shown that 5'-nucleotidase, acid phosphatase, and even glucose-6-phosphatase, "marker" enzymes for plasma membranes, lysosomes, and endoplasmic reticulum, respectively, are present, albeit in small

[62] E. R. Weibel, W. Stäubli, H. R. Gnägi, and F. A. Hess, *J. Cell Biol.* **42**, 68 (1969).
[63] P. Baudhuin, this series, Vol. 32 [1], in press.
[64] A. Leskes, P. Siekevitz, and G. E. Palade, *J. Cell Biol.* **49**, 264 (1971).

and simplifying their separation from mitochondria and peroxisomes in a sucrose gradient [F. Leighton, B. Poole, H. Beaufay, P. Boudhuin, J. W. Coffey, S. Fowler, and C. de Duve, *J. Cell Biol.* **37**, 482 (1968)] (cf. Section VIII, Addendum). The bar denotes 1 μm.

FIG. 12. Electron micrograph of peroxisomes prepared on a sucrose gradient from rats which were injected with Triton (cf. legend of Fig. 11). The bar denotes 1 μm.

FIG. 13. Protein patterns of purified cell fractions. Polyacrylamide gel electrophoresis was carried out using dissociating conditions in phenol–urea–acetic acid (W. L. Zahler, this series, Vol. 32 [7], in press). Each subcellular fraction has a distinctive pattern. Ribonuclease (arrow) was added as an internal standard. The abbreviations are PM, plasma membrane; Mc, microsomes; Mito, mitochondria (R_2); Supt, supernatant. A and B refer to fractions prepared from fresh and frozen livers (293 days), respectively (cf. legend Fig. 4).

amounts, in more than one organelle.[65,66] The "marker" enzyme can still be used but must be used with some caution and understanding. It must also be remembered that the specific activity of the marker enzyme will be in error proportional to contamination. Another implicit assumption is that the specific activity of a marker enzyme is invariant in individual animals, or to perturbing influences on the animal. This is not always so. Starvation causes elevation of glucose-6-phosphatase.[67] There is some variation in the specific activity of enzymes in purified cell fractions. This problem has not been investigated in detail and requires careful study. In the light of possible animal variation, the specific activity of the marker and contaminant fractions is best measured on cell fractions isolated from the same homogenate. The cell fractionation procedure which is described makes this feasible. When the diagnostic enzyme is present in more than one location, the enrichment of a specific activity as compared with the homogenate cannot be applied to estimate the amount of protein referable to a particular organelle in the cell. This is the case for ATPase activity which is present in significant amounts in both mitochondria and plasma membranes. Another assumption is that the enzymatic activity in question can be assayed in the homogenate, which contains many hydrolases and enzymes capable of carrying out many reactions. Misleading conclusions may be drawn if caution is not exercised. In measuring enzymatic activity, it is important to use conditions to measure maximal rate. Some enzymes have "latent activity," which is to say, the activity cannot be assayed because the substrate is inaccessible to the enzyme. The compartment may have to be disrupted to make the substrate accessible to the enzyme. Freezing and thawing is sufficient for lysosomes.[68] Pretreatment with detergents is sometimes useful but may modify the properties of the enzyme. Additional complexities arise as specific systems are studied. Succinate–cytochrome c reductase cannot readily be measured in submitochondrial vesicles because the cytochrome c cannot accept electrons on the matrix face of the membrane, i.e., the face containing the inner membrane particles.[69]

It may seem remarkable considering the potential hazards that many conclusions can be asserted in enzyme localization studies using purified

[65] M. G. Farquhar, J. J. M. Bergeron, and G. E. Palade, *J. Cell Biol.* **60**, 8 (1974).
[66] C. C. Widnell, *J. Cell Biol.* **52**, 542 (1972).
[67] W. J. Arion and R. C. Nordlie, *Biochem. Biophys. Res. Commun.* **20**, 606 (1965).
[68] "Lysosomes, A Laboratory Handbook" (J. T. Dingle, ed.). North-Holland Publ., Amsterdam, 1972.
[69] S. Fleischer, G. Meissner, M. Smigel, and R. Wood, this volume [27].

cell fractions. The investigator has an additional edge in that he brings to bear multiple approaches including morphology, cytochemistry, chemical analysis, the use of different marker and diagnostic enzymes as well as to couple them with the exercise of good judgement.

VII. Fractionation of Liver Homogenate after Prolonged Storage

In article [1] of this volume, we described a procedure for the prolonged storage of liver by preparing a homogenate, which is quick-frozen and stored in a liquid nitrogen refrigerator. Such frozen samples were quick-thawed[13] and then fractionated by the procedure described in this article. The homogenates of frozen and fresh livers were processed in parallel and thereby treated in nearly identical fashion. The homogenate from stored and fresh liver behaved similarly in the fractionation procedure. Purified cell fractions including plasma membrane, mitochondria, nuclei, Golgi complex, rough and smooth microsomes, and the supernatant could be prepared from the stored liver homogenate. The fractions were similar in quality as judged by electron microscopy (Figs. 4–7, 8–10), protein pattern, and as studied by polyacrylamide gel electrophoresis (Fig. 13), chemical composition (phosphorus to protein ratios) and enzymatic criteria (Table III). The electron micrographs are comparisons of cell fractions prepared from fresh liver and liver which was stored for 293 days. The quality of the fractions from fresh and frozen tissue were practically indistinguishable by the criteria used. Furthermore, mitochondria (R_2) from both the fresh and frozen (293 days) liver had optimal P/O ratios of 3 and 2 using D-β-hydroxybutyrate and succinate as substrates. Respiratory control is also preserved by the storage procedure; respiratory control ratios were similar, $\sim$5.5, for mitochondria for both fresh or stored liver.[70]

There were two noticeable differences between fresh and frozen liver in the fractionation procedure. The yield of mitochondria (R_1) was lower by almost a factor of two (cf. legend, Table III). The lysosomes seem to be broken down in large part after quick-freezing and thawing of the homogenate. This was indicated by the appearance of lysosomal activity (aryl sulfatase) in the supernatant and decreased lysosomal specific activity (60–80% less) in the mitochondrial (R_2) fraction, which contains the lysosomes and peroxisomes. The specific activity of the marker enzyme (urate oxidase) for peroxisomes was similar in the mitochondrial (R_2) fraction from fresh and frozen liver. Urate oxidase activity was not detected in the supernatant fraction.[70]

---

[70] S. Fleischer and M. Kervina, in preparation.

VIII. Addendum

Step [D₂]. Use of R₂ for Purification of (4b) Mitochondria,
(5) Lysosomes, and (6) Peroxisomes

Reagents

Triton WR 1339 (Ruger Chemical Co., Irvington, New Jersey)
200 mg/ml in 0.9% NaCl
0.25 M Sucrose–0.1% ethanol, pH 7
0.25 M Sucrose–10 mM HEPES–1 mM EDTA–0.1% ethanol,
pH 7.5
35.0% Sucrose–5% Dextran 10–0.1% EtOH, pH 7 [Dextran-10
(Pharmacia, Uppsala, Sweden)]
57.5% Sucrose–5% Dextran 10–0.1% EtOH, pH 7
65.1% Sucrose–0.1% ethanol

Procedure. These studies were carried out with rats that were previously injected with Triton WR 1339.[10] For each preparation, two male
rats (200–250 g, Sprague-Dawley derived, Holtzman Co., Madison, Wisconsin) were injected intraperitoneally with the nonionic detergent Triton
WR 1339. The amount of Triton used for injection was 75 mg/100 g
weight of rat in a solution of 200 mg/ml Triton in 0.9% NaCl. The rats
were starved overnight (16 hours) prior to sacrifice and killed 3.5 days
after injection of the Triton WR 1339.[71]

A residue 2 fraction is prepared as previously described (cf. [C]). The
amount of contaminating microsomes may be reduced by resuspending the
pellet with buffered sucrose–EDTA containing 0.1% ethanol,[72] sedimenting at 18,000 rpm for 10 minutes in a Spinco type 42.1 rotor and discarding the upper light layer (cf. [D₁]). The lower brown pellet is resuspended
in a final volume of 4 ml of a solution of 0.25 M sucrose containing 0.1%
ethanol and used for separation into tritosomes, mitochondria, and peroxisomes using a continuous sucrose density gradient. The gradient was modified after the procedure of Leighton *et al.*[10] On the bottom of a centrifuge
tube for an SW 50 rotor are added 4 drops of 65.1% sucrose. A linear
gradient (4.0 ml), ranging from 35% to 57.5% sucrose (density 1.15–

---

[71] Lysosomes have a broad density distribution and therefore cannot readily be
separated from mitochondria and peroxisomes by isopycnic gradient centrifugation.
The Triton, which is injected, is taken up selectively and accumulates inside the
lysosomes, decreasing their density. Such lysosomes are loaded with Triton ("tritosomes") and are then readily separable. R. Wattiaux, M. Wibo, and P. Baudhuin,
Lysosomes, Ciba Found. Symp. 1963, p. 176 (1963).

[72] The 0.1% EtOH is used to prevent inactivation of catalase. B. Chance, *Biochem. J.*
46, 387 (1950).

1.27), is then formed in the tube using a gradient former (350052, Beckman Instruments, Palo Alto, California). The sucrose solutions used to form the gradients are fortified with 5% (w/w) Dextran-10[71,73] and 0.1% ethanol. One milliliter of the purified mitochondrial fraction is carefully layered on the top of each tube, and the tubes are centrifuged at 49,000 rpm (199,000 g at R_{av}) in a Spinco type SW 50 rotor for 90 minutes. Three defined bands are obtained. An upper light band at the interface contained the tritosomes which were collected using a syringe. The lower greenish band, the peroxisomes, was collected dropwise through a hole made at the bottom of the tube. The middle brown band is the mitochondria.[38,74] Electron micrographs of the "tritosomes" and peroxisome fractions are shown in Figs. 11 and 12, respectively.

[73] The use of Dextran 10, recommended by Leighton *et al.*,[10] seems to improve the resolution of separation and the preservation of the peroxisomes.

[74] Sucrose gradients can be prepared using D_2O instead of H_2O so that the amount of sucrose required to obtain the same density is less and so is the viscosity of the gradient. In preliminary experiments, D_2O sucrose gradients of the same density range were prepared for the separation of tritosomes, lysosomes, and peroxisomes. The separation time could be decreased to 30 minutes under the same sedimentation conditions, 48,000 rpm in a SW 50 rotor. Again, three bands are obtained; the tritosomes are at the upper interface, the mitochondria form a band in the center and the peroxisomes at the lower portion of the tube. Lysosomes can be prepared using continuous preparative free flow electrophoresis.[75] These "normal" lysosomes have a different lipid composition from those prepared by loading with Triton. (R. Henning and H. G. Heidrich, personal communication.)

[75] K. Hannig and H. G. Heidrich, this volume [75].

[3] Subcellular Fractionation of the Pancreas

By A. M. TARTAKOFF and J. D. JAMIESON

Over the past few years the exocrine pancreatic cell has provided a model system in which to study the synthesis, intracellular transport, concentration, and discharge of exportable proteins. The route and timetable of this secretory process is now reasonably well defined, and the essential cellular components involved have been identified. They include the cisternae of the rough endoplasmic reticulum (RER), the Golgi apparatus, including its condensing vacuoles, and the zymogen granules. These structures are illustrated in Fig. 1.

The isolation, identification, and characterization of subcellular fractions is basic to an understanding of the secretory process. Early cell frac-

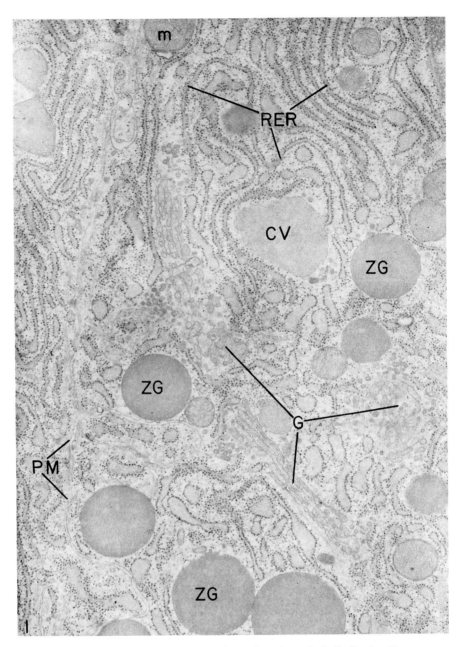

Fig. 1. Electron micrograph of a thin section through OsO₄-fixed calf pancreas. The principal organelles of the exocrine cell are labeled: rough endoplasmic reticulum (RER); smooth membranous elements of the Golgi complex (G); condensing vacuole (CV); zymogen granules (ZG); mitochondria (m); lateral plasmalemma (PM). ×16,000.

42

tionation schemes[1-3] applied to the pancreases of several species produced reasonably pure preparations of zymogen granules and total microsomal fractions, the latter representing primarily elements of the RER. Absent or unidentified in these fractionation schemes, which relied on differential centrifugations with interposed washings of the fractions, were the elements of the Golgi complex, including condensing vacuoles. It is now possible to isolate a smooth microsomal fraction that includes vesicles and cisternae of the Golgi complex and to identify a population of condensing vacuoles that cosediment with the zymogen granules in the zymogen granule fraction.

In the first section of this article, we describe in detail fractionation schemes applied to the guinea pig pancreas. The second section describes comparable procedures for fractionation of the bovine pancreas. While strictly applicable to the pancreas, these schemes should be adaptable, with modification, to other exocrine glands of comparable structure (e.g., the parotid).

Fractionation of the Guinea Pig Pancreas

Isolation of Guinea Pig Pancreas Subcellular Fractions: Rough Microsomes, Smooth Microsomes, Zymogen Granule Fractions[4,5]

The aim of this fractionation scheme is to isolate, at the expense of yield, relatively pure subcellular fractions: (1) rough microsomes, corresponding to elements of the rough endoplasmic reticulum; (2) smooth microsomes, corresponding to elements of the Golgi complex; and (3) a total zymogen granule fraction containing identifiable condensing vacuoles.

Procedure (Fig. 2). Fresh guinea pig pancreas is removed from animals starved overnight with access to water, placed in cold (4°) 0.3 M sucrose, and trimmed free of fat and mesentery. Subsequently all operations are conducted at 4°. Samples (0.5 g) are weighed, minced with scissors, and homogenized in 4.5 ml (1:10 tissue to sucrose, w/v) of 0.3 M sucrose in a motor-driven (3000 rpm) Brendler-type homogenizing apparatus (A. H. Thomas, size A). Three up-and-down strokes suffice. The homogenate is initially subjected to a 600 g, 10-minute centrifugation in 12-ml conical tubes run in a swinging-bucket rotor of the International Centrifuge (size 2, Model K, Yoke No. 240), to eliminate nuclei, unbroken cells, and heavy debris. Further centrifugations of the 600 g supernatant are carried out in

[1] L. E. Hokin, *Biochim. Biophys. Acta* **18**, 379 (1955).
[2] L. J. Greene, C. H. W. Hirs, and G. E. Palade, *J. Biol. Chem.* **238**, 2054 (1963).
[3] P. Siekevitz and G. E. Palade, *J. Cell Biol.* **2**, 671 (1956).
[4] J. D. Jamieson and G. E. Palade, *J. Cell Biol.* **34**, 577 (1967).
[5] J. D. Jamieson and G. E. Palade, *J. Cell Biol.* **34**, 597 (1967).

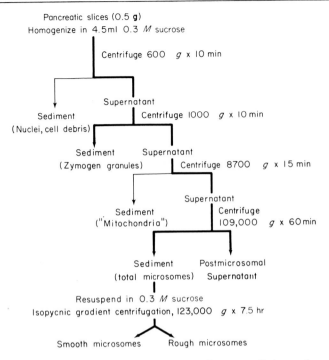

Pancreatic slices (0.5 g)
Homogenize in 4.5ml 0.3 *M* sucrose

Centrifuge 600 *g* x 10 min

Sediment
(Nuclei, cell debris)

Supernatant
Centrifuge 1000 *g* x 10 min

Sediment
(Zymogen granules)

Supernatant
Centrifuge 8700 *g* x 15 min

Sediment
("Mitochondria")

Supernatant
Centrifuge
109,000 *g* x 60 min

Sediment
(total microsomes)

Postmicrosomal
Supernatant

Resuspend in 0.3 *M* sucrose
Isopycnic gradient centrifugation, 123,000 *g* x 7.5 hr

Smooth microsomes   Rough microsomes

Fig. 2. Flow sheet for cell fractionation scheme applied to the guinea pig pancreas. Centrifugal forces (*g*) are average values at the middle of the tube.

a Spinco No. 40.3 angle-head rotor or the appropriate equivalent. Recovery of the supernatant fraction from successive centrifugations is obtained by decantation. The surface of the 1000 *g*, 10-minute sediment (zymogen granule fraction) is rinsed three times with 1.5 ml of 0.3 *M* sucrose by swirling the tube. This rinse removes most of the overlying tan layer of mitochondria. Higher yields of granules may be obtained by resuspending the 600 *g* sediment in 5.0 ml of 0.3 *M* sucrose, repeating the 600 *g* spin, and centrifuging the resulting supernatant again at 1000 *g*. The surface of this second zymogen granule fraction is similarly rinsed with 1.5 ml of 0.3 *M* sucrose. Attempts to wash the granule pellets by resuspension and recentrifugation are unsuccessful since granules tend to aggregate extensively.

While over 75% of the protein in the zymogen granule fraction is secretory protein contained within zymogen granules and condensing vacuoles, for detailed studies on zymogen granule membranes it is important to reduce the residual contamination by microsomes and mitochondria. Accordingly, a modified granule isolation procedure has been developed, which, from the outset, eliminates the bulk of the microsomes and mito-chondria.[6] The procedure is as follows: 5 ml of the original homogenate in 0.3 M sucrose is centrifuged in 12-ml conical tubes for 12 minutes at 1000 g in the swinging-bucket rotor of an International centrifuge (Type 2, Model K, Yoke No. 240). The resulting supernatant (which contains the bulk of microsomes and mitochondria) is discarded. The pellets are resuspended in 5.0 ml of 0.3 M sucrose and centrifuged at 180 g for 12 minutes to resediment debris. This 180 g supernatant is next removed with a disposable pipette and filtered through 110-mesh nylon cloth into 15-ml round-bottomed Corex tubes. Centrifugation is carried out for 3.5 minutes at 1000 g in the same swinging-bucket rotor. The final supernatant is aspirated, and the surface of the white granule pellet is rinsed by swirling with 0.3 M sucrose as above. The process may be repeated by resuspending the 180 g pellet in 0.3 M sucrose and repeating the procedure in order to increase the final yield.

Returning to the standard fractionation procedure, the remaining mito-chondria and zymogen granules in the 1000 g supernatant are pelleted at 10,000 g (cf. Fig. 2), and the total microsomal fraction is subsequently harvested from the 10,000 g supernatant.

The total microsomal fraction is subfractionated into rough and smooth microsomes as follows: After gentle resuspension by hand with a Teflon pestle in 0.25 ml of 0.3 M sucrose, the suspension is layered on top of a linear sucrose density gradient (1.04 to 2.0 M sucrose) formed in the tube of a Spinco SW-39 rotor. Centrifugation at 123,000 g for 7.5 hours results in the formation of three bands in the tube (Fig. 3). Each is collected with a J-shaped needle attached to a syringe, diluted to 0.3 M sucrose, and repelleted at 105,000 g for 90 minutes in a Spinco 40.3 rotor. The pellet from band 1 consists of smooth microsomes (Golgi-derived elements pri-marily); the pellet from band 2 is mixed rough and smooth microsomes with some free ribosomes; and the pellet from band 3 consists of rough microsomes.[7]

The high speed soluble fraction resulting from sedimentation of the total microsomal fraction (the postmicrosomal supernatant) contains the bulk of the nonmembrane-bound proteins of the cytosol (primarily from

---

[6] J. Meldolesi, J. D. Jamieson, and G. E. Palade, *J. Cell Biol.* **49**, 109 (1971).

[7] Alternatively, the total microsomal fraction can be subfractionated on a discon-tinuous "sandwich gradient" similar to the one used for bovine microsomes. The gradient is designed for the Spinco SW 56 rotor. From bottom to top it contains 0.5 ml of 2 M sucrose, 1.25 ml of 1.35 M sucrose, a 0.5 ml load of total micro-somes, 0.5 ml of 1.20 M sucrose, and 1.25 ml of 0.3 M sucrose. It is spun 100 minutes at 50,000 rpm. The smooth microsomes rise to the 0.3/1.2 M interface. The rough microsomes collect at the 1.35/2.0 M interface.

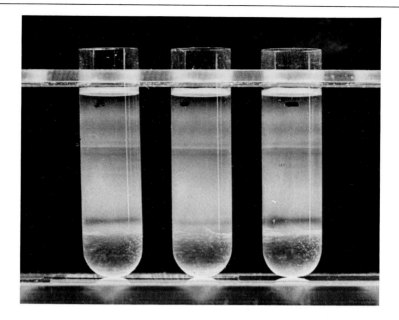

Fig. 3. Photograph of density gradient tubes at the end of 7.5 hours of centrifugation (SW 39L swinging-bucket rotor; 38,000 rpm) showing position of the three major bands. The accessory layer beneath band 3 formed only when the animal had been starved and fed 1 hour before death.

exocrine cells) and, in addition, represents up to 15% of the total content of secretory proteins of the gland. The latter has two main origins: elements of the rough ER, Golgi complex, and zymogen granules ruptured during homogenization, and the soluble content of the duct system of the gland. Each contributes approximately equally to the content of secretory proteins in the postmicrosomal supernatant although the contribution from ruptured cell particulates can be substantially increased (up to ~85% of the gland content of secretory proteins), by prolonging or increasing the vigor of homogenization.

A plasma membrane fraction which contains basal, lateral, and apical plasmalemma can be isolated from the initial low speed sediment.[6]

Chemistry and Identification of the Fractions

Morphology of Fractions

The structural details of the guinea pig pancreas subcellular fractions have been previously published and are similar to fractions obtained from bovine pancreas, as illustrated in Figs. 7–9. Identification of the smooth microsomal fraction as being derived from elements of the Golgi complex, in lieu of biochemical markers, rests on two lines of evidence. First, the main source of smooth intracellular membranous elements within the pancreatic exocrine cell is Golgi vesicles and cisternae with little contribution from other smooth membranes, as in the case of the liver (i.e., the smooth ER tubules); and the second, in *in vitro* experiments on pulse-labeled pancreatic slices,[4] autoradiographic data show that the peak of labeling of Golgi elements during chase intervals corresponds to the peak of labeling of proteins in the content of isolated smooth microsomes. Identification of the cellular source of rough microsomes is unequivocal, based on their morphological characteristics. Finally, electron microscopical autoradiography of thin sections of total zymogen granule fractions isolated from pancreatic slices containing labeled condensing vacuoles enables us to identify in the pellet a small population ($\sim 5\%$) of condensing vacuoles.

Chemistry of Isolated Fractions

The gross chemistry of guinea pig pancreas subcellular fractions is given in Table I. The significance of these data is discussed later in connection with the chemistry of comparable fractions isolated from the bovine gland.

Artifactual Degradation of Subcellular Membrane Components

As mentioned, the postmicrosomal supernatant contains a significant proportion of the gland's secretory proteins and several of these hydrolases are potentially capable of degrading the main membrane components of subcellular fractions (e.g., membrane lipids, proteins, and enzymes). Although the proteases are present as inactive zymogens and do not appear to be activated during fractionation, it has recently been found by Meldolesi et al.[8] that lipase, which is active as synthesized and capable of significant

TABLE I
CHEMISTRY OF GUINEA PIG PANCREAS SUBCELLULAR FRACTIONS

Fraction	mg RNA/mg protein	μg Phospholipid phosphorus/mg protein
Zymogen granule	0.01–0.03	—
Total microsomal	0.28–0.31	4.3
Smooth microsomal (band 1)	0.05–0.09	6.0–6.7
Rough microsomal (band 3)	0.39–0.44	3.6–5.2

[8] J. Meldolesi, J. D. Jamieson, and G. E. Palade, *J. Cell Biol.* 49, 130 (1971).

catalytic activity at 4°, causes extensive degradation of membrane lipids. The data show that membrane phospholipids, particularly lecithin, are converted to lysolecithin by the phospholipase-like activity of lipase in the homogenate (i.e., in the released soluble fraction) and that membranes so altered are capable of scavenging free fatty acids from digested endogenous triglyceride stores. This artifact, in addition to altering the membrane lipid composition, decreases the buoyant density of isolated membrane fractions compared to those from tissues lacking lipase activity. In addition, a number of loosely bound membrane enzymes are either solubilized or inactivated following aging of subcellular fractions, again probably secondary to alteration of membranes by lipase.[9]

Release of Adsorbed Proteins and Content of Pancreatic Subcellular Fractions

Early studies by Hokin[1] had shown that dog zymogen granule fractions could be solubilized by exposure to mild alkaline pH (>7.0). This observation was exploited by Greene et al.[2] in order to obtain the content of bovine zymogen granules and by Jamieson and Palade[4] and Meldolesi et al.[6] to obtain the content of rough and smooth microsomes and zymogen granules from guinea pig pancreas. In the latter cases the fractions were incubated for 45 minutes at 4° with 175 mM NaHCO$_3$, pH 7.8–8.4, with or without 24 mM NaCl, and centrifuged for 60 minutes at 105,000 g to separate soluble released secretory proteins from the pelletable membrane components. Under these conditions, 95% of several representative secretory proteins are soluble, while ~5% are associated with the pellet. The majority of the membrane phospholipids (up to 98%) are recovered in this pellet, although ~50% of the RNA of rough microsomes is solubilized.

Recently Meldolesi and Cova[10] have improved on this method for the release of secretory protein by introducing a treatment with puromycin at high ionic strength. This treatment, originally used by Adelman et al.[11] to release ribosomes from liver microsomal membranes, was found by Meldolesi and Cova to release from guinea pig pancreas subcellular fractions up to 80% of representative secretory proteins and ~70% of the attached ribosomes. Content release was completed by treatment with 0.2 M NaHCO$_3$, pH 7.8. Under these conditions $>98\%$ of the secretory proteins are released, as indicated by recovery of amylase activity. In this way membrane fractions virtually free of residual content and stripped of ribosomes can be isolated for further study.

[9] J. Meldolesi, J. D. Jamieson, and G. E. Palade, *J. Cell Biol.* **49**, 150 (1971).
[10] J. Meldolesi and D. Cova, *Biochem. Biophys. Res. Commun.* **44**, 139 (1971).
[11] M. R. Adelman, G. Blobel, and D. D. Sabatini, *J. Cell Biol.* **47**, 3a (Abst.) (1970).

Fractionation of the Bovine Pancreas

Isolation of Zymogen Granule and Microsomal Fractions

The goal and principle of this subcellular fractionation procedure resemble that used for the guinea pig pancreas.

The general protocol is outlined in Fig. 4. The whole homogenate is centrifuged three times at low speed to eliminate debris, unbroken cells and erythrocytes, after which the supernatant is overlaid on a cushion of 1 M sucrose, and the zymogen granule fraction is collected as a pellet through this cushion. The overlying supernatant and the surface rinse of the granule fraction are combined, centrifuged briefly to eliminate mitochondria, and a total microsomal fraction is sedimented by prolonged

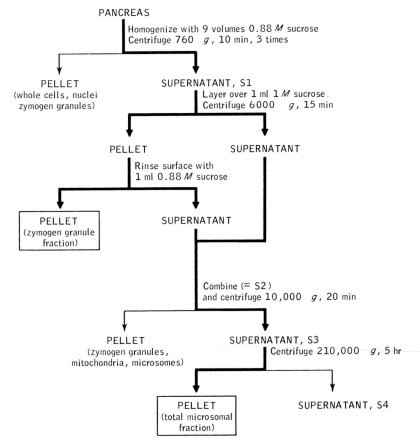

FIG. 4. Flow sheet for cell fractionation scheme applied to the calf pancreas. Centrifugal forces (g) are average values at the middle of the tube.

centrifugation. This total microsomal fraction is later resuspended and subfractionated on a discontinuous sucrose gradient.

Procedure. Calf tissue is placed on ice at the slaughterhouse within minutes of the death of the animals. The processing of the tissue should begin with 1–1.5 hours. All operations are carried out at 0–4°.

Obvious fat and connective tissue are dissected away, and the tissue is minced with scissors and forced through a small tissue press. Nine volumes (per wet weight of tissue) of 0.88 M sucrose are added, and homogenization is carried out with 5–8 strokes in a Brendler-type glass homogenizer fitted with a Teflon pestle (volume 50 ml) operated at 3000 rpm, taking care not to allow the temperature to rise.

Aliquots (10 ml) of the homogenate are centrifuged three times for 10 minutes at 760 g in 12-ml conical glass tubes to eliminate large or especially dense particles. The first pellet is large ($\sim$0.75 ml) and predominantly pink, although obviously heterogeneous. The second and third pellets are much smaller and consist of a tightly packed white button (zymogen granules) and some underlying red blood cells. The third supernatant is designated S1.

The zymogen granule fraction is obtained from S1 by layering it over a 1 ml cushion of 1 M sucrose at the bottom of a No. 40 tube and spinning for 15 minutes at 6000 g in a No. 40 rotor. A well-packed, white pellet collects at the bottom of the tube, and its tan surface (containing the bulk of the mitochondrial and microsomal contamination) is removed by adding 1 ml of 0.88 M sucrose and swirling gently. This 1 ml is added to the supernatant to yield S2, and the tube containing the granule pellet is inverted and allowed to drain for 5 or 10 minutes. The standard mitochondrial spin is 20 minutes at 10,000 g (i.e., 12,000 rpm in the No. 40 rotor). The overlying supernatant, S3, is pipetted off, taking care to avoid disturbing the pellet.

In order to maximize recovery of both smooth and rough microsomes from S3, this supernatant is centrifuged 5 hr at 210,000 g in the A269 rotor of an International B-60 ultracentrifuge to yield a firmly packed pinkish-tan pellet.

Subfractionation of Total Microsomes

Information concerning the isopycnic density of rough and smooth microsomes was used to design a discontinuous "sandwich gradient" in which the load of total bovine microsomes is in the middle of the centrifuge tube. During centrifugation, the smooth microsomes collect at an interface above the load and the rough microsomes collect at an interface beneath the load.

Procedure. Total microsomes from 5 g of tissue are gently resuspended

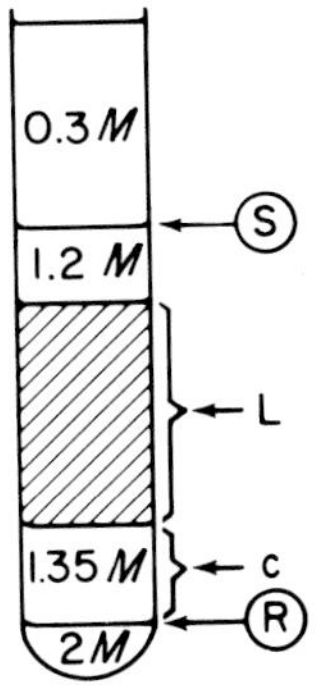

Chemistry of Sandwich Gradient

	% of load recovered in the designated fractions			
	Protein	RNA	PLP-P	RNase
(S)	6	2	22	8
L	30	19	19	21
C	15	16	7	11
(R)	48	62	50	57
Pellet	1	1	1	2
Recovery	106-111%	100-112%	80-116%	

FIG. 5. Design of sandwich density gradient for subfractionation of total microsomal fraction from calf pancreas. Chemistry of the resulting fractions is indicated in the accompanying table.

with a motor-driven homogenizer (Brendler-type: 10 ml) in 3.5 ml of 1.35 *M* sucrose. The resulting suspension is layered in the gradient outlined in Fig. 5, and the tube is spun 12 hours at 40,000 rpm (190,000 g_{avg}) in the SB283 rotor of an International B-60 ultracentrifuge. The volumes of the sucrose layers are as follows: 2 *M* sucrose, 1.5 ml; 1.35 *M* sucrose, 2.25 ml; load, 3.5 ml; 1.20 *M* sucrose, 1.5 ml; 0.3 *M* sucrose, 4.25 ml.

At the end of the run a tan band accumulates at the interface between the 1.35 *M* layer and the 2 *M* cushion, and a white band has risen to the interface between the 1.20 *M* layer and the 0.3 *M* layer. There is a faint annulate pellet at the bottom of the tube, and the region extending from the 0.3/1.2 interface to the 1.35/2.0 interface is somewhat cloudy. The indicated fractions are removed with a Pasteur pipette starting from the top of the tube.

Chemistry

The gross chemical composition of the granule and microsomal fractions is presented in Table II, where it can be compared to the composition of the whole homogenate. Comparable data from subcellular fractions of the guinea pig pancreas are given in Table I.

The recovery of protein in the granule fraction is low but comparable to the value obtained by Greene.[2] The very low phospholipid:protein and RNA:protein ratios indicate that the final pellet is predominantly protein in both the guinea pig and bovine fractionation schemes.

TABLE II
CHEMISTRY OF ZYMOGEN GRANULE FRACTION AND TOTAL
MICROSOMAL FRACTION

	$\dfrac{\text{mg Protein}}{\text{g tissue}^a}$	$\dfrac{\text{mg RNA}}{\text{g tissue}}$	$\dfrac{\text{mg PLP-P}}{\text{g tissue}}$
Homogenate	63	11.4	0.76
	(52–70)	(10.6–12.2)	(0.63–0.89)
Zymogen granule	2.5	0.05	0.0074
fraction	(1–3.6)	(0.045–0.056)	(0.0069–0.0078)
Microsomal fraction	7.6	2.8	0.14
	(6.4–8.8)	(2.6–3.1)	(0.13–0.16)

^{a} Tissue wet weight.

The surface rinses of the bovine granule fraction remove ~70% of both the RNA and the cytochrome oxidase activity of the 6000 g pellet. Only 30% of the protein is removed. If the cushion of 1 M sucrose is omitted, the rinsed pellets are considerably more contaminated, as measured by the RNA:protein ratio and the cytochrome oxidase:protein ratio.

The total g times minutes used to sediment the bovine mitochondrial fraction is one half that used by Siekevitz and Palade.[3] Both procedures sediment the same amount of cytochrome oxidase activity. The shorter spin is routinely used in order to increase the recovery of microsomes in the next centrifugation step. The bovine microsomal spin is much longer than that used by Siekevitz and Palade, in order to increase the recovery of small smooth vesicles; ~12% of the homogenate protein is recovered in the microsomal pellet. In both the guinea pig and bovine cases this fraction has a relatively high phospholipid:protein and RNA:protein ratio.

Figure 5 records the average protein, RNA, and phospholipid content of each of the five subfractions of bovine microsomes, and the results are essentially as expected: (1) relatively little phospholipid remains in the load zone; (2) the RNA content of the rough fraction is much greater than that of the smooth fraction; (3) the phospholipid:protein ratio of the smooth fraction is higher than for the rough fraction; (4) because of the presence of soluble RNase, essentially no polysomes are found in the pellet. The disproportionate amount of protein remaining in the load zone suggests that soluble proteins were entrapped in, or adsorbed to, the total microsomal pellet.

The high amount of phospholipid recovered in the smooth fraction (~25%) relative to the amount recovered in the rough fraction (~47%) is at variance with the morphological picture of the exocrine cell. The

TABLE III

SUBCELLULAR DISTRIBUTION OF SECRETORY ENZYMES OF THE CALF PANCREAS[a,b]

	Trypsin (TAME) (%)	Chymotrypsin (ATEE) (%)	Amylase (%)	Ribonuclease (%)
Debris	19	20	14.5	14.2
Zymogen granule + mitochondrial fraction	46	45	39	35.5
Microsomal fraction	12	11	7.3	18
Postmicrosomal supernatant	25	25	40.5	33
Recovery	102	101	101	102

[a] Data from L. J. Greene (unpublished). Activation of zymogens and measurement of enzyme activity was as in L. J. Greene, C. H. W. Hirs, and G. E. Palade, *J. Biol. Chem.* **238**, 2054 (1963).

[b] A 10% homogenate was prepared in 0.88 M sucrose. The whole homogenate was centrifuged two times for 10 minutes at 650 g to sediment debris. The remaining supernatant was centrifuged 10 minutes at 10,000 g to sediment a common zymogen granule and mitochondrial fraction. The overlying supernatant was centrifuged 80 minutes at 75,000 g to yield a pellet of microsomes and the postmicrosomal supernatant.

smooth fraction may be contaminated with fragments of non-Golgi membranes. Furthermore, since a significant amount of rough microsomes is lost in the differential spins which precede the pelleting of total microsomes, the phospholipid recovered in this fraction is expected to be low.

Table III records the distribution of several secretory enzymes among subcellular fractions, isolated by similar procedures by Greene. The activity in the debris fraction comes primarily from contaminating zymogen granules. The excess RNase relative to the other enzymes in the microsomal fraction may be the result of an adsorption artifact, as discussed in the last section of this article under the heading *Adsorption*. The appreciable amount of activity in the postmicrosomal supernatant suggests how great may be the danger of adsorption artifacts.

Preparation of Fractions for Electron Microscopy

Fixation of Pellets. In our studies we have routinely fixed zymogen granule pellets *in situ* in the centrifuge tube (cellulose nitrate tubes must be used) since fixation by resuspension leads, after recentrifugation, to extensive streaking on the tube walls and unworkable nonrepresentative pellets. In detail, the supernatants over pellets (3–4 mm in diameter) are decanted and replaced by sucrose of the same molarity (either 0.88 M or 0.3 M) containing 1% OsO_4. No buffer is added and the pH should be ~5 to prevent granule lysis or extraction. All operations are at 4° and

a   b

FIG. 6. Cutting of orientable strips during the preparation of pellets for electron microscopy. (a) Cross section of pellet. (b) Individual strip seen in profile.

should be performed with good ventilation, preferably in a fume hood. After a few minutes of fixation to stabilize the pellet surface, the pellets should be cut with a razor blade into small orientable strips (Fig. 6) to facilitate penetration of fixative, except in the case of friable pellets. Leave strips overnight in fixative then dehydrate and embed as described below.

Fixation in Suspension (*0–4°*). We have generally fixed microsomal fractions and subfractions in suspension since microsomal pellets are densely packed and penetration of fixative is very slow. In contrast to zymogen granule fractions, microsomal fractions form coherent pellets following fixation in suspension.

Mix an aliquot of the fraction with an equal volume of 1% OsO_4 dissolved in a solution of the same composition as the suspending fluid, and fix for 30 minutes. If necessary, dilute with water to facilitate pelleting. Centrifuge 30–60 minutes at 100,000 g in cellulose nitrate tubes in a swinging-bucket rotor. Fixed fractions often streak along the walls of tubes when spun in angle-head rotors. The volume of the aliquot should be chosen to yield a pellet 3–4 mm in diameter. Leave overnight. Decant the fixative and replace with 70% ethanol at room temperature. Cut off the bottom of the tube and cut the pellet into recognizable and orientable strips with a clean razor blade (Fig. 6a). Do not let the surface dry. Continue dehydration of the strips. The cellulose nitrate dissolves in propylene oxide. Figure 6b shows, in profile, the appearance of individual strips. Embed in such a way that the left face is parallel to the block face. Thus, sections will display the entire block face and allow survey from bottom to top of the pellet.

If desired, the first fixation can be in 1–2% glutaraldehyde in the appropriate sucrose and postfixation of the pellet can be performed with OsO_4 as above. Similarly, the OsO_4 fixation can be followed by block staining with uranyl acetate.[12]

Observations

There is little stratification in the zymogen granule pellets. A typical field is shown in Fig. 7. Each granule contains a homogeneous dense content and is bounded by a unit membrane. There is some suggestion of

[12] M. G. Farquhar and G. E. Palade, *J. Cell Biol.* **26**, 263 (1965).

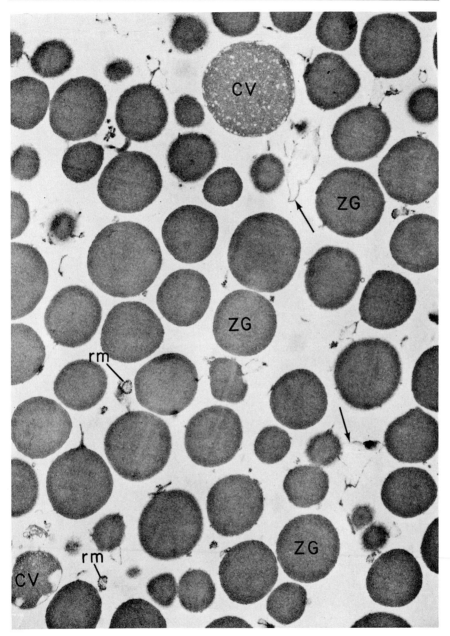

Fɪɢ. 7. Electron micrograph of a thin section through a representative portion of a zymogen granule pellet fixed with OsO₄. The majority of the structures shown are zymogen granules (ZG). Also present are condensing vacuoles (CV), membrane profiles possibly derived from burst zymogen granules (arrows), and a few rough microsomes (rm). ×24,800.

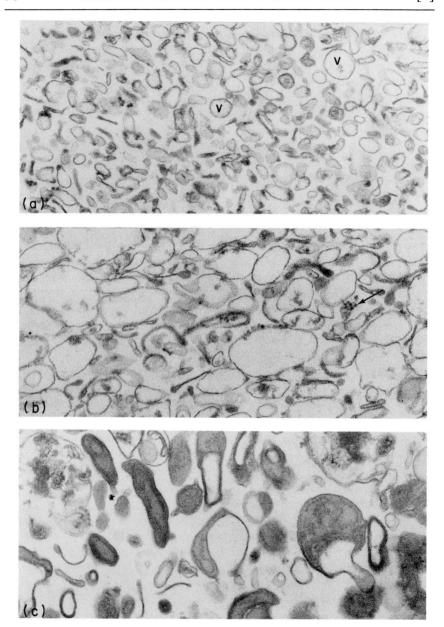

Fig. 8. Electron micrographs of thin sections through the fixed and pelleted are larger closed membrane profiles and a few contaminating ribosomes (arrow). At the top of the pellet (a) are many smooth-surfaced vesicles (v). In the middle (b) are larger closed membrane profiles and a few contaminating ribosomes (arrow). At the bottom of the pellet (c) is a layer of large vacuoles often containing dense heterogeneous material. These are presumably autophagic vacuoles (lysosomes). ×22,750.

adhesion between granules and between them and the principal contaminants: rough microsomes and mitochondria. Such adhesion may explain why attempts to resuspend and repellet granules results in major losses. A small population of condensing vacuoles is also present.

The smooth microsomal fraction (Fig. 8) contains, from bottom to top: a small population of large particles resembling secondary lysosomes, a large population of flattened and distended smooth-surfaced cisternae, and, at the top, a large population of small smooth-surfaced vesicles. Very few ribosomes or rough microsomes are seen at any level.

The rough microsomal fraction (Fig. 9) is more homogeneous. At all levels it consists predominantly of closed, ribosome-studded vesicles, many of which have a visible amorphous content. The larger vesicles are found at the bottom of the pellet.

Limitations of the Fractionation Scheme

The gross chemical and morphological features of the three key bovine pancreas fractions correspond closely to those isolated from the guinea pig pancreas. Furthermore, a variety of experiments involving fractionation of tissue slices labeled *in vitro* with radioactive amino acids indicate that the bovine system should be as manipulable for studies of intracellular transport and secretion as the guinea pig system. On the other hand, in both species, the yield of none of the fractions is quantitative and the granule fraction contains an appreciable admixture of condensing vacuoles.[13]

Greene *et al.* (2) have shown that the spectrum of enzymes in the bovine zymogen granule fraction is very similar to the spectrum of enzymes in pancreatic juice. Therefore it is unlikely that the granule fraction is nonrepresentative of the population of granules in the cell, and it is also unlikely that there is appreciable differential leakage of secretory enzymes from granules or differential adsorption of secretory enzymes to the components of the fraction.

A number of experiments have been performed to find out whether the total microsomal fraction is equally free from leakage and adsorption artifacts, especially with reference to RNase. RNase is the smallest and most basic of the digestive enzymes. Approximately 80% of the RNase in pancreatic juice is RNase A, a nonglycoprotein. The remaining ~20% is a mixture of glycoprotein species, all of which have the same peptide backbone as RNase A.[14,15]

---

[13] Pilot experiments on the guinea pig pancreas show that after homogenization in sucrose, both zymogen granules and condensing vacuoles reach the same equilibrium position in sucrose density gradients.

[14] T. Plummer, Jr., and C. H. W. Hirs, *J. Biol. Chem.* **238**, 1396 (1963).

[15] T. Plummer, Jr., *J. Biol. Chem.* **243**, 5461 (1968).

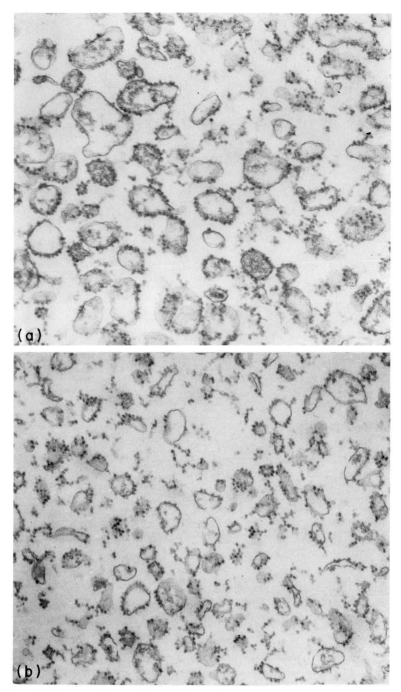

FIG. 9. Electron micrograph of thin sections through fixed and pelleted rough microsomes derived from the sandwich gradient (see Fig. 5). Both at the top of the pellet (a) and at the bottom (b), the primary components are ribosome-studded vesicles derived from fragmented and resealed elements of the RER. ×24,500.

TABLE IV
Leakage and Adsorption of Ribonuclease during Cell Fractionation

	Leakage[a]		
	^{3}H-labeled protein (%)	^{3}H-labeled RNase (%)	Adsorption[b] (%)
Debris, zymogen granules, and mitochondria	34	40	38
Microsomes	46	34	15
Postmicrosomal supernatant	20	26	47

[a] Distribution of ^{3}H-labeled protein and ^{3}H-labeled RNase among subcellular fractions isolated from lobules pulsed 6 minutes with mixed ^{3}H-labeled amino acids. Sum = relative recovery.

[b] Distribution of exogenous RNase among subcellular fractions.

Leakage. Lobules of fresh bovine pancreas were pulse-labeled for 6 minutes with mixed radioactive amino acids[4] and then fractionated. At this time point the bulk of radioactive secretory protein should be in the cisternal space of the rough endoplasmic reticulum. The distribution of total trichloroacetic acid-insoluble radioactive protein and radioactive RNase is given in Table IV. The RNase was recovered from each fraction by affinity chromatography, using a modification of the procedure of Wilchek and Gorecki.[16] Clearly, an appreciable amount of the total protein and a larger percentage (26%) of the radioactive RNase has leaked to the postmicrosomal supernatant.

Adsorption. Radioactive RNase was recovered from *in vitro* labeled lobules and purified by affinity chromatography. Trace amounts of this labeled RNase were added to homogenates of fresh bovine pancreas and the distribution of the exogenous tracer was followed during fractionation. These data are also presented in Table IV. The known specific activity (cpm/RNase activity) of the postmicrosomal supernatant allows an absolute estimation of the percentage of contaminating RNase in each fraction. By this measure, the granule fraction is 4% contaminated, whereas in the microsomal fraction (washed once) 27% of the RNase is exogenous. Comparable experiments concerning total secretory proteins show that they too adsorb to the microsomes, but to a lesser extent than RNase.

[16] M. Wilchek and M. Gorecki, *Eur. J. Biochem.* 11, 491 (1969).

[4] Subcellular Fractionation of Adipocytes

By Leonard Jarett

The ability to study the hormonal regulation of adipose tissue metabolism was significantly advanced by the techniques of Rodbell for isolating adipocytes[1] and preparing "ghost" preparations from the adipocytes.[2] This yielded not only a homogeneous population of cells to study, but a delipidated cellular preparation which retained much of the original metabolic activities of the adipocyte. The development of a rapid and simple procedure for fractionating the adipocyte into purified subcellular components by McKeel and Jarett[3] extended the ability to investigate the hormonal regulation of the fat cell metabolic processes to a subcellular level.

Preparation of Fat Cells

Fat cells isolated from rat epididymal fat pads as described by Rodbell[1] were used as a starting preparation for the fractionation scheme. The choice of the crude collagenase was critical for obtaining metabolically active and hormonally sensitive fat cells from the rat epididymal fat pads. Crude collagenase preparations have been extremely variable from batch to batch regardless of the commercial source, and certain criteria for selecting the enzyme have been established in this laboratory. Trypsin activity was measured by the modification of the technique of Schwert and Takenaka[4] as described by Rick,[5] and collagenase preparations containing activity greater than 4 nmoles of substrate hydrolyzed per milligram of collagenase per minute were unacceptable. A small but detectable amount of trypsin activity must be present or the digestion process will not go to completion. The optimum time for digestion of the minced fat pads from two rats was 25–35 minutes at 37° with gentle shaking in 1.5 ml of Krebs-Ringer bicarbonate buffer, pH 7.4, containing half the usual calcium concentration, 200 mg of glucose per 100 ml, 3% fatty acid-poor bovine serum albumin (BSA) fraction V, and 7.0 mg of the selected collagenase and gassed with 95% O_2–5% CO_2. Treating properly prepared fat cells for 15 minutes with epinephrine (0.125 μg/ml) caused the rate of lipolysis

[1] M. Rodbell, *J. Biol. Chem.* 239, 375 (1964).
[2] M. Rodbell, *J. Biol. Chem.* 242, 5744 (1967).
[3] D. W. McKeel and L. Jarett, *J. Cell Biol.* 44, 417 (1970).
[4] G. W. Schwert and Y. Takenaka, *Biochim. Biophys. Acta* 16, 570 (1955).
[5] W. Rick, *in* "Methods of Enzymatic Analysis" (H. U. Bergmeyer, ed.), p. 815. Academic Press, New York, 1965.

to increase from about 100 to 1600–2000 nmoles of glycerol released per hour per milligram of protein and a 3-fold elevation of cyclic AMP levels. Insulin (500 μunits/ml) suppressed the epinephrine-stimulated lipolysis by over 90%. All these criteria must be satisfied before the collagenase preparation is used for further studies.

Fractionation of Isolated Fat Cells

The procedure described is a slight modification of the original fractionation technique of McKeel and Jarett[3] and based on using 16–24 rats.

Homogenate. The isolated fat cells are washed once with 10 volumes of homogenization medium (medium I: 10 mM Tris·HCl, pH 7.4, 1 mM EDTA, and 0.25 M sucrose) to rid the cells of the Krebs-Ringer buffer. The EDTA in this solution is essential to prevent reaggregation and clumping of membrane components and other subcellular elements in subsequent steps, and the isosmotic sucrose is necessary to minimize damage to cellular organelles, especially mitochondria and nuclei. The cells were resuspended in at least 3.5 volumes of medium I at room temperature and transferred to a glass homogenizer (Arthur H. Thomas, No. 4288-B, type C) fitted with a Teflon pestle (clearance 0.006–0.009 inch). Potter-Elvehjem glass homogenizers fitted with Teflon pestles (clearance 0.004–0.006 inch) were used to resuspend all subsequent pellets. The cell suspension is homogenized with 10 up-and-down strokes by means of a variable-speed motor driven at 2600 rpm. The resulting homogenate is placed in ice, and all subsequent steps are carried out at 4°. The homogenate may be used for assaying by allowing the fat to float to the surface and congeal after 10 minutes on ice and then aspirating needed aliquots of the infranatant material. If the entire homogenate including the endogenous fat is to be used for assay purposes, as in measuring lipase activity, then the homogenates should not be chilled. It is easy to maintain and dispense a homogeneous suspension of the homogenate kept at room temperature.

The homogenization step is critical for a successful fractionation. Too vigorous homogenization or too narrow a clearance of the pestle will cause the plasma membranes to be isolated in the microsomal fraction. In contrast, too gentle homogenization or too wide a clearance of the pestle will cause inadequate rupture of the adipocytes.

Fat Cake. The endogenous lipid liberated from the fat cells during homogenization may be separated from the rest of the homogenate and used for assay purposes, such as measuring lipase activity. The homogenate maintained at room temperature is centrifuged at 1000 g for 1.0–1.5 minutes. The infranatant suspension may be removed with a long needle and syringe, leaving behind the freed lipid, or fat cake. The fat cake is rinsed into the homogenizer with medium I and gently resuspended to the

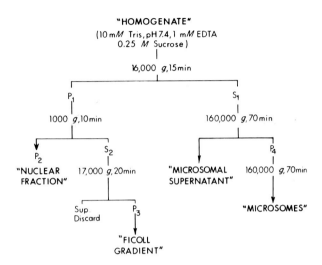

ISOLATED FAT CELLS

"HOMOGENATE"
(10 mM Tris, pH 7.4, 1 mM EDTA
0.25 M Sucrose)

16,000 g, 15 min

P₁

1000 g, 10 min

S₁

160,000 g, 70 min

P₂

"NUCLEAR
FRACTION"

S₂

17,000 g, 20 min

P₄

"MICROSOMAL
SUPERNATANT"

160,000 g, 70 min

Sup
Discard

P₃

"MICROSOMES"

"FICOLL
GRADIENT"

FIG. 1. Schematic representation of the adipocyte fractionation scheme.

original homogenate volume with three strokes of the pestle at 1250 rpm.

Crude Nuclear Fraction. The chilled homogenate is centrifuged at 16,000 g (mid-tube) for 15 minutes, yielding a pellet (P_1), a supernatant solution (S_1), and a congealed fat cake (Fig. 1). S_1 is removed by needle and syringe for subsequent use in preparing the microsomal fraction. The fat cake is discarded, and P_1 is resuspended in 8 ml of medium I by six strokes at 1250 rpm. The suspension is centrifuged at 1000 g for 10 minutes, yielding a pellet (P_2) and a supernatant (S_2). P_2 contains the nuclei as well as cellular fragments similar to the Rodbell "ghosts." The various nonnuclear material may be removed by the various techniques described by Wang[6] to yield an intact and purified nuclear fraction.

Microsomal Supernatant and Microsomal Fractions. Lipid carried over with S_1 from the 16,000 g spin can be removed by placing S_1 in a chilled beaker; this causes the fat to adhere to the glass, and the S_1 material is easily aspirated. S_1 is then centrifuged at 160,000 g for 70 minutes, yielding a pellet. The microsomal supernatant is slowly and gently aspirated from the centrifuge tube to avoid collecting any slight lipid still present. The pellet is gently rinsed with medium I, and the tube is wiped dry. The pellet is resuspended in 1.0 ml of medium I with six strokes at 1700 rpm and a Kel-F Kontes pestle, which fits the Beckman ⅝ × 3 inch polyallomer tubes (No. 326814) used in the high speed centrifugation. The suspension is transferred to a clean polyallomer tube and centrifuged at

[6] T. Y. Wang, this series, Vol. 12A, p. 417.

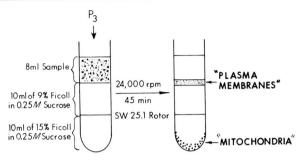

FIG. 2. Schematic representation of the Ficoll gradient fractionation of the pellet designated P_3 to yield plasma membranes and mitochondria.

160,000 g for 70 minutes. The supernatant is discarded, and the pellet is resuspended as before to yield the microsomal fraction.

Plasma Membrane Fraction. S_2 is centrifuged at 17,000 g for 20 minutes to yield a supernatant fluid, which is discarded, and a pellet (P_3). P_3 is resuspended in 6 ml of medium I by 6 strokes at 1700 rpm. The resulting suspension is further diluted with medium I and placed on a discontinuous Ficoll gradient (Fig. 2). If a Beckman SW 25.1 rotor is used, then the entire 8 ml suspension is placed on top of 10 ml of 9% Ficoll in 0.25 M sucrose, which is above 10 ml of 15% Ficoll in 0.25 M sucrose. With a Beckman 27.1 rotor and 17-ml tubes, 12 ml of the P_3 suspension is layered in two tubes over 6 ml and 5 ml of the 9% and 15% Ficoll solutions, respectively, both in 0.25 M sucrose. The gradients are centrifuged at 59,000 g for 45 minutes. The plasma membranes form a sharp band just below the medium I–9% Ficoll interface and can easily be removed with a bent needle and syringe. The aspirate is diluted to 40 ml with medium I to dilute out the Ficoll and centrifuged at 18,000 g for 20 minutes. The supernatant is decanted, the tube is wiped clean, and the pellet is resuspended in 1 ml of medium I by 6 strokes at 1700 rpm, yielding the purified plasma membrane fraction.

The original preparative procedure[3] utilized linear Ficoll and linear sucrose gradients, which yielded comparable results. The Ficoll gradient was chosen because of its shorter centrifugation time requirements and then converted to a discontinuous gradient to sharpen the plasma membrane band and to conserve on the utilization of Ficoll. The Ficoll must be dialyzed 24 hours against three changes of distilled water, lyophilized, and reground into a fine powder before use.

Mitochondrial Fraction. The vast majority of the mitochondria are contained in the P_3 pellet and are purified by the gradient centrifugation used for isolating the plasma membranes. The mitochondria form a pellet at the bottom of the tube under the 15% Ficoll (Fig. 2). The Ficoll is carefully decanted after removal of the plasma membranes, and the tube is wiped clean. The pellet is resuspended in 1 ml of medium I with 6 strokes at 1700 rpm; the suspension is diluted to 40 ml and centrifuged at 18,000 g for 20 minutes. The resulting pellet is the purified mitochondrial fraction; it is resuspended in 1 ml of medium I for subsequent assays. This technique was compared to the differential centrifugation method for preparing mitochondria from the original homogenate and found to be a comparable end product,[3] so that it became the method of choice because of the ease and time-saving factors.

Morphological Characterization. Intact adipose tissue and isolated fat cells were examined by electron microscopy to establish the morphological features characteristic of the major subcellular components. The major difference between the intact fat pad and isolated adipocytes was the loss of the basal lamina from around the adipocyte. More detailed morphological descriptions and micrographs are contained in the original publication.[3]

Plasma Membrane Fraction. This fraction consisted almost entirely of membranous sacs roughly 0.5–2.0 μm in diameter that had numerous invaginations characteristic of the plasma membrane of intact cells.[3,7,8] Microvesicles commonly seen throughout the cytoplasm of adipocytes[3,7,8] were seen trapped inside the plasma membrane sacs as well as occasionally clustered outside the sacs. Other entrapped material consisted of smooth and rough endoplasmic reticular membranes and mitochondria. A rare Golgi tubular system was seen connected to smooth endoplasmic reticular sacs. Interconnections between the plasma membrane and the inner sacs were common. Carter *et al.*[9] reported the plasma membrane fraction prepared by their modifications of the original technique to contain closed vesicles and open sheets of membranes. Brief sonication of this preparation converted the membranes to closed vesicles.

Mitochondrial Fraction. The mitochondria appeared slightly swollen with the occasional loss of the outer membrane and in various energy configurations. The only contamination of this fraction consisted of an occasional plasma membrane vesicle with characteristic invaginations.

Microsomal Fraction. This was the most heterogeneous fraction consisting of altered ribosomal material, smooth and rough endoplasmic reticulum, and microvesicles consistent with the plasma membrane invaginations and microvesicles seen throughout the fat cell cytoplasm.[3,7,8]

Biochemical Characterization

Protein Recovery. The determination of the protein in fat cells requires that the lipid material be removed before assay. In order to circumvent

[7] J. R. Williamson, *J. Cell Biol.* **20**, 57 (1964).
[8] S. W. Cushman, *J. Cell Biol.* **46**, 326 (1970).
[9] J. R. Carter, J. Avruch, and D. B. Martin, *J. Biol. Chem.* **247**, 2682 (1972).

this procedure, two methods have been developed for indirectly determining the protein content of fat cells. The number of fat cells has been correlated with the protein content of the cells and found to be 1.1×10^6 cells per milligram of protein.[10] This value was constant for fat cells from rats weighing 120–200 g and allowed the protein content to be determined by means of cell counts on a fat cell suspension by phase microscopy. The alternative method correlates the packed cell volume (PCV) to protein content. The PCV is measured after centrifugation of a suspension of fat cells at 1000 g for 30 seconds. The PCV per milligram of protein for rats of various weights is: 120–150 g, 0.14 ml/mg; 150–170 g, 0.16 ml/mg; 170–200 g, 0.19 ml/mg. The protein content of all other fractions can be measured by the procedure of Lowry et al.[11] The recovery of protein in the various fractions is illustrated in the table. A 150-g rat yields about 3.0–3.5 mg of fat cell protein.

Nuclear Fraction. Eighty-five percent of the homogenate DNA was recovered in the nuclear fraction. The remaining 15% was not recovered in particulate fractions, indicating that it was solubilized from ruptured nuclei. The plasma membrane fraction contained less than 0.6% of the original homogenate DNA; this could be attributed to nuclei entrapped in the plasma membrane vesicles.[3]

Plasma Membrane Fraction. The plasma membrane fraction was highly purified by the procedure described. Adenyl cyclase and 5′-nucleotidase were used as marker enzymes. Adenyl cyclase specific activity was increased in P_3 to seven times the homogenate with no loss of total activity.[12] The gradient purification increased the adenyl cyclase specific activity by nine to twelve times the homogenate, but with a loss of 70% of the total enzyme activity, which could not be accounted for by the loss of protein. This was supported by the finding that over 90% of the P_3 succinic dehydrogenase (SDH) activity was recovered in the mitochondrial fraction after gradient fractionation.[12] No other fractions had an enrichment in adenyl cyclase activity. Carter et al.[9] found the plasma membrane fraction to be highly enriched in 5′-nucleotidase activity and to contain less than 20% contamination with a microsomal enzyme, NADH oxidase. This was consistent with earlier data[3] that the plasma membrane fraction contained less than 30% contamination with microsomal elements based on NADH–cytochrome c reductase measurements and that the RNA content of the plasma membrane fraction was consistent with native membrane RNA and

[10] L. Jarett and D. W. McKeel, *Arch. Biochem. Biophys.* **140**, 362 (1970).

[11] O. H. Lowry, N. J. Rosebrough, A. L. Farr, and R. J. Randall, *J. Biol. Chem.* **193**, 265 (1951).

[12] L. Jarett, M. Reuter, D. W. McKeel, and R. M. Smith, *Endocrinology* **89**, 1186 (1971).

PROTEIN RECOVERY IN ADIPOCYTE SUBCELLULAR FRACTIONS

	Homogenate	P_1	S_1	Nuclear	P_3	Plasma membrane	Mito-chondrial	Micro-somal	Microsomal supernatant
% Recovery	100	30.0 ± 1.0	66.0 ± 3.1	5.0 ± 0.3	18.0 ± 0.6	5.0 ± 0.3	4.0 ± 0.3	6.0 ± 0.3	68.0 ± 3.4
No. of experiments		43	37	9	37	58	59	58	57

not significant ribosomal contamination. The plasma membrane fraction had less than 10% contamination by mitochondria as measured by SDH activity, which was consistent with the morphological observations.[3] The Mg-ATPase specific activity of the plasma membranes was increased five times over the homogenate but amounted to only 6–7% of the total homogenate activity.[10] This ATPase activity could not be accounted for by mitochondrial contamination since the specific activity of the enzyme in the two fractions was identical, but their patterns of behavior to oligomycin, 2,4-dinitrophenol and ouabain differed.[10] Over 40% of the sialic acid content of the plasma membrane fraction was acessible to neuraminidase. If the sialic acid content of the plasma membranes was on the outer surface of the membranes as was the case for red blood cells,[13] this would indicate the high degree of right-side-outness of this fraction considering the large amount of microvesicles not accessible to the enzyme because of being trapped inside the larger vesicles.[14]

Mitochondrial Fraction. This fraction was biochemically the purest of all the fractions. SDH specific activity was increased about 10-fold over the homogenate with about 50–60% recovery of the total homogenate activity.[3,12] Adenyl cyclase activity was very low and consistent with the morphological contamination by plasma membranes.[3] Endoplasmic reticular contamination was low as indicated by the specific activity of NADH cytochrome *c* reductase as compared to the microsomal fraction. The Mg-ATPase activity was enriched 5-fold over the homogenate, and about 50% of the original homogenate activity was recovered with the mitochondrial fraction.[10]

Microsomal Fraction. This fraction was found to be a mixture of various membrane components. The endoplasmic reticular elements were the predominant component, NADH-cytochrome *c* reductase specific activity being five times that of the homogenate.[3,13] RNA concentration was four times the homogenate, indicating the extent of ribosomal concentration. Plasma membrane content up to 30% was indicated by adenyl cyclase[3] and 5'-nucleotidase[9] activity. Mitochondria were almost nonexistent as determined by SDH measurements.[3]

Applications

Enzyme Distribution. This fractionation scheme has been useful in documenting the subcellular distribution of palmitoyl-CoA synthetase,[15] ATPase,[10] adenyl cyclase,[3,12] cyclic AMP phosphodiesterase,[12] triglyceride

[13] T. L. Steck, R. S. Weinstein, J. H. Straus, and D. F. H. Wallach, *Science* **168**, 255 (1970).

[14] L. Jarett and R. M. Smith, unpublished data (1971).

[15] K. Lippel, A. Llewellyn, and L. Jarett, *Biochim. Biophys. Acta* **231**, 48 (1971).

lipase,[16] phosphorylase,[16] and phosphorylase kinase.[16] Palmitoyl-CoA synthetase, which is involved in the pathway of triglyceride synthesis, was found to be primarily a microsomal enzyme and to be unaffected by a variety of hormones, nucleotides, and other compounds known to influence lipid metabolism.[15] ATPase activity of the fat cell was primarily associated with the mitochondria, and the plasma membranes contained as high a specific activity of the enzyme as the mitochondria.[10] The enzymes could be distinguished in the two fractions by the responses to Na^+ and K^+, oligomycin, 2,4-dinitrophenol, and ouabain.[10] Adenyl cyclase was localized almost entirely with the plasma membranes and showed a unique behavioral pattern during fractionation as discussed above. Cyclic AMP phosphodiesterase was found to be in the microsomal supernatant fraction and not associated with the plasma membranes.[12] Half of the triglyceride lipase activity was associated with the fat cake, the rest being primarily in the microsomal supernatant, but with a small and reproducible amount associated with the microsomes.[16] Phosphorylase and phosphorylase kinase were recovered with the microsomal supernatant.

Hormonal Regulation of Enzyme Activity. The ability of glucagon, ACTH, and epinephrine to stimulate adenyl cyclase during purification of the plasma membranes was altered in patterns unique for each hormone, although these hormones actively stimulated the adenyl acyclase activity of the fat cell homogenate. The ability of glucagon to stimulate the enzyme was completely lost, and that of ACTH and epinephrine was diminished to different degrees.[12] EGTA and propranolol blocked the stimulation of plasma membrane adenyl cyclase by ACTH and epinephrine, respectively.[12] These findings indicated that the hormone receptors were distinct one from the other and from the enzyme itself. Epinephrine treatment of isolated fat cells resulted in stimulation of triglyceride lipase and phosphorylase activity in their various locations.[16] Surprisingly phosphorylase kinase activity could not be shown to be altered by the epinephrine treatment.[16] If this enzyme was already fully activated *in vivo,* then it would not be rate limiting for phosphorylase activation. Alternatively, it could have been activated during fractionation despite attempts to prevent this occurrence. Epinephrine treatment of the fat cells increased the affinity of triglyceride lipase toward endogenous triglyceride substrate to a greater extent than to exogenous triolein substrate.[16] Data showed that the endogenous and exogenous substrates did not mix, but behaved as separate pools of substrate although in the same suspension.[16]

Transport. The plasma membrane preparation has been used success-

[16] J. C. Khoo, L. Jarett, S. E. Mayer, and D. Steinberg, *J. Biol. Chem.* **247**, 4812 (1972).

fully for studying glucose transport[9] and insulin effect on the transport process.[17,18] The glucose transport system displayed expected characteristics showing preference for D-glucose, counter transport of D-glucose, and inhibition of D-glucose transport by phlorizin, 2-deoxyglucose, 3-O-methylglucose and N-ethylmaleimide.[9] Agents such as cyclic AMP, dibutyryl cyclic AMP, other adenine nucleotides and other sugars, such as glucose 6-phosphate, L-glucose, fructose, glucose 1-phosphate and fructose 1,6-diphosphate had no effect on glucose transport.[9] Avruch et al. have data demonstrating that addition of insulin directly to the plasma membrane suspension did not stimulate glucose transport whereas plasma membranes prepared from isolated fat cells incubated with insulin did show stimulated glucose transport.[17,18]

Hormone Binding and Degradation. Crofford and Okayama have utilized the plasma membrane, mitochondrial, nuclear, and microsomal fractions to study insulin binding.[19] The insulin binding was determined indirectly by measuring by immunoassay the disappearance of insulin from the incubation solution. The plasma membranes were found to bind almost twice the amount of insulin per milligram protein as the microsomal fraction while the mitochondrial and nuclear fractions had little binding capacity. Part of the microsomal binding of insulin could be accounted for by the 20% contamination by plasma membranes. Trypsin treatment of the fat cells prior to isolation of the plasma membranes prevented the binding of insulin. Crofford and co-workers, using a dialysis cell, have further studied the binding of insulin to the plasma membranes and the degradation of insulin by the plasma membranes or by a solubilized extract of the membranes.[20] Unlabeled insulin and anti-insulin guinea pig serum prevented the binding and degradation of ^{125}I-insulin in the presence of excess albumin. The degradation products were soluble in 10% trichloroacetic acid, nonreactive with anti-insulin guinea pig serum and nonabsorbable to a ^{125}I-absorbing resin. Trypsin treatment of fat cells prior to isolation of the membranes reduced not only ^{125}I-insulin binding but degradation as well. The data suggested that both binding and degradation of insulin occurred on the external surface of the plasma membranes of fat cells.

[17] J. Avruch, J. R. Carter, and D. B. Martin, "Handbook of Physiology," Section 7, Endocrinology, Vol. I (D. F. Steiner and N. Freinkel, eds.), Chapter 35. Amer. Physiol. Soc., Washington, D. C., 1972.
[18] J. Avruch, J. R. Carter, and D. B. Martin, *Biochim. Biophys. Acta* **288**, 27 (1972).
[19] O. B. Crofford and T. Okayama, *Diabetes* **19**, Suppl. 1, 369 (1970).
[20] O. B. Crofford, N. L. Rogers, and W. G. Russell, *Diabetes* **21**, Suppl. 2, 403 (1972).

The characteristics of the insulin binding to the fat cell plasma membranes was investigated further by Hammond et al.[21] ^{125}I-insulin was found to bind in an identical fashion to adipocytes and the isolated plasma membranes. Two major binding sites were identified, one a high affinity, low capacity site ($K_D = 5 \times 10^{-10} M$) and the other a lower affinity, high capacity site ($K_D = 3 \times 10^{-9} M$). Insulin binding to the membranes was prevented by insulin analogs (desoctapeptide and desalanine insulin) and proinsulin in direct relationship to their biological activities. These studies showed that the lack of stimulation of glucose transport by the plasma membranes by direct incubation with insulin reported previously[9,17,18] could not be due to a loss of insulin receptors. It did suggest that an alteration had occurred in the mechanism that couples the receptor to the effector system and that once the effector or transport system was stimulated it did not require the coupling component.

The binding of epinephrine to adipocytes and plasma membrane, mitochondrial and microsomal fractions has been characterized.[22] The adipocytes and all three fractions were found to contain a high and a low affinity binding site. The dissociation constants derived from the high affinity association constants for adipocytes and the plasma membranes were consistent with the level of hormone necessary to cause half-maximal stimulation of adenylate cyclase, cyclic AMP, and lipolysis. The binding sites on the adipocyte could be accounted for by the sites on the plasma membrane. The high affinity association constants for the mitochondrial and microsomal fractions were slightly less than for the plasma membranes. No significant difference was found between the specific binding of the ^{3}H-epinephrine to adipocyte plasma membranes, mitochondria, or microsomes or liver mitochondria while liver plasma membranes bound twice to three times as much. The use of analogs helped distinguish the epinephrine binding among the three adipocyte fractions. Propranolol at concentrations which prevent epinephrine from stimulating adenylate cyclase and lipolysis prevented only 30% of the hormone binding to plasma membranes. Phentolamine, an α-adrenergic blocking agent, had no effect on epinephrine binding. The significance of the binding sites on the mitochondrial and microsomal fractions is unclear since all of the binding of epinephrine to intact cells can be accounted for by the binding sites available on the plasma membrane fraction. Recent studies on liver have shown epinephrine to affect gluconeogenesis without affecting the adenylate cyclase system suggesting that epinephrine may enter the cell and act

[21] J. M. Hammond, L. Jarett, I. K. Mariz, and W. H. Daughaday, *Biochem. Biophys. Res. Commun.* 49, 1122 (1972).
[22] L. Jarett, R. M. Smith, and S. R. Crespin, unpublished observation (1972).

directly on intracellular systems.[23] Another explanation for epinephrine binding to mitochondria may be for degradation since a propranolol degradation system has been described for mitochondria.[24]

[23] M. E. M. Tolbert and F. R. Butcher, *Endocrinology* **92**, (Supplement, 55th Annual Meeting) A-171, Abstract 246 (1973).
[24] A. Huunan-Seppälä, *Acta. Chem. Scand.* **26**, 2712 (1972).

Section II

Isolation of Purified Subcellular Fractions and Derived Membranes (from Mammalian Tissue Excluding Nerve)

[5] Isolation of Plasma Membranes from Rat and Mouse Livers and Hepatomas

By P. EMMELOT, C. J. BOS, R. P. VAN HOEVEN, and
W. J. VAN BLITTERSWIJK

Erythrocyte ghosts have long since been isolated and studied, but it is only more recently that plasma membranes have been prepared from cells of solid mammalian tissues. The first achievement of this kind was the isolation of plasma membranes from rat liver, reported by Neville[1] in 1960. In subsequent years this method has been modified in various laboratories, and other methods have also been introduced (Table I).

The original Neville procedure, as modified in our laboratory[2-6] and found highly satisfactory for the isolation of plasma membranes from rat and mouse liver and various liver tumors, will be described here.

The principle of the method is the homogenization of the tissue in water buffered with sodium bicarbonate, sedimentation of the plasma membranes followed by a number of washing cycles using low speed differential centrifugation in order to remove the bulk of contaminating materials, and a final flotation in a discontinuous sucrose gradient to remove remaining contaminants. Tissue homogenization forms a crucial step. Success depends on the generation of large plasma-membrane fragments (sheets) which by centrifugation settle upon nuclei and debris, and which can be separated from the bulk of the other organelles by the washings. Therefore, tissue homogenization should not disrupt too much (a) the plasma-membrane skeleton, and (b) the nuclear membrane. Otherwise, (a) small-sized plasma-membrane fragments may not be sedimented by the low speed centrifugation (while increasing speed may introduce too much unwanted material), and (b) a deoxynucleoprotein gel is formed that acts as a glue incorporating the plasma membranes and thus prevents any further effective separation. Different tissues may vary as to the extent to which these requirements can be met. Conditions satisfactory to rat and mouse liver are not necessarily so for liver tumors indigenous to the same hosts. Among the

[1] D. M. Neville, Jr., *J. Biophys. Biochem. Cytol.* **8**, 413 (1960).
[2] P. Emmelot and C. J. Bos, *Biochim. Biophys. Acta* **58**, 374 (1962).
[3] P. Emmelot, C. J. Bos, E. L. Benedetti, and P. Rümke, *Biochim. Biophys. Acta* **90**, 126 (1964).
[4] P. Emmelot and C. J. Bos, *Int. J. Cancer* **4**, 705 (1969).
[5] P. Emmelot and C. J. Bos, *Int. J. Cancer* **4**, 723 (1969).
[6] P. Emmelot and E. L. Benedetti, *in* "Carcinogenesis, a Broad Critique," *Symp. Fundam. Cancer Res.* **20**, 471 (1967).

TABLE I

METHODS USED FOR ISOLATING PLASMA MEMBRANES FROM RAT LIVER[a]

Authors	Homogenization medium	Rate-dependent centrifugation	Isopycnic-zonal centrifugation
Neville[b]; Emmelot and Bos[c]; Emmelot et al.[d]	b	d	SW ↗
Neville[e]	b	r SW	SW ↗
Pfleger et al.[f]	b	r Z^B	Z^B ↙
Song et al.[g]	b	—	A ↗
Evans[h]	b[v]	r Z^A	SW ↙
Ray[i]	b + 0.5 mM $CaCl_2$	d	SW ↗
Newkirk and Waite[j]	b (10 mM)	r Z^B	—
Takeuchi and Terayama[k]	s + 0.5 mM $CaCl_2$	d, r SW	SW ↗
Coleman et al.[l]	s (0.3 M)	d[w]	SW ↙
Stein et al.[m]	s	d	A ↗
Berman et al.[n]	s + 0.5 mM $CaCl_2$	d	SW ↙ , ↗
Touster et al.[o]; Ashworth and Green[p]; Henning et al.[q]; House and Weidemann[r]	s	d	SW ↗ or ↙
El-Aaser et al.[s]	s (0.08 M)	(r Z^A)[x]	Z^A ↙
Weaver and Boyle[t]	s (0.08 M)	r Z^B	Z^B ↙
Hinton et al.[u]	s (0.08–0.25 M)	r Z^A	SW ↗

[a] In all cases plasma membranes were isolated from the crude nuclear fraction, except for studies[o,p,r] in which plasma membranes were also[o,r] or exclusively[p] isolated from the microsomal fraction. Media: bicarbonate (b; 1 mM unless indicated otherwise) or sucrose (s, 0.25 M unless indicated otherwise). Centrifugation methods: rate-dependent centrifugation (d, differential centrifugation with fixed-angle head rotor; r, rate zonal centrifugation) and isopycnic-zonal centrifugation (↙ and ↗, respectively, sedimentation and flotation in sucrose gradient). Rotors: A, fixed-angle head rotor; SW, swing-out rotor; Z^A and Z^B, zonal rotors A and B.

[b–u] Key to references: [b] D. M. Neville, Jr., J. Biophys. Biochem. Cytol. 8, 413 (1960). [c] P. Emmelot and C. J. Bos, Biochim. Biophys. Acta 58, 374 (1962). [d] P. Emmelot, C. J. Bos, E. L. Benedetti, and P. Rümke, Biochim. Biophys. Acta 90, 126 (1964). [e] D. M. Neville, Jr., Biochim. Biophys. Acta 154, 540 (1968). [f] R. C. Pfleger, N. G. Anderson, F. Snyder, Biochemistry 8, 2826 (1968). [g] C. S. Song, W. Rubin, A. B. Rifkind, and A. Kappas, J. Cell Biol. 41, 124 (1969). [h] W. H. Evans, Biochem. J. 116, 833 (1970). [i] T. K. Ray, Biochim. Biophys. Acta 196, 1 (1970). [j] J. D. Newkirk and M. Waite, Biochim. Biophys. Acta 225, 224 (1971). [k] M. Takeuchi and H. Terayama, Exp. Cell Res. 40, 32 (1965). [l] R. Coleman, R. H. Michell, J. B. Finean, and J. N. Hawthorne, Biochim. Biophys. Acta 135, 573 (1967). [m] Y. Stein, C. Widnell, and O. Stein, J. Cell Biol. 39, 185 (1968). [n] H. M. Berman, W. Gram, and M. A. Spirtes, Biochim. Biophys. Acta 183, 10 (1969). [o] O. Touster, N. N. Aronson, Jr., J. T. Dulaney, and H. Hendrickson, J. Cell Biol. 47, 604 (1970). [p] L. A. E. Ashworth and C. Green, Science 151, 210 (1966). [q] R. Henning, H. D. Kaulen, and W. Stoffel, Z. Physiol. Chem. 351, 1191 (1970). [r] P. D. R. House, and M. J. Weidemann, Biochem. Biophys. Res. Commun. 41, 541

liver tumors studied by the authors[4,7] there are rapidly growing, rather un-differentiated rat hepatomas, which contain large nuclei whose membranes are apparently easily disrupted by homogenization in the bicarbonate medium, however carefully performed. In such cases it has been found profitable to change the homogenization medium so as to afford protection to the nuclear membrane. Hardening of the nuclear membrane has been achieved either by replacing the bicarbonate medium by dilute citric acid[4,7] or by fortifying it with calcium ions.[4,8] The citric acid method of homogenization has been found less useful as to yield and properties of the resulting plasma membranes.[4] This is presumably due to the extensive fragmentation of the plasma-membrane skeleton following removal of endogenous Ca^{2+} from the membranes by citric acid.[9] This fragmentation may result in a fractionation of heterogeneous plasma-membrane fragments during preparation. By contrast, the presence of Ca^{2+} during homogenization protects the plasma membranes against breakage as shown by the larger size of the isolated membrane sheets, but it also has certain disadvantages referred to below.

Preparation of Plasma Membranes

Reagents. Analytical grade reagents are used.

Sodium bicarbonate, 1 mM in bidistilled water, pH 7.5, freshly prepared, without or with calcium chloride, 2 mM

Sucrose solutions in bidistilled water of 26.6 (*d* 1.10), 37.6 (*d* 1.14), 42.9 (*d* 1.16), 48.0 (*d* 1.18), 53.4 (*d* 1.20), and 81% (*d* 1.30) (% as w/v; d_4^{20} in parentheses).

---

[7] P. Emmelot, E. L. Benedetti, and P. Rümke, *in* "From Molecule to Cell" (P. Buffa, ed.), p. 253. (Symp. on Electron Microscopy Modena). C.N.R., Rome, 1964.

[8] P. Emmelot and C. J. Bos, *Biochim. Biophys. Acta* **121**, 434 (1966).

[9] E. L. Benedetti and P. Emmelot, *in* "The Membranes" (A. J. Dalton and F. Haguenau, eds.), p. 33. Academic Press, New York, 1968.

(1970). [s] A. A. El-Aaser, J. T. R. Fitzsimons, R. H. Hinton, E. Reid, E. Klucis, and P. Alexander, *Biochim. Biophys. Acta* **127**, 553 (1966). [t] R. A. Weaver and W. Boyle, *Biochim. Biophys. Acta* **173**, 377 (1969). [u] R. H. Hinton, M. Dobrota, J. T. R. Fitzsimons, and E. Reid, *Eur. J. Biochem.* **12**, 349 (1970).

[v] Subfractionation of plasma membrane fragments generated by vigorous rehomogenization in 8% (w/v) sucrose-Tris buffer.

[w] Vigorous rehomogenization of 1000 *g* pellet results in slower sedimentating plasma membrane fragments.

[x] Isolation carried out in a single step; plasma membranes reach equilibrium; particles with sedimentation constant smaller than that of plasma membranes do not.

Tissue Sources. Liver of rats (3-month-old inbred strain R-Amsterdam, and hybrids with other strains) and mice (strain CBA, of various ages). Animals are not fasted prior to sacrifice by decapitation. Gallbladder of the mouse liver is removed.

Hepatomas: Rat hepatoma-484, and its subline 484A, originally induced in a female rat of strain R by 4-dimethylaminoazobenzene, rather anaplastic and rapidly growing hepatoma of the liver-cell type, transplanted intraperitoneally (i.p.), and harvested after 10 days of growth. The peritoneal cavity is opened, then ascites fluid is removed by blotting with paper towel, and tumor nodules are dissected free from adhering, e.g., fat, tissue. (As is frequently found with this type of tumor, subcutaneous (s.c.) transplants cannot be used. The subcutaneous tumors usually have a necrotic core covered by a shell of living tumor cells and many fibrotic elements. The consistency of this tissue is too firm to allow the gentle homogenization required for the purpose of isolating plasma membranes, and the homogenization necessary for breaking up sufficient cells leads to extensive gel formation even in the presence of Ca^{2+}.)

Mouse hepatoma-147042 and -143066 arose spontaneously in old CBA males and were transplanted s.c. on similar young animals. These are well differentiated and slowly growing tumors, containing little necrosis and few fibrotic elements, harvested after 2 months of growth, on the average.[5] The excised tumors are dissected free from their well defined fibrotic capsules.

Plasma Membranes from Liver

All operations are carried out at 0–4° with prechilled materials.

Cell Rupture. Rat liver corresponding to 30–40 g of wet weight of tissue (20 g in the case of mouse liver) is collected in a beaker containing 50 ml of bicarbonate medium and is finely cut with scissors. Portions corresponding to about 5 g of cut liver are each homogenized in 20 ml of bicarbonate medium using an all-glass homogenizer of the Potter-Elvehjem type with a pestle clearance of 0.5–0.6 mm and tube content of about 50 ml. The pestle is driven at about 1400 rpm and should be carefully centered during the 4–6 up-and-down movements of the tube, each of which takes 5 seconds on the average.

The broken cell preparations are poured into a vessel containing 300 ml of bicarbonate medium, and the collected homogenate is diluted with similar medium to 500 ml, vigorously stirred for 2 minutes, and filtered twice through prewetted surgical gauze (18 threads/cm²), first through one layer and then through a double layer.

Low Speed Differential Centrifugations. Centrifugation is carried out in a cooled centrifuge capable of stable low speed. Accordingly, working quantities should be adapted to the volume capacity of the rotor.

The filtered suspension is equally divided among the centrifuge tubes (e.g., 6 of 100 ml), and subjected to 1500 g[10] for 10 minutes. After reaching standstill, the fatty layer (mainly triglycerides) floating at the surface of the supernatant is sucked off with a pipette equipped with a suction bulb, while the tubes are left standing in the rotor so as not to disturb the floating layer.

The supernatant is either decanted or sucked off by a pipette with water-pump aspiration, but, because of the loose packing of the sediment, care should be taken to leave a small layer of the supernatant above the undisturbed sediment. Any fatty material sticking to the wall of the tube is removed with the aid of filter paper. Bicarbonate medium, 5–10 ml, is added to each centrifuge tube, and the precipitate is suspended *in toto* by stirring with a glass rod. Each suspension is transferred, rinsing the tubes with bicarbonate medium, to a 75-ml smooth-walled tube equipped with a loosely fitting smooth Perspex pestle, and made up to a final volume of 35 ml with bicarbonate medium.

Further suspension is carried out with three very gentle strokes of the pestle by hand. The resulting six suspensions are transferred to 35-ml centrifuge tubes (glass or other translucent material). The centrifuge is slowly accelerated by first spinning for 5 minutes at 100 g and then 10 minutes at 1000 g. The centrifuge is also decelerated slowly to avoid disturbance of the precipitate. The precipitated material consists of two easily distinguishable portions: an upper layer of loosely packed membranes, pale tan in appearance, covering a large more consistent bottom layer of dark-red nuclear material and other debris. The bulk of the supernatant is drawn off with a pipette connected with a water pump; the remainder of the supernatant is drawn off without such suction in order not to interfere with the precipitate. Bicarbonate medium, 3–5 ml, is gently layered over the precipitate. The fluffy upper layer is suspended with the aid of a glass rod bent and flattened at one end, while leaving the bottom layer intact. The suspended membranes of two tubes each are carefully transferred with a Pasteur pipette to the hand-driven homogenizer, made up to a final volume of 35 ml of bicarbonate medium and further suspended by three gentle strokes. The three suspensions are recentrifuged for 10 minutes at 1000 g. The supernatant, which still contains many mitochondria, is removed, leaving behind a sediment consisting of a fluffy layer of plasma membranes above a small amount of more tightly packed, mainly nuclear material that sticks to the bottom of the tube. As before, the fluffy layer is carefully suspended, further suspended in the hand-driven homogenizer, and centrifuged (now in 2 tubes) for 10 minutes at 1000 g. The washing procedure is repeated. After the last centrifugation, the supernatant is no longer

---

[10] The centrifugal fields refer to the bottom of the tubes (fixed-angle rotor, 36.5°).

turbid. The precipitate consists of plasma membranes, mitochondria, and some nuclei. This material is used in the flotation step.

Flotation. After removal of the supernatant, the two precipitates are separately suspended and transferred using 3 ml of bicarbonate medium, to a 10-ml tube equipped with a loose Perspex pestle and calibrated at 3.6 ml. Bicarbonate medium is added to a final volume of 3.6 ml, and re-suspension is carried out by three gentle strokes by hand. The two suspensions are transferred with a pipette to two translucent tubes fitting the SW 25-1 swinging-bucket rotor of a Beckman L ultracentrifuge. Ten milliliters of a sucrose solution of d 1.30 is added drop by drop to both tubes with vigorous shaking in order to prevent membrane agglutination (and subsequent contamination of the membranes). This brings the density of the solution to 1.22. This discontinuous sucrose gradient is built up by carefully adding (along the wall of the tube and using a Pasteur pipette with a tip bent at 90°) 4.5 ml of a sucrose solution of d 1.20, 8 ml of d 1.18, and 4.5 ml of d 1.16. The two tubes containing the gradients and a third tube containing sucrose solution are tared, capped, and subjected to 70,000 g for 90 minutes. Plasma membranes gather at the d 1.16/1.18 interface as a compact band. At the d 1.20/1.22 and d 1.18/1.20 interfaces, hazy bands of mitochondria are present. The latter interface also contains some plasma membranes, but these together with the few plasma membranes which are part of the bottom pellet, are discarded.

The plasma membranes are harvested from the d 1.16/1.18 interface using a Pasteur pipette with a tip bent over 90°. The collected membranes are resuspended in the large hand-driven homogenizer by 3 strokes, transferred, and centrifuged in 35 ml of bicarbonate medium for 10 minutes at 2500 g. Routinely two such washings are applied, but for particular experiments (e.g., hexose determination) up to 5 washings have to be carried out. The final membrane precipitate is suspended in bicarbonate medium (routinely) or in any other medium desired for future use.

Comments

1. The method is suitable for the isolation of plasma membranes from rat and mouse[5,6] liver. For routine isolation 1 mM NaHCO$_3$ solution without CaCl$_2$ is used for homogenization and throughout all washing steps.

2. The membranes are gradually concentrated in the various washing cycles. For the last two washings and the flotation, the crude membrane fractions stemming from 30–40 g wet weight of rat liver are divided over two centrifuge tubes, instead of being collected in one. This has been found to improve yield and purity of the membranes, mitochondrial contamination being counteracted. Overloading of the gradient should be avoided, otherwise plasma membranes are trapped in layers of higher

densities containing mitochondria, and some mitochondria may be carried over by the membranes to the position where the latter arrive.

3. To judge the progress of separation, the entire procedure can be followed by phase contrast microscopy. In the final membrane sediment obtained by low speed centrifugation mitochondria are still present. Only if too many mitochondria are present (to be judged from experience) is it advisable to repeat washing. A useful criterion is the turbidity of the supernatant. It has been found that if the wash supernatant is no longer turbid, the mitochondria still present in the membrane sediment can easily be removed in the following flotation step provided the sucrose gradient is composed as indicated. Also the small amounts of nuclei and debris found in the membrane sediment at this stage are easily removed in the flotation step.

4. If, for comparative reasons, liver plasma membranes should be isolated in the manner required for the plasma membranes of certain hepatomas (next section), 2 mM $CaCl_2$ is added to the homogenization medium. Dilution of the homogenate and washings are, however, carried out with the unsupplemented bicarbonate medium. The number of washings should be increased (usually with another two washings) since the calcium ions promote the adherence of mitochondria to the plasma membranes. Under these conditions the first washings remove fewer mitochondria than do later ones, as shown by the turbidity of the supernatant and phase contrast microscopy of both supernatant and membranes. The use of calcium ions does not infrequently lead to the presence of some mitochondria in the final preparations (without $CaCl_2$ none are present), and introduces some cytoplasmic RNA in the membranes.

Plasma Membranes from Hepatomas

Mouse Hepatomas. Conditions for isolation of plasma membranes from slowly growing and well-differentiated hepatomas[5,6] are the same as those described for rat liver. Since a small part of the hepatoma plasma membranes may show a lower buoyant density, 3 ml of a sucrose layer of *d* 1.14 may be added on top of the gradient, at the expense of an equivalent part of all other layers, if this material gathering now at the *d* 1.14/*d* 1.16 interface, is wanted separately.

Rat Hepatomas. Basically, the same procedure is followed except for the following changes. From 4–5 animals 50–60 g of wet weight of tumor tissue is collected in a beaker containing 100 ml of bicarbonate medium fortified with 2 mM $CaCl_2$. After swirling the beaker, the tissue is allowed to settle, and most of the fluid is decanted. The tissue is finely cut and homogenized in 5-g portions each with 20 ml of bidistilled water containing 1 mM sodium bicarbonate and 2 mM $CaCl_2$. Homogenization is carried out very gently, by moving the homogenization tube very slowly up and down, each stroke taking about 8–10 seconds. The collected homogenate is diluted with bicarbonate medium (without $CaCl_2$ in this and the following steps) to 750 ml and stirred for about 2 minutes. For the first centrifugation of the homogenate in translucent tubes, the centrifuge is set at 100 g for 5 minutes, then slowly accelerated and kept for 10 minutes at 2500 g. Much more floating lipid material accumulates at the surface than in the case of rat liver; this and any floating lipid arising later in the procedure is removed as indicated. After the supernatant has been sucked off, the upper fluffy layer of the sediment is harvested as described for liver. Separation between the fluffy membranes and bottom layer is less distinct than in the case of liver because (a) the two layers differ less in color, both appearing slightly tan, and (b) the upper layer sticks somewhat to the bottom one. This first separation should be considered as approximate, visibly carrying over also some of the nuclear material. After resuspension by homogenization by hand, another centrifugation is carried out as before. Separation between membranes and small nuclear sediment is now more pronounced. The membranes, in two portions, are subjected to 2–3 washing cycles with bicarbonate medium during which centrifugation is carried out at 1500 g for 10 minutes. Supernatant containing mitochondria and the sticky bottom pellet are discarded each time. The number of washings is decided by the same criteria as applied to liver. The two membrane preparations are transferred to 2 tubes of the swinging-bucket rotor, as described for liver. Ten milliliters of sucrose solution of d 1.30 is added, followed by 3 ml of d 1.20, 4 ml of d 1.18, 4 ml of d 1.16, 3 ml of d 1.14, and finally 3 ml of d 1.10. Centrifugation is carried out as described. Most of the hepatoma membranes gather at the d 1.14/1.16 interface, and a few at the d 1.10/1.14 interface. At d 1.16/1.18 some membranes heavily contaminated with mitochondria are present; these are discarded. The membranes are collected and washed three times with bicarbonate medium; they are somewhat paler than the liver plasma membranes.

Comments

1. The differential centrifugations are carried out at higher speed than in the case of liver. The lower speed used for the latter is not fast enough to pack the majority of the tumor plasma membranes.

2. The two-step manner of the first and second centrifugation promotes stratification of the sediments, the nuclear material sedimenting faster so that its trapping of plasma membranes is counteracted. This is especially important for the process of hepatoma membrane separation because of the tendency of the nuclear material to gel formation.

3. For the same reason the initial separation of the plasma membranes

from the bulk of the nuclear material is performed already after the first centrifugation, so that during the subsequent resuspension gel formation and mutual exposure is reduced to a minimum.

4. The sucrose layer of d 1.10 is introduced to facilitate collection of the plasma membranes from the gradient. If the d 1.14 layer is the top one, part of the membranes is diffusely floating in this layer and hard to collect. With the d 1.10 layer present, these membranes are packed at the d 1.10/ 1.14 interface. Note that not all hepatomas should necessarily show these differences in buoyant densities of their plasma membranes.

Properties of Isolated Plasma Membranes

Rat-liver plasma membranes will be principally dealt with. In general, the mouse-liver plasma membranes[5,6,9] resemble those of the rat; salient differences will be indicated. The properties of hepatoma plasma membranes are markedly tumor-strain specific, and common characteristics distinguishing these membranes from normal are very scarce.[4-9,11-14] The presentation is limited to findings made in the authors' institute.

Yields. The isolation procedure aims not at quantity nor at rapidity, but at purity. Homogenization of the tissue is not complete in order not to fragment the membrane sheets too much, membranes stick to and are present in the nuclear pellets, membrane fragments may remain in the wash supernatant, and some 10% of the final membranes are lost in the flotation step.

The average yields of plasma membranes in milligrams of protein (biuret) per 10 g, fresh weight, of tissue are 3.5 for rat liver and 6.4 for mouse liver homogenized in bicarbonate medium, 1.8 for rat liver homogenized in the presence of 2 mM CaCl$_2$, and 1.2 for rat hepatomas and about 3.0 for mouse hepatomas. The higher yield of the mouse liver membranes is due to the softness of this tissue which allows more complete homogenization. Lower yield obtained following homogenization in the presence of Ca^{2+}, is due to an increased number of washings, and in the case of the rat hepatoma also by gel formation of the nuclear material trapping plasma membranes.

A rough estimate, based on work of Weibel *et al.*[15] shows the yield of rat-liver plasma membranes (*minus* the protein that is saline soluble, compare below) to be some 15% of the theoretical value. Apart from the aforementioned factors contributing to this restricted yield, the question arises

[11] P. Emmelot and C. J. Bos, *Biochim. Biophys. Acta* **211**, 169 (1970).
[12] P. Emmelot and C. J. Bos, *Biochim. Biophys. Acta* **249**, 285 (1971).
[13] P. Emmelot and C. J. Bos, *J. Membrane Biol.* 9, 83 (1972).
[14] R. P. van Hoeven and P. Emmelot, *J. Membrane Biol.* 9, 105 (1972).
[15] E. R. Weibel, W. Stäubli, H. R. Gnägi, and F. A. Hess, *J. Cell Biol.* **42**, 68 (1969).

as to the extent to which the isolated liver plasma membranes represent the liver cell surface *in situ*. Loosely bound surface materials, such as part of the glycocalyx, could be lost during preparation. Furthermore, the cell surface is composed of heterogeneous elements, for example, membranes lining Disse and bile canalicular spaces, apposed membranes, and structurally differentiated intercellular contacts and contact zones. Disruption by homogenization might create small fragments more easily from the first mentioned than from the other membrane areas, since the former are of relatively long extension and lack structural supports. Such small fragments (vesicles) could behave during the differential centrifugations as mitochondria and microsomes, and subsequently be lost. Plasma membranes have been isolated from the liver microsomal fraction (see Table I).

Buoyant Densities. Liver plasma membranes gather at the d 1.16/d 1.18 interface of the sucrose gradient. If another layer of d 1.17 is included in the gradient, the membranes are approximately equally divided over the two interfaces d 1.16/d 1.17/d 1.18. Thus 1.17 can be considered as the average buoyant density of these membranes. The buoyant density is a relative value dependent on the composition of the medium. The average buoyant density of rat-liver plasma membranes equilibrated in Ficoll, Urografin-[16] and glycerol-water gradients amounts to 1.08, 1.14, and 1.21, respectively, and in corresponding gradients containing ^2H_2O instead of H_2O to 1.14, 1.19, and 1.22, respectively (and 1.21 for a sucrose–^2H_2O gradient). The specific density of the waterless membrane material, calculated from the distance traveled by a drop of membrane-detergent (sodium deoxycholate, 1%, or dodecyl sulfate, 0.8%) solution against that of a drop of detergent solution in a continuous kerosene-bromobenzene density gradient,[17] amounts to 1.33 for fresh rat-liver plasma membranes, and to 1.29 for the saline-insoluble membrane portion.

Purity of Isolated Plasma Membranes. Rat-liver plasma membranes, isolated from plain bicarbonate homogenates, are free from any mitochondrial contamination as shown by electron microscopy[3,9] and the absence of mitochondrial enzymes[3] and cardiolipin.[14] Some (5–10%) microsomal contamination is present on the basis of glucose-6-phosphatase measurements, but the plasma membrane activity could very well stem from an aspecific phosphatase activity.[11] No smooth or rough microsomal membranes are present in the preparations[3,9,18] and certain drug-metabolizing

[16] N,N'-Diacetyl-3,5-diamino-2,4,6-triiodobenzoate, product of Schering A. G., Berlin, Germany.
[17] W. S. Bont, P. Emmelot, and H. Vaz Dias, *Biochim. Biophys. Acta* **173**, 389 (1969).
[18] P. Emmelot and E. L. Benedetti, *in* "Protides of the Biological Fluids" (H. Peeters, ed.), Vol. 15, p. 315. Elsevier, Amsterdam, 1968.

enzymes[18] and cytochromes[13] which are characteristic for liver microsomes (see under Enzymatic Composition below) are absent. Some 25% of the membrane protein is soluble in physiological saline (0.15 M NaCl), and this protein contains several enzymes and antigens of the liver cell cytoplasm.[3-9,13,18] The saline-soluble protein originates mainly, if not exclusively, from the soluble fraction of the homogenate. These proteins which are predominantly positively charged, interact with negatively charged plasma membranes proper under the hypotonic conditions prevailing throughout the isolation procedure. Associated with the saline-soluble protein is hemoglobin derived from lysed erythrocytes. This gives the isolated plasma membranes their faint reddish color. The saline-insoluble membrane portion, which represents the clean plasma membranes, is colorless. Prior perfusion of the liver, e.g., with physiological saline, yields uncolored membranes.

Morphology (as Studied by Electron Microscopy). GENERAL APPEARANCE. The preparations consist of large sheets of membranes interconnected by the various types of junctional complexes found in liver sections, i.e., the gap junction (nexus, formerly called tight junction), desmosome (macula adhaerens), intermediate junction (zonula adhaerens), next to many bile spacelike structures and some vesicles. The preparations are free from any other recognizable organelles. Appropriate illustrations can be found in various publications.[3,5,7,9,18] These give the strong impression that of the liver plasma-membrane skeleton most is preserved, except perhaps some of the blood front lining.

FINE STRUCTURES AND LOCAL SPECIALIZATION OF THE PLASMA MEMBRANES. The isolated membranes reveal[9] the "classical" triple-layered membrane element of an overall width of about 80 Å. Bile spacelike elements exhibit[9,18] an average membrane width of 95–100 Å. As shown by colloidal iron hydroxide staining, sialic acid is located in a rather regular spacing at the extracellular side of the membrane, and is absent from the inner side of the membrane leaflet.[19] Sialic acid in desmosomes and intermediary junctions is occluded by some Ca^{2+}-dependent mechanism, since it can be unmasked and demonstrated at the internal plates of these junctions after treatment of the membranes with EDTA. By these criteria, gap junctions lack sialic acid.

In negatively stained membrane preparations, globular knobs (50–60 Å diameter) are seen projecting from the membrane surface of certain membrane sheets or areas.[9,18] These knobs are released by exposing the isolated membranes to papain, but not to trypsin. The isolated knobs exhibit all the

[19] E. L. Benedetti and P. Emmelot, *J. Cell Sci.* **2**, 499 (1967).

Co^{2+}-activated aminopeptidase activities displayed by fresh membranes.[20,21] With leucyl-β-naphthylamide as substrate, histochemically demonstrable amino-peptidase activity is restricted to the plasma membranes lining the bile spaces. This activity can be considered as a marker enzyme for the globular knobs which coat the bile space-lining plasma membranes. In negative stain a hexagonal subunit (90 Å) pattern of restricted location is outlined.[6,9,22] This lattice is demonstrated by the gap junctions.[23,24] These junctions survive sodium deoxycholate (1%) treatment of the isolated membranes and are thus obtainable by centrifugation, but in an as yet impure form.[23,24]

Only undifferentiated rapidly growing hepatomas may lack both the globular knobs (although not the enzymes contained therein) and the gap junctions.[6] *Structural* continuity between a few ribosome-dotted microsomal vesicles and the isolated plasma membranes has been observed for the various hepatomas examined, but not for liver.[6]

Chemical Composition (Table II). The main components of hepatic plasma membranes are proteins and lipids, amounting to 66 and 30.5%, respectively, of the dry membrane weight in the case of rat liver.[9,13,14] Hepatic plasma membranes are characterized by a high content of cholesterol and protein-bound sialic acid,[9,19] and a particular phospholipid class profile in which the high contents of sphingomyelin and phosphatidylserine are noteworthy.[14] These components may be considered as chemical "markers" which distinguish the liver plasma membranes from intracellular membrane species, except lysosomal membranes.[25] The chemical composition may help to provide criteria for judging the purity of liver plasma membranes preparations (e.g., the absence of cardiolipin).

The various types of hepatoma plasma membranes contained significantly more cholesterol per micromole of phospholipid than did the liver plasma membranes.[14] The former's RNA content was also increased due to ribosomal RNA of the few rough microsomal vesicles structurally connected with the hepatoma plasma membranes.[13] Liver plasma membranes contain about 1% RNA on a protein basis. The presence of calcium ions during homogenization increases this amount by 50%, and this addi-

[20] P. Emmelot, A. Visser, and E. L. Benedetti, *Biochim. Biophys. Acta* 150, 364 (1968).
[21] P. Emmelot and A. Visser, *Biochim. Biophys. Acta* 241, 273 (1971).
[22] E. L. Benedetti and P. Emmelot, *J. Cell Biol.* 26, 299 (1965).
[23] E. L. Benedetti and P. Emmelot, *J. Cell Biol.* 38, 15 (1968).
[24] P. Emmelot, C. A. Feltkamp, and H. Vaz Dias, *Biochim. Biophys. Acta* 211, 43 (1970).
[25] Secondary lysosomes are partly derived from the plasma membranes.

TABLE II
CHEMICAL COMPOSITION OF RAT LIVER PLASMA MEMBRANES

Protein

Percent of dry weight (total membranes)	66 ± 2 (61–69)
Percent of dry weight[a] (saline-insoluble membranes)	58
Soluble in 0.15 M NaCl (% of total protein)	25 ± 4 (18–33)
Total protein per μmole phospholipid-P (mg)	2.83 ± 0.12

Lipids

Percent of dry weight (total membranes)	30.5 ± 0.7
Total lipids per mg protein (mg)	0.46 ± 0.01
Cholesterol/phospholipid-P (molar ratio)	0.65 ± 0.06
Cholesteryl esters (mole-% of total cholesterol)	1–1.5
Free fatty acids (μmoles per mg protein)	0.088 ± 0.039
Triglycerides (μmoles per mg protein)	0.035 ± 0.002
Phospholipids (% of total lipids)	59.9 ± 3.9
Phospholipid composition (% of lipid-P)	
Sphingomyelin[b]	23.2 ± 2.1
Phosphatidylcholine	30.0 ± 2.0
Phosphatidylethanolamine[c]	19.3 ± 1.3
Phosphatidylserine	15.2 ± 0.8
Phosphatidylinositol	6.0 ± 0.5
Phosphatidic acid	2.3 ± 1.1
Lysophosphatidylcholine	3.5 ± 1.1
Lysophosphatidylserine	0.5 (0–1.6)
Cardiolipin	0.0

Nucleic acids

Percent of dry weight (total membranes)	0.6
RNA (μg per mg protein)	8.1 ± 0.4
DNA (μg per mg protein)	1.4 ± 0.1

Bound carbohydrates

Percent of dry weight (total membranes)	2.2
Sialic acids[d] (nmoles per mg protein)	33 ± 2
Percentage neuraminidase-resistant sialic acid	69 ± 3
Hexoses (nmoles glucose per mg protein)	65 ± 2
Hexosamines (nmoles per mg protein)	61 ± 4

[a] Dry weight corrected for saline-soluble protein.
[b] Containing some 1% lysophosphatidylethanolamine.
[c] 17.5% as plasmalogen and 82.5% as diacyl.
[d] Mainly N-acetylneuraminic acid; at least 95% of the sialic acid is protein bound.

tional RNA (very probably transfer RNA) is fully removed by 0.15 M NaCl. DNA is present in trace amounts (rat liver) or below the level of detection (mouse liver). Its increased presence in hepatoma membranes is due to additional contamination.[13]

TABLE III
SPECIFIC ENZYME ACTIVITIES OF ISOLATED RAT-LIVER PLASMA MEMBRANES

Enzyme	μMoles of substrate converted or product formed/mg protein/hour
Mg^{2+} (or Ca^{2+})-ATPase*[a]	46.4 ± 7.8
(Na^+-K^+)ATPase*	11.7 ± 2.3
5'-Mononucleotidase*	51.0 ± 6.7
Glycerolphosphatase	
Alkaline* ♂	3.8 ± 0.4
♀	1.0 ± 0.2
Acid	0.4 ± 0.1
p-Nitrophenylphosphatase	
Alkaline* ♂	3.31 ± 0.2
♀	1.97 ± 0.04
Acid*　　♂	9.1 ± 1.3
♀	7.3 ± 0.33
K^+-nitrophenylphosphatase*	
Alkaline ♂	1.6 ± 0.3
♀	1.6 ± 0.5
Acetylphosphatase*	11.4 ± 2.2
K^+-acetylphosphatase*	10.8 ± 1.9
ADPase*, Mg^{2+}	20.3 ± 1.2
Ca^{2+}	42.2 ± 3.9
IDPase*, Mg^{2+}	30.1 ± 3.2
NAD pyrophosphatase*	5.72 ± 0.2
Inorganic pyrophosphatase*[,b] (P_i)	4.0 ± 0.3
Co^{2+}-Aminopeptidase(s)*	
Leucinamide	6.5 ± 0.7
Triglycine	8.4 ± 0.9
Leucylglycylglycine	11.3 ± 1.2
Leucylglycine	1.4 ± 0.16
Leucyl-β-naphthylamide	3.9 ± 0.2
Adenylcyclase* (nmoles cyclic AMP)	
Basal	2.3 ± 1.0
NaF	25.3 ± 4.2
Glucagon	44.2 ± 8.3
Epinephrine	11.1 ± 2.8
Lipase (μg triolein)	267
Phosphodiesterase (bis-(p-nitrophenyl)phosphate)	
Alkaline*	3.6 ± 0.4
Acid[c]	0.7 ± 0.1
Ribonuclease ($A_{260\ nm}$)	
Alkaline[d]	2.2 ± 0.50
Acid[e]	0.54 ± 0.24
Glucose-6-phosphatase[f]	1.0 ± 0.3
PP_i-glucose transferase	0.18 ± 0.01

TABLE III (*Continued*)

Enzyme	μMoles of substrate converted or product formed/mg protein/hour
Esterase, nonspecific	
α-Naphthyl laurate	0.46 ± 0.02
α-Naphthyl caprylate	16.1 ± 2.3
p-Nitrophenylacetate	2.5 ± 0.5
Triose-3-P dehydrogenase[g]	2.04 ± 0.3
Aspartate aminotransferase[h]	128 ± 15
NADPH–cytochrome c reductase	7.68 ± 0.5
NAD nucleosidase	0.5 ± 0.1

[a] Enzymes marked with an asterisk have been shown, or can reasonably be considered to be intrinsic, though not necessarily exclusive, components of plasma membranes.

[b] Distinct from the microsomal enzyme.[i]

[c] This enzyme appears to be a genuine component of rat hepatoma membranes[i]; some 30% of the liver membrane activity is released by 0.15 M NaCl.

[d] Very probably a genuine plasma-membrane enzyme.[k]

[e] Half of the activity is soluble in 0.15 M NaCl.[l]

[f] Not due to microsomal contamination in the case of rat-hepatoma plasma membranes.[i]

[g] Completely removable by 0.15 M NaCl.[m,n]

[h] Nearly completely removable by 0.15 M NaCl, specific activity of saline-insoluble membrane fractions amounting to 4.2 ± 0.6 μmoles.[n]

[i] P. Emmelot and C. J. Bos, *Biochim. Biophys. Acta* **211**, 169 (1970).

[j] P. Emmelot and C. J. Bos, *Int. J. Cancer* **4**, 705 (1969).

[k] K. A. Norris, M. L. E. Burge, and R. H. Hinton, *Biochem. J.* **122**, 53 P (1971).

[l] P. Emmelot and E. L. Benedetti, *in* "Protides of the Biological Fluids" (H. Peeters, ed.), Vol. 15, p. 315. Elsevier, Amsterdam, 1968.

[m] P. Emmelot and C. J. Bos, *Biochim. Biophys. Acta* **121**, 434 (1966).

[n] P. Emmelot and C. J. Bos, *J. Membrane Biol.* **9**, 83 (1972).

Enzymatic Composition (*see Table III, and references cited in foot-notes 2–9, 11–13, 18, 20, 21, 26–29*). The enzymes marked with an asterisk in Table III either have been shown, or may reasonably be considered, to be intrinsic, though not necessarily exclusive, components of the plasma membranes. Some of the enzymes may provisionally—until more is known of possible isozymes—be used as marker enzymes for hepatic plasma membranes, e.g., 5'-nucleotidase (at least in the case of rat liver; the specific activity of the enzyme in mouse-liver plasma mem-

[26] P. Emmelot and C. J. Bos, *Biochim. Biophys. Acta* **121**, 375 (1966).

[27] P. Emmelot and C. J. Bos, *Biochim. Biophys. Acta* **120**, 369 (1966).

[28] P. Emmelot and C. J. Bos, *Biochim. Biophys. Acta* **150**, 341 (1968).

[29] P. Emmelot and C. J. Bos, *Biochim. Biophys. Acta* **249**, 293 (1971).

branes is one-fourth that of the rat), Co^{2+}-activated aminopeptidase(s), fluoride- and glucagon-stimulated adenylcyclase, NAD-pyrophosphatase.

Enzymes which are also present in the isolated plasma membranes, but whose authenticity is not settled or may stem from (microsomal) contamination, are also included in Table III.

Enzymes below the level of detection of the analytic methods used are: hexokinase and glucokinase (hexokinase is present in and a genuine component of rat-hepatoma plasma membranes), 3'-mononucleotidase, monoamine oxidase, succinate–cytochrome c reductase system, cytochrome c oxidase, phosphoprotein phosphatase, various proteolytic enzymes, microsomal NADPH-oxidase and N-demethylase (substrate: dimethylamino-antipyrine). Cytochromes of liver, including cytochrome c and the microsomal P450 and b_5, could not be detected.

Organ-Specific or Differentiation Antigens Present in Rat-Liver Plasma Membranes. By using a heterologous serum (rabbit) directed against isolated rat-liver plasma membranes freed from their saline-soluble protein, and suitable absorption, three classes of insoluble liver-specific components, protein in nature, have been detected in the liver cell surface and in isolated plasma membranes.[30] One of these is located in the bile space-lining membranes and found to reside in the 50–60 Å globular knobs coating these membranes, thus providing independent evidence for the location of these knobs. Another is present at other regions of the plasma membrane skeleton. The third type of liver-specific component is normally masked, but can be uncovered by very mild trypsin digestion.[31]

[30] K. A. Norris, M. L. E. Burge, and R. H. Hinton, *Biochem. J.* **122**, 53 P (1971).
[31] J. B. Sheffield and P. Emmelot, *Exp. Cell Res.* **71**, 97 (1972).

[6] Isolation of Rat Liver Plasma Membrane Fragments in Isotonic Sucrose

By NATHAN N. ARONSON, JR. and OSCAR TOUSTER

During the differential centrifugation of liver homogenates, certain enzymes are noted to exhibit a bimodal distribution between the nuclear and microsomal fractions. It has been shown that many such enzymes are in actuality components of the plasma membrane.[1,2] Indeed, a good first assumption to be made is that an enzyme is localized on the plasma membrane of rat liver cells if it exhibits such a nuclear-microsomal bimodal

[1] O. Touster, N. N. Aronson, Jr., J. T. Dulaney, and H. Hendrickson, *J. Cell Biol.* **47**, 604 (1970).
[2] C. de Duve, *J. Cell Biol.* **50**, 20D (1971).

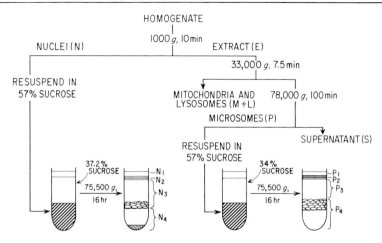

FIG. 1. Scheme showing preparative procedure for liver plasma membranes.

distribution pattern upon differential centrifugation of the homogenate in 0.25 M sucrose, and yet is found to be absent from nuclei purified by a technique such as that of Chauveau et al.[3] Thus, methods for preparing liver plasma membranes are generally based on separating cell-surface membrane fragments from either the nuclear[4] or microsomal[5] portion of the tissue homogenate. The procedure for isolating rat liver plasma membranes described below is based on these principles of tissue fractionation; it is a quantitative method by which plasma membrane fragments are separated from both the nuclear and microsomal fractions of a liver homogenate made in isotonic sucrose. In addition to being a good preparative procedure for liver plasma membranes in terms of yield and purity, it permits a quantitative characterization of most other subcellular organelles during the same experiment. The procedure is shown schematically in Fig. 1.

Assays

Only the assays for measuring 5'-nucleotidase, phosphodiesterase I, and glucose-6-phosphatase will be described. Protein is measured by a modification of the Lowry procedure.[6] Nicotinamide adenine dinucleotide phosphate (NADPH) cytochrome c oxidoreductase,[7] succinate–cytochrome c

[3] J. Chauveau, Y. Moulé, and C. Rouiller, Exp. Cell Res. 11, 317 (1956).
[4] P. Emmelot, this volume [5].
[5] D. F. H. Wallach and V. B. Kamat, this series, Vol. 8, p. 164.
[6] G. L. Miller, Anal. Chem. 31, 964 (1959).
[7] G. L. Sottocasa, B. Kuylenstierna, L. Ernster, and A. Bergstrand, J. Cell Biol. 32, 415 (1967).

oxidoreductase,[7] N-acetyl-β-D-glucosaminidase,[1] and cytochrome oxidase[8] were assayed according to published procedures.

5'-Nucleotidase

Reagents

(a) Na-AMP, 50 mM, pH 7.0
(b) Glycine–NaOH buffer, 0.5 M, pH 9.1
(c) MgCl$_2$, 0.1 M
(d) Trichloroacetic acid (TCA), 8% (w/v)

Procedure. The activity of 5'-nucleotidase is determined by measuring the rate of release of inorganic phosphate from 5'-AMP. A reagent assay mixture is prepared by combining solutions (a), (b), and (c) and water in the volume ratio of 1:2:1:5. This mixture can be stored frozen and used for at least a year. Twenty micrograms of plasma membrane protein in a volume of 0.05 ml are added to 0.45 ml of this reagent mixture in 13 × 100 mm glass tubes; once the enzyme has been added, the sample is incubated for 15–30 minutes at 37° in a shaking incubator bath. The following control tubes are also carried through the assay in a final volume of 0.5 ml: (1) a water blank, (2) 20 μg of enzyme, (3) 0.45 ml of assay mixture, and (4) standard phosphate solution containing 0.5 μmole of sodium phosphate. The reactions and controls are stopped by adding 2.5 ml of 8% TCA at room temperature. The acidified samples are immediately placed in ice and then centrifuged at 3000 g for 5 minutes at 5°. Two-milliliter aliquots of the supernatant solutions are removed from each sample and control, added to an 18 × 150 mm test tube, and diluted with 2.3 ml of water. The inorganic phosphate in each tube is determined by the following modification of the procedure of Fiske and SubbaRow.[9] To the diluted aliquots is added 0.5 ml of a 2.5% (w/v) solution of ammonium molybdate in 5 N H$_2$SO$_4$. The solution is mixed, and 0.2 ml of 1-amino-2-naphthol-4-sulfonic acid reagent (Fischer Gram-Pac, Fischer Scientific, Co.) is added. The solution is mixed, and after 10 minutes the optical density of the solution is read at 660 nm against the water blank. Results are expressed as the rate of release of inorganic phosphate in micromoles per minute per milligram of protein.

[8] F. Appelmans, R. Wattiaux, and C. de Duve, *Biochem. J.* **59**, 438 (1955).
[9] C. H. Fiske and Y. SubbaRow, *J. Biol. Chem.* **66**, 375 (1925).

Phosphodiesterase I[10]

Reagents

(a) Na-*p*-nitrophenyl 5'-thymidylate, 5 mM, stored frozen
(b) Tris·HCl, 0.1 M, pH 9.0
(c) TCA, 8% (w/v)
(d) 0.133 M glycine–83 mM Na_2CO_3–67 mM NaCl (adjusted to pH 10.7 with NaOH)

Procedure. Phosphodiesterase activity is measured by determining the rate of release of *p*-nitrophenol from the synthetic nucleotide substrate, *p*-nitrophenyl-5'-thymidylate (Calbiochem, Los Angeles, California). The reactions are carried out in 13 × 100 mm glass tubes. Each assay contains 0.1 ml each of soltuions (a) and (b) and 0.25 ml of water. Each reaction is started by the addition of 10 μg of plasma membrane protein in a volume of 0.05 ml. The following control tubes are also carried through the assay in a volume of 0.5 ml: (1) a water blank; (2) a standard solution of *p*-nitrophenol (50 nmoles/0.5 ml); (3) substrate (a) alone in buffer (b); and (4) 10 μg of enzyme alone in buffer (b). After incubation for 10–15 minutes at 37° in a shaking bath, the reactions and controls are stopped by the addition of 1.5 ml of 8% TCA at room temperature. These samples are immediately placed on ice and centrifuged at 3000 g for 5 minutes at 5°. A 1-ml aliquot of each supernatant solution is pipetted into 3 ml of basic buffer (d), and the optical density of the resultant solution is read at 400 nm against the water blank. The activity is expressed as micromoles of *p*-nitrophenol released per milligram of protein per minute.

Glucose-6-Phosphatase

Reagents

(a) Na-Glucose 6-phosphate, 0.1 M, pH 6.5
(b) Histidine, 35 mM, pH 6.5
(c) Na-EDTA, 10 mM, pH 7.0

Procedure. Glucose-6-phosphatase is measured by determining the rate of release of inorganic phosphate from glucose 6-phosphate. A reagent assay mixture is made by combining solutions (a), (b), (c) and water in a volume ratio of 2:5:1:1. This mixture may be stored frozen and used

---

[10] This enzymatic activity is easily assayed and is as good a marker enzyme for rat liver plasma membranes as the more commonly used marker, 5'-nucleotidase (see Touster *et al.*[1]).

for at least a year. To begin the reaction 100 μg of plasma-membrane protein in a volume of 0.05 ml are added to 0.45 ml of the above assay mixture. In a volume of 0.5 ml, the following control tubes are also carried through the assay: (1) a water blank; (2) 100 μg of enzyme only; (3) 0.45 ml of the assay mixture; and (4) a standard phosphate solution containing 0.5 μmole of sodium phosphate. The reactants and controls are incubated at 37° for 15–30 minutes in a shaking water bath, 2.5 ml of 8% TCA are added to each tube to stop the reactions, and inorganic phosphate is determined in 2-ml aliquots of the acidified samples by the exact procedure described above for the assay of 5'-nucleotidase. The results are presented as micromoles of inorganic phosphate released from the glucose 6-phosphate substrate per milligram of protein per minute. Generally, the control with substrate only has some inorganic phosphate present.

Preparation of Membranes

Step 1. Preparation of Sucrose Solutions

(a) 1000 ml of 0.25 M sucrose (pH 8.0, Tris·HCl, 5 mM)
(b) 125 g of 37.2% (w/w) sucrose solution (pH 8.0, Tris·HCl, 5 mM)
(c) 125 g of 34% (w/w) sucrose solution (pH 8.0, Tris·HCl, 5 mM)
(d) 200 g of 57% (w/w) sucrose solution (pH 8.0, Tris·HCl, 5 mM)

All sucrose solutions are made 5 mM in Tris and are subsequently adjusted to pH 8.0 with 2 N HCl at room temperature. The pH adjustment is essential for the success of the sucrose density gradient flotation of plasma membranes described in steps 4 and 5. Solutions (b), (c), and (d) are used for the preparation of the gradients. Their sucrose concentrations in percent are based on the total weight of the final solution. In order to calculate the corresponding volumes of these solutions, the total weight of each is divided by its respective density value. The density value is obtained from the table of sucrose densities in water at 5°.[11]

Weigh out the proper amount of sucrose and, knowing the final volume of the solution to be made, add enough 1 M Tris solution to yield a final concentration of 5 mM. Finally, add that amount of water to bring the solution to the desired total weight. After each sucrose solution is completely dissolved by shaking on a mechanical shaker, the pH is adjusted

[11] N. G. Anderson, "Handbook of Biochemistry," p. J-248. Chem. Rubber Publ. Co., Cleveland, Ohio, 1968.

at room temperature to 8.0 by the dropwise addition of 2.0 N HCl. The weight percentage of sucrose is checked at 25° in an Abbe-3L refractometer (Bausch and Lomb Inc., Rochester, New York) and, within 0.5%, should be the calculated weight percentage of sucrose.

Step 2. Preparation and Homogenization of Tissue. Two male rats, of either the Sprague-Dawley or the Wistar strain, each weighing 200–300 g, are fasted for 20 hours to deplete liver glycogen. The animals are killed by decapitation; the livers are removed and placed into cold 0.25 M sucrose. The livers are weighed and then perfused with cold 0.25 M sucrose via the hepatic portal vein until totally blanched.[12] Normally about 20 g of liver are processed, but the procedure may be scaled up or down. The livers (kept cold throughout this processing) are minced on an ice-cold glass plate, and, in order to remove connective tissue, are then passed through a tissue press (screw-type or Arbor tissue press, coarse sieve, Harvard Apparatus Co., Millis, Massachusetts).[13] Three volumes of 0.25 M sucrose per gram of liver are added to the pulverized tissue; the suspension is then mixed with a spatula and homogenized with one up-and-down stroke of the pestle of a Potter-Elvehjem homogenizer mounted in a drill press (Teflon pestle, glass homogenizer tube Size C, A. H. Thomas, Philadelphia, pestle speed 1000–1200 rpm).

Step 3. Preparation of Nuclear (N) and Microsomal (P) Fractions. The nuclear fraction (N) is separated by centrifugation of the homogenate at 2000 rpm (1000 g) for 10 minutes in the International rotor No. 269 (International Equipment Co., Needham Heights, Massachusetts).[14] The supernatant fluid is poured off and saved, while the pellet is resuspended in the same initial volume of 0.25 M sucrose and rehomogenized with one up-and-down stroke of the pestle. The nuclear fraction is again separated as before, and the supernatant liquids are combined. Rehomogenization and recentrifugation of the nuclear pellet are performed a third time. The combined supernatant solutions (E) are saved and processed as described below. The separation of the plasma membranes (N₂) from the washed nuclear pellet is described in step 4.

The postnuclear extract (E) is used to prepare a combined heavy and light mitochondrial fraction (M + L), which contains both mito-

---

[12] One can also perfuse *in situ* prior to excising the liver. Perfusion is important since otherwise red blood cells will be found in the nuclear fraction (N) and some hemoglobin released from disrupted blood cells will be bound to the microsomal membranes (P). Also, perfusion minimizes the possibility of contaminating the membrane fractions with erythrocyte ghosts.

[13] Removal of connective tissue in this manner will minimize its contamination of fraction (N).

[14] The nuclear pellet is best isolated in a centrifuge rotor with swing-out type buckets.

chondria and lysosomes, by centrifugation of this solution at 33,000 g for 7.5 minutes in the No. 30 rotor, Spinco Model L-2 centrifuge (Spinco Division, Beckman Instruments, Inc., Palo Alto, California). The rotor is accelerated at its maximum rate until the speed reaches 25,000 rpm, at which point the speed control is turned down to 25,000 rpm, and finally the rotor is stopped at maximum deceleration with the brake. The total time from start of acceleration to start of braking is 7.5 minutes.[15] The supernatant solution, including the "fluffy pink" layer loosely packed on top of the M + L pellet, is carefully removed by aspiration. This loose layer is most easily discerned by holding the centrifuge tube in such a way that the face of the M + L pellet is horizontal during the removal of the bulk of the supernatant liquid. When most of this liquid has been removed by aspiration, the tube is turned 90° in order to remove the pinkish material which slides off the M + L pellet. In order to wash the M + L pellet, it is resuspended by gentle rubbing with a glass rod into three volumes of 0.25 M sucrose per gram of liver and homogenized with one up-and-down stroke of the pestle. The centrifugation, and the careful removal, of the supernatant liquid from the M + L pellet are repeated as described above. The washed M + L pellet is resuspended in 0.25 M sucrose and saved for enzyme assays.[16]

In order to obtain a microsomal pellet (P), the combined M + L supernatant fractions are centrifuged in the No. 30 rotor at 30,000 rpm (78,000 g) for 100 minutes. The resulting final supernatant fraction (S), including the congealed fat floating on its surface, is removed by aspiration, while care is taken to leave behind a whitish layer which is loosely packed on top of the microsomal pellet. The separation of the plasma membrane fraction (P_2) from the latter material is described in step 5.

Table I shows the intracellular distribution and overall recoveries of all the marker enzymes assayed in preparing the four main subcellular fractions. The dual localization of 5′-nucleotidase and phosphodiesterase I in the microsomal and nuclear fractions is apparent. The isolation of the plasma membranes from each of these fractions is described in steps 4 and 5.

Step 4. Gradient Separation of Plasma Membranes from (N). The nuclear pellet (step 3) is resuspended in three volumes of 0.25 M sucrose per gram of liver with one up-and-down stroke of the homogenizer. In

[15] The M + L fraction may also be prepared in a RC-2 centrifuge (Sorvall, Inc., Norwalk, Connecticut) by centrifugation of extract (E) for 11.5 minutes at 15,000 rpm in the SS-34 rotor.

[16] If the M + L pellet is not washed, a larger portion of the plasma membrane fragments will be found in this portion of the fractionated homogenate and the amount in the microsomal fraction will be reduced.

TABLE I

INTRACELLULAR DISTRIBUTION OF ENZYMES IN RAT LIVER[a]

Enzyme	Liver homogenate (E + N)	Fraction				Recovery
		N (%)	M + L (%)	P (%)	S (%)	
Protein (5)	181 ± 3	11.8 ± 3.6	33.5 ± 2.3	24.1 ± 4.0	28.7 ± 3.0	98.1 ± 5.5
5'-Nucleotidase (5)	13.6 ± 2.4	22.8 ± 4.0	12.9 ± 2.6	54.5 ± 2.1	10.9 ± 4.5	101.1 ± 6.2
Phosphodiesterase I (4)	4.43 ± 0.1	17.4 ± 3.4	13.5 ± 4.8	50.9 ± 6.0	4.5 ± 1.6	86.3 ± 6.5
Glucose-6-phosphatase (5)	15.5 ± 1.2	7.0 ± 2.6	22.0 ± 2.2	68.1 ± 2.5	5.3 ± 1.9	102.4 ± 6.8
NADPH-cytochrome c reductase (2)	3.17 ± 0.1	4.6 ± 0.6	14.9 ± 4.1	68.4 ± 3.0	3.3 ± 0.6	92.2 ± 1.9
Cytochrome oxidase (1)	17.9	7.2	79.8	1.2	0	88.2
Succinate-cytochrome c reductase (2)	14.6 ± 2.5	6.6 ± 0.9	96.0 ± 4.7	5.3 ± 5.6	0	107.9 ± 4.2
N-Acetyl-β-D-glucosaminidase (3)	2.2 ± 0.4	6.3 ± 1.1	83.7 ± 1.6	5.4 ± 1.1	1.5 ± 1.3	96.9 ± 0.8

[a] One unit of enzyme activity equals 1 μmole of substrate changed per minute, except for cytochrome oxidase, which is given in units per minute as described by F. Appelmans, R. Wattiaux, and C. de Duve, *Biochem. J.* **59**, 438 (1955). Protein units are in milligrams. Statistics refer to the means ± the standard deviations, with the number of experiments indicated in parentheses. E = extracts after removal of nuclear fraction, N = nuclear fraction, M + L = combined mitochondrial and lysosomal fraction, P = microsomal fraction, S = final supernatant. For each fraction, % is based on E + N representing 100%. For details, see O. Touster, N. N. Aronson, Jr., J. T. Dulaney, and H. Hendrickson, *J. Cell Biol.* **47**, 604 (1970).

order to pack the nuclear pellet firmly, this homogenate is centrifuged at 25,000 rpm for 7.5 minutes in the No. 30 rotor, and the supernatant fluid is removed by aspiration and discarded. At the bottom of the packed nuclear pellet is a small pellicle of red cells and a tan material. The nuclei can be poured off the latter, which is discarded. To the nuclei, 1.5 volumes of 57% sucrose (w/w) per gram of liver are added, and the resulting suspension is mixed with a spatula and homogenized with two up-and-down strokes of the homogenizer. The homogenizer is rinsed twice with 5 ml of 57% sucrose, and the washings are combined with the homogenate. The final volume is brought to three volumes per gram of liver with 57% sucrose. The weight percentage of sucrose in the thoroughly mixed homogenate is determined with the refractometer and should be at least 47.0%. From 20 to 25 ml of this nuclear suspension in heavy sucrose are placed in the bottom of cellulose nitrate tubes for the Spinco SW 25.2 rotor. Above the sample, 25 ml of 37.2% sucrose are carefully layered from a pipette whose tip is placed against the wall of the centrifuge tube, which is submerged in ice. Finally, enough 0.25 M sucrose to make the total volume 58 ml is slowly layered above the 37.2% sucrose. The discontinuous gradient is centrifuged at 25,000 rpm for 16 hours[17] in the SW 25.2 rotor. After the centrifuge is stopped with the brake on, the tubes are removed and placed in ice. They are then fractionated as shown in Fig. 1 into four fractions. Material is carefully withdrawn from the meniscus at the wall of the centrifuge tube by means of a capillary tube connected to a peristaltic pump.

The first (top) fraction (N_1) contains congealed fat and is free of all enzyme activity. The plasma membrane fraction (N_2) is a band appearing at the interface between the 37.2% sucrose ($\rho_{5^\circ} = 1.17$) and the 0.25 M sucrose overlayer. It consists of small pieces of whitish material. The 37.2% sucrose layer is essentially clear. The material banding at the interface between the original sample and the 37.2% sucrose layer (N_3) is mainly mitochondria and the other cell particulates, including a considerable amount of plasma membrane but excluding the nuclei, which form a pellet (N_4) at the bottom of the centrifuge tube. If the sucrose concentration of the sample fraction is less than 47%, most of the mitochondria and other particulate matter in fraction N_3 will sediment to the bottom of the tube along with the nuclei.

Step 5. Gradient Separation of Plasma Membranes from (P). The microsomal pellet (step 3) is placed in a homogenizer together with 1.5 volumes (per gram of original tissue) of 57% sucrose, thoroughly mixed

---

[17] This time period was chosen for convenience; that is, it is essentially an overnight run in the centrifuge. It is very likely that this time of centrifugation could be reduced considerably.

with a spatula, and homogenized with 2 up-and-down strokes of the pestle. The homogenizer is rinsed twice with 5 ml of 57% sucrose, and the washings are combined with the homogenate. The final volume is brought to 2.5 volumes per gram of liver with 57% sucrose. The weight percentage of sucrose in the thoroughly mixed homogenate is checked by use of the refractometer and should be at least 49.0%. Of this microsomal extract in heavy sucrose, 15–20 ml are placed in the bottom of a cellulose nitrate tube for the Spinco SW 25.2 rotor; 25–30 ml of 34.0% sucrose are layered above the sample; and finally enough 0.25 M sucrose is placed on top of the 34.0% sucrose to make a total volume of 58 ml. Centrifugation and fractionation of the discontinuous gradients are performed exactly as for the isolation of plasma membranes (N_2) from the nuclear fraction as described above (step 4).

The uppermost fraction (P_1) contains a small amount of congealed fat and is free of enzymes. The plasma membrane fraction (P_2) is a thick band of white material appearing between the 34% ($\rho_{5°} = 1.15$) and the 0.25 M sucrose overlayer. Fraction P_3 is cloudy and pale yellow and contains both plasma membrane and endoplasmic reticulum enzymes, while P_4 contains, by enzymatic estimation, the bulk of the membrane derived from the endoplasmic reticulum placed in the gradient, with a considerable amount of plasma membrane also present. (When fraction P_4 is subdivided into the particulate matter at the top of the fraction and the clear, yellow portion at the bottom, lacking any particulate matter, there is very low activity of plasma membrane enzymes in the latter clear solution.) One should avoid adding any of fraction P_3 to fraction P_2 during fractionation of the gradient, since the former fraction contains a considerable amount of internal membrane, which will lower the specific activity of the isolated plasma membranes (P_2).

Step 6. Washing of Membranes N_2 and P_2. In order to wash the plasma membranes in fractions P_2 and N_2 free of sucrose, they are diluted with 4 volumes of either cold distilled water or 5 mM Tris·HCl, pH 8.0, and the diluted fractions are centrifuged at 30,000 rpm for 1 hour in the Spinco No. 30 rotor. The supernatant solutions are carefully removed by aspiration; the membrane pellets are resuspended with a Dounce homogenizer in either distilled water or buffer at an approximate protein concentration of 1–2 mg/ml. When stored frozen at $-20°$, such suspensions of plasma membranes retain the full activity of 5′-nucleotidase and phosphodiesterase I for several months. Depending on the specific biochemical component or membrane function to be investigated, one should independently determine the stability of that particular membrane property or characteristic to washing and freezing. Upon repeated freezing and thawing, the gross physical appearance of the membrane suspension changes.

TABLE II

DISTRIBUTION OF ENZYMES IN RAT LIVER PLASMA MEMBRANE FRACTIONS[a]

Enzyme	Yield						Relative specific activity	
	P_2 (%)	P_3 (%)	P_4 (%)	N_2 (%)	N_3 (%)	N_4 (%)	P_2	N_2
Protein (4)[c]	1.2 ± 0.0	2.1 ± 0.6	19.8 ± 3.1	0.4 ± 0.2	3.4 ± 1.8	10.1 ± 1.5	—	—
5'-Nucleotidase (5)[c]	23.4 ± 2.0	10.0 ± 2.8	21.0 ± 2.6	9.0 ± 4.0	3.2 ± 1.1	10.2 ± 3.1	21.5 ± 2.1 (9)[b]	29.5 ± 8.1 (8)[b]
Phosphodiesterase I (4)[c]	19.1 ± 3.5	10.2 ± 2.9	21.4 ± 4.8	6.2 ± 1.7	3.3 ± 1.6	7.9 ± 2.8	23.9 ± 4.3 (7)[b]	23.7 ± 6.8 (7)[b]
Glucose-6-phosphatase (4)[c]	1.2 ± 0.5	3.9 ± 0.6	62.6 ± 1.3	0.3 ± 0.2	1.6 ± 1.1	6.1 ± 1.0	1.1 ± 0.1	0.7 ± 0.2
NADPH–cytochrome c reductase (2)[c]	1.2 ± 0.1	10.0 ± 1.4	58.1 ± 1.5	0.1 ± 0.0	1.5 ± 0.7	3.0 ± 0.2	1.1 ± 0.2	0.5 ± 0.1
Cytochrome oxidase (1)[c]	0.05	0	1.2	0.2	5.5	1.8	0.04	0.08
Succinate–cytochrome c reductase (1)[c]	0.3	0.27	4.7	0.06	0.9	5.6	0.05	0.16
N-Acetyl-β-D-glucosaminidase (3)[c]	0.2 ± 0.2	0.2 ± 0.2	5.0 ± 0.8	0.1 ± 0.1	1.4 ± 1.7	4.9 ± 0.7	0.2 ± 0.3	0.2 ± 0.3

[a] Statistics are presented as in Table I; % indicates yield based on the original homogenate (E + N). Relative specific activity is the ratio of the specific activity of the fraction to that in the original homogenate (E + N).

[b] Numbers in parentheses indicate the number of experiments in which 5'-nucleotidase and phosphodiesterase activities were measured.

[c] Number of experiments.

Membranes treated thus tend to settle out quickly to the bottom of the solution, while unfrozen membranes will remain dispersed for several days.

Characterization of the Two Isolated Plasma Membrane
    Fractions N_2 and P_2

Table II shows the distribution of the marker enzymes in the nuclear (N_2) and microsomal (P_2) plasma membrane fractions. Several conclusions may be drawn from the table: (a) 5'-Nucleotidase and phosphodiesterase I have very similar distributions. (b) The enzymatic yields (32.4 and 25.3%) and relative specific activities (21.5–28.5) of these two activities in the isolated plasma membranes P_2 and N_2 are high, as is the total yield of membranes (2.9 mg of membrane protein per gram of liver). (c) The contamination by other marker enzymes appears to be very low. Estimates of contamination of the plasma membrane fractions have been made by first calculating the specific activities of the endoplasmic reticulum marker enzyme, glucose-6-phosphatase, and of that for mitochondria, cytochrome oxidase, in their respective organelles. Such calculations indicate that P_2 is composed of endoplasmic reticulum to the extent of 20% and mitochondria to the extent of 2%, and the corresponding values for N_2 are 10 and 4%. (These estimates may be high because of the assumption that the marker enzymes are not normal constituents of the plasma membrane and are present exclusively as contaminants.)

Electron microscopic examination of the plasma membrane fractions indicate that they contain primarily vesicular elements and little nonmembranous material or organelles. In Table III are given the results of analyses for cholesterol, phospholipid, sialic acid, DNA, and RNA. The value for the lipid content is among the highest reported for rat liver plasma membranes, and that for the sialic acid is also high.

TABLE III
CHEMICAL ANALYSES OF PLASMA MEMBRANE FRACTIONS[a]

Component	P_2	N_2
Total cholesterol (μmole/mg protein)	0.78 ± 0.08 (6)	0.76 ± 0.03 (5)
Phospholipid phosphorus (μmole P/mg protein)	0.99 ± 0.3 (5)	0.92 ± 0.12 (5)
Cholesterol/phospholipid (mole/mole)	0.79	0.83
Sialic acid (nmoles/mg protein)	46.4 ± 10 (7)	48.0 ± 4.7 (5)
DNA (μg/mg protein)	12.2 ± 12 (3)[b]	11.2 ± 4.6 (3)[b]
RNA (μg/mg protein)	71.3 ± 5.0 (3)	71.7 ± 4.7 (3)

[a] Statistics are presented as in Table I. For methods see O. Touster, N. N. Aronson, Jr., J. T. Dulaney, and H. Hendrickson, *J. Cell Biol.* **47**, 604 (1970).
[b] Uncertain because of low readings.

Comments

The surfaces of liver cells are exposed to different environments. For example, certain faces of hepatocyte plasma membrane are adjacent to the surface membrane of neighboring cells (e.g., tight junctions and desmosomes), while other faces of the same cell may be exposed to sinusoidal space (space of Disse) or the bile canaliculi. It is therefore to be expected that fragments of plasma membranes derived from such differentiated sections may vary in chemical or enzymatic properties. Indeed, Evans[18] and House, Poulis, and Weidemann[19] have separated from liver two plasma membrane fractions which differ enzymatically, and some methods for preparation of liver plasma membranes yield products with different chemical composition which undoubtedly reflects the presence of desmosomes. Two recent papers have briefly reviewed this problem.[1,2]

Whether there are as yet undetermined differences in the plasma membrane fractions P_2 and N_2 which influence their affinity for the nuclei, or whether the adherence of membrane fragments to nuclei is a purely random matter, is unknown. Until very recently, the two fractions had always been found to be remarkably similar in all biochemical properties that had been measured.[1] We have now found, however, that both P_2 and N_2 contain sialyl[20] and galactosyltransferase[20a] activities, the specific activities in P_2 being much higher than in N_2. It would seem, therefore, that P_2, and to a much lesser extent, N_2, are contaminated by Golgi membranes, the subcellular location reported for these two glycosyltransferases.[21] It should be pointed out, however, that there has been a previous report of the presence of sialyltransferase in rat liver plasma membranes[22] which had been prepared by a modification of the procedure of Neville.[23] Two other reports, on bovine[24] and on rat liver[25] plasma membranes, concluded that galactosyltransferase is absent, but the analyses are complicated by the very high UDP-galactose hydrolase activities in these preparations.

[18] W. H. Evans, *Biochem. J.* **116**, 833 (1970).

[19] D. R. House, P. Poulis, and M. J. Weidemann, *Eur. J. Biochem.* **24**, 429 (1972).

[20] N. N. Aronson, Jr., L. Y. Tan, and B. P. Peters, *Biochem. Biophys. Res. Commun.* **53**, 112 (1973).

[20a] B. Dewald and O. Touster, *J. Biol. Chem.*, **248**, 7223 (1973).

[21] H. Schacter, I. Jabbal, R. L. Hudgin, L. Pinteric, E. J. McGuire, and S. Roseman, *J. Biol. Chem.* **245**, 1090 (1970).

[22] W. E. Pricer and G. Ashwell, *J. Biol. Chem.* **246**, 4825 (1971).

[23] T. K. Ray, *Biochim. Biophys. Acta* **196**, 1 (1970).

[24] B. Fleischer, S. Fleischer, and H. Ozawa, *J. Cell Biol.* **43**, 59 (1969).

[25] D. J. Morré, this series, Vol. 22, p. 130.

[7] Preparation of Isolated Fat Cells and Fat Cell "Ghosts"; Methods for Assaying Adenylate Cyclase Activity and Levels of Cyclic AMP

By M. Rodbell and G. Krishna

A variety of peptide hormones and biogenic amines stimulate the activity of adenylate cyclase, a membrane-bound regulatory enzyme system that catalyzes the production of cyclic AMP from ATP.[1] The role of cyclic AMP in the regulation of cellular metabolism has been reviewed extensively.[2] Two approaches are generally employed to investigate the activity and response of adenylate cyclase to hormones. One approach is to measure the levels of cyclic AMP generated in intact cells or tissues. This approach provides information on the hormonal sensitivity and state of the enzyme system as it exists under physiological conditions. The other approach is to isolate membrane fractions from target cells or tissues and to measure the activity and response of the system under incubation conditions manipulated by the investigator. Both procedures have obvious merit; combined information from these procedures can give valuable insights into the molecular basis for the actions of hormones and into those factors involved in the physiology of hormone action.

Preparations of isolated fat cells have many advantages for investigating adenylate cyclase activity and hormonal response (lipolysis).[3,4] Because of their characteristic high content of fat, and thus lower density than other cell types, fat cells can be easily separated from other cell types in adipose tissue as a homogeneous cell suspension. These fat cell suspensions can be exposed to the incubation medium (and hormones) and can be distributed among a number of incubation flasks to give the identical number of cells. Suspensions of fat cells isolated from rat adipose tissue, the preparation of which is described here, have additionally the distinct advantage of responding to a number of hormones, several of which stimulate the activity of a common adenylate cyclase system.[5,6] A

[1] E. W. Sutherland, T. W. Rall, and T. Menon, *J. Biol. Chem.* **237**, 1220 (1962).
[2] G. A. Robison, R. W. Butcher, and E. W. Sutherland (eds.), "Cyclic AMP." Academic Press, New York, 1971.
[3] M. Rodbell, *J. Biol. Chem.* **239**, 375 (1964).
[4] B. B. Brodie, J. I. Davies, S. Hynie, G. Krishna, and B. Weiss, *Pharmacol. Rev.* **18**, 273 (1966).
[5] L. Birnbaumer and M. Rodbell, *J. Biol. Chem.* **244**, 3477 (1969).
[6] M. Rodbell, L. Birnbaumer, and S. L. Pohl, *J. Biol. Chem.* **245**, 718 (1970).

relatively simple, rapid, and gentle procedure has been developed for isolating from fat cells a plasma membrane-rich fraction, termed "ghosts,"[7] that contains an adenylate cyclase system which responds to the same hormones with many of the characteristics observed in intact fat cells.[5-7]

The principles of the methods to be described here are applicable to other tissues and cells, although each tissue may require slight changes in procedure. The methodology is divided into four categories: a method for preparing isolated fat cells; an assay for cyclic AMP formed in fat cells in response to hormones; a simple method for preparing fat cell "ghosts"; and an assay for adenylate cyclase activity in response to hormones.

Preparation of Isolated Fat Cells

Principle. Adipose tissue is digested with a crude preparation of bacterial collagenase, resulting in release of fat cells. Because of their high fat content, fat cells float and are thereby readily separated from other types of cells in adipose tissue.

Reagents

> Albumin medium: 2% albumin (bovine, fraction V) in a physiological salt solution consisting of 125 mM NaCl, 5 mM KCl, 1.0 mM $CaCl_2$, 2.5 mM $MgCl_2$, 1.0 mM KH_2PO_4, and 25 mM Tris·HCl, pH 7.4. Solution is adjusted, if necessary, to pH 7.4 with 1 M Tris.
>
> Collagenase: Crude collagenase [*Clostridium histolyticum* (125–150 units/mg) obtained from Worthington or Boehringer]. Dissolve 15–30 mg in 1 ml of albumin medium. Solutions can be stored frozen.

Procedure. Rats weighing 150–200 g are killed by decapitation. Epididymal, parametrial, or intrascapular adipose tissue are quickly removed and placed in albumin medium kept at room temperature. Adipose tissue from one animal is placed on Whatman No. 3 MM filter paper wetted with albumin medium and is minced rapidly with a Mickle slicer (H. Mickle, Surrey, England). Alternatively, the wetted tissue can be minced with scissors. Minced tissue from each animal is transferred to a 25-ml plastic vial containing 3 ml of albumin medium plus 0.1 ml of collagenase solution (see Comments). The suspension of tissue is incubated at 37° with shaking (130–150 cycles/min) for 1 hour, at which time most of the tissue should be noticeably disaggregated.

The contents of each vial are expressed at room temperature with the aid of a plunger through fine mesh nylon or silk screening (7×, obtained

[7] M. Rodbell, *J. Biol. Chem.* **242**, 5744 (1967).

from suppliers of art materials) attached at the cut bottom portion of a 25-ml plastic syringe with a band of heavy rubber tubing. This procedure disperses clumps of fat cells and removes most of the stromal-vascular cells and undigested tissue. The expressed cells and those clinging to the screen are collected and centrifuged in a 13-ml plastic centrifuge tube for about 15 seconds at 1000 rpm in a clinical centrifuge. The infranatant fluid and sedimented stromal-vascular cells are aspirated and discarded. The packed floating layer of fat cells is dispersed with 5 ml of albumin medium and centrifuged again as above. The washing procedure is repeated, and the suspension of fat cells is passed through silk or nylon screening, as above, in order to remove final traces of stromal-vascular cells. At this point, the washed cells are ready either for incubation, after suitable dilution, and determination of effects of hormones on cyclic AMP levels or for the preparation of fat cell "ghosts" and determination of adenylate cyclase activity and response to hormones.

The simplest and most rapid method for determining the number of cells used for incubation is to dilute the suspension 10-fold with albumin medium and count under a microscope at low magnification the number of cells in a 5-μl aliquot placed on a siliconized microscope slide.

Comments. The most serious difficulty encountered in the preparation of fat cells is reduction or loss of hormonal response of the cells. This is largely due to contamination of some preparations of crude collagenase with such enzymes as proteases and phospholipases which adversely affect the structure of the hormonally responsive systems in the plasma membrane of fat cells.[8-10] Each preparation of collagenase must be examined, therefore, for its effects on the response of the cells to hormones. In the case of insulin, this is usually accomplished by assaying for its effects on glucose metabolism[3]; in the case of lipolytic hormones, their effects on glycerol or fatty acid production should be assessed.[3] Our usual practice is to obtain small amounts of several lots of the enzyme preparation and, after selecting the best preparation, to purchase and store at 0–4° quantities adequate for a long time.

Assay for Cyclic AMP Formation in Isolated Fat Cells

Principle. The method consists of prelabeling the adenine nucleotides in fat cells by incubating the isolated cells *in vitro* with [³H]adenine.[11] After washing, the prelabeled cells are incubated with various hormones

[8] J. F. Kuo, I. K. Dill, and C. E. Holmlund, *J. Biol. Chem.* **242**, 3659 (1967).
[9] M. Rodbell, *J. Biol. Chem.* **241**, 130 (1966).
[10] M. Rodbell and A. B. Jones, *J. Biol. Chem.* **241**, 140 (1966).
[11] J. Forn, P. S. Schonhofer, I. F. Skidmore, and G. Krishna, *Biochim. Biophys. Acta* **208**, 403 (1970).

in the presence or in the absence of a phosphodiesterase inhibitor. After termination of incubation by addition of trichloroacetic acid, a known amount of ^{14}C-labeled cyclic AMP is added to aid in determining recovery of cyclic AMP. ATP and cyclic AMP are isolated by chromatography on Dowex 50 (H[+]) columns. ATP is purified further by chromatography on Dowex (Cl[-]) based on content of radioactivity and total amount (the latter as determined by the Luciferase method[12]); the specific activity of fat cell ATP is obtained. Cyclic AMP is purified from other contaminating nucleotides by precipitating them with $ZnSO_4 \cdot Ba(OH)_2$ treatment[13]; total cyclic AMP formed is determined by the protein kinase binding assay of Gilman.[14]

Reagents and Materials

Krebs-Ringer phosphate buffer, pH 7.4, containing 4% dialyzed bovine albumin (fraction V)

[3H]Adenine, 0.5 mCi/ml; specific activity, 24 Ci/mmole

[14C]cyclic AMP, 0.1 μCi/ml; specific activity, 50 mCi/mmole

BioRad AG 50 (H[+])-X4, 200–400 mesh

BioRad AG 1 (Cl[-])-X4, 200–400 mesh

7.8% $Ba(OH)_2$ ⎫ Concentrations are adjusted so that mixing of
8% $ZnSO_4 \cdot 7H_2O$ ⎭ equal volumes will give pH 7.6–7.8 or reach a phenolphthalein end point

Glass columns (5 mm id) or disposable plastic columns chromaflex purchased from Kontes Glass

Trichloroacetic acid, 10% w/v

Sodium acetate, 0.5 M, pH 4.0

Protein kinase (beef muscle) 0.2–0.4 mg/ml prepared according to Gilman[14]

Protein kinase inhibitor (beef muscle) 3–8 mg/ml prepared according to Gilman's modification[14] of the Appleman et al.[15] procedure

Cyclic AMP standard solutions (concentration range: 0.04–4.0 μM)

Potassium phosphate buffer, 20 mM, pH 6.1

Methyl Cellosolve (ethyleneglycol monomethylether)

Millipore filter assembly and Millipore filters (0.45 μm, 25 mm)

Preparation of Dowex Resins. The resins obtained from the manufacturer are used as such after they have been washed three times with

[12] B. L. Strehler and W. D. McElroy, this series, Vol. 3, p. 871.

[13] G. Krishna, B. Weiss, and B. B. Brodie, *J. Pharmacol. Exp. Ther.* 163, 379 (1968).

[14] A. G. Gilman, *Proc. Nat. Acad. Sci. U. S.* 67, 305 (1970).

[15] M. M. Appleman, L. Birnbaumer, and H. N. Torres, *Arch. Biochem. Biophys.* 116, 39 (1966).

5 volumes of distilled water. The resins are suspended in distilled water to give 50% resin volume. The columns are prepared by pipetting 2 ml of the uniformly suspended resins and washing the resins with 25 ml of water just prior to use. It is not necessary to maintain liquid on the top of the columns at all times. The columns may be allowed to go dry, and this does not appear to affect the separation.

All operations are performed at room temperature (22°) unless specified.

Prelabeling of Intracellular Adenine Nucleotides with [³H]Adenine.[14] Fat cells isolated from 4 g of adipose tissue are incubated in plastic vials in 10 ml of 4% albumin–Krebs Ringer medium containing 100 μCi of [³H]adenine (0.4 mM) for 30 minutes at 37°. The cells are separated from the medium by centrifugation in plastic tubes for 30 seconds, washed twice with 10 ml of albumin–Krebs-Ringer medium, and suspended in 50 ml of this medium in a plastic container. One-milliliter aliquots of the suspension are transferred to plastic vials using plastic pipettes and are incubated with various concentrations of hormones (for example, norepinephrine, glucagon, ACTH) in the absence and in the presence of theophylline (0.5 mM). Incubations are carried out at 37° and are terminated after 5 minutes by addition of 1 ml of 10% trichloroacetic acid. Two zero-time incubations as well as incubation mixture without fat cells are performed at the same time. [¹⁴C]Cyclic AMP (50 μl of 0.1 μCi/ml) is added to each incubation vial for revolving estimations. The contents are transferred to centrifuge tubes and are centrifuged for 15 minutes. The whole supernatant fluid is chromatographed on Dowex 50 (H⁺) as described below.

Separation of ATP and Cyclic AMP. The trichloroacetic acid supernatant (usually 1.4 ml) is decanted onto Dowex 50 (H⁺) column; the effluent is collected. The column is then washed with 0.6 ml of water; the effluent is also collected in the same tube. This represents the ATP–ADP fraction (fraction I). The column is washed with 2 ml of water, and the eluate is discarded. Cyclic AMP (fraction II) is then collected in 13-ml conical centrifuge tubes by eluting the columns with 2 ml of water.

Separation of ATP from ADP. In order to remove trichloroacetic acid, fraction I is extracted 4 times with 10 ml of diethyl ether saturated with water by mixing for 15 seconds in a vortex mixer. The ether layer is discarded, and the aqueous layer is passed through a Dowex 1 (Cl⁻) column. The column is eluted in the following sequence: (1) 10 ml of water (adenosine, adenine); (2) 10 ml of 0.01 N HCl (5′AMP and cyclic AMP); (3) 20 ml 0.02 N HCl + 0.02 N NH₄Cl (ADP). ATP is then eluted from the column with 5 ml of 0.25 N HCl into tubes containing 2 ml of 1 M Tris. An aliquot is taken for determination of radioactivity. The remaining solution is used for determining the content of ATP by the luciferase method.[12]

Separation of Contaminating Nucleotides from Cyclic AMP. This method is based on the capacity of nascently formed $Zn(OH)_2 \cdot BaSO_4$ to quantitatively absorb all nucleotides except cyclic AMP as well as nucleosides (except adenosine) and most bases. Since adenosine and bases are adsorbed by Dowex 50 (H^+), combination of chromatography with $Zn(OH)_2 \cdot BaSO_4$ adsorption helps to separate cyclic AMP from all other nucleotides, nucleosides, and bases. Fraction II, collected from the Dowex 50 (H^+) column, is treated with 0.2 ml of $ZnSO_4$ and 0.2 ml of $Ba(OH)_2$ solution (this order of additions should not be reversed). The tubes are rapidly mixed and then centrifuged for 5 minutes at 2000 rpm in a clinical centrifuge. The Zn-Ba treatment and centrifugation steps are repeated. Aliquots of the supernatant are taken for determination of radioactivity (Both ^3H and ^{14}C) and for estimation of total cyclic AMP, as described below.

Assay for Total Cyclic AMP. This method is based on the ability of protein kinase, partially purified from beef muscle, to bind specifically cyclic AMP.[12] The method involves the competition between labeled and unlabeled cyclic AMP for binding to protein kinase. The assay is performed at concentrations of cyclic AMP which saturate the binding sites; under these conditions competition between labeled and unlabeled nucleotide is based solely on dilution of the original specific activity of [^3H]cyclic AMP by the added cyclic AMP. An inhibitor of protein kinase is also included in the assay medium since it both increases the affinity of protein kinase for cyclic AMP[16] and stabilizes the binding during isolation of the bound nucleotide.

Of the cyclic AMP fraction purified by Zn-Ba treatment, 100 μl is evaporated to dryness at 60° in a small disposable culture tube (10 × 75 mm). All tubes and the solutions are maintained at 0° (melting ice). The residue is taken up in 10 μl of ice-cold distilled water to which is added 20 μl of an ice-cold mixture containing equal volumes of 0.5 M sodium acetate, pH 4.0, protein kinase inhibitor (6 mg/ml), 0.2 μM [^3H]cyclic AMP (24 Ci/mmole), and water. Twenty microliters of protein kinase (ice-cold) is added, and the tubes are incubated for at least 60 minutes. [Typically, the final incubation mixture contains 2 μg of protein kinase, 20 μg of protein kinase inhibitor, and 1 pmole of [^3H]cyclic AMP (53,000 dpm) in a total volume of 50 μl.] The reaction is stopped by adding 1 ml of ice-cold 20 mM potassium phosphate buffer, pH 6.1. The mixture is filtered at room temperature through a Millipore filter previously washed

[16] D. A. Walsh, J. P. Perkins, and E. G. Krebs, *J. Biol. Chem.* **29**, 470 (1970).

with 5 ml of ice-cold phosphate buffer. The tubes are rinsed with 1 ml of ice-cold phosphate buffer, which is filtered. The filter is washed with 10 ml of ice-cold phosphate buffer, and the filter is transferred to counting vials containing 1 ml of methyl Cellosolve. Fifteen milliliters of scintillation fluor is added after the filter has dissolved, and the amount of radioactivity is determined.

Binding assays of standard solutions of cyclic AMP are carried out in the manner described above. The amount of standard [^{3}H]cyclic AMP bound is plotted on log-log paper against the total amount of cyclic AMP present (note that [^{3}H]cyclic AMP contributes to the quantity added; this quantity must be subtracted when the final concentration is determined).

Calculation of Total Cyclic AMP. The total cyclic AMP determined by the binding assay is adjusted based on [^{14}C]cyclic AMP recovered during the entire procedure. Since the latter contributes significant cyclic AMP to the binding assay, its contribution to the total is determined from the recovery of [^{14}C]cyclic AMP when the isolation and binding assay are carried out in the absence of fat cells; this value is subtracted from the final results.

Calculation of Pool Size of ATP Utilized for Cyclic AMP Production. The above procedures provide an estimate of the total cyclic AMP in the cells. It is possible to calculate the total amount of cyclic AMP formed by dividing the radioactivity in cyclic AMP formed in the cells by the specific activity of ATP. If the specific activities of ATP and cyclic AMP are identical, this calculation will lead to the same values determined by the binding assay. If not, the data may indicate fractional pools of ATP being utilized for cyclic AMP formation. In the brown fat cell, it has been shown that the ATP and cyclic AMP pools are in dynamic equilibrium.[17]

Preparation of Fat Cell "Ghosts"

Principle. The procedure described below is based on swelling of isolated fat cells with a hypotonic medium containing salts that serve to stabilize the membranes and ATP which appears to stabilize adenylate cyclase and its response to hormones. Hypotonic swelling alone does not cause lysis of the cells. Gentle agitation is necessary and is carried out in a manner that produces relatively large sacs or vesicles of membranes that are readily sedimented at low gravitational forces. The resultant "ghosts" appear to be sacs of plasma membrane within which are contained mitochondria, endoplasmic reticulum, and occasionally a nucleus.[7] Fat cell "ghosts" have been used for studying ion transport,[18] amino acid

[17] J. Moskowitz and G. Krishna, *Pharmacology,* in press (1973).
[18] T. Clausen, M. Rodbell, and P. Dunand, *J. Biol. Chem.* **244,** 1252 (1969).

transport,[19] glucose transport and the effects of insulin on this process,[20] the effects of lipolytic hormones on carbohydrate metabolism,[20] and the characteristics of adenylate cyclase and its response to hormones.[5,6,21]

Reagents and Materials

Lysing medium: 50 ml of freshly prepared, ice-cold solution containing 2.5 mM MgCl$_2$, 2.5 mM ATP, 0.1 mM CaCl$_2$, 1.0 mM KHCO$_3$, 2.0 mM Tris·HCl, pH 7.6, and (optional) 1.0 mM mercaptoethanol

Suspending medium: 1.0 mM KHCO$_3$ containing 1.0 mM mercaptoethanol. Solution should be freshly prepared for each experiment and kept in an ice bath.

Procedure. Fat cells are isolated, as described in the section on preparation of fat cells, from epididymal or parametrial adipose tissue obtained from two rats. The washed cells are placed in a 13-ml conical plastic centrifuge tube. The packed fat cells are dispersed gently in 5 ml of lysing medium and centrifuged for 15 seconds at about 1000 rpm (solutions are kept ice-cold, but operations are carried out at room temperature). The infranatant fluid is aspirated and discarded. The swollen fat cells are resuspended in 5 ml of lysing medium and the tubes are inverted in a cyclical fashion 20 times, using a vigorous shaking motion at each downward stroke. The tubes are centrifuged for about 30 seconds at 2000 rpm. With a Pasteur pipette, the infranatant fluid and pellet are transferred to a glass centrifuge tube (conical, 13 ml) chilled in an ice bath, leaving the unlysed fat cells in the same tube. These operations are repeated twice more, each time with 5 ml of lysing medium. The combined infranatants and pellets are centrifuged for 15 minutes at 900 g (2000 rpm in International refrigerated centrifuge, rotor II 269), at 4°. The supernatant fluid is aspirated and discarded; the pellet is suspended, with the aid of a wide-mouthed plastic pipette, in 5 ml of suspending medium. The suspension is again centrifuged as above, the supernatant fluid is discarded, and the pellet is taken up in a volume of suspending medium equal to 5 times the volume of the pellet. In a typical "ghost" preparation isolated from fat cells obtained from 8 rats, the final suspension contains 3.5 mg of protein per milliliter (1.5 ml total volume).

Comments. The adenylate cyclase system and its response to hormones is particularly unstable, decaying at a rate of about 50% in 4 hours even

[19] T. Clausen and M. Rodbell, *J. Biol. Chem.* **244**, 1258 (1969).
[20] M. Rodbell, *J. Biol. Chem.* **242**, 5751 (1967).
[21] L. Birnbaumer, S. L. Pohl, and M. Rodbell, *J. Biol. Chem.* **244**, 3468 (1969).

when "ghosts" are maintained at 0°. For this reason, it is important to prepare "ghosts" and to assay adenylate cyclase activity within as short a time as possible. If frozen in liquid nitrogen as soon as isolated, fat cell "ghosts" can be stored for at least 2 weeks without loss of activity or response to hormones.

For studies of ion, glucose, or amino acid transport, fat cell "ghosts" should be suspended and incubated in Krebs-Ringer solution containing 1–4% albumin.

Method for Assay of Adenylate Cyclase Activity

Adenylate cyclase catalyzes the reaction

$$\text{ATP} \rightarrow \text{cyclic AMP} + \text{P-P}$$

which is essentially irreversible under the reaction conditions usually employed with mammalian systems. In fat cells or fat cell "ghosts" prepared from rat adipose tissue, the adenylate cyclase system is stimulated by at least six hormones (catecholamines, ACTH, glucagon, secretin, luteinizing hormone, and thyroid-stimulating hormone), each of which interacts with distinct receptors that appear to be separate molecular entities from the enzyme.[6] The entire system (receptors and enzyme) seems to be constructed within the plasma membrane so that the receptors are situated on the external surface and the enzyme on the inner surface of the membrane.[6]

A divalent ion (preferably Mg^{2+}) is required for activity; calcium ion inhibits activity. Fluoride ion usually enhances the activity of mammalian adenylate cyclase systems and is frequently used, in lieu of hormones, to determine the activity of the enzyme.[1]

Principle. The assay measures the formation of ^{32}P-cyclic AMP formed from [α-^{32}P]ATP and is based on the finding by Krishna[13] that cyclic AMP, unlike other nucleotides or nucleosides, is not absorbed to nascently formed $Zn(OH)_2$-$BaSO_4$. Because cyclic AMP can be formed nonenzymatically from ATP under slightly alkaline conditions, cyclic AMP is first separated from ATP by chromatography on Dowex 50 (H^+) columns. Since nucleotidases are present in most membrane preparations containing adenylate cyclase, an ATP-regenerating system is included in the incubation medium. Unlabeled cyclic AMP is also included in the medium to reduce the hydrolysis of radioactive cyclic AMP catalyzed by phosphodiesterases present in most enzyme preparations.

The procedures described below permit the assay of 200 samples within 4 hours and with less than 1% variation between replicates.

Procedure

Reagents and Materials

[α-^{32}]ATP, 100–200 mCi/mmole

[^{3}H]Cyclic AMP (2–5 Ci/mmole) diluted with water to contain 300,000 cpm/ml

ATP-regenerating system consisting of solution of 200 mM creatine phosphate containing 10 mg/ml creatine kinase (Sigma)

"Stopping" solution: 1.0% sodium lauryl sulfate containing 0.4% Na cyclic AMP, 5.0% Na$_2$ATP·4H$_2$O adjusted to pH 7.6 with 2 M Tris. The solution is stored frozen.

BioRad AG 50 (H$^+$)-X4, 200–400 mesh. Washed and suspended in distilled water to give 50% bed volume, as described in the section on assay of cyclic AMP

7.8% Ba(OH)$_2$
8.0% ZnSO$_4$·7H$_2$O $\Big\}$ Concentrations are adjusted so that mixing of equal volumes will give pH 7.4–7.6 or reach a phenolphthalein end point.

Glass columns (5 mm id) or disposable plastic columns (Chromaflex, Kontes glass).

Incubation. The incubation conditions described are designed to give linear rates of activity for at least 10 minutes of incubation at 30° with the fat cell "ghost" adenylate cyclase system. Incubations are carried out in 10 × 75 mm glass test tubes containing, in a final volume of 50 μl, the following ingredients: 3.5 mM [α-^{32}P]ATP, 5 mM MgCl$_2$, 2.5 mM cyclic AMP, 20 mM creatine phosphate, 1 mg of creatine kinase per milliliter, 25 mM Tris·HCl, pH 7.6, and either 0.5% bovine serum albumin, 10 mM NaF, or hormones appropriately diluted in 0.5% albumin. Reactions are initiated by addition of fat cell "ghosts" to give final concentrations ranging from 0.5 to 2.0 mg of protein per milliliter. Incubations are terminated by addition of 0.1 ml of "stopping" solution. Zero time (or reaction blank) incubations are carried out by addition of "stopping" solution before addition of "ghosts."

As a means of determining the recovery of [^{32}P]cyclic AMP in the isolation procedure described below, 50 μl of [^{3}H]cyclic AMP solution is added per tube. The reaction mixture is heated in a boiling water bath for 3 minutes, after which 0.9 ml of water is added per tube.

Isolation of Cyclic AMP

Columns of Dowex 50 (H$^+$) are freshly prepared by addition of 2.0 ml of a uniform suspension of the resin to columns containing a small wad of glass wool. The columns are held in a Perspex rack containing suitably spaced holes to hold 50 columns. In the first stage of the procedure, the rack is placed over a tray for collecting the eluates.

The boiled contents of the incubation tubes are decanted into the columns, which are allowed to drain. To the columns are added, with a calibrated syringe, 1.25 ml of distilled water. After allowing the columns to drain, another 1.25 ml of water is added. The eluates are discarded. The collecting tray is replaced with a test tube rack designed to hold 50 (13 × 10 mm) glass test tubes that are positioned directly under the tips of the columns; 3.0 ml of distilled water are added to each column. The eluates, containing cyclic AMP, are collected in the test tubes. To each tube is added, successively, 0.2 ml each of $ZnSO_4$ and $Ba(OH)_2$ solutions (this order of additions should not be reversed). The contents are mixed on a vortex mixer.

As a means of rapidly removing the nascently formed $BaSO_4$ (containing absorbed nucleotides except cyclic AMP), the following procedure is employed: a plug of $BaSO_4$ is formed at the bottom of plastic or glass columns (plugged with glass wool) by the addition of 0.4 ml each of the solutions of $ZnSO_4$ and $Ba(OH)_2$. The columns are placed in the same type of Perspex rack described above and are positioned above a suitable rack holding 50 counting vials filled with 12 ml of scintillation fluor (Aquasol, from New England Nuclear, is usually employed). The contents of the test tubes containing cyclic AMP are decanted into the columns and allowed to drain into the counting vials.

Radioactivity is determined in a three-channel scintillation counter adjusted so that there is little crossover in the channels for ^3H and ^{32}P.

Calculations

$$\frac{[^3H]\text{Cyclic AMP (standard)}}{[^3H]\text{Cyclic AMP (in sample)}} \times \frac{^{32}P(\text{sample})-^{32}P(\text{machine background})}{\text{specific activity } [^{32}P]\text{ATP}}$$
$$= \text{picomoles cyclic AMP}$$

Picomoles cyclic AMP minus zero time cyclic AMP
$$= \text{net formation of cyclic AMP}$$

Specific activity of ATP is obtained by determining both the radioactivity and the optical density at 260 nm of an appropriately dilute sample of the incubation medium. The extinction coefficient of ATP is about 15,000.

Comments. The principles of the method described here can be applied to any adenylate cyclase system. However, it is advisable to check whether the radioactive product is cyclic AMP when the reaction conditions are changed. For example, we have found that the presence of glycerol in the incubation medium containing either a preparation of liver adenylate cyclase or of fat cell "ghosts" gives rise to a product that behaves as cyclic AMP during the assay. The exact chemical nature of the product and the characteristics of the enzyme producing this material have not been determined. It is possible that the product is AMP-glycerol. The same caveat holds when either [^{14}C]ATP or [^{3}H]ATP are used instead of [^{32}P]ATP, and paper or thin-layer chromatography is used instead of the above method for isolation of cyclic AMP. In these cases, radioactive hypoxanthine may be formed which will behave as cyclic AMP when paper or thin-layer chromatography are used instead of the method described here. As a simple method of checking whether the radioactive product is cyclic AMP, the material can be examined for cocrystallization with authentic cyclic AMP.[13] Alternatively, loss of radioactivity in cyclic AMP fraction after treatment with purified cyclic AMP phosphodiesterases can also be employed.[13]

The most common difficulty encountered in assaying cyclic AMP production is loss of linear reaction rates, particularly when the concentration of ATP is reduced and the temperature of incubation is increased. This difficulty is partly due to rapid hydrolysis of ATP by potent nucleotidases present in most membrane preparations, which may not be prevented by addition of an ATP-regenerating system. It should be emphasized, in this regard, that conversion of ATP to cyclic AMP by adenylate cyclase in isolated membrane preparations is small relative to the concentration of ATP available to the system. Loss of ATP to the actions of nucleotide phosphohydrolases can be circumvented by the use of AMP-PNP (5'-adenylylimidodiphosphate), which has been shown to be a substrate for adenylate cyclase and is not hydrolyzed at the terminal phosphate position by the above enzymes.[22,23] This material is supplied in both labeled and unlabeled form by the International Chemical and Nuclear Corp.

[22] M. Rodbell, L. Birnbaumer, S. L. Pohl, and H. M. J. Krans, *J. Biol. Chem.* **246**, 1877 (1971).
[23] G. Krishna, J. Harwood, A. J. Barber, and G. A. Jamieson, *J. Biol. Chem.* **247**, 2253 (1972).

[8] The Isolation of Kidney Brush Border

By David M. Neville, Jr.

This article describes a method for isolating brush border plasma membranes from rat kidney which was developed by Wilfong and Neville[1] and has been used repeatedly in this laboratory for the past three years. The method is reproducible providing that careful attention is paid to the details. Starting with 27 g of kidney, 15 mg of membrane protein is obtained. Morphological and biochemical criteria indicate that 90% of this material is of brush border origin.

The brush border is a specialized portion of the cell surface membrane which occurs in the small intestine and the renal proximal convoluted tubule. It consists of parallel arrays of microvilli, each 1 μm long and 0.1 μm in diameter, which project from the luminal surface of the cell. Purified preparations of brush border were first isolated from hamster intestinal epithelium by Miller and Crane in 1961.[2] The first purified preparations of proximal convoluted tubule were isolated by Thuneberg and Rostgaard from rabbit.[3] This preparation exhibits well-preserved microvilli when examined by electron microscopy. The method has been modified by Berger and Sacktor, who provided morphological and biochemical data for judging purity.[4] Kinne and Kinne-Saffran have described a brush border preparation from rat kidney and provided biochemical and morphological evidence for purity.[5] All these preparations show a 10- to 20-fold enrichment of brush border marker enzymes.

Method

Materials

Water was deionized, and sucrose was reagent grade. All sucrose concentrations, which are reported as percentage of weight by weight, were determined with a Bausch and Lomb Abbe 3L refractometer by reading the "total dissolved solids" scale at room temperature and correcting to

---

[1] R. F. Wilfong and D. M. Neville, Jr., *J. Biol. Chem.* **245**, 6106 (1970).
[2] D. Miller and R. K. Crane, *Biochim. Biophys. Acta* **52**, 293 (1961).
[3] L. Thuneberg and J. Rostgaard, *Exp. Cell Res.* **51**, 123 (1968).
[4] S. J. Berger and B. Sacktor, *J. Cell Biol.* **47**, 637 (1970).
[5] R. Kinne and E. Kinne-Saffran, *Pfluegers Arch. Gesamte Physiol. Menschen Tiere* **308**, 1 (1969).

20° (International Sucrose Tables, 1936). Homogenizations were performed in a large Dounce glass homogenizer (Blaessig Glass, Rochester, New York) equipped with a loose pestle. Fiberglass screening with an opening of 1.5 mm was obtained from Sears, Roebuck and Company, Philadelphia, Pennsylvania (fiberglass insect screen, catalog No. 99K68136C). Nylon gauze with an opening of 0.1 mm (Nytex 114-T) was obtained from Washington Screen Company, Washington, D.C.

All centrifugations were carried out in either a Beckman-Spinco Model L-2B 65 preparative ultracentrifuge or an International Equipment Company (I.E.C., Needham Heights, Massachusetts) B-20 general purpose refrigerated centrifuge. The SW 25.1 or SW 25.2 Spinco rotors were used with cellulose nitrate tubes. The swinging-bucket No. 947 as well as the fixed-angle No. 840 I.E.C. rotors were used with appropriate 250-ml round-bottom glass bottles or 12-ml Lusteroid tubes. The standard 250 ml round-bottom glass centrifuge bottles (Fisher Scientific Co., No. 5-586-5, Pittsburgh, Pennsylvania) can be accommodated in IEC or Sorvall swinging-bucket rotors by cutting off the bottle tops on a Carborundum wheel.

Procedure

All steps were carried out at 4°. Media referred to in the procedure were as follows: Medium A, 20 mM sodium or potassium bicarbonate; Medium B, 10 mM sodium or potassium bicarbonate; Medium C, 4 mM sodium or potassium bicarbonate and 1 mM magnesium chloride. For option II, the concentration of media A and B are increased 3-fold.

Step 1. Twenty-one rats of either sex, 150–200 g body weight, were decapitated, and the kidneys were excised and freed from surrounding connective tissue.

Step 2. Minced kidney, 3 g, was homogenized in a large Dounce homogenizer in 30 ml of medium A. Homogenization was completed with seven downward gentle strokes of the loose pestle. This step was repeated twice.

Step 3. The pooled homogenates, added to 600 ml of medium B, were stirred for approximately 3 minutes.

Step 4. The diluted homogenate was filtered through coarse fiberglass screen and then through nylon gauze.

Step 5. The filtered homogenate, distributed equally among four 250-ml glass bottles (previously chilled), was centrifuged at 500 g for 20 minutes in a swinging-bucket rotor (I.E.C. rotor 947 or Sorvall rotor HS-4, 1500 rpm). This constitutes option I. For option III the rpm are reduced to 1000 and the time is reduced to 10 minutes. The supernatant fluids were carefully poured off, leaving 2–3 mm of thick, loosely packed, light-tan pellets.

Step 6. Steps 2–5 were repeated twice.

Step 7. The combined pellets were poured into a 100-ml glass cylinder. Medium B was added to bring volume up to 28 ml, then 60 ml of 60.0% sucrose were added and mixed well.

Step 8. Adjustments were made with either medium B or 60.0% sucrose until the final percentage was 41.2 ± 0.1%. (The optimal density for flotation varies between 40.4 and 41.5% sucrose depending on the state of the animals. See Comments.)

Step 9. Of the above solution, 32 ml were poured into each of three SW 25.1 tubes. For a double preparation, use 62 ml in each of three SW 25.2 tubes.

Step 10. Centrifugation at 90,000 g_{max} (25,000 rpm) for 75 minutes (brake on) resulted in a thin, light brown floating layer and a dark brown pellet. Increase time to 90 minutes for a double preparation.

Step 11. The floats were removed with a Spoonula (Fisher 14-375-10) and added to 10 ml of medium C.

Step 12. The resuspended floating layers were centrifuged at 3000 g_{max} for 5 minutes in an angle head rotor (I.E.C. rotor 840, 5000 rpm). The supernatant was carefully aspirated and discarded.

Step 13. The loosely packed pellet was gently resuspended in 10 ml of medium C by drawing the pellet up twice into a 2-ml pipette.

Step 14. A 3.0-ml cushion of 41% sucrose was placed into the bottom of two SW 25.1 tubes. Over each cushion, a linear gradient was formed from 20 to 5% sucrose (sucrose solutions for cushions and gradients were made in medium C). Use three gradients for a double preparation. For a Beckman gradient former, use the following solutions and settings: heavy 22%, light 3%, 30-ml syringes containing 20 ml, hold up −8, start 0, end 100. Run on slow.

Step 15. The resuspended pellet from step 13 (1.0 to 1.5 ml) was carefully layered on top of each of the gradients formed in step 14 and centrifuged at 2500 g_{max} (4000 rpm) for 15 minutes (brake off).

Step 16. The brush border fraction, which collects at the gradient–cushion interface, was removed with a syringe and a long, blunt No. 20 gauge needle. The procedure yielded an average of 0.27 g of final brush border membranes, wet weight, when pelleted by centrifugation at 40,000 g_{max} for 20 minutes.

The above volumes can be doubled if a SW 25.2 rotor is available. The final centrifugation step can be carried out in a SW 25.1 rotor with the use of all three buckets. The overlay sample volume will be 2.0 ml.

Comments

Basic Design. The basic design of this method is similar to that of the methods developed in this laboratory to isolate liver cell plasma membranes.[6,7] The homogenization conditions are chosen so that the desired membranes remain in large fragments (5–20 μm) while nuclei are ruptured, resulting in a solution of nucleoprotein. After filtration to remove glomeruli, connective tissue, vessels, tubular basement membrane, and partially disrupted tubules, (step 4) the brush borders are the largest particles in the filtered homogenate, and they may therefore be concentrated

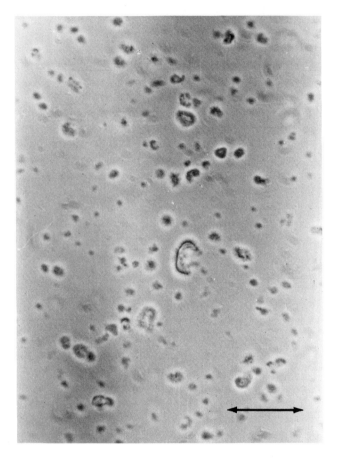

FIG. 1. Phase contrast wet preparation from the final brush border fraction. The large object just below the center shows a dense curved base line with small fine filaments projecting outward. This is an intact brush border. The remaining objects, which are all smaller, are fragments of brush borders. These have rounded up and appear as circular stellate objects. Line indicates 20 μm. Base magnification, ×400. From R. F. Wilfong and D. M. Neville, Jr., *J. Biol. Chem.* **245**, 6106 (1970).

[6] D. M. Neville, Jr., *J. Biophys. Biochem. Cytol.* **8**, 413 (1960).
[7] D. M. Neville, Jr., *Biochim. Biophys. Acta* **154**, 540 (1968).

by a low speed centrifugation (step 5). Step 10 floats brush borders in a sucrose medium, and most of the mitochondria and endoplasmic reticulum are sedimented. The percent sucrose of this medium is chosen as the minimal value that will achieve flotation of 80% of the brush borders in 75 minutes at 90,000 g. A minimal value is chosen in order to sediment as many mitochondria as possible. Mitochondria are more dense than brush borders; however, their density distribution is broad and overlaps

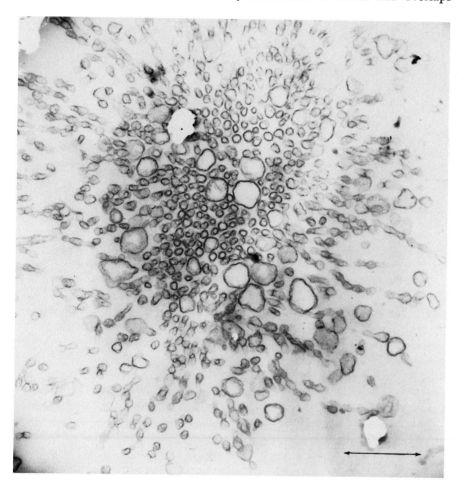

Fig. 2. Electron microscopic section stained with uranyl acetate and lead citrate. Tissue is from the final brush border fraction. The section shows a single brush border fragment. The microvilli are cut perpendicular to their vertical axis in the *center* of the micrograph and more obliquely in the periphery. Line indicates 1 μm. Base magnification, ×7800. From R. F. Wilfong and D. M. Neville, Jr., *J. Biol. Chem.* **245**, 6106 (1970).

that of the brush border. The final zonal centrifugation (step 15) reduces the microsomal and mitochondrial particles which were light enough to be floated by selecting for the larger brush border fragments.

Purity, Morphological Criteria. Examination by phase microscopy shows that the brush borders are concentrated at each stage of the preparation. Brush borders appear in two forms. Fragments of a single brush border round up so that they appear as circular stellate objects 2–10 μm in diameter. Intact sheets of brush borders from single cells or several cells still adherent exhibit a shallow curvature. These structures viewed on edge show a well-defined brush appearance, the base of the brush exhibiting considerable contrast (see Fig. 1). Electron photomicrographs of embedded material show that most of the material exhibits parallel arrays of villous processes typical of brush borders (see Fig. 2).

Purity, Enzymatic Criteria. Alkaline phosphatase, a standard marker enzyme for brush borders,[1] is concentrated at every stage of the preparation and shows a relative specific activity[8] of 15 for purified brush borders (see Table I). The relative specific activity of marker enzymes for kidney endoplasmic reticulum, mitochondria, and lysosomes are all diminished (Table II). From these relative specific activities and the fractionation data on rat kidney by Wattiaux-DeConinck *et al.,*[9] Wilfong and Neville

TABLE I

RECOVERY OF PROTEIN[a,b] AND THE RELATIVE SPECIFIC ACTIVITY (RSA) OF ALKALINE PHOSPHATASE AT VARIOUS STAGES IN THE ISOLATION PROCEDURE FOR OPTIONS I AND III[c]

Sample	Option I, total protein (mg)	Option III, total protein (mg)	RSA
Step 2, homogenate	1750	1750	1
Step 4, homogenate	1000	1000	1
Step 5, 1st sediment	80	50	3.5
Step 10, float	15	10	10
Step 16, brush borders	5	3.3	15

[a] O. H. Lowry, N. J. Rosebrough, A. L. Farr, and R. J. Randall, *J. Biol. Chem.* **193**, 265 (1951).

[b] Protein determinations are referred to a bovine serum albumin standard. Within absorbance ranges between 0.15 and 0.4 per centimeter, homogenate, membrane, and albumin curves are all linear.

[c] Using 9 g of kidney. RSA is identical for options I or III.

[8] Relative specific activity (RSA) is equal to the enzymatic specific activity of the indicated fractionation step divided by the enzymatic activity of the homogenate.

[9] S. Wattiaux-DeConinck, M.-J. Rutgeerts, and R. Wattiaux, *Biochim. Biophys. Acta* **105**, 446 (1965).

TABLE II
SPECIFIC ACTIVITY OF ORGANELLE SPECIFIC ENZYMES IN STARTING
HOMOGENATE AND FINAL BRUSH BORDER FRACTION[a,b]

Enzyme	Homogenate	Brush border	Relative specific activity
Alkaline phosphatase[c]	6.92 ± 0.07	108 ± 4	15.6
Acid phosphatase[d]	1.90 ± 0.20	0.01 ± 0.005	0.005
Glucose-6-phosphatase[d]	1.80 ± 0.30	0.01 ± 0.006	0.006
Succinate–cytochrome c reductase[e,f]	0.10 ± 0.02	0.03 ± 0.01	0.31

[a] From R. F. Wilfong and D. M. Neville, Jr., *J. Biol. Chem.* **245,** 6106 (1970).

[b] Phosphatase data are expressed in micromoles of substrate per milligram of protein per hour at 37°. Data are from two separate brush border preparations (option I, see *Procedure*) using an average of four different time intervals per assay. Standard deviations were calculated from these data.

[c] O. A. Bessey, O. H. Lowry, and M. J. Brock, *J. Biol. Chem.* **164,** 321 (1946).

[d] G. Hubscher and G. R. West, *Nature (London)* **205,** 799 (1965).

[e] H. D. Tisdale, this series Vol. 10, p. 213.

[f] Specific activity of succinate–cytochrome c reductase is reported in units of micromoles per milligram of protein per minute at 25°.

calculated that 7% of the purified membrane preparation was of mitochondrial origin, while lysosomes and endoplasmic reticulum contribute less than 1%.

Optional Steps. These steps are provided for users with varied needs. Option I gives a higher yield of brush border protein and more mitochondrial contamination. This is noted by the presence of a prominent brown button at the bottom of the white brush border pellet when pelleted after step 16. Option III reduced this mitochondrial contamination and the yield. Option II provides the largest fragments of brush borders, and under phase microscopy the brushes appear better defined.

Reproducibility. In this laboratory the relative specific activity of alkaline phosphatase always falls between 14 and 17. With option I or II used in the preparation (designed for high yield), the mitochondrial contamination varies between 5 and 10% (based on RSA values of succinate–cytochrome c reductase). With option III the mitochondrial contamination varies between 2.5 and 5%. The yield of protein is the most variable parameter. For options I or II, yield varies from 10 to 35 mg of protein per 27 g of kidney, with 15 mg as an average value. For option III yield varies between 7 and 13 mg protein per 27 g of kidney. Variation in yield is largely a result of variable recovery after flotation. The variables which affect this step are not fully understood, but it appears that seasonal variations and other environmental variables affect the brush border isopycnic point. In this laboratory, the percentage of sucrose used in the flotation step has been changed several times (see Basic Design).

Sources of Difficulty. The following errors in technique have been noted in this or in other laboratories. When the yield is low, but the brush borders appear well formed and vesicle contamination is not great, check for (a) underspeeding of centrifuge (step 5), (b) insufficient homogenization, which will leave more then one-half of the protein on the filtering screens, (c) density of sucrose too low (steps 8 and 9). When the yield is low and the brush borders do not lie evenly on the cushion (step 16), the gradient is irregular, probably because it was formed too rapidly. When the yield of protein is normal or elevated but the brush borders are generally small and many vesicles without brushes are present, check for (a) overspeeding of centrifuge (step 5), (b) too vigorous homogenization, (c) density of sucrose too high (steps 8 and 9), (d) fine nylon screen clogged or dirty, (e) substitution of angle-head centrifuge for swinging bucket in step 5, (f) all resuspension steps that were too vigorous, resulting in excessive fragmentation of brush borders, (g) relaxation of temperature control resulting in warming and fragmentation of brush borders.

Biochemical Activities. The following enzyme activities have been found in rat kidney brush borders with RSA values between 10 and 20. Alkaline phosphatase,[1,5,10] 5'-nucleotidase,[10] and maltase.[10] Rabbit kidney brush borders have a similar enrichment with trehalase.[4] Brush borders from both animals show a 2- to 4-fold enrichment for NaK+-dependent ATPase[1,4,5]; parathyroid hormone stimulated adenyl cyclase[1] and γ-glutamyltransferase[10] are 2- to 4-fold enriched in rat kidney brush borders. These enzymes either are not specifically concentrated in the brush border or are absent from brush border and represent contamination with cell front or blood plasma membranes.

Recently, high-affinity phlorizin receptors have been found in isolated brush borders,[11,12] and these receptors are enriched 17-fold over the homogenate. Competition experiments with sugars suggest that these receptors are related to the glucose tubular reabsorption system.[12,13]

Storage. Pellets of brush borders can be stored at −70° for over 3 months without loss in alkaline phosphatase activity or distortion of

[10] H. Glossmann and D. M. Neville, Jr., *FEBS (Fed. Eur. Biochem. Soc.) Lett.* **19**, 340 (1972).

[11] W. Frasch, P. P. Frohnert, F. Bode, K. Baumann, and R. Kinne, *Pfluegers Arch. Gesamte Physiol. Menschen Tiere* **320**, 265 (1970).

[12] H. Glossmann and D. M. Neville, Jr., *J. Biol. Chem.* **247**, 7779 (1972).

[13] F. Bode, K. Baumann, and D. F. Diedrich, *Biochim. Biophys. Acta* **290**, 134 (1972).

their complex disc gel electrophoretic pattern in sodium dodecyl sulfate. Storage at 4° for over 24 hours results in detectable proteolysis.[14] Proteolysis of brush borders and liver plasma membranes is very rapid at 37°.[15] When studies utilizing incubations at this temperature are done, degradation of soluble protein incubated with membranes and degradation of many membrane proteins will occur. These phenomena must be corrected for during data analysis.[12,16]

[14] D. M. Neville, Jr. and H. Glossmann, *J. Biol. Chem.* **246**, 6335 (1971).
[15] H. Glossmann and D. M. Neville, Jr., *J. Biol. Chem.* **246**, 6339 (1971).
[16] P. Freychet, R. Kahn, J. Roth, and D. M. Neville, Jr., *J. Biol. Chem.* **247**, 3953 (1972).

[9] Isolation of Plasma Membranes from Intestinal Brush Borders

By Alexander Eichholz and Robert K. Crane

The intestinal epithelial cells form the surface of the small intestine. They are in contact with the digestive chyme and are responsible for the final stages of its digestion as well as for the absorption of its nutrients. Both processes; that is, terminal digestion and absorption, are functions of a specialized substructure at the luminal face of the cells, called the brush border.

The brush border consists of numerous cylindrical, i.e., fingerlike, projections called microvilli. Depending on the species examined, there are about 1700 microvilli in each brush border. They are closely packed and are about 1 μm in length and 0.1 μm in diameter.

The microvilli are bounded by a plasma membrane which appears typically trilaminar under the electron microscope after fixation and staining. The outer surface of the membrane is characteristically covered by a fine filamentous material which extends radially. The extent of this "fuzzy coat," also called the glycocalyx, varies greatly with species and is particularly prominent in the cat, the bat, and man. It contains carbohydrate which presumably represents glycoprotein and mucopolysaccharide. The proteins of the fuzzy coat form an integral part of the brush border membrane since they are not easily removed by mechanical effort or by washing. The outer surface of the membrane contains a number of hydrolytic enzymes, as has been shown by direct enzyme assay of isolated membranes as well as by histochemical techniques. When microvilli are sectioned and observed by transmission electron microscopy, the interior is seen to be filled with long filaments running along their length and comprising an internal core. The presence of such a structure suggests immediately a structural role to support and maintain the microvillus, although other more physiological roles may not be excluded. The core filaments are longer than the microvilli, and extend into the subjacent portion of the cell, where they are seen to embed in another layer of filaments running perpendicular to them. This perpendicular layer of filaments is called the terminal web. It separates the brush border from the main body of the cell. At the level of the terminal web, the lateral plasma membrane forms a tight junction with adjacent epithelial cells, thus forming an effective barrier between the intestinal lumen and the intercellular space.

A detailed description of the morphological organization of this region of the epithelial cell can be found in the review by Trier.[1]

Since there is a significant difference in the purification procedure from different animal species as well as the contaminants encountered, preparation of membranes from each species will be discussed under a separate heading. Only procedures used for the preparation of the purified membrane will be presented. The preparation of the brush border is described where appropriate only as an intermediate step, and no attempt is made to review the methods where the final goal is the isolation of brush borders.

Preparation of Brush Border Membranes from Hamster Small Intestine

The intact brush border structure can be obtained relatively easily from hamster intestine. Hence, preparation of the membrane starts with isolation of brush borders. In this way, contamination by other membranes, such as the nuclear and the lateral membranes of the epithelial cell, is substantially eliminated.

Preparation of Brush Borders

Unlike other membranous structures, the brush border appears to be relatively resistant to treatment with hypotonic EDTA and does not disrupt after swelling. Exposure of mucosal scrapings to 5 mM EDTA, and homogenization results in disruption of the nuclear and lateral plasma membrane, leaving the brush border as the largest structure in the homogenate, which can then be easily removed by differential centrifugation.[2] The degree of contamination of the brush border preparation depends on the species used, hamster intestine giving nearly uncontaminated preparations. The small amounts of contamination routinely observed are tags of lateral

[1] J. Trier, "Handbook of Physiology," Vol. III, p. 1125. Amer. Physiol. Soc., 1968.
[2] D. Miller and R. K. Crane, *Anal. Biochem.* **2**, 284 (1961).

membrane, since the brush borders rarely separate cleanly at the tight junction, and bits of cytoplasmic vacuoles and other material hanging on below the brush border. Other species, such as the rat and guinea pig, show various degrees of contamination, primarily by nuclei, which are removed by additional purification steps.

Method

Reagents

NaCl, 0.15 M
EDTA, 5 mM, pH 7.4

Procedure. Syrian Golden Hamsters 8–10 weeks old are fasted for 24 hours and sacrificed; the small intestine is excised and rinsed through with 0.15 M NaCl. The intestine is everted with a stainless steel wire and washed again in three successive beakers of ice-cold NaCl. The intestine is then gently blotted and laid out on a cold glass plate. Using the edge of a glass slide, the mucosa is removed by light scraping. Collected mucosal scrapings from four animals are homogenized in cold 5 mM EDTA, pH 7.4, in a Waring Blendor. The speed of the Waring Blendor is controlled by the use of a Powerstat variable transformer. Usually, with a setting of 90 V, the homogenization is carried out for 15 seconds. The best time and speed should be determined for each individual case by monitoring the completeness of the disruption under the phase microscope. The homogenate is then passed through bolting silk No. 25[3] or fine nylon, to remove muscle fibers and larger contaminations. The filtrate is subjected to centrifugation at 1500 rpm (500 g) for 10 minutes in an International PR-2 refrigerated centrifuge. The supernatant is decanted, and the small amount of sediment is resuspended in EDTA to the original volume. The cycle of centrifugation and washing is repeated several times until most of the small-sized contaminating material has been removed as determined by phase microscopy. Larger contaminations are removed by centrifugation in an International Clinical Centrifuge at its lowest setting for 1 minute. By careful monitoring with the microscope and additional centrifugations in the PR-2 centrifuge to remove small contaminants, or in the Clinical Centrifuge for larger-sized particles, brush borders substantially free of other particles can be prepared. Such preparations contain 60–70% of the alkaline phosphatase, disaccharidase, and leucyl-naphthyl amidase activities of the original homogenate. About a 10-fold increase in specific activities over those in the homogenate are obtained.

[3] This may be obtained from Tobler, Ernst and Tobler, New York, New York.

Subfractionation of Brush Borders

After exposure to 1 M Tris, the isolated brush border disintegrates into what appears under phase contrast microscope as an array of barely distinguishable particles. Fractionation by density gradient centrifugation yields five separate bands. Two of these contain the membrane of the microvilli, and the precipitate fraction contains, in addition to some undisrupted brush borders and some unidentified fragments, fibrillar material forming a rod of proper dimensions to be the core of the microvilli.[4,5]

Method

Reagents

Tris, 1 M pH 7.0
20%, 30%, 40%, 50%, 60% Glycerol in 50 mM MgCl$_2$
EDTA, 5 mM

Disruption of Isolated Brush Borders. Washed brush borders are disrupted by addition of cold 1 M Tris buffer, pH 7.0, in a volume equal to

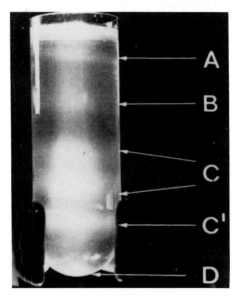

Fig. 1. Distribution of visible brush border subfractions on a glycerol 20 to 60% gradient after Tris disruption. From J. Overton, A. Eichholz, and R. K. Crane, *J. Cell Biol.* **26**, 692 (1965).

[4] A. Eichholz and R. K. Crane, *J. Cell Biol.* **26**, 687 (1965).
[5] J. Overton, A. Eichholz, and R. K. Crane, *J. Cell Biol.* **26**, 692 (1965).

four times the weight of the original mucosa. The suspension is vigorously agitated on a Vortex mixer for several minutes.

Separation of the Membrane Fraction by Density Gradient Centrifugation. Tris-disrupted brush borders are subfractionated on a discon-

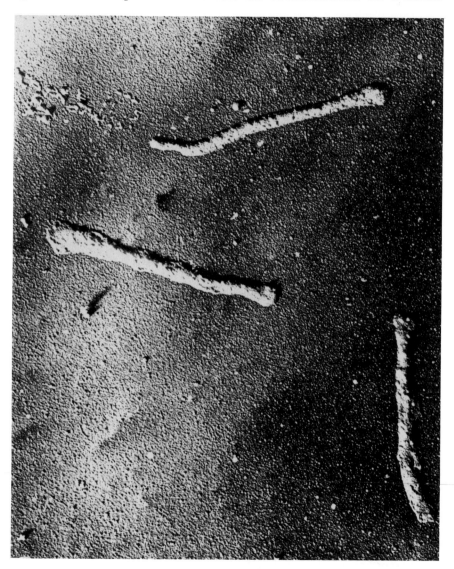

FIG. 2. Shadowed preparation of membrane fraction. From J. Overton, A. Eichholz, and R. K. Crane, *J. Cell Biol.* **26**, 692 (1965).

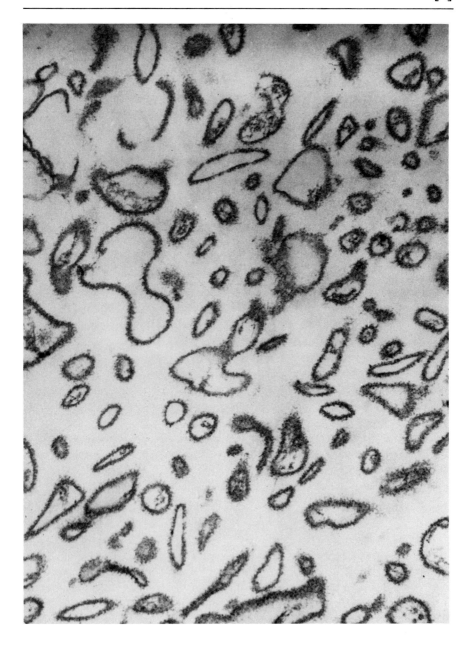

FIG. 3. Sectioned pellet of the membrane fraction after fixation in gluteraldehyde and osmium.

tinuous gradient made by successive layering of equal volumes of 20, 30, 40, 50, and 60% glycerol containing 0.05% $MgCl_2$. The centrifugation is carried out for 10 minutes at 63,000 g in the swinging-bucket rotor of the Spinco Model L ultracentrifuge. Figure 1 is a photograph of a separated gradient. The fractions are visualized by the Tyndall effect. Fractions C and C' contain the microvillous membrane. As can be seen from the electron micrograph of a shadowed preparation in Fig. 2, the membranes are obtained in the form of hollow tubes closed at one end, and open at the other, like cut fingers of a glove. When sectioned, the membrane appears empty and still has pronounced fuzz on its outer surface, as seen in Fig. 3. Fractions are collected by any one of the standard techniques used in density gradient centrifugation. Because of the precipitate at the bottom of the tube, we use a device similar to that employed by Kahler and Lloyd.[6] The collected membrane fractions are diluted 3- to 5-fold with distilled water and subjected to centrifugation at 48,000 g for 40 minutes, a Sorvall angle-head rotor being used. The sediment contains a translucent precipitate containing all of the protein and enzymatic activities of the glycerol fraction.

Properties of the Purified Microvillous Membranes

Yield and Enzyme Activities. All the enzymatic activities known to be associated with the brush border membrane are found to be localized in the membrane fraction of Tris-disrupted brush borders. Starting with brush border, recoveries of most of the enzymatic activities are close to 100% except for maltase, isomaltase, and trehalase, which often show activation. Generally, we have found the specific activity of the membrane fraction to be four times that of the brush border. Fraction C' exhibits much lower specific activities of enzymes and about one-tenth the total activity of the C fraction. The number of hydrolytic enzyme activities found to be associated with the membrane fraction is growing continuously as new enzymes are tested. The variety and number of such enzymes found contribute to the uniqueness of its specialized structure. The table lists enzymes currently shown to be localized in the membrane. Although with many membranes it is said that one cannot place great reliance on the specific activity of enzyme markers as a criterion of membrane purity, owing to possible redistribution during homogenization, in the brush border the danger of such changes is minimized because of the large number of enzymes that can be monitored.

Chemical Composition. While a variable amount of DNA contamination is found to be associated with the whole isolated brush border, the separated membrane fraction is virtually devoid of DNA and any con-

---

[6] H. Kahler and B. J. Lloyd, Jr., *J. Phys. Colloid Chem.* **55**, 1344 (1951).

Enzymes of the Microvillous Membrane

Enzymes	Reference or source
A. Enzymes found in the microvillous membrane	
Sucrase	a
Maltase	a
Isomaltase	a
Lactase	a
Trehalase	b
Alkaline phosphatase	a
ATPase	a
Cholesterol ester hydrolase	c
Retinyl ester hydrolase	c
Phlorizin hydrolase (β-glucosidase activity)	d
Leucylnaphthyl amidase	e
Leucylglycine hydrolase	e
γ-Glutamyl transpeptidase	f
B. Enzymes probably localized to the membrane, but as yet demonstrated only in the brush border	
Enterokinase	g
γ-Amylase	h
Oligopeptidase	i
Folate deconjugase	j

[a] Obtained from Tobler, Ernst, and Tobler, New York, New York.
[b] H. Kohler and B. J. Lloyd, Jr., *J. Phys. Colloid Chem.* **55**, 1344 (1951).
[c] P. Malathi and R. K. Crane, *Gastroenterology* **52**, 1106 (1967).
[d] P. Malathi and R. K. Crane, *Biochim. Biophys. Acta* **173**, 245 (1969).
[e] J. Rhodes, A. Eichholz, and R. K. Crane, *Biochim. Biophys. Acta* **135**, 956 (1967).
[f] J. Cerda, H. Preiser, J. Woodley, and R. K. Crane, *J. Clin. Invest.* **19**, 655 (1971).
[g] A. Dahlqvist, *Biochim. Biophys. Acta* **198**, 621 (1970).
[h] *Abstracts Meeting Amer. Gastroenterol. Asso.* **56**, 1182 (1969).
[i] T. J. Peters, K. Modha, and C. N. C. Drey, *Biochem. J.* **119**, 20P (1970).
[j] I. H. Rosenberg, R. A. Streiff, H. A. Fodwin, W. B. Castle, *N. Engl. J. Med.* **280**, 985 (1969).

tamination would be found in fractions A and D.[7] We were also successful in removing DNA contamination from brush borders by pretreatment with DNase in the presence of activating Mg ion concentrations and subsequent washing. The membrane contains approximately 50% protein by weight, similar to the erythrocyte membrane. The ratio of cholesterol to phospholipid is around 0.8. The membrane also contains about 5–10% carbohydrate on dry weight basis.

[7] A. Eichholz, *Biochim. Biophys. Acta* **135**, 475 (1967).

Preparation of Membranes from Human Small Intestine

Human brush borders seem to be relatively fragile, and if the tissue is frozen, as is practical for most tissue samples obtained by peroral biopsy from human intestine, the yield of brush borders becomes very small. In this case, therefore, it is more practical to proceed directly to the preparation of microvillous membranes from frozen intestinal tissue.[8]

Method

Reagents

NaCl, 0.15 M
EDTA, 5 mM
EDTA, 5 mM in Tris, 5 mM, pH 7.3
Tris, 1.25 M pH 7.4

Procedure. Full sections of the human small intestine are removed at surgery, sealed in foil, and frozen immediately on dry-ice. For the preparation, the cut portion is thawed and washed in cold 0.15 M NaCl. The tissue is cleaned of extraneous fat, a 0.2–4.2 g specimen is placed in 40–90 ml of 5 mM EDTA and 5 mM Tris buffer, pH 7.3, and homogenized in a Waring Blendor for 10–15 seconds using a Powerstat variable transformer at a setting of 90. The homogenate is filtered through No. 25 bolting silk, a suction flask being used. If the original specimen is particularly fatty or large, the volume is brought up to 90–135 ml. The filtered homogenate is then centrifuged in a Sorvall RC-2 centrifuge using a SS-3 rotor and 19,200 g for 15 minutes. The supernatant is decanted, and the pellet is resuspended in 1.5–2 ml of the EDTA–Tris buffer or in distilled water. Freshly prepared 1.25 M Tris, pH 7.4, is then added to the resuspended precipitate in a 1:1 proportion, stirred repeatedly on a Vortex mixer and kept in the cold for 1 hour. The mixture is applied to a discontinuous gradient of 20% to 60% glycerol containing 50 mM $MgCl_2$, as with the hamster brush border. Centrifugation is carried out for 10 minutes at 63,000 g in a Spinco Model L centrifuge using an SW 25.1 swinging-bucket type rotor.

The gradient obtained by this procedure is similar to that obtained from hamster intestine. The largest fraction obtained is about half way through the gradient and is labeled fraction III. This fraction contains the microvillous membrane as identified by its enzymatic content and by elec-

[8] J. D. Welsh, H. Preiser, J. F. Woodley, and R. K. Crane, *Gastroenterology* **62**, 10 (1972).

tron microscopic observation. Probably owing to the greater fragility of the brush border, somewhat lower recoveries of the enzymatic activities are obtained. They range between 42% for trehalase and 69% for nitrophenyl-β-galactosidase (a synthetic substrate for β-galactosidase). Increase in specific activities for sucrase, trehalase, and alkaline phosphatase range from 12- to 37-fold over the homogenate. The preparation is to some extent contaminated by other membranous fractions of the cell.

A new procedure for the preparation of microvillous membranes from human tissue is now available.[8a] This procedure is applicable to fresh or frozen intestine and is suitable for the preparation of membranes from preoral jejunal biopsies. The method employs calcium for the removal of microsomal contaminants and uses Tris and differential centrifugation for release of membranes and for their further purification.

Preparation of Membranes from Rat Intestine

Unlike hamster brush borders, which can be obtained by the simple purification procedure outlined above, brush borders obtained by the same procedure from other species show various degrees of contamination primarily by nuclei. Additional purification steps have been incorporated for such species, as can be seen by the procedures developed for the rat.[9]

Method

Reagents

NaCl, 0.15 M
EDTA, 5 mM
EDTA, 2.5 mM
EDTA, 90 mM, in 0.8 mM EDTA
15% Sucrose and 30% sucrose

Procedure. Rats are killed, and the intestine is removed and irrigated with cold 0.15 M NaCl. The intestine is everted, and the mucosa is removed with a glass slide. The mucosa is weighed, placed in 75 volumes of 5 mM EDTA, adjusted to pH 7.4, and homogenized in a Waring Blendor for 25 seconds. All operations are performed at 4°. After centrifugation at 450 g for 10 minutes, the sediment is washed three times in 5 volumes of EDTA buffer. The washed sediment is then suspended in two volumes of 90 mM NaCl containing 0.8 mM EDTA, mixed, and kept until a well defined sediment has developed (usually 20–30 minutes). A similar alter-

[8a] J. Schmitz, H. Priser, D. Maestracci, B. Ghosh, and R. K. Crane, *Biochim. Biophys. Acta* 323, 98 (1973).
[9] G. G. Forstner, S. M. Sabesin, and K. J. Isselbacher, *Biochem. J.* 106, 338 (1968).

nate procedure for the flocculation of nuclei and debris is suggested by Porteous.[10,11] See alternate procedure below.

The supernatant and sediment are then filtered through glass wool, as had been suggested by Harrison and Webster in an alternative procedure for brush border preparation.[12] About 300 mg dry weight of Pyrex brand Corning No. 3950 wool is used. After filtration, the glass wool pad is washed with 20 ml of 5 mM EDTA and the combined filtrate is again subjected to centrifugation for 10 minutes at 450 g.

Isolation of the membrane from the rest of the brush border structure can be achieved either by Tris disruption using the technique described for hamsters above, or by the following procedure using distilled water for disruption.[9]

Brush borders are resuspended in 150 volumes of cold water to bring the concentration of EDTA to 17 μmoles, and the mixture is allowed to stand at 4° for exactly 12 minutes. Sufficient EDTA buffer is then added with rapid swirling to bring the concentration to 4.5 mM. The suspension is then centrifuged at 1800 g for 20 minutes, the precipitate containing undisrupted brush borders is removed, and the supernatant is subjected to another centrifugation at 30,000 g for 30 minutes to sediment the membrane. The precipitate from the second centrifugation, labeled membrane fraction I, contains 34% of brush border protein and 54% of the total sucrase activity. Therefore, only a portion of the total microvillar membrane was released from the brush border by hypotonic disruption. An additional 16% of sucrase activity associated with the membrane can be recovered by collection of another membrane fraction from undisrupted brush borders on a sucrose gradient. To accomplish this, the low speed precipitate is applied to a sucrose gradient consisting of 3 ml of 30% sucrose and 3 ml of 15% sucrose in a 2.5 mM EDTA buffer. After centrifugation for 20 minutes at 1000 g, the upper aqueous phase contains membrane fraction II and can be collected by recentrifugation at 30,000 g for 30 minutes. Membrane fraction I and II appear to be identical in all respects.

Alternate Flocculation Procedure.[10,11] After disruption of the mucosa by homogenization, filtration through nylon and centrifugation, the precipitate is resuspended in a buffer containing 2.5 mM EDTA, 50 mM potassium phosphate, 25 mM potassium citrate, and 77 mM potassium chloride. It is left to stand in this solution for 20–30 minutes, and the supernatant is decanted through a fine nylon cloth to remove the flocculant

[10] J. W. Porteous, *in* "Subcellular Components" (G. D. Birnie and S. M. Box, eds.), p. 57. Butterworth, London, 1969.
[11] J. W. Porteous, *FEBS (Fed. Eur. Biochem. Soc.) Lett.* 1, 46 (1968).
[12] D. D. Harrison and H. L. Webster, *Biochim. Biophys. Acta* 93, 662 (1964).

sediment. *A priori* there appears to be no reason why removal of DNA from rat brush border preparations should not be accomplished by treatment with DNase as described in the procedure used for hamster intestine.

Preparation of Membranes from Guinea Pig Intestine

Similar to the rat brush border, DNA and nuclei are the major contaminants in the guinea pig brush border and can be removed by flocculation with a high ionic strength buffer.[13]

Method

Reagents

NaCl, 0.15 M
EDTA, 5 mM, pH 7.4
Krebs Ringer bicarbonate reagent

Procedure. The initial procedure followed is identical to that used for hamster brush borders. After centrifugation in the International Centrifuge, the pellet is resuspended in Krebs-Ringer bicarbonate solution, and the clumped material is removed by light centrifugation at 500 rpm for 1 minute. The sediment is discarded. Although this step substantially reduced the yield of the brush borders, it provides a purer preparation as evaluated by light microscopy. The supernatant is then centrifuged at 2500 rpm for 10 minutes in the International PR-2 centrifuge to yield a pellet of relatively pure brush borders.

Further steps for disruption of brush borders and isolation of the membrane are identical to those used in the preparation from hamster intestine described above.

[13] R. M. Donaldson, Jr., I. L. Mackenzie, and J. S. Trier, *J. Clin. Invest.* **46**, 1215 (1967).

[10] Isolation of Plasma Membrane from Smooth, Skeletal, and Heart Muscle

By A. M. KIDWAI

For studying the characteristics of membranes, it is important to have a pure preparation that can be obtained in a short time and under mild conditions to preserve the enzymatic activities and chemical composition as close as possible to the ones present *in situ*.

Differential centrifugation leads to a heterogeneous preparation, e.g., microsomes. In the past, a microsomal preparation was used whenever a membrane preparation was required irrespective of the nature of the membrane in question. For example, protein biosynthesis expected to be the property of rough ER[1] and $Na^+ + K^+$-ATPase present in plasma membrane[2] for transport of ions were studied in the same preparation. Methods to isolate sarcolemma from skeletal muscle involve extraction of homogenate with salt solutions[3] to remove contractile proteins or incubation in dilute salt solutions at 37°.[4] These treatments might make the preparation unsuitable for some studies, for example, enzymatic, protein, and ionic composition of the membrane.

Density gradient centrifugation is a useful technique for separation of components with different densities. The difference in density between the plasma membrane and intracellular membranes is not great, but it is possible to separate the two membranes on a sucrose density gradient prepared in ISCO density gradient former at neutral pH and low ionic strength. The present method describes a density gradient which enables one to isolate all the subcellular fractions including plasma membrane in one step.[5] Use of rather drastic homogenization condition helps in emptying the cells of viscous intracellular myofibrils.

Sucrose used in the homogenizing medium reduces the contamination of subcellular fractions by contractile proteins. Most of the contractile proteins were filtered first, and the remaining ones were centrifuged down to the bottom of the gradient, as they were heavier than the ($\bar{v}$ = approximately 0.7) plasma membrane[6] and other fractions except the nuclear fraction.

Method

Reagents. Prepare solutions in glass-distilled water and store between 0 and 5°.

Sucrose, 0.25 M, d 1.040
Sucrose, 2.00 M, d 1.269

[1] A. Van Der Decken, *in* "Techniques in Protein Biosynthesis" (P. N. Campbell and J. R. Sargent, ed.), Vol. I, p. 88. Academic Press, New York, 1967.
[2] J. C. Skou, *Physiol. Rev.* **45**, 596 (1965).
[3] T. Kono and S. P. Colowick, *Arch. Biochem. Biophys.* **93**, 520 (1961).
[4] D. L. McCollester, *Biochim. Biophys. Acta* **57**, 427 (1962).
[5] A. M. Kidwai, M. A. Radcliffe, and E. E. Daniel, *Biochim. Biophys. Acta* **233**, 538 (1971).
[6] A. M. Katz, D. I. Repke, J. E. Upshaw, and M. A. Polascik, *Biochim. Biophys. Acta* **205**, 473 (1970).

Equipment

Polytron Generator PT20, Kinematica G.m.b.H., 6000 Lucerne, Switzerland (supplied by Brinkman Instruments Inc., Westbury, New York)

Generators other than PT20 can be used depending upon the amount of tissue to be processed.

ISCO Density gradient former Model 570 (Instrumentation Specialities Co. Inc., Lincoln, Nebraska) or Beckman Gradient Former (Spinco Division of Beckman Instruments, Inc., Palo Alto, California)

Filtration device (as described in the text)

Ultracentrifuge either a or b.

(a) Beckman ultracentrifuge L or L2-65B (Spinco Division of Beckman Instruments, Palo Alto, California) Rotors: Swinging-bucket, SW 36 for centrifuge L or SW 40 for centrifuge L2-65B; fixed-angle rotors, 40 or 65.

(b) IEC ultracentrifuge B-50 or B-60, International Equipment Co., Needham Heights, Massachusetts. Rotors: Swinging-bucket, SB-283; fixed-angle rotors, 494 or 480.

Procedure

Preparation of the Gradient. Adjust volumes of the delivery syringe (ISCO density gradient former) so that the left-side syringe delivers twice as much sucrose solution as that of the right side. Fill the left side with heavy sucrose (2 *M*) solution and the right side with the lighter solution (0.25 *M*). If rotor SW 40 or SW 36 is used, settings of delivery syringe

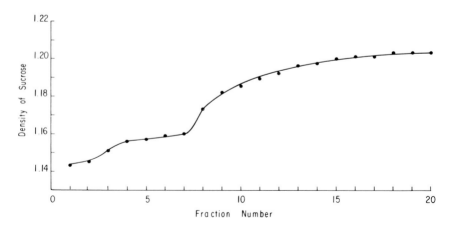

FIG. 1. Sucrose density gradient. Density measured in 0.5-ml fractions collected in ISCO density gradient fractionator.

7 ml on left side and 3.5 ml on the right side should be used. At the start, the right syringe moves faster and delivers approximately 2–3 ml of light gradient in the bottom of the tube; the heavier gradient coming later pushes this initial volume delivered to the bottom of the tube to the top of the gradient, resulting into sharp breaks in the upper part of the gradient. It is advisable to leave the gradient in the refrigerator for equilibration for at least 12 hours before use. Freshly prepared gradient can also be used, but for the sake of reproducibility, the gradient should be used at a fixed time after formation of the gradient. A representative density profile of the gradient is shown in Fig. 1.

Homogenization. Homogenize the tissue in 0.25 *M* sucrose using the Polytron PT20 for a short time at half-maximum speed. The time of homogenization is different for different tissues. The time required is in the order heart < skeletal muscle < myometrium < aorta.

The time of homogenization should be worked out so that the minimum

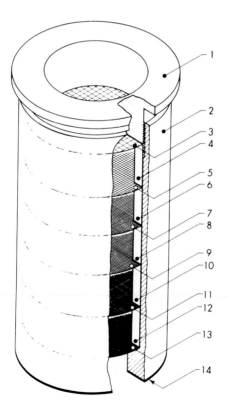

FIG. 2. Filtration device. 1, Screw top; 2, acrylic cylinder; 4, 6, 8, 10, and 12, removable O-ring spacings; 3, mesh 16; 5, mesh 20; 7, mesh 30; 9, mesh 40; 11, mesh 60; 13, mesh 80; 14, RTV-112 pourable silicone rubber.

activity of cytochrome *c* oxidase can be detected in the plasma membrane preparation. As the time increases, so does the cytochrome *c* oxidase activity in fractions other than mitochondrial fraction. All operations are carried out at 4°.

Filtration. The filtration step is an important one since this replaces the low speed centrifugation, which usually removes the cell debris. A filtration device is described below which is essential for skeletal muscle plasma membrane preparation but is helpful in smooth and heart muscle as well.

An acrylic cylinder of 75 mm i.d. can be used as a support. O rings of the same material, 17 mm long and outer diameter 75 mm, are fused in the bottom using plastic solvent. A circular stainless steel cloth is layered on this support and sandwiched with another ring from the top, but not fused. Five different mesh sizes of wire cloths are used, one over the other as shown in the Fig. 2. Start with mesh 16 from the top and go down to mesh 80 at the bottom. High mesh cloths tend to collapse under vacuum and therefore need support of other, lower mesh cloths. The size of the filtration device can be made according to the amount of tissue to be processed. The

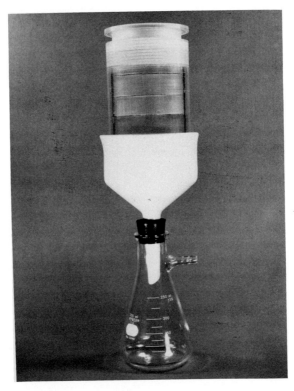

Fig. 3. Filtration assembly: filtration device, Büchner funnel, and filtration flask.

above-mentioned dimensions are suitable for processing 50 g of skeletal muscle.

This assembly with a rubber gasket at the bottom can be placed on a Büchner funnel, and the Büchner flask can be connected to a vacuum line. A mild vacuum was necessary for fast filtration (Fig. 3).

Preparation from Smooth Muscle

Homogenize about 2.0 g of smooth muscle either from rat uterus or rabbit aorta in 0.25 M sucrose in Polytron PT20 for 10 seconds and 30 seconds, respectively, at half maximum speed. Filter the homogenate in the cold room or cool the filtration assembly before use and immerse the flask in an ice bath and rehomogenize lumps at the top of the filter for the same period of time and filter again. Pool the filtrates and make 10% with respect to the wet weight of the original tissue. Centrifuge this homogenate at 100,000 g for 1 hour. Save supernatant for studies if needed. Resuspend the sediment in 2 ml of 0.25 M sucrose and layer on the top of the gradient and centrifuge for 2 hours at 111,688 g. The plasma membrane forms a band at the interphase of the loading medium and the gradient. The endoplasmic reticulum is spread more diffused next below the plasma membrane and extends to the middle of the tube. A band of mitochondria can be seen around the middle of the tube. The nuclear fraction settles down at the bottom of the tube. A few mitochondria remain trapped in the plasma membrane band and can be removed by low speed centrifugation (13,000 g 15 minutes), if extremely pure plasma membrane is required.

Collect the bands on the gradient either in fractionator or by Pasteur pipette, dilute to reduce the concentration of sucrose to 0.25 M, and centrifuge at 105,000 g for 30 minutes to get the fractions in pellet. Resuspend the pellet in buffer as desired.

Preparation from Heart Muscle

Chop about 5 g of heart muscle with scissors and wash with sucrose repeatedly to free it of blood as much as possible. Homogenize in 0.25 M sucrose in Polytron PT20 for 15 seconds at half maximum speed.[7] Prepare 10% homogenate and filter through wire cloths to remove any connective tissue and some of the contractile proteins. Centrifuge the filtrate for 60 minutes at 100,000 g. Resuspend the sediment in a small volume of 0.25 sucrose and load on the top of the gradient and centrifuge for 90 minutes at 111,688 g.

A thick band of plasma membrane collects at the interphase of the loading medium and the sucrose gradient. Sarcoplasmic reticulum is spread

[7] A. M. Kidwai, M. A. Radcliffe, G. Duchon, and E. E. Daniel, *Biochim. Biophys. Res. Commun.* **45**, 901 (1971).

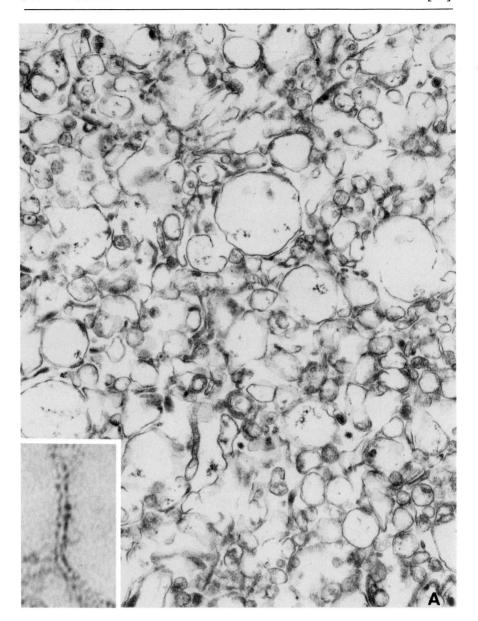

FIG. 4. (A) Electron micrograph of plasma membrane preparation of smooth muscle (rat myometrium). ×40,000. Inset shows the triple layer of plasma membrane. ×350,000.

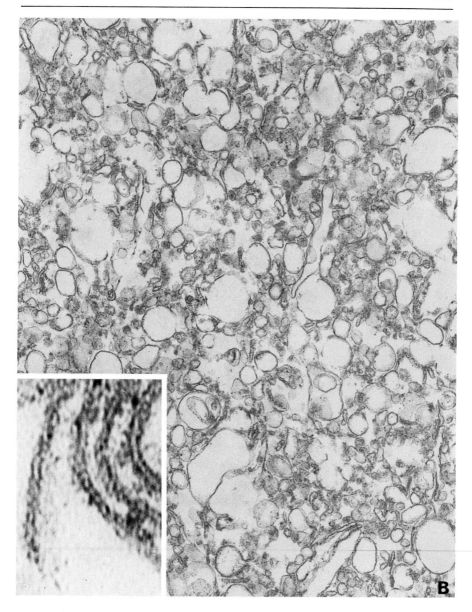

(B) Electron micrograph of rat heart muscle plasma membrane. ×40,000. Inset shows the triple layer of plasma membrane. ×350,000. From A. M. Kidwai, M. A. Radcliffe, G. Duchon, and E. E. Daniel, *Biochim. Biophys. Res. Commun.* **45**, 901 (1971).

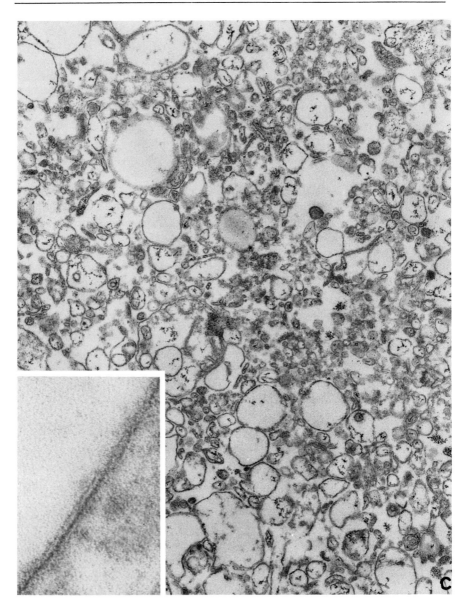

(C) Electron micrograph of rat skeletal muscle plasma membrane. ×30,000.
Inset shows the triple layer of plasma membrane. ×250,000.

more widely in a band just below the plasma membrane which extends to the middle of the tube. There a thick mitochondrial band can be seen, followed by a dark red band of blood cells; the contractile proteins, nuclei and cell debris settle down at the bottom of the gradient.

The heart mitochondria moves down the gradient faster than do mitochondria of smooth muscle. The homogenization conditions should be properly controlled so as to minimize the damage to mitochondria.

Preparation from Skeletal Muscle

Skeletal muscle plasma membrane preparation is essentially the same as that of smooth and heart muscle, except that special attention has to be paid to reduce the amount of contractile proteins before the density gradient step. This can be achieved by the use of a special filter as mentioned earlier. Prepare a 10% homogenate in 0.25 M sucrose using Polytron PT20. A homogenization time of 30 seconds at half maximum speed for each batch of 6 g of tissue is sufficient to break most of the cells.

The translucent filtrate can be centrifuged to reduce the volume. The density gradient centrifugation is then carried out as usual.

Applications

The absence of any salt, alkali and other drastic treatment makes this general procedure suitable for chemical and biological studies of plasma membrane, endoplasmic reticulum, and mitochondria. There is minimum damage to the enzymes associated with the plasma membrane and other subcellular fractions. We were able to measure many enzymes in this preparation including adenyl cyclase and $Na^+ + K^+$-ATPase. Active Ca^{2+} uptake, drug-binding studies, protein biosynthesis, and fractionation of structural proteins are some of the applications of this preparation.

Limitations

Since there is no clear line of demarcation between the bands on the gradient, there is a possibility of cross contamination of fraction which can be remedied by sacrificing part of the yield and discarding the interphase of the bands.

Use of the Polytron homogenizer requires special attention and the homogenization should be properly controlled.

Effect of isopycnic centrifugation in our sucrose gradient on subcellular fractions is not known. It has been shown that mitochondria deteriorate during centrifugation in a sucrose gradient.[8]

---

[8] R. Wattiaux, S. Wattiaux-De Coninck, and M.-F. Ronveaux-Dupal, *Eur. J. Biochem.* **22**, 31 (1971).

Properties

The plasma membrane isolated by this method is vesicular in nature as revealed by electron microscopy (Fig. 4). A triple-layered structure of plasma membrane can be seen at high magnifications.

The preparation of plasma membrane is heterogeneous in density and can be further subdivided if centrifuged on a discontinuous gradient of densities 1.072, 1.108, 1.136, 1.16, and 1.172. All the subfractions are vesicular in electron micrographs.

$Na^+ + K^+$-ATPase and 5'-nucleotidase activities are concentrated in the plasma membrane fraction, protein biosynthetic activity in the endoplasmic reticulum, and cytochrome c-oxidase activity in mitochondrial fraction.

[11] Isolation of the Plasma Membrane from the Thyroid

By KAMEJIRO YAMASHITA and JAMES B. FIELD

Recently, methods for the isolation of plasma membranes have been developed for several tissues, such as liver,[1,2] kidney,[3] intestinal mucosa,[3] cultured fibroblasts,[4] Ehrlich ascites carcinoma cells,[5] HeLa cells,[6] and myeloma cells.[6] The plasma membranes of HeLa cells and cells in suspension are obtained from the supernatant fluid after a light centrifugation of disrupted cells.[6] In contrast, the plasma membranes from liver,[1,2] kidney[3] and intestine[3] are obtained from the nuclear pellet. This paper describes a method for preparing the plasma membranes of thyroid cells from the nuclear pellet.

Histologically, the thyroid gland consists of follicles enmeshed in thick, tough connective tissue. The follicles are composed of epithelial cells in the periphery and colloid containing thyroglobulin in the center. Mammalian thyroid tissue also contains parafollicular cells which are the source of calcitonin but are not concerned with thyroid hormone formation. Therefore, it is very hard to obtain homogenates from thyroid glands containing only a single cell type. Furthermore, disruption of the tissue by motor-driven homogenizers requires high shear forces, which damage the plasma mem-

[1] D. M. Neville, Jr., *J. Biophys. Biochem. Cytol.* **8**, 413 (1960).

[2] P. Emmelot, C. J. Bos, E. L. Benedetti, and P. Rumke, *Biochim. Biophys. Acta* **90**, 126 (1964).

[3] R. Coleman and J. B. Finean, *Biochim. Biophys. Acta* **125**, 197 (1966).

[4] L. Warren, M. C. Glick, and M. K. Nass, *J. Cell. Physiol.* **68**, 269 (1966).

[5] V. B. Kamat and D. F. H. Wallach, *Science* **148**, 1343 (1965).

[6] E. H. Eylar and A. Hagopian, this series, Vol. 22, p. 123.

branes. Since the main objective in the preparation of plasma membranes is to preserve the membrane integrity and biological properties, the procedure should be done rapidly and gently. A loose-fitting Dounce homogenizer is preferable to disrupt the thyroid tissue for this purpose.[7] The plasma membranes of the thyroid contain receptor sites for thyrotropin (TSH), and the interaction of TSH with these is regarded as the first step in the TSH stimulation of the thyroid. Although adenylate cyclase activity has been found in the plasma membranes of various tissues,[8,9] TSH stimulation of adenylate cyclase occurs only after binding of the hormone to the plasma membrane. Therefore, measurement of both basal and TSH-stimulated activity of adenylate cyclase provides information concerning the integrity of two components of the plasma membranes obtained. The integrity and purity of the preparation can also be monitored by electron microscopic appearance and assessment of other enzymatic activities and chemical composition.

Preparation of Plasma Membranes

Reagents

NaHCO$_3$ solution, 1 mM, pH 7.5
Sucrose solutions in NaHCO$_3$, 1 mM, pH 7.5, at 63%, 45%, 41%, and 37% (w/w) on refractometer

Disruption of Cells. All operations are carried out at 0–4°. Bovine or calf thyroid glands (25–100 g) are trimmed free of fat and connective tissue, sliced with a Stadie-Riggs microtome at about 1 mm thickness, and then minced into small pieces with scissors or a razor blade. These steps are necessary since homogenization of large pieces of thyroid tissue would require too much force. Six to eight grams of minced tissue are homogenized in 50 ml of NaHCO$_3$ solution in a loose-fitting Dounce homogenizer using 30–40 strokes. The number of homogenizer strokes is important. If too few, the yield may be low; if too many, the plasma membranes will be damaged. Homogenized tissues (100–200 ml) are pooled and diluted to 250 ml using NaHCO$_3$ solution. The homogenate is kept on ice for several minutes to allow cell lysis.

Removal of Connective Tissue and Large Fragments. The supernatant of the homogenate is filtered through cheesecloth using two fine layers (No. 120) and four coarse layers (No. 40). The yield of plasma membranes can be increased by homogenizing the residue again in 50 ml of

[7] K. Yamashita and J. B. Field, *Biochem. Biophys. Res. Commun.* **40**, 171 (1970).
[8] P. R. Davoren and E. W. Sutherland, *J. Biol. Chem.* **238**, 3016 (1963).
[9] S. L. Pohl, L. Birnbaumer, and M. Rodbell, *Science* **164**, 566 (1969).

NaHCO$_3$ solution using 10–20 strokes and filtering through cheesecloth. The filtrates are then combined.

Removal of Most of the Mitochondria and Microsomes. The filtrate is centrifuged in 50-ml conical tubes at 2900 rpm (1860 g) for 25 minutes in an International PR2 centrifuge. The resulting pellet contains most of the plasma membranes and nuclei. The mitochondria, microsomes, and some plasma membranes are in the supernatant. After the supernatant is aspirated, the pellets from several tubes are combined and then resuspended in 30 ml of NaHCO$_3$, homogenized using a few strokes of a loose-fitting Dounce homogenizer, and centrifuged at 2900 rpm for 15 minutes. The supernatant is aspirated. The pellet (usually 5–7 ml) is mixed with 2–3 ml of NaHCO$_3$ solution and homogenized gently in a 20-ml Dounce homogenizer using 5 strokes with a loose-fitting pestle. The homogenate is folded slowly into 10–12 ml of 63% sucrose to provide a final concentration of sucrose of 48% determined on a refractometer.

Sucrose Gradient. The membrane suspension in sucrose solution (30–38 ml) is divided equally into three centrifuge tubes used with the Beckman SW 25.1 rotor. Sucrose solutions of 45% (7 ml), 41% (10 ml), and 37% (2–3 ml) are carefully layered over the membrane suspension. The tubes are centrifuged at 25,000 rpm (63,600 g) for 120 minutes. After centrifugation, material is present at each interface of the layers and a pellet is at the bottom of the tube. Plasma membranes with a minimum contamination are obtained at the interface between the layers of 37% and 41% sucrose solutions. This band is collected with a pipette, diluted 1:1 with NaHCO$_3$ solution, and centrifuged at 16,000 rpm for 30 minutes in a Sorvall angle-head rotor (30,800 g_{max}). The precipitate is suspended in 0.25–0.3 ml of NaHCO$_3$ solution per 6 g of starting tissue. The protein content of the plasma membrane preparation is approximately 0.8–1.2 mg/ml. This preparation can be stored at −20°C for as long as 2 weeks with no loss in adenylate cyclase activity or TSH responsiveness.

Properties of the Purified Plasma Membrane

Electron Microscopy. An electron micrograph of the material at the interface between the layers of 37% and 41% revealed that the main structures consisted of vesicular membranous materials and a continuous layer of plasma membranes with a minimum amount of other cellular components, primarily rough reticulum. Other layers and the pellet at the bottom consisted of nuclei, rough reticulum, mitochondrial fragments, plasma membranes, and amorphous debris.

Enzymatic Characterization. Adenylate cyclase activity is associated with the plasma membrane fraction in pigeon erythrocytes,[8] rat liver,[9] bovine thyroid,[7] and other tissues. Therefore, the specific activity of

TABLE I

SPECIFIC ACTIVITIES OF ADENYLATE CYCLASE, Na^+-K^+-DEPENDENT ATPASE,
AND 5'-NUCLEOTIDASE IN WHOLE HOMOGENATE AND PLASMA
MEMBRANE FRACTION[a]

Fraction	Adenylate cyclase (nmoles/10 min/mg protein)	Na^+-K^+-dependent ATPase (μmoles/hr/mg protein)	5'-Nucleotidase (μmoles/15 min/mg protein)
Whole homogenate	Control, 0.012	0.83	0.39
	TSH (10 mU), 0.025	—	—
	NaF (10^{-2} M), 0.134	—	—
Plasma membranes	Control, 0.248	8.71	5.61
	TSH (10 mU), 1.185	—	—
	NaF (10^{-2} M), 5.143	—	—

[a] The results are the average of two preparations.

adenylate cyclase can be measured as an indicator for the purification of plasma membranes. Adenylate cyclase was assayed by the method of Krishna et al.[10,11] The specific activity in the plasma membrane fraction was 15–25 times that in the whole homogenate (Table I). In the presence of TSH (10 mU), it was more than 40 times that produced by the hormone using whole homogenates. TSH (10 mU) increased the adenylate cyclase activity 5-fold or more in the plasma membrane fraction as compared to only 2-fold in the whole homogenate. TSH showed a dose-response relationship in the activation of adenylate cyclase, and the activity using 50 mU of TSH was 13–15 times greater than the control level.[11] Although the purification in terms of adenylate cyclase activity was somewhat lower than that observed by Wolff and Jones using different methods for the membrane preparation,[12] the stimulation by TSH of adenylate cyclase was much higher in our preparation. In addition, stimulation was obtained with smaller concentrations of TSH. This indicates that some components in plasma membranes, such as TSH receptors, may be more susceptible to damage or loss during the preparative procedure. NaF is a more potent stimulator of adenylate cyclase in the thyroid than is TSH. In contrast, very little adenylate cyclase activity was demonstrable in the material obtained from the interface of the other layers or in the pellet, and the effects of TSH and NaF were either minimal or absent.[7] Although long-acting thyroid stimulator (LATS) had no effect on the whole homogenate, it stimulated adenylate cyclase activity 2-fold in purified plasma membranes.[11] The specific ac-

[10] G. Krisna, B. Weiss, and B. B. Brodie, J. Pharmacol. Exp. Ther. 163, 379 (1968).
[11] K. Yamashita and J. B. Field, J. Clin. Invest. 51, 463 (1972).
[12] J. Wolff and A. B. Jones, J. Biol. Chem. 246, 3939 (1971).

tivity of other enzymes considered to be present in plasma membranes was increased along with adenylate cyclase activity. Na^+-K^+-dependent ATPase and 5′-nucleotidase which have been demonstrated in plasma membranes in other tissues[3,5,13] showed a 10- to 20-fold increase in the plasma membrane fraction. Wolff and Jones also reported an equivalent purification of 5′-nucleotidase, Mg^{2+}-activated ATPase and K^+-activated p-nitrophenyl phosphatase in their thyroid plasma membrane preparation but a greater purification of Na^+-K^+-activated ATPase.[12] Cytochrome c oxidase activity was not detected in the plasma membrane fraction indicating very little, if any, contamination with mitochondria.[7] However, DPNH–cytochrome c reductase activity was detectable in the plasma membrane fraction indicating the presence of some microsomal elements.[7] Similar findings were reported in preparations of liver plasma membranes[2] and thyroid plasma membranes.[12] It is possible that plasma membranes may have a continuity with microsomes at certain sites.

Chemical Composition. The molar ratio of cholesterol:phospholipid has been reported to be a characteristic feature of plasma membranes.[3,6,14] A high ratio is a general property of plasma membranes obtained from various tissues, such as myelin,[3] erythrocyte,[3,14] intestinal mucosa,[3,14] and HeLa cells.[6] However, this ratio in rat liver plasma membranes was reported as 0.26–0.8.[2,15-17] Coleman and Finean suggested that this low ratio may be due to contamination of the preparations.[3] The ratio in bovine

TABLE II

CHEMICAL COMPOSITION OF PLASMA MEMBRANES AND TOTAL MEMBRANES FROM BOVINE THYROID GLANDS

	Protein (mg): P(μmole)	Cholesterol (μmole): P(μmole)	Phospholipid (μmole): P(μmole)	Cholesterol: phospholipid
Plasma membranes	1.32 ± 0.07	0.56 ± 0.02	0.71 ± 0.03	0.79 ± 0.02
Total membranes[a]	1.20 ± 0.03	0.07 ± 0.01	0.22 ± 0.05	0.32 ± 0.03

[a] Total membranes were prepared by centrifuging the whole homogenate at 110,000 g_{max} for 60 minutes. The results are the average $\pm$ SEM of four preparations.

[13] H. B. Bosmann, A. Hagopian, and E. H. Eylar, *Arch. Biochem. Biophys.* **128**, 51 (1968).

[14] J. B. Finean, R. Coleman, and W. A. Green, *Ann. N.Y. Acad. Sci.* **137**, 414 (1966).

[15] M. Takeuchi and H. Terayama, *Exp. Cell Res.* **40**, 32 (1965).

[16] V. P. Skipski, M. Barclay, F. M. Archibald, O. Terebus-Kekish, E. S. Reichman, and J. J. Good, *Life Sci.* **4**, 1673 (1965).

[17] L. A. E. Ashworth and C. Green, *Science* **151**, 210 (1966).

thyroid plasma membranes was 0.79 as compared to 0.32 in total membranes (Table II). The total cellular membranes were prepared by centrifuging the whole homogenate at 110,000 g_{max} for 60 minutes. Therefore, this criterion may also be used as an indicator of the degree of purity of the preparation from thyroid glands. The molar ratio of phospholipid:membrane P was more than 0.7. This indicates that phospholipid comprises most of the membrane P. Phospholipid (e.g., lecithin) is an important component for TSH stimulation of adenylate cyclase (and other metabolic effects) in thyroid slices[18,19] and plasma membranes.[20]

Yield. The yield of plasma membrane protein was 0.32–0.50 mg per gram of protein of the original homogenate while Wolff and Jones reported 0.1–0.2 mg per gram of protein in the crude material in their preparation.[12]

[18] V. Macchia, O. Tamburrini, and I. Pastan, *Endocrinology* **86**, 787 (1970).
[19] K. Yamashita, G. Bloom, B. Rainard, U. Zor, and J. B. Field, *Metabolism* **19**, 1109 (1970).
[20] K. Yamashita and J. B. Field, unpublished observations, 1972.

[12] Isolation of Human Platelets and Platelet Surface Membranes

By Nancy L. Baenziger and Philip W. Majerus

Because erythrocytes, leukocytes, and platelets occur as free-floating cells in blood, their surface membranes can be isolated relatively intact, avoiding the damage suffered by plasma membranes of organ-based cells during cell isolation. The plasma membranes of platelets are of particular interest because of their role in platelet function in hemostasis. Human platelets contribute to the process of blood coagulation by providing a physical barrier across a damaged vessel wall and a catalytic lipoprotein surface upon which the interaction of certain clotting factors is thought to occur. During hemostasis platelets rapidly change from isolated discoid cells to a compact aggregated mass of spherical cells with pseudopodia and liberate adenine nucleotides, sulfated mucopolysaccharides, and certain proteins into the surrounding medium, a phenomenon termed the "release reaction." These cellular effects can be induced *in vitro* by thrombin and other agents.[1]

Certain properties of platelets which seem closely related to their function in hemostasis can make them difficult to isolate and study. However, several simple precautions help to yield a pure platelet preparation and are

[1] A. J. Marcus, *N. Engl. J. Med.* **280**, 1213 (1969).

important for obtaining meaningful metabolic and functional studies in washed platelets:

1. Platelets show a marked intolerance to cold, tending to aggregate spontaneously. Hence all preparative procedures should be carried out at room temperature, and blood banks should be requested not to refrigerate the blood prior to platelet isolation.

2. Platelets will stick to glass surfaces and aggregate. They should be handled only in siliconized glassware or in the plastic equipment described below, as platelets will also stick to some plastic centrifuge tubes. Wire brushes should not be used to clean plastic tubes; the resultant scratching peels away shreds of plastic from the tube walls, to which platelets aggregate.

3. Platelets utilize the glycolytic and tricarboxylic acid pathways for energy production and possess glycogen stores. However, once separated from plasma they rapidly become depleted of adenosine triphosphate during washing unless 1 mg/ml (5.5 mM) glucose is added to the buffers. Platelet potassium levels also become depleted during washing, so that potassium should be included in washing buffers.

4. Centrifugation of platelet suspensions at greater than 2500 g makes them more difficult to resuspend.

5. The use of pooled platelets isolated from multiple donors will decrease the variability of response in metabolic studies that occurs from one batch of platelets to another.

Aggregation problems can often be circumvented by using an Na$_2$EDTA anticoagulant in place of the acid-citrate-dextrose formula used by blood banks and by adding 1–5 mM Na$_2$EDTA to washing buffers. However, EDTA has been shown to solubilize proteins from a variety of cell membranes, and thus the following procedures avoid the extensive use of EDTA.

Isolation of Human Platelets

Reagents[2]

> Washing buffer: pH 6.5 buffer containing 0.113 M NaCl, 4.3 mM K$_2$HPO$_4$, 4.3 mM Na$_2$HPO$_4$, 24.4 mM NaH$_2$PO$_4$, 5.5 mM glucose
> Resuspension buffer: pH 7.5 buffer containing 0.14 M NaCl, 15 mM Tris·HCl, and 5.5 mM glucose; *or* pH 7.5 buffer containing 0.109 M NaCl, 4.3 mM K$_2$HPO$_4$, 16 mM Na$_2$HPO$_4$, 8.3 mM NaH$_2$PO$_4$, and 5.5 mM glucose

[2] N. L. Baenziger, G. N. Brodie, and P. W. Majerus, *Proc. Nat. Acad. Sci. U.S.* **68**, 240 (1971).

Blood banks can provide large numbers of platelets either in the form of platelet-rich plasma or as platelet concentrates containing the cells in a small volume of plasma. Concentrates are not recommended as a source for isolation of washed platelets since the platelets tend to be aggregated, and the erythrocyte and leukocyte contamination is harder to remove than from platelet-rich plasma.

Method I

Blood banks prepare platelet-rich plasma by centrifuging whole blood (450 ml = 1 unit) in plastic bags at 4200 g for 2–3 minutes and drawing off the supernatant containing platelets plus some contaminating erythrocytes and leukocytes. To isolate platelets the platelet-rich plasma is divided into 50-ml plastic centrifuge tubes (Nalgene polypropylene, 29 × 102 mm) and is centrifuged for 25 minutes at 100 g. (Platelet-rich plasma from donors of different ABO blood groups should not be mixed in this initial spin, but they can be pooled for all subsequent steps.) The supernatant plasma then is centrifuged for 17 minutes at 1000 g to sediment the platelets. It is advisable to use an accurate tachometer to ensure proper g forces during centrifugation. Platelet yield can be increased by adding the platelet-poor plasma from the 1000 g centrifugation to the cell pellet from the 100 g centrifugation, resuspending the cells, and repeating both 100 g and 1000 g spins. The platelet pellets are resuspended in 2 ml of washing buffer by washing the latter in and out of a plastic Pasteur-type pipette (Nalgene, 15.5 cm). Platelets from 1–2 units of platelet-rich plasma are pooled into a single centrifuge tube in 30–40 ml final volume of washing buffer. After two washes the platelets can be resuspended in either of the pH 7.5 buffers described above. The platelet pellet from the 1000 g centrifugation of platelet-rich plasma should not be resuspended directly in Tris·saline buffer, as the platelets will aggregate spontaneously.

Platelet pellets should always be resuspended in a small volume (2–5 ml) and then further diluted. This evenly disperses the platelets, breaks up aggregates which occasionally form even during room temperature isolation, and facilitates the removal from the suspension of any irreversibly aggregated platelets and residual erythrocytes and leukocytes. These all appear at the bottom of the cell pellet, so that the unaggregated platelets may be gently agitated loose from them and transferred to a clean 50-ml plastic centrifuge tube.

Platelets can be enumerated by spinning an aliquot of a suspension for 5 minutes in a standard micro hematocrit apparatus (1% packed cell volume equals approximately 10^9 platelets/ml) or by phase microscopy in a standard hemacytometer after dilution and thorough mixing in 1%

ammonium oxalate to lyse erythrocytes in the preparation.[3] Alternatively, they can be counted with an electronic particle counter (Coulter counter, Model F). The average yield of platelets by method I is 40 to 60×10^9 cells from 1 unit of platelet-rich plasma (250–300 ml).

Method II

This platelet isolation procedure is particularly useful for small amounts of whole blood.[4] Whole blood (40 ml) is added to a 50-ml plastic centrifuge tube containing 0.75 ml of 0.25 M Na_2EDTA as anticoagulant and centrifuged at 1400 g for 3 minutes. The supernatant platelet-rich plasma is centrifuged at 2250 g for 15 minutes, and the resulting platelet pellet is resuspended in washing buffer as described above (method I). The platelet-poor plasma is then mixed with the original pellet from the 1400 g centrifugation, and the 1400 g and 2250 g spins are repeated. Platelets from these spins are resuspended in washing buffer and added to those from the initial 2250 g spin, and the entire mixture is diluted to 40 ml in washing buffer. The platelet suspension is centrifuged at 120 g for 7 minutes to remove contaminating leukocytes and erythrocytes. The pellet of erythrocytes and leukocytes is then resuspended in 40 ml of washing buffer and centrifuged at 120 g for 7 minutes again to recover platelets which sedimented in the first 120 g spin. The platelets are pelleted by centrifugation at 2000 g for 15 minutes and are washed once and finally resuspended in pH 7.5 buffer as described in method I. Both methods give a final contamination of <10 erythrocytes and <1 leukocyte per 10^5 platelets; the platelet yields are also similar (50–75%).

Other Methods

Two other methods for isolating platelets are available which are less widely used than the foregoing. One involves centrifuging whole blood or platelet-rich plasma in siliconized oil bottles (Corning Glass Co.), which are funnel-shaped bottles with 4 cm-long stems, 0.5 ml in volume at the bottom, for 30 minutes at 3000 rpm in No. 395 adapters in the International centrifuge at room temperature. The erythrocytes sediment to the bottom of the stem, followed by leukocytes and then platelets.[5,6] The other method has been used for studies of coagulation factors, since washing by methods I and II has been found by some investigators to activate a platelet coagulant activity.[7] Platelet-rich plasma (9 ml) is layered over 1 ml of

[3] G. Brecher and E. P. Cronkite, *J. Appl. Physiol.* 3, 365 (1950).
[4] S. M. Wolfe and N. R. Shulman, *Biochem. Biophys. Res. Commun.* 35, 265 (1969).
[5] I. Green and W. Solomon, *J. Clin. Pathol.* 16, 180 (1963).
[6] R. Nachman, *Blood* 25, 703 (1965).
[7] P. N. Walsh, *Brit. J. Haematol.* 22, 205 (1972).

40% bovine albumin in a 10-ml plastic tube, the interface is gently stirred for 30 seconds to produce a 1 cm-wide density gradient, and the tube is centrifuged for 15 minutes at 2700 g. The platelets appear in a 0.5 cm-broad band in the albumin. No leukocyte and erythrocyte contamination data are available for either technique.

Disruption of Platelets

Platelets have been difficult to disrupt because of their small size and resistance to shear. Methods of platelet disruption have been compared by Barber et al.[8] Simple osmotic lysis in distilled water or freeze-thawing is not satisfactory, and decompression in a nitrogen bomb yields variable results. Sonication is the simplest means of disrupting platelets. It is satisfactory for obtaining soluble platelet proteins and the total platelet particulate fraction. Sonication of platelets suspended at 2 to 8×10^9/ml for 15 seconds at 70% intensity with a Biosonik sonifier, using a microprobe for 0.05–2 ml and a standard probe for 3–10 ml samples, disrupts all the platelets, but releases only 10% of their lysosomal enzymes. After 2 minutes of sonication in 15-second intervals with cooling periods between them, all the lysosomal granules have been disrupted.[9] The total platelet particulate fraction may be separated from the soluble proteins by centrifugation for 30 minutes at 48,000 g.

Marcus et al. have described the use of a "no-clearance" homogenizer for disrupting platelets.[10] Platelets are disrupted at 20×10^9/ml in 0.44 M sucrose containing 1 mM EDTA by homogenization in a Potter-Elvehjem-type homogenizer at 1700 rpm for 5–10 minutes at 4° in a Kontes size D homogenizing vessel. The no-clearance Teflon pestle (Kontes No. 12-152-A) does not fit into the vessel at 20°, but will when it contracts slightly after cooling in ice. As homogenization proceeds, the pestle warms and expands, providing firm shearing forces. This method gives satisfactory rupture of the platelets. It is technically difficult in that small volumes of platelets and of suspending buffer must be manipulated in a large vessel, causing appreciable losses of material and foaming of the homogenate. Even when operated in an ice bath in a cold room, the Teflon pestle tends to overexpand and seize the vessel, particularly when volumes of 4–5 ml or more are homogenized.

[8] A. J. Barber, D. S. Pepper, and G. A. Jamieson, Thromb. Diath. Haemorrh. 26, 38 (1971).

[9] N. L. Baenziger, G. N. Brodie, and P. W. Majerus, J. Biol. Chem. 247, 2723 (1972).

[10] A. J. Marcus, D. Zucker-Franklin, L. B. Safier, and H. L. Ullman, J. Clin. Invest. 45, 14 (1966).

Barber and Jamieson[11] have described a technique for lysis of platelets using glycerol loading. Washed platelets (30 to 150 × 10^9) are resuspended in 5–10 ml of Tris resuspension buffer and layered on a 30-ml continuous gradient of 0–40% glycerol in the same buffer in a 50-ml plastic centrifuge tube. The gradients are centrifuged in either an angle rotor (e.g., Sorvall SS-34) or a swinging-bucket rotor (e.g., Sorvall HB-4) at 4° at 1465 g for 30 minutes followed by 5860 g for 10 minutes to pellet the platelets. As the platelets sediment through the gradient they slowly take up glycerol so that its intracellular concentration reaches 4.3 M. The platelet pellet is then rapidly resuspended in 4–5 volumes of cold isotonic buffer containing 0.25 M sucrose and 0.01 M Tris·HCl, pH 7.5 (or in Tris·saline resuspension buffer) by Vortexing or agitation with a pipette, causing 80–90% lysis within 1–2 minutes. Since intracellular organelles do not take up high concentrations of glycerol, they are not ruptured when the pellet is restored to isotonicity.

Isolation of Platelet Plasma Membranes

The most widely used method of subcellular fractionation of platelet organelles is that of Marcus et al.[10] Washed platelets are disrupted by "no-clearance" homogenization of 15 to 20 × 10^9 platelets in 1 ml of 0.44 M sucrose containing 1 mM EDTA. The homogenate is washed into a centrifuge tube with more buffer to make a final volume of 10 ml per 30 to 40 × 10^9 platelets, and is centrifuged 30 minutes at 2000 g to remove debris. The supernatant is layered in 1.2-ml aliquots on 4 ml of continuous sucrose gradients (30–60%, containing 1 mM EDTA), which are centrifuged 2 hours at 3° at 39,400 rpm in the Spinco SW 39L rotor (130,576 g). Two particulate bands result, a wide one (d 1.21–1.17) containing granules, mitochondria, and large membrane fragments, and a narrower one at a density of 1.13–1.12 containing plasma membrane vesicles. The latter can be sedimented at 130,576 g, washed in 0.25 M sucrose, and stored at −85°C. Marcus's procedure is useful chiefly for preparations of 30 to 60 × 10^9 platelets.

Barber and Jamieson[11] have developed a method that can accommodate 30 to 900 × 10^9 platelets in one run. A platelet lysate (30 to 150 × 10^9) prepared by the 0 to 40% glycerol gradient loading technique is layered over a 10-ml cushion of 27% sucrose (d 1.106) and centrifuged at 4° at 63,500 g in the Spinco SW 25.1 rotor for 3 hours. This density of sucrose has been chosen such that granules and intracellular debris rapidly sediment away from the plasma membranes, which themselves only slightly penetrate the gradient in 3 hours. Debris and granules sediment

[11] A. J. Barber and G. A. Jamieson, J. Biol. Chem. 245, 6357 (1970).

to the bottom of the tube, and the plasma membranes remain in a narrow band at the sucrose interface. They can be diluted 1:10 in a 0.25 M sucrose 0.01 M Tris·HCl, pH 7.4, and collected by centrifugation at 105,000 g for 1 hour. Further centrifugation of the plasma membrane fraction on a 15–40% (w/v) sucrose gradient for 18 hours at 63,500 g resolves two membrane bands. The upper one (d 1.090) contains double membrane vesicles averaging 1750 Å in diameter, and the lower (d 1.120) contains single-layered vesicles of 700 Å diameter. Protein to phospholipid ratios vary between the two bands, although the bands show no difference with respect to marker enzymes. The reason for the separation of platelet plasma membranes into two bands is not clear.

Platelet plasma membranes prepared by these two techniques are comparable in purity based on recovery of the marker enzyme acid phosphatase. We have found more variation in the gradient bands using the Marcus procedure, which may reflect the variable damage of no-clearance homogenization rather than difficulty with the 30–60% density gradient fractionation. The 27% sucrose cushion procedure of Barber and Jamieson can be used to isolate the plasma membrane and granules from platelets disrupted by limited sonication as well as by glycerol loading and hypotonic lysis.

Whether the membranes isolated by these procedures represent the total intact plasma membrane of platelets is not certain. In platelets the surface membrane is deeply invaginated to form a network of channels which penetrate into the interior of the cell, giving the platelet a sponge-like structure. This internal network continuous with the plasma membrane is called the "surface-connected system." Its fate upon disruption of the cell is not known. The yield of platelet membranes by either technique described above is 20–30%, which is similar to that for plasma membranes of liver and kidney cells; however, the low yield may possibly be due to loss of the surface-connected system in the isolation procedure. Furthermore, proteolytic activity is present in platelets and may give rise to artifacts during isolation and handling of platelet subcellular fractions. Some of the protein associated with the platelet particulate fraction is especially susceptible to proteolysis,[9] which can occur in spite of storage at $-20°$ if the preparations contain high concentrations of glycerol or sucrose. Proteolytic alterations in proteins of the platelet particulate fraction have been noted by gel electrophoresis in sodium dodecyl sulfate even during the short centrifugations of the two membrane isolation procedures described above. The incorporation of suitable inhibitors of proteolysis into platelet membrane isolation techniques may circumvent this problem in part.

[13] Isolation of Plasma Membrane from Tissue Culture—L Cells[1]

By Leonard Warren

Methods will be described for the isolation of purified fractions of the surface membranes of L cells. The starting material must be free, rounded cells, i.e., cells grown in suspension culture or cells grown on glass or plastic surfaces which have been freed from these surfaces by trypsin. The underlying principles of the methods are simple: surface membranes are stabilized by heavy metal ions (Zn^{2+}), by blocking of -SH groups with fluorescein mercuric acetate (FMA)[2] or with Tris base. The molecular basis for stabilization is unknown. Further, stabilization takes place in a hypotonic medium, so the cells swell and the surface structure rises off the underlying cytoplasm, leaving a clear area under the surface membrane. The cells are then broken in a Dounce homogenizer, which acts by building up and suddenly releasing pressure. A bag with a large hole through which nucleus and cytoplasm were ejected remains. Membranes are then isolated by differential centrifugation on gradients of sucrose solution.

Materials

Dounce homogenizers can be purchased from the Kontes Glass Co., Vineland, New Jersey. Three sizes can be used, 7, 15, and 40 ml, all with tight-fitting pestles (type B). All chemicals used are of reagent grade. Solutions of fluorescein mercuric acetate (Nutritional Biochemical Corp., Cleveland, Ohio) are made to a saturation concentration of 2.2 mM with 20 mM Tris·HCl buffer, pH 8.1, by stirring for 1 hour at room temperature. Final adjustment to pH 8.1 is made with a solution of 2 M Tris base. Undissolved material is removed by filtration. The solution seems to be best when fresh but can be used for 3 weeks. Sucrose solutions are made according to the "Handbook of Chemistry and Physics," 27th edition, 1943, p. 1556.

[1] Much of the work described in this section was done with the help of grants from the American Cancer Society BC-16A, PRP-28 and the U.S. Public Health Service 5 PO1 AI-0700507.
[2] F. Karush, N. R. Klinman, and R. Marks, *Anal. Biochem.* 9, 100 (1964).

Methods

General[3]

It is of the utmost importance that every step be carefully followed and evaluated by phase contrast microscopy. These methods of isolation of membranes can never be so standardized that they can be carried out "blind." All procedures are carried out at 4° unless otherwise stated. Harvesting of fractions can be accomplished on a small scale with a syringe fitted with a No. 15 needle whose final 6 mm is bent at a right angle with the bevel facing upward. On a larger scale, fractions can be harvested by using a Pasteur pipette on a plastic tube attached to a side-arm suction flask under moderate negative pressure.

In this laboratory the L cells used are grown in suspension culture in a modified Eagle's medium containing 5% fetal calf serum. However, other media can be employed. The cells are washed twice in physiological saline (0.16 M NaCl) and then suspended in saline at a concentration of 5×10^7 cells/ml.

Fluorescein Mercuric Acetate Method

Small Scale.[4,5] L cells (3×10^6 to 10^8 cells) in 1–1.5 ml of saline are transferred to a small Dounce homogenization tube. Three milliliters of a solution of FMA are added per milliliter of cells. After 5 minutes of occasional stirring at room temperature, the cells are chilled in ice until the temperature falls below 5°, at which time the cells are broken by 20–30 gentle strokes of a tight Dounce pestle (type B). The preparation is viewed every 5–8 strokes by phase contrast microscopy. Empty ghosts appear which have a hole in them bordered by a dark lip formed by a rolling up of the free edge. A point is reached at which more ghosts are being fragmented, torn, and destroyed than are being formed, and it is at this time that homogenization should cease. There are always some cells (perhaps one-fourth of the total) which are small and dense and do not yield a surface membrane. These are probably leaky, damaged cells which allowed FMA to enter and fix the interior of the cell.

Four milliliters of 60% sucrose solution are added to the homogenate with vigorous stirring. The homogenate is then placed on 10 ml of 45% sucrose solution in a pointed 40-ml centrifuge tube with straight sides.

[3] L. Warren and M. C. Glick, *in* "Biomembranes" (L. A. Manson, ed.), Vol. 1, p. 257. Plenum, New York.
[4] L. Warren and M. C. Glick, *J. Cell Biol.* **37**, 728 (1968).
[5] L. Warren and M. C. Glick, *in* "Fundamental Techniques in Virology" (K. Habel and N. P. Salzman, eds.). Academic Press, New York, 1969.

The tube is centrifuged for 1 hour at 800 rpm (150 g) in a PR2 refrigerated International centrifuge. The upper orange layer is removed with a syringe down to within 1 mm of the interface. This upper fraction contains whole cell membranes and fine debris. A second fraction is removed, the interface region, containing some ghosts and a few cells and nuclei. The lower phase contains virtually all the nuclei and whole, unbroken cells. Water is added to the interface material (0.1 volume), and this fraction is recentrifuged for 1 hour on a solution of 45% sucrose at 800 rpm (150 g). The upper fraction of this run is combined with that of the first. This fraction, containing cell ghosts and fine debris, is diluted with 0.1 volume of water and is layered over 10 ml of 35% sucrose solution in a 40-ml centrifuge tube with straight sides. The tube is centrifuged for 1 hour at 1800 g (2900 rpm). Fine debris remains in the upper phase while the denser cell ghosts form an orange pellet. Both upper and lower layers in the tube are removed through a Pasteur pipette connected to a sidearm flask under suction. The pellet is left intact, and to it are added a few drops of 35% sucrose solution (volume of approximately 0.3 ml). The pellet is suspended and is transferred to a small tube containing sucrose solution in a linear gradient of 45 to 65% concentration. The tube is centrifuged in the SW 39 head of a Spinco L-2 ultracentrifuge for 1 hour at 33,000 g. The membranes form an orange band about halfway down the tube and are harvested with a 2-ml syringe fitted with a No. 18 needle bent at its tip. A discontinuous gradient of sucrose (65, 60, 55, 50, and 45%) has also been used successfully. The suspension of membrane is diluted with distilled water to 12 ml and is then centrifuged at 4000 g for 30 minutes in an RC-2 centrifuge with an HB-4 rotor. The supernatant solution is removed, the pellet is suspended in distilled water, and the cell ghosts are counted in a hemacytometer. The preparation contains few or no nuclei or mitochondria and little debris. The yield is variable—usually 20–55%.

Large Scale.[6] The procedure is essentially the same as in the small-scale method. Thirty milliliters of a solution of FMA are added to 10 ml of cells (5×10^8 cells) in a 40-ml Dounce homogenizer tube. After 10 minutes at room temperature, the tube is cooled to 5° in ice. The cells are broken with a type B pestle (20–40 strokes). Forty milliliters of 60% sucrose solution are added with stirring, and this is layered over 30 ml of a solution of 45% sucrose in two 250-ml centrifuge bottles. The bottles are centrifuged at 550 g for 1 hour. The top part of the upper layer and the interface layers are harvested separately. The upper layer contains cell ghosts and fine debris while the interface material contains cells and

[6] L. Warren, M. C. Glick, and M. K. Nass, *J. Cell Physiol.* **68,** 269 (1966).

nuclei as well as surface membranes. This layer is recentrifuged at 1200 g for 20 minutes on a solution of 45% sucrose as in the first centrifugation. The upper layers of the two centrifugations are combined. The membranes diluted to 30% concentration in sucrose are placed on a solution of 35% sucrose (40 ml in a 250-ml centrifuge bottle) and are centrifuged at 1600 g for 1 hour in the PR-2 International Centrifuge. The membranes form a pellet at the bottom of the bottle. The overlying solutions are removed by suction. This step removes fine debris and concentrates the membranes. The pellet is suspended in 7 ml of 35% sucrose solution and is transferred to a tube containing sucrose solution in a linear gradient of 45 to 65% concentrations. The membranes are then centrifuged in the SW 25.1 head of a Spinco ultracentrifuge at 90,000 g for 1 hour. An orange band of membranes is found halfway down the gradient. The band of membranes is harvested, and the preparation is dialyzed to remove sucrose. Discontinuous gradients of 45, 55, 60, and 65% sucrose solution can substitute for the continuous gradient.

Tris Method[5]

L cells are washed twice with cold Tris·HCl buffer (50 mM, pH 7.4) and are suspended at a concentration of 5×10^7 cells per milliliter in this buffer. To 10 ml of this suspension in a 40-ml Dounce homogenizer tube are added 10 ml more of buffer and 1 ml of a solution of 50 mM MgCl$_2$. The cells are allowed to stand for 8 minutes at 4° and are then broken by approximately 20 gentle strokes with a tight, type B pestle. Whole, intact cell ghosts are formed. A pH of 7–8 can be used. The yield is higher at the higher pH, but the membranes obtained are less stable. All sucrose solutions used in the Tris method contain 5 mM MgCl$_2$. An equal volume of 20% sucrose solution is added to the homogenate. Five milliliters of the homogenate-sucrose solution are layered onto a discontinuous gradient of sucrose solutions in a 40-ml centrifuge tube. Each tube contains 5 ml each of 65, 50, 40, 30, and 20% and 10 ml of 35% sucrose solutions. The tubes are centrifuged at 1380 g for 15 minutes in a PR-2 International centrifuge (Rotor No. 269). Debris remains in the upper layer, and the membranes are found in the layers of 35 and 40% sucrose solution. These are harvested with a syringe and are then centrifuged at 6000 g for 15 minutes in the HB-rotor of a Sorvall RC-2 centrifuge. The supernatant solution containing debris is removed, and the pellet, which consists of cell ghosts, is dispersed in 2 ml of 30% sucrose solution. This suspension is placed on a gradient consisting of 5 ml each of 60, 55, 50, and 10 ml each of 48 and 45% and 4 ml each of 43 and 40% sucrose solutions in a centrifuge tube of an SW 25.2 rotor of a Spinco L-2 ultracentrifuge. After centrifuging for 90 minutes at 90,000 g, the surface membranes come to rest at the top of the 48 and 45% sucrose solutions. The membranes are harvested with a syringe, diluted with distilled water to a sucrose concentration of approximately 35%, and centrifuged at 5000 g for 20 minutes. The pellet of cell ghosts is suspended in a few milliliters of 35% sucrose solution and is counted in a hemacytometer. The yield of membranes is usually 15–25%. It is possible to increase the yield by combining and processing layers of the first gradient that contain cells and nuclei as well as cell ghosts. This layer is diluted and recentrifuged on a gradient similar to that used in the first centrifugation, followed by a high speed centrifugation on a gradient similar to the second gradient previously used except that 10 ml of 50% and 5 ml of 48% sucrose solutions are used.

Zinc Ion Method[5]

L cells are washed twice with cold saline and are suspended in saline at 5×10^7 cells/ml. Thirty ml of 1 mM ZnCl$_2$ solution are added to 10 ml of a suspension of L cells (5×10^8 cells) in a 40-ml Dounce homogenizing tube, and the suspension is allowed to stand at room temperature for 10 minutes. The suspension, is then cooled to 5° and is gently homogenized with a type B, tight pestle. After 100–200 strokes, a maximum number of intact cell ghosts can be seen in the phase contrast microscope. Forty milliliters of 60% sucrose solution are added to the homogenate, and this is divided between two gradients in 150-ml Corex centrifuge bottles. The gradient consists of 15, 10, 25, 20, 10, and 10 ml, respectively, of 50, 48, 45, 43, 40, and 35% sucrose solutions. The bottles are centrifuged at 1500 g for 30 minutes in the HG-4 rotor in a Sorvall RC3 centrifuge. Surface membranes are found in the 40 and 43% sucrose solutions and are removed in three parts by suction. The upper fraction contains whole ghosts, fragments of membranes, and particles. The middle area contains mainly cell ghosts, and the lowed part (lower 43% sucrose solution) contains cell ghosts and nuclei. All three fractions are diluted to a sucrose concentration of approximately 35% and are centrifuged at 6000 g for 15 minutes in the RC3 centrifuge. The pellets are suspended separately in 35% sucrose solution and are centrifuged for 1 hour at 90,000 g in the SW 25.1 head over discontinuous gradients of sucrose solutions. The gradient consists of 3 ml each of 65 and 60%, 4 ml of 55%, 6 ml of 30%, and 5 ml each of 45 and 40% sucrose solution. The surface membranes, found on top of the 50 and 55% sucrose solutions are harvested with a syringe, diluted to a concentration of approximately 35% in sucrose, and centrifuged at 6000 g for 15 minutes. The pellet is dispersed in a solution of 35% sucrose and is recentrifuged. The yield of cell ghosts is about 20–30%. They are stable for several months when kept at $-20°$ in 35% sucrose solution. Membranes can be prepared on a smaller scale by the Zn^{2+} method by centrifugation essentially as described for the small-scale FMA method.

Comments

Although the methods were devised for the L cell, they have been successfully used on a number of other cells—KB, BHK21/C_{13}, CHO, CV-1, primary chick and human fibroblasts, MCIM, 3T3, WI38, WI26, a number of cells transformed by Rous sarcoma, polyoma, and SV40 viruses, Ehrlich ascites cells, MT80 (a lymphoma), 70-F (squamous cell carcinoma) as well as amoeba (*Chaos chaos* and *Amoeba proteus*). Lymphoid and other cells which have a large nucleus and only a thin rim of cytoplasm are not good candidates for these methods since there is little swelling and rising of the surface membrane in hypotonic media. Optimal conditions for each type of cell can be established by trial and error.

The yield and quality of some cell membranes can be improved by preliminary treatment of cells (e.g., chick embryo fibroblasts) with an equal volume of distilled water for 5 minutes at room temperature. This procedure seems to "soften" the cell so that it swells more when the hypotonic solutions of toughening reagents are added.

The FMA method is the most versatile, but the -SH groups of the isolated membranes are blocked, and so they are not suitable for enzyme studies. However, membranes isolated by the FMA method contain H-2 antigen.[7] The Tris method is the least predictable, but the product is best for enzyme studies. Membranes isolated by the zinc ion method are enzymatically active[8] and are capable of incorporating amino acids into hot trichloroacetic acid-precipitable material.[9] They also contain sialyltransferases.[10]

For unknown reasons the Zn^{2+} and FMA methods work best when cells are washed with saline rather than phosphate-buffered or Tris-buffered saline. The cells should be processed as soon after washing as possible. Homogenization in a small Dounce tube (7 ml) seems to result in a better preparation with a higher yield than homogenization carried out in larger tubes. There is considerable variation in the clearances between

[7] L. A. Manson, C. Hickey, and J. Palm, in "Biological Properties of the Mammalian Surface Membrane" (L. A. Manson, ed.). Wistar Inst. Press, Philadelphia, Pennsylvania, 1969.

[8] J. F. Perdue and J. Sneider, *Biochim. Biophys. Acta* 196, 125 (1970).

[9] M. C. Glick and L. Warren, *Proc. Nat. Acad. Sci. U.S.* 63, 563 (1969).

[10] L. Warren, J. P. Fuhrer, and C. A. Buck, *Proc. Nat. Acad. Sci. U.S.* 69, 1838 (1972).

various tubes and pestles; the quality of the preparation obtained may vary with the particular combination used. It has also been found that, in purification, centrifugation at lower speeds for longer times results in better separations than when higher speeds for shorter periods are employed.

A number of variations of these methods have been devised for various lines of tissue culture cells.[8,11–15] Brunette and Till[11] have successfully used a dextran-polyethylene glycol aqueous two-phase system to purify surface membranes of L cells prepared by the Zn^{2+} ion method. Heine and Schnaitman[16] have isolated surface membrane material of L cells by allowing them to ingest polystyrene latex beads, isolating the membrane-covered beads, and then removing the membranes from the beads by sonication.

[11] D. M. Brunette and J. E. Till, *J. Membrane Biol.* **5**, 215 (1971).
[12] I. Scher and P. Barland, *Biochim. Biophys. Acta* **255**, 580 (1972).
[13] P. Barland and E. A. Schroeder, *J. Cell Biol.* **45**, 662 (1970).
[14] E. D. Kiehn and J. J. Holland, *Biochemistry* **9**, 1716 (1970).
[15] P. H. Atkinson and D. F. Summers, *J. Biol. Chem.* **246**, 3762 (1971).
[16] J. W. Heine and C. A. Schnaitman, *J. Cell Biol.* **48**, 703 (1971).

[14] The Isolation and Characterization of Plasma Membrane from Cultured Chicken Embryo Fibroblasts

By JAMES F. PERDUE

The current widespread interest in the genetically expressed functions of the cell surface is reflected by numerous studies in the areas of immunology, virology, developmental biology, and oncology. In particular, cells, which can be grown in culture under specified growth conditions, which can be synchronized with respect to the cell cycle, and which can be chemically and virally transformed to the malignant state, are invaluable for studies on the mechanism of oncogenesis. Many of these studies have employed isolated plasma membrane. Steck and Wallach[1] and Warren and Glick[2] have recently reviewed the methods used to isolate and characterize the plasma membrane from animal cells. Perdue and Sneider[3]

[1] T. L. Steck and D. F. H. Wallach, *in* "Methods in Cancer Research" (H. Busch, ed.), Vol. 5, pp. 109, 137. Academic Press, New York, 1970.
[2] L. Warren and M. C. Glick, *in* "Biomembranes" (L. Manson, ed.), Vol. 1, p. 272. Plenum, New York, 1971.
[3] J. F. Perdue and J. Sneider, *Biochim. Biophys. Acta* **196**, 125 (1970).

have developed a method for the isolation of plasma membrane from cultured chick embryo fibroblasts by flotation equilibrium centrifugation of cellular homogenates on continuous density gradients of sucrose. The method is reproducible, permits the isolation of pure plasma membrane, and has been found applicable to the isolation of membrane from other cell types and tissues. With these procedures, plasma membrane has been isolated from uninfected and oncogenic virus-converted cells, and differences in the carbohydrate composition of these membranes have been found.[4] Plasma membrane was also isolated from a parenchyma-like rat liver cell culture and from such cells transformed by murine sarcoma virus.[5] Recently,[6] plasma membrane was isolated from virus-induced chicken tumors.

Harvesting of Cells

Confluent cultures of cells, grown on Falcon plates (150 mm in diameter), are rinsed twice with $0.16 M$ NaCl (saline). Five milliliters of saline is then added to the plates, and they are scraped with a rubber policeman. The cell sheets are collected, centrifuged at 1100 rpm for 10 minutes on a clinical centrifuge, and washed twice by resuspension and centrifugation. The fibroblasts can also be released by incubating the saline-rinsed plates with 6 ml of 0.1% collagenase in 0.16 NaCl at 37° for about 8 minutes in a CO_2 incubator. The cells are scraped from the plates with a rubber policeman, collected, and centrifuged. The cells are separated from the collagenase (which can be reused three to four times) and washed three times by resuspension and centrifugation. In work with oncogenic virus-converted cells, the medium and washes are saved for treatment with virus-denaturing compounds. Release of the cells with collagenase gives a preparation of viable, single cells with no detectable loss of surface sialic acids.[3] These cells can be used in plasma membrane isolation procedures which employ a toughening of the cell surface with Zn^{2+}, fluorescein mercuric acetate, acetic acid, etc.[2] prior to the homogenization. With these procedures, plasma membrane can be isolated as intact cell membranes or as large fragments. An advantage of this method is that the fractionation procedure can be followed by phase microscopy because of the distinctive morphology of the cell membrane. Disadvantages of this procedure include denaturation of some marker enzymes and the selective loss of some membrane components.[2]

---

[4] J. F. Perdue, R. Kletzien, and K. Miller, *Biochim. Biophys. Acta* **249**, 419 (1971).

[5] J. F. Perdue, R. Kletzien, K. Miller, G. Pridmore, and V. L. Wray, *Biochim. Biophys. Acta* **249**, 435 (1971).

[6] J. F. Perdue, D. Warner, and K. Miller, *Biochim. Biophys. Acta* **298**, 817 (1973).

Isolation of Plasma Membrane

The pellet of washed cells is resuspended in about 10 ml of 0.16 *M* NaCl for every 36 plates of harvested cells. These cells are homogenized in a glass homogenizer with a Teflon pestle; one with a clearance of 0.08 mm works well for breaking up chick fibroblasts. Homogenization is carried out for intervals of 5 or 8 minutes, the samples are cooled in ice, and the degree of breakage is estimated by examination with a phase microscope. When the cells are well broken up, they are sedimented at 200,000 *g* for 20 minutes, and the pellet is resuspended in 65% sucrose. This resuspended pellet, termed the particulate homogenate, since it is free of most of the cell's soluble proteins, must be thoroughly homogenized. It is brought to a refractive index of 1.430 (68% w/v) with cold, saturated sucrose. The volume of the resuspended pellet, which is equated to 100% is measured, an aliquot is removed as a control sample, and 1 to 1.5 ml of this material is placed on the bottom of 15 × 90 mm cellulose nitrate tubes. Continuous 25–65% (w/v) density gradients of sucrose are formed above the particulate homogenate. These are about 10 ml in volume, and with certain gradient-forming apparatus, two to six gradients can be formed simultaneously. A procedure used in this work was to estimate the volume of the lighter sucrose needed to make the gradients, determine the weight of this sucrose, and form the gradients with an equal weight of the denser sucrose. The gradients are made slowly at 4°, and about 1 ml of 10% (w/v) sucrose is added to the tops of the tubes. The gradients are then centrifuged; we have used 90,000 *g* for about 16 hours in an SW 41 rotor (Spinco). This length of time is convenient, being an overnight run, and a centrifugation period of at least 10 hours is required. The

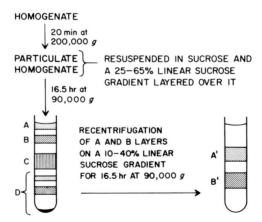

FIG. 1. Procedures used to isolate plasma membrane from chick embryo fibroblasts, fibroblasts infected with oncogenic viruses, or virus-induced tumors.

centrifugation of the gradients is terminated with the brake in the off position.

The membranes and organelles in the homogenate are separated into bands of differing density and a pellet, designated bands A to D (Fig. 1). The layers can be removed, beginning with the A band, with a syringe affixed to a bent needle. The A band is generally distinct from the B, but in some poorly formed gradients this distinction is not clear. Similarly, the break between the B and C bands is not always sharp. It has been found that the plasma membrane is concentrated in the A and B bands. Therefore, although the lower bands in the gradient can be separated, they are combined with the pellet, designated D, for chemical and enzymatic analysis.

The material removed from these layers is diluted with cold H_2O or $0.16 M$ NaCl and sedimented at 200,000 g for 40 minutes. The pellets are resuspended by homogenization in saline, water, or $0.25 M$ sucrose. The latter medium is used when enzyme assays are to be run immediately. The sucrose-suspended particles remain more finely dispersed, and higher levels of enzyme activity are found in these preparations than in the water- or saline-washed fractions. If the chemical compositions of the fractions other than enzymes and protein are to be determined (e.g., phospholipid, neutral sugar), then the material in the respective layers is washed three times by sedimentation and resuspension in saline or water. The volume of the final resuspended pellet is determined so that the recovery and concentration of enzymes, sialic acid, etc., in the fraction can be equated to the quantities present in the homogenate.

In any isolation procedure it is important to select enzymes and chemical components indicative of the presence of specific cellular organelles or membranes. As "markers" for the plasma membrane, we have determined CTPase activity[3,7] (EC 3.6.1.4) and sialic acid[8]; for mitochondria, succinic dehydrogenase (EC 1.3.99.1)[9]; and for endoplasmic reticulum, the antimycin-insensitive NADH–cytochrome c reductase (EC 1.6.2.1)[10] and RNA.[11] The distribution of components which may be common to membranes, such as cholesterol,[12] phospholipid,[13] and neutral[14] and amino sugars,[15] was also determined.

[7] J. F. Perdue, *Biochim. Biophys. Acta* **211**, 184 (1971).
[8] L. Warren, *J. Biol. Chem.* **234**, 1971 (1959).
[9] D. Ziegler and J. S. Rieske, this series, Vol. 10, p. 231.
[10] B. Mackler and D. E. Green, *Biochim. Biophys. Acta* **21**, 1 (1956).
[11] A. Fleck and H. N. Munro, *Biochim. Biophys. Acta* **55**, 571 (1962).
[12] D. Glick, B. F. Fell, and K. Sjølin, *Anal. Chem.* **36**, 1119 (1964).
[13] P. S. Chen, T. Y. Toribara, and H. Warner, *Anal. Chem.* **28**, 1756 (1956).
[14] M. Dubois, K. A. Gilles, J. K. Hamilton, P. A. Rebers, and F. Smith, *Anal. Chem.* **28**, 350 (1956).
[15] D. J. Allison and Q. T. Smith, *Anal. Biochem.* **13**, 510 (1965).

TABLE I

SPECIFIC ACTIVITY AND CONCENTRATION OF MEMBRANE-ASSOCIATED ENZYMES
AND CHEMICAL COMPONENTS AMONG BANDS A, B, C, AND D AFTER
CENTRIFUGATION OF A HOMOGENATE ON CONTINUOUS DENSITY
GRADIENTS OF SUCROSE[a]

Enzyme or component	Units/mg protein				
	Homog-enate	Band A	Band B	Band C	Band D
CTPase (μmoles P_i/30 min)	2.85	10:84	9.53	3.92	1.07
Cytochrome c reductase (μmoles reduced/min)	0.046	0.026	0.045	0.088	0.018
Succinate dehydrogenase (μmoles oxidized/min)	0.006	0.00	0.006	0.030	0.005
Sialic acid (μg)	7.5	17.8	15.2	7.1	2.2
Phospholipid (μg)	368	2186	1057	499	194
Cholesterol (μg)	95	573	345	117	53
RNA (μg)	89	18	48	54	104

[a] Data from J. F. Perdue and J. Sneider, *Biochim. Biophys. Acta* **196**, 125 (1970).

On the basis of these enzymatic and chemical criteria, it is apparent that the plasma membrane is concentrated in the A and B bands (Table I). This material can be recentrifuged a second time by resuspending the sedimented pellets in sucrose to a refractive index of 1.391 (42% w/v) and layering above them a continuous density gradient, 10–40% (w/v), of sucrose. These gradients were usually centrifuged overnight at 90,000 g, but we recently found that the membranes reach equilibrium in 3 hours. The position of the A′ band is clearly different from that of the B′ band (Fig. 1). With care, these bands can be removed and the mean density can be determined by measuring the refractive index of the sucrose with which they are in equilibrium. For plasma membrane isolated from different cell types and tissues, a characteristic density is obtained (Table II). The material in the bands is washed by sedimentation and resuspension.

The membranes in both the A′ and B′ bands are plasma membranes. Both fractions contain sialic acid and CTPase activity (Table III), but the latter band has more endoplasmic reticulum contamination. The major difference between these preparations is in their amounts of phospholipid and cholesterol. The A′ band contains almost twice as much of these components as does the membrane in the B′ band.

Plasma membrane was isolated from oncogenic virus-transformed cells, and major differences in the carbohydrate content of these mem-

TABLE II

Density of Membranes in the A' and B' Bands from Uninfected and
Virus-Converted Cells or Virus-Induced Tumors

Cells	A' Band	B' Band
Uninfected	1.060 ± 0.001[a]	1.108 ± 0.006
RAV-49-infected	1.064 (2)[a]	1.106 ± 0.017
RBA-converted	1.066 ± 0.003	1.108 ± 0.007
morph[r] Fujinami-converted	1.074 ± 0.004	1.110 ± 0.006
morph[f] Fujinami-converted	1.058 ± 0.001	1.111 ± 0.002
RBA-induced tumor	1.097 (2)	1.119 (2)
morph[r] Fujinami-induced tumor	1.104 ± 0.004	1.129 ± 0.003

[a] Standard deviation shown, except that, when the number of experiments was less than three; this number is shown in parentheses.

branes were found (Table III). To determine the applicability of flotation equilibrium centrifugation as a general method for the isolation of plasma membrane from tissues, this membrane was isolated from an RBA virus-induced chicken tumor. The isolation procedures were modified in that a differential centrifugation step was employed after homogenization of the tissues in 0.25 M sucrose to remove nuclei and whole cells, and the centrifugation of the A and B bands was carried out in 15–45% (w/v)

TABLE III

Specific Activity and Concentration of Membrane-Associated Enzymes
and Chemical Components in Isolated Plasma Membrane from
Uninfected Fibroblasts, Fibroblasts Converted by RBA
Virus, and RBA Virus-Induced Tumors

Enzyme or component	Units/mg protein		RBA virus-converted fibroblasts A' Band	RBA virus-induced tumor A' Band
	Uninfected fibroblasts			
	A' Band	B' Band		
CTPase[a]	1.78 ± 0.28	2.32 ± 0.24	2.67 ± 0.19	8.1 ± 1.2
Sialic acid	33 ± 1	25 ± 2	25 ± 3	23 ± 3
Neutral sugar	243 ± 8	153 ± 6	358 ± 12	322 ± 27
Amino sugar	33 ± 2	30 ± 1	34 ± 2	ND[b]
Phospholipid	1998 ± 63	916 ± 43	1979 ± 98	780 ± 73
Cholesterol	613 ± 46	403 ± 23	619 ± 12	273 ± 43
RNA	12 ± 6	39 ± 8	23 ± 9	30 ± 16

[a] The units of enzyme activity and concentrations are the same as listed in Table I.
[b] Not determined.

sucrose gradients. The carbohydrate composition of this isolated plasma membrane was similar to that of RBA virus-transformed cells (Table III). The most striking differences were found in the lipid content of these membranes. The tumor cell membranes contained one-half the amount of phospholipid and cholesterol found in the membranes isolated from the cultured cells.

[15] The Preparation of Red Cell Ghosts (Membranes)

By DONALD J. HANAHAN and JANICE E. EKHOLM

The mammalian erythrocyte, or red cell, represents a readily available source of plasma membranes. In addition to the rather unlimited amounts of starting material, these cells contain no demonstrable subcellular organelles, thus providing a unique supply of pure plasma membranes.

In the preparative method described here, the membranes (ghosts) are obtained through the use of osmotic lysis. It is well accepted that a normal human erythrocyte will behave as a good osmometer. Hence, when placed in a hypotonic medium, these cells will swell and change initially from a biconcave disk to an almost perfect sphere. The latter form contains the greatest volume per minimum surface area. Further stress, i.e., decreasing ionic strength or osmolarity, will allow release of hemoglobin and cell contents, and hence lysis will occur. Seeman[1] has provided evidence that the permeable state of the human cells was observable only within 15–25 seconds after initiation of hemolysis in a low osmolarity solution. Coldman *et al.*[2] published data on the osmotic fragility of mammalian erythrocytes; they concluded that this fragility is highly species specific.

The technique outlined here is a modification of the procedure of Dodge *et al.*[3] and uses the human erythrocyte as the starting material. It is a highly reproducible, high yield method, which has been in use in this laboratory for a number of years.

Experimental

Reagents. Two buffer solutions are routinely used.

a) Isotonic Tris buffer (310 imOsM[4]) pH 7.6; due to ionization

---

[1] P. Seeman, *J. Cell. Biol.* 32, 55 (1967).
[2] M. F. Coldman, M. Gent, and W. Good, *Comp. Biochem. Physiol.* 31, 605 (1969).
[3] J. T. Dodge, C. D. Mitchell, and D. J. Hanahan, *Biochim. Biophys. Acta* 104, 348 (1965).

characteristics of this buffer, 0.172 M Tris instead of 0.155 M is prepared. It is brought to pH 7.6 with 12 N HCl prior to diluting to volume.

b) 20 imOsM Tris buffer, pH 7.6 prepared by dilution of an aliquot of the 0.172 M (310 imOsM) buffer with water

Collection of Blood and Washing of Cells. Blood is collected directly into heparinized (Becton-Dickinson) Vacutainer tubes and centrifuged immediately in these tubes at 1000 g for 30 minutes at 4°.

Plasma and buffy coat are removed by careful suction, and the cells are resuspended in pH 7.6, 310 imOsM Tris buffer. After mixing well by inversion, the sample is centrifuged again at 1000 g for 30 minutes at 4°. The supernatant is removed by careful suction, and a few red cells are sacrificed to remove any remaining buffy layer. This washing procedure is repeated twice more. Washed cells are kept on ice after the final washing.

The washed cells are suspended in isotonic Tris buffer pH 7.6 to an approximate hematocrit of 50%. The sample is mixed gently by inversion for approximately 1 minute before a hematocrit is taken or before aliquots are measured for membrane preparation.

Preparation of Membranes. In a typical run, 5-ml aliquots of the above 50% cell suspension are transferred to 50-ml polyethylene tubes. Thirty milliliters of 20 imOsM, pH 7.6, Tris buffer at 4° are forcefully blown into the cell suspension. The tubes are allowed to stand approximately 5 minutes prior to centrifuging at 20,000 g for 40 minutes at 4°. An index of the progress of hemolysis process is the development of a deep red color, due to hemolysis, in the soluble portion. The membranes are very light weight and tend to be lost more easily in the initial supernatant: thus the first decantation of supernatant is a little more difficult than with subsequent washes. If the sample is held so that light can come up through the tube as it is slanted, the membranes can be more easily seen. When as much supernatant is decanted as is possible without losing membranes, the tube is placed on a vibrating mixer to loosen and resuspend the membranes in the remaining buffer. A small fibrinlike skin often attaches to the bottom of the tube; an attempt is made to dislodge it, since it may trap membranes. An additional 30 ml of 20 imOsM, pH 7.6, buffer is blown into the tube, and the suspension is centrifuged for 40 minutes at 20,000 g at 4°. A total of four washes is necessary before the membranes are colorless.

---

[4] imOsM, ideal milliosmolarity; the use of this term assumes that there is complete ionization of a particular buffer. For example, 310 imOsM Tris buffer would indicate complete dissociation into two ions.

After removal of the last buffer wash, the membranes, suspended in pH 7.6, 20 imOsM Tris buffer are transferred quantitatively to a 10-ml volumetric flask by use of a Pasteur pipette. The fibrinlike clot is left behind at this point, but is washed well with transferring buffer to remove any remaining membranes. The contents are diluted to 10 ml with buffer, and this then represents the erythrocyte membrane preparation. Recoveries can most conveniently be determined by assay for lipid content,[5] and also by acetylcholine esterase levels.[6] Morphological characteristics can be checked by phase and electron microscopy.

Comments on Operational Details of Procedure

The preparation of membranes by the above techniques is quite straightforward and usually presents few difficulties in execution. However, one should be cognizant of certain possible technical problems in handling these membranes and of possible changes in the characteristics of the membrane that may arise during preparation.

General. The removal of hemoglobin is dependent not only on the pH of the hemolyzing buffer, but also on the osmolarity of the buffer. Hemoglobin release is near maximal at pH 7.6 and in a 20–25 imOsM buffer. At lower pH values and higher osmolarity a significant amount of hemoglobin is retained; the presence of divalent cations, for example, Ca^{2+}, Mg^{2+} in concentrations in the range of 1–5 mM can cause significant retention of hemoglobin.

The volume ratio of cells to hemolyzing buffer is approximately 1:7 as described in Methods. An increased ratio may allow the cells to be washed fewer times. It is important to note that nonhemoglobin protein is lost during the preparation of the membranes; a lower osmolar buffer at pH 7.6, for example, 10 imOsM, not only leads to greater loss of nonhemoglobin protein, but to distinct fragmentation of the membranes.[7] Thus, in order to prepare "intact" membranes, free of hemoglobin, careful attention must be paid to the osmolarity and pH of the hemolyzing buffer and to the volume ratio of buffer to cells.

Storage of Cells and Membranes. The length of time that the cells and/or membranes may be stored is dependent on the type of subsequent analysis planned. In any event, the cells should be separated from the plasma as soon as possible and washed with isotonic buffer. After the cells have been suspended to approximately 50% hematocrit, they may be stored overnight at 4°, but should be used within 24 hours. The pres-

[5] D. J. Hanahan and J. Ekholm, *Biochim. Biophys. Acta* **255**, 413 (1972).
[6] M. Heller and D. J. Hanahan, *Biochim. Biophys. Acta* **255**, 251 (1971).
[7] T. A. Bramley, R. Coleman, and J. B. Finean, *Biochim. Biophys. Acta* **241**, 752 (1971).

ence of leukocytes, with their high level of protease activity, can cause difficulty.

Membrane preparations should be completed on the same day. Storage of the membranes at one of the hemolyzing stages, followed by centrifugation and removal of the supernatant, could cause removal of more nonhemoglobin protein than would occur during a shorter exposure time. Membranes should be stored in the same low ionic strength medium as used in their final wash. Isotonic solutions tend to decrease the permeability of the membrane in a rather irreversible manner. Freezing of the membranes should be avoided unless it is desired to produce a specific effect. Occasionally certain preparations contain a small hard, red pellet under the white membranes. This may be due to the particular characteristics of the donor's blood, and the membranes should be carefully removed to prevent any contamination by this residue.

Source of Cells

In general, other mammalian erythrocytes, such as the cow, pig, dog, rat, guinea pig, respond to hypotonic hemolysis in a manner comparable to that of the human cell. However, it is an oversimplification to state that all these cells exhibit morphologically and biochemically similar behavior on hemolysis. Even though a "plasma" membrane, devoid of hemoglobin, can be obtained in each case, the procedure is not without some trauma or alteration to the membrane structure. Thus each investigator must evaluate each preparation in terms of the type of information being sought. Certain of the alterations encountered in these preparations are outlined in the following section.

Alterations to Membranes during Preparation. Two specific examples, i.e., the human and the cow erythrocyte, will serve to illustrate the types of alterations that can occur in these membranes during preparation.

One criterion used in evaluating the character of the human erythrocyte membranes has been the level and nature of the Na^+-, K^+-, Mg^+-activated ATPase activity. In the intact human erythrocyte this activity cannot be elicited unless the cell is subjected to a treatment such as freeze thaw or sonication. However, upon preparation of hemoglobin-free membranes, this activity is expressed without the use of a treatment such as freeze thaw or sonication.[5,7] Thus a latent ATPase activity[7] or more permeable membrane[5] has developed during preparation, and the active site of the ATPase is made available to the ATP substrate. In part the development of this overt ATPase activity in the hemoglobin-free membranes is pH dependent, since in membranes prepared at pH 5.8 such changes are not noted.[4] Also, on the basis of preliminary evidence it would appear that addition of Ca^{2+} (but not Mg^{2+}) in the initial hemolyzing buffer may control the degree to which this ATPase activity is made available in the final membrane preparation.[8] In very low osmolarity buffers, e.g., less than 10 imOsM at pH 7.6, the human erythrocyte membrane can undergo distinct fragmentation; however, at higher osmolarities, e.g., 25–30 imOsM and above, this type of phenomenon is not noted and the cells appear morphologically intact.[7]

In the case of the adult cow erythrocyte, it is necessary to include Ca^{2+} or Mg^{2+} at a level of 5 mM in the initial hemolyzing buffer to prevent disruption of the membrane per se. This approach prevents loss of considerable lipid and enzyme (protein).[9]

[8] D. J. Hanahan and J. Ekholm, unpublished observations.
[9] S. P. Burger, T. Fujii, and D. J. Hanahan, *Biochemistry* **7**, 3682 (1968).

[16] Preparation of Impermeable Ghosts and Inside-out Vesicles from Human Erythrocyte Membranes[1,2]

By THEODORE L. STECK and JEFFREY A. KANT

A central feature in a membrane's organization is the asymmetrical distribution of functions between its two surfaces. In order to chemically probe each face selectively, closed membranous sacs can be prepared that expose either the original external side of the membrane or the internal side, i.e., sealed right-side-out and inside-out vesicles.

Right-side-out membranes are found, of course, on intact red blood cells, but they may also be prepared relatively free of cytoplasm by "resealing" erythrocyte membrane "ghosts" after hemolysis. Impermeable inside-out vesicles are formed when unsealed erythrocyte ghosts are incubated in a chilled alkaline buffer of very low ionic strength. These vesicles can be separated from most of the contaminating ghosts and right-side-out vesicles by suitable equilibrium density gradient centrifugation. The quality of each preparation may be ascertained by measuring the extent to which intrinsic markers affixed to the unexposed side of the membrane are inaccessible to probes.

[1] T. L. Steck, R. S. Weinstein, J. H. Straus, and D. F. H. Wallach, *Science* **168**, 255 (1970).
[2] JAK was supported during this work by Training Grant No. 5T05GM0193904 from the U.S. Public Health Service; TLS was supported by Grant No. P-578 from the American Cancer Society and by a fellowship from the Schweppe Foundation.

Membrane Preparations

Preparation of Unsealed Ghosts[1,3]

Reagents

PBS (phosphate-buffered saline): 150 mM NaCl, 5 mM sodium phosphate, pH 8.0
5P8: 5 mM sodium phosphate, pH 8.0

Procedure. Human blood can be obtained fresh or out-dated through a blood bank; heparin, EDTA, or citrate may be used for anticoagulation. All procedures are performed at 0–4°. Red blood cells are washed three times by sedimenting in a swinging-bucket rotor (2300 g_{max} for 10 minutes) and resuspending the pellet in five volumes of PBS. The buffy coat is carefully aspirated each time from the surface of the pellet.

Hemolysis is initiated by rapidly and thoroughly mixing 1 ml of packed cells with approximately 40 ml of 5P8. The membranous ghosts are pelleted by centrifugation at 22,000 g_{max} for 10 minutes in an angle-head rotor. The supernatant is removed with a tap aspirator. The tube is tipped and rotated on its axis so that the loosely packed ghosts slide away from a small hard button, rich in contaminating proteases, which may then be aspirated. After two more identical wash cycles, the ghost membranes are creamy white and morphologically intact.

Preparation of Sealed Inside-out Vesicles

Reagents

0.5P8: 0.5 mM sodium phosphate, pH 8.0–8.2. A 0.1 M sodium phosphate (pH 10.0) stock solution is diluted 200-fold.
Dextran barrier solution (density 1.03 g/ml) in 0.5P8: A density 1.10 g/ml Dextran T-110 (Pharmacia) stock solution (26.7 g/100 ml, stored frozen) is diluted 3 parts to 10.

Procedure. To initiate vesiculation, each milliliter of pelleted unsealed ghosts (containing approximately 4 mg of protein) is diluted to about 40 ml with 0.5P8. After 0.5–18 hours on ice, the membranes are pelleted at 28,000 g_{max} for 30 minutes. The membranes are resuspended to 1 ml in 0.5P8 by vortex mixing and passed 3–5 times through a No. 27 gauge needle on a 1-ml syringe to complete vesiculation. Phase contrast microscopy at this point should demonstrate nearly complete breakdown to small vesicles.

[3] G. Fairbanks, T. L. Steck, and D. F. H. Wallach, *Biochemistry* **10**, 2606 (1971).

The homogenates are diluted 2- to 4-fold in 0.5P8 and layered upon an equal volume of the Dextran barrier solution. For example, homogenates from 2 ml of ghosts may be diluted to 6.2 ml and layered on a 6.2-ml barrier in a Spinco SW 41 rotor tube. Centrifugation for 2 hours at 40,000 rpm resolves the mixture into a pellet and a band floating on top of the barrier. The top band material is collected and washed by diluting to 40 ml in buffer (e.g., 0.5P8) and pelleting at 28,000 g_{max} for 30 minutes. The pellet, containing unsealed ghosts and right-side-out vesicles, is discarded.

Comments. Sealed inside-out vesicles arise by endocytosis in fresh, well-washed ghosts following incubation in an alkaline buffer which is low in ionic strength and lacking in divalent cations. Various modifications of the original procedure[1] have been explored. In the method outlined above, speed and simplicity were favored by the shortened centrifugation times and use of a density barrier rather than a continuous gradient; the quality of the preparations equals that of the original method (see the table).

The pH of the 0.5 mM phosphate vesiculation medium may be varied between 7.7 and 8.7. While the more alkaline pH tends to increase the yield of inside-out vesicles, it favors the loss of certain easily eluted polypeptides from the membrane.[3,4] Other alkaline buffers, such as NH_4HCO_3 or Tris·HCl, can be substituted for the sodium phosphate; Na^+ can also be replaced by K^+. Greater yields of inside-out vesicles may be obtained by washing the membranes a second time in the 0.5 mM phosphate medium prior to homogenization. In that case, the incubation in 0.5 mM phosphate can be reduced to 15 minutes.

The 1.03 g/ml density shelf described above can lead to decreased purity of the inside-out vesicle fraction, seemingly because mixed aggregates form when concentrated vesicle homogenates are rapidly sedimented against this barrier. The use of a continuous[1] or discontinuous density gradient (e.g., 1.005–1.05 g/ml) minimizes this source of contamination.

The protocol may be interrupted at any point after vesiculation. The sealed inside-out vesicles are stable at 4° for a week or two. However, freezing the preparation may disrupt the integrity of the membrane. Once formed, the vesicles are stable in pH 5.5–9.0 buffers. They may become "unsealed" during prolonged incubation at elevated ionic strength (≥ 0.15) and should be monitored for this possibility (see below). They also may aggregate when pelleted from buffers of acidic pH or high ionic strength.

Divalent cations critically affect both the sidedness and sealing of the vesicles. Including 10^{-4} M $MgSO_4$ in the 0.5 mM phosphate medium from the outset suppresses both sealing and endocytosis, so that almost all the

[4] T. L. Steck, G. Fairbanks, and D. F. H. Wallach, *Biochemistry* 10, 2617 (1971).

vesicles formed are unsealed and right-side-out. If divalent cations are omitted entirely, about half of the vesicles formed upon incubation and homogenization will be both inside-out and sealed (to Na^+, K^+, sucrose, etc.).[5] If 10^{-4} M $MgSO_4$ is added *after* incubation in the low ionic strength buffer but *before* homogenization, sealed right-side-out vesicles form instead of inside-out.[5] Ca^{2+} can replace Mg^{2+} throughout. It appears that the resealing process requires the transient absence of divalent cations while their presence during homogenization predisposes toward exocytic rather than endocytic vesiculation.

We originally supposed[1] that the inside-out vesicles achieved a lower equilibrium density than ghosts and right-side-out vesicles because unequal fixed charge densities at the two membrane faces produces asymmetrical Donnan-osmotic effects.[6] We now know that ghosts and inside-out and right-side-out vesicles equilibrate in the 1.005–1.03 g/ml density region when they are impermeable to small solutes, while unsealed ghosts and vesicles band in the 1.05–1.065 g/ml zone.[1,5] Thus, the dextran gradient principally separates sealed from unsealed species. The buoyancy of the sealed species suggests that they behave as osmometers which have trapped ions or other solutes in their interior aqueous compartment. The precise equilibrium density point of individual sealed vesicles also appears to be limited by their size, as though each were osmotically expanded to its elastic limits.[6] Vigorous homogenization, for example, produces small vesicles that equilibrate near 1.03 g/ml, while gentle shearing yields larger vesicles that are recovered in the 1.005 g/ml range.

Preparation of Sealed Ghosts

Reagents

> PBS (phosphate-buffered saline): 150 mM NaCl, 5 mM sodium phosphate, pH 8.0
> 5P8: Sodium phosphate, 5 mM, pH 8.0
> 5P8-1Mg: sodium phosphate, 5 mM, pH 8.0 containing 1 mM $MgSO_4$

Procedure 1. Erythrocytes are prepared and hemolyzed as in the preparation of unsealed ghosts except that 5P8-1Mg is substituted for 5P8 throughout. These ghosts retain significant amounts of hemoglobin (1–2% of initial) because they spontaneously "reseal" shortly after lysis and trap cytoplasmic contents. For this reason, higher dilutions than 40-fold may be desirable in the initial hemolysis step.

[5] J. A. Kant and T. L. Steck, *Nature (London) New Biol.* **240**, 26 (1972).
[6] T. L. Steck, J. H. Straus, and D. F. H. Wallach, *Biochim. Biophys. Acta* **203**, 385 (1970).

Procedure 2. Ghosts may also be resealed by warming in saline.[7,8] The erythrocytes are initially hemolyzed in 40 volumes of 5P8 (as in the preparation of unsealed ghosts) and pelleted. The membranes are resuspended in 40 volumes of PBS and incubated 40 minutes at 37° to induce resealing. The ghosts are then pelleted and washed twice more in PBS.

Comments. The 5P8-1Mg resealing protocol is simple and avoids high ionic strength and warming. However, the PBS procedure affords cleaner preparations, presumably because the ghosts remain permeable until the second dilution (into PBS). These preparations are stable for at least a week at 4° if maintained in 5P8-1Mg and PBS, respectively.

Sidedness Assays

Membrane impermeability is operationally defined by the exclusion of specific probes, such as substrates, to the membrane enzymes. The quality of the sealed ghosts is measured by the latency of enzymes presumed to reside on the cytoplasmic face of the membrane (NADH–cytochrome *c* oxidoreductase[9] and glyceraldehyde 3-phosphate dehydrogenase[10]). Similarly, the inside-out vesicles are assessed by the inaccessibility of markers found on the exterior of the intact erythrocyte (e.g., sialic acid[1,11] and acetylcholinesterase[12]). Surfactants are used to disrupt the permeability barrier and expose the latent marker. Protein is estimated by the procedure of Lowry *et al.,*[13] using bovine serum albumin as standard.

Sialic Acid Accessibility to Sialidase

Reagents

Sialidase stock. Type VI neuraminidase from *Clostridium perfringens* (Sigma Chemical Co.) is stored at 4° as a 1 mg/ml solution in 0.3 mg/ml bovine serum albumin[14]

Sialidase reagent: 0.1 mg/ml sialidase in 0.1 *M* Tris·acetate buffer, pH 5.7

Sialidase and Triton reagent: 0.1 mg/ml sialidase in 0.1 *M* Tris·acetate buffer, pH 5.7–0.2% (v/v) Triton X-100

[7] J. F. Hoffman, D. C. Tosteson, and R. Wittam, *Nature (London)* **185**, 186 (1960).
[8] H. Bodemann and H. Passow, *J. Membrane Biol.* **8**, 1 (1972).
[9] I. Zamudio, M. Cellino, and M. Canessa-Fischer, *Arch. Biochem. Biophys.* **129**, 336 (1969).
[10] G. Duchon and H. B. Collier, *J. Membrane Biol.* **6**, 138 (1971).
[11] G. M. W. Cook and E. H. Eylar, *Biochim. Biophys. Acta* **101**, 57 (1965).
[12] W. W. Bender, H. Garan, and H. C. Berg, *J. Mol. Biol.* **58**, 783 (1971).
[13] O. H. Lowry, N. J. Rosebrough, A. L. Farr, and R. J. Randall, *J. Biol. Chem.* **193**, 265 (1951).
[14] J. T. Cassidy, G. W. Jourdian, and S. Roseman, *J. Biol. Chem.* **240**, 3501 (1965).

Procedure. Aliquots of membranes containing 100–300 μg protein in 0.05 ml are incubated at room temperature (22°) for 30 minutes with an equal volume of sialidase reagent containing or lacking Titron X-100. The sialic acid released is determined directly by the method of Warren.[15] The total sialic acid determined following incubation of membranes at 80° for 1 hour with an equal volume of 0.2 N H_2SO_4 agreed with the sialidase and Triton values within $\pm 10\%$, indicating that this detergent disrupts the membrane permeability barrier completely.

Acetylcholinesterase Accessibility

Reagents

Sodium phosphate, 100 mM, pH 7.5

DTNB [5,5'-dithiobis-(2-nitrobenzoic acid)], 10 mM in 100 mM sodium phosphate, pH 7.0. This stock contains 3 mg of $NaHCO_3$ per 8 mg of DTNB

Acetylthiocholine chloride, 12.5 mM

5P8: 5 mM sodium phosphate, pH 8.0

Triton X-100, 0.2% (v/v) in 5P8

Procedure.[16] Aliquots of 2–10 μg of membrane protein in 0.01–0.10 ml are briefly preincubated in semimicro cuvettes with an equal volume of 5P8 or Triton X-100. Sodium phosphate (100 mM, pH 7.5) is added to make 0.7 ml; 0.05 ml each of DTNB and acetylthiocholine chloride are then mixed in, and the reaction is followed spectrophotometrically at 412 nm at room temperature.

NADH-Cytochrome c Oxidoreductase Accessibility

Reagents

PBS (phosphate-buffered saline): 150 mM NaCl, 5 mM sodium phosphate, pH 8.0

Saponin, 0.1% (w/v) in PBS

β-NADH, 2 mM in PBS

Cytochrome c, 5 mg/ml in PBS

Procedure. Aliquots of 50–200 μg of membrane protein in 0.025–0.05 ml are brought to 0.375 ml with PBS. PBS or saponin (0.025 ml) is then mixed in. β-NADH and cytochrome c (0.05 ml each) are added sequentially, and the reaction is followed spectrophotometrically in microcuvettes at 550 nm at room temperature.

---

[15] L. Warren, *J. Biol. Chem.* **234**, 1971 (1959).

[16] G. L. Ellman, K. D. Courtney, A. Valentino, Jr., and R. M. Featherstone, *Biochem. Pharmacol.* **7**, 88 (1961).

Glyceraldehyde 3-Phosphate Dehydrogenase Accessibility

Reagents

Sodium pyrophosphate, 30 mM, adjusted to pH 8.4 with HCl and made 4 mM in cysteine-HCl just before use
Sodium arsenate, 0.4 M
β-NAD, 20 mM
DL-Glyceraldehyde 3-phosphate, 15 mM, adjusted to pH 7.0 with 1 N NaOH
5P8: 5 mM sodium phosphate, pH 8.0
Triton X-100, 0.2% (v/v) in 5P8

Procedure.[17] Aliquots of 2–10 μg of membrane protein in 0.01–0.10 ml are preincubated for 1 minute in semimicrocuvettes with an equal volume of 5P8 or Triton X-100. Sodium pyrophosphate is added to 0.82 ml, followed by 0.03 ml of sodium arsenate and 0.05 ml of β-NAD. Then 0.1 ml of glyceraldehyde 3-phosphate is added, the solution is mixed, and the reaction is followed spectrophotometrically at 340 nm at room temperature. Taking the addition of glyceraldehyde 3-phosphate as the starting point of the reaction, the increase in absorbance between the first and second minute of reaction is taken as a measure of enzyme activity. This increment is not linear with time but is linear with respect to enzyme concentration in the range indicated.

Discussion

The table summarizes a representative experiment. It is seen that the unsealed ghosts used in preparing the inside-out vesicles appear permeable to all of the sidedness probes tested. In contrast, the inside-out vesicles exhibit great inaccessibility in the sialidase and acetylcholinesterase assays, while sealed ghosts (and right-side-out vesicles) show a converse latency of the NADH–cytochrome c oxidoreductase and glyceraldehyde 3-phosphate dehydrogenase activities. These data confirm the asymmetrical orientation of four membrane proteins deduced previously.[1,9-12] All the major polypeptides in this membrane have similarly been found to be asymmetrical in disposition.[18] The table supports the crucial premise that the membrane constituents are not randomized or scrambled during the preparation of the ghosts and vesicles. Furthermore, it indicates that sealed ghosts and vesicles are impermeable to small substrate molecules; recent studies indicate that they are even rather impermeable to Na$^+$ and K$^+$.[5]

The accessible sialic acid and acetylcholinesterase in the top band fraction has ranged between 2 and 20% over many preparations. This

[17] G. T. Cori, M. W. Slein, and C. F. Cori, *J. Biol. Chem.* **173**, 605 (1948).
[18] T. L. Steck, *in* "Membrane Research" (C. F. Fox, ed.), p. 71. Academic Press, New York, 1972.

ACCESSIBILITY OF MARKERS IN UNSEALED GHOSTS, Mg-SEALED
GHOSTS, AND INSIDE-OUT VESICLES

Marker	Unsealed ghosts	Mg-sealed ghosts	Inside-out vesicles
Sialic acid released			
− Triton X-100[a]	101	71.7	16.4
+ Triton X-100[a]	102	72.2	162.3
% Accessible[c]	99	99	10
Acetylcholinesterase			
− Triton X-100[b]	2.02	1.59	0.29
+ Triton X-100[b]	2.11	1.52	3.18
% Accessible[c]	96	105	9
NADH–cytochrome c Oxidoreductase			
− Saponin[b]	0.0156	0.0005	0.0178
+ Saponin[b]	0.0202	0.0123	0.0128
% Accessible[c]	77	4	139
Glyceraldehyde 3-phosphate dehydrogenase			
− Triton X-100[b]	2.32	0.15	1.99
+ Triton X-100[b]	2.50	1.48	2.56
% Accessible[c]	93	10	78

[a] Expressed as nanomoles of N-acetylneuraminic acid per milligram of protein.

[b] Expressed as micromoles of product per milligram of protein per minute.

[c] (Value minus detergent/value plus detergent) × 100.

value is taken to represent the contamination of sealed inside-out vesicles by both sealed right-side-out and permeable membranes. Every vesicle preparation should be assessed in this way. This was not done in a previous study, so that sealed right-side-out vesicles banding in the low density Dextran fraction were apparently mistaken for inside-out vesicles.[4]

NADH–cytochrome c oxidoreductase has frequently shown a minor degree of latency in unsealed ghosts, perhaps reflecting partial membrane impermeability to the cytochrome acceptor. In addition, saponin causes mild inactivation of this enzyme (e.g., see the inside-out vesicles in the table) but it is preferred to Triton X-100, which seriously interferes with the reduction of the cytochrome c (but not with the reduction of ferricyanide).

The reduced specific activities found in the resealed ghosts reflect the persistence of trapped cytoplasmic proteins, while the increased specific activities of the inside-out vesicle preparations is consonant with their loss of certain elutable polypeptides.[1,3,4]

Recently, these methods were used to prepare inside-out vesicles from rat fat cell plasma membranes,[19] suggesting that their applicability may be more general than previously supposed.

[19] V. Bennett and P. Cuatrecasas, *Biochim. Biophys. Acta* **311**, 362 (1973).

There have been several studies indicating the sidedness of certain protein components in human erythrocyte membrane. However, questions have arisen concerning the fidelity with which the native orientation and reactivity of membrane constituents are conserved following hemolysis and the impermeability of the intact cell to certain small probe molecules. This topic was recently reviewed.[20] It was concluded that proteins in the purified ghosts and vesicles under discussion thus far exhibit the same asymmetric disposition inferred from studies on intact red blood cells.

[20] T. L. Steck, *J. Cell Biol.* **62** (1974), in press.

[17] Isolation and Characterization of Golgi Apparatus and Membranes from Rat Liver

By Becca Fleischer

Until very recently, methods for the isolation of Golgi membranes depended primarily on morphological criteria. Although considerable progress was made by early workers in this field using such criteria,[1,2] critical evaluation of the purification procedures used was not possible since no unique enzymatic activity characteristic of this organelle was known at that time. The observation that UDP galactose:*N*-acetylglucosamine, galactosyltransferase activity is localized exclusively in Golgi-derived vesicles of bovine liver[3,4] paved the way for the isolation of this organelle in large quantities so that chemical and enzymatic characterization could be carried out. Subsequently galactosyltransferase was shown to also be a marker enzyme for Golgi vesicles in rat liver.[5]

Two methods will be described for the isolation of Golgi membranes from rat liver since this tissue is more suitable for metabolic study than bovine liver. One method involves isolation of a Golgi-rich fraction by differential centrifugation followed by purification of the Golgi in a sucrose step-gradient. The other method is a direct isolation of Golgi apparatus from a homogenate of liver using a sucrose step-gradient. The Golgi fractions obtained by the two procedures differ somewhat in the amount and kind of contamination by other cell organelles as well as in yield. Each procedure has obvious advantages in certain situations which

[1] E. L. Kuff and A. J. Dalton, "Biochemical studies of isolated Golgi membranes," *in* "Subcellular Particles" (T. Hayashi, ed.), p. 114. Ronald Press, New York, 1959.
[2] D. J. Morré and H. H. Mollenhauer, *J. Cell Biol.* **23**, 295 (1964).
[3] B. Fleischer, *Fed. Proc., Fed. Amer. Soc. Exp. Biol.* **28**, 404 (1969).
[4] B. Fleischer, S. Fleischer, and H. Ozawa, *J. Cell Biol.* **43**, 59 (1969).
[5] B. Fleischer and S. Fleischer, *Biochim. Biophys. Acta* **219**, 301 (1970).

will be pointed out in the last section (Comments). Each procedure can be applied to the fractionation of small amounts of liver (3–10 g) for metabolic studies, as well as to large amounts (100–300 g) which may be useful for isolation of individual enzymes or other components of Golgi membranes.

Preparation of Golgi Fractions

Male rats, Sprague-Dawley or Holtzman 200–250 g each, are used. They are fed ad libitum before use. They are killed by decapitation and exsanguinated before the livers are removed. All sucrose solutions are prepared using "Ultra-Pure" grade (Mann Research Laboratories), and percent sucrose (w/w solution) is adjusted to the desired concentration $\pm 0.1\%$ using a Bausch and Lomb Abbe 3L refractometer thermostatically controlled at 25°. Dextran 500 is obtained from Pharmacia Fine Chemicals, New Market, New Jersey. All centrifugations are carried out in a Spinco L-50 or L-2 65B ultracentrifuge, and all operations are carried out at 4°. Centrifugation times given are settings used on the Spinco time dial.

Method A: Differential Centrifugation Method.[5] The livers of 25–30 rats are collected in cold 0.25 M sucrose, blotted, weighed and minced slightly with scissors. The minced liver (200–300 g) is suspended in 0.5 M sucrose containing 0.1 M sodium phosphate (pH 7.2) and 1% Dextran 500 (w/v solution). The mixture is homogenized (3 full strokes) with a Potter-Elvehjem homogenizer (50 ml) and a Teflon pestle with a clearance of 0.026 inch[6] at 1000 rpm. The homogenization step is repeated, once again with a pestle having a clearance of 0.018 inch. The final pH of the mixture should be 7.1–7.2. The mixture is centrifuged for 30 minutes at 10,000 rpm (8720 g_{av}) in a No. 30 rotor. The supernatant is poured through 4 layers of cheesecloth to remove coagulated fat and centifuged as before for 30 minutes at 20,000 rpm (34,880 g_{av}). The supernatant is discarded, and any lipid adhering to the tubes is wiped away. The pellet (R_2) is suspended in the small amount of liquid remaining in each tube and is transferred with a Pasteur pipette to a large Dounce homogenizer. To the pellet is added 53% sucrose containing 0.1 M sodium phosphate (pH 7.1), and the mixture is made homogeneous by hand homogenization with a type A pestle. The final concentration of sucrose is adjusted to 43.7% using 0.25 M sucrose. Fractionation of R_2 is carried out by means of zonal centrifugation in a Ti 14 rotor. A step-gradient is used which can be prepared with 17.8% sucrose and 38.7%

---

[6] Glass–Teflon homogenizers, 50 ml in size, as purchased from Arthur H. Thomas Co. have a clearance of 0.008–0.009 inch and must be modified by machining of the Teflon pestle to give the desired clearance.

sucrose and a Spinco gradient maker programmed to give the following series of sucrose concentrations in the rotor: 230 ml of 17.8%, 120 ml of 30.8%, 120 ml of 36.2%, and 180 ml of 38.8%. The gradient is placed in the rotor while it is spinning at 4000 rpm. A simpler and more reproducible method of preparing the gradient is to pump sequentially each concentration of sucrose into the rotor using a large capacity nonpulsing pump [for example a Sage Model 375 tubing pump (Sage Instruments, Inc.)]. The R_2 fraction is introduced into the periphery of the rotor, followed by 100 ml of 53% sucrose containing 0.1 M sodium phosphate (pH 7.1). The sample (about 50 ml) is introduced using the Sage pump or a syringe-type infusion pump (Model No. 975, Harvard Apparatus Co., Inc.) at the rate of 6–8 ml per minute, followed by 100 ml of 53% sucrose containing 0.1 M sodium phosphate (pH 7.1). The gradient is centrifuged for 45 minutes at 35,000 rpm (91,300 g_{max}). When the rotor has decelerated to 4000 rpm, the rotor is unloaded from the center by pumping 55% sucrose into the periphery. Fractions of 20 ml each are collected at a rate of 10–20 ml per minute. The effluent may be conveniently monitored using an ISCO UV analyzer (Model 222, Instrumentation Specialties Co.). Tubes containing the first two protein peaks (Fig. 1) are combined (generally tubes 10–19) diluted with 0.5 volume of cold distilled water and centrifuged 60 minutes at 30,000 rpm (78,480 g_{av}) in a No. 30 rotor. The pellets are combined and suspended finally in 10 ml of 0.25 M sucrose using a 10-ml Potter-Elvehjem homogenizer.

Method B: Single Step-Gradient Method.[7] The livers of 10 rats are col-

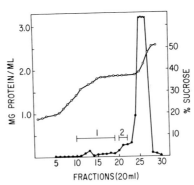

FIG. 1. Fractionation of crude Golgi fraction (R_2) from rat liver by zonal centrifugation. Fraction 1, tubes 10–19; fraction 2, tubes 20–22; O—O, % sucrose; ●—●, mg of protein per milliliter.

[7] L. M. G. van Golde, B. Fleischer, and S. Fleischer, *Biochim. Biophys. Acta* **249**, 318 (1971).

lected, washed briefly in cold 0.25 M sucrose, blotted, weighed (approximately 100 g), and minced with scissors directly into 3 volumes of 53% sucrose containing 0.1 M sodium phosphate (pH 7.1). The mixture is homogenized as in method A and filtered through two layers of cheesecloth. Filtration of the viscous homogenate is aided by stirring the liquid on the cheesecloth during the filtration with a thick, blunt stirring rod. With 0.25 M sucrose, the filtrate is adjusted to 43.7% sucrose. The homogenate (approximately 300 ml) is pumped into the periphery of a Ti 15 zonal rotor previously filled with a step-gradient consisting of the following sucrose concentrations: 350 ml of 29%, 260 ml of 33%, 300 ml of 36%, and 425 ml of 38.7%. After loading the sample, the rotor is completely filled using the homogenizing medium. The gradient is centrifuged at 35,000 rpm (121,750 g_{max}) for 50 minutes. When the rotor has decelerated to 4000 rpm, 40-ml fractions are collected from the core by displacement of the rotor contents from the periphery with 55% sucrose at the rate of 15–20 ml per minute. Fractions corresponding to the interface between 20 and 33% sucrose (usually tubes 7–12) are combined as one fraction. A second fraction at the interface between 33 and 36% sucrose (usually tubes 13–18) is also collected. Each fraction is then diluted with 0.5 volume cold distilled water, and the vesicles are recovered by centrifugation at 30,000 rpm (78,480 g_{av}) for 1 hour. The Golgi fractions are suspended finally in 10 ml of 0.25 M sucrose each as in method A.

Preparation of Membranes from Golgi Vesicles

To the Golgi fraction in 0.25 M sucrose (2–3 mg protein per milliliter) is added 2 volumes of 2 N NaCl followed by 7 volumes of 0.2 N NaHCO$_3$. The mixture is equilibrated in a Parr Model No. 4635 Cell Disruption Bomb (Parr Instrument Co.) at about 1800 psi of N_2 for 15 minutes. The mixture is released dropwise from the discharge tube. The mixture is then centrifuged at 30,000 rpm (78,480 g_{av}) in a No. 30 rotor for 90 minutes. The pellet is taken up in the original volume of 0.25 M sucrose using a small Potter-Elvehjem homogenizer. The contents of the Golgi vesicles may be concentrated from the supernatant using an Amicon ultrafiltration device with a PM-30 filter.

Characterization of Golgi Fractions

The Golgi-rich fractions described above have at least two characteristics which may be used to define them. They have a unique protein profile when the proteins are resolved by polyacrylamide disc gel electrophoresis, and they contain galactosyltransferase, an enzyme important in the biosynthesis of glycoproteins. Electrophoresis on polyacrylamide gels

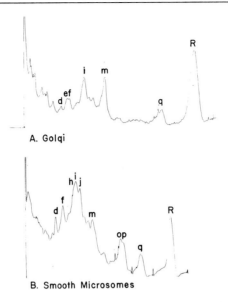

FIG. 2. Densitometer tracings of electrophoresis patterns of organelles from rat liver. Electrophoresis was carried out on polyacrylamide gels using phenol–acetic acid–urea [W. L. Zahler, B. Fleischer, and S. Fleischer, *Biochim. Biophys. Acta* **203**, 283 (1970)]. R = ribonuclease A added to each gel to serve as a marker protein for calculation of relative mobilities. The gels were stained with Coomassie Blue.

can be carried out using the phenol–acetic acid–urea method of Takayama *et al.*[8] as modified by Zahler *et al.*[9] This method is described in detail elsewhere in this series, Volume 32 [7].

Using this procedure, we have shown that Golgi preparations have a characteristic protein profile compared to other purified organelles of rat liver[5] (Fig. 2). They can easily be distinguished from smooth microsomes by their lack of bands h, j, o, and p and by their large and characteristic major band m. Both rat serum albumin and very low density lipoprotein probably contribute to band m in Golgi.[5]

UDP-galactose:*N*-acetylglucosamine, galactosyltransferase activity is determined by a modification of the method of Babad and Hassid[10] as follows:

[8] K. Takayama, D. H. MacLennan, A. Tzagoloff, and C. D. Stoner, *Arch. Biochem. Biophys.* **114**, 223 (1966).
[9] W. L. Zahler, B. Fleischer, and S. Fleischer, *Biochim. Biophys. Acta* **203**, 283 (1970).
[10] H. Babad and W. Z. Hassid, this series, Vol. 8, p. 346.

Reagents

Sodium cacodylate, 0.2 *M*, adjusted to pH 6.5 with HCl
MnCl$_2$, 1 *M*
Mercaptoethanol, 1 *M*
Triton X-100 (Rohm & Haas), 10% w/v
N-Acetylglucosamine, 1 *M*
UDP-galactose, 10 m*M*, containing UDP-C^{14} galactose (New England Nuclear), 1 μCi/μmole
EDTA, 0.3 *M*, neutralized to pH 7.4 with NaOH

Procedure. Mix 0.02 ml of sodium cacodylate, 0.003 ml of MnCl$_2$, 0.003 ml of mercaptoethanol, and 0.005 ml of Triton X-100. Add 0.003 ml of *N*-acetylglucosamine, 0.02 ml of enzyme (20–100 μg of protein), and start the reaction with 0.015 ml of UDP-galactose. Incubate at 37° for 60 minutes. Stop the reaction with 0.017 ml of EDTA, and place tubes on ice. The mixture is passed through a column of Dowex 2 X8, 200–400 mesh in the Cl$^-$ form, 0.5 cm in diameter and 2 cm high; the columns have been previously washed in distilled water. Unreacted UDP-galactose remains bound to the column while galactose which has been transferred to *N*-acetylglucosamine to form lactosamine, as well as free galactose, is washed directly onto a tared planchet with two washes of 0.5 ml of distilled water. The planchets are dried with an infrared lamp and radioactivity determined with a Nuclear-Chicago gas-flow counter Model No. 4312 equipped with a micromil window. The dried planchets are weighed and corrections are made for self-absorption when necessary. For each assay, a control tube is run in which all ingredients are present except *N*-acetylglucosamine. This value represents galactose released. The difference between the tubes in which acceptor is present or absent represents transferase activity. For each set of assays a control tube without enzyme is included to correct for nonenzymatic hydrolysis or for contamination of the substrate with free radioactive sugar. In purified Golgi preparations, the ratio of galactose hydrolyzed to that transferred is usually less than 20%.

Contamination of the Golgi preparations with other organelles can best be estimated using the following assays: for endoplasmic reticulum, Rotenone-insensitive NADH–cytochrome *c* reductase[11] or glucose-6-phosphatase[7]; for plasma membrane, Mg^{2+}-stimulated ATPase[11] or galactosidase[7]; for mitochondria, succinate cytochrome *c* reductase[11]; and for

---

[11] S. Fleischer and B. Fleischer, this series, Vol. 10, p. 406.

TABLE I

Enzyme Profiles of Golgi Fractions Prepared by the Differential Centrifugation Method (A) and by the Single Step-Gradient Method (B)[a]

Assay	PM[b]	Golgi[c]			Microsomes[b]		Homogenate
		A	B-1	B-2	Smooth	Rough	
Glucose-6-phosphatase	0.036	0.007	0.019	0.035	0.21	0.15	0.085
NADH–cytochrome c reductase[d]	0.244	0.008	0.175	0.286	2.2	1.4	0.358
Succinate–cytochrome c reductase[d]	0.039	0.002	0.006	0.009	0.006	0.004	0.130
Acid phosphatase	—	0.054	0.060	0.083	0.035	0.010	0.031
Aryl sulfatase	—	0.014	0.027	—	0.005	0.009	—
ATPase[d]	1.4	0.154	0.153	0.253	0.088	0.065	0.191
Galactosidase	300	61	41	55	36	6.1	—
Galactosyltransferase	0.0	190	564	413	19	0.4	—
μg P/mg protein	22.0	23.3	23.9	24.8	28.3	38.5	6.7
Yield (mg protein per gram liver)	—	0.05	0.18	0.21	—	—	—

[a] Enzyme activities expressed as micromoles per minute per milligram of protein except for galactosidase and galactosyltransferase, which are expressed as nanomoles per hour per milligram of protein.

[b] Plasma membranes were prepared by a modification (see S. Fleischer and M. Kervina, this series Vol. 32 [2]) of the method of Y. Stein, C. Widnell, and O. Stein [J. Cell Biol. 39, 185 (1968)]; and microsomes by a modification [B. Fleischer and S. Fleischer, Biochem. Biophys. Acta 183, 265 (1969)] of the method of G. Dallner [Acta Pathol. Microbiol. Scand. Suppl. 166, 1 (1963)].

[c] B-1 refers to the fraction obtained from the 29/33% sucrose interface, and B-2 refers to the fraction obtained at the 33/36% sucrose interface.

[d] At 32°; all others at 37°.

lysosomes, acid phosphatase[12] or aryl sulfatase.[13] In order to evaluate the contamination, however, it is also necessary to have purified and characterized preparations of the other organelles on hand so that the enzymatic activity of these may be compared directly with the fraction under study. For example, an estimate of the percent contamination of the Golgi fraction with endoplasmic reticulum on a protein basis can be obtained from the specific activity of NADH–cytochrome c reductase in the Golgi fraction divided by the specific activity of NADH–cytochrome c reductase in purified smooth endoplasmic reticulum multiplied by 100. Thus, the contamination of Golgi fraction B with smooth endoplasmic reticulum (Table I) would be $(0.175/2.2) \times 100 = 8\%$. Table I summarizes the

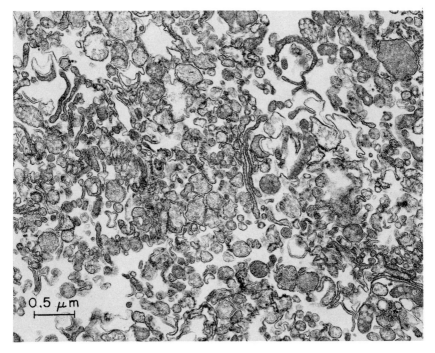

0.5 μm

FIG. 3A. Electron micrographs of Golgi fraction isolated by method A. Samples were fixed in suspension in 2.5% glutaraldehyde in 0.1 M cacodylate buffer, pH 7.4, overnight in the refrigerator. After centrifugation at 10,000 rpm for 10 minutes in a Spinco Ti 50 rotor they were washed free of glutaraldehyde, treated with 1% osmium tetroxide, dehydrated, embedded, and sectioned as described previously [B. Fleischer and S. Fleischer, *Biochim. Biophys. Acta* **219**, 301 (1970)]. ×35,000.

[12] C. B. de Duve, B. C. Pressman, R. Gianetto, R. Wattiaux, and F. Appelmans, *Biochem. J.* **60**, 604 (1955).

[13] A. B. Roy, *Biochem. J.* **53**, 12 (1953).

enzymatic characteristics of a number of Golgi fractions prepared as described under methods A and B.

The Golgi-rich fractions obtained by both methods A and B have characteristic morphologies, as illustrated in Fig. 3. Prepared by method A, the Golgi fraction consists of a heterogeneous collection of sacs, often with attached tubules which appear in cross sections as associated small vesicles[5] (Fig. 3A). Prepared by method B, the Golgi appear most often as flattened cisternae filled with particles (Fig. 3B).

When a membrane fraction is prepared from the Golgi vesicles using the Parr bomb technique (Table II), about 37% of the protein and 30% of the phosphorus becomes nonsedimentable. Galactosyltransferase remains bound to the membrane whereas serum albumin, a secretory protein presumably contained in the lumen of the Golgi vesicles,[14] is released to

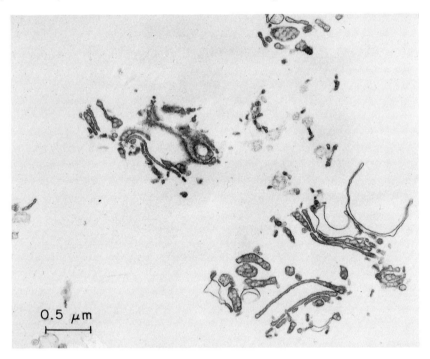

0.5 μm

Fig. 3B. Electron micrographs of Golgi fraction isolated by method B. Samples obtained directly from the gradient were fixed in suspension by the addition of an equal volume of 5% glutaraldehyde in 0.1 M cacodylate buffer, pH 7.4, overnight in the refrigerator; they were then processed for electron microscopy as described for Fig. 3A. ×35,000.

[14] T. Peters, Jr., B. Fleischer, and S. Fleischer, *J. Biol. Chem.* **246**, 240 (1971).

TABLE II

RECOVERY OF GOLGI MEMBRANES AFTER DISRUPTION OF GOLGI VESICLES
WITH A PARR BOMB

Component	Orig.	Pellet	Supernatant	Recovery (%)
Total protein (mg)	25	13.7	7.5	85
μg P/mg protein	24.7	34.0	14.4	93
Galactosyltransferase[a]	594	1059	157	105
Rat serum albumin[a] (μg)	0.90	0.27	0.65	102

[a] Galactosyltransferase is expressed as nanomoles per hour per milligram of protein at 37°. Rat serum albumin was determined by precipitation with specific antibody [T. Peters, Jr., B. Fleischer, and S. Fleischer, *J. Biol. Chem.* **246**, 240 (1971)].

the extent of 72%. The phosphorus released is probably a measure of the amount of serum lipoproteins also present in the lumen of the Golgi vesicles.[15]

Comments

The methods described for the preparation of Golgi vesicles are designed to yield large amounts of this fraction for further analysis. Both methods have also been adapted for the fractionation of a portion of a rat liver (for method A see ref. 16; for method B, ref. 5). Method A gives lower yields of Golgi but is the method of choice when other organelles must be isolated from the same homogenate. Method B on the other hand gives excellent yields but tends to give higher contamination with smooth endoplasmic reticulum (10–15% for B as compared to less than 5% in A).

The preparations are very reproducible; only two difficulties have been encountered. Erratic separations occur if (a) rats are starved before use and (b) gradient sucrose solutions are not prepared less than 1 day before use. It should also be noted that additions of any salts, buffers, or EDTA to the gradient sucrose solutions makes the separations impossible.

The best morphological preservation of the Golgi structure has been obtained using method B. It is important, however, that fixation of the sample be carried out on the fractions as they are isolated from the zonal gradient, before dilution and centrifugation to recover the fractions. The procedure we have found to be best is to take 5 ml of the gradient fraction (approximately 0.1 mg of protein per milliliter) and add 0.5 ml of 25% glutaraldehyde in 0.1 M cacodylate, pH 7.4. After it has stood overnight in the refrigerator, the sample is diluted with 2.5 ml 0.25 M sucrose and

[15] R. L. Hamilton, D. M. Regen, M. E. Gray, and V. S. LeQuire, *Lab. Invest.* **16**, 305 (1967).
[16] S. Fleischer and M. Kervina, this series, Volume 32 [2].

centrifuged at 10,000 rpm (6037 g_{av}) for 5–10 minutes, and the supernatant is discarded. The pellet is washed with 0.25 M sucrose containing 0.1 M cacodylate by overlaying the pellet with the solution, letting the mixture stand for 10 minutes in the cold and decanting the wash. This is repeated 4–5 times over a period of an hour in the cold. The pellet is then treated with 1% osmium tetroxide, dehydrated, and sectioned as described previously.[5]

For storage of the samples for future analysis, it is recommended that they be quick-frozen in liquid nitrogen and stored at $-70°$ in a low temperature freezer.

It may be noted that the specific activity of galactosyltransferase for Golgi vesicles obtained by the one-step procedure, although variable, is consistently higher than that obtained by the zonal procedure. Yet the level of known contaminants, as judged by assay of known marker enzymes, is about the same for the two preparations. One possible explanation is that the Golgi apparatus is somewhat heterogeneous in its content of transferase activity and that the two isolation procedures used select slightly different populations of particles derived from the Golgi. There is considerable evidence of heterogeneity in the Golgi membranes *in situ* in staining for complex carbohydrates,[17] in ordinary osmium staining,[18] and in heavy osmication.[19] We have observed that Golgi vesicles isolated by the zonal procedure have a higher content of serum albumin than those prepared by the flotation procedure,[14] which lends support to the idea of a somewhat heterogeneous population of vesicles.

D. J. Morré has described a different isolation procedure for obtaining Golgi rich fractions from rat liver.[20] His method of preparation differs from that presented here in that his fraction is sedimented from the homogenate at very low g forces (2000 g for 30 minutes). This is probably due to the high concentration of divalent cations used in the homogenizing medium (5 mM MgCl$_2$). It has been shown previously by Dallner and Nilsson[21] that 10 mM Mg^{2+} causes extensive aggregation of isolated smooth microsome fractions from rat liver and that this aggregation is essentially irreversible. Added to the initial homogenate, however, the effect of Mg^{2+} appears to be more specific. Galactosyltransferase was shown to be present in this type of Golgi preparation at about the same level as we have found in our preparations. Further, Schachter *et al.*[22] have recently shown that such

[17] A. Rambourg, W. Hernandez, and C. P. Leblond, *J. Cell Biol.* **40**, 395 (1969).
[18] S. N. Grove, C. E. Bracker, and D. J. Morré, *Science* **161**, 171 (1968).
[19] D. S. Friend and M. J. Murray, *Amer. J. Anat.* **117**, 135 (1965).
[20] D. J. Morré, this series, Vol. 22, p. 130.
[21] G. Dallner and R. Nilsson, *J. Cell Biol.* **31**, 181 (1966).
[22] H. Schachter, I. Jabbal, R. L. Hudgin, L. Pinteric, E. J. McGuire, and S. Roseman, *J. Biol. Chem.* **245**, 1090 (1970).

preparations contain other sugar transferases, such as *N*-acetylglucosaminyl and sialyltransferase, and suggested that the Golgi apparatus in liver is equipped with a multienzyme system for the synthesis of the terminal sugars in glycoproteins.

Acknowledgments

The author would like to thank Mr. Akitsugu Saito for the electron micrographs used in this paper and Dr. Theodore Peters, Jr., of The Mary Imogene Bassett Hospital, Cooperstown, New York, for the determination of rat serum albumin in the Golgi fractions. This work was supported in part by Grant AM-14632 of the U.S. Public Health Service.

[18] Isolation of Rough and Smooth Microsomes—General

By GUSTAV DALLNER

Definition. The isolated total microsomal fraction in most tissues consists of two major components: rough and smooth microsomes, the former having attached ribosomes on the outer surface. The definition of this fraction is based not on its content but on the result of the centrifugation procedure employed to isolate it. By this definition, the microsomes are those elements of a tissue homogenate that can be sedimented by ultracentrifugation from the mitochondrial supernatant. Whereas the rough microsomes originate from the rough-surfaced endoplasmic reticulum (ER), the smooth microsomes, depending on the tissue and isolation conditions, vary in origin. Smooth microsomes isolated from the liver derive either almost exclusively or to a great extent from the smooth ER[1]; when isolated from the pancreas[2] and testis,[3] the same fraction contains mainly Golgi elements; and at subfractionation of kidney,[4] Ehrlich ascites tumor cells,[5] or fibrocytes,[6] parts of the plasma membranes are recovered as smooth microsomes. In most of the other organs, the microsomes may still contain additional elements.

Another complication facing those wishing to isolate rough and smooth microsomes is the variation in size and density even among particles deriving from the same origin. Unlike many other cellular organelles, none of the lamellar and cisternal structures of the rough and smooth ER can be

---

[1] G. Dallner and L. Ernster, *J. Histochem. Cytochem.* **16,** 611 (1968).

[2] J. D. Jamieson and G. E. Palade, *J. Cell Biol.* **34,** 597 (1967).

[3] E. L. Kuff and A. J. Dalton, *in* "Subcellular Particles" (T. Hayashi, ed.), p. 114. Ronald Press, New York, 1959.

[4] J. Rostgaard and O. J. Møller, *Exp. Cell Res.* **68,** 356 (1971).

[5] D. F. H. Wallach and V. B. Kamat, this series, Vol. 8, p. 164.

[6] C. G. Gahmberg and K. Simons, *Acta Pathol. Microbiol. Scand. B,* **78,** 176 (1970).

separated in the intact form.[7] The new morphological structure appearing after homogenization acquires a spectrum of physical properties which to a great extent mirror the conditions employed. There is a consequence of this which is unfortunately not always recognized: the method used for the separation of rough and smooth microsomes in one tissue almost invariably cannot be used for another tissue.[8] Indeed, when a procedure found successful for one type of animal is used on another type of animal, even in the case of the same organ, radical modifications are often necessary.

Recovery and Purity. The clash between high yield and absence of contamination is greatly accentuated during the isolation of rough and smooth microsomes. These two basic requirements of all subfractionation work cannot be reconciled at present. In simple terms, this clash means that we must choose between high quantitative recovery plus high impurity, and partial recovery plus high purity. Since extensive contamination of a subfraction runs counter to the very purpose of the procedure, the investigator in most cases must choose the second alternative. A high recovery would be desirable, certainly, but at present such an objective belongs in the realm of wishful thinking. The basic condition for a high yield of both rough and smooth microsomes is effective homogenization. The extensive homogenization of liver, for example, results in the breakage of plasma membranes[9] and Golgi cisternae,[10] it releases outer mitochondrial membranes,[11] and ruptures lysosomes,[12] all of which contribute to a high degree of impurity of smooth and in some instances even of rough microsomes. Furthermore, connective tissue cells and cells of erythropoiesis, particularly in the newborn—cell types that possess plasma membranes more resistant to shearing forces—also break up and contribute to a false heterogeneity.[13,14] Prolonged homogenization leads to an increased release of bound ribosomes, modified vesicle structure and composition, and also interference with enzyme activities. It does not seem, at least in the case of the liver, that the ER portion lost in the first centrifugation represents a specific and unique part of the system. It is very probable that the loss is caused by the presence of unbroken or incompletely broken cells of random origin.

[7] G. E. Palade and P. Siekevitz, *J. Biophys. Biochem. Cytol.* **2**, 671 (1956).

[8] Y. S. Kim, J. Perdomo, and J. Nordberg, *J. Biol. Chem.* **246**, 5466 (1971).

[9] R. Coleman, R. H. Michell, J. B. Finean, and J. N. Hawthorne, *Biochim. Biophys. Acta* **135**, 573 (1967).

[10] B. Fleischer, S. Fleischer, and H. Ozawa, *J. Cell Biol.* **43**, 59 (1969).

[11] W. E. Criss, *J. Biol. Chem.* **245**, 6352 (1970).

[12] R. L. Deter and C. de Duve, *J. Cell Biol.* **33**, 437 (1967).

[13] I. T. Oliver, W. F. C. Blumer, and I. J. Witham, *Comp. Biochem. Physiol.* **10**, 33 (1963).

[14] G. Dallner, P. Siekevitz, and G. E. Palade, *J. Cell Biol.* **30**, 73 (1966).

Consequently, the isolated rough and smooth microsomes, at least in qualitative respects, represent the true picture.

Medium. Practically all fractionation procedures are performed in sucrose solutions.[15] For homogenization, 0.25 M or 0.44 M solutions are used in most cases; 0.88 M sucrose is better able to keep an elongated, tubulelike structure among the rough microsomes,[7] but this increases centrifugation time, which is a drawback for the separation procedure itself. Under the conditions of ultracentrifugation, the liver microsomal vesicle is completely permeable to sucrose,[16] a property that seems to be shared by peroxisomes alone.[17] Thanks to this property, sucrose has a less damaging effect on microsomal vesicles than many other suspension media. The vesicles do not display any osmotic response, and their high equilibrium density—mainly provided by the solid membrane material plus hydration water—is advantageous in the separation procedure. High polymer dextran, glycogen, and Ficoll, being nonpermeable solutes, lend low density to both rough and smooth microsomes to such an extent that their isolation in a discontinuous gradient is more difficult.

Influence of Surface Charge. The microsomal membranes possess a high net negative surface charge density, rough microsomes being more negative than the smooth variety.[18] This property can be utilized for effective and rapid separation, and it is also important for a number of other reasons. During the isolation procedure, a significant amount of cytoplasmic basic proteins becomes attached to the particle surfaces, contributing in this way about 30% of the total microsomal protein.[19] The extremely common finding that microsomal vesicles aggregate during the isolation procedure, at times as early as during homogenization, is a very unfortunate consequence of the surface charge, invariably leading to incomplete separation. Purity of all glassware as well as the absence of cations in the distilled water are of basic importance in the subfractionation of microsomes. Foreign charged substances ingested with food or drinking water accumulate in the liver, particularly on the surface of the ER. A well-documented finding is the influence of detergents used, for example, for cleaning cages.

---

[15] The major disadvantage of sucrose is the steep rise in viscosity with increasing concentration and the very sensitive relationship between the viscosity of sucrose and temperature. To give an example: an increase in temperature from 5° to 10° (which frequently occurs during long centrifugations) will decrease the viscosity by 20% and consequently increase the sedimentation velocity by 20%.

[16] R. Nilsson, E. Peterson, and G. Dallner, *J. Cell Biol.* **56**, 762 (1973).

[17] F. Leighton, B. Poole, H. Beaufay, P. Baudhuin, J. W. Coffey, S. Fowler and C. de Duve, *J. Cell Biol.* **37**, 482 (1968).

[18] G. Dallner and A. Azzi, *Biochim. Biophys. Acta* **255**, 589 (1972).

[19] G. Dallner, *in* "Proceedings of the 4th International Congress of Pharmacology" (R. Eigenmann, ed.), Vol. 4, p. 70. Schwabe, Basel, 1970.

It is no exaggeration to say that all these factors are decisive for the ultimate physicochemical properties of the microsomal membranes and certainly constitute one of the explanations why different laboratories obtain different results. It has also been suggested that the inner surfaces of several membranes have a different structure and consequently a different charge,[20] which may contribute toward a selective change of particle density when divalent cations are present in low amounts in a polymer gradient. All these considerations indicate how important it is not to use buffers in the gradient medium without controlling side effects. In most cases, the proteins present during subfractionation have a large enough buffering capacity to keep the pH around 7.

Stability. Microsomes have a great tendency to aggregate, often even in the absence of cations. This statement is valid for both subfractions, but smooth microsomes are decidedly more sensitive and unstable.[21]

The stability involving both an existence in nonaggregated form and an uninfluenced enzyme activity requires either high protein or high sucrose concentration or, in the most ideal case, both. It is not advisable to use a particle concentration of less than 5 mg of protein per milliliter or a sucrose concentration of less than 0.25 M. If it is difficult to obtain a high protein concentration, stabilization can be achieved by adding protecting colloids, such as albumin,[22] spermine, or spermidine.[23] Stability may be increased by using a high sucrose concentration (30–40%), but this is not often practicable. Isolated subfractions tolerate sedimentation badly and aggregate in many cases. If smooth microsomes are pelleted, they always aggregate independently of the protein or sucrose concentration.

If isolated subfractions must be concentrated and aggregation is deleterious (for example, in the case of further subfractionation) a suitable concentration may be obtained by layering the suspension above a 2–4 ml sucrose cushion ($\sim$1.6 M) in an angle-head or swinging-bucket rotor and collecting the particles at the interface after short centrifugation ($\sim$30 minutes) at 60,000–80,000 g. A high sucrose concentration and high centrifugal force lead to aggregation of the particles at the interface and are no more advantageous than pelleting.

The simplest way to test for aggregation is to filter microsomes (4–5 mg of protein per milliliter) by suction pump through a filter of 0.45 μm (Millipore Corp., Bedford, Massachusetts) followed by an additional volume of sucrose for washing.[22] In the nonaggregated state, about 90% of both rough and smooth microsomes appear in the filtrate.

[20] T. L. Steck and D. F. H. Wallach, *Methods Cancer Res.* **5**, 93 (1970).
[21] H. Glaumann and G. Dallner, *J. Cell Biol.* **47**, 34 (1970).
[22] G. Dallner and R. Nilsson, *J. Cell Biol.* **31**, 181 (1966).
[23] P. H. Jellinck and G. Perry, *Biochim. Biophys. Acta* **137**, 367 (1967).

In the isolated state, microsomes can be damaged by their own enzymes in the membrane (e.g., PLPase A,[24,25] enzymatic lipid peroxidation[26]) or in the lumen (e.g., trypsin, lipase[27]). Specific inhibitors may counteract this: absence of cations and low temperature (PLPase A), small amounts of antioxidants or chelating agents (lipid peroxidation), trypsin inhibitor (trypsin), Zn^{2+} or Ag^+ (pancreatic lipase), etc.

Centrifugal Force and Rotors. It is very tempting to use high centrifugal forces to sediment microsomes, since their small size often requires a long centrifugation time, but this should be avoided. Microsomal vesicles in the usual concentration and in 0.25 M sucrose centrifuged in an angle-head rotor (26°) in the Spinco-Beckman ultracentrifuge at more than 40,000 rpm show signs of damage such as increased permeability and an oval rather than round form in the electron microscopical picture. Naturally, much higher speeds may be employed when the particle is moving in a high concentration of medium and does not hit the bottom of the tube. In most gradient centrifugation, swinging-bucket rotors are required. In the case of microsomal fractions, particularly when the sedimentation velocity of rough microsomes is increased by the addition of monovalent cations, angle-head rotors with a large tube angle of inclination to the axis of rotation may be used for the discontinuous sucrose gradient.[28] Actually, this can be advantageous, since the convections set up during centrifugation decrease centrifugation time without causing mixing of the two main components in the surface layer and in the pellet.

Isolation of Rough and Smooth Microsomes

Rat Liver

Rats weighing around 180–200 g give the most reproducible preparations. Starvation for 20 hours before sacrifice decreases the liver glycogen content to about 1 mg/g (wet weight). Starvation induces increased glucose-6-phosphatase (G6Pase) activity,[29] but in nonfasted rats the loss of both rough and smooth microsomes during the preliminary centrifugation is increased by 30–40%. Constituents from broken erythrocytes (e.g., hemoglobin) are easily adsorbed on the surface of microsomes,[30] and, if this is undesirable, perfusion with 0.25 M cation-free sucrose may be

[24] P. Bjørnstad, *Biochim. Biophys. Acta* 116, 500 (1966).
[25] M. Waite and L. L. M. van Deenen, *Biochim. Biophys. Acta* 137, 498 (1967).
[26] P. Hochstein and L. Ernster, *Cell. Inj., Ciba Found. Symp., 1963* p. 123 (1964).
[27] J. Meldolesi, J. D. Jamieson, and G. E. Palade, *J. Cell Biol.* 49, 130 (1971).
[28] A. Bergstrand and G. Dallner, *Anal. Biochem.* 29, 351 (1969).
[29] W. J. Arion and R. C. Nordlie, *Biochem. Biophys. Res. Commun.* 20, 606 (1965).
[30] M. L. Petermann and A. Pavlovec, *J. Biol. Chem.* 236, 3235 (1961).

performed. Livers are minced into small pieces with a pair of scissors and rinsed several times with cold sucrose solution. Homogenization is performed in the commercially available Teflon homogenizer in a volume that roughly corresponds to the volume of liver. In the motor-driven homogenizer, run at a speed not exceeding 400–500 rpm, four strokes are used. Complete homogeneity of the suspension is not obtained and is, in fact, not to be desired.

Cation-Containing Sucrose Gradient

Principle. Rough microsomes display a high binding affinity for Cs^+ in comparison with smooth microsomes.[31] In the presence of a limited amount of Cs^+, the net negative surface charge density selective for rough membranes decreases, with consequent abolishment of repelling forces. Extensive aggregation of rough microsomes follows,[32] and the mean radius of the sedimenting particles is at least doubled or trebled; this increases the sedimentation velocity by 4–9-fold.[22]

Beckman-Spinco Centrifuge. Practical. 40.2 ROTOR.[28] Homogenization of the liver is performed in 0.25 M sucrose followed by dilution to a tissue concentration of 25% (w/v). The homogenate is centrifuged at 10,000 g for 20 minutes, either in an ultracentrifuge or in a low-speed cooled centrifuge. The supernatant (10,000 g supernatant) is decanted to the last drop into a graded cylinder and is not diluted. In order to obtain a final CsCl concentration of 15 mM, the suspension is mixed with 0.15 ml of 1 M CsCl per 10 ml; 4.5 ml of CsCl-containing "10,000 g supernatant" is layered over 2 ml of 1.3 M sucrose–15 mM CsCl.[33,34] After centrifugation in a 40.2 rotor (tube angle 40°) at 102,000 g for 120 minutes,[35] the clear reddish upper phase can be removed either with a pipette provided with a rubber aspirator or with a water pump. This portion is discarded. The fluffy double layer (upper: reddish, lower: light brown) at and under the gradient boundary is collected.[36] This fraction, *the total smooth microsomes,* contains some soluble cytoplasmic protein. If no further subfractionation is planned and aggregation has no importance, the fraction can

---

[31] G. Dallner, L. Ernster, and A. Azzi, *Chem.-Biol. Interactions* 3, 254 (1971).

[32] G. Dallner, *Acta Pathol. Microbiol. Scand.,* Suppl. 166 (1963).

[33] Layering may be done either with a pipette or with a peristaltic pump.

[34] For standardization, all sucrose solutions should be made up first by warming followed by cooling in an ice water bath with subsequent volume adjustment with cool water.

[35] Starting from a 20% homogenate, 90 minutes of centrifugation is sufficient.

[36] The intermediate layer may be removed with an ordinary or Pasteur pipette, but ideal is a syringe with a long needle whose tip has been bent into an L shape. The syringe is clamped above the tube, which rests on a variable stand that can raise it to the desired level.

be recentrifuged at 105,000 g for 90 minutes in order to obtain a pellet. The 1.3 M sucrose layer above the pellet is removed with a pipette or a syringe, except for the last few drops, which contain the fluffy layer. This layer with 15% of the total microsomal protein is discarded except when rough III microsomes are prepared (see below). The pellet with the few drops immediately covering it is resuspended by hand homogenization after the addition of water (*total rough microsomes*). The yield is ~12 and 5 mg of protein per gram of liver for rough and smooth microsomes, respectively.

SW 39 AND SW 40 ROTORS. Swinging-bucket rotors giving high *g* values can also be utilized for subfractionation.[14] The decrease of convection in these rotors, however, increases centrifugation time in spite of the high centrifugal forces employed. Using a SW 39 rotor, 3 ml of Cs⁺-containing 10,000 g supernatant is layered above 2 ml of 1.3 M sucrose–15 mM CsCl; after 3 hours of centrifugation with the highest speed for this rotor (39,000 rpm), the two microsomal subfractions are separated. In the case of SW 40 rotor, 9 ml is layered above a 4-ml cushion. Centrifugation for 3 hours is still required for separation of rough and smooth microsomes in spite of the high centrifugal force employed (198,000 g).

40 ROTOR. Low tube angle to the axis of rotation increases convections to an extent incompatible with the successful separation of microsomal subfractions. The 40 rotor of the Spinco has a tube angle of 26°, but if 7 ml of Cs⁺-containing 10,000 g supernatant are layered above a 4-ml Cs⁺-containing cushion, after 140 minutes of centrifugation rough and smooth microsomes are separated. The result of this, however, probably because of side wall impaction, is that about 15% of smooth microsomes are recovered in the pellet.[37]

Christ Omega and International B-60 Ultracentrifuge. Angle-head rotors with large tube angles (34° in the 60 rotor of Christ Omega and 35° in the A-321 rotor of International B-60 ultracentrifuge) can be spun at 60,000 rpm, which in both cases provides a force of 250,000 g. The conditions created in these rotors are close to ideal for the separation of rough and smooth microsomes by the method described above.[32] In both centrifuges, 3.5 ml of 1.3 M sucrose–15 mM CsCl in the lower layer and 10,000 g supernatant with 15 mM CsCl in the full upper layer gives after 60 minutes of centrifugation at 60,000 rpm in both centrifuges an almost complete separation of rough and smooth microsomes with only limited amounts of particulate material in the 1.3 M sucrose phase.

Separation of Rough I, II, and III Microsomes. Rough vesicles with

[37] T. E. Gram, L. A. Rogers, and J. R. Fouts, *J. Pharmacol. Exp. Ther.* **155**, 479 (1967).

few ribosomes remain in an intermediate position on a Cs^+-containing discontinuous sucrose gradient and can be separated with a simple one-step centrifugation procedure.[38] Liver from starved rat is homogenized in 0.44 M sucrose at a tissue concentration of 20% (w/v). After centrifugation at 10,000 g for 20 minutes, the supernatant is mixed with 0.15 ml of 1 M CsCl/10 ml; 3.5 ml of the suspension is layered over 3 ml of 1.3 M sucrose–15 mM CsCl and centrifuged in a 40.2 rotor (Spinco) at 102,000 g for 90 minutes. The clear upper phase and the smooth microsomes at the gradient boundary are removed, and the 1.3 M sucrose layer down to the upper edge of the pellet is sucked off (*rough III microsomes*). The remaining sucrose solution above the pellet is taken up with a Pasteur pipette (*rough II microsomes*), and the pellet is suspended in 0.25 M sucrose (*rough I microsomes*). The two former subfractions may be pelleted after dilution with water or sucrose.

Separation of Smooth I and II Microsomes.[28] After centrifugation in the 40.2 rotor, total smooth microsomes from two tubes are combined, diluted dropwise with distilled water to 4.5 ml, mixed with 1 M $MgCl_2$ to give a final concentration of 7 mM, and layered over 2 ml of 1.15 M sucrose–7 mM $MgCl_2$. After centrifugation at 102,000 g for 45 minutes in the 40.2 rotor, the clear upper phase is removed and discarded. The fluffy layer at the gradient boundary is collected, diluted with 0.25 M sucrose, and recentrifuged at 105,000 g for 90 minutes (*smooth II microsomes*). The pellet formed on the Mg^{2+}–sucrose gradient represents the *smooth I microsomes.*

Discontinuous Sucrose Gradient

Principle. The majority of rough microsomes possess a higher equilibrium density than smooth microsomes and on a suitably designed discontinuous sucrose gradient they will sediment, while the smooth microsomes remain in the vicinity of the interface.[39] The attainment of isopycnic equilibrium is not necessary; in fact, it is not even possible because of the type of gradient and the continuous density spectrum of the microsomal vesicles.

Practical. The following procedure[40] is a modification of the original one,[39] designed to abolish aggregation of the vesicles in the subfractions. Homogenization is performed in 0.44 M sucrose (2 g/10 ml). After centrifugation at 10,000 g for 20 minutes, the supernatant is diluted with 0.44 M sucrose to restore the original volume. Of this suspension, 8 ml is layered over 3 ml of 1.3 M sucrose and centrifuged at 105,000 g for 7

[38] L. C. Eriksson and G. Dallner, *FEBS (Fed. Eur. Biochem. Soc.) Lett.* **19**, 163 (1971).

[39] J. Rothschild, *Biochem. Soc. Symp.* **22**, 4 (1963).

[40] G. Dallner, A. Bergstrand, and R. Nilsson, *J. Cell Biol.* **38**, 257 (1968).

hours 40 minutes in a 40 rotor (Spinco). The upper 0.44 M sucrose phase is sucked off. The milky layer localized in the upper part of the 1.3 M sucrose layer, which is in the process of sedimenting down, is removed. If the fraction is not used for further subfractionation, it may be recentrifuged after dilution with 0.25 M sucrose (*smooth microsomes*). The 1.3 M sucrose layer is removed and discarded, but the fluffy layer just above the pellet is left behind and rehomogenized by hand together with the pellet, after the addition of a few drops of water (*rough microsomes*). The 1.3 M sucrose layer, which is discarded, contains a mixture of rough and smooth microsomes ($\sim$10–15% of both). The yield after recentrifugation is $\sim$10 and 4 mg of protein per gram of liver for rough and smooth microsomes, respectively.

Mouse, Rabbit, and Guinea Pig Liver

The isolation of rough, smooth I, and II microsomes from mouse liver, as well as rough and total smooth microsomes from rabbit liver, is carried out in the same way as described for rat liver. In the case of the rabbit, however, the density gradient with Mg^{2+} for subfractionation of smooth microsomes is composed of 0.25 M/1.00 M sucrose rather than 0.25 M/1.15 M sucrose. Because of the frequent aggregation of smooth microsomes from guinea pig liver, the soluble supernatant must be removed during the isolation procedure by a three-layered gradient. Over 2 ml of 1.3 M sucrose–15 mM Cs$^+$, and 1 ml of 0.5 M sucrose–15 mM Cs$^+$, 3.5 ml of 15 mM CsCl containing 10,000 g supernatant is layered and centrifuged at 102,000 g for 90 minutes in a 40.2 rotor. The double-layered total smooth microsomes in the 0.5 M sucrose may be further subfractionated by centrifugation after dilution with 0.25 M sucrose and layering over 1.00 M sucrose, both containing 7 mM $MgCl_2$.

Guinea Pig Pancreas

Continuous Gradient[41,42]

Pancreas from 20-hour-starved guinea pig is passed through a stainless steel tissue press and homogenized in 0.3 M sucrose (1 g/10 ml) by 30 strokes in a hand homogenizer fitted with a rubber pestle. After filtration through 110-mesh nylon cloth, the homogenate is centrifuged at 11,000 g for 15 minutes in a Spinco 40.3 rotor. The upper two-thirds of the supernatant is removed, centrifuged at 115,000 g for 60 minutes, and the pellet is resuspended in 0.3 M sucrose to a concentration of $\sim$6 mg of protein per milliliter. Of this suspension, 0.2–0.3 ml is layered in a Spinco SW 39

[41] J. D. Jamieson and G. E. Palade, *J. Cell Biol.* **34**, 577 (1967).
[42] J. Meldolesi, J. D. Jamieson, and G. E. Palade, *J. Cell Biol.* **49**, 109 (1971).

tube on a 5-ml linear sucrose density gradient (1.04 to 2.00 M) and centrifuged for 5 hours at 115,000 g. This results in the appearance of 3 bands on the gradient: the upper band close to the top represents smooth microsomes; the band in the middle contains both rough and smooth microsomes as well as free ribosomes; and the third band, in the lower part, is made up of rough microsomes.

Cation-Containing Sucrose Gradient[43]

Cleaned pancrease from 48-hour-starved guinea pig is homogenized in 0.44 M sucrose (2 g/10 ml) with a Teflon–glass homogenizer (400–500 rpm, 10 strokes). Twice during homogenization the adherent tissue clumps are removed from the pestle. After centrifugation at 15,000 g for 20 minutes, the supernatant is supplied with CsCl (final concentration 10 mM). Of this suspension, 3.5 ml is placed on a double layer consisting of 2 ml 1.05 M sucrose–15 mM CsCl and 1 ml of 0.6 M sucrose–10 mM CsCl. Centrifugation is performed in a 40.2 rotor (Spinco) at 102,000 g for 120 minutes. The double layer in the 0.6 M sucrose represents smooth microsomes (in this case deriving from the Golgi complex). The pellet together with the adherent fluffy layer consists of rough microsomes.

Remarks

Choice of Method. The simplicity of a method is often decisive, and it is understandable that many laboratories still employ differential centrifugation techniques for the isolation of the two main subfractions[44] in spite of the apparent cross contamination. Cation-containing gradients are effective and also fulfill one of the most important requirements in work on membrane structure: rapid, time-saving separation. On the other hand, they require a cation-free laboratory environment as well as relatively high quality laboratory animals, which are not readily available everywhere. Because of the heterogeneity of both rough and smooth microsomes, further subfractionation employing gradient technique is of great importance in studying microsomal membranes. Subfractionation of total microsomes by continuous gradients in general gives poor results, which may be explained by electrostatic interaction among differently charged particles. On the other hand, appreciable heterogeneity can be demonstrated by using isolated rough or smooth microsomes as starting fractions. Rough microsomes isolated in the presence of monovalent cations are in aggregated form and therefore cannot be used for further subfractionation.

Chemical Markers. The isolated microsomal particle contains not only membranous proteins but also adsorbed (~30% of the total) and

---

[43] G. Dallner and H. Glaumann, *Abstr. Int. Congr. Biochem., 7th* **5**, 933 (1967).
[44] Y. Moulé, C. Rouiller, and J. Chauveau, *J. Biophys. Biochem. Cytol.* **7**, 547 (1960).

luminal ($\sim$15%) proteins, and in many studies these must be removed. Adsorbed basic proteins are removed most simply by suspending the pellet in a sucrose-free medium consisting of 0.15 M Tris-buffer, pH 8.0, followed by recentrifugation.[30,45] The majority, if not all, of excretory blood proteins in the lumen can be extracted by incubation in distilled water ($\sim$0.5 mg of protein per milliliter) at 30°C for 15 minutes followed by chilling in an ice-water bath and recentrifugation.[46,47] In the case of pancreas, the suspension of microsomes in alkaline buffers also extracts most of the secretory proteins.[43] The consecutive buffer and water treatments also remove most of the bound ribosomes. The PLP:protein ratio in nonwashed smooth microsomes is $\sim$0.34 and in rough microsomes $\sim$0.26. The corresponding ratios for RNA:protein are $\sim$0.07 and 0.25, respectively.[48] The cholesterol content on a PLP basis also differs, being $\sim$12% in smooth, and 7% in rough, microsomes.

Enzymatic Markers. Most of the enzymes studied are more or less evenly distributed during steady-state conditions in total rough and total smooth microsomes[1] (but not in later subfractions of these two main divisions). One of the few exceptions is GDP–mannosyltransferase, which is present mainly or exclusively in rough microsomes.[49,50] This is true only if GDP-mannose is used as substrate and endogenous protein as acceptor. On the other hand, during dynamic conditions, such as in newborn[45] and drug-induced animals,[51] the enzymatic composition of the two subfractions differs greatly.

[45] G. Dallner, P. Siekevitz, and G. E. Palade, *J. Cell Biol.* **30**, 97 (1966).
[46] M. Schramm, B. Eisenkraft, and E. Barkai, *Biochim. Biophys. Acta* **135**, 44 (1967).
[47] H. Glaumann and G. Dallner, *J. Lipid Res.* **9**, 720 (1968).
[48] A part of the free ribosomes, present in the 10,000 g supernatant, remain in the smooth microsomal fraction and contribute to its RNA content.
[49] J. Molnar, M. Tetas, and H. Chao, *Biochem. Biophys. Res. Commun.* **37**, 684 (1969).
[50] C. M. Redman and M. G. Cherian, *J. Cell Biol.* **52**, 231 (1972).
[51] S. Orrenius, *J. Cell Biol.* **26**, 725 (1965).

[19] Nondestructive Separation of Rat Liver Rough Microsomes into Ribosomal and Membranous Components

By M. R. Adelman, G. Blobel, and D. D. Sabatini

Preparation of Rough and Smooth Microsomes

Most fractionation procedures designed to obtain purified microsomes involve the preparation of a postmitochondrial supernatant (PMS) from which smooth (SM) and rough microsomes (RM) and free ribosomes are prepared, discontinuous sucrose density gradients being used. It is known that the RNA of membrane-bound ribosomes amounts to ~60% of the total RNA of rat liver.[1] Usually, 50% or more of the total RNA is sedimented with the nuclei and mitochondria, and most of this reflects the loss of RM.[1-3] Furthermore, the bound ribosomes in the PMS are only partially recovered in the purified RM fraction. Such losses of RM elements during cell fractionation should be avoidable, however. It is known that mitochondria can be washed relatively free of contaminating RM if simple sucrose solutions are used,[4,5] but that the addition of mono- or divalent cations must be avoided since these cause clumping and aggregation of membranous organelles.[6,7] Nuclei, on the other hand, cannot be washed extensively in salt-free media, since they tend to swell and produce a nucleoprotein gel.[8,9] Rather pure nuclei can be separated from total homogenates, however, by sedimentation through dense sucrose solutions.[9,10] The following fractionation scheme,[11] which yields RM preparations representing nearly 50% of the membrane-bound ribosomes of rat liver, has been devised following the above considerations. The method is in no way a radical departure from previously published ones. Procedures in common use involve sedimentation through concentrated sucrose media to purify nuclei,[9,10] washes with simple sucrose media to reduce the microsomal contamination of mitochondria,[4,5] and sedimentation through discontinuous sucrose density gradients to separate SM, RM, and free ribosome fractions.[12-15] The scheme described here simply represents one convenient

[1] G. Blobel and V. R. Potter, *J. Mol. Biol.* 26, 279 (1968).
[2] R. R. Howell, J. N. Loeb, and G. M. Tomkins, *Proc. Nat. Acad. Sci. U.S.* 52, 1241 (1964).
[3] A. Bergstrand and G. Dallner, *Anal. Biochem.* 29, 351 (1969).
[4] C. de Duve, B. C. Pressman, R. Gianetto, R. Wattiaux, and F. Appelmans, *Biochem. J.* 66, 1955 (1955).
[5] A. Amar-Costesec, H. Beaufay, E. Feytmans, D. Thinès-Semploux, and J. Berthet, *in* "Microsomes and Drug Oxidations" (J. R. Gillette, A. H. Conney, G. J. Cosmides, R. W. Estabrook, J. R. Fouts, and G. J. Mannering, eds.). Academic Press, New York, 1969.
[6] W. C. Schneider and G. H. Hogeboom, *Cancer Res.* 11, 1 (1951).
[7] G. H. Hogeboom, W. C. Schneider, and G. E. Palade, *J. Biol. Chem.* 172, 619 (1948).
[8] R. M. Schneider and M. L. Peterman, *Cancer Res.* 10, 751 (1950).
[9] M. Muramatsu, *in* "Methods of Cell Physiology" (D. M. Prescott, ed.), Vol. IV, p. 195. Academic Press, New York, 1970.
[10] G. Blobel and V. R. Potter, *Science* 154, 1662 (1966).
[11] M. R. Adelman, G. Blobel, and D. D. Sabatini, *J. Cell Biol.* 56, 191 (1973a).
[12] E. Reid, *in* "Enzyme Cytology" (D. B. Roodyn, ed.), p. 321. Academic Press, New York, 1967.
[13] G. Dallner and L. Ernster, *J. Histochem. Cytochem.* 16, 611 (1968).
[14] G. Blobel and D. D. Sabatini, *J. Cell Biol.* 45, 130 (1970).
[15] D. D. Sabatini and G. Blobel, *J. Cell Biol.* 45, 146 (1970).

combination of these techniques designed especially to maximize the yield of membrane bound ribosomes.

All solutions are prepared using deionized distilled water, Millipore filtered (0.45 μm for most, 1.2 μm for concentrated sucrose stock solutions), and stored in the cold. All operations, unless otherwise specified, are carried out in an IEC B-60 centrifuge (International Equipment Co., Needham Heights, Massachusetts). In the figures, the notation "30 min-44K-A211 (200,000)" is used to denote a 30-minute centrifugation at 44,000 rpm in the A211 rotor under which conditions g_{max} ~200,000. In Spinco centrifuges (Beckman Instruments, etc.) rotors A211 and SB110 can be replaced with Spinco rotors 42.1 or Ti 60 and SW 27, respectively. All pH's are those measured at room temperature.

Male Sprague-Dawley rats (~120–150 g) starved for about 18 hours are sacrificed between 9:00 and 10:00 AM by decapitation with guillotine (Harvard Apparatus Co., Inc., Millis, Massachusetts). Livers are quickly excised, immersed into ice-cold 0.25 M sucrose and cut into three to five large pieces. All subsequent operations are carried out in the cold room. Pieces of liver are blotted on absorbent paper and forced through a tissue press consisting of a piston and a stainless steel plate with 1-mm perforations. The pulp is weighed, slurried with 2 ml of 1.0 M sucrose solution per gram, and homogenized (8–10 passes) with a Teflon pestle, motor-driven tissue grinder (Arthur H. Thomas Co., Philadelphia, Pennsylvania size C, pestle rotating at 1000–2000 rpm). Slightly more or less vigorous homogenization does not affect the fractionation significantly.

The fractionation scheme is described below with reference to the numbered steps in flow diagrams I and II (Figs. 1 and 2). The scheme essentially involves first removing nuclei from a liver homogenate, the density of which is adjusted so that most other membranous organelles either float, are isopycnic, or sediment very slowly during the appropriate centrifugation. No ionic components (KCl, Tris, $MgCl_2$, etc.) are introduced into the homogenate. The postnuclear supernatant is then diluted to a density low enough to allow sedimentation of mitochondria which are washed with relatively ion-free sucrose solutions. The combined PMS is then fractionated on a discontinuous sucrose density gradient (which essentially contains no ions) into smooth and rough microsome and free ribosome fractions.

Flow Diagram I (Fig. 1)

Step a. The liver pulp is homogenized in 2 volumes of 1.0 M sucrose (2 ml/g): a higher sucrose concentration makes homogenization more difficult, while a lower concentration necessitates a greater dilution of the homogenate in the subsequent density adjustment step. The homogenate is filtered through a single layer of Nytex cloth (No. 130, Tobler, Ernest

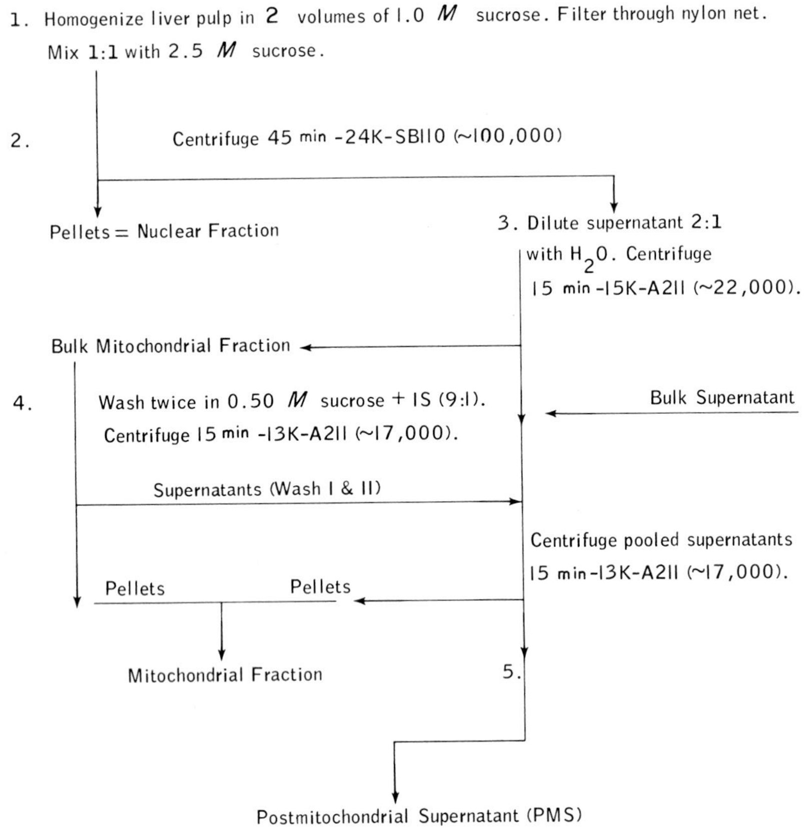

1. Homogenize liver pulp in 2 volumes of 1.0 *M* sucrose. Filter through nylon net. Mix 1:1 with 2.5 *M* sucrose.

2. Centrifuge 45 min -24K-SB110 (~100,000)

Pellets = Nuclear Fraction

3. Dilute supernatant 2:1 with H_2O. Centrifuge 15 min -15K-A211 (~22,000).

Bulk Mitochondrial Fraction

4. Wash twice in 0.50 *M* sucrose + IS (9:1). Centrifuge 15 min -13K-A211 (~17,000).

Bulk Supernatant

Supernatants (Wash I & II)

Centrifuge pooled supernatants 15 min -13K-A211 (~17,000).

Pellets   Pellets

Mitochondrial Fraction   5.

Postmitochondrial Supernatant (PMS)

FIG. 1. Flow diagram I.

and Traber, Inc., New York); this removes connective tissue debris and improves the subsequent separation of nuclei. To the filtered homogenate, an equal volume of 2.5 *M* sucrose is added, followed by thorough mixing (repeated inversion in a stoppered measuring cylinder) thus producing a mixture of suitable density. Normally, 50 ml of filtered homogenate (representing ~17 g of liver pulp derived from four or five rats) are processed in this manner.

Step b. The 100 ml of density-adjusted homogenate are transferred to four SB 110 tubes and each is overlaid with 1 ml of 1.0 *M* sucrose to assure that material which floats to the top during the ensuing centrifugation is not exposed to an air–water interface. After centrifugation [45 minutes at 24,000 rpm in the IEC rotor SB 110 (~100,000 g)], each tube contains a well-packed, mottled, pinkish-gray pellet, a tan supernatant, and a thick, reddish-tan pellicle. Shorter centrifugation (15–30 minutes) often fails to give well packed pellets; longer centrifugation does not increase the yield of nuclei. Recentrifugation of the rehomogenized supernatant (see below) gives a small pellet with only marginal increase in the overall yield of nuclei. Separation of nuclei under these conditions of minimal ion content does lead to some swelling and gelation. However, since the elimination of ions lessens the aggregation of membranous organelles, this disadvantage is deemed acceptable. Attempts to remove nuclei by step gradient centrifugation of the entire homogenate[15] have been unsuccessful in that, with the large volumes used, considerable DNA is trapped at the lower interface and does not sediment further, even after prolonged centrifugation. This trapping occurs whether or not the underlay contains ion components (e.g., TKM: $0.050 M$ Tris·HCl, pH 7.5, 25 mM KCl, 5 mM MgCl$_2$).

Step c. The pellicle is dislodged from the walls of the centrifuge tube with a metal spatula, and, together with the viscous supernatant, carefully transferred into the tissue grinder and homogenized (two or three passes) to disperse all clumps. To the ~100 ml of this dispersed postnuclear supernatant 50 ml of water are added, with thorough mixing, to achieve a dilution sufficient to allow subsequent sedimentation of the mitochondria. The mixture is divided into six portions and centrifuged for 15 minutes at 15,000 rpm in the IEC rotor A211 (~22,000 g).

Step d. The pink, turbid supernatant is decanted and stored in a beaker (to which the mitochondrial washes are subsequently added). The pellets, tan with a small red bottom layer (presumably erythrocytes), are suspended and homogenized in 25 ml of a mixture of 9 parts $0.50 M$ sucrose and 1 part inhibitory supernatant (IS).[16] The inclusion of IS at this point is a precaution to minimize nuclease attack on bound polysomes. The mitochondrial suspension is placed in two centrifuge tubes and sedimented for 15 minutes at 13,000 rpm in the IEC rotor A211 (~17,000 g). The supernatant is decanted and saved and the entire mitochondrial washing is repeated. The pellets obtained after each wash show a dark tan, tightly packed lower layer and a lighter tan, less tightly packed upper layer. In addition, an appreciable amount of pinkish, fluffy material is found which is only lightly packed and is decanted with the supernatants. Further washes of the mitochondrial fraction are ineffective in removing residual RNA. If the two washes are carried out with $0.25 M$ STKM, ($0.25 M$

---

[16] IS: Inhibitory supernatant. Homogenize rat liver in 2 ml of $0.25 M$ per gram of sucrose. Centrifuge for 15 minutes at 20,000 rpm in the IEC rotor A211 to remove large debris. Recentrifuge the supernatant 2 hours at 44,000 rpm in the IEC rotor A211 (200,000 g). Remove the clear red supernatant (avoiding the floating milky scum as well as the loosely pelleted material. This supernatant, referred to as IS, stored frozen until use is a source of RNase inhibitor.[17]

[17] G. Blobel and V. R. Potter, *Proc. Nat. Acad. Sci. U.S.* **55**, 1283 (1966).

sucrose with the ionic composition of TKM) the contamination of the mitochondrial fraction with RNA is even greater.

Step e. The combined mitochondrial supernatants are put into 8 tubes and centrifuged for 15 minutes at 13,000 rpm in the IEC rotor A211 (~17,000 g). Each tube contains a tiny two-layer pellet (as in step d above) and a large layer of loosely packed, pinkish, fluffy material. The supernatants are decanted with gentle swirling to assure transfer of the pinkish fluff, and then gently homogenize to disperse any clumps. The pooled supernatants constitute the final postmitochondrial supernatant (PMS). All the two-layer pellets (steps d and e) are combined and suspended for analysis as the "mitochondrial" fraction.

Flow Diagram II (Fig. 2)

The PMS (~180–190 ml, derived from ~17 g of liver) is used to separate free ribosomes, rough, and smooth microsomes on a discontinuous sucrose density gradient.

Step a. The total PMS is distributed evenly in 8 centrifuge tubes, and, with a syringe and large steel cannula, underlaid with: (i) 4 ml of a mixture of 3 parts 2.0 M sucrose plus 1 part IS, and (ii) 1 ml of 2.0 M STKM. Layer i serves to separate rough (RM) from relatively smooth (SM) microsomes, and is approximately equivalent in density to 1.5 M sucrose. While the choice of this density cutoff results in an appreciable loss of

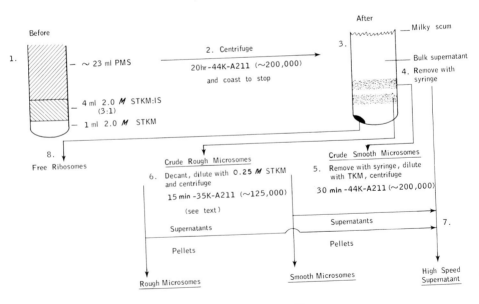

FIG. 2. Flow diagram II.

membrane-bound RNA to the SM fraction, it serves to minimize the extent to which mitochondrial fragments contaminate the RM. The addition of ions (e.g., TKM) to the underlay is avoided, since this leads to poorer separation of RM from SM (as shown by the RNA distribution). If IS is not present in the 1.5 M sucrose underlay, free polysomes are more extensively degraded, sediment more slowly, and therefore heavily contaminate the RM fraction.

Layer ii serves to separate RM from free ribosomes, which sediment through the 2.0 M STKM into a pellet. Addition of IS to the 2.0 M STKM layer does not improve the yield of preservation of free polysomes, so long as IS is present in layer i, above. Use of 2.0 M sucrose (without TKM) gives low yields of free ribosomes.

Step b. The step gradients are centrifuged 20 hours at 44,000 rpm in the IEC rotor A211 ($\sim$200,000 g), and the rotor is allowed to coast to a stop. Shorter centrifugation times greatly decrease the yield of free ribosomes, and, if short enough ($>$4–8 hours), result in poor separation of RM from SM. Even after 20 hours' centrifugation, sedimentation of free ribosomes is only two-thirds to three-fourths complete. These incompletely sedimented free ribosomes (mostly monomers) are easily removed from the RM during the subsequent differential centrifugation (step f, below).

Step c. As indicated in flow diagram II (Fig. 2), after centrifugation, each tube contains a clear, pink-to-red supernatant above which floats a thin, milky scum. Membranous material accumulates in the lower part of the tube. Upon close examination, it can be seen that there are two reddish-brown membranous bands, one at each interface, with a small, relatively clear zone between them. Under these conditions of separation, the upper band (crude smooth microsomes) is uniform, with no sign of clumping or adherence to the tube walls. If TKM is present in the $\sim$1.5 M sucrose underlay (or if the microsomes in the PMS are first sedimented and resuspended to allow application of a more concentrated sample to the discontinuous gradient), the upper membrane band is not uniform. Clumping of material in this band was found to be always associated with higher contamination of SM with RM and lower yields of purified RM. The lower membrane band contains crude rough microsomes which, having been in contact with the 2.0 M STKM, are somewhat clumped. Free ribosomes sediment through the 2.0 M STKM form a small, pale-orange pellet, the orange color being due to contaminating ferritin.

Step d. A syringe with a large steel cannula is used to remove the bulk clear supernatant (including the floating scum) from each tube and to transfer this to a beaker.

Step e. Using the same syringe and cannula, the upper membrane layer is removed and transferred to a graduated cylinder, care being taken not to disturb the lower membrane layer. Including the residual superantant fluid removed with this layer, a total of 50–60 ml of crude SM suspension is obtained. This is diluted with TKM to ~150 ml, distributed in six tubes, and centrifuged 30 minutes at 44,000 rpm in the IEC rotor A211 (~200,000 g). The clear supernatants are then added to the bulk super- natant (step d) while the pellets constitute the SM fraction.

Step f. The residual fluid contents of the step gradient tubes are de- canted into a graduated cylinder. Each tube is then gently rinsed with ~5 ml of 0.25 M STKM, and the rinses added to the same cylinder, care being taken to maximize transfer of the turbid fluid while minimizing dis- turbance of the ribosome pellets. This crude RM suspension is brought to a volume of ~100 ml with 0.25 M STKM, gently homogenized, and cen- trifuged (in six tubes) 15 minutes at 35,000 rpm in the IEC rotor A211 (~125,000 g). The supernatant is added to that stored from steps d and e. The pellets are homogenized in 100 ml of 0.25 M STKM and recentrifuged (6 tubes) 15 minutes at 30,000 rpm in the IEC A211 (~95,000 g). The supernatants are saved as above, while the pellets constitute the fraction.

Step g. The combined supernatants (steps d–f) constitute the high-speed supernatant.

Step h. The small, pale-orange, slightly opalescent pellets left in the step gradient tubes constitute the free ribosome fraction. The pellets of free ribosomes, RM and SM can be frozen and stored at −20°C. Since RM and SM stored in this way form aggregates it is better to store the samples after resuspension in 0.25 M STKM (one pellet in ~2.0 ml) to which 4–6 ml of glycerol are added. The suspensions are kept at −20°C. RM and SM can be recovered from such glycerol suspensions by dilution (at least 5-fold) with the appropriate solution and centrifugation.

Analytical data on the various cell fractions are presented in Table I. The nuclear fraction, which accounts for ~80% of the total DNA, is only slightly contaminated with mitochondria, as judged by the low cytochrome oxidase activity. The small amount of RNA present, being not much higher than that found in nuclei purified by other procedures,[18] suggests minimal trapping of ribosomal and/or rough microsomal elements. The mitochon- drial fraction, which accounts for ~85% of the recovered cytochrome oxidase, contains the bulk of the residual DNA. In addition, this fraction contains ~20% of the total RNA. The assumption that this RNA reflects the presence of RM is supported by the observation that treatment of the mitochondrial fraction with puromycin, under appropriate ionic condi- tions, leads to release of ribosomal subunits and "stripped" membranes

[18] A. Fleck and H. N. Munro, Biochim. Biophys. Acta 55, 571 (1960).

TABLE I

ANALYTICAL DATA ON RAT LIVER CELL FRACTIONS

Fraction	% RNA	% DNA	% Protein	% Cytochrome oxidase	$\dfrac{\text{Mg PLP}}{\text{mg protein}}$	Catalase (% units recovered)	Acid phosphatase (% units recovered)
Homogenate	[a]	[a]	[a]	[a]	0.248	[b]	[c]
Nuclei	7.7	80.5	9.2	4.0	0.170	3.3	4.3
Mitochondria	20.4	16.2	33.9	85.9	0.219	7.6	32.9
SM	5.6	0.6	7.8	4.8	0.660	3.4	18.4
RM	28.3	1.3	7.3	4.5	0.679	5.4	10.8
Ribosome	17.8	0.3	1.4	0.05	0.078	0.9	0.3
Supernatant	19.8	1.1	40.5	0.81	0.071	80.5	33.4

[a] The amount in each fraction is expressed as a percent of the sum of the amounts recovered in all fractions. Expressed as a percent of the contents of the homogenate, recoveries of RNA, DNA, protein, and cytochrome oxidase were 92, 103, 97, and 141%, respectively. The apparent overrecovery of cytochrome oxidase presumably reflects the difficulty of accurately assaying the activity in the homogenate.

[b] Homogenate contained 810 U. Total recovered = 1033.

[c] Homogenate contained 167 U. Total recovered = 196.

(see next section). The significance of and the reasons for the persistent contamination of the mitochondrial fraction with RM remain unclear; numerous modifications of the fractionation scheme have failed to minimize the contamination. About 28% of the RNA in rat liver, is recovered in the RM fraction while the SM and RM fractions jointly contain ~35% of the total RNA. Virtually all the RNA in these RM is due to ribosomes tightly bound to the membranes (i.e., contamination with free ribosomes is negligible). Assuming that 60% of all liver RNA is the ribosomal RNA of membrane-bound ribosomes,[1] the RM fraction contains nearly 50% of all the membrane-bound ribosomes. The RNA:protein ratio of these RM (0.20–0.25 mg/mg) is in good agreement with results of others. The RM contain ~5% of the cytochrome oxidase and a small amount of DNA, the significance of which remains obscure.[19]

The free ribosome fraction contains slightly less than the expected 20% of the total RNA,[1] because some of the free ribosomes fail to sediment through the 2.0 M STKM layer of the step gradient and are left in the crude RM layer. However, during the washing of the RM these free ribosomes, along with some small RM elements, are transferred to the combined high-speed supernatant fraction. Prolonged centrifugation of the supernatant leads to sedimentation of one-third to one-half of the RNA in this fraction, primarily as a mixture of free and bound ribosomes.

Phospholipid phosphorus analysis indicates the expected distribution of lipids.[12,13] Both SM and RM fractions contain 0.6–0.7 mg of PLP per milligram of protein. Analysis of catalase, as a peroxisomal marker,[20] reveals ~80% of the activity in the supernatant fraction, which suggests (as might be expected in view of the repeated homogenization involved in this procedure) extensive rupture of peroxisomes. However, since a rather large fraction of rat liver catalase may exist in nonparticulate form[21] the exact extent of peroxisome rupture is difficult to assess. The RM fraction contains ~5% of the recovered catalase activity. Damage to lysosomes seems less extensive, since only ~33% of the acid phosphatase activity is released to the supernatant, while an equal amount was found in the mitochondrial fraction (which is equivalent to the M + L fraction of de Duve et al.[4] The RM account for ~10% of the recovered acid phosphatase activity. Electron microscopic examination of the SM and RM fractions corroborates these biochemical analyses. The SM consist of a fairly heterogeneous population of membranous vesicles most of which are smooth-surfaced, although occasional ribosome-studded vesicles are found. In addition,

[19] W. C. Schneider and E. L. Kuff, J. Biol. Chem. 244, 4843 (1969).
[20] F. Leighton, B. Poole, H. Beaufay, P. Baudhuin, J. W. Coffey, S. Fowler, and C. de Duve, J. Cell Biol. 37, 482 (1968).
[21] R. S. Holmes and C. J. Masters, Arch. Biochem. Biophys. 148, 217 (1972).

mitochondrial fragments, presumptive lysosomes, and large, flattened sheets (presumably plasma membrane) are present. The RM fraction is considerably more homogeneous, consisting primarily of ribosome-studded vesicles, which, when sectioned tangentially show a fairly high density of attached ribosomes.

The ribosomes in the RM and free ribosome fractions have been shown to be active in an *in vitro* amino acid incorporation mixture with endogenous messenger and to be stimulated by the addition of polyuridylic acid.

Separation of Ribosomes from Membranes of Rough Microsomes

It is known that ribosomes interact with microsomal membranes via the large (60 S) subunit[22] and that the nascent polypeptide chain, which grows within a protected region in this subunit,[14,23] enters into close relationship with the membrane immediately upon emerging from the ribosome.[15] Chelating agents[22] and concentrated salt solutions[24] devoid of magnesium ions can be used to release some or most of the bound ribosomes from rough microsomes (RM), but such treatments produce a mixture of intact membranes and damaged or denatured ribosomes. On the other hand, detergents, which have been extensively used to release functional ribosomes from RM,[25] are effective only by greatly altering or destroying membrane structure. Recent developments[26] have made it possible to nondestructively separate RM into the component parts, viz., ribosomes and membranes. These developments are based on the finding[27] that treatment of free polysomes with puromycin in solutions of high ionic strength leads to the disassembly of the polysomes into functionally viable ribosomal subunits.

Examination of the effect of puromycin on the stability of ribosome-membrane interaction has shown[26] that the combined action of this aminoacyl tRNA analog and appropriate high KCl conditions can be used to produce an efficient release of almost all the bound ribosomes from RM. Exposure of RM to a solution containing high KCl but no puromycin, results in release of some ribosomes (Fig. 3), the exact extent of release and state of the ribosomes being a function of the KCl and $MgCl_2$ concentrations, as well as of the time and temperature of treatment. It does not appear possible, however, to remove all ribosomes from RM by alterations in ionic constituents alone, unless conditions are used which lead to unfolding

[22] D. D. Sabatini, Y. Tashiro, and G. E. Palade, *J. Mol. Biol.* 19, 503 (1966).
[23] L. I. Malkin and A. Rich, *J. Mol. Biol.* 26, 329 (1967).
[24] T. Scott-Burden and A. O. Hawtrey, *Biochem. J.* 115, 1063 (1969).
[25] J. K. Kirsh, P. Siekevitz, and G. E. Palade, *J. Biol. Chem.* 235, 1419 (1960).
[26] M. R. Adelman, D. D. Sabatini, and G. Blobel, *J. Cell Biol.* 56, 206 (1973).
[27] G. Blobel and D. Sabatini, *Proc. Nat. Acad. Sci. U.S.* 68, 390 (1971).

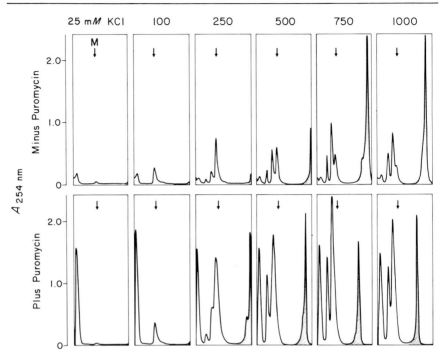

Fig. 3. Release of ribosomes from rough microsomes (RM) by combined action of high KCl and puromycin. Equal amounts of RM were incubated in S 250, Mg 5, T 50 plus 25, 100, 250, 500, 750, or 1000 mM KCl in the absence (upper panels) or in the presence (lower panels) of 0.79 mM puromycin. Samples were incubated for 70 minutes at 0°C, 10 minutes at 37°C, and then 10 minutes at room temperature. Samples (0.45 ml, 0.37 mg RNA) were applied to 15–40% sucrose gradients containing the appropriate K, Mg, T buffer. Sedimentation was at 20°C: 1¼ hours 40K-SB283 (~270,000). All profiles are presented with top of gradient at left, and the direction of sedimentation is from left to right. The small arrow (M) indicates the position of the ribosomal monomer (80 S). Gradients run at [KCl] ≥ 250 mM also show small (40 S) and large (60 S) subunit peaks. The shading in the lower portions of these gradients is added to indicate that the UV absorption corresponded to a region in which membranes were visible (turbidity) but no attempt is made to indicate the exact extent of the turbidity. Membranous bands are designated with the symbol Mb. From ref. 26.

—hence denaturation—of the ribosomal subunits. But, under ionic conditions where partial release of functional ribosomes (or subunits) occurs, puromycin strongly enhances the extent of release. It is possible (see below) to remove up to 85% of all bound ribosomes by the combined action of puromycin and high KCl under conditions where ribosome, as well as membrane, integrity is maintained.

Those membrane bound ribosomes which are released solely in response to elevated ionic strength appear to be a mixture of inactive ribosomes and of ones bearing relatively short nascent chains. Experiments with [³H]puromycin and/or [³H]leucine pulse-labeled RM have shown that the puromycin releasable ribosomes react with the drug at low KCl and release their nascent chains vectorially to the microsomal membrane. However, at low ionic strength, some as-yet-unspecified interaction between the large ribosomal subunit and the membrane is maintained and only when the KCl concentration is elevated are the ribosomes released. Thus all membrane-bound ribosomes are attached via a large subunit–membrane interaction. In addition, some, but not all, ribosomes are tightly held to the RM by an interaction involving the nascent polypeptide chain. Only when both interactions are destabilized, by the combined action of high KCl and puromycin can the bound ribosomes be released in a nondestructive manner.

For routine disassembly RM (prepared as described in the preceding section) are suspended in 0.25 M sucrose, 0.75 M KCl, 5 mM MgCl$_2$, 50 mM Tris·HCl, pH 7.5 plus 1 mM puromycin at a final concentration of 1–2 mg RNA/ml (5–10 mg protein/ml). The suspension is then incubated for 1–2 hours at room temperature. A shorter incubation time (15–30 minutes) at 37°C is also effective but variable results, some suggestive of large subunit breakdown, have been found using the high temperature. The reaction mixture can be directly separated into large and small subunits and stripped membrane fractions by zone sedimentation in sucrose density gradients (see legend to Fig. 3). The appropriate zones are collected separately, diluted with a suitable buffer (e.g., TKM) and centrifuged: 15 minutes at 30,000 rpm in the IEC rotor A211 (~90,000 g) for the stripped membranes; 3 hours at 44,000 rpm in the IEC rotor A211 (~200,000 g) for the ribosomal subunits. It should be noted that density gradient analyses such as those in Fig. 3 are considerably facilitated by the tendency to aggregate of RM (or stripped membranes therefrom) which have been stored frozen in pellet form. When RM which are freshly prepared or have been stored after suspension in glycerol are used, the membranes aggregate less, form broader bands in the gradients (because they take longer to reach their isopycnic position) and are thus less well resolved from the ribosomal subunits. When stripped membranes, but not ribosomal subunits, are required the reaction mixtures can be subjected directly to differential centrifugation 15 minutes at 30,000 rpm in the IEC rotor A211 (~90,000 g) and the membrane pellets may be washed as desired by repeated suspension (gentle homogenization) in and resedimentation from a suitable washing medium, such as 0.25 M sucrose containing 50 mM Tris·HCl pH 7.5, 25 mM KCl, and 5 mM MgCl$_2$.

By subjecting RM to the above protocol, it is possible to prepare stripped membranes which have released ~85% of their bound RNA (Table II). The exact nature of the residual 15% is not clear; these resid-

TABLE II

CHEMICAL ANALYSIS OF ROUGH (RM) AND SMOOTH (SM) MICROSOMES BEFORE AND AFTER
STRIPPING OF RIBOSOMES WITH KCl AND PUROMYCIN[a]

Sample	Protein (mg/ml)	RNA (mg/ml)	PLP (mg/ml)	[³H]RNA (cpm/mg RNA)	RNA/protein (mg/mg)	RNA/PLP (mg/mg)	PLP/protein (mg/mg)
RM	9.66	2.07	5.31	25,400	0.214	0.389	0.550
Stripped RM	6.57	0.358	5.20	21,600	0.054	0.069	0.791
SM	12.2	0.498	7.10	24,500	0.041	0.070	0.581
Striped SM	7.95	0.063	5.46	12,600	0.008	0.012	0.687

[a] Both rough microsomes and smooth microsomes were prepared at the same time from rats which had received an injection of [³H]-orotic acid (200 µCi, 2.5 mCi/mol) ~40 hours before sacrifice. RM and SM pellets were resuspended in S 250, K 25, Mg 5, T 50. Samples of each were incubated for 2 hours at room temperature in the presence of (final concentrations) S 250, K 750, Mg 5, T 50, and 0.63 mM puromycin. The samples were diluted with an excess of room temperature K 750, Mg 5, T 50 and centrifuged at room temperature 15 minutes, 30K-A211 (~90,000). The pellets were resuspended in cold buffer and recentrifuged. Each pellet was then resuspended in S 250, K 25, Mg 5, T 50 to the same volume as the original sample. Samples of the untreated RM and SM, as well as the KCl-puromycin-treated microsomes, were analyzed for protein, phospholipid phosphorus, RNA, and ³H cpm in the RNA hydrolysate.

ual ribosomes may be removed (in denatured form) by washing the membranes with 1.0 M KCl (no $MgCl_2$). It must be reemphasized, however, that total removal of all bound ribosomes has been achieved only under conditions leading to extensive unfolding of the subunits. With the nondestructive procedure presented here, ~85% of all bound ribosomes are recovered as subunits which are at least partially active in translating polyuridylic acid in an *in vitro* amino acid incorporating system. The stripped membranes are recovered as closed, apparently intact vesicles. The data of other workers indicate that the ionic conditions employed do not markedly affect characteristic membrane enzymes.

[20] Procedure for the Selective Release of Content from Microsomal Vesicles without Membrane Disassembly

By G. KREIBICH and D. D. SABATINI

Preparation of Microsomes

Rough microsomes (RM) and smooth microsomes (SM) are prepared by the procedure of Adelman *et al.*[1] (see this volume, 19a) from male albino rats of the Sprague-Dawley strain. Microsome pellets (10–30 mg protein per fraction) are stored at $-20°C$ to $-40°C$ for up to 2 months. Alternatively, and to reduce aggregation,[2] freshly prepared microsomes can be resuspended in 0.25 M sucrose, 25 mM KCl, 50 mM Tris · HCl pH 7.4, 5 mM $MgCl_2$, mixed with 2 volumes of glycerol and kept at $-20°$ to $-40°C$. Before use, the microsomes stored in glycerol are diluted four times with a solution containing 0.25 M sucrose, 0.5 M KCl, 50 mM Tris · HCl pH 7.5, 10 mM $MgCl_2$, and recovered by sedimentation (15 minutes at 40,000 rpm in the IEC rotor A321). Microsomes stored as pellets are washed once (20 minutes at 20,000 rpm in the IEC A211 or in the Ti 60 Spinco rotors) in a medium of high ionic strength (HSB) (25 mM sucrose, 50 mM Tris · HCl pH 7.5, 0.5 M KCl, 10 mM $MgCl_2$).

From 22 g of rat liver (4–5, 120–150 g rats) an average of 180 mg protein is recovered in RM and 190 mg protein in SM.

Detergent Treatment

Table I lists five detergents of different polarity which can be used to release the content of microsomes without producing membrane disassem-

[1] M. R. Adelman, G. Blobel, and D. D. Sabatini, *J. Cell Biol.* **56**, 191 (1973).
[2] D. Borgese, G. Blobel, and D. D. Sabatini, *J. Mol. Biol.* **74**, 415 (1973).

TABLE I

DIFFERENT DETERGENTS USED TO RELEASE THE VESICULAR CONTENT FROM ROUGH MICROSOMES (RM)

Detergent	Formula	MW	Detergent concentration to release microsomal content[a]	
			%	mM
Brij 35[b]	$CH_3-(CH_2)_{11}-O-(C_2H_4O)_{23}H$	~1200	0.15	1.3
Triton X-100[c]	$CH_3-\overset{CH_3}{\underset{CH_3}{C}}-CH_2-\overset{CH_3}{\underset{CH_3}{C}}-\underset{}{}$⟨aromatic ring⟩$-(OC_2H_4)_{9-10}OH$	~625	0.08	1.3
Sodium deoxycholate[d] (DOC)	$C_{23}H_{37}(OH)_2-COO^-Na^+$	414.6	0.05	1.3
Sodium dodecyl sulfate (SDS)[e]	$CH_3-(CH_2)_{11}-O-SO_3^-Na^+$	288.4	0.07	2.5[f]
Hexadecyltrimethyl-ammonium bromide[e] (HTA Br)	$CH_3(CH_2)_{15}-\overset{+}{N}(CH_3)_3Br^-$	364.5	0.091	2.5[g]

[a] RM are resuspended in low-salt buffer at a concentration of 3–4 mg of protein per milliliter.
[b] Atlas Chemicals, Inc., Wilmington, Delaware.
[c] Eastman, Rochester, New York.
[d] Schwarz/Mann, Orangeburg, New York
[e] DDH Chemicals Ltd., Poole, England.
[f] Low-salt buffer contains Na+ instead of K+.
[g] HTABr causes protein aggregation.

bly. Stock solutions of deoxycholate (DOC) (5% and 7.84%, 0.2 M), are prepared by dissolving NaDOC and adjusting the pH at 20°C to 7.5 with 0.1 N HCl. Washed microsomes are resuspended in LSB (50 mM Tris· HCl, pH 7.5, 50 mM KCl, 5 mM MgCl$_2$) to a final concentration of 3–4 mg of protein per milliliter using a Potter-Elvehjem homogenizer operated by hand. For rough microsomes the protein concentration of suspensions can be estimated from the OD$_{260}$ of samples treated with 0.5% DOC (ribosomes $E_{260}^{1\%} = 135$).[3] For RM a value of 6.75 OD$_{260}$ units/ml corresponds to approximately 1 mg of protein per milliliter. For smooth or for stripped rough microsomes, the protein concentration should be measured chemically. RM suspensions kept in the cold receive detergent solutions of 10× the desired final concentration. During addition of detergent the suspension is vigorously mixed by a Vortex tube mixer. Samples are kept at 0°C for 30 minutes before loading onto gradients, or separating subfractions by differential centrifugation.

In Vivo Labeling of Specific Microsomal Components

Table II gives a list of radioactive precursors which can be used to label specifically different components of microsomes (see also Kreibich et al.[4]). The specific radioactivity of each and the time of in vivo labeling elapsed before sacrifice, are given together with the final specific radioactivity of the microsomes. Volumes of solutions injected vary from 0.2 to 0.5 ml. When radioactive leucine is administered, different time periods of in vivo labeling can be chosen to obtain incorporation of this amino acid into nascent polypeptide chains still bound to ribosomes (2 minutes) or into completed but recently synthesized proteins which are already discharged into the cisternal cavity (30 minutes). Proteins that are stable and remain in the microsomes a minimum of one day after synthesis are labeled by injecting the precursor once every day for 3 days and waiting for 24 hours after the last injection before sacrifice. Most of these proteins are part of the microsomal membranes. For 2 minutes of labeling with [^{3}H]-or [^{14}C]leucine, the amino acid is injected into the portal vein under ether anesthesia; all other injections are given intraperitoneally (ip). For convenience, proteins labeled for 2 minutes will be called "pulse-labeled" proteins, whereas proteins labeled for 30 minutes before sacrifice will be called "short-term" labeled proteins, in contrast to stable, or "long-term" labeled, proteins (labeled for 3 days).

Table III shows a typical distribution of [^{3}H]- and [^{14}C]leucine radioactivity in rat liver cell fractions and in serum obtained after 2-minute,

[3] Y. Tashiro and P. Siekevitz, J. Mol. Biol. 11, 149 (1965).
[4] G. Kreibich, P. Debey, and D. D. Sabatini, J. Cell Biol. 58, 436 (1973).

TABLE II

In Vivo Incorporation of Radioactively Labeled Precursors into Rough (RM) and Smooth (SM) Rat Liver Microsomes

Radioactive precursor	Specific activity (SA) (Ci/mmole)	μCi/g rat injected	Time of labeling	SM		RM		Labeled microsomal component
				SA (dpm/mg protein)	% of injected radioactivity	SA (dpm/mg protein)	% of injected radioactivity	
Choline-methyl [³H]chloride[a]	0.55	1.6	4 Hours	—	—	5.2×10^4	0.48	Membrane phospholipids[d]
Orotic-[5-³H]acid[a]	2.81	1.0	40 Hours	—	—	5.0×10^4	0.8	RNA of bound ribosomes
L-[¹⁴C]Leucine	0.312	3 × 0.83	3 Days	1.9×10^4	0.17	2.5×10^4	0.23	Stable microsomal proteins
L-[4,5-³H]Leucine[b]	40.0	3.2	30 Min	1.6×10^5	0.7	1.5×10^5	0.63	Secretory proteins; proteins of the cisternal content
L-[4,5-³H]Leucine[b]	40.0	4.0	2 Min	2.8×10^5	2.3	5.1×10^5	3.0	Nascent polypeptides
D-[GL-³H]Glucosamine[c]	1.9	3.5	30 Min	8.4×10^4	0.39	6.0×10^3	0.024	Newly synthesized glycoproteins

[a] New England Nuclear Corp., Boston, Massachusetts.
[b] Schwarz BioResearch, Orangeburg, New York.
[c] International Chemical and Nuclear Corp., Irvine, California.
[d] After Folch extraction [J. Folch, M. Lees, and G. H. J. Stanley, J. Biol. Chem. 226, 497 (1957)] 92% of the radioactivity in the rough microsomes was found in the chloroform phase.

TABLE III

DISTRIBUTION OF ACID INSOLUBLE [³H]- AND [¹⁴C]LEUCINE RADIOACTIVITY
IN RAT LIVER CELL FRACTIONS[a]

	2 Min (pulse)[c]	30 Min (short term)[c]	3 Days (long term)[c]
Total homogenate[b]	100% (3.34×10^8 dpm)	100% (3.34×10^8 dpm)	100% (2.1×10^7 dpm)
% of injected isotope recovered in homogenate	17.8%	5.0%	3.2%
Postnuclear supernatant	92.1%	84%	81%
Postmitochondrial supernatant	78.2%	73%	67%
Final supernatant	21.2%	45%	53%
Smooth microsomes	20.8%	14%	4.8%
Rough microsomes	26.6%	13%	6.6%
TCA precipitable radioactivity in serum (dpm/ml)	0.0%	3.21×10^6	2.4×10^6

[a] See Table II for labeling conditions.

[b] TCA-precipitable dpm in ~25 g of liver.

[c] Preparations are those listed in Table II.

30-minute, or long-term administration (1–3 days) of the labeled amino acids.

Separation of Content from Membrane-Bound Microsomal Components

After detergent treatment, soluble and sedimentable (membrane bound) microsomal subfractions can be separated preparatively by differential centrifugation or by using sucrose density gradients. For differential centrifugation, 8-ml samples are placed in tubes of the A321 IEC rotor or of the No. 60 Spinco rotor and underlayered with 2 ml of 20% sucrose LSB (20% sucrose containing 50 mM Tris·HCl pH 7.5, 50 mM KCl, 5 mM MgCl₂). After centrifugation at 4°C (90 minutes at 50,000 rpm) membranes are recovered in the sediment and the solubilized material in the top 8 ml.

For sucrose density gradient separation up to 0.5-ml aliquots are loaded onto linear sucrose gradients (10–60% sucrose) containing the same salt composition as LSB, prepared in tubes for the SB283 IEC rotor or for the SW 41 Spinco rotor, which are centrifuged for 2–3 hours at 40,000 rpm at 4°C.

Comparison of the absorbance and radioactivity profiles in gradients of detergent-treated RM containing labeled phospholipids has shown[4] that concentrations of **DOC** lower than 0.098% have no detectable effect on the sedimentation properties of the membranes or on the distribution of labeled phospholipids. The joint appearance of radioactive [³H]choline in the top part and absorbance profiles of ribosomes in the middle part of the gradients indicates membrane breakdown which begins at 0.098%

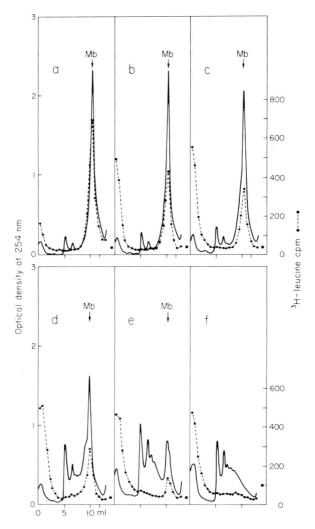

Fig. 1. Effect of deoxycholate (DOC) on the release of short-term labeled proteins from rough microsomes. RM containing labeled content ([³H]leucine, 30 minutes, see Table II) resuspended in low salt buffer were treated with a series of DOC concentrations: (a) control, 0%; (b) 0.025%; (c) 0.049%; (d) 0.098%; (e) 0.196%; (f) 0.392%. The treatment and analysis are described in the text. ———, OD₂₅₄, ●- -●, [³H]leucine radioactivity.

DOC. Total solubilization of the RM membranes only occurs after treatment with DOC concentrations higher than 0.392%. Above this DOC concentration labeled polysome patterns appear within the gradients. A selective release of the microsomal content proteins occurs at DOC concentrations between 0.025% and 0.049%.

To demonstrate the release of content, microsomal proteins are labeled with [³H]leucine administered 30 minutes before sacrifice (Fig. 1, short-term labeled proteins: [³H]leucine, 30 minutes). After this time of *in vivo* incorporation, secretory proteins, which are manufactured in the ER, are extensively labeled. On the other hand, polypeptide chains, growing in bound ribosomes, are no longer significantly radioactive[4] (Table III). The top 2.5 ml from gradients of RM treated with 0.049% DOC are collected and pooled. A large fraction (~50–60%) of the total radioactivity no longer sediments with the membranes and is found in this region (Fig. 1).

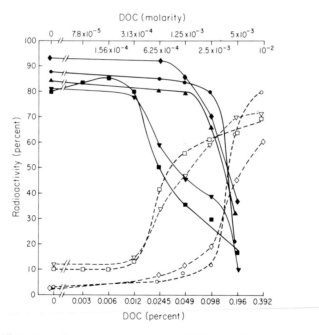

FIG. 2. Differential effect of deoxycholate (DOC) on the release of labeled microsomal constituents in a solution of low ionic strength (LSB). RM labeled *in vivo* in different constituents (see Table II) were resuspended in LSB, and treated with a series of DOC concentrations. Subfractions were separated by sucrose density gradients as described in the text. - - -, Percent of total radioactivity in first five fractions near the top of the gradient. ———, Percent of total radioactivity in seven fractions corresponding to the membrane band. ○, ●, [³H]choline, 4 hours; ▲, [³H]orotic acid, 40 hours; □, ■, [³H]leucine, 30 minutes; ◇, ◆, [¹⁴C] leucine, long term; ▽, ▼, [³H]glucosamine, 30 minutes.

Figure 2 summarizes the effect of a series of DOC concentrations on the release of several labeled components from rough microsomes. For each marker and for each DOC concentration the fraction of label associated with the membrane band or found in the upper regions of the gradient was obtained from analyses similar to those illustrated in Fig. 1. Content proteins and glycoproteins are released at lower concentrations than structural components associated with the sedimentable membranes (phospholipids, membrane proteins, and ribosomes).

Differential centrifugation is better suited for the preparation of larger amounts of membrane and content fractions for electrophoretic or enzymatic analysis. It can be seen in Table IV that membrane enzymes

TABLE IV

EFFECT OF DEOXYCHOLATE (DOC) ON THE RELEASE OF SEVERAL CONSTITUENTS FROM ROUGH MICROSOMES[a]

Approximate DOC conc. (%)	Phospho-lipid[b] (%)	Protein[c] (%)	TCA[d] precip-itable radio-activity	NADPH[e] Cyt c reductase	NADH[e] Cyt c reductase	Cyto-chrome b_5[f]
0	2	4	7	3	3	<2
0.025	3	9	28	4	5	<2
0.05	3	15	42	5	6	3
0.10	4	20	45	6	7	13
0.20	96	35	50	43	32	70
0.40	98	65	69	88	92	93

[a] RM ([³H]leucine, 30 minutes, Table II) resuspended in a modified LSB (10 mM Tris·HCl, pH 7.5) to a final concentration of 3–4 mg of protein per milliliter; 1–2 ml aliquots are incubated at 0°C for 30 minutes with DOC concentrations ranging from 0.025 to 0.392%. Sedimentable membranes and ribosomes are separated from the supernatant by differential centrifugation (50,000 rpm, 60 min in the A269 IEC rotor). Phospholipid, protein, and TCA-insoluble radioactivity, NADPH, and NADH-cytochrome c reductase activities and cytochrome b_5 concentrations are determined before fractionation as well as in the fractions. For each sample, the 100% value is the sum of the values in supernatant and pellet.

[b] Determined according to B. N. Ames and D. T. Dubin [J. Biol. Chem. 235, 769 (1960)] after Folch extraction [J. Folch, M. Lees, and G. H. J. Stanley, J. Biol. Chem. 226, 497 (1957)].

[c] Determined according to J. F. Foster, M. Sogami, H. A. Peterson, and W. J. Leonard, Jr., J. Biol. Chem. 240, 2495 (1965).

[d] Measured on filter disks [R. J. Mans and G. B. Novelli, Arch. Biochem. Biophys. 94, 48 (1961)].

[e] Determined according to T. Omura, P. Siekevitz, and G. E. Palade, J. Biol. Chem. 242, 2389 (1967).

[f] Measured according to D. Garfinkel, Arch. Biochem. Biophys. 71, 100 (1957).

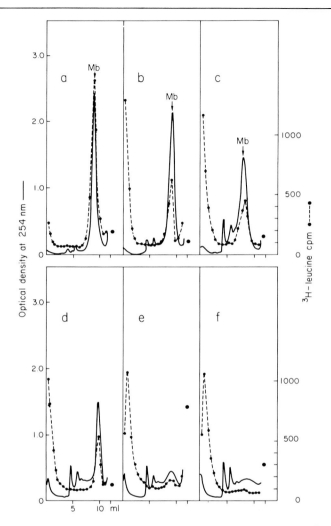

FIG. 3. Effect of the ratio of detergent to microsome concentration. Rough microsomes containing short-term labeled content proteins ([³H]leucine, 30 minutes; Table II) were resuspended in low-salt buffer and adjusted to various concentrations (expressed as milligrams of protein per milliliter): (a) control, 6.25 mg; (b) 25 mg; (c) 12.5 mg; (d) 6.25 mg; (e) 3.13 mg; (f) 1.6 mg. All samples except the control (a) received DOC to a final concentration of 0.196%. After 30 minutes at 0°C, 50–800-μl aliquots were loaded on top of sucrose gradients, all of which received approximately equal amounts of microsomes. Composition of the gradients, centrifugation, and analysis are described in the text.

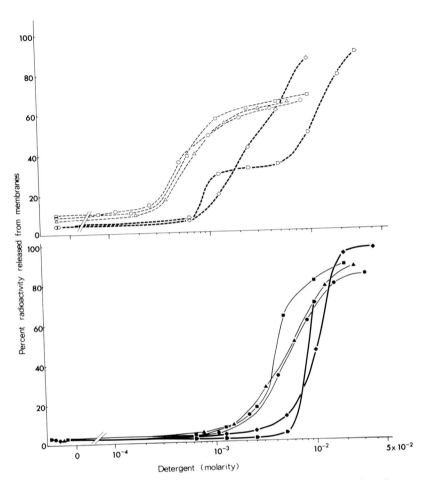

FIG. 4. Effect of detergents of varying polarity on the release of the microsomal content and on the integrity of microsomal membranes in a solution of low ionic strength. Rough microsomes containing tritiated phospholipids ([³H]choline, 4 hours, bottom panel) and tritiated content proteins ([³H]leucine, 30 minutes, upper panel) were resuspended in low-salt buffer, treated with increasing concentrations of different detergents and analyzed by sucrose density gradient as described in the text. In this graph, only the percentage of radioactivity released from the membranes and found in the five top fractions of the gradients is shown. ●, ○, Brij 35; ▲, △, Triton X-100; ■, □, deoxycholate; ◆, ◇, sodium dodecyl sulfate; ▶, ▷, hexa-decyltrimethylammonium bromide.

(NADPH– and NADH–cytochrome c reductases and cytochrome b_5), which are proteins tightly bound to microsomal membranes,[5] behave as the phospholipids and are released only at DOC concentrations higher than 0.098%. This release behavior contrast with that of short-term labeled proteins of the content ([³H]leucine, 30 minutes).

The ratio of detergent to the amount of phospholipid present in membranes determines the level of release of membrane constituents. The dependency of the DOC effect on membrane concentration, at a fixed level (0.196% = 5 mM) of detergent is shown in Fig. 3. Using high microsome concentrations (25 mg of protein per milliliter) short-term labeled proteins ([³H]leucine, 30 minutes) can be largely released without affecting the integrity of the membranes. The complete sequence of release can be made to proceed at a single DOC concentration by decreasing the microsomal concentration (see the height of the absorbancy peak at the membrane region and the release of ribosomes in Fig. 3).

The lowest detergent concentration which first induces a selective release of short term-labeled proteins and glycoproteins from microsomes also depends on the ionic strength of the medium of incubation. Values given here are for LSB, a buffer of moderate ionic strength. If microsomes are resuspended in HSB (0.5 M KCl, 50 mM Tris·HCl, 5 mM MgCl$_2$) the DOC concentration needed to produce maximum release of content is about half of that needed in LSB.[4]

Other detergents of varying polarity (Table I) can also be used to selectively release microsomal content. From Fig. 4 it can be seen that nonpolar detergents are more effective in inducing leaking of [³H]leucine (30 minutes) labeled proteins from microsomes. The effectiveness in solubilizing microsomal membrane phospholipids ([³H]choline, 4 hours) also correlates with the polarity.

[5] T. Omura, P. Siekevitz, and G. E. Palade, *J. Biol. Chem.* **242**, 2389 (1967).

[21] Separation of Hepatic Smooth and Rough Microsomes Associated with Drug-Metabolizing Enzymes

By THEODORE E. GRAM

The endoplasmic reticulum (ER) of the parenchymal cells of mammalian liver has two forms which are distinguishable by electron microscopy: (1) rough (granular) endoplasmic reticulum (RER), characterized by the presence of ribosomes bound to its cytoplasmic surface; (2) smooth (agranular) endoplasmic reticulum (SER), conspicuous by its lack of (membrane-bound) ribosomes.

RER and SER exist in hepatocytes both as tubules and as vesicles. Numerous morphological investigations have demonstrated RER and SER to be in direct physical continuity; RER is thought to give rise to SER by a process of cisternal "budding." Physical disruption of hepatocytes (homogenization) results in the conversion of both forms of ER into spherical vesicles which can be isolated as the microsomal fraction by differential centrifugation.[1]

The predominant elements of the microsomal fraction of liver are rough-surfaced vesicles, smooth-surfaced vesicles, and free ribosomes, unattached to any membranous structure; quantitatively minor yet nearly ubiquitous constituents of this fraction are fragments of mitochondria, Golgi apparatus, and plasma membrane.[1] In contrast to mitochondrial or lysosomal fractions which have discrete counterparts in the hepatocyte, the microsomal fraction was originally described by Claude[2] in operational terms, namely the total particulate material present in a postmitochondrial supernatant. The question whether the presence in microsomal fractions of fragments of plasma membrane or Golgi apparatus constitutes "contamination" or "heterogeneity" is moot because of the imprecision of the term "microsomal" and the intrinsic heterogeneity of the ER itself.[3]

Principle

The phospholipid-rich ER membranes have a density of ~ 1.1 whereas the density of ribosomal particles (80 S particle) is ~ 1.6. Thus the presence of ribosomes confers upon rough vesicles a slightly greater mean density relative to smooth vesicles[4] (Table I). This difference permits resolution of the two populations by isopycnic density gradient centrifugation. Both smooth and rough vesicles exhibit rather broad variations in particle diameter, smooth being somewhat more variable than rough[1,4] (Table I).

The *rate* of sedimentation of a spherical particle is a function of both its density and its diameter. Thus, the sedimentation rate of a rough vesicle 50 nm in diameter may be the same as that of a smooth vesicle 150 nm in diameter. Accordingly, simple differential centrifugation, which relies on differences in sedimentation rates, cannot provide adequate resolution of

[1] G. E. Palade and P. Siekevitz, *J. Biophys. Biochem. Cytol.* **2**, 171–198 (1956).

[2] A. Claude, *Science* **97**, 451 (1943).

[3] M. Wibo, A. Amar-Costesec, J. Berthet, and H. Beaufay, *J. Cell Biol.* **51**, 52 (1971).

[4] J. Rothschild, *Biochem. Soc. Symp.* **22**, 4–28 (1963).

TABLE I
Some Physical Properties of Liver Microsomal Components

Component	Density	Diameter (nm)
Smooth microsomes	1.05–1.18	50–300
Rough microsomes	1.18–1.27	50–150
Ribosomes	~1.6	~15

smooth and rough microsomal membranes, whereas the density gradient technique has been more successful.[4]

The Binding of Ribosomes to ER Membranes: the Relationship between Rough and Smooth ER. RER and SER are parts of a membranous continuum within hepatocytes; RER is generally recognized to be the morphological and biochemical precursor of SER. The transformation of RER to SER apparently occurs through "budding" of the cisternal ER membranes and the concomitant loss of ribosomes from the membrane surface. In liver and other organs, ribosomes exist both free in the cytoplasm and bound to the membranes of the ER. About 95% of the total RNA of rat liver is cytoplasmic; of this, about 75% is membrane bound and 25% free in the cytoplasm.[5] Although little is known about the precise nature of the ribosome-ER membrane binding, it is known that the ratio of free to membrane-bound ribosomes is not constant but is subject to modification both *in vivo* and *in vitro*. Physiological states such as acute starvation, neonatal growth, and partial hepatectomy evoke significant changes in the relative proportions of free and bound ribosomes in rat liver.[6] However, this should not be construed as evidence that the binding of ribosomes to ER membranes is generally weak or readily disrupted; for example, prolonged homogenization of rat liver (up to 200 strokes of a Potter-type homogenizer) did not change the ratio of free to membrane-bound ribosomes.[5,7]

On the other hand, membrane-bound ribosomes can be removed ("stripped") from microsomal vesicles under certain conditions *in vitro* without influencing the composition or morphology of the membranous vesicles. Thus, incubation of microsomes with EDTA,[1] pyrophosphate,[8] citrate,[9] ribonuclease,[1] 4 M lithium,[10] or trypsin[11] results in a nearly com-

[5] G. Blobel and V. R. Potter, *J. Mol. Biol.* **26**, 279 (1967).
[6] E. A. Smuckler and M. Arcasoy, *Int. Rev. Exp. Pathol.* **1**, 305 (1969).
[7] J. N. Loeb, R. R. Howell, and G. M. Tomkins, *J. Biol. Chem.* **242**, 2069 (1967).
[8] H. Sachs, *J. Biol. Chem.* **233**, 650 (1958).
[9] R. Suss, G. Blobel, and H. C. Pitot, *Biochem. Biophys. Res. Commun.* **23**, 299 (1966).

plete loss of membrane-bound RNA with the concomitant formation of smooth-surfaced vesicles. Smooth vesicles prepared in this manner are indistinguishable from endogenous smooth microsomes[1,9,10]; however, the two membrane fractions appear to differ appreciably in their capacity to bind ribosomes. Incubation of "stripped" RER membrane fractions with freshly isolated ribosomes reveals that only this fraction possesses significant ribosomal rebinding capacity as indicated by electron microscopy and by RNA:protein ratios.[12] By contrast, similarly treated ("stripped") SER membrane fractions or untreated microsomes exhibit negligible ribosomal binding capacity[12] suggesting that smooth microsomes are not simply rough membranes lacking ribosomes. It is possible that the transition of rough microsomes (or RER) to smooth microsomes (or SER) is associated with conformational changes in the ER membrane which bury certain ribosome-binding and enzyme sites and expose others.

In liver cells, the distribution of ribosomes on the ER surface is highly irregular and variable; for example, in transition zones between SER and RER are commonly found mixed vesicles or tubules which have only a partial ribosome complement.[3] Thus, there exist in hepatocytes and in microsomal fractions spectra of ER membranes and vesicles ranging from maximum ribosome "loading" to the ribosome-free (smooth) state.

It is therefore apparent that a rigid classification of ER membranes into smooth and rough types may be neither accurate nor useful. In hepatocytes there may be extensive interconversion of RER and SER by detachment and possibly reattachment of ribosomes.

General Procedures

Methods

Many techniques have been described for separating smooth and rough microsomal membranes, but only two have gained general acceptance and documented reproducibility.

Method of Rothschild. Utilizing discontinuous density gradient centrifugation, Rothschild[4] achieved resolution of rough and smooth microsomal membranes in relatively high purity. In principle, the method is quite simple; smooth and rough membranes differ sufficiently in density to permit the construction of a two-phase gradient which allows penetration of the heavier rough vesicles through the dense phase but excludes or greatly retards passage of smooth vesicles.

[10] T. Scott-Burden and A. O. Hawtrey, *Biochem. J.* **115**, 1063 (1969).
[11] J. Lust and P. Drochmans, *J. Cell Biol.* **16**, 81 (1963).
[12] W. L. Ragland, T. K. Shires, and H. C. Pitot, *Biochem. J.* **121**, 271 (1971).

Finely minced liver is homogenized[13] with three volumes (w/v) of 0.88 M sucrose solution, and the homogenate is centrifuged at 10,000 g_{av} for 25 minutes. The supernatant is aspirated carefully and recentrifuged at 10,000 g for 25 minutes.

One volume of the 10,000 g supernatant fraction is mixed thoroughly with 5 volumes of cold, distilled water; 12 ml of this preparation, now approximately 0.15 M in sucrose, are pipetted into a centrifuge tube. Using a syringe fitted with a long needle,[14] 2 ml of 1.31 M sucrose (density = 1.215 at 0°) are carefully layered *under* the diluted supernatant without significant mixing of the two phases. The tubes are gently placed in a pre-cooled Spinco No. 40 rotor, and centrifuged (0–2°) at 40,000 rpm (105,000 g_{av}) for 8 hours. Acceleration and deceleration time are not included. At the conclusion of this step, rough microsomes are recovered as a pellet at the bottom of the centrifuge tube, and smooth microsomes form a band at the gradient interphase (Fig. 1).[15] The 1.31 M sucrose phase between the smooth vesicle band and the pellet is transparent. The smooth microsomes may be collected by aspiration, diluted suitably with water or 0.15 M KCl (a small volume of the 1.31 M sucrose phase is unavoidably removed with the band of smooth microsomes), and recovered as a pellet after centrifugation (105,000 g for 90 minutes).

In the Spinco number 40 rotor, the duration of centrifugation may be

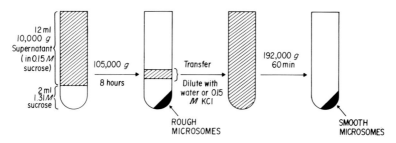

FIG. 1. A schematic representation of the technique described by J. Rothschild [*Biochem. Soc. Symp.* **22**, 4 (1963)], for the subfractionation of microsomal membranes of liver.

[13] We routinely employ a smooth-walled Potter-type homogenizer with a Teflon pestle (A. H. Thomas & Co., Philadelphia, Pennsylvania; size C clearance 0.006–0.009 inch). The pestle is driven by an electric hand drill which generates 2500 rpm. With liver from common laboratory animal species, about five complete excursions of the pestle through the medium give satisfactory results.

[14] A 4-inch 14-gauge unbeveled laboratory cannula is advanced until its tip reaches the bottom of the tube.

[15] Preparations from fed animals exhibit a translucent glycogen pellet underlying the rough microsomal pellet.

critical; 8 hours yielded an excellent resolution of smooth and rough membranes whereas 7 hours or 8.5 hours seemed less satisfactory.[16] As might be expected, the time required for optimal resolution of the two fractions was not as critical with a swinging-bucket rotor (Spinco SW 39L). The sucrose concentration in the upper phase (0.15 M) can be critical and modifications upward to 0.25 M may interfere with satisfactory fractionation.[4] This may account for the poor resolution, turbid lower phase, and cross-contamination reported by some investigators.[17,18]

Method of Dallner. In 1963, Dallner described a method for microsomal fractionation based upon selective aggregation of rough vesicles in the presence of CsCl.[19] This method provides excellent and rapid resolution of smooth and rough membranes and permits handling of relatively large tissue samples for large-scale analysis. Exposure of postmitochondrial supernatants to Cs^+ (10–15 mM) produces aggregation of rough microsomes, a 2- to 3-fold increase in their effective particle size and a 4- to 9-fold increase in their sedimentation velocity.[17] In principle, this method closely resembles that of Rothschild, but utilizes the increased effective particle size of the rough vesicles produced by Cs^+ to decrease the centrifugation time from 8 hours to about 2 hours.

As originally described,[19] livers were homogenized[13] in 0.25 M sucrose (25% homogenate), and the homogenate was centrifuged at 10,000 g_{av} for 20 minutes. The supernatant was aspirated and to each 9.85 ml was added 0.15 ml of 1 M CsCl giving a final Cs^+ concentration of 15 mM. Seven milliliters of this mixture were layered over 4.5 ml 1.30 M sucrose containing 15 mM CsCl in centrifuge tubes,[14] and the gradients were centrifuged at 250,000 g_{av} for 60 minutes (Christ Omega I ultracentrifuge) in a fixed-angle rotor. After centrifugation, the rough microsomes form a pellet at the bottom of the tube and the smooth microsomes accumulate as a well-delineated band at the gradient interphase[15] (Fig. 2). The 1.31 M sucrose phase is transparent. The band of smooth microsomes is removed by aspiration, diluted with water or 0.15 M KCl, and collected as a pellet by centrifugation (250,000 g for 60 minutes). This method has been modified for use with other centrifuges and rotors.[20,21]

Additional Methods. Other methods have been devised for subfractionating hepatic microsomes. Many of these fail to achieve adequate sepa-

[16] J. R. Fouts, L. A. Rogers, and T. E. Gram, *Exp. Mol. Pathol.* **5**, 475 (1966).
[17] G. Dallner and R. Nilsson, *J. Cell Biol.* **31**, 181 (1966).
[18] H. Glaumann and G. Dallnner, *J. Cell Biol.* **47**, 34 (1970).
[19] G. Dallner, *Acta Pathol. Microbiol. Scand.,* Suppl. 166 (1963).
[20] T. E. Gram, L. A. Rogers, and J. R. Fouts, *J. Pharmacol. Exp. Ther.* **155**, 479 (1967).
[21] A. Bergstrand and G. Dallner, *Anal. Biochem.* **29**, 351 (1969).

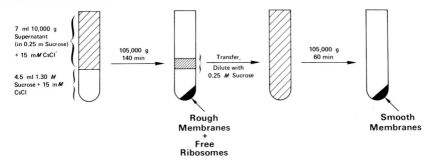

7 ml 10,000 g
Supernatant
(in 0.25 m Sucrose)
+ 15 mM CsCl

4.5 ml 1.30 M
Sucrose + 15 mM
CsCl

105,000 g
140 min

Transfer,
Dilute with
0.25 M Sucrose

105,000 g
60 min

**Rough
Membranes
+
Free
Ribosomes**

**Smooth
Membranes**

FIG. 2. A schematic representation of the technique described by G. Dallner [*Acta Pathol. Microbiol. Scand.,* Suppl. 166 (1963)] for the subfractionation of microsomal membranes of liver.

ration of smooth and rough vesicles resulting in fractions which are cross-contaminated. One such method, devised by Moulé *et al.,*[22,23] attempted to achieve microsomal fractionation by differential centrifugation, a technique not admirably suited to the purpose. From a postmitochondrial supernatant of rat liver in 0.88 M sucrose, two pellets were collected, the first (heavy) fraction by centrifugation at 40,000 g_{max} for 2 hours, the second (light) fraction by centrifuging the supernatant at 145,000 g_{max} for three additional hours. Chemical analysis of the two fractions revealed that RNA was evenly distributed between them. Electron microscopy indicated that although the heavy and light fractions were relatively enriched in rough and smooth vesicles, respectively, distinct separation of smooth and rough vesicles had not been attained and significant cross-contamination of the fractions existed.

Unfortunately, a number of subsequent investigations adopted this method to study subtle interrelationships between RER and SER. One such often-cited investigation[24] attempted to examine smooth and rough microsomes of liver vis-à-vis amino acid incorporation, phospholipid and amino acid composition, and enzyme distribution. Although no morphological evidence of the purity of the subfractions was provided, the authors themselves estimated that their rough-surfaced microsomal fraction contained nearly equal amounts of rough and smooth vesicles. In preparations so grossly impure, the authors' inability to demonstrate significant differences between the two membrane fractions might have been expected but was considered by them to have important implications with respect to the dynamic properties of intracellular membranes. A modification of the method of Moulé *et al.,* further complicated the problem by exposing the

[22] Y. Moulé, C. Rouiller, and J. Chauveau, *J. Biophys. Biochem. Cytol.* **1,** 547 (1960).
[23] J. Chauveau, Y. Moulé, C. Rouiller, and J. Schneebeli, *J. Cell Biol.* **12,** 17 (1962).
[24] V. C. Manganiello and A. H. Phillips, *J. Biol. Chem.* **240,** 3951 (1965).

TABLE II

ESTIMATED MORPHOLOGICAL PURITY OF SMOOTH AND ROUGH
MICROSOMAL FRACTIONS PREPARED FROM LIVER[a]

Species	Microsomal fraction	Method of preparation	
		Rothschild	Dallner
Rat	Rough	20–25%	25–30%
	Smooth	<5%	<1%
Rabbit	Rough	~10%	15–20%
	Smooth	<1%	<1%

[a] Values represent estimated contamination of the indicated fraction by vesicles of the opposite type.

membrane fractions to isooctane during isolation.[25] Unfortunately, isooctane causes detachment of ribosomes from ER membranes, extracts membrane phospholipids, and inactivates many microsomal enzymes. This method has been applied in studies of microsomal RNA and phospholipid turnover.

Morphological Purity of Microsomal Fractions. We have estimated the purity of smooth and rough microsomal preparations by random vesicle counts made on numerous electron microscopic fields. These estimates were made on fractions prepared from livers[20] of two animal species by two different methods in a Spinco No. 40 fixed-angle rotor (Table II). It is possible that higher purities could be achieved through the use of swinging-bucket rotors. In our preparations, the smooth vesicle fractions were invariably of higher purity than the rough vesicle fractions (Tables II and III). In the rabbit, the extent to which enzymes which catalyze drug oxidation were concentrated in smooth microsomes was directly related to the purity of the rough microsomal fraction. Thus, as the relative number of smooth vesicles contaminating the rough vesicle fraction decreased, the enzyme activity of the rough vesicle fraction declined in a roughly proportional manner.

Centrifuge Rotors and Rotor Speeds

A discussion of the attributes of swinging-bucket and fixed-angle rotors in density gradient centrifugation is not possible here. The theoretical advantages of swinging-bucket rotors tend to be offset by their cost, inconvenience, and limited volume capacity. Thus, it is common practice to sacrifice the generally superior resolution provided by bucket rotors for the practical advantages offered by fixed-angle rotors. This compromise

[25] T. Hallinan and H. N. Munro, *Quart. J. Physiol.* **50**, 93 (1965).

TABLE III

DISTRIBUTION OF COMPONENTS OF THE HEPATIC MICROSOMAL MIXED-FUNCTION
OXIDASE SYSTEM BETWEEN SMOOTH AND ROUGH VESICLES[a]

Method of separation	Substrate or component	Animal species		Ref- erence
		Rabbit	Rat	
Rothschild	Aminopyrine	5.4	—	b
	Hexobarbital	4.1	—	
	Codeine	3.5	—	
	Aniline	4.7	2.8	c
	Aminopyrine	26.1	2.5	
	Codeine	4.1	1.8	
	Benzopyrene	1.5	0.9	
	Hexobarbital	7.3	1.4	
Dallner	Aminopyrine	5.2	3.0	c
	Hexobarbital	3.4	1.5	
	Codeine	2.9	2.7	
	Aniline	2.2	2.0	
	Benzopyrene	0.8	2.0	
Dallner	Ethylmorphine	5.7	1.8	d
	NADPH-cytochrome *c* reductase	3.5	—	
	Cytochrome P-450	2.1	—	
	NADPH-cytochrome P-450 reductase	5.6	—	
	Progesterone (total hydroxylation)	3.2	—	e
	Cytochrome P-450	3.3	—	
	Aminopyrine		2.5	b
	o-Chloroaniline	2.8	—	g
	p-Nitrobenzoic acid	—	3.6	h
Glucuronyltransferase	*p*-Nitrophenol	0.5	—	i
	o-Aminophenol	0.5	—	
	Phenolphthalein	1.0	—	
	Bilirubin	—	0.5	h

[a] Data are expressed as a ratio of specific activities, smooth:rough.

[b] J. R. Fouts, *Biochem. Biophys. Res. Commun.* **6**, 373 (1961).

[c] T. E. Gram, L. A. Rogers, and J. R. Fouts, *J. Pharmacol. Exp. Ther.* **155**, 479 (1967).

[d] J. L. Holtzman, T. E. Gram, P. L. Gigon, and J. R. Gillette, *Biochem. J.* **110**, 407 (1968).

[e] G. Lange and K. J. Thun, *Naunyn-Schmiedebergs Arch. Pharmakol. Exp. Pathol.* **267**, 265 (1970).

[f] G. W. Lucier and O. S. McDaniel, *Biochim. Biophys. Acta* **261**, 168–176 (1972).

[g] Y. Ichikawa and T. Yamano, *Biochem. Biophys. Res. Commun.* **40**, 297 (1970).

[h] R. F. Potrepka and J. L. Spratt, *Biochem. Pharmacol.* **20**, 2247 (1971).

[i] T. E. Gram, A. R. Hansen, and J. R. Fouts, *Biochem. J.* **106**, 587 (1968).

introduces the complications of convection, mixing, and impaction of sub-cellular particles against the outer wall. Using the Rothschild procedure, it was found[26] that the large volume Spinco No. 30 rotor (tube angle 26°) did not allow resolution of microsomal subfractions comparable to that achieved with the Spinco No. 40 rotor (tube angle 26°). Thus, fractions prepared in the No. 30 rotor exhibited greater cross-contamination both at equivalent time–velocity (g–minute) values (78,000 g_{av} for 10.7 hours vs 105,000 g_{av} for 8 hours) and at several empirically chosen times bracketing the 10.7 hours. It is of interest in this regard that isopycnic centrifugation of rat liver mitochondria in sucrose gradients has been reported to produce, at different rotor speeds, markedly different enzyme distribution patterns, even though the integrated time velocity values were held constant.[27]

With the Cs[+]-aggregation technique Dallner[19] achieved microsomal fractionation by centrifuging the discontinuous gradients for 60 minutes at 60,000 rpm (250,000 g_{av}) in a fixed angle (34°) rotor. Subsequent investigation, including electron microscopy of the subfractions, revealed that the method could be adapted for use in a Spinco No. 40 rotor.[22] Cross-contamination was sufficiently low and estimatable as to permit the modification. Curiously, our attempts to employ the Dallner technique in an International B-60 ultracentrifuge equipped with an A-321 rotor (tube angle 35°) were not rewarding.[27a] The discontinuous gradients were centrifuged at 60,000 rpm (220,000 g_{av}) for 68 minutes, thus closely approximating the original conditions; resolution of smooth and rough microsomal fractions was unsatisfactory as judged by turbid, unpelleted material in the lower sucrose phase, changes in enzyme distribution (relative to those obtained with a Spinco No. 40 rotor) and increased cross-contamination.

Fixed-angle rotors predispose to particle impaction against the outer wall; this results in spuriously high sedimentation properties. Applied to microsomal fractionation, this tendency would be manifested as contamination of the heavy particle fraction (rough) by light particles (smooth). On this basis, Bergstrand and Dallner[21] argued that the Cs[+] aggregation method should not be used in rotors, such as the Spinco No. 40, having a steep (26°) tube angle. In lieu of the Christ Omega 60 rotor used in earlier experiments[19] (tube angle 34°), these workers utilized a more generally available rotor (Spinco No. 40.2, tube angle 40°) and obtained microsomal subfractions described as comparable in purity to those obtained with the

---

[26] J. R. Fouts, *Biochem. Biophys. Res. Commun.* 6, 373 (1961).

[27] R. Wattiaux, S. Wattiaux-De Conick, and M. F. Ronveaux-Dupol, *Eur. J. Biochem.* 22, 31–39 (1971).

[27a] J. R. Fouts, in "Methods in Pharmacology" (A. Schwartz, ed.), Vol. 1, pp. 287–325. Appleton-Century-Crofts, New York (1971).

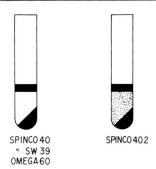

SPINCO 40   SPINCO 40.2
" SW 39
OMEGA 60

FIG. 3. The appearance of the discontinuous sucrose gradients after centrifugation in different rotors.

Omega 60 rotor.[21] However, they acknowledged the "disadvantage," confirmed by us,[28] that the Spinco 40.2 rotor does not effect resolution into two discrete fractions. Instead of a pellet at the bottom of the tube and a discrete band at the gradient interphase, an additional "intermediate" fraction occurs (Fig. 3) which consists of both rough and smooth vesicles. If one is interested in obtaining fractions with the lowest possible cross-contamination, the intermediate fraction must be discarded, thus compromising recovery. However, a much more serious complication arises from the homogeneous turbidity of the lower phase and the difficulty in delineating the boundaries between the smooth microsomes and the intermediate phase (Fig. 3) and between the intermediate phase and rough microsomes. In practical terms, this poses a serious and highly subjective sampling problem. Inasmuch as the intermediate fraction is composed of both rough and smooth microsomes, the prospect of obtaining "pure subfractions" under these conditions seems remote indeed.

Effects of Aggregation and Pelleting on Subfractionation

In his early work, Rothschild[4] observed that smooth microsomes suspended in dilute sucrose solutions undergo extensive aggregation which can interfere with subsequent fractionation. This tendency was prevented by inclusion of KCl (20–50 mM) in all solutions. The predisposition of smooth microsomes to undergo spontaneous aggregation was confirmed by Dallner and Nilsson,[17] who found that it could be prevented by bovine serum albumin. Storage of suspensions of smooth vesicles under conditions of either dilute protein or dilute sucrose concentration favored aggregation which could be prevented by increasing the concentration of either or both.[18] The aggregation of rough vesicles evoked by Cs$^+$ is readily

[28] T. E. Gram, D. H. Schroeder, D. C. Davis, R. L. Reagan, and A. M. Guarino, Biochem. Pharmacol. 20, 1371 (1971).

reversible whereas the aggregation of smooth I vesicles in the presence of Mg^{2+} is essentially irreversible.[17]

Pelleting of smooth vesicles and subsequent resuspension in 0.6 M sucrose results in intense aggregation, which cannot be prevented by serum albumin.[4,29] Moreover, it is now generally recognized that the sedimentation of total microsomes as a pellet seriously compromises any subsequent attempts at subfractionation.[4,30] It is for this reason as well as expediency that postmitochondrial supernatant fractions in preference to resuspended microsomes are routinely used in microsomal subfractionation procedures.

Distribution of the Mixed Oxidase System and Its Components in Microsomal Subfractions

The specific activities of many mixed-function oxidase components are significantly higher in smooth microsomes than in rough microsomes (Table III) prepared from the livers of rabbits, rats, guinea pigs, and monkeys.[20,26,28,31]

Differences appear to exist between animal species in the degree to which activity is concentrated in smooth vesicles (Table III); this uneven distribution is most prominent in rabbits and least prominent in mice. In rabbits, specific activities are higher in smooth microsomes whether they are expressed per unit of protein or per unit of phospholipid.[31]

Animal Species and Extrahepatic Organs

In general, methods for subfractionating microsomes have been developed with rat liver, and there is an obvious potential risk in extrapolating these methods to other animal strains, species, or even organs. It is therefore incumbent upon each investigator to validate his extrapolations by biochemical, morphological, or other criteria.

We have used the Cs^+ aggregation method to fractionate microsomes from the livers of rats, rabbits, guinea pigs, monkeys (rhesus), and mice. No marked differences have been found in the separability or purity of smooth and rough microsomal fractions based on electron microscopical evaluation and on RNA distribution.[20,28] It is conceivable that livers of exotic animals might pose special problems. For example, the livers of certain fish, such as sharks, have uncommonly high levels of total lipids

[29] G. Dallner and L. Ernster, *J. Histochem. Cytochem.* **16**, 611 (1968).
[30] J. R. Gillette, *in* "Fundamentals of Drug Metabolism and Drug Disposition" (B. N. LaDu, ed.), p. 403. Williams & Wilkins, Baltimore, Maryland, 1971.
[31] J. L. Holtzman, T. E. Gram, P. L. Gigon, and J. R. Gillette, *Biochem. J.* **110**, 407 (1968).

which might influence the sedimentation characteristics of subcellular particles.

The Cs^+ aggregation technique has been employed without modification to subfractionate microsomes from thyroid[32,33] and testes[34,35] and with minor modifications with intestinal mucosa,[36] kidney, and pancreas.[29] Unfortunately, in many instances, successful fractionation is termed "adequate" or "satisfactory" with varying degrees of documentation whereas when modifications are deemed necessary to achieve or improve the separation, the basis of and criteria for adequacy of the modifications frequently are not provided.

Assay and Properties

Electron Microscopy

The fundamental distinction between smooth and rough microsomes is morphological, and thus the primary criterion for determining the completeness of separation and purity of fractions should be electron microscopy. Obviously, whenever possible, additional evidence, such as chemical analysis, is highly desirable.

Chemical Analysis

RNA. Since >95% of microsomal RNA is ribosomal, measurements of RNA in smooth and rough membrane fractions should reveal marked concentration of RNA in rough membranes, assuming that free ribosomes are not randomly distributed between the two fractions. Thus, RNA distribution commonly normalized as RNA:protein would seem to be a reliable index of the completeness of separation of smooth and rough vesicles. However, published ratios of RNA:protein in rough and smooth fractions range from about unity to >10. This variability can usually be attributed to inadequate separation of rough vesicles from smooth vesicles or to contamination of smooth vesicles by free ribosomes; in either case electron microscopy may be very useful. It must be emphasized that the exclusive reliance upon chemical estimates of RNA distribution can be misleading because this approach (even when normalized to phospholipid) does not differentiate free and membrane-bound RNA.

[32] C. Cheftel, S. Bouchilloux, and O. Chabaud, *Biochem. Biophys. Acta* **170**, 29 (1968).
[33] T. Hosoya, S. Matsokawa, and Y. Nagai, *Biochemistry* **10**, 3086 (1971).
[34] H. Inano, A. Inano, and B. Tamaoki, *J. Steroid Biochem.* **1**, 83 (1970).
[35] H. Inano and B. Tamaoki, *Biochemistry* **10**, 1503 (1971).
[36] Y. S. Kim, J. Perdomo, and J. Nordberg, *J. Biol. Chem.* **246**, 5466 (1971).

[22] Isolation of Sarcoplasmic Reticulum from Skeletal Muscle

By GERHARD MEISSNER[1]

Sarcoplasmic reticulum (SR) of skeletal muscle forms an extensive network of membranous tubules and cisternae surrounding the muscle fibrils in a sleevelike fashion. During homogenization, part of this SR structure is disrupted and forms vesicular structures that make up most of the "microsomal" fraction of the homogenate. A crude SR vesicle fraction can be isolated in the form of a "microsomal" pellet from the homogenate by differential centrifugation. The crude SR preparation is then further purified by sucrose gradient centrifugation[2-5] or by washing with 0.6 M KCl to remove contaminating muscle protein.[6] In the best preparations, contamination with other cell organelles is low; however, appreciable amounts of extraneous muscle proteins and neutral lipids may be associated with the SR preparation.

The isolation procedure to be described is distinct in that it avoids the formation of a "microsomal" pellet and extensive washing with salt. After a low speed centrifugation to remove cell debris, nuclei, mitochondria and myofibrils, SR vesicles are isolated from the supernatant fraction by two successive sucrose gradient centrifugations using zonal rotors. The first zonal rotor centrifugation serves both to reduce the volume of the SR preparation as well as to effect partial purification. The crude SR preparation is further purified in a second zonal run containing a density gradient from 26 to 37% sucrose. Only small amounts of extraneous muscle proteins and neutral lipids remain with the purified SR vesicles.[7]

[1] The encouragement and advice of Dr. S. Fleischer during the course of this work is gratefully acknowledged. This work was done in part during the tenure of an Established Investigatorship of the American Heart Association. The studies were also supported by a Special Fellowship of the United States Public Health Service (CA 41263), a United States Public Health Service Grant (AM-14632), and Grant-in-Aids from the Tennessee Heart and Middle Tennessee Heart Associations.
[2] W. Hasselbach and M. Makinose, *Biochem. Z.* 339, 94 (1963).
[3] K. Seraydarian and W. F. H. M. Mommaerts, *J. Cell Biol.* 26, 641 (1965).
[4] B. P. Yu, F. D. DeMartinis, and E. J. Masoro, *Anal. Biochem.* 24, 523 (1968).
[5] G. Meissner and S. Fleischer, *Biochim. Biophys. Acta* 241, 356 (1971).
[6] A. Martonosi, *J. Biol. Chem.* 243, 71 (1968).
[7] G. Meissner, G. E. Conner, and S. Fleischer, *Biochim. Biophys. Acta* 298, 246 (1973).

Preparation of Sarcoplasmic Reticulum Vesicles

Albino rabbits, about 3 kg each, are killed by injecting an overdose of Nembutal (60 mg/kg rabbit). The leg and back muscles are excised, chilled in ice, freed of fat and connective tissue, and passed through a meat grinder. All operations are carried out at 0–4°. Ground muscle, 63 g, is homogenized in 180 ml of 0.3 M sucrose (Grade Ultrapure, Schwarz/ Mann, Orangeburg, New York 10962) containing 10 mM HEPES (N-2-hydroxyethylpiperazine-N'-2-ethanesulfonic acid) (pH 7.5) for 30 seconds in a Waring Blendor. The homogenate of 9 blendings is centrifuged in six bottles for 20 minutes at 8000 rpm (7020 g, average force) in a JA 10 rotor in a Beckman Model J-21 centrifuge. The supernatant is poured through five layers of cheesecloth to remove various amounts of floating fat particles. A Spinco Ti 15 zonal rotor equipped with a B-29 type core plus liner to allow loading and unloading of the gradient from the edge of the rotor is filled with 1200 ml of the filtrate. After acceleration of the rotor to 4000 rpm, a steep, narrow sucrose gradient is established at the outer edge of the rotor. Three sucrose solutions consisting of 80 ml of 24%, 30 ml of 40%, and 50 ml of 50% sucrose in 5 mM HEPES (pH 7.4) are pumped in at a speed of 10 ml per minute with a Sage tubing pump Model 375. The "percent sucrose" (w/w) is adjusted using a Bausch and Lomb refractometer at 25°. The rotor is then accelerated to 34,000 rpm and maintained at this speed for 45 minutes. For unloading, the rotor is decelerated to 3000 rpm, and distilled water is pumped into the center of the rotor (20 ml per minute) to unload a crude SR preparation from the edge of the rotor. The material sedimenting in the gradient between 40 and 23% sucrose is collected. The contents of two zonal runs, run either in parallel or sequentially, are combined and diluted with 5 mM HEPES (pH 7.4) to give about 250 ml of a crude SR preparation adjusted to 18% sucrose.

For further purification the SR preparation is placed into a Spinco Ti 14 zonal rotor, and a sucrose gradient is established as the rotor spins at 4000 rpm by successively pumping in at the outer edge 70 ml each of 26.5%, 29.1%, 31.6%, 33.9%, and 37.1% sucrose containing 5 mM HEPES (pH 7.4) and about 50 ml of 50% sucrose (at a rate of 8 ml per minute). The rotor is accelerated to and centrifuged at 43,000 rpm for a total of 60 minutes. This time of centrifugation has been found to be optimal. The majority of the SR vesicles have entered the gradient (26–30% sucrose) and are well separated from mitochondrial fragments and extraneous muscle proteins which are present in the gradient ranging from 31% sucrose and higher. The rotor is then decelerated to 4000 rpm and the sample is unloaded through the center of the rotor by pumping in a solution of 55% sucrose. The first 150 ml are discarded. The following 100 ml and then fractions of 25 ml are collected so that the material sedimenting in the gradient in up to 26.5% (fraction 1) and in 26.5–30.5% (fraction 2) sucrose can be combined. The material sedimenting between 30.5–33% and 33–40% sucrose contains in addition to SR appreciable amounts of mitochondrial fragments and other muscle proteins, respectively, and is therefore generally discarded. Fraction 1 is diluted with an equal volume of 1.2 M KCl in 2.5 mM HEPES (pH 7.5) and kept in ice overnight to remove contaminating muscle proteins. Fraction 2 may be stored in ice overnight and then diluted with an equal volume of 2.5 mM HEPES (pH 7.5) before centrifugation. Both fractions are centrifuged at 30,000 rpm for 90 minutes in a Spinco 30 rotor. The yellowish translucent pellets are resuspended in 0.3 M sucrose in 1 mM HEPES (pH 7.5) to a protein concentration of 10–20 mg/ml and are then quick-frozen using liquid N_2. The preparations are stable for many months when stored at $-70°$ with regard to energized Ca^{2+} accumulation and [^{32}P]phosphoenzyme formation.

The yield is about 300 mg of protein of a highly purified SR vesicle preparation using two rabbits which give 1000 g of skeletal muscle. An approximately equal amount of SR of somewhat lesser purity is obtained in addition (see the table). These yields correspond to a recovery of 10–15% of the SR of skeletal muscle as judged by the [^{32}P]phosphoenzyme levels of total muscle (~ 0.15 nmole of ^{32}P per milligram of protein)[7] and the two purified fractions together. The low recovery raises the question as to whether the isolated SR vesicles originate from limited areas of the SR network; e.g., they could be derived from the longitudinal sections of SR which are believed to be mainly responsible for reabsorbing Ca^{2+}, after it has been released by the cisternae of SR during the excitation event.[8] The yield may be somewhat increased by prolonging the time of homogenization to 45 seconds or 1 minute in the Waring Blendor. However, under these conditions, fraction 2 is slightly contaminated with extraneous muscle proteins. These can be removed by washing in 0.6 M KCl using identical conditions as described above for fraction 1. The crude SR preparation may be stored for many hours in a solution of sucrose of 25% or higher without loss of energized Ca^{2+} loading capacity or [^{32}P]phosphoenzyme concentration. The procedure may therefore be scaled up for 2000 g of muscle tissue, four Ti 15 zonal rotors (two times two in parallel) being used for the first separation. The SR preparation is then further purified in a Ti 15 zonal rotor using 140 ml each of the five sucrose gradient solutions mentioned above. In this case optimal

[8] S. Winegrad, *J. Gen. Physiol.* **55**, 77 (1970).

separation of SR vesicles and contaminants is obtained by accelerating the rotor to 34,000 rpm and maintaining it at this speed for 60 minutes.

Characterization of Sarcoplasmic Reticulum Vesicles

Vesicles derived from SR of skeletal muscle rapidly accumulate Ca^{2+} from the medium through the action of a Ca^{2+}-stimulated ATPase.[2,9] A covalently linked phosphoenzyme intermediate mediates the splitting of ATP.[10,11] Ca^{2+}-dependent [^{32}P]phosphoenzyme formation and Ca^{2+} loading capacity (in the presence of 5 mM oxalate) have been found in our experience to be the two most reliable and informative assays for characterizing the purity and quality of a SR preparation. Ca^{2+}-stimulated ATPase activity, which is often used by others to estimate the purity of SR preparations, has been less useful in our hands. This is understandable since the ATPase activity is also dependent on the leakiness of the membrane.

PROPERTIES OF SARCOPLASMIC RETICULUM VESICLES[a]

		Purified SR	
	Crude SR	Fraction 1	Fraction 2
Yield (mg protein/1000 g muscle)	1700 ± 200	320 ± 50	300 ± 50
μg bound P/mg protein	—	21.5 ± 2.0	24.5 ± 1.0
Ca^{2+} loading (μmoles Ca^{2+}/mg protein)	2.7 ± 0.4	3.0 ± 0.5	5.5 ± 0.7
^{32}P-phosphoenzyme (nmoles ^{32}P/mg protein)	3.0 ± 0.3	5.4 ± 0.4	6.4 ± 0.3
ATPase (μmoles P_i/mg protein per min)			
With 1 mM EGTA	0.15 ± 0.05	0.20 ± 0.05	0.10 ± 0.05
With 0.4 mM Ca^{2+}-0.5 mM EGTA[b]	0.7 ± 0.1	1.3 ± 0.2	1.4 ± 0.3
Bound Ca^{2+} (nmoles Ca^{2+}/mg protein)	35 ± 5	30 ± 5	70 ± 5
Bound Mg^{2+} (nmoles Mg^{2+}/mg protein)	65 ± 10	17 ± 4	60 ± 4

[a] Crude and purified SR vesicles are obtained from the first and subsequent zonal rotor separations, respectively. Fraction 1 of the second zonal rotor separation is extracted with 0.6 M KCl overnight at 0°. Fraction 2 contains only small amounts of KCl extractable protein and is not extracted with salt. Bound phosphorus is measured as an estimate of lipid phosphorus [G. Meissner and S. Fleischer, *Biochim. Biophys. Acta* **241**, 356 (1971)] using a modification (G. Rouser and S. Fleischer, this series, Vol. 10, p. 385) of the method of P. S. Chen, T. Y. Toribara, and H. Warner [*Anal. Chem.* **28**, 1756 (1956)]. Protein is determined by the procedure of O. H. Lowry, N. J. Rosebrough, A. L. Farr, and R. J. Randall, [*J. Biol. Chem.* **193**, 265 (1951)] using bovine serum albumin as a standard. Each value represents the mean of 10 determinations ± SE.

[b] EGTA, ethyleneglycol-bis-(β-aminoethylether)-N,N'-tetraacetic acid.

[9] S. Ebashi and F. Lipmann, *J. Cell Biol.* **14**, 389 (1962).
[10] T. Yamamoto and Y. Tonomura, *J. Biochem.* **64**, 137 (1968).
[11] M. Makinose, *Eur. J. Biochem.* **10**, 74 (1969).

All three assays are described below in detail, and the activities of the crude and the two purified SR vesicle preparations are summarized in the table. The ratio of total phosphorus to protein may be determined as a preliminary test for the presence of extraneous muscle proteins.

A more detailed characterization of the SR preparation includes determination of the protein profile by polyacrylamide gel electrophoresis

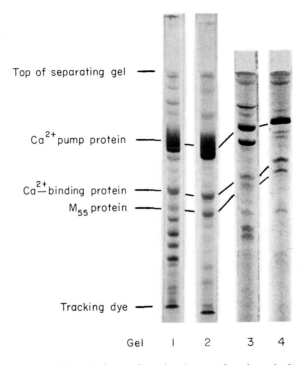

Top of separating gel —

Ca^{2+} pump protein —

Ca^{2+}—binding protein —

M_{55} protein —

Tracking dye —

Gel   1   2   3   4

Fig. 1. Polyacrylamide gel electrophoresis of sarcoplasmic reticulum proteins on sodium dodecyl sulfate (SDS) and soaked-acid gels. SDS gels (gels 1 and 2) consisting of a stacking (3% polyacrylamide) and separating gel (7.5% polyacrylamide) are prepared and run as described by U. K. Laemmli [*Nature* (*London*) **227**, 680 (1970)]. Samples are reduced in $4 M$ urea, 2% SDS, 60 mM Tris·HCl (pH 6.8), and 3% 2-mercaptoethanol by heating for 4 minutes at 100°. The sample mix (approximately 0.05 ml) contains also 0.001% bromophenol blue as a tracking dye. Soaked-acid gels (7.5% polyacrylamide, gels 3 and 4) are prepared and run as described previously [G. Meissner and S. Fleischer, *Biochim. Biophys. Acta* **241**, 356 (1971)]. The samples are not reduced prior to electrophoresis. SDS and soaked-acid gels contain 50 and 20 μg protein, respectively, and are stained with 1% Amido Schwartz as described previously [Meissner and Fleischer, *loc. cit.;* G. Meissner, G. E. Conner, and S. Fleischer, *Biochim. Biophys. Acta* **298**, 246 (1973).

Gels 1 and 3 are from crude sarcoplasmic reticulum; 2 and 4, from purified sarcoplasmic reticulum, fraction 2.

(Fig. 1), estimation of the content of neutral lipids which may be present as contaminants, and measurement of endogenous Ca^{2+} and Mg^{2+} by atomic absorption spectroscopy.[7,12] A relatively simple protein and lipid composition is found, in accordance with the apparent restriction of SR to one function, i.e., Ca^{2+} release and uptake. There are altogether three major proteins as observed by polyacrylamide gel electrophoresis in soaked-acid and SDS gels.[7] The Ca^{2+}-pump protein has a molecular weight of about 105,000 and, based on a phosphoenzyme level of 6.4 nmoles of ^{32}P per milligram of protein (see table), accounts to about 70% of the total protein of SR vesicles. Two other bands, designated Ca^{2+}-binding and M_{55} proteins, account each for about 5–10% of the SR protein. MacLennan[13] describes the purification of the three major proteins using SR vesicles prepared by differential centrifugation. The purification of the Ca^{2+}-pump and Ca^{2+}-binding proteins may be simplified using fraction 2 of the zonal rotor separation as a starting material.[7] Phospholipids account for more than 95% of the total lipid[7,14] with phosphatidylcholine, phosphatidylethanolamine, and phosphatidylinositol, accounting for about 95% of the total phospholipid.[5,14]

Contamination with other organelles is low in the preparation described. Sarcolemma and lysosomes are practically absent as judged by nonexistent or insignificant 5′-nucleotidase[15] and acid phosphatase[16] activities, respectively. Contamination with inner mitochondrial fragments is less than 1% as indicated by a succinate–cytochrome c reductase activity[17] of less than 0.005 μmole of cytochrome c reduced per minute per milligram of protein.

Assays for Characterization of Sarcoplasmic Reticulum Vesicles

Ca^{2+}-Dependent [^{32}P]Phosphoenzyme Formation

Ca^{2+}- and ATP-binding studies have shown that the [^{32}P]phosphoenzyme concentration under steady-state conditions agrees closely with the number of active Ca^{2+}-pump protein sites of SR[18] and can thus be used to test the purity and activity of a SR vesicle preparation.

[12] J. B. Willis, *Methods Biochem. Anal.* **11**, 1 (1963).

[13] D. H. MacLennan, this series, Vol. 32.

[14] W. R. Sanslone, H. A. Bertrand, B. P. Yu, and E. J. Masoro, *J. Cell. Physiol.* **79**, 97 (1972).

[15] R. H. Michell and J. N. Hawthorne, *Biochem. Biophys. Res. Commun.* **21**, 333 (1965).

[16] R. Wattiaux and C. de Duve, *Biochem. J.* **63**, 606 (1956).

[17] S. Fleischer and B. Fleischer, this series, Vol. 10, p. 406.

[18] G. Meissner, *Biochim. Biophys. Acta* **298**, 906 (1973).

Reagents

 HEPES buffer, 40 mM, pH 7.5 at 0°, containing 400 mM KCl and
  20 mM MgCl$_2$

 Calcium chloride, 2 mM

 EGTA (ethyleneglycol-bis-(β-aminoethylether)-N,N'-tetraacetic acid),
  20 mM, pH 7.5

 SR vesicles, approximately 10 mg of protein per milliliter (cf. section on preparation of vesicles)

 [γ-^{32}P]ATP, 5 mM, pH 7.0, 10^3–10^4 cpm per nanomole of ATP

 Trichloroacetic acid, 2% and 5%. The solutions are prepared 1 day
  before use and are kept at 4°. Before the assay they are placed
  into ice and a third solution of trichloroacetic acid is prepared
  by diluting 4 volumes of 5% trichloroacetic acid with 1 volume
  of a solution containing 2.5 mM ATP and 5 mM P$_i$

 Sodium hydroxide, 0.1 M, containing 2% Na$_2$CO$_3$ and 0.5%
  sodium dodecyl sulfate

 Scintillation fluid: 60 g of naphthalene, 4.2 g of PPO (2,5-diphenyl-
  oxazole), 180 mg of POPOP {1,4-bis[2-(5-phenyloxazolyl)]}
  benzene, and 70 ml of water in 900 ml of dioxane

Procedure. The phosphorylation reaction and subsequent washing steps
are carried out at 0°. The reagents are added to a 40-ml centrifuge tube
sequentially: (1) distilled water to a final volume of 0.98 ml; (2) 0.25 ml
of HEPES buffer; (3) 0.05 ml of CaCl$_2$ to a sample tube or 0.05 ml of
EGTA to a control sample tube; and (4) SR vesicles, to give a final
concentration of 1 mg of protein per milliliter. Phosphoenzyme formation
is initiated by the addition of 20 μl of [^{32}P]ATP with rapid stirring and
is stopped after 6 seconds by addition of 25 ml of ice cold 4% trichloroacetic acid containing 0.5 mM ATP and 1 mM P$_i$.[19] The suspension is
centrifuged at 20,000 g for 15 minutes and the pellet is washed twice
with 25 ml of 2% trichloroacetic acid and once with 10 ml of 5% trichloroacetic acid by resuspension and centrifugation. The final centrifugation step is carried out at 5000 g in a glass centrifuge tube. The pellet
is then resuspended in 1 ml of 0.1 M NaOH containing 2% Na$_2$CO$_3$ and
0.5% SDS and heated at 100° for 5 minutes. Aliquots are taken for
protein and radioactivity determination. For the latter a 0.1-ml sample
is added to 5 ml of scintillation fluid.

Measurement of Ca^{2+} Loading

 Ca^{2+} uptake by isolated SR vesicles is indicative of a functional Ca^{2+}
transport system. Upon energization with ATP, SR vesicles have the ability

[19] R. L. Post and A. K. Sen, this series, Vol. 10, p. 773.

to sequester 0.1–0.2 μmole Ca^{2+} per milligram of protein. Ca^{2+} accumulation is greatly increased (>1 μmole Ca^{2+} per milligram of protein) when carried out in the presence of 5 mM oxalate (Ca^{2+} loading). A precipitate of calcium oxalate is formed inside the vesicles, and Ca^{2+} loading may be easily and rapidly measured using $^{45}Ca^{2+}$ and the Millipore filtration technique.

Reagents

HEPES buffer, 20 mM, pH 7.2, containing 200 mM KCl, 10 mM $MgCl_2$ and 0.2 mM $^{45}CaCl_2$ (approximately 10^4 cpm/ml)

Potassium oxalate, 50 mM

ATP, 100 mM, pH 7.0

SR vesicles, approximately 10 mg of protein per milliliter (cf. section on preparation of SR vesicles)

0.3 M sucrose–1 mM HEPES, pH 7.5

Scintillation fluid: 60 g of naphthalene, 4.2 g of PPO, 180 mg of POPOP, and 70 ml of water in 900 ml of dioxane

Procedure. The reagents are added sequentially as follows: (1) distilled water to a final volume of 2.5 ml; (2) 1.25 ml of HEPES buffer; (3) 0.1 ml of ATP; (4) 0.25 ml of potassium oxalate; and (5) SR vesicles (0.02–0.05 mg protein), or a corresponding volume of 0.3 M sucrose–1 mM HEPES to a control tube. The reaction is carried out at room temperature for 8 minutes and is terminated by pressing the mixture through a Millipore filter (GS 0.22 μm) in a syringe adapter. The control mixture without added SR is used to determine the zero time concentration of $^{45}Ca^{2+}$. To obtain the percentage of $^{45}Ca^{2+}$ which is accumulated by the SR vesicles, the radioactivity of the filtrates (0.1 ml) is counted in 4 ml of scintillation fluid.

Measurement of Ca^{2+}-Stimulated ATPase Activity

Ca^{2+}-stimulated ATPase activity is estimated as the difference between total and Ca^{2+}-independent ("basic") ATPase activity. Total ATPase activity is measured in a medium containing 0.4 mM $CaCl_2$ and 0.5 mM EGTA. "Basic" ATPase activity is determined as total ATPase activity except that the free Ca^{2+} concentration is kept about 10^{-8} M. No Ca^{2+} is added to the solution and the EGTA concentration is increased to 1 mM.

Reagents

HEPES buffer, 20 mM, pH 7.2, containing 200 mM KCl and 10 mM $MgCl_2$

4 mM $CaCl_2$–5 mM EGTA

EGTA, 10 mM
SR vesicles, approximately 10 mg protein per milliliter (cf. section
    on preparation of SR vesicles)
ATP, 100 mM, pH 7.0
Perchloric acid, 1.5 M

Procedure. The reagents are added sequentially as follows: (1) distilled water to a final volume of 2.0 ml; (2) 1.0 ml of HEPES buffer; (3) 0.2 ml of 4 mM CaCl$_2$–5 mM EGTA (for total ATPase activity), or 0.2 ml of 10 mM EGTA (for basic ATPase activity); and (4) SR vesicles (0.03–0.10 mg of protein). The samples are preincubated for 5 minutes at 32°. The reaction is then started by the addition of 0.08 ml of ATP and stopped after 2.5, 5, and 10 minutes with 0.7 ml of 1.5 M HClO$_4$. Inorganic phosphate is determined on 1 ml of the protein-free supernatant[20] using Elon Developing Agent (0.2 g/100 ml, Eastman Kodak Co.) as a reducing agent.

[20] C. H. Fiske and Y. SubbaRow, *J. Biol. Chem.* **66**, 375 (1925).

[23] Isolation of Nuclei from the Thymus

By VINCENT G. ALLFREY

The lymphocyte population of the thymus gland is particularly well suited for nuclear isolations because the nuclei comprise such a large proportion of the total cell mass. In the calf thymus, for example, nuclei comprise about 61% of the tissue by weight. This has made possible a variety of isolation procedures which yield nuclear fractions of high purity in amounts sufficient for extensive biochemical study. The choice of isolation procedure depends largely upon the aims of the investigator because the purity, physical state, enzyme composition, and biosynthetic activity of the isolated nuclei are greatly influenced by the fractionation procedure employed. For studies of nuclear composition, particularly with regard to low-molecular-weight diffusible constituents (such as nucleotide or amino acid "pools"), isolations in *nonaqueous* media are necessary, because they preclude any exchange of water-soluble components between nucleus and cytoplasm when the cells are broken.[1,2] Such "nonaqueous" methods, however, would not be recommended for studies of the nuclear envelope,

[1] V. Allfrey, H. Stern, A. E. Mirsky, and H. Saetren, *J. Gen. Physiol.* **35**, 529 (1952).
[2] V. G. Allfrey, *in* "The Cell" (J. Brachet and A. E. Mirsky, eds.), Vol. 1, p. 193. Academic Press, New York, 1959.

because the organic solvents employed in grinding and fractionating the lyophilized tissue are known to extract some of the lipid components of the membranes. For this reason, recourse is made to cell fractionations in aqueous media, usually sucrose solutions, which permit the isolation of thymocyte nuclei under relatively gentle conditions. Such "sucrose" nuclei are capable of extensive metabolic and biosynthetic activity.[2-5] The advantages of purity and high yield recommend the isolation procedures now to be described.

Isolation of Calf Thymus Nuclei in Isotonic Sucrose: Method I

In outline, the procedure is as follows: minced calf thymus tissue is suspended in 0.25 M sucrose containing small amounts of divalent cations, and the cells are broken by shearing forces, usually in a blender running at low speed. The nuclei are sedimented by differential centrifugation and washed to remove loosely adhering cytoplasmic debris.

Calf thymus glands from 12- to 20-week-old calves are obtained *as quickly as possible* after the death of the animal and transported to the laboratory in ice cold 0.25 M sucrose solution. It is important that the gland be excised quickly, otherwise autolysis will inactivate the nuclei and destroy nuclear components. Postmortem changes in the thymus are very rapid. In the rat, for example, ATP concentrations drop by over 50% within 5 minutes after the death of the animal.[6] Proteolysis of nuclear proteins, particularly the F3 and F1 histone fractions, also occurs unless special precautions are taken.

Fifty grams of fresh calf thymus tissue is finely minced with scissors and placed in a Waring Blendor vessel with 50 ml of cold 0.50 M sucrose and 400 ml of cold 0.25 M sucrose containing 3 mM CaCl$_2$. (Other divalent cations, such as Mg^{2+}, can also be used. However, calcium has been preferred for studies of nuclear energy metabolism because it uncouples oxidative phosphorylation in mitochondria without affecting nuclear ATP synthesis.[7]) The presence of divalent cations in the sucrose medium acts to stabilize nuclear structure. In their absence, the nucleohistone gel within the nuclei tends to expand—especially if the pH is allowed to exceed neutrality—and the nuclei rupture to form an unmanageable gel. If divalent cations must be avoided, 1 mM spermidine may be used instead.[8]

[3] V. G. Allfrey, V. C. Littau, and A. E. Mirsky, *J. Cell Biol.* **21**, 213 (1964).

[4] V. G. Allfrey and A. E. Mirsky, *Proc. Nat. Acad. Sci. U.S.* **48**, 1590 (1962).

[5] B. S. McEwen, V. G. Allfrey, and A. E. Mirsky, *J. Biol. Chem.* **238**, 758, 2571, 2579 (1963).

[6] S. Osawa, V. G. Allfrey, and A. E. Mirsky, *J. Gen. Physiol.* **40**, 491 (1957).

[7] V. G. Allfrey, *Exp. Cell Res.,* Suppl. 9, 418 (1963).

[8] E. L. Gershey, G. Vidali, and V. G. Allfrey, *J. Biol. Chem.* **243**, 5018 (1968).

The cells are broken by homogenization for 4 minutes at 700 rpm. This procedure breaks over 90% of the cells without appreciable destruction of the nuclei. More prolonged blending, or the use of higher speeds, must be avoided because the nuclei are fragile and cannot withstand very rigorous treatment. The effects of prolonged homogenization or high speed mixing are to destroy nuclei and leave many small thymocytes intact. As a result, nuclear yields are low and cell contamination becomes appreciable.[3]

All the following operations are carried out at 2°. The tissue homogenate is filtered through a double layer of gauze (Johnson & Johnson type I) and then through a single-thickness of double-napped flannelette. The filtrate is centrifuged at 700 g for 10 minutes and the supernatant is discarded. The nuclear sediment is resuspended in two volumes (100 ml) of 0.25 M sucrose–3 mM CaCl$_2$, transferred to a 100-ml cylinder and allowed to settle for 10 minutes. The supernatant is carefully decanted through a double-thickness of gauze and a double-thickness of flannelette, and the clumps of nuclei, fiber, and whole cells that remain at the base of the cylinder are discarded. The filtrate is centrifuged at 400 g for 7 minutes, and the supernatant phase is discarded. The nuclear sediment is washed twice more with 120-ml portions of 0.25 M sucrose–3 mM CaCl$_2$.

Nuclear fractions obtained in this way are better than 90% pure as judged by tests for cytoplasmic enzymes and by cytological examination.[3,5] The contamination consists mainly of small cytoplasmic tabs or strands attached to some of the nuclei, plus a small proportion of intact cells (red cells plus very small thymocytes). The whole cell contamination, as judged by electron microscopy of the nuclear pellet, ranges from 29 to 77 cells per 1000 nuclei. These figures are in accord with estimates based on differential staining of cells and free nuclei by basic dyes.[3] The cells can be removed by centrifugation of the nuclei through density barriers, as described below, or alternatively, the number of intact cells present in the original homogenate can be lowered by the use of hypotonic conditions during blending, as described in Method II, below.

It should be stressed that the fragility of thymocytes varies from one species to another and with the physiological state of the animal. The above procedure applies to calf thymus and is not as successful with rat thymus because of the resistance of the rat thymocytes to shearing under the conditions of low-speed blending.

Isolation of Calf Thymus Nuclei: Method II

Cell Disruption in Hypotonic Sucrose Solution.[3] In this method cells are swollen in hypotonic sucrose briefly to increase their fragility during blending. The homogenate is then readjusted to isotonicity before isolating the nuclei. All operations are carried out at 0–2°. Fifty grams of fresh calf thymus tissue is finely minced with scissors and transferred to a Waring Blendor vessel of 1-liter capacity containing 450 ml of 0.20 M sucrose–3 mM CaCl$_2$. The cells are broken by shearing at 700 rpm for 3 minutes, using a high-torque motor. At that time, sufficient 0.5 M sucrose–3 mM CaCl$_2$ is added to bring the sucrose concentration to 0.25 M, and the mixture is filtered through two layers of gauze (Johnson & Johnson type I) and through two layers of prewashed, double-napped flannelette. The filtrate is blended for an additional 2 minutes at 700 rpm and filtered again through a double thickness of flannelette.

Sedimentation of the Nuclear Fraction. The filtrate is centrifuged at 400 g_{av} for 7 minutes, and the supernatant phase is discarded. The nuclear sediment is washed twice more with 120-ml portions of 0.25 M sucrose– 3 mM CaCl$_2$.

At this stage, the nuclei are better than 90% pure, as judged by the absence of cytoplasmic marker enzymes, such as succinoxidase or cytochrome c oxidase. Electron microscopic examination shows the presence of some cytoplasmic tabs attached to the outer nuclear membrane, and a small percentage of intact cells (about 44 cells per 1000 nuclei, on the average). These estimates of cell contamination are in good agreement with other estimates based on staining the nuclear suspension with 0.2% crystal violet in 0.25 M sucrose; in such tests 2–6% of the suspended particles resisted immediate penetration by the dye and are presumed to be intact cells. The remaining 94–98% were nuclei and stained darkly within a few seconds.

Further purification of the nuclei can be achieved by centrifugation through density barriers.

Purification of Thymus Nuclei by Centrifugation in Dense Media

Procedures which involve sedimentation of the nuclei through media of high density depend on the fact that free nuclei, because of their high nucleic acid content, have a higher average density than do whole cells or nuclei contaminated by cytoplasm.

The sucrose-layering procedure now to be described is similar in principle to the original method of Chauveau *et al.,*[9] which utilizes 2.2 M sucrose to float whole cells and cytoplasmic debris away from the denser nuclei of rat liver cells.

Suspensions of calf thymus nuclei in 0.25 M sucrose–3 mM CaCl$_2$ (containing about 40 mg of nuclei, dry weight, per milliliter) are carefully layered over a two-step barrier consisting of a 1.6 M sucrose layer above a 2.0 M sucrose layer. This operation is conveniently carried out

[9] J. Chauveau, Y. Moulé, and C. Rouiller, *Exp. Cell Res.* **11**, 317 (1956).

by placing 10 ml of 2.0 M sucrose–3 mM CaCl$_2$ in the bottom of each centrifuge tube (1 × 3 inch Lusteroid tubes of 30-ml capacity) and then carefully layering 10 ml of 1.6 M sucrose–3 mM CaCl$_2$ over this base. The nuclear suspension in isotonic sucrose–3 mM CaCl$_2$ is applied next, adding 10 ml per tube, and the tubes are centrifuged at 45,000 g_{av} for 30 minutes in the swinging-bucket SW 25 rotor of the Beckman Model L-2 ultracentrifuge. The upper clear layer is removed by aspiration, and the interphase zone, which contains light nuclei and intact cells, is removed with a syringe. The dense sucrose solution is then decanted, leaving the purified nuclear pellet.

Nuclear sediments obtained in this way show less than 2% contamination by intact thymocytes, as judged by electron microscopy and staining tests. About 60–70% of the nuclei in the original isotonic sucrose suspension are recovered in the sediment.

Thymus nuclei isolated by methods I and II contain a complete nuclear envelope, i.e., an inner membrane and an outer membrane studded with ribosomes. The latter can be removed by exposing the nuclei to detergents, such as sodium deoxycholate, Triton X-100, or Cemulsol NPT 6. Removal of the outer membrane in 0.25% Triton X-100 has been employed as a purification step in studies of cytochrome pigments attached to the inner nuclear membrane.[10]

Isolation of Calf Thymus Nuclei in Nonaqueous Media

The nonaqueous isolation procedures are the most reliable methods for the study of intracellular enzyme distributions, and they are the only methods that allow quantitative measurements of low-molecular-weight, water-soluble constituents of the nucleus.[1,2] The steps in the isolation procedure are as follows: (1) the tissue is quickly frozen, lyophilized and minced; (2) the dry powder so obtained is suspended in petroleum ether and ground in a ball mill to fragment the cells and liberate the nuclei; (3) the nuclei are then separated from both lighter and heavier components of the ground tissue suspension by alternate sedimentations and flotations in cyclohexane–carbon tetrachloride mixtures of varying specific gravity.[1] In almost all cases, the nuclei have a characteristic and relatively high density which permits their separation from blocks of free cytoplasm and the partly ground cells which form a large part of the ground tissue suspension. For if the suspension is adjusted to a specific gravity slightly lower than that of the nuclei, centrifugation will sediment the nuclei and float the lighter (cytoplasmic) components. And in organic media of

[10] K. Ueda, T. Matsuura, N. Date, and K. Kawai, *Biochem. Biophys. Res. Commun.* **34**, 322 (1969).

higher specific gravities, the nuclei can be floated away from heavier cellular debris.

Preparation of Tissue. Fresh calf thymus tissue is trimmed and cut into thin slices that are dropped into liquid nitrogen. After submersion for 10–20 minutes, the tissue chips are transferred to wide-mouth flasks and lyophilized. The flasks are maintained under vacuum for 60 hours. (To avoid thawing, the drying is not accelerated by heat.) The frozen dry tissue is pulverized in a Waring Blendor, and the resulting dry powder is stored frozen at −20° until needed. All further operations are at 2°.

Grinding. A 100-g portion of the dried tissue powder is transferred to a porcelain milling jar (type A, 15 cm diameter; Paul O. Abbe, Inc., Little Falls, New Jersey) containing 450 ml of petroleum ether (bp 60°) and 1400 g of irregular grinding stones (2–3 cm in diameter). The jar cover is locked into position and the jar is mounted on its side on motor-driven rollers and rotated at 110 rpm for 44–48 hours.[1]

To remove coarse fiber, the ground suspension is passed either through a wire screen (35 mesh, 0.5 mm apertures) or through a single layer of gauze (Johnson & Johnson type I). Grinding stones and filter pads are washed twice with about 100 ml of petroleum ether and the washings and filtrate are combined.

Isolation of Nuclei. The suspension is transferred to glass centrifuge tubes of 100-ml capacity and centrifuged at 900 g for 20 minutes. The supernatant is discarded. The sediment is resuspended in 800 ml of 1:1 cyclohexane–CCl_4 mixture and centrifuged at 900 g for 20 minutes. (In this and following operations with the organic solvents, good ventilation should be provided.) The supernatant is discarded. The residue is washed twice more with the 1:1 solvent mixture as described, discarding the supernatants.

The sediment is next suspended in 700 ml of 1:1.5 cyclohexane–CCl_4, the specific gravity is adjusted to 1.312 with the aid of a hydrometer, and the suspension is centrifuged at 2000 g for 20 minutes. The supernatant contains most of the free cytoplasm and some slightly ground cells. It is discarded. The residue is again taken up in 700 ml of cyclohexane–CCl_4 at specific gravity 1.312 and centrifuged as described. The supernatant is again discarded.

The sediment is resuspended in about 700 ml of cyclohexane–CCl_4, brought to specific gravity 1.345 and centrifuged at 2000 g for 60 minutes. The supernatant contains many partly ground cells and nuclei with large cytoplasmic tabs. It is discarded.

The sediment at specific gravity 1.345 is taken up in about 700 ml of cyclohexane–CCl_4, brought to specific gravity 1.358, and centrifuged at 2000 g for 60 minutes. The supernatant is discarded.

The sediment at specific gravity 1.358 is next suspended in about 700 ml of 1:2 cyclohexane–CCl_4; the suspension is adjusted to specific gravity 1.410 and centrifuged at 2000 g for 60 minutes. The supernatant, which contains the nuclei, is carefully decanted and saved. The sediment is discarded.

Cyclohexane is added to the supernatant to bring the specific gravity to 1.368 and the suspension is centrifuged at 2000 g for 60 minutes. The supernatant, which contains tabbed nuclei and fiber, is discarded.

The sediment at specific gravity 1.368 is suspended in about 500 ml of 1:2 cyclohexane–CCl_4, brought to specific gravity 1.410 and centrifuged at 2000 g for 1 hour. The supernatant which contains the nuclei is carefully decanted and the sediment is discarded.

A little cyclohexane is added to the supernatant to lower the specific gravity to 1.373 and the suspension is centrifuged at 2000 g for 80 minutes. The supernatant is discarded. The sediment is washed twice more at specific gravity 1.373, using about 400 ml of solvent mixture for each washing and discarding the supernatants.

The sediment at specific gravity 1.373 is taken up in about 400 ml of 1:2 cyclohexane–CCl_4, adjusted to specific gravity 1.378, and filtered through 140-mesh wire screen. The filtrate is then centrifuged at 2000 g for 80 minutes. The supernatant contains many tabbed nuclei and is discarded. The sediment is washed thrice more at specific gravity 1.378 using about 400 ml of solvent for each washing. The supernatants are discarded.

The sedimented nuclei are finally taken up in about 200 ml of petroleum ether, filtered twice through 325-mesh wire screen, and centrifuged at 900 g for 20 minutes. They are dried *in vacuo* at room temperature and weighed.

About 23–25 g of nuclei are obtained from 100 g of lyophilized tissue. These nuclei show negligible cytoplasmic contamination. Their utility for studies of nuclear enzymatic composition has been stressed repeatedly.[1,2,11] Some indication of their stability is provided by the fact that a 20-year-old preparation of calf thymus nuclei—stored frozen and dry after isolation as described—is as active in the histone deacetylase reaction as is the corresponding amount of fresh tissue.

[11] V. G. Allfrey, *in* "Aspects of Protein Biosynthesis" (C. B. Anfinsen, ed.), Part A, p. 247. Academic Press, New York, 1970.

[24] Isolation of Nuclei from Liver and Other Tissues

By J. R. TATA

As with other methods of subcellular fractionation, the most suitable procedure for the isolation of nuclei is determined by the purpose for which the organelle is required. For example, an isolation procedure which yields good nuclei for ultrastructural studies may be useless for studying RNA synthesis *in vitro*. An ideal method of isolation should satisfy any requirement but in practice one is forced to compromise in view of the diversity of conditions, often mutually incompatible, that have to be met.[1,2] The methods described below are based on a compromise between good recovery, little ultrastructural damage, and low contamination with cytoplasmic material. They are particularly recommended for nuclei that are highly active in RNA synthesis *in vitro* and retain NAD pyrophosphorylase activity and from which chromatin with reproducible template activity can be derived.

Although methods for separating nuclei from cytoplasm have been described over the last one hundred years, it was Dounce[3] who first described a procedure for isolating "clean" nuclei in reasonable amounts. The method was based on homogenization of liver in a medium containing citric acid to lower the pH. However, freezing of the tissue and homogenization in a Waring Blendor, suggested in this method, causes considerable damage to the nuclei, as well as the loss of a large part of soluble nuclear enzymes. In order to reduce such a loss, methods based on sequential extraction with organic solvents were developed.[4,5] But these methods are now rarely used because of the damage caused by organic solvents to nuclear and cytoplasmic membranes.[6] Other methods for nuclear isolation involved the use of bivalent cations[7-10] in the homoge-

[1] D. B. Roodyn, *Biochem. Soc. Symp.* **23**, 20 (1963).

[2] D. B. Roodyn, *in* "Subcellular Components" (G. D. Birnie and S. M. Fox, eds.), p. 15. Butterworth, London, 1969.

[3] A. L. Dounce, *J. Biol. Chem.* **147**, 685 (1943).

[4] A. L. Dounce, G. H. Tishkoff, S. R. Barnett, and R. M. Freer, *J. Gen. Physiol.* **33**, 629 (1950).

[5] G. P. Georgiev, *in* "Enzyme Cytology" (D. B. Roodyn, ed.), p. 27. Academic Press, New York, 1967.

[6] G. Siebert, *Proc. Int. Congr. Biochem. 5th 1961*, Vol. 2, p. 93 (1963).

[7] M. E. Maver, E. A. Greco, E. Løvtrup, and A. J. Dalton, *J. Nat. Cancer Inst.* **13**, 687 (1952).

[8] G. H. Hogeboom, W. C. Schneider, and M. J. Striebich, *J. Biol. Chem.* **196**, 111 (1952).

nization medium to preserve nuclear structure. The cations most frequently used are Ca^{2+}, which is particularly effective in preventing nuclear fragmentation and clumping, and Mg^{2+}, which is particularly beneficial for preserving nucleolar structure and also causes less damage to mitochondria than does Ca^{2+}.

In all the above isolation procedures in aqueous media, the major drawback is the substantial contamination of nuclei with cytoplasmic particles, cellular debris, and erythrocytes. This is partially alleviated by centrifuging the homogenate in 0.25 M sucrose over a layer of 0.32–0.35 M sucrose.[10] A real advance was therefore made in 1956 by Chaveau et al.[11] when they introduced the use of hypertonic sucrose to homogenize the tissue. In their procedure for rat liver, the tissue homogenized in 2.2 M sucrose was centrifuged at 40,000 g for 60 minutes. The pellet obtained consists of virtually pure nuclei[12] since all the other cytoplasmic organelles, plasma membranes, and erythrocytes float up to the top of the centrifuge tube. The disadvantage of the method of Chauveau et al.[11] is that the high shear forces required to homogenize the tissue in such a viscous sucrose solution causes considerable breakage of nuclei and thus leads to low yields. The use of detergents, such as Tween 80, Triton X-100, BRIJ, etc., to break cells in isotonic or hypotonic media has been introduced for tissue culture cells[13,14] which also allows the isolation of nuclei with low cytoplasmic contamination. However, it is often difficult to find a concentration of detergent sufficient to break the cell and lyse cytoplasmic membranes but not damage the nuclei. On the other hand, treatment of nuclei, isolated by other methods, with small amounts of the nonionic detergent, Triton X-100, has been found to be particularly effective in eliminating the nuclear envelope.[15,16]

The methods most commonly used for nuclear isolation at the present time, and described below, are based on a combination of homogenization in an isotonic or slightly hypertonic medium containing divalent cations followed by centrifugation of the 600 g pellet or the homogenate through

[9] V. G. Allfrey, A. E. Mirsky, and S. Osawa, J. Gen. Physiol. 40, 451 (1957).
[10] F. Stirpe and W. N. Aldridge, Biochem. J. 80, 481 (1961).
[11] J. Chauveau, Y. Moulé, and C. H. Rouiller, Exp. Cell Res. 11, 317 (1956).
[12] W. D. Currie, N. M. Davidian, W. B. Elliott, N. F. Rodman, and R. Penniall, Arch. Biochem. Biophys. 113, 156 (1966).
[13] W. C. Hymer and E. L. Kuff, J. Histochem. 12, 359 (1964).
[14] M. H. Vaughan, J. R. Warner, and J. E. Darnell, J. Mol. Biol. 25, 235 (1967).
[15] E. D. Whittle, D. E. Bushnell, and V. R. Potter, Biochim. Biophys. Acta 161, 41 (1968).
[16] J. R. Tata, M. J. Hamilton, and R. D. Cole, J. Mol. Biol. 67, 231 (1972).

2.2 M sucrose.[17-21] Maggio et al.[18] homogenized rat liver in 0.88 M sucrose containing $CaCl_2$, but since Ca^{2+} inhibits nuclear RNA polymerase[22] the method does not yield enzymatically active nuclei. In 1964 Widnell and Tata[19] described for rat liver a method in which the tissue was homogenized in isotonic sucrose containing 3 mM $MgCl_2$, suspending the pellet obtained at 600 g in 2.2 M sucrose and then centrifuging the suspension at 50,000 g. This method was later extended to hard or fibrous tissues[20] and yields nuclei of high purity with good recoveries, which actively synthesize RNA in vitro. The inconvenience of this method was the extra time required in two centrifugation steps and suspension of a crude nuclear pellet in 2.2 M sucrose. This was overcome by Blobel and Potter,[21] who adjusted a homogenate made in 0.25 M sucrose to 1.66 M sucrose in a single step.

Factors Important in Isolation Procedures

Widnell and Tata[19] have considered in detail the most important technical problems of isolation of nuclei.

The first step in any isolation procedure involves complete breakage of the cell with the minimum amount of damage to the nucleus. For liver, kidney, and other soft tissues, it is best to use a glass–Teflon Potter-Elvehjem homogenizer with a wide clearance of 0.12–0.16 mm. For hard or fibrous tissues or cultured cells, it is necessary to use other methods of disruption, such as detergents or disintegration with rotating blades or pressure homogenization. The ratio of tissue or cells to the homogenization medium is critical. For example, in the method of Widnell and Tata[19] a higher yield of rat liver nuclei was obtained if a high ratio of tissue:medium was used for homogenization followed by subsequent dilution of the homogenate rather than homogenize the tissue initially at a low tissue:medium ratio. The number of strokes and speed of rotation of the pestle for homogenization with a Potter-Elvehjem homogenizer is critical for the yield and morphological state of the nuclei and the optimal speed and number should be determined for each preparation. It is also advisable to filter the homogenate through a gauze (100–120 mesh) made of synthetic material, such as nylon or Dacron, to free it from connective tissue, unbroken cells, etc.

The composition of homogenization medium is also an important

[17] M. B. Sporn, T. Wanko, and C. W. Dingman, J. Cell Biol. 15, 109 (1962).
[18] R. Maggio, P. Siekevitz, and G. E. Palade, J. Cell Biol. 18, 267 (1963).
[19] C. C. Widnell and J. R. Tata, Biochem. J. 92, 313 (1964).
[20] C. C. Widnell, T. H. Hamilton, and J. R. Tata, J. Cell Biol. 32, 766 (1967).
[21] G. Blobel and V. R. Potter, Science 154, 1662 (1966).
[22] C. C. Widnell and J. R. Tata, Biochim. Biophys. Acta 87, 531 (1966).

factor and is ultimately a question of the use to which isolated nuclei are to be put. Small changes in tonicity or divalent ion concentration may have large repercussions. For example, homogenization in 0.32 M sucrose instead of 0.25 M sucrose causes less damage to liver nuclei; inclusion of low levels of Mg^{2+} (<0.5 mM), in hypertonic sucrose (2.2 M) causes nuclei to agglutinate whereas concentrations above 2 mM Mg^{2+} may lead to contamination with whole cells. It is also preferable to suspend isolated nuclei in isotonic sucrose at pH 7.4 with low salt concentrations. All procedures should be carried out at 2°–4°, and freezing and thawing should be avoided for studies on nucleic acid synthesis as it leads to activation of nuclease activity.

Isolation of Liver Nuclei

Two methods are recommended, the first one described below has the advantage of multiple sample handling as it can be carried out in an angle-head ultracentrifuge rotor; the second method requires a swing-out rotor (and therefore limited sample capacity) but is less time-consuming.

Method of Widnell and Tata[19]

Reagents

0.32 M Sucrose–3 mM $MgCl_2$
0.25 M Sucrose–1 mM $MgCl_2$
2.4 M Sucrose–1 mM $MgCl_2$

Procedure. Livers are dropped into chilled 0.32 M sucrose–3 mM $MgCl_2$, weighed, finely chopped with a pair of curved scissors, rinsed, and suspended in 3 volumes (3 ml of medium per 1 g of liver) of the same solution. The tissue is homogenized in a loose-fitting Potter-Elvehjem glass–Teflon homogenizer (clearance 0.15 mm) with 20 slow up- and-down strokes at speed of 800–1200 rpm. (For liver from embryonic or immature animals or for small-scale work, <1 g total tissue, the speed should be reduced to about 500 rpm and 5–6 strokes to be applied.[23]) The homogenate is filtered through two layers of nylon bolting cloth (110 mesh) and the filtrate is diluted with 0.6 volume of 0.32 M sucrose–3 mM $MgCl_2$ and 0.22 volume of water to lower the tissue and sucrose content of the homogenate. Of 0.32 M sucrose–3 mM $MgCl_2$, 0.8 volume is carefully inserted with a fine long-tipped pipette below the diluted homogenate in a centrifuge tube, and the contents are centrifuged at 700 g for 10 minutes.

The pellet of crude nuclei is suspended in 7–12 ml of 2.4 M sucrose–

[23] J. R. Tata, *Biochem. J.* **105**, 783 (1967).

1 mM MgCl$_2$ (the volume depending on size of the centrifuge tube) with 4–6 strokes in a tight-fitting ground-glass homogenizer. The suspension is centrifuged in an angle-head rotor of Beckman, MSE, or other preparative ultracentrifuge at 50,000 g for 1 hour. The top plug, which floats up and contains whole cells, erythrocytes, and some mitochondria, is scooped out with a spatula, the dense sucrose medium is siphoned off, and the walls of the tube are wiped with nonfluff tissue paper. The pellet of nuclei is suspended with 2–3 strokes in a small all-glass, hand-operated homogenizer in 0.25 M sucrose–1 mM MgCl$_2$ in a ratio of 1 ml per 2–3 g equivalent of liver.

Procedure of Blobel and Potter[21]

The principle of this method involves homogenization of liver in a small volume of an isotonic medium, raising the concentration of the sucrose in the homogenate sufficiently to float off the endoplasmic reticulum and mitochondria upon ultracentrifugation and allowing the nuclei to sediment through an interface with an even denser medium.

Reagents

Sucrose, 0.25 M, in TKM (5 mM Tris·HCl, pH 7.5, at 20°C, 2.5 mM KCl and 5 mM MgCl$_2$)
Sucrose, 2.3 M, in TKM
TKM without sucrose

Procedure. Livers are chilled in 0.25 M sucrose–TKM, blotted, weighed, and minced in 2 volumes of 0.25 M sucrose–TKM. The mince is homogenized in a glass–Teflon homogenizer (clearance 0.25 mm) with 10–15 strokes at 1700 rpm (or at a lower speed if the homogenizer clearance is around 0.15 mm). The homogenate is filtered and its volume is measured and thoroughly mixed with 2 volumes of 2.3 M sucrose to bring the sucrose concentration to 1.62 M. In the original method the fractionation of the homogenate was carried out in a Spinco SW 39 rotor, which has the disadvantage of a very small volume. In fact any of the modern large-capacity rotors of the swing-out type (Beckman SW 27, SW 41, MSE 6 × 15 ml, etc.) of high length:diameter ratio are even more suitable. If using an older model swing-out rotor, such as the SW 25.1 or SW 25.2, it is preferable to use only the bottom two thirds of the tube and topping up the top one-third with 0.25 M sucrose–TKM. About 5/7 of the usable volume of the swing-out rotor tube is filled with homogenate now made up to 1.62 M sucrose, and 2/7 of the volume of 2.3 M sucrose–TKM is then underlaid with a syringe and a long needle (or a rigid plastic tubing). The sample is centrifuged at 60,000 g for 60 minutes (or for less time at higher speeds). The supernatant is removed preferably by siphoning off, the walls of the tube are wiped with tissue paper, and the pellet of nuclei is suspended in TKM buffer without any sucrose, using an all-glass homogenizer.

Isolation of Nuclei from Fibrous and Hard Tissues

The above two procedures are suitable for tissues that can be relatively easily disintegrated (liver, kidney, spleen thymus, brain, etc.), but for material with a high content of connective tissue (i.e., skin) or tissues in which nuclei tend to get trapped in fibers (muscle, heart, uterus, etc.), it is necessary to modify the process of homogenization.

Edelman et al.[24] were the first to describe a procedure for isolating reasonably "clean" nuclei from skeletal muscle. They suspended a crude nuclear pellet obtained at 800 g in 2.15 M sucrose and collected the pellet after centrifugation at relatively low speed (16,000 g for 2 hours). However, their recommendation of pretreating muscle with Ca^{2+} and inclusion of 0.7 mM ATP in 2.15 M sucrose would be a major disadvantage in assaying RNA polymerase activity in the nuclei. Breuer and Florini[25] avoided the use of Ca^{2+} and ATP, but their use of a VirTis disintegrator and centrifugation through a solution of low density (1.0 M sucrose) would tend to yield nuclei that are damaged or contaminated with cytoplasmic particles. Widnell et al.[20] adapted for heart and uterus an earlier method used for obtaining liver nuclei with a high RNA polymerase activity.[19] This method is recommended for fibrous or connective tissue-rich material.

Procedure of Widnell, Hamilton, and Tata[20]

RAT HEART NUCLEI

Reagents

> 0.32 M sucrose–3 mM $MgCl_2$ (homogenizing medium)
> 2.4 M sucrose–1 mM $MgCl_2$
> 0.25 M sucrose–1 mM $MgCl_2$

Procedure. Hearts from 5 or more rats (150 ± 20 g body weight) are removed and weighed in ice-cold homogenizing medium. The tissue is minced finely first with a pair of scissors and then by passing it successively through a tissue press with sieves of pore diameter of 1.27 and 0.95 mm. The tissue is further disintegrated in 3 volumes of homogenizing medium

[24] J. C. Edelman, P. M. Edelman, K. M. Knigge, and I. L. Schwartz, *J. Cell Biol.* **27**, 365 (1965).
[25] C. B. Breuer and J. R. Florini, *Biochemistry* **5**, 3857 (1966).

for 2.5 minutes with an Ultraturrax TP 18/2 tissue disintegrator (Janke and Kunkel, KG, Staufen 1 BR, Germany) run at 25% of maximum voltage. The homogenate is filtered through two layers of nylon bolting cloth, which was then rinsed with an equal volume of the homogenizing medium. The filtered homogenate is diluted with 28% of its volume of water, 60% of its volume of homogenizing medium was layered under the diluted homogenate in a swing-out centrifuge tube and the tube centrifuged at 700 g for 10 minutes. Each pellet is suspended in a ground glass homogenizer in sufficient volume of 2.4 M sucrose–1 mM $MgCl_2$ to fill a suitable ultracentrifuge tube and centrifuged at 50,000 g for 45 minutes. The supernatant is siphoned off, the walls of the tube are wiped with tissue paper, and the nuclear pellet suspended in 0.25 M sucrose–1 mM $MgCl_2$.

NUCLEI FROM UTERUS

Reagents. Same as for heart muscle.

Procedure. Uteri from 10 or more female rats (100–200 g) are pooled, weighed in the homogenizing medium, chopped finely with a pair of scissors and passed successively through a tissue press with sieve pore sizes of 1.27, 0.95, and 0.63 mm. The fragmented tissue is homogenized in 3 volumes of homogenizing medium in the Ultraturrax disintegrator for 80 seconds at 25% of the normal voltage. The homogenate is filtered through nylon bolting cloth and the filter rinsed with an equal volume of homogenizing medium. Substantial amounts of nuclei are left in the residue on the bolting cloth which has to be scraped with a spatula and rehomogenized as for the whole tissue. The combined filtrate is diluted with 26.5% of its volume of water and 30% of its volume of homogenizing medium is underlayered. The rest of the procedure is as described above for heart nuclei.

Preparation of Nuclei by Zonal Centrifugation

Johnston *et al.*[26] have successfully applied the zonal centrifuge to the isolation of liver nuclei. This method has the advantage that it allows the fractionation of nuclei according to the degree of ploidy or nuclei originating from different cell types.

Chemical and Enzymatic Characterization of Nuclei

The suspension of isolated nuclei is analyzed for DNA, protein, RNA, and phospholipid. They are enzymatically characterized by measuring their DNA-dependent RNA polymerase and NAD-pyrophosphorylase content.

[26] I. R. Johnston, A. P. Mathias, F. Pennington, and D. Ridge, *Nature (London)* **220**, 668 (1968).

TABLE I

Composition of Nuclei from Liver, Heart, and Uterus, Isolated by the Techniques Described Above[a]

Tissue	Method used	Protein: DNA	RNA: DNA	Phospholipid: DNA	Recovery (as % of tissue DNA)
Liver	Widnell and Tata[c]	3.74 ± 0.21	0.20 ± 0.01	0.06 ± 0.02	70.4 ± 4.9
	Blobel and Potter[d]	2.61 ± 0.32	0.23 ± 0.02	0.10 ± 0.02	74.8 ± 4.3
Liver (Triton-treated)[b]	Tata et al.[e]	1.48 ± 0.17	0.16 ± 0.03	0.02 ± 0.004	69.8 ± 7.3
Heart	Widnell et al.[f]	4.30 ± 0.60	0.23 ± 0.05	—	43.0 ± 6.5
Uterus	Widnell et al.[f]	3.90 ± 0.55	0.26 ± 0.04	—	47.0 ± 5.7

[a] All preparations made in the author's laboratory.
[b] Isolated nuclei were treated with 1% Triton X-100 for extraction of chromatin, as described by Tata et al.[e]
[c] C. C. Widnell and J. R. Tata, *Biochem. J.* **92**, 313 (1964).
[d] G. Blobel and V. R. Potter, *Science* **154**, 1662 (1966).
[e] J. R. Tata, M. J. Hamilton, and R. D. Cole, *J. Mol. Biol.* **67**, 231 (1972).
[f] C. C. Widnell, T. H. Hamilton, and J. R. Tata, *J. Cell Biol.* **32**, 766 (1967).

TABLE II
ACTIVITY OF NUCLEAR ENZYMES ASSOCIATED WITH NUCLEI FROM RAT
LIVER, HEART, AND UTERUS[a]

Tissue	Method used	RNA polymerase[b]		NAD pyrophosphorylase[c]
		With Mg^{2+}	With Mn^{2+} + 0.4 M $(NH_4)_2SO_4$	
Liver	d	1034	2980	425
	e	560	1575	378
Heart	f	456	1355	99
Uterus	f	377	1235	69

[a] Preparations correspond to those described in Table I. Values given are the average of a minimum of three determinations in duplicate.

[b] Picomoles of AMP incorporated per milligram of DNA per 10 minutes for the Mg-activated reaction and per 20 minutes for the $Mn^{2+}/(NH_4)_2SO_4$-activated reaction.

[c] Nanomoles of NAD synthesized per milligram of DNA per 20 minutes.

[d] C. C. Widnell and J. R. Tata, *Biochem. J.* **92**, 313 (1964).

[e] G. Blobel and V. R. Potter, *Science* **154**, 1662 (1966).

[f] C. C. Widnell, T. H. Hamilton, and J. R. Tata, *J. Cell Biol.* **32**, 766 (1967).

Cytoplasmic contamination of nuclei is conveniently determined by assaying for cytochrome oxidase as a mitochondrial marker enzyme and glucose-6-phosphatase or NADH–cytochrome c reductase as microsomal markers.

Methods

Chemical Analysis. Protein is determined by the method of Lowry et al.,[27] DNA according to Burton,[28] RNA according to Fleck and Munro,[29] and phospholipid as described by Tata[30] and based on the phosphorus determination method of Ames and Dubin.[31]

Enzyme Determinations. DNA-dependent RNA polymerase is assayed with Mg^{2+} or with Mn^{2+} ions in the presence of 0.3 M $(NH_4)_2SO_4$ as described by Widnell and Tata.[19,32]

NAD pyrophosphorylase is determined as described by Branster and Morton.[33]

[27] O. H. Lowry, N. J. Rosebrough, A. L. Farr, and R. J. Randall, *J. Biol. Chem.* **193**, 265 (1951).

[28] K. Burton, *Biochem. J.* **62**, 315 (1956).

[29] A. Fleck and H. N. Munro, *Biochim. Biophys. Acta* **55**, 571 (1962).

[30] J. R. Tata, *Biochem. J.* **116**, 617 (1970).

[31] B. N. Ames and D. T. Dubin, *J. Biol. Chem.* **235**, 769 (1960).

[32] C. C. Widnell and J. R. Tata, *Biochim. Biophys. Acta* **123**, 478 (1966).

[33] M. V. Branster and R. K. Morton, *Biochem. J.* **63**, 640 (1956).

TABLE III

RECOVERY OF SOME CYTOPLASMIC MARKER ENZYMES IN NUCLEI PREPARED
FROM RAT LIVER, HEART, AND UTERUS[a]

Tissue	Method	Enzyme activity recovered (% of homogenate)[b]			
		Cytochrome oxidase	Glucose-6-phosphatase	NADH–cytochrome c reductase	5'-Nucleotidase
Liver	c	0.27	1.60	0.62	—
	d	0.25	3.15	0.95	2.35
Liver (Triton-treated)	e	0.07	0.72	0.40	1.04
Heart	d	0.68	—	0.63	—
Uterus	d	0.62	—	1.85	—

[a] Preparations as in Table I.

[b] Values are the average of at least three separate determinations in duplicate.

[c] C. C. Widnell and J. R. Tata, Biochem. J. 92, 313 (1964).

[d] G. Blobel and V. R. Potter, Science 154, 1662 (1966).

[e] J. R. Tata, M. J. Hamilton, and R. D. Cole, J. Mol. Biol. 67, 231 (1972).

For cytoplasmic marker enzymes, cytochrome oxidase is determined by the method of Cooperstein and Lazarow,[34] glucose-6-phosphatase and NADH–cytochrome c reductase as described by Tata et al.[35]

Results

Table I summarizes the chemical composition of nuclei and recovery of total tissue DNA in the nuclear fraction from the three rat tissues described above. The effect of washing nuclei with 1% Triton X-100 is also included in Table I because detergent treatment of nuclei yields more stable chromatin preparations.[16] Electron microscopic studies[16,21,36] have shown that 1% Triton X-100 strips nuclei of their outer as well as inner envelope. This is reflected in a marked drop in phospholipid content (Table I) as well as a significant reduction in the activity of membrane-associated extranuclear enzymes (Table III). Table II shows that nuclei prepared by the methods described here have active RNA polymerase (assayed at low and high salt concentrations) and NAD pyrophosphorylase. At the same time they contain only traces of enzymes associated with mitochondria, microsomes, and plasma membranes.

[34] S. J. Cooperstein and A. Lazarow, J. Biol. Chem. 189, 665 (1951).

[35] J. R. Tata, L. Ernster, O. Lindberg, E. Arrhenius, E. Pedersen, and R. Hedman, Biochem. J. 86, 408 (1963).

[36] E. D. Whittle, D. E. Bushnell, and V. R. Potter, Biochim. Biophys. Acta 161, 41 (1968).

[25] Specific Methods for the Isolation of Nuclei from Chick Oviduct

By THOMAS C. SPELSBERG, JOHN T. KNOWLER, and HAROLD L. MOSES

For several years, we have utilized many reported techniques in the isolation of nuclei from the oviduct of immature chicks. This organ represents not only a hormone target tissue, but also an excellent model system for studies into hormone-induced cytodifferentiation.[1] We have examined these nuclei for chemical composition, species of RNA, endogenous RNA polymerases of both nucleolar and nucleoplasmic origin, and cleanliness and ultrastructure by electron microscopy. Further, the effects of the various procedures on the chromatin composition, template capacity, and histone species have been studied. The procedures described below represent the preferred methods in this laboratory for nuclear isolation for the stated objectives.

Source of Tissue. Oviducts were obtained from immature chicks (Rhode Island Reds) which had received daily injections (subcutaneous) of 5 mg of diethylstilbestrol for 17 days. At this time, the oviduct is fully developed and weighs about 1–2 g per chick. Oviduct is quickly trimmed of connective tissue and either used immediately to minimize autolysis or quickly frozen in dry-ice and stored at $-70°$ until needed. (Those oviducts to be processed without freezing were processed as soon as possible after removal from the chicks.) All steps were performed at $0°-4°$ unless otherwise specified.

Method I: Nuclei for RNA Polymerase Measurements
(Hypertonic Sucrose)

This method is designed to yield nuclei which retain maximum RNA polymerase activity. Methods that utilize concentrated solutions of sucrose have essentially four purposes: (1) to maintain nuclear integrity; (2) to provide a medium of specific gravity higher than that of most cytoplasmic constituents (except for ribosomes and other ribonucleoproteins) so that nuclei can be freed of this material; (3) to strip nuclei from the attached endoplasmic reticulum during sedimentation through concentrated sucrose solutions; and (4) to help prevent loss of material from the nuclei. There are often problems in using hypertonic sucrose methods for isolating nuclei. Important suggestions and safeguards for good yield and purity are there-

[1] B. W. O'Malley, W. L. McGuire, P. O. Kohler, and S. G. Korenman, *Rec. Progr. Hormone Res.* **25**, 105 (1969).

TABLE I
Suggestions for Optimal Results in Nuclei Isolations Using Hypertonic Sucrose Solutions at pH 7.0 to 8.0 Applicable to Most Tissues

Yield

A. *Homogenization*

1. Avoid sucrose solutions below 0.3–0.4 M, since nuclear damage increases and final yield thus decreases.

2. Presence of divalent cations (Mg or Ca)[a] at 1–3 mM in sucrose solutions is important in preventing nuclei fragmentation, clumping, as well as excessive swelling of nuclei. At later steps, less concentration is needed since less cation binding protein is present.

3. Ratio of homogenizing solutions to tissue amount is critical when using solutions of high specific gravity (>1.28 g/ml such as >1.67 M sucrose) where cytoplasmic debris float. At 25 volumes (v/w) yield is best; with 10 volumes (v/w) and less the yield begins to decrease rapidly owing to trapping of nuclei in rising (floating) cell debris. At 5 volumes (v/w) and less, a complete loss of nuclei is often observed. Longer periods of centrifugation at higher gravitational force does not increase the yield. Rehomogenization of supernatant with the floating debris and the addition of more solvent followed by centrifugation recovers much of the trapped nuclei.

4. Each resuspension of nuclei should be thorough, otherwise groups of nuclei with cell debris may float in succeeding centrifugations.

5. The number of strokes on pestle-type homogenizers (or length of homogenization in blade-type) should be limited to maximize cell breakage; excess homogenizations damage nuclei and lowers yield.

6. The clearance on the pestle-type homogenizer is critical and should be close enough for sufficient cell breakage but not too close so as to damage nuclei nor increase temperature by friction. (0.003 to 0.010 inch clearance, the closer for tissues with

Purity

A. *Homogenization*

1. Sucrose solutions greater than 1.67 M which gives a specific gravity above 1.28 g/ml are recommended at some stage of the nuclei isolation. Most cytoplasmic organelles and debris will not sediment with the nuclei during the centrifugations but will float to the top; this helps to separate nuclei from debris. Sucrose solutions of 2.0 M or greater are recommended for such centrifugation steps, since it yields preparations with less debris and whole cells, and nuclei with minimal attached cytoplasmic particles are found. The choice of the molarity of the sucrose solutions in any case will depend on the tissue in use. Different tissues require different concentrations of sucrose solutions for optimal yield and purity; e.g., the cleanest preparations of rat nuclei are achieved by sedimentation through 2.2 or 2.3 M sucrose[b], whereas chick oviduct nuclei are difficult to sediment in this medium.

2. Use of high ratios of volume of media to amount of nuclear material or repetitions of steps can assist in cleaner nuclei preparations.

3. Neutral detergents remove the outer membranes of nuclei with the attached ribosomes and generally destroy practically all cytoplasmic membrane systems.[c–f] No other solvents (including heavy sucrose solutions) remove the outer membrane of the nuclear envelope so thoroughly as these detergents. Dilute solutions (0.2–1%) of these detergents have a less drastic effect on nuclei than citric acid solutions, based on nuclear composition, RNA polymerase activity, and template activity and composition of the isolated chromatin. The detergents of the octylphenoxyethanol series (Triton X-100 and Triton X-114) as well as the nonylphenoxyethanol series (Triton N-101) are all effective.[f] They should not be used with initial homogenization of tissues

since they can enhance lysis of lysozymes and expose the nuclei to the released lytic enzymes.[g] Solutions of sucrose containing 0.2% Triton X-100 have been found to enhance nuclear ribonuclease activity in chick oviducts and rat liver. Triton X-45, Triton X-165, Tween-20, Lubrolux, Pluronic L-64, and Tetronic 707 have been found not to be as effective in cytoplasmic membrane solubilization as the Triton X-100.[f]

4. Overhomogenization (especially by blade-type homogenizers) should be prevented since such finely ground cell debris tend to accompany nuclei in filtrates through 100–200-mesh cloth and as pellets through heavy sucrose solutions during centrifugations.

B. Filtration

The use of screens with extremely small pores (>than 200–300 mesh) for the filtration may cause excessive trapping, and thus loss of nuclei. Use of screens with rather large pores (<100 mesh) fail to remove undesirable debris.

C. Centrifugations

1. The layering of tissue homogenates or crude nuclear preparations over fresh solutions of higher sucrose concentration followed by centrifugation yields cleaner nuclear preparations than when such layering is omitted.

2. Complete draining of tubes containing pellets of nuclei from sedimentation in sucrose solutions and the thorough wiping of walls with tissue paper reduce contamination.

3. Resuspensions of nuclei in succeeding solvents should be thorough throughout the procedure so that attached or trapped debris can better be removed.

4. See also under Yield, the Centrifugation section, item No. 5.

(Continued)

smaller nuclei.) (For fibrous tissue, rotating blade or pressure-type homogenizers must be used in the initial homogenization.)

7. The pressure and vacuum created in pestle-type homogenizers as well as the rate of pestle rotation can help break cells but can also cause damage to nuclei if excessive. Individual expertise should be analyzed in considering pestle-type homogenizers. A rotation of 300–600 rpm of the pestle is usually optimal. The shape of blades and vessel in blade-type homogenizers have a marked effect on the extent of shear and should be considered.

B. Filtration

1. A significant (≃50% or more) loss of nuclei can occur if the debris is allowed to accumulate on the screen, cloth, or gauze; the filtering should be spread over a rather wide area since the cell debris can trap many nuclei.

2. Filtering of initial tissue homogenates should be performed with 10–40-mesh material; later filterings of solutions with less debris should involve 100–200-mesh screens.

C. Centrifugations

1. The exact amount of sucrose used in making the more concentrated solutions (>1.8 M) is critical since the viscosity of sucrose solutions increases practically logarithmically between 1.8 and 2.4 M.[h] At 5°C, an additional error of 0.1 M in making up a 1.6 M sucrose solution can cause an increase of only 10% in the viscosity of the solution; however, such an error in making up a 2.2 M sucrose solution causes a 60% increase in viscosity.[h] Such errors in these solutions can result in a marked loss of nuclei through lack of complete sedimentation during centrifugation.

2. Similarly, a shift in rotor temperature from 5°C to 0°C will cause an increase in viscosity of a 2.2 M sucrose solution

TABLE I (Continued)

Yield	Purity

Yield

by 60%; solutions with less than 1.8 *M* sucrose show minimal or no change in viscosity during this temperature change. Temperature control is therefore critical during the centrifugation period in heavy sucrose solutions.

3. Losses of nuclei from 20 to 100% can occur with sucrose solutions of >1.7 *M* if the ratio of solvent to amount of material is low (see No. 3 under Homogenization). When losses are significant at this step, the supernatant and floating debris can be rehomogenized with added solvent followed by centrifugation to retrieve the lost nuclei.

4. Another cause of heavy loss of nuclei is the layering of concentrated tissue homogenates (or any solutions with cell debris present) over heavy sucrose solutions (>1.7 *M* sucrose and specific gravity >1.28 g/ml). Centrifugation of this system results in trapping of nuclei by the cell debris at the interphase. Longer periods of centrifugations even at higher rotor speeds do not increase the yield of nuclei. The only solution is to dilute the homogenate with more solvent and more importantly, mix the interphase well to form a crude gradient.

5. The length and speed of a centrifugation step should be considered; undercentrifugation results in loss of nuclei, overcentrifugation often results in sedimentation of undesired cell debris. The distance the nuclei must travel which is greater in larger tubes should be considered; the time for nuclei to sediment in a small tube is shorter than the time required for sedimentation in larger tubes.

D. General

All quantitative methods demand a minimum of transfers from vessel to vessel; rinsing of vessels during transfers to reduce loss is required.

D. General

It is obvious that individual expertise plays a major role in determining the effectiveness of any procedure for obtaining pure nuclei.

Biological properties

Of all the biological parameters one can measure, one cannot be sure that measurements in isolated nuclei represent exact cellular states. The most sensitive studies of biological properties of nuclei are probably the transfer of nuclei from one cell to another with maintenance of the vital functions.[i] Comments on subjects of interest are made for consideration:

1. Small contaminants in materials used in the media—especially sucrose, which is used in high concentrations—may alter such properties (enzyme activities, etc.).

2. The presence of low levels of salts, such as KCl, has been shown to be essential in maintaining biologically active nuclei in studies involving nuclear transplants among amoeba.[j]

3. The use of solutions of high osmolarity, e.g., 2.0 M sucrose or 25% glycerol, during isolation and purification of nuclei generally yield nuclei with maximum enzyme activities (RNA polymerase, etc.).

4. The most important suggestion that this laboratory can make concerning maintenance of biological properties in isolated nuclei is "speed of isolation." Long periods of isolation of nuclei often result in nuclei with altered morphology, loss of enzyme activities and large species of RNA, reduced protein and RNA content, and altered chromatin (composition and template capacity for DNA-dependent RNA synthesis). Whatever the technique of isolation employed, the faster the nuclei are prepared, the better the results obtained with all the parameters mentioned above.

5. The use of freshly excised tissue to obtain nuclei with maximum biological properties is recommended. The immediate freezing of tissues on dry-ice and storage for several days had no major effect on the ultrastructure of subsequently isolated nuclei. The endogenous nuclear RNA polymerase activity decreased only 10-15% compared to nuclei from unfrozen tissue, and the chromatin isolated from nuclei from frozen tissue appeared unchanged (in composition and template capacity for RNA synthesis) from that from unfrozen tissue. However, more critical studies of the effects of freezing tissue need to be performed, such as the viability of nuclei when transplanted in amoeba.

[a] B. D. Roodyn, in "Subcellular Components" (G. D. Birnie and S. M. Fox, eds.), p. 15. Butterworth, London, 1972.

[b] G. Blobel and V. R. Potter, *Science* **154**, 1662 (1966).

[c] E. Holtzman, I. Smith, and S. Penman, *J. Mol. Biol.* **17**, 131 (1966).

[d] M. T. Hubert, P. Favard, N. Carasso, G. R. Rozencwaj, and J. P. Zalta, *J. Microsc.* **1**, 435 (1962).

[e] D. A. Rappoport, R. R. Fritz, and A. Moraczewiski, *Biochim. Biophys. Acta* **74**, 42 (1963).

[f] W. C. Hymer and E. L. Kuff, *J. Histochem. Cytochem.* **12,** 359 (1964).

[g] M. L. E. Burge and R. H. Hinton, *in* "Separations with Zonal Rotors" (E. Reid, ed.), p. S-5.1. Guildford, 1971.

[h] C. de Duve, J. Berthet, and H. Beaufay, *Progr. Biophys. Chem.* **9**, 325 (1959).

[i] N. Burnstock and J. L. Philpot, *Exp. Cell Res.* **16**, 657 (1959).

[j] M. J. Ord and L. G. E. Bell, *Nature (London)* **226**, 854 (1970).

fore summarized in Table I. This table was derived from both the literature (for an excellent review see Roodyn[2]) and experience gained in this laboratory.

Reagents

TKM Buffer: 50 mM Tris·HCl (pH 7.5) containing 25 mM KCl and 5 mM MgCl$_2$
Solution A: 2.0 M sucrose in TKM buffer
Solution B: 1.7 M sucrose in TKM buffer
Solution B + Triton: solution B containing 0.2% Triton X-100
Solution C: 1.8 M sucrose in TKM buffer
Solution D: 25% glycerol in 10 mM Tris·HCl (pH 7.9) and 1 mM MgCl$_2$

Procedure. This procedure is a modification of that described by Chaveau *et al.*[3] In this method 2.0 M sucrose (specific gravity ~1.35 at 0°) has been substituted for the 2.2 M sucrose (specific gravity ~1.39 at 0°) to improve the yield. Either fresh or frozen oviduct (10 g wet weight) is cut into small pieces with scissors and placed in a Thomas Teflon pestle–glass homogenizer (0.010 inch clearance) with 20 volumes of solution A. Homogenization with a pestle rotation of about 300–400 rpm (0.5 horsepower hand drill mounted on a drill stand—Sears, Roebuck & Co.) requires four to six complete strokes. The homogenate is then poured through 4 layers of cheesecloth, taking care not to allow filtration through excessive amounts of debris, which trap nuclei and cause low yields. The homogenizer and beakers are rinsed with 4 volumes (milliliters per gram of tissue) of solution A. For the Beckman 35 rotor, 50 ml of homogenate is layered over 25 ml of rinse in each tube. The tubes are centrifuged for 1 hour at 70,000 g. The floating debris is loosened with a rubber policeman and discarded along with the supernatant. The tubes are rinsed, drained well and wiped with tissue paper, taking care not to remove nuclei pelleted on the side of the tube. The nuclear pellet is resuspended gently but thoroughly (with a loose-fitting Teflon homogenizer) in either 2 volumes (ml/g tissue) of solution B or solution B containing Triton X-100. The homogenate is filtered through organza cloth (100 mesh), care being taken to prevent excessive accumulation of debris in any one area of the cloth. About 15 ml of the filtrate is layered over 15 ml of solution C in a 30-ml glass Sorvall centrifuge tube and then centrifuged for 20 minutes at 15,000 g in a HB-4 Sorvall rotor. The supernatant is discarded,

[2] D. B. Roodyn, *in* "Subcellular Components" (G. D. Birnie and S. M. Fox, eds.), 2nd ed., p. 15. Butterworth, London, 1972.
[3] J. Chaveau, Y. Moulé, and C. H. Rouiller, *Exp. Cell Res.* **11**, 317 (1956).

and the pellet of purified nuclei is resuspended in 0.5 volume (milliliters per gram of tissue) of solution D and either analyzed immediately by the techniques described below or stored at −20°C until needed. The total time required for the isolation of these nuclei is about 2 hours.

Remarks on the Procedure. Based on electron micrographs, the nuclei prepared by this method appear relatively free of contamination (Fig. 1A), although round dense bodies thought to be secretory granules from oviduct glandular cells are sometimes seen. Very few intact cells or cytoplasmic

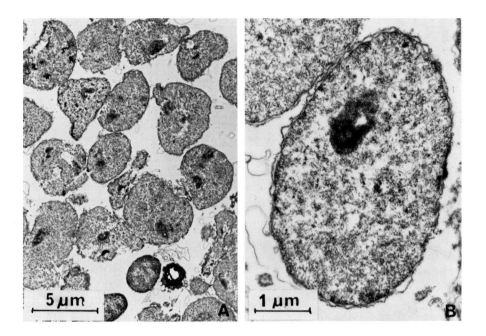

FIG. 1. Electron micrographs of nuclei isolated by the hypertonic sucrose method (method I) for polymerase assay. Pellets of nuclei were fixed with 2% glutaraldehyde in a 0.1 M phosphate buffer, pH 7.4, for 2 hours at room temperature, rinsed twice with 7.5% sucrose in a similar phosphate buffer, and stored overnight at 4° in the sucrose solution. After postfixation in 1% osmium tetroxide in Millonig's phosphate buffer for 1 hour at room temperature, nuclear pellets were rinsed three times with deionized water, stained en bloc for 2 hours with 1% uranyl acetate, dehydrated with graded solutions of ethanol, passed through propylene oxide, and embedded in an Araldite. After heat polymerization, 1.0–2.0 sections including a full thickness of the pellet were cut with glass knives on an LKB Pyramitome and examined by phase-contrast microscopy. Selected areas were then thin-sectioned on an LKB ultrotome III ultramicrotome equipped with a diamond knife. Sections 30–45 nm thick were collected on Formvar-carbon coated grids, stained with uranyl acetate and lead citrate, and examined in a Phillips EM-300 electron microscope. (A) ×3500; (B) ×15,000.

debris are observed. Use of the neutral detergent Triton X-100 eliminates the outer nuclear membrane and attached ribosomes, which remain intact in the absence of detergent (Fig. 1B).[4-7] Otherwise, the general morphology of nuclei prepared with and without Triton are similar and resemble the nuclei of intact oviduct cells. Nucleoli showing both fibrillar and granular components are present in the nuclei. Euchromatin and heterochromatin areas are usually distinguishable, although areas of the latter are less discernible in the Triton preparations. Few disrupted nuclei are observed; however, frequent small breaks in the inner nuclear membrane are present in Triton-treated nuclei. The ratio of protein to DNA in the non-Triton preparations of nuclei range from 2.5 to 3.0, while the ratio of RNA to DNA ranges from 0.30 to 0.40. Triton treatment of nuclei lowers the protein and RNA ratios by about 20% as compared to untreated preparations. The best yields of nuclei range from 40 to 60%.

Assay for Nucleolar and Nucleoplasmic RNA Polymerase Activity

DNA is initially determined by the diphenylamine reaction described by Burton.[8] As discussed later, the nuclei are stored in a glycerol containing medium. In order to avoid interference with color development, 25–50-μl aliquots of nuclei in solution D are suspended in 3 ml of 0.3 N HClO$_4$ and sedimented in a clinical centrifuge at 2000 g for 10 minutes. The pellet is then taken for DNA determination.[9] The assay for the endogenous RNA polymerase activity is a modification of that described by Roeder and Rutter[10] as described elsewhere.[11,12] The reactions, started by the addition of nuclei, are incubated for 10 minutes at 15°. At this temperature, ribonuclease activity has been shown to be minimal.[10,12,13] Assays in low salt conditions [50 mM (NH$_4$)$_2$SO$_4$] contain 0.2 μg of the polymerase II inhibitor, α-amanitin,[14,15] while those in high salt [0.25 M (NH$_4$)$_2$SO$_4$] con-

[4] M. T. Hubert, P. Favard, N. Carasso, G. R. Rozencwaj, and J. P. Zalta, *J. Microsc. (Paris)* 1, 435 (1962).
[5] D. A. Rappoport, R. R. Fritz, and A. Moraczewski, *Biochim. Biophys. Acta* **74**, 42 (1963).
[6] W. C. Hymer and E. L. Kuff, *J. Histochem. Cytochem.* **12**, 359 (1964).
[7] E. Holtzman, I. Smith, and S. Penman, *J. Mol. Biol.* **17**, 131 (1966).
[8] K. Burton, *Biochem. J.* **62**, 315 (1956).
[9] T. C. Spelsberg and L. S. Hnilica, *Biochim. Biophys. Acta* **228**, 202 (1971).
[10] R. G. Roeder and W. J. Rutter, *Biochemistry* 9, 2543 (1970).
[11] S. R. Glasser and T. C. Spelsberg, *Biochem. Biophys. Res. Commun.* 47, 951 (1972).
[12] S. R. Glasser, F. Chytil, and T. C. Spelsberg, *Biochem. J.* **130**, 947 (1972).
[13] P. Chambon, M. Ramuz, P. Mandel, and J. Doly, *Biochim. Biophys. Acta* **157**, 504 (1968).
[14] F. Stirpe and L. Fiume, *Biochem. J.* **105**, 779 (1967).
[15] C. Kedinger, M. Gniazdowski, J. L. Mandel, F. Gissinger, and P. Chambon, *Biochem. Biophys. Res. Commun.* 38, 165 (1970).

tain no inhibitor. The low salt reactions with α-amanitin synthesize RNA with a U:G value of 0.75 while reactions containing high salt synthesize RNA with a U:G value of 1.10. These ratios demonstrate that the low-salt conditions are synthesizing ribosomal RNA, and the high-salt conditions DNA-like RNA.

We have found that a variety of factors affect polymerase activity. Triton X-100 treatment results in lowered activities. Concentrations of sucrose of 0.5 M or less caused reduced nucleoplasmic polymerase activity and variable activity for the nucleolar polymerase. For optimal polymerase activity, it seems advantageous to forego long purification washes of nuclei in order to minimize the time required for isolation. Nuclei can be stored in a buffered glycerol solution without loss of enzyme activity, as discussed below. Interestingly, the quick freezing of freshly excised oviducts on dry-ice followed by storage for several days at $-70°$ before isolation of nuclei results in only a 10–20% loss in activity of both polymerases.

Preservation of RNA Polymerase Activity in Stored Nuclei

It is often advantageous to be able to store nuclei at low temperatures for later use. We have found that only minimal loss of RNA polymerase activity occurs at $-20°$ using the following storage buffer: 25% glycerol, 1 mM $MgCl_2$, and 10 mM Tris·HCl (pH 7.9).[12] In contrast to the other buffers and temperatures studied (Fig. 2), this glycerol medium demonstrated preservation of both nucleolar (low salt with α-amanitin) and nucleoplasmic (high salt) polymerase activities for months.[12,16] Also,

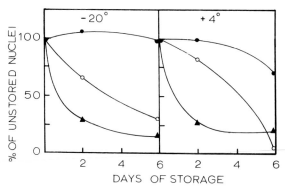

FIG. 2. Levels of RNA polymerase II activity (assayed in high salt) in isolated nuclei after storage. Nuclei, isolated by method I (hypertonic sucrose method) were resuspended in (▲) 10 mM Tris·HCl (pH 7.5), (○) 0.5 M sucrose in TKM buffer, or (●) 25% glycerol in 10 mM Tris·HCl (pH 7.5) + 1 mM $MgCl_2$. The solutions were stored at 4° or at −20°C for various periods of time.

[16] T. C. Spelsberg, unpublished results, 1972.

nuclear morphology is maintained[17] and chromatin isolated 3 months after storage shows no alteration in template capacity for RNA synthesis, DNA, RNA, and protein composition, or acidic protein and histone profiles on polyacrylamide gels.[16] Thus, it seems that glycerol helps maintain the native structure and biological activity of nuclei which are stored for more than 3 months at $-20°$.

Method II: Nuclei for Isolating High-Molecular-Weight RNA (Citric Acid)

This method is designed to provide nuclei with the least amount of RNA degradation and involves the inclusion of citric acid in the isolation solutions. Citric acid is known to reduce nuclear fragility,[18,19] and Higashi *et al.*[20] reported that, in Walker carcinoma tissue, citric acid permitted the isolation of nuclei which retained high-molecular-weight RNA despite considerable changes in their ultrastructure. In our hands, the method of Higashi *et al.*,[20] when applied to oviduct, did preserve much of the nuclear RNA. The results were found, however, to be very variable, and there was always some decay of heterogeneous RNA (HnRNA). The method described here is much faster and gives cleaner preparations and a reproducible preservation of nuclear RNA. The preparation is not suitable, however, for other purposes. It is known that citric acid alters nuclear morphology,[21] removes some DNA[22] and 18–55% of nuclear protein,[23] including lysine rich histones.[24–26] Our observations confirm these findings.

Reagents

Solution A: 1 mM MgCl$_2$
Solution B: 0.1 M citric acid + 1 mM MgCl$_2$
Solution C: 50 mM citric acid + 1 mM MgCl$_2$ + 1.0% Triton X-100
Solution D: 0.5 M sucrose in solution C

[17] R. S. D. Read and C. M. Mauritzen, *Can. J. Biochem.* **48**, 559 (1970).
[18] A. L. Dounce, *Ann. N.Y. Acad. Sci.* **50**, 982 (1948).
[19] A. L. Dounce, *in* "The Nucleic Acids" (E. Chargaff and J. N. Davidson, eds.), Vol. II, p. 93. Academic Press, New York, 1955.
[20] K. Higashi, K. S. Narayanan, H. R. Adams, and H. Busch, *Cancer Res.* **26**, 1582 (1966).
[21] H. Busch, this series, Vol. 12, p. 421.
[22] A. L. Dounce, *Exp. Cell Res.,* Suppl. 9, 126 (1963).
[23] V. G. Allfrey, H. Stern, A. E. Mirsky, and H. Saetren, *J. Gen. Physiol.* **35**, 529 (1952).
[24] A. L. Dounce, F. Seaman, and M. MacKay, *Arch. Biochem. Biophys.* **117**, 550 (1966).
[25] K. Murray, G. Vidali, and J. M. Neelin, *Biochem. J.* **107**, 207 (1968).
[26] K. Murray, *J. Mol. Biol.* **39**, 125 (1969).

Solution E: 0.32 *M* sucrose in solution C
Solution F: 25% glycerol + 10 m*M* Tris·HCl (pH 7.5) + 1 m*M* MgCl$_2$

Procedure. This method has been used successfully in the preparation of uterine nuclei with retention of high-molecular-weight RNA[27] and has been slightly modified for the preparation of oviduct nuclei. All steps are at 0°–4°. The method gives better yields and higher degrees of purity with small amounts of tissue. Finely chopped oviduct, 50 mg–5 g, is homogenized in solution A with a chilled Polytron PT-10 homogenizer (Kinematica GMBH, Lucerne, Switzerland) for 30 seconds at 36 V. The volume for homogenization is kept as low as possible; 2 ml of 1 m*M* MgCl$_2$ is added for amounts of tissue up to 800 mg and 2.5 ml/g for larger amounts. An equal volume (ml/ml of homogenate) of solution B is rapidly mixed with the homogenate which is filtered through 4 layers of cheesecloth. After centrifugation at 300 *g* for 5 minutes, the crude nuclear pellet is resuspended in 5 ml of solution C and is made 0.25 *M* with respect to sucrose by mixing with an equal volume of solution D. The 10-ml suspension is gently layered on 10 ml of solution E in 30-ml glass centrifuge tubes and centrifuged for 5 minutes at 800 *g* in a Sorvall HB-4 rotor. The pellet of purified nuclei is characterized as prepared, or is resuspended in 0.5–1 ml of solution F and stored at −20°C. The use of a discontinuous sucrose gradient is essential in removing cell fragments which otherwise accompany the nuclei. After centrifugation in a discontinuous gradient, the supernatant is removed with care to avoid losing part of the nuclear pellet. Gentle removal of the supernatant with a water aspirator is recommended. The extent of homogenization of the tissues is to be kept to a minimum since this will generate fibrous elements which tend to accompany the nuclei even through the sucrose gradients. All steps are performed as quickly as possible. The total time for this procedure is 30 minutes.

Remarks on the Procedure. Nuclei prepared by this citric acid method show no intact cells and only few fragments of cytoplasmic debris in electron micrographs (Fig. 3A and B). Small strips of outer nuclear membrane remain attached with interspersed denuded areas (Fig. 3B). Most of the nuclei show a homogeneous coarse reticulation formed by interanastomosing beaded strands of chromatin material. Many nuclei have large irregular areas devoid of structure, and others contain peripherally clumped chromatin. Nucleolar areas can be identified, but the ultrastructure is distorted by the same pattern of reticulation observed in the rest of the nucleus. The ratios of protein to DNA in these nuclear preparations range from 2.0 to 2.5 and that of RNA to DNA range from 0.20 to 0.30. The

[27] J. T. Knowler and R. M. S. Smellie, *Biochem. J.* **131**, 689 (1973).

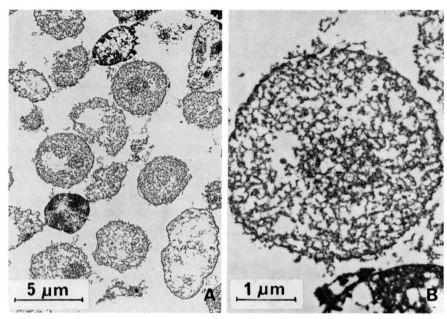

FIG. 3. Electron micrographs of nuclei prepared by the citric acid method (method II) for isolating high-molecular-weight RNA. See the legend of Fig. 1 for methods for preparing the micrographs. (A) ×3500; (B) ×15,000.

best yields of nuclei (DNA as nuclei versus DNA as starting tissue) in this procedure range from 40 to 50%.

Analysis of the Species of RNA in Isolated Nuclei

In order to obtain sufficient specific activity in the nuclear RNA for polyacrylamide gel electrophoresis, tissue slices were incubated with radioactive uridine *in vitro*. Chicks were killed by cervical dislocation; the oviduct was removed, divided in half longitudinally and cut into small sections. Each chopped oviduct was incubated in a 25-ml conical flask in 4 ml of Eagle's medium containing 25 μCi/ml of [5-^{3}H]uridine. Incubations were for 1 hour at 37° under an atmosphere of 95% O_2 and 5% CO_2 in a shaking water bath. The tissue was then washed twice in saline at 4°, and the nuclei were isolated as described above (method II).

RNA was prepared by a modification of the method of Joel and Hagerman[28] the extraction process being modified for isolation from chick oviduct nuclei.[29] Nuclei are lysed in 50 mM sodium acetate (pH 5.2) containing 1% SDS and 1 mg of bentonite per milliliter before extraction

[28] P. B. Joel and D. D. Hagerman, *Biochim. Biophys. Acta* **195**, 328 (1969).
[29] J. T. Knowler, unpublished results, 1973.

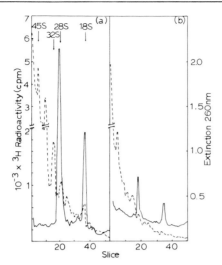

FIG. 4. Retention of RNA during nuclear isolation. Excised chick oviducts were incubated *in vitro*, in the presence of 25 μmCi per milliliter of [5-³H]uridine as described in the Methods section. RNA was extracted either from the incubated tissue or from isolated nuclei. Purified RNA was separated on 2.7% polyacrylamide gels for 5 hours at 5 mA per gel. (a) Total oviduct RNA; (b) RNA of nuclei prepared by the citric acid method (method II); ————, extinction at 260 nm; - - - -, radioactivity per slice.

with phenol. Contaminating oligonucleotides are not exhaustively removed from the preparation since they are run off the end of the polyacrylamide gels. The separation of RNA on polyacrylamide gels has been previously described.[30,31]

Figure 4A shows the incorporation of precursor into total tissue RNA. Peaks of labeling occur in the preribosomal and ribosomal RNA species, and these are superimposed on a background of HnRNA. The identity of the pre-rRNA has been confirmed by methylation studies; in the very similar profile in the labeling of rat uterine RNA, the various species have been characterized by methylation, base composition, rate of synthesis and decay, and intracellular location.[27,30]

Figure 4B shows the profile of RNA species detected in nuclei prepared by method II.[27] The preparation has a distribution of radioactivity in HnRNA very like that of total tissue as well as distinct peaks of 45 S and 32 S pre-rRNA but only very small peaks associated with the ribosomal subunit species (28 S and 18 S). The extinction at 260 nm of 28 S and 18 S RNA is also markedly reduced. One would expect these species to

[30] J. T. Knowler and R. M. S. Smellie, *Biochem. J.* **125**, 605 (1971).
[31] U. E. Loening, *Biochem. J.* **102**, 251 (1967).

be a less conspicuous feature of a nuclear RNA preparation but to what extent they should disappear is not entirely clear. RNA prepared from purified HeLa cell nuclei contains some 28 S RNA but very little of the 18 S species and it has been suggested that the absence of 18 S RNA should be a criterion of nuclear purity.[32] However, otherwise apparently clean preparations of nuclei from other cell types contain both ribosomal RNA species, often detectable both by their extinction at 260 nm and the incorporation of radioactive precursor.[20,27,33]

The distribution in the polyacrylamide gels of labeled RNA from nuclei which have been isolated by other methods, e.g., hypertonic sucrose, reveal that the preribosomal RNA species are still strongly labeled, but there is marked decay of HnRNA. The distribution of label in RNA in nuclei isolated in organic solutions or solutions containing isotonic sucrose with detergents (Triton X-100) all show extensive degradation with little or no high molecular weight RNA. In conclusion, the citric acid procedure described above[27] gives more consistent results and the least loss of the high-molecular-weight RNA of all procedures attempted in this laboratory.

Method III: Isolation of Nuclei for Bulk Yields and for Chromatin Preparation (Bulk Preparation)

This method has been reported previously[34] and is essentially a hypertonic sucrose method which avoids long periods of ultracentrifugation. It utilizes the high specific gravity of concentrated sucrose solutions, filtration, and a nonionic detergent to purify nuclei. The same problems and pitfalls as described in Table I apply to this method. The hypertonic sucrose (method I) used to assay nuclear RNA polymerase activity is excellent for preparing nuclei for chromatin isolation but is limited to small amounts ($\simeq$10 to 20 g) of tissue. The method (III) described here allows the use of bulk quantities of tissue ($\simeq$50 to 100 g) with relatively good yield of nuclei, is more rapid, and is applicable to all tissues studied, e.g., liver, spleen, heart, kidney, erythrocyte, brain, and lung of chickens and rats.

Reagents

TKM buffer: 50 mM Tris·HCl (pH 7.5) + 25 mM KCl + 5 mM MgCl$_2$

Solution A: 0.5 M sucrose in the TKM buffer

Solution B: 1.7 M sucrose in the TKM buffer

Solution C: solution A containing 0.2% Triton X-100

[32] S. Penman, *J. Mol. Biol.* **17**, 117 (1966).

[33] W. J. Steele, N. Okamura, and H. Busch, *J. Biol. Chem.* **240**, 1742 (1965).

[34] T. C. Spelsberg, A. W. Steggles, and B. W. O'Malley, *J. Biol. Chem.* **246**, 4188 (1971).

Solution D: 25% glycerol in 10 mM Tris·HCl (pH 7.5) + 1 mM MgCl$_2$

Procedure. Fifty grams of oviduct are minced into small pieces and homogenized in 6 volumes of solution A using a Thomas Teflon–glass homogenizer (0.010 inch clearance) with about 400 rpm of the power-driven pestle. Usually, 4–5 complete strokes is sufficient. The homogenate is filtered through 4 layers (1-fold) of cheesecloth taking care to prevent excessive accumulation of debris at any one area on the cloth (see Table I, No. 1 under Filtrations), and centrifuged for 5 minutes at 10,000 *g*. The supernatant is discarded and the pellet is gently resuspended in 6–10 volumes (milliliters per gram of tissue) of solution B, using the same Teflon–glass homogenizer with slow ($\simeq$200–400 rpm) rotation of the pestle. One or two strokes usually suffice. After centrifugation for 10 minutes at 20,000 *g*, the floating debris is loosened with a rubber policeman and poured along with the supernatant back into the homogenizer. The tubes with the nuclear pellets are retained in an ice bath. The supernatant with debris is rehomogenized and recentrifuged, and the debris and supernatant are discarded. The combined nuclear pellets are rehomogenized in solution B (6 volumes), recentrifuged as described above, and the supernatant and floating debris discarded. The pellets are resuspended in 5 volumes of solution C using the homogenizer, filtered through organza cloth (100 mesh) and centrifuged for 10 minutes at 10,000 *g*. The tubes are drained and the walls are wiped with tissue paper. The pellets are resuspended in 0.5 volume (milliliters per gram of tissue) of solution D and stored at −20° until needed. This procedure requires about 1.5 hours to complete.

Remarks on the Procedure. Electron micrographs of nuclear preparations obtained by this method closely resemble those of intact oviduct cells, with some artifacts. The outer nuclear membrane is totally removed, and occasional breaks in the inner membrane are evident. Euchromatin and heterochromatin are clearly distinguishable, and nucleoli display their characteristic fibrillar and granular components (Fig. 5A and B). Minor visible contaminants usually include strips of intact smooth muscle cells and round dense bodies 1–2 μm in diameter. The ratio of protein to DNA in these nuclear preparations ranges from 2.2 to 2.6, and the ratio of RNA to DNA ranges from 0.30 to 0.40. The optimal yield of nuclei is 40–60%, depending on the quantity of starting material.

Preparation and Characterization of Chromatin

Isolation of Chromatin. Chromatin is isolated and purified by methods described previously.[9] Nuclei are homogenized by hand in 50 volumes of

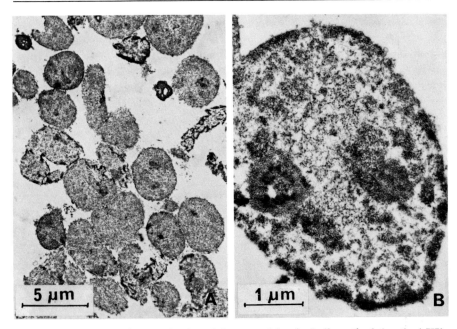

FIG. 5. Electron micrograph of nuclei prepared by the bulk method (method III) for isolating chromatin. See the legend of Fig. 1 for details on preparing the micrographs. (A) ×3500; (B) ×15,000.

cold 80 mM NaCl, 20 mM EDTA, pH 6.3, using a Teflon pestle homogenizer. Homogenates are centrifuged for 10 minutes at 10,000 g. The pellets are rehomogenized and centrifuged twice more in the same buffer, once in cold 0.35 M NaCl, and twice in cold 0.1 mM EDTA and 2 mM Tris·HCl (pH 7.5). Before sedimentation in the latter buffer, the lysate is filtered through organza cloth (100 mesh). Chromatin can be stored at −20° in the latter buffer for long periods without noticeable changes in its composition and template capacity.

Characterization of Chromatin. The chemical analysis of chromatin was reported previously.[9,34] It is based on selective solubilization by hydrolysis of DNA and RNA[35,36] and the chemical analysis of the hydrolyzates.[8,37] Histones and acidic chromatin proteins are selectively isolated and quantitated as described elsewhere.[34,38] Electrophoresis of histones in polyacrylamide gels is readily performed with the method of Panyim and

[35] H. N. Munro and A. Fleck, *Methods Biochem. Anal.* **14**, 114 (1966).
[36] J. M. Webb and H. V. Lindstrom, *Arch. Biochem. Biophys.* **112**, 273 (1965).
[37] G. Ceriotti, *J. Biol. Chem.* **214**, 59 (1955).
[38] O. H. Lowry, N. J. Rosebrough, A. L. Farr, and R. J. Randall, *J. Biol. Chem.* **193**, 265 (1951).

Chalkley.[39] The measurement of the template capacity (for DNA-dependent RNA synthesis) of the isolated chromatins using bacterial RNA polymerase[40] has also been reported.[9,34]

We have found that 2.5–3% of the DNA of chick oviduct chromatin isolated from nuclei of method III can serve as template for *in vitro* DNA dependent RNA synthesis. Analysis of chromatin composition resulted in the following average values: histone:DNA = 1.20, acidic protein/DNA = 0.71, and RNA:DNA = 0.08. The citric acid procedure (method II) proved inadequate for chromatin isolation, for it revealed high open template activity found to be due to the loss of all of lysine rich F_1 (Ia and Ib) and partially F_{2b} histones as reported by other investigators.[24,41,42]

[39] S. Panyim and R. Chalkley, *Arch. Biochem. Biophys.* **130**, 337 (1969).
[40] R. R. Burgess, *J. Biol. Chem.* **244**, 6160 (1969).
[41] J. T. Knowler, H. L. Moses, and T. C. Spelsberg, *J. Cell Biol.,* in press.
[42] A. J. MacGillivray, A. Cameron, R. J. Krauze, D. Richwood, and J. Paul, *Biochim. Biophys. Acta* **277**, 384 (1972).

[26] Isolation and Properties of the Nuclear Envelope

By CHARLES B. KASPER

Furthering our knowledge of membrane structure and function depends upon the acquisition of detailed chemical and biochemical information on purified membrane systems. In the case of the nuclear envelope, the purity of the nuclei is of crucial importance, since the presence of other membranous organelles will result in serious contamination of the nuclear membrane fraction. Consequently, the method of preparing nuclei must be carefully selected and the degree of homogeneity rigorously established. In recent years, procedures for the isolation of the bileaflet nuclear membrane in quantities suitable for chemical and biochemical studies have been reported for rat liver,[1-4] pig liver,[4] bovine liver,[5] and hen erythrocytes.[6] A major concern in the isolation procedure is the preservation of the morpho-

[1] M. Bornens, *C.R. Acad. Sci. Ser. D* **266**, 596 (1968).
[2] D. M. Kashnig and C. B. Kasper, *J. Biol. Chem.* **244**, 3786 (1969).
[3] I. B. Zbarsky, K. A. Perevoshchikova, L. N. Delektorskaya, and V. V. Delektorsky, *Nature (London)* **221**, 257 (1969).
[4] W. W. Franke, B. Deumling, B. Ermen, E. Jarasch, and H. Kleinig, *J. Cell Biol.* **46**, 379 (1970).
[5] R. Berezney, L. Funk, and F. L. Crane, *Biochim. Biophys. Acta* **203**, 531 (1970).
[6] H. Zentgraf, B. Deumling, E. Jarasch, and W. W. Franke, *J. Biol. Chem.* **246**, 2986 (1971).

logical and functional integrity of the membrane. In the method to be described, the envelope is released by sonic disruption of the nuclei, the nucleoplasm is solubilized by the addition of appropriate amounts of potassium citrate, and the final purification is performed by density gradient centrifugation. This procedure permits the isolation of nuclear ghosts and large fragments of nuclear envelope with intact pore complexes.

Isolation of the Nuclear Envelope

Preparation of Nuclei

Reagents

Tris·HCl, 0.5 M, containing 50 mM MgCl$_2$ and 25 mM KCl, pH 7.50 (10X TKM buffer)
Sucrose, 2.3 M, in 50 mM TKM buffer, pH 7.5 (filter)
Sucrose, 1.0 M, in 50 mM TKM buffer, pH 7.5
Sucrose, 0.25 M, in 50 mM TKM buffer, pH 7.5

Procedure. The procedure used for the isolation of rat liver nuclei is basically that of Blobel and Potter[7] with several minor innovations as described by Kashing and Kasper.[2] Male Holtzman rats are starved for 15–20 hours before being sacrificed. After decapitation, the animals are allowed to bleed freely and the excised livers are rinsed in cold distilled water and then immersed in ice slush. Three animals at a time are processed in this manner. The livers are freed of connective tissue, blotted on filter paper, weighed into 12-g packets, wrapped in Parafilm, and stored in crushed ice. The livers (in 12-g batches) are finely minced with scissors and transferred to the grinding chamber of a Potter-Elvehjem homogenizer equipped with a Teflon pestle (A. H. Thomas Co., Cat. No. 3431-E25). Then 24 ml of 0.25 M sucrose–50 mM TKM buffer is added, and homogenization of the tissue is accomplished with the aid of a power-driven pestle connected to a variable transformer to regulate the speed at approximately 1700 rpm. A motor that has proved to be quite satisfactory is the Thomas Power Drive Assembly No. 8590-H30, A. H. Thomas Co. Approximately 15 up-and-down strokes are satisfactory to release the nuclei. The homogenate is filtered into a beaker through four layers of cheesecloth. Stirring with a glass rod facilitates filtration, and finally the cheesecloth is closed at the top to form a bag and the excess liquid is expressed by applying gentle pressure with the glass stirring rod. Two 12-g packets of liver provide sufficient filtrate for one run with the Beckman SW 25.2 rotor. With a

[7] G. Blobel and V. R. Potter, *Science* **154**, 1662 (1966).

pipette, 16 ml of filtrate are transferred to each of three cellulose nitrate tubes that fit the SW 25.2 rotor. Next, 32 ml of 2.3 M sucrose–50 mM TKM are added with a 50-ml syringe (no needle) to each tube and the contents are thoroughly mixed by inversion after the tubes are covered with Parafilm. The homogenate, which is close to 1.64 M in sucrose, is carefully underlayered with 10 ml of 2.3 M sucrose–50 mM TKM; a 10-ml syringe equipped with a 13-gauge needle is used to deliver the viscous sucrose solution. The nuclei are forced through the 2.3 M sucrose–50 mM TKM layer and collected in the bottom of the tube by centrifugation at 24,500 rpm (104,000 g_{max}) for 70 minutes (from start to brake) at 3°. The heavy brownish-red plaque at the top of the tube is loosened by rimming with a small spatula and removed. The supernatant is removed by quickly inverting the tube and allowing it to drain thoroughly. Residual sucrose solution is removed by gently directing a stream of distilled water from a plastic squeeze bottle against the wall of the inverted tube. Care

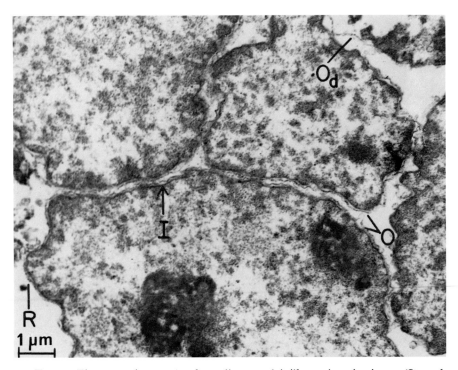

FIG. 1. Electron micrograph of rat liver nuclei illustrating the inner (I) and outer (O) nuclear membranes. Ribosomes (R) attached to the outer leaflet are visible as electron-dense granules. The designation O_d refers to a segment of outer membrane which has become partially detached from the nucleus.

must be taken to prevent the wash solution from contacting the nuclear pellet. After draining, the inside wall of the tube is carefully cleaned with an absorbent tissue and the nuclear pellet suspended by vortex mixing in approximately 3–5 ml of 1.0 M sucrose–50 mM TKM. The nuclei from one SW 25.2 tube are transferred with rinsing to a 30-ml Corex centrifuge tube (No. 8445, Corning Glass Works, Corning, New York) and centrifuged at 3400 g_{max} for 10 minutes in a Sorval RC-3 equipped with the HG-4 rotor. The supernatant is discarded, and a final washing is performed by suspending the nuclei in 0.25 M sucrose–50 mM TKM and centrifuging at 750 g_{max} for 10 minutes at 3°. The yield of nuclei based on the DNA recovery is 50–70%. An electron micrograph of intact nuclei illustrating the inner and outer leaflets of the nuclear envelope is shown in Fig. 1. Available evidence indicates that hepatocyte nuclei are selectively separated from nonhepatocyte nuclei by this procedure.[8,9]

Preparation of Nuclear Membrane

Reagents

60% (w/v) potassium citrate–H$_2$O in 50 mM TKM buffer

2 M sucrose–50 mM TKM containing 10% (w/v) potassium citrate·H$_2$O

Discontinuous density gradient solutions. Each solution contains (1) 10 ml of 0.5 M TKM buffer, (2) 10.0 g of K$_3$ citrate·H$_2$O, and (3) the following amounts of sucrose per 100 ml of final solution: $d^{24} = 1.16$ g/cc, 26.4 g; $d^{24} = 1.18$ g/cc, 31.3 g; $d^{24} = 1.20$ g/cc, 37.0 g.

Procedure. Only freshly prepared nuclei should be used. Nuclear preparations that have been frozen and thawed are unsatisfactory, and nuclei stored at 3° are not reliable in terms of biochemical alterations occurring during storage. To each washed nuclear pellet contained in a 30-ml Corex centrifuge tube, 10 ml of cold 50 mM TKM buffer (pH 7.50) are added. The tube is positioned vertically in an ice bath, and the standard microtip of the Sonifer Cell Disrupter W 140-D (Heat Systems–Ultrasonics, Inc., Plainview, New York) is placed approximately 1 cm above the bottom of the tube. The output control of the sonifier is placed at a setting of 6.5, and the nuclei are disrupted by sonic treatment for 15 seconds. Immediately after this step, 2 ml of 60% (w/v) potassium citrate–H$_2$O in 50 mM TKM buffer are added with vortex mixing. This solubilizes much of the nucleoplasm, which otherwise adheres to the nuclear membrane

[8] T. W. Sneider, D. E. Bushnell, and V. R. Potter, *Cancer Res.* **32**, 1867 (1970).
[9] H. M. Rabes, R. T. Hartenstein, and W. Ringelman, *Cancer Res.* **32**, 83 (1972).

and contaminates the final product. The sonication is a most crucial step, and the proper balance between nuclear disruption and membrane disintegration must be achieved. Since uniform performance will not be obtained from one sonifier to another, the optimal conditions for a particular sonifier and microtip must be individually determined. Care should be taken to properly tune the sonifier, and this setting should be checked periodically. Also, the microtip will develop, through use, a pitted surface which markedly decreases its efficiency. The optimal conditions for sonication are best determined by examining the sonicate (after addition of potassium citrate to a final concentration of 10% w/v) under the phase microscope after different periods of sonication, e.g., 10, 15, and 20 seconds. The proper end point is one at which there is a maximum number of nuclear ghosts and large membrane fragments and a minimum number of intact nuclei. Care must be taken to avoid excessive sonication. Figure 2A–D illustrates the various structures observed in a sonically treated nuclear suspension to which potassium citrate was added. Intact nuclei appear as dense, well-defined, spherical bodies (Fig. 2A), whereas nuclei with a broken envelope and exuding nucleoplasm appear swollen and considerably less dense (Fig. 2B and C). A nucleus that has lost the bulk of its intranuclear contents appears as a collapsed irregular structure and is termed a nuclear ghost (Fig. 2D). The nuclear sonicate is transferred to nylon tubes to fit the SS-34 rotor of the Sorvall RC-2 and centrifuged at 39,000 g_{max} for 45 minutes at 3°. The semigelatinous pellets of crude nuclear membrane from 100–200 g of liver are transferred to a 15-ml conical Corex centrifuge tube (Cat. No. 806A, Corning Glass Works, Corning, New York) containing 1–3 ml of 50 mM TKM buffer–10%

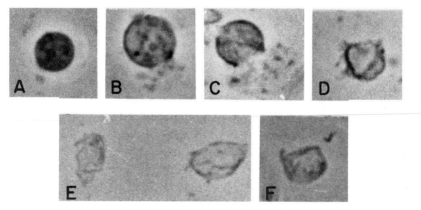

FIG. 2. Phase-contrast micrographs depicting different stages in the isolation of the nuclear envelop [D. M. Kashnig and C. B. Kasper, *J. Biol. Chem.* **244**, 3786 (1969)]. See text for explanation. Final magnification in each case, ×1250.

citrate. The pellet is evenly suspended by gentle sonication while immersed in an ice bath. The subsequent density gradient centrifugation is carried out in either the Beckman SW 25.1 or SW 27 rotor. The membrane suspension is transferred to the appropriate cellulose nitrate centrifuge tube, and 2 *M* sucrose–50 m*M* TKM containing 10% potassium citrate is added to adjust the density to approximately 1.21–1.22 g/cc. The suspension is overlayered, using a peristaltic pump, in a stepwise manner with appropriate volumes of each of the sucrose–TKM–citrate solutions having densities of 1.20, 1.18, and 1.16 g/cc. The discontinuous gradient is developed by centrifugation at 88,000 g_{max} for 10–14 hours at 3° in the case of the SW 25.1 rotor and at approximately 120,000 g_{max} when the SW 27 is used. A typical gradient is pictured in Fig. 3. The bileaflet nuclear membrane collects at both the *d* 1.16–1.18 and 1.18–1.20 g/cc interfaces; however, most of the membrane is usually found at the upper interface. Nuclear ghosts representative of those isolated at both interfaces are shown in Fig. 2, E and F. Occasionally small amounts of material will appear at the lowest interface (between the bottom layer and the *d* 1.20 g/cc layer). Electron microscopic examination of this fraction indicates variable amounts of the bileaflet nuclear membrane, with large amounts of nonmembrane nucleoplasmic material. Reprocessing of this grossly contaminated fraction has not yielded significant levels of purified nuclear membrane. The purified membrane is isolated from the gradient by lateral puncture with a syringe

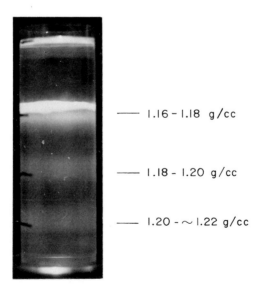

— 1.16 - 1.18 g/cc

— 1.18 - 1.20 g/cc

— 1.20 - ~1.22 g/cc

Fig. 3. Separation of nuclear membrane on a discontinuous sucrose-citrate density gradient. The interfaces are indicated.

and 18-gauge needle, diluted appropriately with 50 mM TKM buffer–10% citrate, and collected by centrifugation for 30 minutes at 314,000 g_{max} in the Beckman type 65 rotor. For most studies, the two nuclear membrane fractions are combined. The membrane is finally washed with water to remove residual sucrose and salts and is either used directly or suspended in H_2O and stored at $-18°$. The yield of nuclear membrane is approximately 5% of the total nuclear protein and in the range of 30–40% based on the recovery of nuclear phospholipid phosphorus.

Comments. Discussions of various procedures used for the isolation of nuclei may be found in the literature.[10-12] From a practical as well as theoretical point of view, the one-step isolation of nuclei from a tissue homogenate has several distinct advantages over those involving the intermediate isolation of a crude nuclear pellet which is subsequently purified. The crude pellet contains, in addition to nuclei, whole cells, cellular debris, and numerous cytoplasmic structures. This increases the likelihood of adsorption onto the nuclear surface of cytoplasmic contaminants that may not be easily removed by rehomogenization and further purification. The fact that purified nuclear membrane can form a hybrid complex with microsomal membrane[13] suggests an additional reason why collecting nuclei in a pellet along with other membranous organelles should be avoided.

The exposure of the nuclear membrane to citrate and sonication results in the removal of ribosomes from the outer nuclear leaflet. The main function of the citrate is to disperse and solubilize the nucleoplasm, which otherwise adheres to an occludes the nulear envelope. Since citrate may be inhibitory to certain enzymes or enzyme complexes, careful controls are necessary when performing biochemical studies.

Because of the bileaflet nature of the nuclear envelope, it is logical to think in terms of separating the inner and outer membranes into entities that can be meaningfully characterized with respect to morphology, biochemistry, and composition. To date, techniques suitable for this task are lacking. Although methods have been described for the selective removal of the outer nuclear membrane with such agents as Triton X-100[7,14] and citric acid,[15] the resulting modification of the membrane composition and structure limits their application.

[10] A. L. Dounce, *in* "The Nuclei Acids" (E. Chargaff and J. N. Davidson, eds.), p. 93. Academic Press, New York, 1955.
[11] T. Y. Wang, this series, Vol. 12a, p. 417.
[12] H. Busch, this series, Vol. 12a, p. 421.
[13] C. B. Kasper and H. Kubinski, *Nature (London)* **231**, 124 (1971).
[14] P. D. Sadowski and J. A. Howden, *J. Cell Biol.* **37**, 163 (1968).
[15] S. J. Smith, H. R. Adams, K. Smetana, and H. Busch, *Exp. Cell Res.* **55**, 185 (1969).

Properties of the Nuclear Envelope

Morphology. The nuclear envelope isolated at the d 1.16–1.18 and d 1.18–1.20 g/cc interfaces is characterized by the usual double membrane structure with the accompanying nuclear pores (Figs. 4 and 5). Morphologically, the membranes obtained at the two interfaces are indistinguishable.[16] Both the inner and outer nuclear membranes measure approximately 75 Å thick and are joined together at frequent intervals by

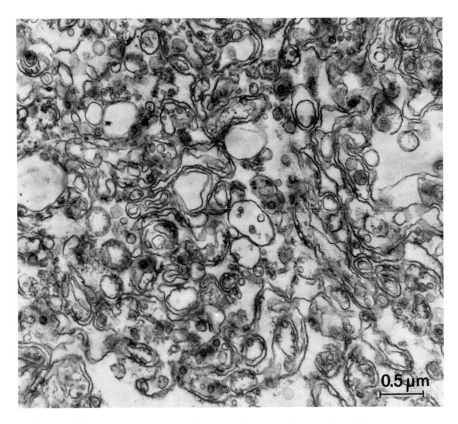

Fig. 4. Electron micrograph of the nuclear envelope isolated from a sucrose-citrate gradient.

[16] The claim of N. S. Mizuno, C. E. Stoops, and A. A. Sinha, *Nature* (*London*) **229**, 22 (1971) and N. S. Mizuno, C. E. Stoops, and R. L. Peiffer, Jr., *J. Mol. Biol.* **59**, 517 (1971) that the inner and outer nuclear membranes are localized at the d 1.16–1.18 and d 1.18–1.20 g/cc interfaces, respectively, is not supported by a detailed morphological, chemical, and biochemical examination of the two membrane zones.

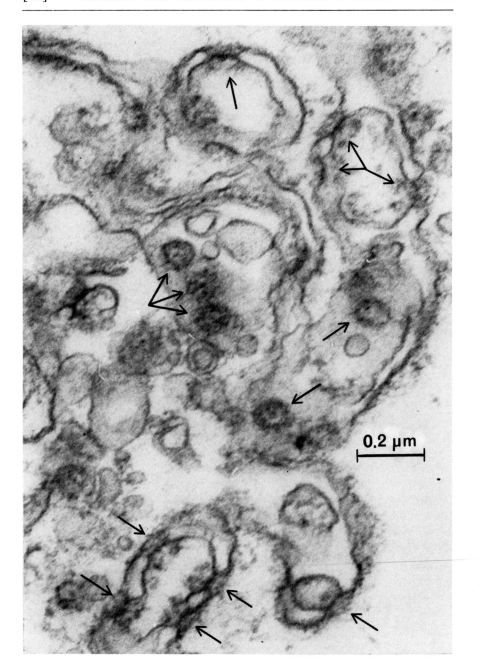

FIG. 5. Enlargement of the lower center section of Fig. 4. Arrows direct attention to transverse and tangential sections of pore complexes.

pore complexes. The diameter of the pore complex is in the range of 800–1000 Å. In tangential view, pores appear as annuli with particularly electron-dense periphery, whereas, in transverse section, the inner and outer membranes appear to converge to form a single electron-dense diaphragm-like structure bordered by electron-opaque material. Negative staining techniques have demonstrated that nuclear pores have 8-fold symmetry.[17]

Chemical Composition. Analysis of the two nuclear envelope fractions indicates no major variation in composition (Table I). Protein and lipid account for approximately 90% of the membrane, the remainder being contributed by carbohydrate and nonribosomal RNA. The proteins of the bileaflet nuclear membrane have been fractionated and characterized with respect to molecular weight and other physicochemical parameters.[18] A minimum of 23 different molecular weight classes of proteins can be detected in the nuclear envelope, ranging from 160,000 to 16,000. The major polypeptide chains, as determined by SDS gel electrophoresis, have molecular weights of 74,000, 70,000, 64,000, and 53,000 (doublet). The three proteins in the range of 64,000 to 74,000 account for 25% of the total envelope protein while the doublet at 53,000 represents an additional 20%. The carbohydrate of the membrane is in the form of glycoprotein. Com-

TABLE I
COMPOSITION OF NUCLEAR MEMBRANE FROM RAT LIVER[a]

Component	Composition at interface	
	d 1.16–1.18 g/cc (%)	d 1.18–1.20 g/cc (%)
Protein	61.3	63.7
Carbohydrate	3.0	4.0
Neutral sugars	2.7	3.7
Hexosamine	0.24	0.21
Sialic acid	0.06	0.09
Lipid	32.4	26.0
Phospholipid	30.1	25.0
Cholesterol	2.3	1.0
RNA	3.3	6.3
DNA	0.0	0.0

[a] Taken in part from D. M. Kashnig and C. B. Kasper, *J. Biol. Chem.* **244**, 3786 (1969). The values in this table have been adjusted to incorporate recent analytical data obtained with chromatographically purified cholesterol fractions (A. S. Khandwala and C. B. Kasper, unpublished results).

[17] B. J. Stevens and J. Andre, *in* "Handbook of Molecular Cytology" (A. Lima-De-Faria, ed.), p. 837. North-Holland Publ., Amsterdam, 1969.
[18] M. Bornens and C. B. Kasper, *J. Biol. Chem.* **248**, 571 (1973).

positional studies indicate that mannose is the predominant neutral sugar while galactose is present in only trace quantities.[19] According to column chromatographic analysis of acid hydrolyzates of delipidated nuclear membrane, at least 95% of the hexosamine is accounted for by glucosamine.[2] Sialic acid has consistently been found at a low level.

No DNA has been detected as an integral part of the nuclear envelope; however, loosely associated DNA could easily be removed by this isolation procedure. The fact that nuclear envelopes with numerous intact pore complexes can be obtained free of DNA offers direct chemical evidence that DNA is not a discrete structural component of the complex as has been claimed by Agutter.[20] Small amounts of DNA have been found attached to nuclear envelope that has not been subjected to citrate treatment[4,5]; these same preparations also have a low phospholipid to protein ratio (approximately 0.22).

Phosphorus and lipid values for purified nuclear membrane appear in Table II. The phospholipid to protein ratio (w/w) ranges from 0.4 to 0.5 for the two fractions, and lipid phosphorus accounts for approximately 75% of the total membrane phosphorus. No significant differences are noted in the phospholipid composition of the nuclear membrane and the nucleus (Table III). Differences between the nucleus and nuclear membrane were detected, however, in the fatty acid composition of the sphingomyelin, lysophosphatidylcholine, and phosphatidylserine plus inositol fractions. These differences were confined largely to variations in the levels of fatty acids with carbon chain lengths of C_{20} and above.

TABLE II

PHOSPHORUS AND LIPID VALUES FOR ISOLATED RAT LIVER NUCLEAR MEMBRANE[a]

| | Relative composition at interface | |
Quantity measured	d 1.16–1.18 g/cc	d 1.18–1.20 g/cc
Total phosphorus per mg of protein	25.9 ± 3.0 µg	21.4 ± 4.1 µg
Lipid phosphorus per mg of protein	19.8 ± 3.1 µg	15.5 ± 2.5 µg
Lipid phosphorus per µg of total phosphorus	0.76 ± 0.05 µg	0.73 ± 0.05 µg
Phospholipid per mg of protein	0.50 ± 0.07 mg	0.39 ± 0.06 mg
Cholesterol per µmole of lipid phosphorus	0.16 ± 0.03 µmole	0.08 ± 0.01 µmole
Protein per µmole of lipid phosphorus	1.57 ± 0.22 mg	2.00 ± 0.27 mg

[a] From D. M. Kashnig and C. B. Kasper, *J. Biol. Chem.* **244**, 3786 (1969).

[19] C. B. Kasper, unpublished results.
[20] P. S. Agutter, *Biochim. Biophys. Acta* **255**, 397 (1972).

TABLE III

PHOSPHOLIPID COMPOSITION OF RAT LIVER NUCLEI AND NUCLEAR MEMBRANE[a]

	Percent of total phosphorus	
Phospholipid	Nuclei	Nuclear membrane
Cardiolipin + phosphatidic acid	1.5 ± 0.4	1.4 ± 0.3
Phosphatidylethanolamine	18.8 ± 1.0	18.3 ± 1.0
Phosphatidylserine + phosphatidylinositol	14.6 ± 0.2	13.9 ± 0.2
Phosphatidylcholine	59.7 ± 2.6	61.8 ± 1.0
Sphingomyelin	2.7 ± 0.3	2.5 ± 0.3
Lysophosphatidylcholine	2.3 ± 1.3	1.4 ± 0.4

[a] A. S. Khandwala and C. B. Kasper, *J. Biol. Chem.* **246**, 6242 (1971).

Enzymatic Composition. Various enzymatic activities have been found associated with the purified nuclear envelope (Table IV); however, to date no enzyme has been studied that could be considered unique to this bileaflet structure. Quite significant is the finding that glucose-6-phos-

TABLE IV

ENZYMATIC ACTIVITIES ASSOCIATED WITH THE NUCLEAR AND MICROSOMAL MEMBRANES[a]

Enzyme	Nuclear membrane	Microsomal membrane
Glucose-6-phosphatase[e]	$3.82,$[b] 5.32[c]	7.79
Mg^{2+}-Adenosine triphosphatase[e]	$0.433,$[b] 0.591[c]	2.67
DPNH–cytochrome c reductase[f]	$0.381,$[b] 0.379[c]	0.902
Cytochrome b_5[g]	0.183[d]	0.492
TPNH–cytochrome c reductase[f]	0.104[d]	0.332
Cytochrome P-450[g]	0.16[d]	0.62
N-Demethylase[h]	2.6[d]	24.1
Aryl hydroxylase[i]	259[d]	3640

[a] Taken in part from C. B. Kasper, *J. Biol. Chem.* **246**, 577 (1971) and from D. M. Kashnig and C. B. Kasper, *J. Biol. Chem.* **244**, 3786 (1969).
[b] Value for membrane isolated at d 1.16 to 1.18 g/cc interface.
[c] Value for membrane isolated at d 1.18 to 1.20 g/cc interface.
[d] Values obtained for the combined nuclear membrane fractions.
[e] Micromoles of P_i released per 15 minutes per milligram of protein.
[f] Micromoles of cytochrome c reduced per minute per milligram of protein.
[g] Millimicromoles per milligram of protein.
[h] Millimicromoles of 3-methyl-4-aminoazobenzene formed per 30 minutes per milligram of protein.
[i] Micromicromoles of hydroxylated benzo[a]pyrene formed per 30 minutes per milligram of protein.

phatase is an intrinsic nuclear envelope enzyme[2]; this activity, which has generally been considered a marker for the endoplasmic reticulum, is present in the nuclear envelope at approximately 50% the level noted for the microsomal membrane. Recent biochemical studies dealing with rat[21] and bovine[22] liver nuclear membranes confirm the presence of glucose-6-phosphatase. In addition, histochemical data further demonstrate the localization of glucose-6-phosphatase activity around the periphery of the nucleus.[23-26] Failure of other laboratories[20,27-29] to account for the nuclear glucose-6-phosphatase activity in their envelope preparations is difficult to explain. It seems well established, however, that glucose-6-phosphatase must be considered a marker enzyme for both the nuclear envelope and the microsomal membrane, at least in the case of rat and beef liver.

The DPNH and TPNH electron transport systems similar to those found in the endoplasmic reticulum are also present in the nuclear envelope. Nuclear DPNH cytochrome c reductase is rotenone-insensitive and in this regard resembles the microsomal enzyme as well as the enzyme located in the outer mitochondrial membrane. Although variable results have been obtained for the cytochrome P-450 content of the nuclear envelope,[30] results from the analysis of a large number of preparations yield an average value of 0.16 ± 0.04 nmole per milligram of protein (A. S. Khandwala and C. B. Kasper, unpublished results). Other components of the TPNH pathway include TPNH–cytochrome c reductase and various drug metabolizing activities (benzo[a]pyrene hydroxylase and an aminoazo dye N-demethylase). In contrast to the microsomal TPNH electron transport chain, the chain associated with the nuclear envelope is not induced by the administration of phenobarbital.[30] However, when 3-methylcholanthrene is used as the inducing agent, the specific activity of nuclear envelope aryl hydroxylase is increased 17-fold.[31] This preferen-

[21] R. R. Kay, D. Fraser, and I. R. Johnston, *Eur. J. Biochem.* **30**, 145 (1972).

[22] R. Berezney, L. K. Macaulay, and F. L. Crane, *J. Biol. Chem.* **247**, 5549 (1972).

[23] S. Orrenius and J. Ericsson, *J. Cell Biol.* **31**, 243 (1966).

[24] S. I. Rosen, *J. Anat.* **105**, 579 (1969).

[25] S. I. Rosen, *Experientia* **26**, 839 (1970).

[26] A. Leskes, P. Siekevitz, and G. E. Palade, *J. Cell Biol.* **49**, 264 (1971).

[27] I. B. Zbarsky, K. A. Perevoshchikova, L. N. Delektorskaya, and V. V. Delektorsky, *Nature (London)* **221**, 257 (1969).

[28] W. W. Franke, B. Deumling, B. Ermen, E. Jarasch, and H. Kleinig, *J. Cell Biol.* **46**, 379 (1970).

[29] I. B. Zbarsky *in* "Methods in Cell Physiology" (D. M. Prescott, ed.), Vol. V, p. 167. Academic Press, New York, 1972.

[30] C. B. Kasper, *J. Biol. Chem.* **246**, 577 (1971).

[31] A. S. Khandwala and C. B. Kasper, *Biochim. Biophys. Res. Comm.,* in press.

tial inductive effect indicates that the biosynthetic regulation of the nuclear enzyme differs from that observed for the endoplasmic reticulum.

The occurrence of certain microsomelike enzymes (TPNH and DPNH cytochrome c reductases, glucose-6-phosphatase, and cytochrome b_5) in the nuclear envelope at a level of approximately 30–60% that found for the microsomal membrane suggests that these activities may reside on only one of the nuclear leaflets, presumably the outer. Thus, a partial biochemical as well as morphological continuum may exist between the endoplasmic reticulum and the outer nuclear leaflet, whereas the inner leaflet may be biochemically and functionally distinct.

[27] Preparation of Submitochondrial Vesicles Using Nitrogen Decompression

By SIDNEY FLEISCHER, GERHARD MEISSNER, MURRAY SMIGEL, and ROBERT WOOD

A procedure is described for disrupting mitochondria to yield submitochondrial vesicles. The mitochondria are subjected to high nitrogen pressure (2000 psi) in a sealed chamber and are then rapidly decompressed to atmospheric pressure by passing them through a needle valve. The combined effect of nitrogen decompression and the mechanical shear due to flow is the disruption of the mitochondria. This treatment results in various degrees of disruption and ultimately gives small single membrane "inside-out" vesicles. The treatment is repeated to improve the efficiency of disruption, and the vesicles are separated from the unbroken or partially broken mitochondria by the use of differential centrifugation.

Reagents and Preparations

Beef heart mitochondria were isolated and purified according to Hatefi and Lester[1] with some modifications.[2,3] The mitochondria (25–40 mg of protein per milliliter) in 0.25 M sucrose were stored frozen in 20–40 ml aliquots in plastic bottles at $-80°$.

Sucrose solution, 0.25 M, neutralized to pH 7.4 with a solution of potassium hydroxide

Sucrose–HEPES solution: 0.25 M–1 mM, neutralized to pH 7.4 with

[1] Y. Hatefi and R. L. Lester, *Biochim. Biophys. Acta* **27**, 83 (1958).
[2] S. Fleischer, G. Rouser, B. Fleischer, A. Casu, and G. Kritchevsky, *J. Lipid Res.* **8**, 170 (1967).
[3] H. G. Bock and S. Fleischer, this series, Vol. 32 [35] in press.

a solution of potassium hydroxide. [*N*-2-Hydroxyethylpiperazine-*N'*-2-ethanesulfonic acid (HEPES) A grade, Calbiochem, Lajolla, Calif.]

Equipment

Potter-Elvehjem type glass homogenizer equipped with a Teflon pestle (A. H. Thomas "tissue grinder"). The homogenizer is used to resuspend sedimented mitochondria or submitochondrial vesicles.

Cell Disruption Bomb[4]

The 4635 cell disruption bomb (Parr Instrument Co., 211 Fifty-third Street, Moline, Illinois) is a stainless steel chamber that may be used to subject samples to high pressure (Fig. 1). It is readily disassembled and reassembled. The bomb head (upper portion) is equipped with a pressure gauge, an inlet (gas filling) valve for charging the bomb with nitrogen and a throttling (release) valve for releasing the sample. A dip tube, attached to the release valve, extends close to the bottom of the sample container inside the bomb cylinder when the bomb is assembled. The dip tube and the sample exit tubing which extends beyond the release valve are usually of stainless steel. We have specified Teflon in place of steel tubing so that the sample contacts metal only when it flows through the release valve. The sample is placed into the bomb cylinder either in a plastic beaker or tube. The bomb head is then seated on the bomb cylinder, and the bomb is sealed (cf. Fig. 1). An 1831 nitrogen filling connection is provided

---

[4] A pressure bomb is also available from Yeda, Research and Development Co. Ltd., P.O. Box 95, Rehovot, Israel. This unit is smaller in capacity (1–28 ml) and is capable of sustaining pressures up to 6000 psi. We have not had experience with this unit. It has been used for disrupting the alga *Euglena gracilis* to prepare chloroplasts or chloroplast fragments of high photosynthetic activity.[5] Nitrogen decompression and shear have been used for the controlled disruption of bacteria, mammalian tissues, and tissue culture cells.[6,7,8] The amount of shear can be modulated by varying the nitrogen pressure in the bomb.

Other procedures, such as sonic oscillation or the use of a French pressure cell, have been used for disruption of biological materials. These methods have also been used to disrupt heart mitochondria.[9-11] The use of nitrogen decompression is a simple, readily controllable procedure which is carried out under inert atmosphere and is free from heating problems.

[5] A. Shneyour and M. Avron, *FEBS (Fed. Eur. Biochem. Soc.) Lett.* **8**, 164 (1970).
[6] D. Fraser, *Nature (London)* **167**, 33 (1951).
[7] M. J. Hunter and S. L. Commerford, *Biochim. Biophys. Acta* **47**, 580 (1961).
[8] D. F. H. Wallach, J. Soderberg, and L. Bricker, *Cancer Res.* **20**, 397 (1960).
[9] A. W. Linnane and D. M. Ziegler, *Biochim. Biophys. Acta* **29**, 630 (1958).
[10] C. P. Lee and L. Ernster, *Eur. J. Biochem.* **3**, 391 (1968).
[11] S. Fleischer, B. Fleischer, and W. Stoeckenius, *J. Cell Biol.* **32**, 193 (1967).

Fig. 1. Photograph of Parr cell disruption bomb. The upper and lower photos show the unit assembled and disassembled, respectively. The inset shows, in top view, the plastic filler (on the right) which fits into the Parr bomb base (on the left). The plastic filler is used to decrease nitrogen consumption as well as to support the tubes containing the sample (cf. text).

which connects the inlet valve to the nitrogen tank. Instructions for assembly and disassembly are provided by the manufacturer.

For disruption of small amounts of sample (5–50 ml), we have fashioned a solid plastic filler which occupies most of the inner volume of the bomb (Fig. 1, inset). The filler contains two vertical cylindrical holes of 1.5 cm and 4.0 cm in diameter. The tube containing the sample is inserted into the appropriate size hole, which would otherwise be filled with a plexiglass rod. The filler serves both as a support for the sample tube as well as to decrease the empty space so that less nitrogen is required to pressurize the bomb.

Procedure for Preparation of Submitochondrial Vesicles Using the Parr Bomb

All the following operations are carried out in the cold (0–4°). The mitochondria (5–500 ml of a suspension of 10 mg of protein per milliliter in aqueous 0.25 M sucrose–1 mM HEPES, pH 7.4) are placed into the Parr bomb in a plastic container. We find it convenient to fill the dip tube, the release valve, and the exit tubing with sucrose solution prior to assembly of the bomb, so that later the adjustment of sample flow can be made without waiting for passage of the sample through the exit tubing. This is accomplished by inverting the bomb head, filling the tubes with liquid, and then closing the release valve. The nitrogen inlet is also closed. The bomb is then sealed and connected to the nitrogen tank with the 1831 nitrogen filling connection. The inlet valve is cautiously opened, and the bomb is filled with nitrogen to a pressure of 2000 psi from a tank of pre-purified nitrogen. Nitrogen pressure is maintained for 20 minutes.[12] The discharge valve is then carefully opened allowing the sample to come out dropwise into a collecting vessel. The rate is approximately 10–15 ml per minute, i.e., it takes about 20–30 minutes for 300 ml to empty. When the bomb is emptied of sample contents, the pressure must be released before

TABLE I

PREPARATION OF SUBMITOCHONDRIAL VESICLES

Preparation	ml	Protein (mg/ml)	Total protein	
			mg	%
Beef heart mitochondria	266	9.8	2606	100
Supernatant	266	1.4	372	14.3
Submitochondrial vesicles				
Fraction 1	25	18.5	462	17.7
Fraction 2	21	28.0	588	22.6
Unbroken or partially disrupted mitochondria	20	33.1	662	25.4

Protein recovery 80%

[12] It takes time for the nitrogen gas to saturate the sample, as can be observed by the decrease in pressure if the inlet valve is closed soon after the bomb is filled with nitrogen gas to 2000 psi. Nitrogen decompression has a disruptive effect since dissolved nitrogen expands at atmospheric pressure giving the sample "the bends." However, it is our experience that rapid decompression is not sufficient for optimal disruption. The sample must flow slowly (dropwise) through the valve to effect sufficient shear for disruption.

the bomb can be reopened. A bleed valve supplied on the nitrogen filling connection is opened, and the bomb pressure is allowed to decrease to atmospheric pressure. Before reassembly, the discharge valve, the dip and exit tubing should be carefully rinsed to remove any ice that may have formed during the nitrogen discharge. The sample is then placed back into the bomb, and the disruption process is repeated one more time.

The disrupted mitochondria are centrifuged for 10 minutes at 20,000

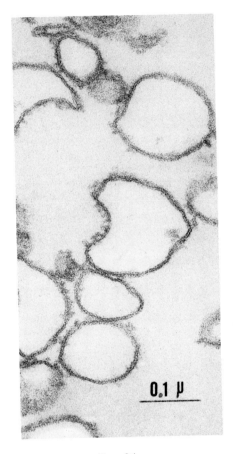

FIG. 2A.

FIG. 2. Electron micrographs of submitochondrial vesicles prepared using the Parr cell disruption bomb. The vesicles were prepared for electron microscopy by conventional sample preparation, visualized in thin sections (Fig. 2A) or by negative staining (Fig. 2B). Note that these vesicles are matrix side out, in that the inner membrane particles are on the outer face of the membrane. The magnification of the two electron micrographs is the same.

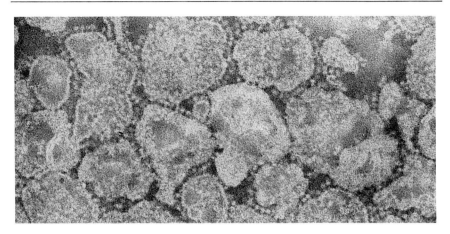

FIG. 2B.

rpm in a Spinco 42.1 fixed-angle rotor (44,000 *g* at maximum radius 9.9 cm). This first sediment consisting of mitochondria or partially disrupted mitochondria appears opaque brown. The turbid supernatant is carefully decanted and recentrifuged for 1 hour at 40,000 rpm in the same rotor (177,000 *g* at maximum radius). The pellet consisting of submitochondrial vesicles is brownish and translucent. It is resuspended in 0.25 *M* sucrose with the aid of a Potter-Elvehjem type homogenizer (submitochondrial vesicles fraction 1; cf. Table I). If the bottom part of the pellet is of opaque appearance, it is separated from the remaining pellet and combined with the first sediment. The first sediment usually contains a dense bluish-black lipofuscin-containing material at the very bottom of the pellet. This is left behind in combining the mitochondrial pellets. The combined sediments are resuspended to about 10 mg of protein per milliliter in 0.25 *M* sucrose–1 m*M* HEPES, pH 7.4. The suspension is then disrupted twice more in the Parr bomb and centrifuged as described above. The sedimented submitochondrial vesicles are resuspended in 0.25 *M* sucrose (submitochondrial vesicles, fraction 2; cf. Table I). The submitochondrial vesicles are placed in plastic tubes and quick-frozen using liquid nitrogen. They can be stored at $-70°$ for prolonged periods of time (>half year).

Properties of Submitochondrial Vesicles

Submitochondrial vesicles, obtained from the first and second disruption cycle, do not differ appreciably from one another and are usually combined. The yield is approximately 1050 mg of protein starting with 2.6 g of protein of bovine heart mitochondria (Table I).

TABLE II

DPNH AND SUCCINATE OXIDATION BY MITOCHONDRIA AND SUBMITOCHONDRIAL
VESICLES WITH CYTOCHROME c AND OXYGEN AS ELECTRON ACCEPTORS[a]

		Electron transfer activity			
Preparation	Substrate	Micromoles Cyt. c reduced/ (min · mg^{-1})	Microatoms O consumed/(min · mg^{-1})		Ratio activities[b] (oxidase/ reductase)
			−Cyt. c	+Cyt. c	
Mitochondria	DPNH	0.81	0.70	1.7	4.2
Submitochondrial vesicles	DPNH	0.21	2.8	2.8	26.7
Mitochondria	Succinate	0.96	0.51	0.96	2.0
Submitochondrial vesicles	Succinate	0.12	0.96	1.15	19.2

[a] The bound phosphorus of mitochondria and submitochondrial vesicles was 17.6 and 19.9 μg of P per milligram of protein, respectively. Cytochrome c reductase activity was measured at 32° as described previously by S. Fleischer and B. Fleischer (this series, Vol. 10, p. 406). Oxygen consumption was measured polarographically at 37° using a Clark oxygen electrode (R. W. Estabrook, this series, Vol. 10, p. 41).

[b] The ratio of electron transport rates of oxidase/reductase is calculated taking into account that cytochrome reductase c and oxidase activities involve the transfer of one and two electrons, respectively.

The submitochondrial vesicles are composed of single-layered vesicles (Fig. 2). These are inside-out in the sense that the inner membrane particles are on the outer face of the vesicles as visualized by negative staining. The vesicles are derived mainly from the mitochondrial inner membrane (~95%) in accordance with the high content of inner membranes in heart mitochondria.

The submitochondrial vesicles are quite distinct from mitochondria in that they have a high respiratory activity from DPNH or succinate to oxygen but do not readily transfer electrons to cytochrome c. The low activities are due to the inaccessibility of cytochrome c to the electron donor, since normal rates can be obtained by "opening" the vesicles with bile acids. Electron transport is accompanied by proton uptake by the vesicles, confirming the inverted nature of the membrane. The proton pulses are sensitive to uncouplers, such as carbonyl cyanide m-chlorophenylhydrazone (m-Cl-CCP). The low conductivity of the membrane to protons is comparable with that of the intact mitochondria, indicating that proton gradients are available to energize oxidative phosphorylation and that the membrane behaves in an anisotropic manner.[13,14] The vesicles have an

[13] P. Mitchell and J. Moyle, *Biochem. J.* **105**, 1147 (1967).
[14] V. P. Skulachev, *Curr. Top. Bioenerg.* **4**, 127 (1971).

"inner" volume of about 2 μl per milligram of protein that is very slowly accessible (1–2 days) to radioactive sucrose, but not to bovine serum albumin. These results indicate that the vesicles form a sealed compartment and are well suited for studies concerned with sidedness of the membrane.

[28] Rapid Isolation Techniques for Mitochondria: Technique for Rat Liver Mitochondria

By F. Carvalho Guerra

Isolation of mitochondria implies homogenization of the organ or tissue in order to disrupt the cells and release its subcellular components. Proteolytic enzymes may be helpful.[1] For homogenization, several techniques are described. Methods for different tissues have been revised in this series.[2] Disruption of cells may require grinding in a mortar with glass beads,[3,4] alumina,[5] sand,[6,7] or even electrical blending apparatus.[8,9] For soft material, motor-driven pestle homogenizers, the Potter-Elvehjem[10] or similar type, are frequently referred to in the common methodology.[11,12] Apart from the need for accurate clearance between the pestle and the tube, a major objection raised to the use of pestle homogenizers has been the heating of the material during the operation. This is usually counterbalanced by immersion of the tube in ice cold water and by compromising between completeness of the homogenization and the final activity of the preparation. Since the homogenate must still be subjected to centrifugal fractionation, the overall process is relatively long, and is lengthened further if multiple washings are introduced in order to remove micro-

[1] B. Chance and B. Hagihara, *Proc. Int. Congr. Biochem., 5th, 1961,* Vol. 5, p. 3 (1963).

[2] See this series, Vol. 10, Section II and Vol. 1 [2] and [4].

[3] J. R. Mattoon and F. Sherman, *J. Biol. Chem.* **241**, 4330 (1966).

[4] R. Wu and L. A. Sauer, this series, Vol. 10 [19].

[5] G. C. Hill and D. C. White, *J. Bacteriol.* **95**, 2151 (1968).

[6] A. Millerd, J. Bonner, B. Axelrod, and R. S. Bandurski, *Proc. Nat. Acad. Sci. U.S.* **37**, 855 (1951).

[7] J. T. Wiskich, R. E. Young, and J. B. Biale, *Plant Physiol.* **39**, 312 (1964).

[8] J. K. Grant, *Biochem. J.* **64**, 559 (1956).

[9] S. Ichii, E. Forchielli, and R. I. Dorfman, *Steroids* **2**, 631 (1963).

[10] V. R. Potter and C. A. Elvehjem, *J. Biol. Chem.* **114**, 495 (1936).

[11] C. de Duve, *Harvey Lect.* **59**, 49 (1965).

[12] D. Johnson and H. Lardy, this series, Vol. 10 [15].

somes. To obviate these inconveniences, we have introduced lately[13] a simplified procedure, for rat liver mitochondria, where the preparation time is markedly reduced, while still yielding structurally and functionally intact mitochondria.

Method

Principle. Minced rat livers are ground in a mortar with sand in order to disrupt the cells. Removal of the nuclei, whole cells, cell debris, and red blood cells is accomplished by filtration through glass wool followed by centrifugation of the filtrate at low centrifugal forces. The mitochondrial fraction is obtained from the resulting supernatant by centrifugation at higher speeds.

Reagents

HCl, conc.
Na-EDTA (disodium salt), 20 mM
Sand, specially treated (see below)
Isolation medium: sucrose 0.33 M, Na-EDTA (disodium salt) 0.025 mM, Tris 15 mM, pH 7.4
Washing medium: sucrose 0.33 M

All the solutions are prepared with all glass, CO_2-free redistilled water.

Preparation of the Sand. Quartzitic sand, sifted through a 500-mesh sieve, is treated with conc. HCl for 2 hours at 60°. Excess of acid is subsequently removed by washing with tap water under shaking. The packed sand is then mixed with three parts (w/v) of 20 mM Na-EDTA and heated for 30 minutes at 80°. This step is repeated twice. Finally the sand is washed six times by shaking with deionized and distilled water for 30 minutes each time. A large quantity of the prepared sand is stored at $-20°$. Immediately prior to commencing the homogenization procedure, an aliquot is transferred to a chilled mortar.

Preparation of the Homogenate. Rats are killed by decapitation, and the livers are removed, minced with scissors and immediately placed in ice cold isolation medium. All subsequent operations are performed in the outer cold room at 0–4°. The liver tissue is gently ground for about 5 minutes in a mortar with twelve times its weight of sand, which had been precooled to $-20°$. Amounts ranging from 5 to 30 g of liver have been used. The resulting pulp is placed over glass wool in a column and percolated with 10 volumes of the isolation medium. The filtration may be accelerated by a very slight reduction of pressure.

[13] F. C. Guerra, A. Albuquerque, J. M. Santos Mota, and L. M. G. van Golde, *Biochim. Biophys. Acta* **234**, 222 (1971).

Isolation of the Mitochondria. The filtrate is distributed into centrifuge cups and spun at 800 *g* in a Lourdes 9RA rotor, for 10 minutes. The supernatant fraction is carefully collected and centrifuged at 8200 *g* for 10 minutes. The resultant supernatant is discarded. The pellets are resuspended in the washing medium, using the side of a stirring rod against the wall of the cup, and quantitatively collected in one cup. A further centrifugation at 8200 *g* for 10 minutes should be done immediately thereafter. The yield of mitochondria thus obtained is 15 mg of

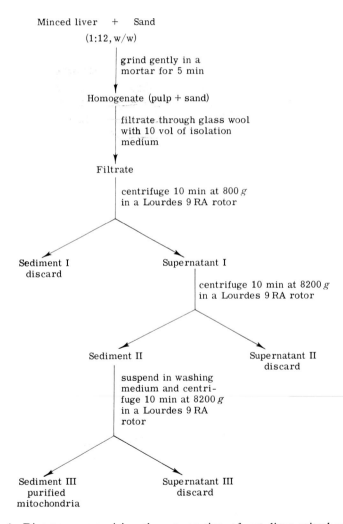

FIG. 1. Diagram summarizing the preparation of rat liver mitochondria.

PHOSPHOLIPID:PROTEIN RATIO AND PHOSPHOLIPID DISTRIBUTION (CARDIOLIPIN P VALUE TAKEN AS 1) IN RAT LIVER MITOCHONDRIA ACCORDING TO DIFFERENT AUTHORS

	Present method	Newman et al.[a]	Caplan and Greenawalt[b]	Schnaitman et al.[c]	Bartley et al.[d]	Wirtz and Zilversmit[e]
Total phospholipid (ng-atom of P/mg protein)	0.182	0.163	0.178	0.154	—	—
Relative distribution of phospholipids:						
Cardiolipin P	1	1	—	—	1	1
Phosphatidylcholine P	3.8	3.2	—	—	4.6	3.1
Phosphatidylethanolamine P	3.6	2.9	—	—	5.0	2.9

[a] H. A. J. Newman, S. E. Gordesky, C. Hoppel, and C. Cooper, *Biochem. J.* **107**, 381 (1968).
[b] A. I. Caplan and J. W. Greenawalt, *J. Cell Biol.* **31**, 455 (1966).
[c] C. Schnaitman, V. G. Erwin, and J. W. Greenawalt, *J. Cell Biol.* **32**, 719 (1967).
[d] W. Bartley, G. S. Getz, B. M. Notton, and A. Renshaw, *Biochem. J.* **82**, 540 (1962).
[e] K. W. A. Wirtz and D. B. Zilversmit, *J. Biol. Chem.* **243**, 3596 (1968).

protein per gram of liver, as assessed by the procedure of Lowry *et al.*[14]

Testing the Quality of the Mitochondria. The quality of the mitochondria prepared according to this procedure, which is summarized in Fig. 1, has been so far assessed through their phospholipid:protein ratio and phospholipid composition, respiratory activity, ability to swell in the presence of phosphate, and ultrastructure. Contamination has been evaluated by study of the electron micrographs and assay of enzymatic activities.

The phospholipid content of the preparations examined was of 0.182 ± 0.021 ng-atom of phospholipid P per milligram of protein, which is within the range of the values found by others with preparations obtained through different methods (see the table).[15-17] The same applies for the molar proportions cardiolipin P–phosphatidylcholine P–phosphatidylethanolamine P (1:3.8:3.6).[15,18,19] The respiratory activity assessed by the P:O ratios (2.96 for glutamate, 3.89 for α-ketoglutarate, 2.67 for β-hydroxybutyrate, and 1.71 for succinate) approached the theoretical values. Besides, good agreement was found with the ADP:O ratios. Phosphate induced noticeable swelling, a result which suggests the presence of a rich endogenous substrate.[20]

Ultrastructural study showed (Figs. 2 and 3), besides good preservation of the mitochondria, with intact membranes and a dense matrix, a low contamination with microsomes, glycogen, and lysosomes. Assay of glucose-6-phosphatase activity pointed to a 5% contamination with microsomes but the use of the more reliable marker cholinephosphotransferase yielded a lower value (2.3%) comparing favorably with those obtained for mitochondria isolated through other techniques.[21,22]

[14] O. H. Lowry, N. J. Rosebrough, A. C. Farr, and R. J. Randall, *J. Biol. Chem.* **193**, 265 (1951).

[15] H. A. J. Newman, S. E. Gordesky, C. Hoppel, and C. Cooper, *Biochem. J.* **107**, 381 (1968).

[16] A. I. Caplan and J. W. Greenawalt, *J. Cell Biol.* **31**, 455 (1966).

[17] C. Schnaitman, V. G. Erwin, and J. W. Greenawalt, *J. Cell Biol.* **32**, 719 (1967).

[18] W. Bartley, G. S. Getz, B. M. Notton, and A. Renshaw, *Biochem. J.* **82**, 540 (1962).

[19] K. W. A. Wirtz and D. B. Zilversmit, *J. Biol. Chem.* **243**, 3596 (1968).

[20] F. E. Hunter, Jr., J. F. Levy, J. Fink, B. Schutz, F. Guerra, and A. Hurwitz, *J. Biol. Chem.* **234**, 2176 (1959).

[21] M. G. Sarzala, L. M. G. van Golde, G. de Kruyff, and L. L. M. van Deenen, *Biochim. Biophys. Acta* **202**, 106 (1970).

[22] G. L. Sottocasa, L. Ernster, B. Kuylenstierna, and A. Bergstrand, *in* "Mitochondrial Structure and Compartmentation" (E. Quagliariello, S. Papa, E. C. Slater, and J. M. Tager, eds.), p. 74. Adriatica Editrice, Bari, 1967.

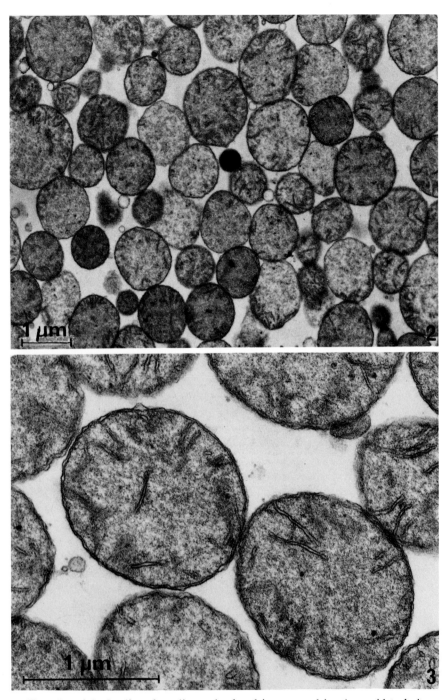

Figs. 2 and 3. Quality of rat liver mitochondria prepared by the rapid technique as revealed by ultrastructural study.

Comments

The procedure herewith described seems to yield rather well-preserved and quite pure rat liver mitochondrial preparations. The method seems also to be adaptable to rat brain mitochondria. It appears that the filtration step "clears" the homogenate, making multiple centrifugation steps unnecessary, therefore reducing the processing time and allowing experiments to be carried out with "fresher" mitochondria. However, the need for a quite exhaustive preparation of the sand is still an inconvenience. Hence we have looked lately for a substitute for the sand requiring a less demanding treatment before use. From the tested materials, the best results so far have been obtained with polyethylene (0.98 density) previously ground to 250 mesh in a stainless steel mill. Preparation in this case is reduced to three washings, once with 0.1 mM sodium EDTA, once with redistilled water, and finally with the isolation medium. The results of the analytical assays and the ultrastructural studies carried out in mitochondrial preparations obtained with this procedure have so far been at least comparable to those reported above.

[29] Isolation of Mitochondria from Mouse Mammary Gland

By CHARLES W. MEHARD

The mammary gland has different physiological states, i.e., nonpregnant state, pregnant (prelactating), lactating, and postlactating states and abnormal (tumor), and each tissue state of the gland must be considered as a different entity for the isolation of mitochondria. It must also be considered that isolated mitochondria derived from these tissues are composed of a mixed cell population containing both adipose and parenchymal elements. The parenchymal cell population increases during pregnancy and becomes most abundant during the lactational period which is the most advanced stage of the mammary gland's development. Each state of the gland's development as well as species-specific differences present unique problems which must be dealt with in order to isolate functional mammary gland mitochondria.

A method previously reported for isolation of mitochondria from guinea pig mammary gland[1] was found inadequate for the mouse system. Coupled mitochondria were obtained from lactating glands of the guinea

---

[1] W. L. Nelson and R. A. Butow, this series, Vol. 10, p. 103.

pig, however, when the same isolation medium was used with the lactating mouse mammary gland, then isolated mitochondria were uncoupled.[2]

In the mouse system, milk in even small amounts, if present in the preparation during isolation, alters both the structure and function of the isolated mitochondria,[2] resulting in a lack of volume changes in hypotonic solutions as well as failure to demonstrate active ion transport, suggesting that membrane permeability alteration occurs as a result of contact with the milk. This detrimental effect may be due in part to the release of free fatty acids by the lipase activity, which occurs in both the tissue and milk.[3,4] The presence of bovine serum albumin (BSA) in the isolation medium is essential to obtaining functional mitochondria from mouse mammary glands in all stages of development, because, when BSA is not used, the mitochondrial ultrastructure is greatly altered even with non-lactating tissue.[2]

The structural intactness and the functional capacity of mouse mammary gland mitochondria are then dependent on the absence of milk and the use of protective agents, such as BSA, which will adsorb the free fatty acids.

Tissue Preparation for Homogenization

Procedures to prepare tissue for homogenization are outlined below. These take into account the different stages of tissue development and will supply the largest amount of tissue per animal.

A tissue dissociation procedure for obtaining parenchymal cell suspensions is also given for both normal and tumor tissue. This technique provides mitochondria from the parenchymal cell population. These cell suspensions are much easier to homogenize than are the tissues, and the respiratory rates of the mitochondria isolated from cell suspensions are higher than those isolated from tissues.[5] However, dissociated parenchymal cell suspensions yield far fewer mitochondria per gram of original wet weight tissue than does the tissue itself.

Nonlactating Tissue from Pregnant Mice. During pregnancy, the mammary gland parenchymal cell population increases. The largest yields per animal of the prelactating gland are obtained during days 15–20 of gestation. However, just prior to parturition, milk may be found in the gland and tissues containing small amounts of milklike fluid should be avoided.

Lactating Tissue. When studying lactiferous glands, the maximum amount of tissue may be obtained from the animal between day 15 and

[2] C. W. Mehard, L. Packer, and S. Abraham, *Cancer Res.* 31, 2148 (1971).
[3] M. Hamosh and R. O. Scow, *Biochim. Biophys. Acta* 231, 283 (1971).
[4] O. W. McBride and E. D. Korn, *J. Lipid Res.* 4, 17 (1963).
[5] C. W. Mehard and L. Young, unpublished observations, 1972.

day 20 of lactation. The functional capacity of mitochondria from these glands can be improved by first removing as much milk from the tissue as possible. This can be accomplished by separating the mother and the pups for a period of several hours or overnight and then placing them together for an hour or two just prior to sacrifice of the dam. If preferred, the animals may be milked by hand with a small suction device. To facilitate the release of milk from the gland, 1–2 units of oxytocin may be injected intraperitoneally per animal prior to milk removal. This procedure removes most of the milk from the gland; however, a small amount still remains in the ducts, which will be released during homogenization of the tissue. Thus, the tissue should be sliced (0.4 mm thick) with a razor blade or motorized tissue chopper[6] and washed free of the lacteal residue. This is accomplished by agitating the sliced tissue (1 part) in cold mitochondrial isolation medium (10 parts). The solution should be changed periodically until it no longer becomes lactescent and the tissue color changes from milky tan to pink. The washing technique does not appear to be detrimental since coupled mitochondria have been isolated from tissue washed up to 4 hours.[2]

Milk-Free Lactating Tissue. Closing the nipples on only one side of the mouse by electrocautery before parturition makes it possible to obtain mitochondria from milk-free tissue of lactating mice. Since the pups are unable to suckle, the glands become inactive to the production of milk. The glands with uncauterized nipples will be able to supply sufficient milk to allow the pups to grow at a normal rate. When the animals are sacrificed (day 14 to 18 of lactation), the glands with cauterized nipples are fully developed, devoid of milklike fluid, and somewhat smaller than the undisturbed lactating glands.

Mammary Gland Tumor Tissue. Mitochondria can be successfully isolated from mammary gland tumor tissue (spontaneous or those propagated by subcutaneous transplantation). Functional mitochondrial preparations are more easily and consistently obtained from tumors of 1–2 g weight than from larger ones since larger tumors are usually more cartilaginous and contain more necrotic tissue. All necrotic regions should be removed and the tissue be cut into small pieces with scissors prior to homogenization.

Some transplanted tumors are more difficult to disrupt than others owing to larger amounts of fibrous tissue. After initial homogenization, it may be necessary to rehomogenize some undisrupted tissue pieces after they are resuspended in fresh medium. Care should be taken that frictional heat generated in the homogenization process does not warm the

---

[6] J. McIlwain and H. L. Buddle, *Biochem. J.* **53**, 412 (1953).

homogenate. Therefore the process should be carried out in an ice bath.

Cell Suspensions—Normal Mammary Gland. The excised mammary glands from mice 15–18 days pregnant are washed in saline A solution[7] and finely minced with a razor blade. The cellular dissociation of 10–12 g of tissue is carried out with 0.1% collagenase in 24 ml of Hanks' BSS medium,[7] pH 7.4, in a gyratory water bath shaker at 37° for 60 minutes. The cell suspension is then centrifuged at 200 *g* for 5 minutes (23°), resuspended in Waymouth's medium,[7] centrifuged, resuspended, and centrifuged again (200 *g*, 5 minutes, 23°). The washed pellet is suspended in 10 ml of Waymouth's, left to stand for 1–2 minutes to permit the larger parenchymal cells to begin sedimenting and then centrifuged by bringing the *g* force up to 50 *g*. When the *g* force reaches 50 *g*, the centrifuge is turned off. The resulting pellet is resuspended in 10 ml of Waymouth's, left again to stand for 1–2 minutes and again brought to 50 *g*. The final pellet is resuspended in cold (0°) mitochondrial isolation medium and is ready for homogenization.

Cell Suspension—Tumor. Tumors (10–12 g) should be finely minced with scissors and quickly rinsed twice with STV medium (saline A solution containing 0.05% trypsin and 0.025% EDTA) at 23°, pH 7.4. Three separate dissociations of 10 minutes each in 35 ml of STV (23°) should be carried out as follows: The STV preparation is agitated with a magnetic stirrer and stirring bar for 10 minutes and allowed to stand for 15–20 seconds to sediment the remaining tissue. The suspension is then filtered through four layers of cheesecloth and pelleted immediately at 200 *g* for 5 minutes; the pellet is resuspended in 1–2 ml of Waymouth's containing 2% fetal calf serum and set aside at room temperature to be pooled later. The sediment is resuspended in STV as above and agitated for 10 minutes; the entire process repeated until three 200 *g* pellets have been prepared. All three resuspended pellets of dissociated cells may now be pooled and processed in the same manner as the 200 *g* pellet described in the normal mammary gland cell suspension procedure. The resulting pellet is resuspended in cold (0°) mitochondrial isolation medium in preparation for homogenization.

Cell Suspension Media

> Saline A, concentrated stock solution: dilute 1 part of stock solution with 9 parts of water. Components: NaCl, 80 g/l; KCl, 4 g; glucose, 10 g; NaHCO$_3$, 3.5 g; penicillin, 100 units; streptomycin, 100 μg; phenol red, 9 ml of 1% solution. Make up to 1 liter with distilled water.

[7] P. I. Marcus, S. J. Cieciura, and T. T. Puck, *J. Exp. Med.* **104**, 615 (1956).

STV medium: 100 ml of saline A (concentrate solution); 825 ml of sterile distilled H_2O; 50 ml of 1% trypsin; 25 ml of 1% EDTA (Versene). Adjust the pH to 7.4 with $NaHCO_3$, and filter through a Millipore filter into a sterile container.

Hank's BSS medium, concentrated stock solution: dilute 1 part medium with 9 parts water. Components: NaCl, 80.0 g; KCl, 4.0 g; $MgSO_4 \cdot 7H_2O$, 2.0 g; $CaCl_2$, 1.4 g; $Na_2HPO_4 \cdot 7H_2O$, 0.94 g; KH_2PO_4, 0.6 g; glucose, 10.0 g; phenol red, 0.2%; penicillin, 100 units/ml; streptomycin, 100 $\mu g/ml$. Make up to 1 l of solution and filter through a Millipore filter into a sterile container. Available commercially.

Waymouth's medium: available commercially—Gibco; Berkeley, California.

Isolation of Mitochondria

As soon as possible after removal from the animal or completion of the cell suspension, the tissue or preparation should be placed in mitochondrial isolation medium ($0°$) composed of 0.33 M sucrose, 1 mM EDTA, 1 mM Tris·HCl, pH 7.4, and 1% bovine serum albumin (BSA). The tissues or cell suspensions (1:10, tissue:isolation medium) are homogenized in a motor-driven (high speed) Potter-Elvehjem homogenizer with a loose-fitting (0.015-inch clearance) Teflon pestle (for cell suspension, use a standard 0.0075-inch clearance). The number of passes of the pestle through the homogenate should be kept to a minimum (3 to 4).

Homogenates should be immediately filtered through two layers of 50-mesh nylon cloth and centrifuged at 4500 g for 3 minutes to sediment cell debris, unbroken cells, and nuclei; the resulting lipid upper layer is carefully removed by aspiration and the supernatant fraction centrifuged at 10,000 g for 10 minutes to sediment the mitochondria. The mitochondrial pellet is then resuspended in 50 ml of isolation medium per 10 g of original tissue and centrifuged at 750 g for 5 minutes. The supernatant fraction containing the mitochondria is centrifuged at 8700 g for 10 minutes; the resulting supernatant fraction is discarded. The mitochondrial pellet may have a fluffy upper layer of light-colored protein material, which should be removed by adding a few drops of isolation medium to the tube. A quick swirling action of the tube should dislodge the layer, which may then be aspirated and the pellet surface rinsed with a few drops of isolation medium. The mitochondrial pellet is then resuspended in 25 ml of isolation medium and centrifuged at 8700 g for 10 minutes. The twice-washed mitochondrial pellet is suspended in a minimal amount of mitochondrial isolation medium and is ready for assay.

MOUSE MAMMARY GLAND MITOCHONDRIAL YIELDS

State of:		Yield (mg mitochondrial protein/g wet wt. tissue)
Mouse	Tissue	
Pregnant	Nonlactating	6.0–8.0
Pregnant	Nonlactating, cell suspension	0.3–0.7
Lactating	Lactating	4.0–8.0
Lactating	Lactating, washed	1.0–2.0
Lactating	Nonlactating[a]	6.0–8.0
Nonpregnant	Tumor	2.0–5.0
Nonpregnant	Tumor, cell suspension	0.1–0.3

[a] Nonsuckled mammary glands taken from lactating mice, 14–18 days postpartum, which had their nipples closed by electrocautery.

Results

Mammary mitochondria obtained by this process are coupled, intact, and demonstrate energized ion transport and are maintained during storage by including BSA (0.1–1.0%) in the holding medium. BSA in the reaction medium also enhances the mitochondrial functions.

Mitochondrial yields for each tissue state and treatment are listed in the table. The variation in the yields reflects the wide variation within the tissues. More mitochondria may be obtained by scaling up the procedures without any loss in activity or yield as long as procedures are carried out quickly at 0°. Although the amount of mitochondria obtained is much less and the procedure more demanding when the technique is used for dissociation of the tissue into cell suspension, more active mitochondrial preparations are obtained.

[30] The Isolation of Outer and Inner Mitochondrial Membranes

By JOHN W. GREENAWALT

Within the last several years, procedures have been described for the isolation of submitochondrial fractions enriched in the inner or in the outer mitochondrial membrane. The procedures vary with regard to the principles upon which membrane disruption is based, the rapidity with which fractions can be isolated, and the degree of enrichment and/or yield of a particular mitochondrial component. Some of these procedures have been discussed in recent reviews.[1-3]

The method presented here is that described by Schnaitman and Greenawalt[4] and extended by Chan *et al.*[5] and Pedersen and Schnaitman.[6] This procedure utilizes digitonin to break the outer mitochondrial membrane; after separation of the outer membrane, the inner membrane is disrupted by treatment with Lubrol WX.

General Principles. An essential step in the subfractionation of isolated mitochondria is the preferential disruption of the outer membrane by means of physical, chemical, or enzymatic treatments. The two membranes are then separated, usually by either gradient or rate (differential) centrifugation in a single or multistep process. In most cases the inner membrane-containing fraction is treated further, prior to isolation of the membrane, to remove matrix proteins.

The procedure described here has a number of features not found in some of the other methods described in the literature. In principle, it utilizes a specified ratio of digitonin to mitochondrial protein under controlled conditions so that the outer membrane is preferentially broken. Digitonin shows a fair degree of specificity toward the outer membrane in its disruptive effects[7]; however, above certain concentrations it, at least partially, solubilizes and/or disrupts the inner membrane.[7,8] Therefore, since the time of treatment and the concentration of digitonin can be varied, this method provides considerable flexibility for the isolation of specific subfractions. For example, with rat liver mitochondria, the ratio of digitonin to milligrams of mitochondrial protein can be manipulated, i.e., raised or lowered within limits, to yield, upon fractionation, especially clean preparations of either the inner or the outer membrane, respectively. As a result it is possible to follow kinetically the release of soluble enzymes from the intermembrane space and/or the mitochondrial matrix as a means of monitoring the disruptive action of this reagent.

With rat liver mitochondria this procedure permits the systematic and sequential isolation of submitochondrial fractions which correspond to each of the physical regions of the intact mitochondrion as revealed in the electron microscope.[4,5] In fact, from the results of numerous investigations utilizing the various techniques of subfractionating the mitochondrion,

[1] J. M. Smoly, B. Kuylenstierna, and L. Ernster, *Proc. Nat. Acad. Sci. U.S.* **66,** 125 (1970).
[2] M. Ashwell and T. S. Work, *Annu. Rev. Biochem.* **39,** 251 (1970).
[3] See this series, Vol. 10 [72a], [72b], [72c].
[4] C. A. Schnaitman and J. W. Greenawalt, *J. Cell Biol.* **38,** 158 (1968).
[5] T. L. Chan, J. W. Greenawalt, and P. L. Pedersen, *J. Cell Biol.* **45,** 291 (1970).
[6] P. L. Pedersen and C. A. Schnaitman, *J. Biol. Chem.* **244,** 5065 (1969).
[7] C. Schnaitman, V. Gene Erwin, and J. W. Greenawalt, *J. Cell Biol.* **32,** 719 (1967).
[8] W. E. Cassady, E. H. Leiter, A. Berquist, and R. P. Wagner, *J. Cell Biol.* **53,** 66 (1972).

it has become increasingly possible to relate the enzyme topology of the mitochondrion to the regional anatomy of this organelle. Thus, fractions can be isolated which are highly enriched in (a) the outer membrane, (b) components of the intermembrane space, (c) the inner membrane, and (d) components of the matrix.

A unique and important feature of the procedure described here is that a "mitoplast" fraction (inner membrane plus matrix), which is essentially devoid of outer membrane and intermembrane proteins, can be isolated.[9] The mitoplasts retain the major integrated functions of intact mitochondria including oxidative phosphorylation, acceptor control, and configurational changes concomitant with shifts from state 4 to state 3 respiration.[4,9,10] The retention of these functions is due, in part, to the fact that fractions are isolated by differential centrifugation and, thus, are not exposed to the high concentrations of solutes, e.g., sucrose, used in gradient separations. Furthermore, there is little doubt that the inclusion of bovine serum albumin in the isolation and disruption media is essential for satisfactory results although its specific contributions to the success of this procedure have not been systematically evaluated.

As with all fractionation procedures, the quality of the subfractions can be no better than that of the initial starting material (in this case, the mitochondrial preparation). A number of criteria can be used to estimate the quality of the mitochondrial preparation; this information is useful to determine at the outset whether further subfractionation is warranted, to provide a means of possibly determining why a particular fractionation attempt may not have been successful, and to establish criteria that help guarantee reproducible results. In this laboratory, measurement of acceptor control ratio, electron microscopic examination, and assay(s) for microsomal contamination are used on a routine basis for this purpose. Only mitochondrial fractions exhibiting acceptor control ratios equal to 5.0 or better (with succinate or D-β-hydroxybutyrate as substrate) are used. Experience has shown that clean fractionations are less frequently obtained when mitochondria having lower acceptor control ratios are subfractionated. Electron microscopic examination of thin sections of the initial mitochondrial preparation aids in monitoring for gross contamination and shows whether or not the mitochondria have the condensed configuration of freshly isolated preparations. Assays for G-6-Pase or NADPH–cytochrome c reductase provide quantitative estimations of microsomal contamination in the mitochondrial preparation. The specific activity of a

[9] J. W. Greenawalt, in "Monoamine Oxidases—New Vistas" (E. Costa and M. Sandler, eds.) (*Advan. Biochem. Psychopharmacol.* 5). Raven Press, New York.
[10] J. W. Greenawalt, *Fed. Proc., Fed. Amer. Soc. Exp. Biol.* 28, 663 (1969).

microsomal enzyme in the mitochondrial fraction should be less than 5% of that in a microsomal fraction itself. This is an important point since the optimal concentration of digitonin is determined on the basis of the ratio of digitonin to what is assumed to be *mitochondrial protein.*

The present procedure was developed in this laboratory for the subfractionation of rat liver mitochondria and has been applied primarily to mitochondria from this source. However, trials with rat kidney, bovine liver, and bovine kidney mitochondria indicate that it can be applied successfully to them. Partial success has been achieved in attempting to subfractionate rat heart and bovine heart mitochondria. It seems likely that this lack of success is related to differences in the relative proportion of outer membrane vs. inner membrane and possibly also to differences in the composition of the outer membranes. It has been shown for example, that application of this procedure to isolated *Neurospora* mitochondria requires much higher ratios of digitonin per milligram of mitochondrial protein than with rat liver mitochondria.[8] It seems essential, therefore, that the optimal ratio of digitonin per milligram of mitochondrial protein be determined for specific kinds of mitochondria and for mitochondrial preparations obtained by different isolation procedures. By "titrating" the liberation of the outer membrane and its associated marker enzymes with increasing concentrations of digitonin as described earlier,[7] the optimal ratios for specific purposes can be determined.

Correlation of ultrastructural changes with changes in the distribution of enzymatic markers or other biochemical activities is an extremely important, if not essential, criterion in trying to establish the sequential events taking place in a multistep fractionation procedure. However, the purpose to which a particular method is put may determine the extent to which such correlation is necessary. In some circumstances an enzymatic activity may serve as a valid *marker* for an ultrastructural component of an organelle or cell even though it may not be localized exclusively in that component. Our present knowledge and available methodology rarely permit the unequivocal localization of a given enzymatic activity to a single site within the cell or to a given subfraction of an organelle. Reaching conclusions regarding enzyme localization generally requires that certain assumptions be made; for example, that optimal assay conditions for a certain activity are the same in all fractions. Such conclusions are largely dependent upon the degree to which ultrastructural studies can be used to correlate the distribution of the ultrastructural entity with the biochemical activity and are based on compromise.

Three criteria in addition to ultrastructural correlations must be applied and adhered to strictly before a given activity can even be considered to have a localized site on the basis of its distribution in subfractions. These are (1) that the *specific activity* of the marker is increased significantly in the subfraction in which the component is distributed, (2) that the subfraction also contains a high percentage of the *total activity* measured in the starting material from which the subfractions are derived, and (3) that quantitative recoveries of the marker activity are obtained. Even when these three conditions are stringently met, it is almost impossible to decide simply on the basis of this type of study whether a small percentage of an activity is distributed differently from the major portion because it is truly localized differently, or whether such results are due to the limitations of the methodology.

Reagents and Solutions

Isolation Medium

Sucrose (Baker Analyzed Reagent), 70 mM
D-Mannitol (Mann Research Lab.), 220 mM
HEPES (N-2-hydroxyethyl piperazine-N-2-ethanesulfonic acid) buffer, 2.0 mM
Bovine serum albumin (BSA), (defatted, Pentex, Inc., Kankakee, Illinois), 0.5 mg/ml
Adjust to pH 7.4 with KOH just prior to use.

Digitonin (A Grade Calbiochem) Solution. In this laboratory the ratio of milligrams of digitonin per milligram of mitochondrial protein has been varied from about 0.10 to 0.12 to suit the purpose of a given experiment. Therefore, the digitonin stock solution is varied, correspondingly, from 1.0% to 1.2%. For example, to prepare very clean mitoplasts, essentially devoid of outer membrane, for subsequent isolation of the inner membrane, a relatively high ratio of digitonin to protein is desirable, i.e., 0.12 mg of digitonin per milligram of protein. The stock solution used in this case (1.2% digitonin) is prepared by dissolving 60 mg of digitonin in 5.0 ml of hot (almost boiling) isolation medium to which the BSA has *not been added.* The hot medium and digitonin are swirled on a hot plate until the digitonin is completely dissolved (ca. 1 minute). The solution is allowed to cool slowly to room temperature, and then 0.05 ml of a stock solution of BSA (containing 50 mg/ml) is added. The digitonin solution prepared in this manner will remain stable at 0–4°C for several hours. In early studies the commercial digitonin was recrystallized before use, but it has been found that recrystallization generally is not necessary.

Lubrol WX (ICI America Co., Stamford, Connecticut 06904) Solution. A stock solution is made in isolation medium containing 19 mg of Lubrol WX per milliliter of isolation medium.

Respiration Medium

Sucrose, 70 mM
D-Mannitol, 220 mM
EDTA, 0.5–1.0 mM
Na succinate (or D-β-hydroxybutyrate), 5–20 mM
K phosphate, pH 7.4, 2.5 mM
HEPES, 2.0 mM
BSA, 0.5–1.0 mg/ml
± ADP, 150–200 μM
± MgCl$_2$, 1.0–2.5 mM
The medium should have a pH of 7.4.

Procedure for Mitochondrial Isolation and Treatment with Digitonin

Isolation of Mitochondria. Many procedures and media are described in the literature for the isolation of biochemically and ultrastructurally intact mitochondria. However, the medium and procedure described here have proved especially useful in reproducibly obtaining clean mitochondrial fractions from rat liver which are suitable for subsequent separation of submitochondrial fractions. It should be noted, for example, that removal of the outer membrane from rat liver mitochondria isolated and treated with digitonin in 0.25 M sucrose results in significant loss of matrix proteins; furthermore, functional mitoplasts are not obtained.

Four to six rats ranging in weight from 200 to 250 g are fasted overnight and sacrificed by guillotine. After bleeding has largely subsided (3–4 minutes), the livers are removed, trimmed of fat, and placed in approximately 60 ml of ice-cold isolation medium per 30 g of liver. It is recommended that a minimum of 30 g of liver be utilized for initial attempts at subfractionation. With experience this can be reduced according to need. All subsequent steps are carried out at 0–4°. Livers are chopped into small pieces and rinsed twice with isolation medium; blood and extraneous material are decanted carefully so that the liver tissue is not lost. A suspension is made consisting of one volume of liver mince plus two volumes of isolation medium. The suspension is homogenized in a glass homogenizer (55-ml capacity) with a pestle containing multiple radial serrations. Four complete cycles up and down with the pestle are made. The pestle is motor-driven and operates at approximately 1500 rpm.

The homogenate is diluted approximately 1:3 with isolation medium and centrifuged at 660 g in the International swinging-bucket rotor or in the Sorvall GSA rotor (RC2-B centrifuge) for 10 minutes. The supernatant fluid is carefully removed so that no loosely packed material is removed; this is conveniently done with a Pasteur pipette. The supernatant fluid is transferred into ice-cold Sorvall tubes and centrifuged for 15 minutes at 6800–7000 g in the Sorvall SS-34 rotor. Resulting supernatant fluids are discarded. Sediments are gently muddled into a paste with a "cold finger," i.e., a test tube filled with ice, or a precooled small homogenizer pestle with the addition of a very small volume of isolation medium. The sediment is suspended in one half of the original volume of the homogenate prior to the initial centrifugation. This suspension is then centrifuged in the Sorvall SS-34 rotor for 15 minutes at 7000 g. The resulting supernatant fluid is decanted, and any resulting fluffy layer on the sediment is carefully rinsed off. The sediment is suspended in one-fourth of the original volume of the homogenate prior to initial centrifugation and again centrifuged for 15 minutes at the above speed. The supernatant fluid is again decanted. The sedimented fraction contains the mitochondria to be used for subsequent subfractionation. Speeds as high as 9770 g have been used in obtaining the mitochondria, and the yield is somewhat greater when the higher speed is used.

The mitochondrial fraction is suspended in the centrifuge tube using a loose-fitting pestle with the addition of a minimal volume of isolation medium. The volume is adjusted to give 100 mg of mitochondrial protein per milliliter as measured by the biuret protein method. The Lowry procedure for measuring protein has also been used, but some difficulty may arise with this method because of its sensitivity and the very high concentration of protein being analyzed. Also, it has been shown recently that HEPES buffer interferes with the Lowry protein procedure. At the concentration of HEPES in the isolation medium used here, estimation of protein concentration by the Lowry method may be only 83% of that indicated by the biuret procedure. It has been possible upon occasion to get good separation of the outer membrane and the mitoplast fraction when the concentration of mitochondrial protein (by the biuret method) has been as low as 90 mg/ml. With concentrations lower than this, however, separations have been uniformly unsuccessful.

By this procedure approximately 600 mg of mitochondrial protein is obtained from 30 g (wet weight) of rat liver from fasted animals or about 15–20 mg of mitochondrial protein per gram of wet tissue.

Digitonin Treatment. A volume of the mitochondrial suspension (100 mg of protein per milliliter), predetermined to give the quantities of subfractions desired, is added to a precooled cylindrical vial having a capacity of about 20 ml; a scintillation vial used for counting radioisotopes is a convenient size. Four to 5 ml of the mitochondrial suspension have been adequate for most studies. The vial containing the mitochondria is surrounded by ice in a beaker and gently stirred with a magnetic flea. An equal volume of the digitonin stock solution (1.0–1.2%) is added, and the gentle stirring is continued for 15 minutes after complete mixing has taken place. It appears that the rate of stirring is *not* extremely critical; however, to avoid frothing of the suspension, stirring is regulated so that a vortex on the surface of the suspension can barely be detected. The importance of two quantitative aspects of the procedure should be emphasized. The concentration of the mitochondria during digitonin treatment (50 mg of protein per milliliter) is crucial to obtain the fractions described here. Although the ratio of digitonin:mitochondrial protein may be varied to suit specific purposes, a ratio of 0.11–0.12 mg of digitonin per milligram of protein is optimal for studies designed to recover each of the 4 major submitochondrial fractions and for enzyme localization studies using rat liver mitochondria.

Separation of Mitoplasts, Outer Membrane, and
  Intermembrane Components

Mitoplasts. The overall procedure for the subfractionation of isolated rat liver mitochondria is outlined in Fig. 1. The digitonin-treated mitochondrial suspension is diluted (1:4) by the addition of 3 volumes of isolation medium. The diluted suspension is homogenized briefly by hand to ensure complete mixing, to produce a uniform suspension, and, possibly, to aid in the vesiculation of the outer membrane. The suspension

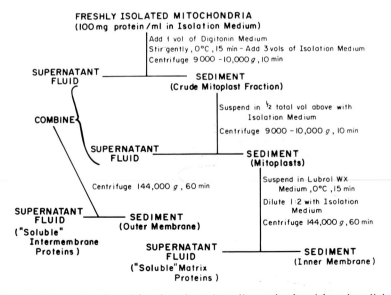

FIG. 1. Scheme for the subfractionation of rat liver mitochondria using digitonin and Lubrol WX to disrupt the outer and inner membranes, respectively.

then is centrifuged at 9000–10,000 g in the Sorvall SS-34 rotor for 10 minutes. The supernatant fluid is carefully removed so that any "fluffy" material is left with the sediment. It has been shown by electron microscopic examination that this fluffy layer consists primarily of broken inner mitochondrial membranes and that the sediment is largely mitoplasts. When carefully prepared mitochondria are used and when the optimal ratio of digitonin to protein is applied, little fluffy layer is formed. The supernatant fluid which contains both outer membrane vesicles and intermembrane components is pipetted off carefully and saved for the separation of outer membrane vesicles. The sedimented crude mitoplast fraction is suspended with a cold finger in isolation medium to one-half the volume of the diluted (1:4) digitonin-treated mitochondria. The mitoplasts are sedimented again at 9000–10,000 g for 10 minutes in the Sorvall SS-34 head. This supernatant fluid also is removed with a Pasteur pipette and combined with the supernatant fluid obtained previously from the first mitoplast sedimentation. The mitoplast fraction is then suspended in about 8 ml of isolation medium to give a concentration of about 30–40 mg of protein per milliliter.

Outer Membrane and Intermembrane Components. The combined supernatant fluids resulting from the sedimentation and washing of the mitoplasts contain the outer membrane and the soluble proteins, which, in the intact mitochondrion, are housed between the two membranes. The outer membrane fraction consists of small vesicles which are collected by centrifugation at 144,000 g for 1 hour in a Spinco Model L centrifuge using the No. 40 rotor. The resulting supernatant fluid which forms the soluble, intermembrane fraction is removed with a Pasteur pipette leaving the outer membrane vesicles in the sediment. The vesicles are suspended in a minimal volume of isolation medium.

The total protein of the initial mitochondrial suspension should be recovered quantitatively in these three fractions, the mitoplasts, the outer membrane, and the soluble intermembrane fraction. Approximately 5% and 13% of the total mitochondrial protein are recovered in the outer membrane and in the intermembrane fractions, respectively. The rest of the protein is recovered in the mitoplast fraction. The outer mitochondrial membrane appears in the electron microscope as small, smooth vesicles containing no projected 9 nm particles (ATPase molecules) which mark the inner membrane.

Since the outer membrane itself constitutes such a small percentage of the total protein, even small amounts of contamination from the inner membrane or other fractions may be of major significance, especially in experiments designed for certain purposes. The outer membrane fraction can be purified further by centrifugation on a 25–38% (w/w) sucrose density gradient in 10 mM HEPES, pH 7.4, for 90 minutes. When a Spinco SW 25.1 rotor at 25,000 rpm is used under these conditions, the outer membrane bands at about 30% sucrose on the gradient.

A number of enzymatic activities have been determined to be components of the outer mitochondrial membranes[1–3]; however, monoamine oxidase, MAO (measured with benzylamine as substrate), and kynurenine hydroxylase seem to be localized exclusively on the outer membrane.[4,11] The former activity provides an extremely useful marker because it is very stable and easy to assay. However, MAO is not a suitable marker for the outer membrane of mitochondria from all sources nor for rat liver mitochondria under all physiological conditions, e.g., during liver regeneration.[9] On the other hand, kynurenine hydroxylase has proved useful as a marker for the outer membranes of *Neurospora* mitochondria of which MAO is not a component.[8,12]

The mitochondrial adenylate kinase activity seems to be localized exclusively in the intermembrane space of rat liver mitochondria. Therefore, this enzyme serves as a marker for this mitochondrial compartment; in addition, since it is soluble, the loss of this activity from the mitochondrion can be used to monitor the disruption of the outer membrane without the loss of the outer membrane. By monitoring in the election microscope the digitonin treatment of rat liver mitochondria with time or as a function of concentration, it can be shown that the initial breakage of the outer membrane is accompanied by the loss of adenylate kinase activity into the soluble fraction before the outer membrane is converted into vesicles which can be separated from the low speed (9000–10,000 g) sediment.[12] The separation of the vesicles accompanies the liberation of MAO activity from the mitochondria.

Properties of the Mitoplast Fraction. Compared to other submitochondrial preparations reported in the literature, the mitoplast fraction isolated by the present procedure is unique in both its biochemical and ultrastructural characteristics. Because the mitoplasts represent a distinctive fraction, their properties are listed here in some detail. These properties also serve as valuable indicators in evaluating the success of the fractionation procedure. Mitoplasts from normal, adult rat liver mitochondria are characterized by:

1. a distinctive morphology in the election microscope due to pseudopodlike extensions of the inner membrane and matrix (see Fig. 2).

[11] J. W. Greenawalt and C. Schnaitman, *J. Cell Biol.* **46**, 173 (1970).
[12] J. W. Greenawalt, unpublished results (1971).

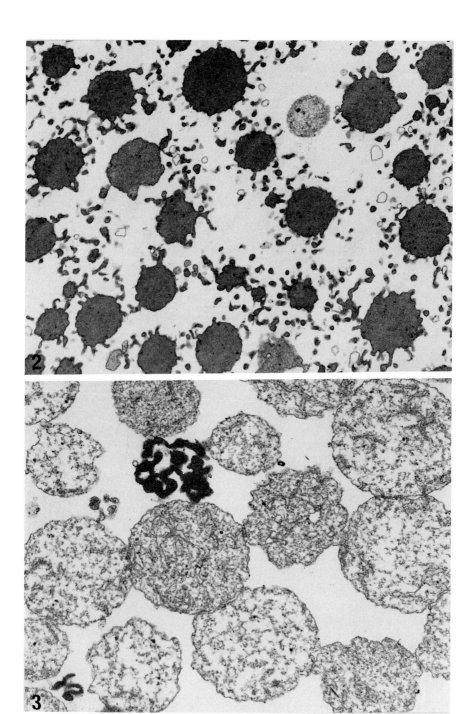

2. the absence of essentially all ($<2\%$ of the total) monoamine oxidase and adenylate kinase activities,

3. the ability to undergo respiration-dependent configurational changes including orthodox (expanded) to condensed (contracted) changes concomitant with state 4 to state 3 respiratory transitions (see Figs. 3 and 4),

4. the oxidization NAD-linked substrates coupled to phosphorylation,

5. the quantitative presence of all citric acid cycle enzymes and an intact electron transport chain,

6. acceptor control ratios ranging from 1.7 to 2.5 with succinate or D-β-hydroxybutyrate as substrates *in the absence of added* Mg^{2+} to the medium described by Schnaitman and Greenawalt.[4] The addition of Mg^{2+} causes rapid and apparently irreversible uncoupling of acceptor control responses and conversion to the expanded configuration (see Fig. 5),

7. the presence of ADP–ATP and P_i–ATP exchange reactions, and

8. the ability to accumulate Ca^{2+} with the ejection of H^+ (in the absence of added Mg^{2+}).

Separation of the Inner Membrane and Matrix Proteins

The mitoplasts can be subfractionated into separate fractions enriched in the inner mitochondrial membrane and the matrix proteins. To a volume of mitoplast suspension (approximately 30–35 mg of protein per milliliter) in a calibrated centrifuge tube is added a volume of Lubrol WX (approximately 19 mg/ml) to give a final concentration of 0.16 mg of Lubrol WX per milligram of mitoplast protein. This mixture is allowed to stand at 0°C for 15 minutes, and then it is diluted (1:2) by the addition of an equal volume of isolation medium. The suspension is then centrifuged for 1 hour at 144,000 g in the Spinco Model L centrifuge using the No. 40 rotor. The supernatant fluid which contains the matrix proteins is removed with a Pasteur pipette and used for subsequent studies. The pellet constituting the inner membrane fraction is washed twice as a sediment with a small amount of isolation medium. This sediment is then suspended in isolation medium to give a concentration of approximately 10 mg of protein per milliliter.

The matrix proteins are characterized by containing essentially all of the Krebs' cycle dehydrogenase activities including malate dehydrogenase

---

FIG. 2. Electron micrograph of freshly prepared mitoplasts illustrating the homogeneity of the fraction and its characteristic pseudopodlike extensions of the inner membranes and matrix. Fixed with 3% glutaraldehyde followed by 1% OsO₄, stained with uranyl acetate and lead citrate. Markers in all the micrographs represents 1 μm. ×19,500.

FIG. 3. Electron micrograph of mitoplasts fixed (as above) during state 4 respiration; orthodox (expanded) configuration is apparent in most mitoplasts. ×19,500.

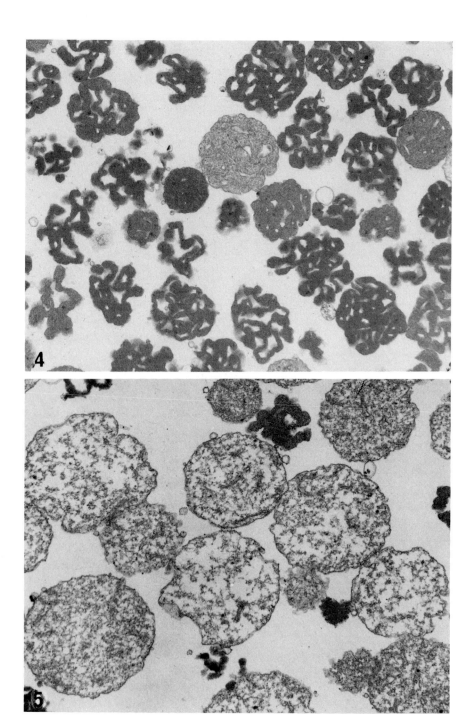

322

and glutamate dehydrogenase. Succinate dehydrogenase is membrane-bound and is associated with the inner membrane fraction. The biochemical and morphological characteristics of the inner membrane prepared in this manner are as follows:

1. This fraction consists of small vesicles less than 0.4 μm in diameter which, when negatively stained with phosphotungstate, exhibit projected 9 nm particles.

2. The vesicles exhibit no acceptor control.

3. The preparation respires in the presence of either succinate or TMPD plus ascorbate, but not in the presence of either D-β-hydroxybutyrate or NADH.

4. The inner membrane fraction will phosphorylate in the presence of succinate as substrate at low protein concentration.

5. The inner membrane is essentially free of those proteins and enzymatic activities associated with the matrix, with the intermembrane space, and with the outer membrane.

6. The particles also show the ability to accumulate calcium ions with the concomitant release of protons.[5]

FIG. 4. Mitoplasts from the same reaction vessel as those in Fig. 3 but after ADP has been added and conversion to state 3 respiration has occurred. Condensed (contracted) configuration marks the morphology of about 70% of the mitoplasts. ×19,500.

FIG. 5. Electron micrograph of mitoplasts uncoupled from state 4 respiration by the addition of 1.0 mm Mg^{2+}. The biochemical and ultrastructural changes are rapid and irreversible. ×19,500.

[31] Isolation of Modified Liver Lysosomes

By ANDRÉ TROUET

The complete purification of lysosomes from normal liver by centrifugation methods raises very great difficulties.[1] The mean rate of sedimentation of lysosomes is very close to that of peroxisomes and only somewhat lower than that of mitochondria. Besides, the distribution of lysosomes obtained by density equilibration in sucrose gradient is very broad (median density: 1.22) and is overlapped by the distribution of mitochondria (median density: 1.19) on its light side and by that of peroxisomes (median density: 1.23–1.25) on its dense side.

[1] H. Beaufay, in "Lysosomes, a Laboratory Handbook" (G. T. Dingle, ed.), p. 1. North-Holland Publ., Amsterdam, 1972.

The size and the density of lysosomes are however selectively modified by undigestible substances which accumulate inside the lysosomal apparatus after injection to the animal and cellular capture by endocytosis. Such induced modifications of lysosomes may serve to separate these organelles from mitochondria and peroxisomes. The density of lysosomes in sucrose gradient increases following intraperitoneal administration of Dextran 500. Thinès-Sempoux[1,2] has worked out a purification method taking advantage of this effect. The most widely used compound, however, is Triton WR-1339, which accumulates inside the lysosomes of the liver and decreases their equilibrium density to about 1.11.[3,4]

The purification procedure we will describe here is based on the use of this compound. It includes two main steps. First, a large granule fraction containing mainly lysosomes, mitochondria, and peroxisomes is prepared by differential centrifugation in 0.25 M sucrose. Nuclei, microsomal components, and the cell sap are eliminated to a large extent at this stage. The large granule fraction is obtained under the conditions used for fractionating normal liver. Indeed Triton WR-1339-filled lysosomes sediment at the same rate as normal lysosomes, the increase of their size compensating for the lowering of their density. Afterward, the large granule fraction is suspended in a sucrose solution of high density (1.21), and the lysosomes are separated from mitochondria, peroxisomes, and other contaminants by flotation through a solution of 34.5% sucrose (density: 1.155).

It must be stressed that the modified lysosomes prepared by this method contain large amounts of Triton WR-1339 and several plasma proteins.[3-5] This should be kept in mind during experimental work on these preparations.

The procedure described for the rat liver has been used successfully without modification for obtaining lysosomes from mouse liver. It can probably be adapted to many other mammalian species.

Purification Procedure

Reagents

Triton WR-1339 (Rohm and Haas, Philadelphia) at 20% (w/v) in 0.15 M NaCl

[2] D. Thinès-Sempoux, personal communication.
[3] R. Wattiaux, M. Wibo, and P. Baudhuin, *Lysosomes, Ciba Found. Symp. 1963,* p. 176.
[4] R. Wattiaux, *in* "Etude expérimentale de la surcharge des lysosomes." Duculot, Gembloux, Belgium, 1966.
[5] A. Trouet, *Arch. Int. Physiol. Biochim.* **72,** 698 (1964).

Sucrose 8.55% (w/v); 0.25 M
Sucrose 45.0% (w/w); density 1.21
Sucrose 34.5% (w/w); density 1.155
Sucrose 14.3% (w/w); density 1.06

Procedure. Twenty rats (150–250 g) are injected intraperitoneally with a Triton WR-1339 solution at a dosage of 1 ml per 100 g of body weight. Four days later, they are killed by decapitation after overnight fasting. The livers (90 g) are quickly removed, weighed in ice-cold 0.25 M sucrose, minced rapidly with scissors, and homogenized in 0.25 M sucrose (5 ml of solution per gram of liver) using a glass tube fitted with a Teflon pestle (A. H. Thomas, Co., Philadelphia, Pennsylvania) rotating at 1000 rpm. The homogenate is centrifuged at 1700 rpm for 10 minutes in a refrigerated centrifuge (International, Model PR-J, head No. 253). This corresponds to 6500 g-min at an average radial distance (R_{av}) of 20 cm. The supernatants are removed and the sediments are homogenized again in 0.25 M sucrose (3 ml of solution per gram of liver). This suspension is centrifuged at 1500 rpm for 10 minutes (5000 g-min). The pooled supernatants are brought to a volume of 10 ml per gram of liver by the addition of 0.25 M sucrose and constitute the cytoplasmic extract (fraction E). The sediments are resuspended in 0.25 M sucrose (fraction N). They contain mainly nuclei and large cell debris.

A large granule fraction is subsequently prepared by centrifuging the cytoplasmic extract in a No. 30 rotor (Beckman-Spinco) at 25,000 rpm during 10 minutes, time of acceleration included (340,000 g-min given $R_{av} = 7.8$ cm). The supernatant and the pink fluffy layer which covers the tight brown pellet are removed. The pellet is suspended in 0.25 M sucrose (5 ml of solution per gram of liver) using a glass or a plastic rod, and centrifuged again in a No. 30 rotor for 340,000 g-min. The supernatant and the fluffy layer are removed and pooled with the material decanted after the preceding centrifugation. These combined supernatants contain the microsomes and the cell sap and form the PS fraction. The pellets constituted by the large cytoplasmic granules (mitochondria, lysosomes, and peroxisomes) are carefully suspended in 45% sucrose (1 ml of solution per gram of liver) by one up-and-down stroke in the homogenizer. Thirty milliliters of this suspension are deposited in a tube of the SW 25.2 rotor (Beckman-Spinco) and covered successively with 20 ml of 34.5% sucrose and 10 ml of 14.3% sucrose, by gently pipetting one solution on top of another. The tubes are centrifuged for 2 hours at 25,000 rpm. The lysosomes float during this centrifugation and band at the interface between the 34.5% and 14.3% sucrose solutions. The corresponding layer (fraction T) is carefully removed with a 25-ml pipette fitted with a rubber aspirator. After removal of the lysosomes, the remainder of the gradient and the pellet are homogenized and make up the fraction R.

Comments. The procedure described here uses the SW 25.2 Beckman Spinco rotor and allows the purification of lysosomes from 90 g of liver with a yield of ±110 mg of proteins. It can be adapted to other types of swinging-bucket rotor provided the amount of processed liver is adapted to the capacity of the centrifuge tubes and the time integral of the centrifugal force to the geometrical specifications of the rotor. It must also be remembered that high centrifugal pressures generated in high speed rotors may cause extensive alternations of lysosomes and of other subcellular organelles.[6]

The processing of the sediment obtained by centrifuging the cytoplasmic extract is a crucial step. The tightly packed brownish pellet is covered by a fluffy pink sediment. This fluffy sediment, consisting mainly in microsomes, is sucked with the supernatant as completely as possible. The sediment is then carefully resuspended by working it into a smooth paste with a thick rod and adding 0.25 *M* sucrose dropwise, until mixing can be achieved by gentle agitation.

Biochemical Criteria of Purity

Distribution of Reference Enzymes

The outcome of the purification is established by determining the protein content of the T fraction and the activity of marker enzymes for lysosomes and for the other cell components. The constituents should be assayed in all the fractions (E, N, PS, R, and T) to establish the balance sheet of the fractionation process. This point is of great importance, because the enzyme activities may serve to estimate the cytological composition of subcellular fractions only if no extensive loss or gain of activity has occurred.

The most widely measured enzymes are acid phosphatase for lysosomes, cytochrome oxidase[7] for mitochondria, urate oxidase[8] or catalase[9] for peroxisomes, and glucose-6-phosphatase[7] for the endoplasmic reticulum.

As an example, the results of an experiment are shown in Table I.

[6] R. Wattiaux, S. Wattiaux-De Coninck, and M. Collot, *Arch. Int. Physiol. Biochim.* **79**, 1050 (1971).
[7] F. Appelmans, R. Wattiaux, and C. de Duve, *Biochem. J.* **59**, 438 (1955).
[8] C. de Duve, B. C. Pressman, R. Gianetto, R. Wattiaux, and F. Appelmans, *Biochem. J.* **60**, 604 (1955).
[9] P. Baudhuin, H. Beaufay, Y. Rahman-Li, O. Z. Sellinger, R. Wattiaux. P. Jacques, and C. de Duve, *Biochem. J.* **92**, 179 (1964).

TABLE I

DISTRIBUTION OF MARKER ENZYMES IN THE FRACTIONS OBTAINED
DURING A TYPICAL PURIFICATION PROCEDURE

| Marker | E + N | Percentage in fractions | | | | Recovery |
		N	PS	R	T	
Protein	100	26.3	50.0	20.8	0.65	97.75
Acid phosphatase	100	22.1	28.5	23.5	22.0	96.10
Cytochrome oxidase	100	30.7	0.8	60.3	0.13	91.93
Urate oxidase	100	26.0	5.2	75.8	0.17	107.17
Glucose-6-phosphatase	100	29.4	51.9	21.0	0.3	102.60

The content of the fractions is expressed in percent of the amount of enzyme or protein present in the reconstructed homogenate (sum of fractions E and N), taken as 100%. The recovery, given by adding the percentage values of fractions N, PS, R, and T, is satisfactory for all constituents. By applying the method described by Leighton et al.,[10] the contamination of fraction T is easily estimated from the activities of enzymes associated with various subcellular components. It can be deduced that 5% of the protein in fraction T are contributed by mitochondria, 1% by peroxisomes, and 11% by elements derived from the endoplasmic reticulum. This latter value is undoubtedly overestimated since the "glucose-6-phosphatase" activity found in the T fraction is largely due to the hydrolysis of glucose 6-phosphate by the lysosomal acid phosphatase. In a similar way, in other experiments, it has been found, by assaying alkaline phosphodiesterase I and galactosyltransferase, that 3% and 1% of the protein in the T fraction belong to plasma membranes and Golgi elements, respectively.

Such a detailed analysis of the T fraction is absolutely indispensable when the purification of lysosomes is attempted in a species other than the rat or the mouse. It is also advisable when the procedure is applied for the first time, especially if other equipment is used for centrifugation.

Routine Check of Purity

In routine work, the determination of the relative specific activity (RSA) of acid phosphatase in fraction T is often a sufficient check of the degree of purification of lysosomes. The average distribution of protein and acid phosphatase in 22 experiments is presented in Table II. The RSA is given by the ratio of the percent of enzyme activity to the percent

[10] F. Leighton, B. Poole, H. Beaufay, P. Baudhuin, J. W. Coffey, S. Fowler, and C. de Duve, J. Cell Biol. 37, 482 (1968).

TABLE II

DISTRIBUTION OF PROTEINS AND ACID PHOSPHATASE DURING THE PURIFICATION PROCEDURE[a]

Marker	E + N	Percentage in fractions				Recovery
		N	PS	R	T	
Protein	100	33.3 ± 6.4	47.4 ± 4.2	16.9 ± 3.2	0.6 ± 0.1	98.2 ± 2.8
Acid phosphatase	100	21.8 ± 4.8	30.4 ± 4.9	25.5 ± 5.5	18.6 ± 4.0	96.3 ± 2.8

[a] Results are expressed as means ± SD of 22 experiments. Relative specific activity of acid phosphatase in fraction T: 32 ± 3.5.

of protein in a given fraction. To make the RSA independent of the recoveries, the percentages values used better refer to the sum of activities or protein in the fractions (N + PS + R + T). In fraction T of the experiments reported here (Table II), the RSA of acid phosphatase exhibited little deviation from the average value: 32 ± 3.5. A 30–35-fold purification of acid phosphatase over the homogenate may thus be taken as a satisfactory result.

Proteins are determined by the Lowry procedure[11] with bovine serum albumin as standard.

Acid Phosphatase

Reagents

Substrate: β-glycerophosphate, 0.5 M
Buffer: acetate buffer, 1 M, pH 5
Detergent: Triton X-100, 2% (w/v)
Stopping reagent: trichloroacetic acid 8% (w/v)
Molybdate–H_2SO_4: 2.5% (w/v) ammonium molybdate in 5 N H_2SO_4
Reducing agent: 10 mM aminonaphtholsulfonic acid in 1.5 M $NaHSO_3$ and 10 mM Na_2SO_3

Procedure. One milliliter of a suitably diluted cell fraction (about 25 mg liver for the homogenate) is added to a mixture containing 0.2 ml of β-glycerophosphate solution, 0.1 ml of acetate buffer, 0.1 ml of Triton X-100, and 0.6 ml of water. After 30 minutes of incubation at 37° the reaction is stopped by 10 ml of trichloroacetic acid. Blanks are run in a similar way, except that β-glycerophosphate is added after the stopping reagent. The denatured proteins are eliminated by filtration through Whatman No. 42 paper, and the inorganic phosphate of 1 ml of filtrate is measured by adding 1 ml of molybdate solution, 0.2 ml of solution of reducing agent, and distilled water to a final volume of 10 ml. Absorbance is read after 10 minutes at 660 nm.

[11] O. H. Lowry, N. J. Rosebrough, A. L. Farr, and R. J. Randall, *J. Biol. Chem.* 193, 265 (1951).

[32] Isolation of Kidney Lysosomes[1]

By ARVID B. MAUNSBACH

Kidney lysosomes were first isolated by Straus from homogenates of rat kidney cortex by means of a combination of differential centrifugation and filtration through cotton.[2,3] The isolated lysosome fractions were enriched in acid phosphatase, cathepsin, β-glucuronidase, acid deoxyribonuclease, and acid ribonuclease, enzymes now recognized as typical lysosomal markers. More recently kidney lysosomes have been isolated by other procedures or modified fractionation schemes.[4-9]

The procedure described here is a slight modification of our previous method[5] and results in the isolation of kidney lysosomes with a minimum of contamination. The procedure has two steps: (i) the isolation of a semipurified lysosomal fraction by differential centrifugation and (ii) the purification of the semipurified lysosomal fraction by isopycnic centrifugation in a linear sucrose gradient.

Isolation Procedure

Reagents

0.3 M Sucrose, 1 mM EDTA, pH 7.0
1.1 M Sucrose, 1 mM EDTA, pH 7.0
2.1 M Sucrose, 1 mM EDTA, pH 7.0
Glycogen (1%) may be added to the above sucrose solutions to counteract aggregation of the lysosomes, but it is not necessary for maximal purification of the lysosomes.

Gradients. Linear sucrose gradients are formed in 38.5-ml cellulose

[1] Supported in part by Research Grants Nos. 512-542, 512-727, and 512-1545 from the Danish Medical Research Council.
[2] W. Straus, *J. Biol. Chem.* **207**, 745 (1954).
[3] W. Straus, *J. Biophys. Biochem. Cytol.* **2**, 513 (1956).
[4] A. B. Maunsbach, *Nature (London)* **202**, 1131 (1964).
[5] A. B. Maunsbach, *J. Ultrastruct. Res.* **16**, 13 (1966).
[6] S. Shibko and A. L. Tappel, *Biochem. J.* **95**, 731 (1965).
[7] S. Wattiaux-De Coninck, M.-J. Rutgeerts, and R. Wattiaux, *Biochim. Biophys. Acta* **105**, 446 (1965).
[8] P. Baudhuin, M. Müller, B. Poole, and C. de Duve, *Biochem. Biophys. Res. Commun.* **20**, 53 (1965).
[9] J. T. Dingle and A. J. Barrett, *in* "Lysosomes in Biology and Pathology" (J. T. Dingle and H. B. Fell, eds.), Vol. 2, p. 555. North-Holland Publ., Amsterdam, 1969.

nitrate tubes held in the swinging-bucket rotor SW 27 of a Beckman L2-65 B ultracentrifuge. The gradients are formed with a gradient mixer consisting of two communicating chambers.[10] The mixing chamber is filled with 11.7 ml of 2.1 M sucrose, and the other chamber with 13 ml of 1.1 M sucrose.

Preparation of Homogenate

Three adult male rats are anesthetized by intraperitoneal injections of sodium pentobarbital (35 mg/kg body weight) and perfused transcardially with ice-cooled 0.3 M sucrose for about 1 minute to remove the blood from the vascular system. The kidneys are excised and the renal cortex, distinguished by its light brown color, is cut with a razor blade from the outer stripe of the outer zone of the renal medulla, which is yellow-brown. The cortical tissue is immediately weighed, minced with scissors and diluted 1:8 (w/v) with 0.3 M sucrose. It is homogenized in a Potter-Elvehjem glass homogenizer by 10 complete strokes with a loosely fitting Teflon pestle rotating at about 1000 rpm.

Step 1. Isolation of Semipurified Lysosomal Fraction. All tissue fractions are maintained at 0–4° throughout the isolation procedure. The homogenate is first centrifuged at 800 rpm (about 143 g at the bottom of the tubes) for 10 minutes (MSE Multex centrifuge, universal swing-out head) to sediment nuclei and unbroken tissue as a nuclear fraction. The supernatant is decanted and centrifuged once again in the same way. The two sediments are combined to form the nuclear fraction. The nuclear supernatant is divided into 12-ml portions which are placed in cellulose nitrate tubes fitting the angle rotor Ti 50 of a Beckman L2-65 B ultracentrifuge. They are centrifuged at 10,000 rpm (9000 g at the bottom of the tubes) for 3 minutes, excluding times for acceleration and deceleration.

The pellet formed during this centrifugation is macroscopically layered and consists of three differently colored layers. The bottom layer is dark brown and is clearly distinguishable from the middle layer, which is yellow-brown. The bottom layer is the semipurified lysosomal fraction and the middle layer the mitochondrial fraction. The top layer which is almost white contains a mixture of membrane structures, prominent among which are the brush border fragments, and it is referred to as the brush border fraction.

The separation of these three fractions is a critical step in the procedure and is carried out with a Pasteur pipette and guided by the color difference between the layers. First, the supernatant is carefully removed.

[10] R. J. Britten and R. B. Roberts, *Science* 131, 32 (1960).

Then a few milliliters of 0.3 M sucrose is added along the wall of the tube and the brush border fraction resuspended by slowly swirling the tube. After the removal of this suspension, a few milliliters of 0.3 M sucrose is added over the mitochondrial layer of the pellet and repeatedly jetted, with the aid of the pipette, against this layer, which thereby becomes resuspended. The dark brown bottom layer containing lysosomes is more resistant to resuspension than the mitochondrial layer and remains at the bottom of the tube if the resuspension of the mitochondrial layer is carefully performed. The resuspended mitochondrial layer is removed, and the brown pellet and the walls of the tube are rinsed once with 0.3 M sucrose. This residual semipurified lysosomal fraction is finally resuspended in about 2.5 ml of 0.3 M sucrose.

Samples of the resuspended semipurified lysosomal fraction are used for enzyme and protein analyses. For electron microscopic analysis the semipurified lysosomal fraction or the other sedimented fractions are fixed *in situ* in the centrifuge tube with a solution containing 3% glutaraldehyde, 0.3 M sucrose, and 25 mM sodium cacodylate buffer, pH 7.2, and then further prepared for electron microscopy.[5]

Step 2. Isolation of Purified Lysosomal Fraction. The semipurified lysosomal fraction is purified by isopycnic centrifugation in the sucrose gradient specified above. Two milliliters of the resuspended semipurified lysosomal fraction are layered on top of the gradient. The sharp interface between the gradient and the resuspended fraction is broken by very cautiously stirring the uppermost layer of the gradient with a glass rod. The gradient is centrifuged at 25,000 rpm (about 113,000 g at the bottom of the tube) in rotor SW 27 of a Beckman L2-65B ultracentrifuge for 150 minutes (excluding 10 minutes for acceleration and 30 minutes for deceleration). After centrifugation the gradient has three macroscopically visible bands and a small sediment. The lowest band, which is the purified lysosomal fraction, is dark yellow to brown. The middle band, which is the purified mitochondrial fraction, is light yellow. The uppermost band, which is located immediately above the middle band, is gray and usually very faint.

The density gradient is fractionated according to a principle described by Hogeboom and Kuff.[11] The centrifuge tube is secured in a Lucite holder,[12] and heavy sucrose (2.1 M) is slowly pumped into the tube close to its bottom. The gradient is thereby displaced upward and through the outlet of the holder and is divided by an automatic fraction collector (LKB Ultrorac) into 1-ml fractions. Samples of these fractions are pre-

[11] G. H. Hogeboom and E. L. Kuff, *J. Biol. Chem.* **210**, 733 (1954).
[12] R. J. Romani and L. K. Fisher, *Anal. Biochem.* **21**, 333 (1967).

pared for electron microscopic analysis and analyzed for enzymes, protein, and refractive index.

In those experiments where the fractions are subsequently found to have a high degree of biochemical and ultrastructural purity, the purified lysosomal fraction and the purified mitochondrial fraction were always separated by a visible clear band in the gradient. Conversely, when the semipurified lysosomal fraction contains more than 5–10 mg of protein or is extensively contaminated with mitochondria, the fractions appear less well separated in the gradient.

Samples from the purified lysosomal or mitochondrial fractions are fixed for electron microscopy by dilution 1:1 with a solution containing 5% glutaraldehyde, 0.3 M sucrose, and 0.1 M sodium cacodylate buffer. The cell fractions are fixed for at least 30 minutes and then sedimented by centrifugation for 50 minutes at 30,000 rpm (about 84,500 g) in 0.8-ml tubes held in commercial adapters for the swinging buckets on the rotor SW Ti 65 of a Beckman L2-65 B ultracentrifuge. For quantitative electron microscopic analysis, the glutaraldehyde-fixed purified lysosomal fraction is centrifuged onto a flat disk oriented so as to be perpendicular to the centrifugal force.[5,13] Fixed and sedimented fractions are processed for electron microscopy in the same manner as small tissue block. Ultrathin sections, perpendicular to the surface of the pellet, are cut through its entire depth.

Properties

The Semipurified Lysosomal Fraction. When analyzed by electron microscopy, the semipurified lysosomal fraction contains predominantly membrane-limited granules with densely stained contents. The granules vary somewhat in size and shape, but the majority are round and 0.5–1.5 μm in diameter. Cytochemical methods show them to contain acid phosphatase at the electron microscope level and they are identified as lysosomes. Mitochondria are also present but are considerably less numerous than the lysosomes. In survey sections there are 5–20 lysosomes for each mitochondrial profile. A few vesicles are also present. Some of them have attached ribosomes and are apparently derived from the rough-surfaced endoplasmic reticulum while others are smooth-surfaced. Microbodies are only very rarely observed.

Biochemically, the semipurified lysosomal fraction contains 7.5% of the acid phosphatase activity of the homogenate and the specific activity of this enzyme is 9 times greater than in the homogenate (Table I). By contrast, only 0.5% of the cytochrome oxidase activity of the homogenate

---

[13] F. S. Sjöstrand, this series, Vol. 4, p. 391.

TABLE I

DISTRIBUTION OF REFERENCE ENZYMES AFTER ISOLATION OF RAT KIDNEY LYSOSOMES BY DIFFERENTIAL CENTRIFUGATION[a]

	Number of experiments	Homogenate (absolute values[b])	Yield[c] (% of homogenate)					Recovery[d] (%)	RSA[e] in SL
			N	SL	M	BB	S		
Protein[f]	5	720 ±82	38.0 ±4.1	0.83 ±0.20	16.6 ±1.4	4.5 ±0.9	37.6 ±3.7	97.5 ±7.8	1.00
Acid phosphatase[g]	5	2.08 ±0.26	32.4 ±8.9	7.5 ±1.8	17.2 ±3.8	3.3 ±0.7	43.6 ±10.8	104.0 ±12.8	9.1 ±0.8
Cytochrome oxidase[h]	3	0.119 ±0.035	37.2 ±4.3	0.52 ±0.30	42.0 ±11.6	4.9 ±1.4	16.5 ±3.8	101.1 ±17.1	0.59 ±0.20
Glucose-6-phosphatase[i]	3	8.50 ±1.80	44.3 ±1.5	0.36 ±0.16	10.9 ±3.4	5.4 ±0.4	40.9 ±7.8	101.9 ±9.6	0.40 ±0.17

[a] Results are given as means ± standard deviations.

[b] Absolute values are given in milligrams for proteins, micromoles inorganic phosphate released per hour per milligram of protein for acid phosphatase and glucose-6-phosphatase, and in micromoles of cytochrome c oxidized per minute per milligram of protein for cytochrome c oxidase.

[c] N, nuclear fraction; SL, semipurified lysosomal fraction; M, mitochondrial fraction; BB, brush border fraction; S, low speed supernatant.

[d] Recovery is defined as the percentage ratio of the sum of the activities recovered in all fractions to the activity of the homogenate.

[e] Relative specific activity (RSA) is the percentage of activity in the fraction: percentage of protein in the same fraction.

[f] Protein concentrations were determined by the method of O. H. Lowry, N. J. Rosebrough, A. L. Farr, and R. J. Randall, J. Biol. Chem. 193, 265 (1951).

[g] Acid phosphatase (orthophosphoric monoester phosphohydrolase, EC 3.1.3.2) was measured by the method of R. Wattiaux and C. de Duve, Biochem. J. 63, 606 (1956). Released P_i was determined by the method of C. H. Fiske and Y. SubbaRow, J. Biol. Chem. 66, 375 (1925).

[h] Cytochrome c oxidase (EC 1.9.3.1) was measured by the method of D. C. Wharton and A. Tzagoloff, this series, Vol. 10, p. 245.

[i] Glucose-6-phosphatase (EC 3.1.3.9) was measured according to C. de Duve, B. C. Pressman, R. Gianetto, R. Wattiaux, and F. Appelmans, Biochem. J. 60, 604 (1955). Released P_i was determined by the method of C. H. Fiske and Y. SubbaRow, J. Biol. Chem. 66, 375 (1925).

is present in this fraction, and it has a low relative specific activity. In addition glucose-6-phosphatase, a marker of the endoplasmic reticulum, has a low relative specific activity.

The biochemical and ultrastructural data are thus in good agreement and show that the semipurified lysosome fraction is quite rich in lysosomes although still contaminated by other cell components. Such semipurified lysosomal fractions may be useful in studies in which highest

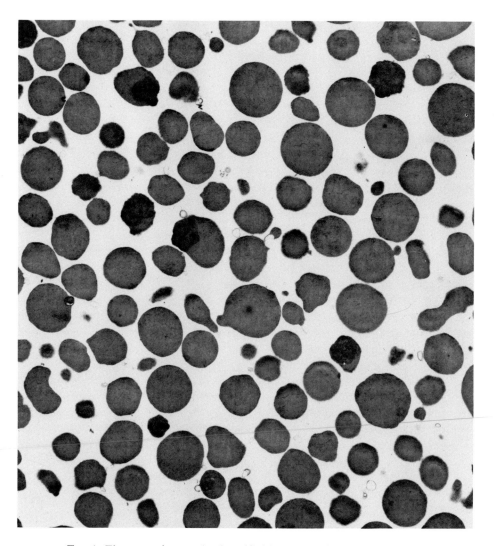

FIG. 1. Electron micrograph of purified lysosomal fraction. ×10,900.

possible purification of the lysosomes is not necessary but ease and speed in preparation is important.

The Purified Lysosomal Fraction. As seen with the electron microscope, the purified lysosomal fraction contains almost exclusively membrane-limited granules approximately 0.5–1.5 μm in diameter (Fig. 1). The granules are identical in ultrastructure to the types of lysosomes most frequently observed in sections of proximal tubule cells, and they contain acid phosphatase and other lysosomal enzymes in histochemical preparations.[14] Mitochondria are very rare in this fraction, and there are

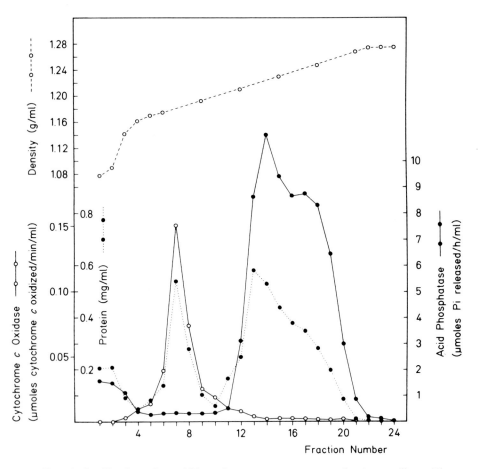

Fig. 2. Purification of rat kidney lysosomes on sucrose density gradient. The equilibrium density of lysosomes (1.21–1.26) is distinctly different from that of mitochondria (1.17–1.19).

[14] A. B. Maunsbach, *in* "Lysosomes in Biology and Pathology" (J. T. Dingle and H. B. Fell, eds.), Vol. 1, p. 115. North-Holland Publ., Amsterdam, 1969.

TABLE II

DISTRIBUTION OF REFERENCE ENZYMES AFTER PURIFICATION OF RAT KIDNEY LYSOSOMES BY ISOPYCNIC CENTRIFUGATIONS[a]

	Number of experiments	Purified mitochondrial fraction		Purified lysosomal fraction (whole fraction)		Purified lysosomal fraction (lower half of fraction)		
		Yield (% of homogenate)	RSA[b]	Yield (% of homogenate)	RSA	Yield (% of homogenate)	RSA	Recovery (% of load on gradient)
Protein	5	0.12 ± 0.04		0.227 ± 0.042		0.090 ± 0.057		89.5
Cytochrome oxidase	3	0.37 ± 0.20	3.1	0.019 ± 0.010	0.085	0.0058 ± 0.0061	0.036	102.5
Acid phosphatase	5	0.28 ± 0.11	2.7	3.90 ± 0.98	17.5	1.73 ± 0.68	22.2	90.3

[a] The experiments were performed with the semipurified lysosomal fractions from the fractionations presented in Table I. Results are given as means ± standard deviations.
[b] RSA, relative specific activity.

usually 50–200 lysosomal profiles for each mitochondrial profile in complete cross sections of the pellet. Vesicles with or without attached ribosomes are virtually absent, and microbodies are never observed. The fraction is ultrastructurally identical in composition to the purified granule fraction obtained with the first modification of this procedure; this fraction, in a quantitative electron microscopic analysis, contained by volume 97.7% lysosomes (dense granules) 1.2% mitochondria, and 1.1% unidentified material.[5]

Samples of the purified lysosomal fraction from different density levels in the gradient have similar contents except that the lysosomes in the lower half of the gradient are somewhat smaller and more irregular in shape than in the upper half and that mitochondria are somewhat less frequent in the lower half of the gradient.

The purified mitochondrial fraction contains almost exclusively mitochondria and only occasional lysosomes are observed.

Biochemically, the purified lysosome fraction is very rich in acid phosphatase and nearly devoid of cytochrome oxidase (Fig. 2). It contains 3.9% of the acid phosphatase of the homogenate, and the relative specific activity calculated for the whole fraction is 17.5 (Table II). In the lower half of the fraction the acid phosphatase is somewhat more concentrated and shows a specific activity of 22.5. The relative specific activity for cytochrome oxidase in the purified lysosomal fraction is 37 times less than in the purified mitochondrial fraction despite the fact that the latter fraction is contaminated with lysosomal protein. This indicates that 2–3% or less of the protein in the purified lysosomal fraction represents mitochondrial contamination. In the lower half of the purified lysosomal fraction the mitochondrial contamination approaches 1%.

Glucose-6-phosphatase is virtually absent in the purified lysosomal fraction but is present in the faint band that is located in the gradient immediately above the band of the purified mitochondrial fraction.

The biochemical observations are in good agreement with the electron microscope observations and demonstrate a very high degree of purification of the lysosomes.

Comments

When kidney lysosomes are isolated, it should be remembered that they constitute a heterogeneous group.[14] Therefore even seemingly minor differences between isolation methods may result in the isolation of lysosomes with slightly different characteristics. The lysosomes isolated by the present procedure are mainly derived from cells in the convoluted part of the proximal tubule.[5] Albumin and other proteins, which are absorbed through endocytosis by the proximal tubule cells, accumulate and are presumably digested in this type of lysosome,[14] which therefore corresponds to a secondary lysosome.[15] It should be noted that the low speed supernatant (Table I) contains an appreciable amount of the total acid phosphatase activity of the homogenate. Some of this activity is sedimentable at higher speeds and represents other types of lysosomes in proximal tubule cells or other kidney cells.

[15] C. de Duve and R. Wattiaux, *Annu. Rev. Physiol.* **28**, 435 (1966).

[33] The Preparation of Macrophage Lysosomes and Phagolysosomes

By ZENA WERB and ZANVIL A. COHN

Macrophage Lysosomes

Sources of Cells and Homogenization. Most reported experiments have employed alveolar macrophages,[1,2] induced peritoneal cells[3] or populations of unstimulated mouse peritoneal cells[4] cultivated *in vitro*. The preparation and properties of these populations may be found in another chapter of this volume.

Cells suspended in aqueous sucrose solutions are best homogenized at concentrations approaching 5 to 6×10^7/ml. This high cell density accelerates cell breakage and reduces lysosome disruption by allowing a reduced number of strokes with the pestle. Because of the variability of the process, cell breakage should be continually monitored by means of phase contrast microscopy. In most instances disruption of 75–90% of the cells is acceptable. Adequate breakage occurs with the use of a machined motor-driven Teflon pestle for three 2-minute periods with equal intervals in an ice bath. Tight-fitting Dounce, all-glass homogenizers will disrupt macrophages with 15–20 strokes, but damage to organelles is more prominent with prolonged homogenization.

Differential Velocity Centrifugation

For many purposes the separation of a "large granule" fraction is adequate for preliminary studies of hydrolase localization, latency, and redistribution.[1]

[1] Z. A. Cohn and E. Wiener, *J. Exp. Med.* **118**, 991 (1963).
[2] Z. A. Cohn and E. Wiener, *J. Exp. Med.* **118**, 1009 (1963).
[3] E. Kolsch and N. A. Mitchison, *J. Exp. Med.* **128**, 1059 (1968).
[4] Z. Werb and Z. A. Cohn, *J. Exp. Med.* **134**, 1545 (1971).

An example of the method and results with BCG-induced alveolar macrophages is presented.[1]

After homogenization at 5×10^7 cells/ml the contents of the homogenization tube are diluted 1:2 with cold 8.5% sucrose. After thorough mixing the homogenate was centrifuged at 500 g for 12 minutes at 0°C. The milky supernatant fluid was removed by means of a fine-tipped Pasteur pipette and transferred to lusteroid centrifuge tubes. The pellet containing primarily nuclei and unbroken cells was taken to volume with sucrose and constituted the "nuclear" fraction. The postnuclear supernatant was then centrifuged at 12,000–15,000 g for 12 minutes at 0° in an angle-head rotor (Lourdes, rotor 9RA). This resulted in a firm, tan pellet and a slightly opalescent "supernatant" fraction. The pellet was resuspended in sucrose; it constituted the "15 g" fraction and contained the majority of lysosomes as well as mitochondria. The "supernatant" fraction contained soluble hydrolases, microsomes, and other soluble proteins.

The table illustrates the distribution of enzymes in three populations of macrophages from rabbit and mouse. In almost all cases 70% or more of the hydrolases were concentrated in the 15 g fraction. The presence of acid hydrolases in the nuclear pellet was correlated with contamination of this low speed fraction with unbroken cells and variable numbers of

PERCENT DISTRIBUTION OF ACID HYDROLASES IN MACROPHAGE FRACTIONS SEPARATED BY DIFFERENTIAL CENTRIFUGATION[a]

Acid hydrolase	BCG-induced rabbit alveolar macrophage			Oil-induced rabbit peritoneal macrophage			Cultivated mouse peritoneal macrophage		
	N	15 g	S	N	15 g	S	N	15 g	S
Acid β-glycerophosphatase	12	78	10	4	72	14	12	80	10
β-Glucuronidase	16	69	15	13	74	9	13	78	9
Cathepsin D	14	74	12	18	70	12	13	76	11
Acid ribonuclease	21	72	7	17	69	14	11	76	13
Acid deoxyribonuclease	11	78	11	16	72	12	16	74	10
N-Acetyl β-glucosaminidase	9	82	9	14	76	10	14	77	9
β-Galactosidase	8	80	12	10	75	15	11	79	10
Cholesterol esterase	12	78	10	—	—	—	16	70	14
Aryl sulfatase	10	83	7	10	78	12	13	78	9

[a] Recoveries for all enzymes varied from 90 to 110% of original homogenate activity. Values are means from three experiments. N = Nuclear, 15 g = 15,000 g pellet, S = 15,000 g supernatant.

large lysosomes. The soluble enzyme in the 15 g supernatant fraction was in part the result of lysosome breakage during homogenization. If vigorous homogenization was continued, a gradual increase in soluble hydrolases occurred at the expense of the large granule fraction. Lysosomes isolated in this fashion exhibit latency but are heavily contaminated with mitochondria. Following phagocytosis and degranulation large amounts of the lysosomal enzymes are transferred to the phagocytic vacuole. These can be accounted for in isolated phagolysosomes[5] or, in the usual fractionation procedures in which the phagolysomes are disrupted, are found in the 15 g supernatant fraction.[2]

The Separation of Macrophage Lysosomes by Isopycnic Sucrose Gradient Centrifugation

The large granule fraction obtained by differential centrifugation contains both secondary lysosomes and 80% of the cell's mitochondria. The separation of these organelles can be accomplished in the rabbit alveolar macrophage since its lysosomes are considerably more dense than mitochondria.[1] The separation of lysosomes and mitochondria from cultivated or induced peritoneal cells is more difficult because of the similar densities of the two cytoplasmic organelles.[3]

Alveolar macrophages are harvested and homogenized as described previously. The postnuclear supernatant fluid is the starting point. This may be layered directly over the gradient but then contains both particle-bound and soluble enzyme. An alternative procedure is to sediment the lysosomes and mitochondria at 15,000 g for minutes at 0°, resuspend in 10% sucrose, and utilize this for the gradient. The latter procedure is sometimes complicated by particle aggregation. Linear sucrose gradients are prepared between 30 and 70% sucrose (w/v) in tubes fitting the SW 25 rotor of the Spinco Model L ultracentrifuge. Three milliliters of either the postnuclear supernatant or resuspended large granule fraction are then carefully layered over the 30-ml gradient. After 6 hours of centrifugation at 53,000 g, two visible bands are present in the gradient. One-milliliter fractions are collected from the bottom of the gradient into calibrated tubes and assayed for protein and enzymatic activity after six cycles of freezing and thawing.

Two protein peaks are present in such gradients. An upper band with an equilibrium density of 1.19 corresponds to the position of mitochondria as evidenced by the presence of cytochrome oxidase activity and mitochondrial profiles by electron microscopy. Between 50 and 65% of the

[5] T. P. Stossel, R. J. Mason, T. D. Pollard, and M. Vaughn, *J. Clin. Invest.* **51**, 604 (1972).

total lysosomal enzyme content of the initial homogenate is present in the second, denser band equilibrating at a density of 1.25. This fraction contains a morphologically heterogeneous population of electron dense membrane-bounded granules similar to those found in the intact cell. The remainder of the lysosomal enzymes (15%) applied to the gradient are present in the load volume. In the majority of experiments there was little or no overlap between the mitochondrial and lysosomal bands. Each of seven acid hydrolases gave similar distribution with sharp peaks in the lysosomal band. In addition, large quantities of muramidase were also localized to this fraction. When the intact macrophages were vitally stained with neutral red prior to homogenization, an intense red band corresponded to the position of the acid hydrolases. The specific activity of the acid hydrolases was 4–5 times that present in the initial homogenate.

The density of the lysosomes present in alveolar macrophages is presumably the result of large stores of nonenzymatic protein taken up by the cell as a consequence of endocytosis. The lysosomes are strongly PAS positive and contain large amounts of carbohydrate–protein complexes—similar in composition to the secretory products of the tracheobronchial tree.[6] It should be pointed out that the alveolar macrophage is particularly enriched with respect to lysosomes and these organelles comprise a major portion of the cytoplasmic mass.

Preparation of Macrophage Phagolysosomes

Principle. Isolated phagocytic vacuoles provide a useful system for studying secondary lysosomes, plasma membrane, and membrane fusion in the macrophage. The phagolysosomes may be isolated from the macrophages at various times after phagocytosis of polystyrene latex particles using a modification of the method of Wetzel and Korn.[7] After disruption of the cells the latex particles with their adherent membranes and contents are collected by flotation through a discontinuous sucrose gradient.

Phagolysosomes from Mouse Peritoneal Macrophages[8]

Cell Sources. Macrophages are harvested from mice and cultivated in 100-mm petri dishes in medium 199 supplemented with newborn calf serum.[8]

Ingestion of Latex Particles. Latex particles (1.1 μm in diameter, Dow Chemical) are washed in 50–100 volumes of medium 199, collected by centrifugation, then resuspended in medium 199 supplemented with 20% newborn calf serum, at a final concentration of 500 μg latex/ml

---

[6] E. Hawrylko and Z. A. Cohn, *Lab. Invest.* **19**, 421 (1968).
[7] M. G. Wetzel and E. D. Korn, *J. Cell Biol.* **43**, 90 (1969).
[8] Z. Werb and Z. A. Cohn, *J. Biol. Chem.* **247**, 2439 (1972).

(approximately 10^8/ml). The beads are then added to the macrophage cultures and incubated together for 1 hour at 37° in an atmosphere of 5% CO_2, balance air. The macrophage monolayer is rinsed twice with medium 199, to remove extracellular beads, and the cells can be harvested immediately or incubated for an additional period of time in fresh serum medium.

Cell Rupture. Macrophages are harvested by scraping with a rubber policeman into phosphate-buffered saline (PBS) (about 2 ml per 100-mm dish), then concentrated by centrifugation at 500 g for 5 minutes for 1 to 5 × 10^7 macrophages. The cell pellet is resuspended in 3 ml of 30% sucrose and homogenized at 4° in a small Dounce homogenizer with tight-fitting pestle. More than 95% cell breakage is achieved in 20–30 strokes, as judged by phase contrast.

Sucrose Gradient. Homogenate, 2.5 ml, is placed in the bottom of a 12-ml cellulose nitrate centrifuge tube, overlaid with 6 ml of 20% sucrose, then 2.5 ml of 10% sucrose. After centrifugation at 105,000 g for 60 minutes in the SW 41 rotor of a Beckman L-2 ultracentrifuge, the latex phagolysosomes are concentrated at the 10% to 20% sucrose interface. Sometimes phagolysosomes are also found at the 20% to 30% sucrose interface. The fractions can be removed by suction with a Pasteur pipette.

Phagolysosomes from Rabbit Alveolar Macrophages[9]

Macrophages from rabbit lungs[1] can also be used for preparing phagolysosomes. Alveolar macrophages (1 × 10^9) are suspended in minimal essential medium supplemented with 10% rabbit serum and equilibrated with 5% CO_2–air. About 1 × 10^{11} washed latex particles (100 mg) are added, and the cells and beads are incubated for 30 minutes at 37° with constant shaking. The cells are harvested by centrifugation at 400 g for 10 minutes and washed by resuspending in PBS and centrifuging, twice to remove uningested beads. The macrophages are resuspended in 12.5 ml of PBS, and 12.5 ml of 60% sucrose are added. Homogenization is performed in a large Dounce homogenizer with a tight-fitting pestle (about 35 strokes), then 4 ml of the homogenate is placed in the bottom of each of 6, 12-ml centrifuge tubes. Over the homogenate, 4 ml of 20% sucrose is carefully layered, followed by 4 ml of 10% sucrose. After centrifugation at 105,000 g for 60 minutes the phagolysosomes are found in a layer at the 10% to 20% sucrose interface.

Phagolysosomes have also been isolated in a similar fashion from alveolar macrophages after ingestion of emulsified paraffin oil containing Oil Red O.[5]

[9] R. L. Nachman, B. Ferris, and J. G. Hirsch, *J. Exp. Med.* **133**, 785 (1971).

Phagolysosomal Membrane

Limited success has been achieved in the isolation of membrane from the phagocytic vacuoles. Sonication releases small pieces of membrane in low yield.[9] Destruction of latency of the phagocytic vacuoles by hypotonic shock, followed by reisolation of the latex particles with their adherent membranes,[8] preserves enzymatic activity and lipid composition, but varying amounts of vacuolar contents remain.

Properties of Macrophage Phagolysosomes

Yield. The yield of phagolysosomes is measured by assaying the amount of polystyrene latex recovered in the 10% to 20% interface fraction,[8,10] and depends on the efficacy of the homogenization procedure. With peritoneal macrophages, homogenization proceeds more smoothly; yields are greater when the cells are cultivated for a few hours after the phagocytic pulse; 30% to 90% of the latex may be recovered in the 10% to 20% interface fraction.

Microscopic Examination. By phase contrast microscopy the 10% to 20% sucrose interface fraction appears as single beads, and the 20% to 30% sucrose interface fraction contains small clumps of beads. Electron microscopic examination of the phagolysosomes from the 10% to 20% sucrose interface shows single polystyrene latex particles surrounded by a unit membrane. In addition, some electron-opaque material is often seen within the membrane sac. This appearance is similar to that of the phagolysosomes of intact cells.[11]

Chemical Composition. Membrane lipid components are enriched relative to protein in the phagolysosomal fractions. In mouse peritoneal macrophages the cholesterol is recovered in the phagolysosomes with a 4-fold enrichment.[8] Phospholipids are also increased to give a cholesterol to phospholipid molar ratio of about 0.9 in this fraction. There are at least five membrane proteins discernible by polyacrylamide gel electrophoresis.[9]

Enzymatic Characteristics. Lysosomal marker enzymes are enriched 2- to 5-fold in the phagolysosomal fractions; 35% to 80% of the acid phosphatase is recovered in this fraction. Acid cholesterol esterase,[12] β-glucuronidase, and cathepsin are also enriched in parallel. Essentially no succinic dehydrogenase, a mitochondrial marker enzyme, and lactate dehydrogenase, a cytoplasmic marker, are recovered in latex-enriched fractions. 5'-Nucleotidase, which is usually considered to be a plasma

[10] R. A. Weisman and E. D. Korn, *Biochemistry* **6**, 485 (1967).
[11] S. G. Axline and Z. A. Cohn, *J. Exp. Med.* **131**, 1239 (1970).
[12] Z. Werb and Z. A. Cohn, *J. Exp. Med.* **135**, 21 (1972).

membrane marker enzyme, is enriched up to 4-fold in phagolysosomes isolated immediately after phagocytosis.[8] The amount of 5'-nucleotidase recovered with the latex phagolysosomes is proportional to the amount of latex ingested.[8] Phagolysosomes isolated 2–24 hours after phagocytosis contain about the same percentage of the cholesterol and acid hydrolases but decreasing levels of 5'-nucleotidase, indicating a change in the plasma membrane-derived membrane of the phagolysosomes.[8]

[34] The Isolation of Granules from Neutrophile Polymorphonuclear Leukocytes (PMN's)

By Marco Baggiolini

In recent years, electron microscopical and cytochemical investigations have clearly established that rabbit PMN's contain two types of granules— azurophiles and specifics—differing in size, biochemical composition, and mode of formation in the maturing PMN's.[1-5] These results have been substantiated by fractionation data and biochemical studies on highly purified granule fractions.[6-8] The azurophile granules represent a special type of primary lysosomes containing acid hydrolases as well as peroxidase. The specifics are not lysosomal in nature—they contain alkaline phosphatase, lysozyme, and lactoferrin but lack common lysosomal hydrolases.

Morphological and cytochemical studies strongly suggest that PMN's of other species also contain two types of granules, and that the concept established for the rabbit may well be generally valid. Unfortunately, the biochemical data backing this hypothesis are so far rather more suggestive than conclusive.[9-11]

[1] D. F. Bainton and M. G. Farquhar, *J. Cell Biol.* **28**, 277 (1966).

[2] B. K. Wetzel, R. G. Horn, and S. S. Spicer, *Lab. Invest.* **16**, 349 (1967).

[3] B. K. Wetzel, S. S. Spicer, and R. G. Horn, *J. Histochem. Cytochem.* **15**, 311 (1967).

[4] D. F. Bainton and M. G. Farquhar, *J. Cell Biol.* **39**, 286 (1968).

[5] D. F. Bainton and M. G. Farquhar, *J. Cell Biol.* **39**, 299 (1968).

[6] M. Baggiolini, J. G. Hirsch, and C. de Duve, *J. Cell Biol.* **40**, 529 (1969).

[7] M. Baggiolini, J. G. Hirsch, and C. de Duve, *J. Cell Biol.* **45**, 586 (1970).

[8] M. Baggiolini, C. de Duve, P. L. Masson, and J. F. Heremans, *J. Exp. Med.* **131**, 559 (1970).

[9] D. F. Bainton, J. L. Ullyot, and M. G. Farquhar, *J. Exp. Med.* **134**, 907 (1971).

[10] B. K. Wetzel, *in* "Regulation of Hematopoiesis" (A. S. Gordon, ed.), p. 769. Appleton-Century-Crofts, New York, 1970.

[11] M. Baggiolini, *in* "Metabolism of Leucocytes in Health and Disease" (J. Frei, ed.), *Enzyme* **13**, 132 (1972).

Peritoneal Exudates

PMN's are collected from peritoneal exudates induced in the rabbit by the infusion of isotonic NaCl solution containing glycogen.[12] This solution is prepared as follows: 250-ml portions of 0.9% NaCl are sterilized in glass infusion bottles and kept at room temperature until they are used. Just before infusion into the animals, unsterilized shellfish glycogen (Amend Drug Co., New York) is dissolved into the saline solution at a concentration of 1 mg/ml. Glycogen from other sources proved less satisfactory. Although the cell yield was normal, cells appeared to be more resistant to homogenization.

Female New Zealand White rabbits, weighing 3.5–5.0 kg, are used. Each animal is fastened in position lying on its back; the abdominal area is clipped, and the skin is wiped with ethanol. Then 250 ml of saline containing glycogen are infused intraperitoneally using a hypodermic needle introduced through the skin in the midline above the umbilicus. The exudate is collected 3–4 hours later into a glass bottle containing 3 mg of heparin dissolved in 1 ml of isotonic NaCl using an 8–10 cm long and 1.5–1.8 mm thick (od) needle with four perforations 2 mm in diameter in its shaft within 2 cm from the bevel. The rabbit is fastened in the prone position on a board (75 × 30 cm) with an oval opening (diameters: 19 and 14 cm) in the middle. The abdomen of the animal fits into the opening and can be reached from underneath. The needle is connected to a transparent plastic tubing. It is introduced in the midline 5 cm below the umbilicus for about 2 cm in the subcutaneous tissue and then plunged into the peritoneal cavity. The tip of the needle thus reaches the lowest area of the abdominal cavity. In most cases, fluid will drain spontaneously through the plastic tubing. Flow can be improved by gently kneading the abdomen in order to pool the exudate about the tip of the needle, and by cautiously moving the needle accordingly. On average, 150 ml of exudate are collected. If the yield is not satisfactory, 50 ml of sterile isotonic saline are injected through the plastic tubing, and the abdomen is kneaded before restarting collection. Suitable exudates contain 5 to 10 × 10⁶ cells per milliliter. PMN's account for 96–99% of the total cell count, lymphocytes and monocytes total together 1–3%, and eosinophiles about 1%.

Rabbits should rest for at least 7–10 days after each collection of exudate.

Biochemical Assays

Assaying for myeloperoxidase and alkaline phosphatase, which are selectively localized in azurophile and specific granules, respectively[5–7]

---

[12] J. G. Hirsch, *J. Exp. Med.* **103**, 589 (1956).

is the most suitable way of assessing the purity of the granule fractions. Enzyme activities should be measured in all fractions and in the starting material, thus allowing computation of recoveries, and conveniently plotted as relative concentration versus cumulative volume of fractions.[13]

Myeloperoxidase; Automated Assay with a Technicon Autoanalyzer Unit

Samples in 0.34 M sucrose, pumped at a rate of 0.42 ml/minute, and segmented by air (1.20 ml/minute) are mixed with an equal amount (0.42 ml/minute) of a 0.2% (w/v) solution of N-hexadecyltrimethyl-ammonium bromide by means of a HO connector followed by a 105-88 glass mixing coil. Substrate solution (2.00 ml/minute) containing 10 parts of 0.1 M citric acid–sodium citrate buffer, pH 5, one part of 0.2% o-tolidine in ethanol, one part of 1.5 mM H_2O_2, and 0.05% BRIJ 35 is added through a DO connector. The reaction mixture passes through a 105-89 glass mixing coil (kept at room temperature or at 25°), enters a C5 debubbler, and is pumped through the flow cell of a colorimeter (10 mm light path) at the rate of 2.00–2.50 ml/minute. Transmission at 550 nm is recorded on log paper. Washing is done with 0.34 M sucrose (1.2 ml/minute). The sampling rate is 20–30 per hour with a 1:1 sampling: washing ratio. The substrate mixture is kept in a dark glass bottle. Glass tubing is used for all connections up to the flow cell.

Myeloperoxidase; Hand Assay

Samples of 0.1–0.2 ml are rapidly mixed (Vortex mixer) with 1.0 ml of substrate solution containing 10 parts of 0.1 M citric acid–sodium citrate buffer pH 5.5, one part of 0.1% o-dianisidine in ethanol, one part of 1.5 mM H_2O_2, and 0.05% Triton X-100, and incubated for 1.5 minutes at 25°. Incubation is stopped with 1.0 ml of 40% perchloric acid, and the OD is read at 560 nm.

Since no standard is available for the described assays, myeloper-oxidase activity of the preparations may be compared with known amounts of purified horseradish peroxidase.

Alkaline Phosphatase; Automated Assay with a Technicon Autoanalyzer Unit

Samples in 0.34 M sucrose (0.32 ml/minute) are mixed with sub-strate solution (0.80 ml/minute), segmented by air (1.20 ml/minute) by means of a HO connector followed by a 105-89 glass mixing coil. The substrate solution contains 1.5 mM p-nitrophenyl phosphate (disodium

[13] C. de Duve, in "Enzyme Cytology" (D. B. Roodyn, ed.), p. 1. Academic Press, New York, 1967.

salt) and 0.05% BRIJ 35 in 0.1 M diethanolamine·HCl buffer pH 9.75. Incubation (at room temperature or at 25°) is extended to 5–15 minutes with a glass delay coil. Incubation is stopped by 2 N NaOH (0.80 ml/minute) added through a D1 connector. After debubbling (C5), the mixture is pumped through the flow cell of the colorimeter (10 mm light path) at the rate of 1.60 ml/minute. Transmission at 400 nm is recorded on log paper. Washing (1.20 ml/minute) is done with 0.34 M sucrose. A 5 μM solution of p-nitrophenol in 10 mM NaOH is used as a standard. Glass tubing is used for all connections up to the flow cell. With minor modifications, this method can be carried out by hand.

In both automated assays, samples are kept at approximately 4°, the sampler being in a refrigerator. Depending on the density, fractions may be diluted with fresh 0.34 M sucrose or with water. Fractions from isopycnic equilibration runs are conveniently diluted 5–10 times with water.

β-Acetylglucosaminidase; Hand Assay

Samples of 0.2 ml are rapidly mixed (Vortex mixer) with 0.3 ml of substrate solution containing 8 mM p-nitrophenyl-N-acetyl-β-D-glucosaminide in 0.2 M citric acid–sodium citrate buffer pH 4.2 and 0.5% Triton X-100, and incubated for 1–3 hours in a stoppered glass tube at 37°. Incubation is stopped by the addition of 2 ml of a solution containing 133 mM glycine, 83 mM Na$_2$CO$_3$, and 67 mM NaCl at pH 10.7. OD is read at 400 nm. A 20 μM p-nitrophenol solution in water is used as a standard.

Fractionation Techniques[14]

Homogenization. The exudates from single rabbits are filtered through six layers of gauze into one graduated cylinder, and the total volume is recorded. After mixing, samples are taken for total and differential cell count. Exudates with less than 3×10^6 cells per milliliter or which are contaminated by large numbers of erythrocytes, are discarded. The cell suspension is spun at approximately 2000 g-min in a refrigerated centrifuge set at 4°, using a swing-out rotor with 50–100 ml glass tubes. The supernatant is thoroughly decanted, the cells are resuspended in about half the original volume of 0.34 M sucrose kept at room temperature, and spun under the above conditions. The washed cell button is cooled in ice, and resuspended finely to a concentration of 0.7–1.0×10^8 cells per milliliter in cold (0–4°) 0.34 M sucrose. Ten- to fifteen-milliliter

[14] The methods described here have been developed on a Spinco Model L ultracentrifuge (Spinco Division, Beckman Instruments, Inc., Fullerton, California). Any ultracentrifuge suitable for zonal operation, however, may be used.

portions of this suspension are then vigorously shaken at short intervals (3–5 seconds) for 1 minute in a stoppered 50-ml glass centrifuge tube using a Vortex mixer set at maximum speed. By this procedure, 50–70% of the PMN's are broken up almost instantaneously. Other exudate cells are also likely to be disrupted as suggested by our recent observation in monocytes.[15] The homogenate is centrifuged under the above conditions at exactly 4000 g-min, and the milky "postnuclear supernatant" is collected by aspiration and mixed.

Zonal Differential Sedimentation in a B-XIV Rotor.[16] Sedimentation is carried out through a discontinuous stabilizing gradient resting on a cushion of 100 ml of 60% (w/w) sucrose, and extending between 4.4 and 6.15 cm radial distance, and from 0.45 to 0.80 M sucrose. The gradient is established by 35-ml portions of 0.45, 0.50, 0.55, 0.60, 0.65, and 0.70 M sucrose, followed by 45 ml each of 0.75, and 0.80 M sucrose, introduced in this order through the edge line. Twenty milliliters of the "postnuclear supernatant" are loaded on top of the gradient through the core line and displaced outward by 230 ml of overlayer consisting of 0.25 M sucrose. Before loading, the rotor is refrigerated at approximately 4°. During operation, all sucrose solutions are kept in ice and the temperature control of the centrifuge is set at 4–6°. The rotor is filled and emptied at a rate of 25–30 ml/minute while rotating at 2000–2200 rpm. Under these conditions, no sedimentation of particles occurs within the loading period.[7] When the initial conditions are established, centrifugation is continued for 15 minutes at 7000 rpm. The rotor is then rapidly decelerated to 2200 rpm and emptied by centripetal displacement with 60% sucrose. Two hundred milliliters of overlayer are collected first in 50–100-ml fractions while the material which follows is collected in fractions of 20–25 ml into graduated tubes.

At the end of this type of sedimentation run, azurophile granules have accumulated against the cushion while the specifics are distributed in a rather broad band covering the middle of the gradient (Fig. 1). A third group of quite heterogeneous particles of lysosomal character (C particles[6,7]), which most likely arise from the mononuclear cells present in the exudate,[15] are distributed between the starting zone and the middle of the gradient, where they partially overlap with the specific granules.

Isopycnic Equilibration in Beaufay's Rotor. The characteristics and the operation of this rotor are described in detail in previous publica-

[15] M. G. Farquhar, D. F. Bainton, M. Baggiolini, and C. de Duve, *J. Cell Biol.* **54**, 141 (1972).
[16] N. G. Anderson, D. A. Waters, W. D. Fisher, G. B. Cline, C. E. Nunley, L. H. Elrod, and C. T. Rankin, Jr., *Anal. Biochem.* **21**, 235 (1967).

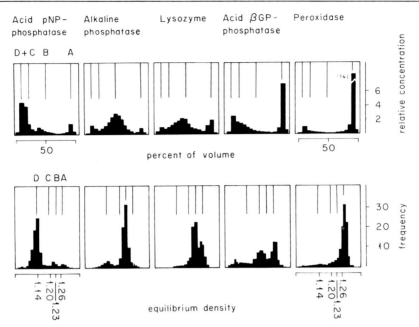

Fig. 1. Enzyme distribution patterns after fractionation of subcellular components of rabbit polymorphonuclear leukocytes by zonal sedimentation at 6500 rpm for 15 minutes, and by zonal isopycnic equilibration (35,000 rpm for 60 minutes). In both fractionation systems, four particulate bands (A to D, in order of decreasing sedimentation coefficient and modal equilibrium density) have been resolved. The distribution of β-acetylglucosaminidase (Table II) is almost identical to that of acid β-glycerophosphatase [see M. Baggiolini, J. G. Hirsch, and C. de Duve, *J. Cell Biol.* **40**, 529 (1969); *ibid.* **45**, 586 (1970)] represented here as the traditional lysosomal marker enzyme. In the bottom histograms, it should be noted that the abscissa represents equilibrium density (in Tables I and II, calculations for both types of fractionations are based on relative volume distribution data). Both experiments have been taken into account for calculation of the data given in Tables I and II.

tions.[17,18] For this type of fractionation, 14 ml of the "postnuclear supernatant" brought to a density of approximately 1.10 by the addition of 60% sucrose are layered over a 19-ml sucrose gradient extending linearly with respect to volume between the densities 1.18 and 1.32 and resting on 6 ml of a sucrose cushion of density 1.32. Centrifugation is carried out at 35,000 rpm for 60 minutes. After recovery of the clear sucrose cushion, 15–18 fractions of 1.5–2.2 ml followed by few larger ones are collected

[17] H. Beaufay, "La centrifugation en gradient de densité." Ceuterick S. A., Louvain, 1966.

[18] F. Leighton, B. Poole, H. Beaufay, P. Baudhuin, J. W. Coffey, S. Fowler, and C. de Duve, *J. Cell Biol.* **37**, 482 (1968).

in preweighed glass tubes. In these experiments, azurophile and specific granules equilibrate in two well-resolved, sharp bands with modal density of 1.26, and 1.23 g/ml, respectively. The C particles band in a somewhat broader peak with modal density of 1.19–1.20 (see Fig. 1).

Isolation of the Granules

After fractionation, the volume, or the weight, of each fraction, which is needed for computation of the relative enzyme distributions, is recorded. The fractions are then mixed, and their myeloperoxidase, alkaline phosphatase, and β-acetylglucosaminidase activity is measured.

From sedimentation runs, 1–2 fractions (25–50 ml) adjacent to the cushion containing as much as 70–80% of the myeloperoxidase of the preparations are processed for the isolation of azurophile granules, while specific granules are prepared from an average of 5 fractions (100–125 ml) with highest alkaline phosphatase activity accounting for 50–60% of the total amount of this enzyme. After isopycnic equilibration, azurophiles and specifics are prepared from the two, or possibly three fractions having the highest myeloperoxidase and alkaline phosphatase content, respectively. The granules are conveniently pelleted from known amounts of the selected fractions at a total centrifugal force of approximately 500,000 g-min using an angle rotor.[19] (Before centrifuging, the azurophile granule fraction from sedimentation experiments which is regularly contaminated by the 60% sucrose of the cushion, and the fractions obtained by isopycnic equilibration, are diluted 1:2 with 0.25 M sucrose.) The supernatant fluid is decanted thoroughly, and the granules are resuspended in the desired medium with a glass rod smoothed at the tip.

Purity of the Azurophile and Specific Granule Preparations

Isolation by Zonal Differential Sedimentation. Resolution between myeloperoxidase and alkaline phosphatase is here almost complete. As shown in Tables I and II, up to 80% of the azurophile, and 50–60% of the specific granules can therefore be recovered in fractions which are only minimally (5–6%) cross contaminated. The specific granule preparations are also contaminated by C particles (β-acetylglucosaminidase) up to approximately 20%. However, if only the fast sedimenting half of the alkaline phosphatase peak is considered, the contamination drops to an average of 12%. Although extremely clean under normal conditions, the azurophile granule preparations are likely to be seriously contaminated by heterogeneous particle aggregates present in the starting material. This has been shown to occur sometimes, at least in minor propor-

[19] Rotor No. 30 (Beckman Instruments), or rotor No. 59595, 8 × 50 ml (M.S.E.).

TABLE I

PURITY OF AZUROPHILE GRANULE PREPARATIONS FROM RABBIT
POLYMORPHONUCLEAR LEUKOCYTES OBTAINED BY ZONAL
SEDIMENTATION, AND ZONAL ISOPYCNIC EQUILIBRATION

	Azurophile granule preparation			Percent contamination[d] by alkaline phosphatase
Type of fractionation	Volume (ml)	Percent of total myeloperoxidase	Percent of total protein[c]	
Zonal sedimentation[a]	33.9 ± 11.6	76.4 ± 6.6	21.4 ± 3.0	4.9 ± 1.1
Zonal isopycnic equilibration[b]	—	65.2 ± 5.9	12.3 ± 1.3	9.6 ± 3.1

[a] Mean value ± standard deviation (SD) from 9 experiments. After loading, sedimentation was carried out at an average angular velocity of 7050 (6500–7700) rpm. Percentage recoveries (mean ± SD) were 101.6 ± 11.9 for protein, 96.8 ± 7.9 for myeloperoxidase, 90.6 ± 8.0 for alkaline phosphatase, and 99.3 ± 11.4 for β-acetylglucosaminidase.

[b] Mean values ± SD from 4 experiments carried out exactly under the conditions described in the text. Percentage recoveries (mean ± SD) were 105.3 ± 13.1 for protein, 79.6 ± 8.0 for myeloperoxidase, 83.6 ± 8.0 for alkaline phosphatase, and 91.2 ± 7.1 for β-acetylglucosaminidase. The output of the computation used for these experiments does not give the actual fraction volume. The percentage volume is directly calculated from weight and density of the fractions.

[c] Protein was determined by the automated method described by F. Leighton, B. Poole, H. Beaufay, M. Baudhuin, J. W. Coffey, S. Fowler, and C. de Duve, *J. Cell Biol.* **37**, 482 (1968).

[d] Percent contamination is calculated from the relative amounts of contaminant (c), and marker (m) enzyme present in the preparations, according the formula: (c/m) · 100.

TABLE II

PURITY OF SPECIFIC GRANULE PREPARATIONS FROM RABBIT POLYMORPHONUCLEAR
LEUKOCYTES OBTAINED BY ZONAL SEDIMENTATION, AND ZONAL
ISOPYCNIC EQUILIBRATION[a]

	Specific granule preparation			Percent contamination by	
Type of fractionation	Volume (ml)	Percent of total alkaline phosphatase	Percent of total protein	Myeloperoxidase	β-Acetylglucosaminidase
Zonal sedimentation	113.8 ± 11.3	59.6 ± 2.3	12.2 ± 1.4	6.1 ± 1.5	18.6 ± 2.7
Zonal isopycnic equilibration	—	54.8 ± 7.3	14.0 ± 0.5	14.2 ± 4.7	24.9 ± 7.1

[a] See Table I for explanations.

tions, by low-speed sedimentation experiments.[7] Agglutinated material may also be responsible for the somewhat high average protein content of the azurophil granule fraction obtained by sedimentation.

Isolation by Isopycnic Equilibration. As shown in Tables I and II, in isopycnic centrifugation the yield in azurophile and specific granules is slightly lower, and the cross contamination approximately twice as high as in sedimentation. Also somewhat higher is the contamination of the specific granules by C particles. Therefore, for preparative purposes, isopycnic equilibration appears to be less convenient than sedimentation.

[35] Isolation of Lysosomes from Lymphoid Tissues

By WILLIAM E. BOWERS

Two important points must be considered before one attempts to isolate lysosomes from lymphoid tissues: (1) Lymphoid tissues have a very heterogeneous cellular composition, and thus fractionation of whole tissue results in a preparation containing lysosomes from many cell types. (2) Although centrifugation methods have been most successful in isolating lysosomes from lymphoid tissues, the maximum relative specific activities attained for lysosomal enzymes are only 2–3 for rat spleen and 5–6 in the case of rat thymus.

Homogenization and Fractionation of Solid Lymphoid Tissues

Medium. A satisfactory homogenization medium for lymphoid tissues is $0.20 M$ KCl. The use of isotonic sucrose can result in considerable agglutination.

Homogenization

All operations are carried out at $0°$. The tissue to be homogenized is cut into small pieces 1–2 mm on a side; these are transferred to an all-glass Dounce homogenizer (Kontes Glass Company, Vineland, New Jersey). The volume of medium added is twice that of the tissue.

The loose-fitting A pestle is used first to prepare a crude homogenate. On the first downward stroke it is necessary to disrupt completely the tissue fragments. Homogenization is continued until the pestle can be moved freely through the homogenate; usually a total of 7–10 up-and-down strokes suffices.

The A pestle is removed and rinsed with another volume of homogenization medium. The final homogenate is prepared with three up-and-down strokes of the tight-fitting B pestle. Excessive homogenization only produces further rupture of lymphoid tissue lysosomes. In any case, 30–50% of the total lysosomal enzyme activity is released during the homogenization procedure.

The homogenate is poured into a graduate, and the B pestle and Dounce homogenizer are rinsed with one volume of homogenization medium. The homogenate is diluted to the desired volume, mixed, and then filtered through gauze three single layers thick to remove connective tissue debris.

Fractionation

Differential Centrifugation. The fractionation scheme devised by de Duve *et al.*[1] has been applied to rat lymphoid tissues. A 1:5 (w/v) homogenate in 0.2 M KCl is centrifuged at 1000 g for 10 minutes at 2° in a 15-ml conical centrifuge tube. The supernatant (postnuclear extract) is carefully removed with a Pasteur pipette and transferred to a graduate. Aliquots, 10 ml, of the supernatant are pipetted into high speed centrifuge tubes and spun at 20,000 g for 5 minutes at 2°. The pellet is resuspended in a total volume of 10 ml of 0.2 M KCl and centrifuged again at 20,000 g for 5 minutes at 2°. The pellet is then resuspended to the desired volume. This fraction corresponds to the combined M and L fractions in the scheme of de Duve *et al.,*[1] and it shows a relative specific activity for lysosomal enzymes of 2–3 for rat spleen[2] and 5–6 for rat thymus.[3]

Density Gradient Isopycnic Centrifugation. The procedure to be described has been worked out for the SW 39 swinging-bucket rotor of the Beckman Model L series of preparative ultracentrifuges, but the scheme can be suitably modified for other types of rotors.   ·

Whole homogenate or a postnuclear extract, 0.5 ml, is layered above 4 ml of a linear sucrose density gradient extending between density values of 1.10 and 1.30; the gradient itself rests on a cushion of 0.3 ml of a sucrose solution having a density of 1.35. KCl at a uniform concentration of 0.2 M is included throughout the gradient and underlying cushion.

[1] C. de Duve, B. C. Pressman, R. Gianetto, R. Wattiaux, and F. Appelmans, *Biochem. J.* **60**, 604 (1955).

[2] W. Bowers and C. de Duve, *J. Cell Biol.* **32**, 339 (1967).

[3] W. Bowers. Distribution of tissue proteinases in lymphoid tissues, *in* "Tissue Proteinases" (A. Barrett and J. Dingle, eds.), p. 221. North-Holland Publ., Amsterdam, 1971.

The tubes are centrifuged for 2.5 hours at 39,000 rpm at 2°. After centrifugation the tube contents are divided into 10–12 fractions by means of either a sectioning device or a collecting device which displaces fractions upward with fluorocarbon (FC-43, Fluorochemical, 3M Chemical Co., St. Paul, Minnesota).[4]

In the case of rat spleen, isopycnic centrifugation yields three distinct lysosomal populations having the following characteristics: (1) a major population with a modal density around 1.19. This population sequesters intravenously injected materials; available evidence suggests that it derives chiefly from macrophages, although other cell types undoubtedly contribute to the lysosomal enzyme activities found in this region of the gradient; (2) a population showing a modal density around 1.15. This population contains cathepsin D and a few other acid hydrolases, but it is poor in, or devoid of, many other lysosomal enzymes. The fact that it is not involved in segregating endocytosed material and that it disappears after corticosteroid treatment suggests that this population resides in lymphocytes; (3) a dense population appearing at a density of 1.30. Its significance is not clear at the present time.

Although the purification of rat spleen lysosomes by this method yields relative specific activities no higher than those obtained by differential centrifugation, it provides the possibility of preparing fractions relatively enriched in lysosomes from specific cell types.

Homogenization and Fractionation of Lymphocytes

Homogenization

Lymphocytes cannot be homogenized by the same procedure described for solid lymphoid tissues. It is necessary to expose the cells to hypotonic conditions in order to facilitate adequate cell breakage.

Lymphocytes are washed 3 times in Hanks' balanced salt solution by centrifuging the resuspended cells each time at 375 g for 10 minutes. After the last centrifugation, the supernatant is removed and the cell pellet is resuspended in 40 mM KCl–10 mM ethylenediaminetetraacetate (EDTA), pH 7, at a concentration of 150 to 300 × 10^6 cells per milliliter. Four minutes after suspending the cells in 40 mM KCl–10 mM EDTA, they are homogenized in an all-glass Dounce homogenizer (Kontes Glass Co., Vineland, New Jersey) with 30 up-and-down strokes of the tight-fitting B pestle. Isotonicity is restored by mixing the homogenate with an equal volume of 0.24 M KCl–10 mM EDTA, pH 7.

[4] F. Leighton, B. Poole, H. Beaufay, P. Baudhuin, J. Coffey, S. Fowler, and C. de Duve, *J. Cell Biol.* **37**, 482 (1968).

Fractionation

Differential Centrifugation. The homogenate is centrifuged at 650 g for 10 minutes, the supernatant is removed, and the pellet is rehomogenized according to the above procedures. After a second centrifugation at 650 g for 10 minutes, the supernatants are combined to form the postnuclear extract.

Ten-milliliter aliquots of the postnuclear extract are centrifuged at 100,000 g for 30 minutes to yield a high speed pellet that shows a relative specific activity of 3 for most lysosomal enzymes.

Density Gradient Isopycnic Centrifugation. Using conditions identical to those described for the isopycnic centrifugation of lymphoid tissue homogenates, a suitable volume of a postnuclear extract is layered over a sucrose gradient and centrifuged until the lysosomes attain their equilibrium density position. In the SW 39 rotor, 39,000 rpm for 2.5 hours suffices. The lysosomes equilibrate around a density of 1.18, and the enzymes display a relative specific activity of 5–6.[5]

[5] W. Bowers, *J. Exp. Med.* **136**, 1394 (1972).

[36] Isolation of Rat Liver Peroxisomes

By Pierre Baudhuin

Principle of the Method

In higher organisms, peroxisomes have so far been fully characterized biochemically in liver and kidney. Methods for purification of these organelles are based on their equilibrium density in a sucrose gradient; the most purified peroxisomal preparations have been obtained from rat liver using the technique described by Leighton *et al.*[1]

In the method developed by these authors, the fractionation of the homogenate by differential centrifugation yields first a pellet enriched in peroxisomes. In this preparation most of the protein, however, is still associated with mitochondria. Moreover, there is appreciable contamination by lysosomes and endoplasmic reticulum fragments.

Complete removal of mitochondria is the major difficulty for further purification by density equilibration. In the case of rat liver at least, this problem has been satisfactorily solved only by performing the isopycnic

[1] F. Leighton, B. Poole, H. Beaufay, P. Baudhuin, J. W. Coffey, S. Fowler, and C. de Duve, *J. Cell Biol.* **37**, 482 (1968).

centrifugation in the rotor designed by Beaufay.[2,3] As discussed in more detail below, this is probably due to the lower hydrostatic pressure generated during centrifugation in Beaufay's rotor, as compared to conventional rotors. However, since Beaufay's device is not yet commercially available, a conventional rotor with swinging buckets is used in the method described here, although the purification is significantly lower than with Beaufay's rotor and the amount of material processed is much smaller. For purposes of comparison, however, results obtained with the Beaufay equipment will be presented.

Contaminants other than mitochondria are also removed by the density equilibration procedure. Animals are treated with Triton WR-1339 to ensure separation of lysosomes from peroxisomes[4]; in normal liver considerable overlapping occurs between the density distribution of the two types of organelles. For fragments of the endoplasmic reticulum, contamination is considerably reduced by the density equilibration step.

Peroxisomes from kidney cannot be purified as extensively as in the case of liver; peroxisomal markers are purified only about 15 times with respect to the homogenate.[5] Moreover, separation of peroxisomes from lysosomes cannot be achieved in kidney, since captation of Triton WR-1339 is much less effective than in liver.[6] No data are presently available on the use of Beaufay's rotor for density equilibration of kidney peroxisomes.

Fractionation Procedure

Injection of Triton WR-1339 into the Animals

Female rats (200 g) are preferred because of the lower content of their liver in soluble catalase.[7] They are injected, either intravenously or intraperitonally, with Triton WR-1339 (Rohm and Haas, Philadelphia), at the dosage of 85 mg per 100 g body weight. The injection should be given at least 3.5 days and at most 8 days before sacrifice of the animals.

Preliminary Purification by Differential Centrifugation

Cytoplasmic Extract. When a Spinco SW 39 rotor (Beckman Instruments, Palo Alto, California) is used for the density equilibration, about

[2] H. Beaufay, "La centrifugation en gradient de densité," Thesis, University of Louvain. Ceuterick, Louvain, Belgium (1966).

[3] H. Beaufay, Spectra 2000, 1 (1973).

[4] R. Wattiaux, M. Wibo, and P. Baudhuin, Lysosomes Ciba Found. Symp. 1963, p. 176 (1963).

[5] P. Baudhuin, M. Müller, B. Poole, and C. de Duve, Biochem. Biophys. Res. Commun. 20, 53 (1965).

[6] S. Wattiaux-De Coninck, M.-J. Rutgeerts, and R. Wattiaux, Biochim. Biophys. Acta 105, 466 (1965).

[7] D. H. Adams and E. A. Burgess, Brit. J. Cancer 11, 310 (1957).

6 g of liver will be sufficient as starting material. Animals are fasted over-night and killed by decapitation, and the liver is quickly removed and immersed in a preweighed beaker containing ice-cold 0.25 M sucrose with 0.1% ethanol. Ethanol is included in the homogenization medium to prevent inactivation of catalase.[8] Homogenization is performed in 3 ml of 0.25 M sucrose with 0.1% ethanol per gram of liver, with a glass tube fitted with a Teflon pestle (A. H. Thomas, Co., Philadelphia) rotating at about 1000 rpm. The tube of the homogenizer is maintained in crushed ice during homogenization, and all subsequent steps are performed at 2°. A single up-and-down stroke is used for the homogenization, and the homogenate (± 25 ml) is centrifuged at 6500 g-min. With the centrifuge used in this laboratory (International refrigerated centrifuge, PR-J, rotor 253, radius at the meniscus = 17.5 cm, at the bottom 22.5 cm), this cor-responds to a centrifugation at 1700 rpm for 10 minutes.

The supernatant is poured off and the pellet is transferred to the homogenizer, where it is again homogenized in the sucrose ethanol medium (3 ml of medium per gram of liver), by a single up-and-down stroke of the pestle rotating at about 1000 rpm. A second centrifugation at 4500 g-min is then performed (10 minutes at 1400 rpm with the cen-trifuge mentioned above). The supernatant is combined with the previous one, and the pellet is again homogenized and centrifuged at 4500 g-min. The combined supernatants are brought to a volume of 10 ml per gram of liver by addition of the sucrose–ethanol medium. They will be desig-nated as the 1:10 cytoplasmic extract.

Large-Granule Fraction. Sixty milliliters of cytoplasmic extract are distributed in six tubes of the No. 40 Spinco rotor and centrifuged at 250,000 g-min. With the acceleration and deceleration of the L2 Spinco centrifuges used in our laboratory this corresponds to 6¾ minutes (ac-celeration time included) at 25,000 rpm. The supernatant is sucked off, with the loosely packed material above the pellets. The latter are resus-pended in about 2 ml of sucrose and combined quantitatively in 2 centrif-ugation tubes. The volume of fluid is adjusted to about 10 ml in each tube and the tubes are centrifuged again at 250,000 g-min. A second washing is eventually performed under the same conditions.

The washed pellets are resuspended by means of a loosely fitted Teflon ball rotating at a few hundred rpm at most. Alternatively, a homogenizer of the Dounce type (with loosely fitting pestle) may be used. The prepara-tion is brought to a volume equivalent to the initial liver weight and is used for density equilibration.

Light Mitochondrial Fraction. If a large yield is sought, the quantity

[8] B. Chance, *Biochem. J.* **46**, 387 (1950).

of material processed with the density equilibration in a SW 39 rotor can be increased approximately by a factor of 2 by preparing a light mitochondrial fraction[9] rather than the large-granule fraction described above. The centrifugation of the cytoplasmic extract is then first performed at 33,000 g-min (3 minutes at 12,500 rpm, with the No. 40 Spinco rotor), and the pellets are washed twice, combining the pellets of 3 tubes after the first centrifugation. The combined supernatants are then centrifuged at 250,000 g-min as described above. The light mitochondrial fraction is brought to a volume equal to 1 ml for 3 g of liver.

A method for preparing a similar fraction from larger quantities of liver (80–100 g) is described by Leighton et al.[1]

Isopycnic Centrifugation in a SW 39 Spinco Rotor

Preparation of the Gradient. A linear sucrose gradient is used, with addition of 0.1% ethanol and Dextran-10. Some observations suggest that this last substance may improve resolution of the gradient and preservation of peroxisomes.[1] In the experiment presented here as an example, the density limits of the gradient are, respectively, 1.12 and 1.28. For 100 g of light solution, 24.24 g of sucrose, 3.79 g of Dextran-10 are dissolved in 71.97 g of 0.1% ethanol in H_2O; for the heavy solution the corresponding values are 54.51 g, 2.275 g, and 56.78 g, respectively. The pH of the solutions should be checked and eventually adjusted with NaOH to neutrality.

Any kind of device may be used for forming the linear gradient, provided the total volume of the gradient can be as low as 4.8 ml (this supposes a mixing device with a very small volume) and that thorough mixing of the solutions is ensured, since the viscosity of the heavy solution is quite high. We have used the gradient maker described by de Duve et al.,[10] but a similar apparatus is commercially available (Beckman Instruments, Palo Alto, California). The gradient should be prepared at 2°, directly in one or more tubes of the Spinco SW 39 rotor.

The particulate fraction (0.8 ml per tube) is then carefully layered on the gradient and the rotor is spun at 39,000 rpm for 2.5 hours. Although it is less critical than for sedimentation analysis, the initial acceleration and the final deceleration should be progressive, and the rotor should be stabilized when started or stopped.

For collection of fractions, the tube used for centrifugation is sectioned. Collection by puncture of the bottom of the tube is not suitable here, since particles have a tendency to stick on the wall of the tube and

[9] C. de Duve, B. C. Pressman, R. Gianetto, R. Wattiaux, and F. Appelmans, *Biochem. J.* **60**, 604 (1955).

[10] C. de Duve, J. Berthet, and H. Beaufay, *Progr. Biophys. Biophys. Chem.* **9**, 325 (1959).

clog the needle. As tube sectioning device, the system described by de Duve *et al.*[10] is used in our laboratory; a tube slicer is presently commercially available (Beckman Instruments, Palo Alto, California). It is somewhat difficult to control the thickness of the slices with the latter equipment, and, since it does not have a cooling jacket, it has to be operated in a cold room.

Before sectioning, the position of the meniscus with respect to the top of the tube is accurately measured with a caliper. Up to 10 or 12 slices can be cut; their content is transferred in preweighed tubes and their thickness is measured with the caliper. If several tubes are processed, they are cut in identical fashion, and homologous fractions are pooled.

The number of slices, as well as the levels at which they are cut depends on the aim of the experiment. This point is discussed in more detail in the section dealing with the properties of peroxisomal fractions.

The density of each fraction is measured. This can be done by refractometry, but high concentration of particles may raise difficulties for the determination in some fractions. A cold, thermostatically controlled gradient of *O*-dichlorobenzene and light petroleum (bp 80–100°) can also be used.[11] A small droplet (10 μl) is allowed to settle in the gradient, and its position with respect to KBr solutions used as standards is measured with a cathetometer.

The weight of the fractions is then determined and an adequate dilution of each is made for enzyme assays. A solution containing Triton X-100 (0.1 g/liter), 1 mM EDTA and sodium bicarbonate (90 mg/liter) is convenient as a diluent.

Analysis of the Fractions Isolated

Biochemical Assays

In addition to the determination of peroxisomal reference enzymes, the complete characterization of the peroxisomal fraction would require the determination of marker enzymes for all possible contaminants. However, in view of the existing data on peroxisomal liver fractions, and more specifically in view of the observations made by Leighton *et al.,*[1] measurement of the distribution of catalase, acid phosphatase, and protein will provide most of the information necessary to assess the purity of the preparation.

It should be stressed that omission of the determination of markers for major contaminants rests on the assumption that the relative specific

[11] H. Beaufay, P. Jacques, P. Baudhuin, O. Z. Sellinger, J. Berthet, and C. de Duve, *Biochem. J.* **92,** 184 (1964).

activity of pure peroxisomes is approximately 40, with respect to the homogenate, as evaluated by Leighton *et al.*[1] Hence this simplification of the enzyme determinations is valid only for the fractionation of the liver of Triton WR-1339 injected female rats. It is implied that it can reasonably be assumed that the maximal relative specific activity for liver peroxisomes will not vary significantly according to the strain of rat used. Should any doubt arise as to the purity of the fraction isolated, determination of a mitochondrial (cytochrome oxidase) and a endoplasmic reticulum (glucose-6-phosphatase) marker enzyme should be included.

Catalase. For catalase determination, the residual H_2O_2 is measured, after incubation with the enzyme, using the yellow titanyl sulfate–H_2O_2 complex.[12] A solution (500 ml) containing bovine serum albumine (1 g/liter) is made in imidazole buffer, pH 7, 20 mM. An aliquot is set aside for blanks or dilutions of the samples, and H_2O_2 is added to the rest, which will be used for incubation with the enzyme. Addition of H_2O_2 is done so that 5 ml of the solution yields an optical density of about 0.75 at 405 nm, after addition of 3 ml of titanyl sulfate (2.25 g/liter) in 2 N sulfuric acid.

Add 0.1 ml of the enzyme preparation, suitably diluted, to 0.1 ml of Triton X-100 (20 g/liter). Incubation is performed at 0° and started by addition of 5 ml of substrate. After at most 10 minutes, the reaction is stopped by addition of 3 ml of titanyl sulfate. Since in the case of liver there is no significant contribution from the enzyme preparation to the blank, the latter is estimated by adding 0.1 ml of Triton X-100 (20 g/liter) to 0.1 ml of water, 5 ml of bovine serum albumin in imidazole buffer, and 3 ml of titanyl sulfate. The initial H_2O_2 concentration in the substrate is estimated by replacing the enzyme preparation by water.

The reaction follows first-order kinetics. One unit of enzyme is defined as the amount destroying 90% of the H_2O_2 present in a 50 ml reaction volume, in 1 minute.

Acid Phosphatase. The substrate used is sodium β-glycerophosphate (50 mM) in acetate buffer pH 5 (50 mM). The incubation mixture contains also thiomersalate (0.1 g/liter) to prevent bacterial contamination, since a long incubation time (20 hours)[13] is used. The total volume of the incubation mixture is 2 ml. Owing the long incubation time, the test can be performed on a small amount of material (0.2 ml of a dilution 1/1600 for a homogenate) and no deproteinization is necessary at the end of the incubation. The reaction is stopped by addition of acid ammonium molyb-

[12] P. Baudhuin, H. Beaufay, Y. Rahman-Li, O. Z. Sellinger, R. Wattiaux, P. Jacques, and C. de Duve, *Biochem. J.* **92**, 170 (1964).

[13] H. Beaufay, D. S. Bendall, P. Baudhuin, and C. de Duve, *Biochem. J.* **73**, 628 (1959).

date, and the phosphate determination is carried directly on the incubation mixture by addition of aminonaphtholsulfonic acid. Dilution in hypotonic medium and incubation at pH 5 are sufficient to rupture lysosomes. Triton X-100 is present only at low concentration in the diluted fractions, since, at higher concentration, it could interfere when the whole incubation medium is used for the phosphate determination. A common blank is obtained by replacing the enzyme by the diluent (Triton X-100, EDTA, bicarbonate). The unit of activity is defined as the amount of enzyme liberating 1 μmole of phosphate per minute.

Although other methods can be used, it should be emphasized that β-glycerophosphate should be used as substrate. Acid phosphatase determinations with phenylphosphate or phenylphosphate derivatives are much easier, but this substrate is also hydrolyzed by other liver enzymes, which are not lysosomal.[14]

Protein. Protein is measured according to the method of Lowry *et al.,*[15] using bovine serum albumin as standard.

Calculations

As will be clear from the examples given below, it is essential that both yield and specific activity are calculated for each fraction. Furthermore, the validity of the enzyme distributions should be checked by verifying that the sum of the enzyme content of all fractions is compatible with the amount present in the initial homogenate.

For the fractions isolated by differential centrifugation catalase activity and protein content are calculated per milliliter of the preparation and multiplied by the dilution factor to obtain activity per gram of liver. Since the homogenate is used completely for the separation of the nuclear fraction from the cytoplasmic extract, the recovery is computed with respect to the sum of activities found in those two fractions; this value will be referred to herein as the E + N value.

For the fractions of the isopycnic gradient, the dilution factor is first computed taking into account the dilution, the amount (in equivalent of grams of liver) layered on the gradient and the volume of each fraction. This last parameter is obtained by multiplying the cross section of the tube by the thickness measured for the corresponding slice. The calculations are then independent of losses that may have occurred during the cutting of the tube. Thus, if 0.8 ml of a large granule fraction 1:2 was layered

[14] H. Beaufay, *in* "Lysosomes, a Laboratory Handbook" (V. T. Dingle, ed.), p. 1. North-Holland Publ., Amsterdam, 1972.
[15] O. H. Lowry, N. J. Rosebrough, A. L. Farr, and R. J. Randall, *J. Biol. Chem.* **195**, 265 (1951).

and if a slice of 0.4 cm was cut in a tube of 1.25 cm^2 section, the corresponding fraction would be 1 : 1.25.

A difficulty is encountered in the evaluation of the volume for the top and bottom fractions. For the first one, the height of fluid is deduced from the slice thickness and from the position measured for the meniscus. For the bottom fraction, the sphericity of the tube is taken into account by subtracting one-third of the tube radius from the thickness measured with the caliper.

The recovery is then computed by adding the amount of enzyme or protein, per gram of liver processed, found in all fractions, including those which are not subjected to gradient analysis. The value obtained is compared to the E + N value. The yield is calculated in all fractions, the sum of activities recovered in all fractions being taken as 100%, rather than the E + N values. The same computation is performed for protein. With this procedure, distributions will be normalized for variations in recoveries.

Relative specific activities, with respect to the initial homogenate, are obtained by dividing the percent of enzyme in a fraction by the percent protein.

In order to have a graphical representation of distributions in the gradient, the relative concentration is plotted versus the cumulated volume, the latter being expressed in percent of the total volume. The relative concentration is defined as the concentration with respect to the concentration of the sample uniformly distributed in the tube. Otherwise stated, if the centrifuge tube had been thoroughly mixed after layering, the distribution of all components would be represented as rectangles of unit height and of base equal to 100%. With such representation for each fraction, the ordinate will correspond to the concentration of components, while the surface will be a measure of the yield.

Properties of the Purified Peroxisomal Fractions

Preparation Obtained with a Spinco SW 39 Rotor

Figure 1 summarizes the results of the isopycnic centrifugation on a large granule fraction, which contained respectively 73.8% of the catalase present in the homogenate, 36.4% of the acid phosphatase, and 25.8% of the protein. As already mentioned, in such preparations most of the protein is mitochondrial and the distribution of protein is thus very similar to that observed for cytochrome oxidase. In this experiment, the recoveries, with respect to the E + N value were 105% for protein, 95% for catalase, and 107% for acid phosphatase.

The distributions presented in Fig. 1 show clearly the concentration of peroxisomes in the lower part of the gradient; Table I gives the properties

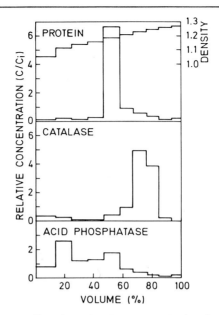

FIG. 1. Isopycnic centrifugation of a large granule fraction in a Spinco SW 39 rotor. In the graph representing the protein distribution, the upper line corresponds to the density of the fractions (right ordinate). Animals were injected with Triton WR-1339, 8 days before sacrifice.

TABLE I

PROPERTIES OF A PEROXISOMAL FRACTION OBTAINED WITH A SPINCO
SW 39 ROTOR[a]

	Percent of the homogenate	Relative specific activity
Protein	2.1	—
Catalase	56.7	27.0
Urate oxidase	56.2	26.8
D-Amino acid oxidase	40.6	19.3
L-α-Hydroxyacid oxidase	32.1	15.3
Cytochrome oxidase	0.9	0.4
Acid phosphatase	1.8	0.9

[a] The preparation here described was obtained by pooling the 2 fractions of the catalase peak in the gradient shown in Fig. 1. This corresponds to the part of the distribution between densities of 1.216 and 1.25. Data from P. Baudhuin, M. Müller, B. Poole, and C. de Duve, *Biochim. Biophys. Res. Commun.* **20**, 53, 1965.

of the preparation obtained by pooling the two fractions corresponding to the catalase peak. From the data of Leighton *et al.,*[1] the relative specific activity of catalase in pure peroxisomes is approximately 40. The fraction here described is thus only 67% pure.

The distribution of acid phosphatase in Fig. 1, and the low specific activity of the enzyme in the peroximal preparation (Table I) shows that good separation of peroxisomes from lysosomes was achieved and that the equilibrium density of the latter particles was adequately modified by Triton WR-1339. Since the amount of cellular protein associated with lysosomes is small, measurement of a marker for these particles is necessary, since their presence would not affect seriously the specific activity of catalase.

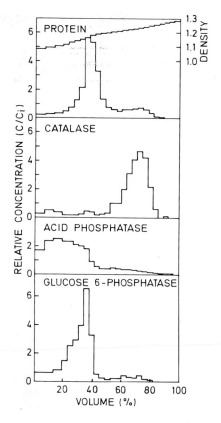

FIG. 2. Isopycnic centrifugation of a light mitochondrial fraction, using Beaufay's automatic rotor. In the graph representing the protein distribution, the upper line corresponds to the density of the fractions (right ordinate). Animals were injected with Triton WR-1339, 8 days before sacrifice.

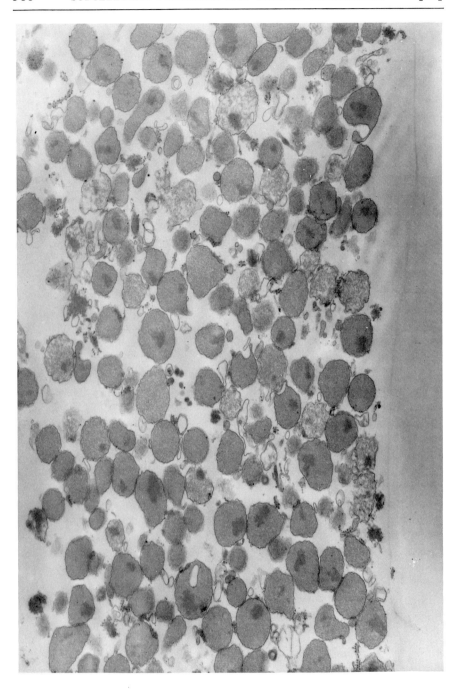

TABLE II

PROPERTIES OF A PEROXISOMAL FRACTION OBTAINED WITH BEAUFAY'S
AUTOMATIC ROTOR[a]

	Percent of the homogenate	Relative specific activity
Protein	0.47	—
Catalase	17.10	36.3
Urate oxidase	18.40	50.0
D-Amino acid oxidase	12.40	30.0
L-α-Hydroxy acid oxidase	15.90	35.8
Cytochrome oxidase	0.05	0.11
Acid phosphatase	0.13	0.27
Glucose-6-phosphatase	0.04	0.09

[a] The preparations were obtained by pooling the material which equilibrated below a density of 1.22 after isopycnic centrifugation. Data from F. Leighton, B. Poole, H. Beaufay, P. Baudhuin, J. W. Coffey, S. Fowler, and C. de Duve, *J. Cell Biol.* **37**, 482 (1968).

Values found for other peroxisomal enzymes or for other contaminants have also been listed in Table I in order to provide a complete description of the preparation. The lower specific activities observed for D-amino acid oxidase and for L-α-hydroxyacid oxidase are due to the occurrence of these enzymes in larger amount in the nuclear fraction or in the postmitochondrial supernatant.

Since the cumulated volume of the gradient is proportional to the height of the fluid in the tube, the graph presented in Fig. 1 can be used as a guide for the sectioning of the tube after isopycnic centrifugation. For example, if the interest is solely in obtaining purified peroxisomes, the upper part of the gradient, including the easily identifiable protein peak, may be collected in a first slice. Through the lower part, 3 or 4 slices (0.4 cm thick) can then be cut. If a greater number of fractions is made at the top of the gradient, preparations enriched in lysosomes or mitochondria can be obtained.

Comparison with a Preparation Obtained with Beaufay's Automatic Rotor

The distribution of catalase and of various markers obtained by equilibrating a light mitochondrial fraction on a sucrose gradient in Beaufay's

FIG. 3. Electron micrograph of a peroxisomal preparation isolated by isopycnic centrifugation using Beaufay's automatic rotor. This field shows clearly the abundance of peroxisomes, although some of them are damaged. Some fragments of rough endoplasmic reticulum can be seen; mitochondria and lysosomes are absent. ×22,000.

automatic rotor is shown in Fig. 2. Biochemical characteristics of the peroxisomal fraction isolated are summarized in Table II. The fractions were examined at the electron microscope and Fig. 3 shows the morphology of the preparation corresponding to the peak of catalase distribution.

Both the biochemical analysis and quantitative morphological analysis showed that more than 90% of the protein of such preparations is associated with peroxisomes. The original article of Leighton et al.[1] should be consulted for more detail.

As already mentioned, the better purification obtained here is essentially due to a more efficient separation of peroxisomes from mitochondria. As pointed out by Beaufay,[3] the distribution of mitochondria is more symmetrical with his rotor than with conventional rotors. At 39,000 rpm, in a swinging-bucket rotor, some mitochondria are altered by the high hydrostatic pressure and their equilibrium density is increased; hence overlapping between mitochondrial and peroxisomal distribution profiles is observed.

It is worthwhile mentioning here that other cell particles, and more specifically peroxisomes and lysosomes, are also sensitive to hydrostatic pressure (Wattiaux et al.[16]; Bronfman, unpublished results). For peroxisomes, this point deserves careful consideration, since they are damaged when conventional swinging-bucket rotors are used at 50,000 and 60,000 rpm,[16] and that even at 39,000 rpm some damage to these particles is not excluded. Recent experiments (Bronfman, unpublished results) show that in fact peroxisomes are much more sensitive to hydrostatic pressure than are lysosomes or mitochondria.

[16] R. Wattiaux, S. Wattiaux-De Coninck, and M. Collot, *Arch. Int. Physiol. Biochim.* **79**, 1050 (1971).

[37] Isolation of Nucleoids (Cores) from Peroxisomes

By HIDEYUKI TSUKADA

In contrast to the fact that cytoplasmic membraneous structures are solubilized readily by treatment with detergents, crystalloid nucleoids (crystalloid cores) of peroxisomes are resistant to detergents, retaining their ultrastructural details after the treatment.[1-4] Furthermore, urate oxidase,

[1] H. Tsukada, Y. Mochizuki, and S. Fujiwara, *J. Cell Biol.* **28**, 449 (1966).
[2] H. Tsukada, Y. Mochizuki, and T. Konishi, *J. Cell Biol.* **37**, 231 (1968).
[3] H. Tsukada, Y. Mochizuki, and M. Gotoh, *Symp. Cell Biol.* **21**, 263 (1970).
[4] H. Tsukada, S. Koyama, M. Gotoh, and H. Tadano, *J. Ultrastruct. Res.* **36**, 159 (1971).

the only enzyme in peroxisomes which is found to be firmly bound to the nucleoids so far as reported,[1-8] is neither released therefrom nor inactivated by treatment with detergents. In these aspects, the principle of isolation and purification of the nucleoids is based on the treatment of cytoplasmic particulate fractions rich in peroxisomes with detergents, using activity of urate oxidase as a parameter for assessing the magnitude of purity of the isolated nucleoids.

Isolation of Nucleoids from Rat Liver

Rats of both sexes weighing between 150 and 200 g are used. The animals are fasted overnight before use in order to reduce glycogen content of the liver, since glycogen particles are contaminants usually found in the isolated nucleoid fraction. Livers are perfused *in vivo* through the portal vein with 10–15 ml of cold Mg- and Ca-free PBS or saline solution under light anesthesia with ether.

The tissue homogenate, 10% (w/v), is prepared with an ice-cold 0.25 M sucrose solution containing 1 mM ethylenediamine tetraacetate. Homogenization is made with about 10 hand strokes in a glass homogenizer fitted with a Teflon pestle. After filtered through double layers of gauze, the homogenate is centrifuged at 700 g for 10 minutes to remove unbroken cells, nuclei, erythrocytes, and fragments of collagen fibers. The sediment is washed once with the sucrose solution of about half the volume of the original homogenate. The supernatant and the washing are combined and centrifuged at 12,000 g for 15 minutes. The sediment is washed once with the sucrose solution.

The sediment thus obtained from 10 ml of 10% homogenate is suspended in about 10 ml of 0.1% Triton X-100–0.25 M sucrose solution, stirred vigorously by pumping with the aid of a 2-ml syringe equipped with a long needle of about 20 gauge, and then centrifuged at 50,000 g for 15 minutes. The pH of the Triton X-100–sucrose solution is adjusted to 9.5–9.8 by adding with 1/100 N NaOH solution (about 2–3 ml per 100 ml of the detergent solution) immediately prior to use.

This Triton treatment is repeated three times; three treatments with 0.1% Triton X-100–sucrose solution are necessary and suffice for isolation of the nucleoids. Four or more treatments result in a decrease in the recovery of the nucleoids, particularly in the case of the isolation of the band-shaped nucleoids from mouse liver, probably as a result of mechanical breakdown of the nucleoids. Furthermore, the use of Triton X-100 in concentrations higher than 0.2% does not improve the isolation and puri-

[5] P. Baudhuin, H. Beaufay, and C. de Duve, *J. Cell Biol.* **26**, 219 (1965).
[6] Z. Hruban and H. Swift, *Science* **146**, 1316 (1966).
[7] C. de Duve and P. Baudhuin, *Physiol. Rev.* **46**, 323 (1966).
[8] F. Leighton, B. Poole, P. B. Lazarow, and C. de Duve, *J. Cell Biol.* **41**, 521 (1969).

fication of the nucleoids. The resulting sediment is suspended in 0.25 M sucrose solution or distilled water and centrifuged at 50,000 g for 15 minutes to remove the detergent. These operations are made at about 4°. The sediment thus obtained is the nucleoid fraction, showing a light grayish brown color. It is more or less contaminated with glycogen particles which can be eliminated by digestion with amylase.

The yield is about 0.05 mg as protein from 100 mg of wet liver tissues (25–28 mg on protein basis); the recovery of the nucleoids from the original homogenate is about 80% in terms of urate oxidase activity. Protein is determined according to the method of Lowry *et al.*,[9] with crystalline bovine serum albumin as the standard. Urate oxidase activity is measured according to Praetorius' method[10] with modification.[2,4] The specific activity of urate oxidase in the nucleoid fraction is increased about 380-fold on an average (300- to 500-fold) over the homogenate. Electron microscopic examinations of the thin sections of Epon-embedded osmium-fixed specimens (Fig. 1) and on the negatively stained dispersed preparations (Fig. 2) reveal that the purified nucleoid fraction consists of the crystalloid nucleoids of compact polytubular type (1:10 pattern) with the same fine structures as those observed in peroxisomes in hepatocytes and of a small number of glycogen particles.[1,2,6]

Nucleoid Fraction Isolated from Livers of Guinea Pigs and Mice

Nucleoids in hepatocyte peroxisomes of guinea pigs are crystalloid nucleoids of compact polytubular type (1:6 pattern), and those in mice are nucleoids of band-shaped type. According to the isolation procedure mentioned above, about 0.05 mg of protein is obtained as the nucleoid fraction from 100 mg of the wet tissues in both cases of guinea pigs and mice. The recovery of urate oxidase activity in the nucleoid fraction is about 80% in guinea pigs, while it is 40–50% in mice. This relatively low recovery in mice may be due to a breakdown and a subsequent loss of the band-shaped nucleoids during the repeated treatments with the detergent. The fine structure of the nucleoids isolated from guinea pig liver is the same as that of the nucleoids observed on the cells (Fig. 3); however, the nucleoids obtained from mouse liver are more or less fragmented as compared with their profiles in peroxisomes in hepatocytes (Figs. 4 and 5).[3,4] The specific activity of urate oxidase in the nucleoid fraction is raised about 300- to 500-fold in guinea pigs, and 200- to 300-fold in mice over the homogenate.

[9] O. H. Lowry, N. J. Rosebrough, A. L. Farr, and R. J. Randall, *J. Biol. Chem.* **193**, 265 (1951).
[10] E. Praetorius, *Biochim. Biophys. Acta* **2**, 590 (1948).

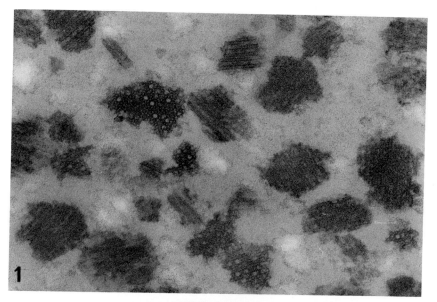

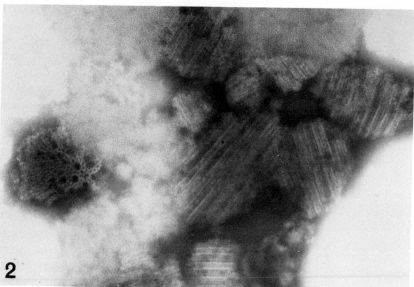

Fig. 1. Thin section of the nucleoids isolated from rat liver. Osmium-fixed nucleoid fraction is embedded in Epon 812. The section is stained doubly with uranyl and lead. Some nucleoids are sectioned longitudinally, and some are cut transversely. ×69,000.

Fig. 2. Negatively stained dispersed preparation of the nucleoids isolated from rat liver. A nucleoid seen at the left-hand side of this figure is observed on the surface falling at a right angle to its longitudinal axis, while other nucleoids are seen on their longitudinal surfaces. Glycogen particles are also observed. ×120,000.

371

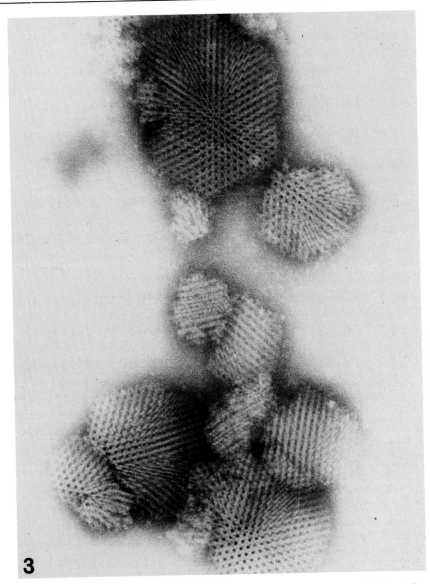

3

FIG. 3. Negatively stained dispersed preparation of the nucleoids isolated from guinea pig liver. The nucleoids are seen on their surfaces which fall at a right angle to the longitudinal axis, on their longitudinal and on their oblique surfaces. ×150,000.

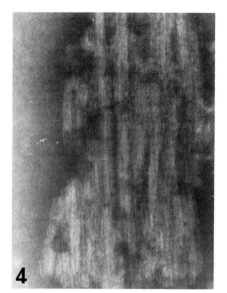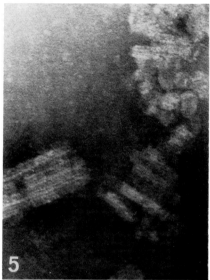

FIGS. 4 and 5. Negatively stained dispersed preparations of the nucleoids isolated from mouse liver. The nucleoids are seen on the surfaces parallel to the longitudinal axis of the unit tubules. Fragmentation of the nucleoids is noted at the upper-right corner of Fig. 5. ×120,000.

Selection of Detergents[11]

Among several detergents tested, Triton X-100 is most satisfactory for the isolation and purification of the nucleoids; 0.5% sodium deoxycholate is less active than 0.1% Triton X-100, and 1% Tween 20 is far less active. Sodium lauryl sulfate must not be used, since it inactivates urate oxidase completely in concentrations capable of yielding the nucleoid fraction containing protein of comparable amounts to that isolated with 0.1% Triton X-100.

Effects of pH of Triton X-100-Sucrose Solution[11]

Notwithstanding that Triton X-100 is a nonionic detergent, the effectiveness in isolation of the nucleoids depends on pH of the isolation medium. Since borate does not inhibit urate oxidase activity, 0.1% Triton X-100–0.25 M sucrose–2.5 mM borate buffer solution is used as the isolation medium for examining the effects of pH of the medium on purity of the isolated nucleoid fraction. The specific activity of urate oxidase in the isolated specimens decreases as the pH of the medium is lowered at the pH range between 8.5 and 7.2. The pH between 9.0 and 9.5 is most effec-

[11] H. Tsukada, Y. Mochizuki, and M. Gotoh, to be published (1973).

tive in this case. The specific activity of urate oxidase in the nucleoid fraction increases more than 300-fold over the homogenate, when the fraction is isolated with medium of pH 9.0–9.5, whereas it increases only about 100-fold when medium of pH 7.2 is used. The concentrations of borate buffer higher than 3 mM are not adequate for purification of the nucleoids. When NaOH solution is used for adjusting the pH of the isolation medium, the pH falls more rapidly during the treatment than when borate buffer is used. Therefore, a relatively high initial pH is required in this case for successful purification. However, 2.5 mM borate buffer is not necessarily a concentration sufficient to keep the pH of the isolation medium during the treatment. A pH level of the medium higher than 10 results in a considerable decrease in the recovery of the nucleoid fraction in terms of urate oxidase activity as well as of protein.

[38] Fractionation of Islet Tissue in Processing Proinsulin to Insulin

By WOLFGANG KEMMLER and DONALD F. STEINER

Isolation of Islets of Langerhans from Rat Pancreas

This procedure is based on a method originally devised by Moskalewski[1] and later modified slightly by Lacy and Kostianovsky.[2] Two fed normal rats weighing 150–250 g are anesthetized with Nembutal and exsanguinated; the entire pancreas is excised and cleared of adherent fat and connective tissue with the aid of a dissecting microscope. The pancreas is quickly and finely minced with a small scissors, and the bits of tissue are suspended in 8 ml of Hanks' salt solution buffered with bicarbonate[3] containing per milliliter 1 mg of bovine serum albumin (BSA), 3 mg of glucose, and 5–10 mg of crude collagenase or 1–2 mg of purified collagenase.[4] The amount of collagenase needed varies with each batch, and some lots are better than others for unknown reasons. The suspension is transferred to a 20-ml Erlenmeyer flask fitted with a large stirring bar. The flask is tightly stoppered after gassing with 95% O_2–5% CO_2 for 20 seconds and

[1] S. Moskalewski, *Gen. Comp. Endocrinol.* **5**, 342 (1965).
[2] P. E. Lacy and M. Kostianovsky, *Diabetes* **16**, 35 (1967).
[3] J. H. Hanks and R. E. Wallace, *Proc. Soc. Exp. Biol. Med.* **71**, 196 (1949).
[4] Worthington Biochem. Corporation (Freehold, New Jersey). Since various batches of crude collagenase vary in their efficacy in producing good islet preparations (yield and freedom from contamination), it is useful to purchase small amounts of several lots and compare them in use before buying a larger supply.

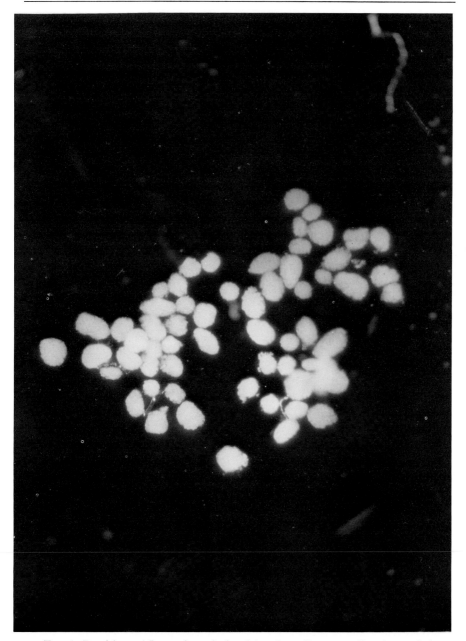

FIG. 1. Rat islets of Langerhans isolated by the collagenase digestion technique. ×30.

is placed over a magnetic stirrer unit[5] submerged in a 37° water bath and rapidly stirred with declining speed over a period of 15–30 minutes. Individual preparations of pancreas and different lots of collagenase vary in the time required for dispersion of the acinar pancreas. Each digest must be monitored and the operator must judge from the texture of the suspension when digestion of the acinar tissue is complete; considerable experience is required to determine this end point. Insufficient digestion yields fewer islets contaminated by acinar remnants whereas excessive digestion may yield partially digested islets (moth-eaten appearance). Normal islets are shown in Fig. 1.

After incubation the digest is diluted 5-fold in a 50-ml graduated cylinder with cold Hanks' medium and is allowed to stand on ice for about 3 minutes. The larger islets settle out during this time, and the supernatant suspension of acinar cells is carefully drawn off and discarded. The residue is resuspended with fresh buffer solution, and the islets are allowed to sediment for about 3 minutes before decanting the supernatant suspension. This operation is repeated 5 times. Small aliquots of the sediment are placed in plastic petri dishes under a dissecting microscope and diluted with Hanks' solution. The free islets are drawn up into Lang-Levy micropipettes and transferred to a drop of cold medium. If free acinar cells are still present at this stage the islets can be picked up individually and transferred again to a fresh drop of fluid. Although somewhat tedious and time consuming, this method yields islets that have only a few percent contamination with exocrine cells (Fig. 1). From 100 to 500 islets (about 1–2 mg of wet tissue) usually can be obtained by this method. Total time required is about 3 hours (2 persons).

Incubation of Isolated Islets

Freshly isolated rat islets are transferred to 50 or 100 μl of modified Hanks' medium, pH 7.4, containing all the essential amino acids in the same amounts as in Eagle's basal medium.[6] The amino acid to be used as labeled precursor (usually leucine) is omitted from the medium and 5–20 μCi or ^{3}H-amino acid (dissolved in medium) is added to the small incubate and thoroughly mixed by blowing obliquely against the fluid drop from a micropipette. The incubation can be carried out in a small plastic culture dish or in a small tapered centrifuge tube in a heated enclosure or bath at 37° under 95% O_2/5% CO_2. To incorporate labeled proinsulin into Golgi structures and secretion granules, the islets are preincubated for 10 minutes and then pulse-labeled with ^{3}H-labeled amino acid for 30 min-

[5] Tri R, MS-7 Stirrer (Tri R Instruments, Inc., 48 Merrick Road, Rockville Centre, New York, 11570).
[6] H. Eagle, *Science* 122, 501 (1955).

utes.[7,8] The labeled islets are washed several times with medium containing 100-fold excess of unlabeled leucine and then incubated 15 minutes at 37° in this medium.

Fractionation of Islets[9]

Prelabeled or unlabeled islets are homogenized manually in a Duall homogenizer, size 20, with Teflon pestle (Kontes Ltd., Vineland, New Jersey) in 0.3 ml of buffer containing 0.25 sucrose, 1% Ficoll (Pharmacia Co., Uppsala, Sweden), 2% BSA, and 10 mM potassium glycerophosphate, pH 6.0. The homogenate is centrifuged at 800 g for 10 minutes to sediment nuclei, debris, and intact cells. The sediment is rehomogenized in 0.2 ml of the same buffer and centrifuged. The combined supernatants are then transferred to a $3/16 \times 1\frac{5}{8}$-inch cellulose nitrate ultracentrifuge tube containing a lower layer of 0.1 ml of 2.0 M sucrose, 5% Ficoll, 2% BSA, 10 mM potassium glycerophosphate, pH 6.0, and an intermediate layer of 0.15 ml of 0.6 M sucrose, 1% Ficoll, 2% BSA, and 10 mM potassium glycerophosphate, pH 6.0. After centrifugation at 5×10^5 g-min a visible layer of subcellular particles can be seen on the top of the 2.0 M sucrose layer. This zone is transferred to a small tapered centrifuge tube coated with silicone (Siliclad, Clay Adams, Parsippany, New Jersey) containing 0.5 ml of the following buffer: 160 mM KCl, 5 mM NaCl, 5 mM MgCl$_2$, 1 mM CaCl$_2$, 4 mM cysteine, 2% BSA, 10 mM potassium glycerophosphate, pH 6.3. During incubation at 37° of this suspension for 1–5 hours transformation of proinsulin to insulin occurs at rates initially similar to those observed *in vivo*, i.e., half-time about 1 hour.[10] After the incubation, 1 mg of carrier insulin is added and sufficient glacial acetic acid to give a final concentration of 3 M. Any precipitate is centrifuged off and discarded. Proinsulin and insulin are separated by gel filtration of the acid extract over a 1×50 cm column of Biogel P-30 (Bio-Rad Laboratories, Richmond, California) equilibrated with 3 M acetic acid at a rate of 10–20 ml/hour. The fractions (1.2–1.5 ml) are collected in siliconized or albumin-coated[11] tubes to minimize absorption of radioactive proteins to the glass. Radioactivity is determined in a liquid scintillation counter using

[7] D. F. Steiner, J. L. Clark, C. Nolan, A. H. Rubenstein, E. Margoliash, F. Melani, and P. E. Oyer, *in* "The Pathogenesis of Diabetes Mellitus. Proceedings of the Thirteenth Nobel Symposium" (E. Cerasi and R. Luft, eds.), p. 57. Almqvist & Wiksell, Stockholm, 1970.

[8] S. L. Howell, M. Kostianovsky, and P. E. Lacy, *J. Cell Biol.* **42**, 695 (1969).

[9] W. Kemmler and D. F. Steiner, *Biochem. Biophys. Res. Commun.* **41**, 1223 (1970).

[10] D. F. Steiner, *Trans. N.Y. Acad. Sci.* **30**, 60 (1967).

[11] D. F. Steiner, D. D. Cunningham, L. Spigelman, and B. Aten, *Science* **157**, 697 (1967).

a Triton-toluene scintillation fluid made up as follows: Triton X-100, 875 ml; POPOP, 175 mg; PPO, 14 g; volume to 3.5 liters with toluene.

Comments

The crude particulate fraction separated in the above procedure contains approximately 65% of the total radioactive proinsulin and insulin of the postnuclear supernatant fraction. By electron microscopy this fraction is seen to contain large numbers of intact secretion granules, as well as some vesicles of rough and smooth endoplasmic reticulum, mitochondria, and presumably lysosomes, but not intact cells. Studies in other laboratories have shown that isolated beta granules are most stable at pH 6.0 in media of low ionic strength.[12-14] Labeled proinsulin added to this system during incubation is not converted to insulin, indicating that the conversion process is localized within the particulate elements, presumably mainly in newly formed secretion granules or in Golgi vesicles. Repeated freezing and thawing, addition of detergents such as 1% Triton X-100 or 0.2% sodium deoxycholate, or sonication all strongly inhibited the conversion, in keeping with the above conclusion.[9]

Further fractionation of this crude granule fraction by ultrafiltration or by zonal and isopycnic density gradient centrifugation have been only partially successful,[12-15] because of the similar density and sedimentation characteristics of beta granules, rough endoplasmic reticulum vesicles and other subcellular organelles. Separation of lysosomes from beta granules by means of phase distribution in polyethylene glycol and dextran systems buffered with lithum phosphate has been reported.[13] Using repeated freeze thawing as a method of lysis of the granule fraction, it has been possible to show the presence of weak trypsinlike as well as carboxypeptidase B-like activity in these preparations.[15,16] The very low residual tryptic activity and the marked inhibition of conversion produced by lysis of the granule fractions suggests that this activity may be membrane bound. More precise localization of the conversion mechanism and further characterization of these proteolytic activities will require the development of more adequate islet fractionation techniques.

[12] A. W. Lindall, Jr., G. E. Bauer, P. K. Dixit, and A. Lazarow. *J. Cell Biol.* **19,** 317 (1963).

[13] H. G. Coore, B. Hellman, E. Pihl, and I.-B. Täljedal, *Biochem. J.* **111,** 107 (1969).

[14] S. L. Howell, C. J. Fink, and P. E. Lacy. *J. Cell Biol.* **41,** 154 (1969).

[15] W. Kemmler, D. F. Steiner, and J. Borg, *J. Biol. Chem.* **248,** 4544 (1973).

[16] W. Kemmler, J. D. Peterson, and D. F. Steiner, *J. Biol. Chem.* **246,** 6786 (1971).

[39] Adrenal Chromaffin Granules: Isolation and Disassembly

By S. F. BARTLETT and A. D. SMITH

The catecholamines of the adrenal medulla are stored in a specific cell particle, the chromaffin granule. This granule also stores the other secretory products of the gland: adenine nucleotides and chromogranins, a group of soluble proteins. In electron micrographs, chromaffin granules appear as spherical, membrane-limited particles containing a dense core. The diameter of the whole particle ranges from 150 nm to 350 nm (ox adrenal).

Isolation of Chromaffin Granules

Between 20 and 60 bovine adrenal glands are obtained from the abattoir, where they are placed on ice within 20 minutes of death. On arrival at the laboratory the glands are dissected quickly at room temperature, to remove the cortex. The medullas are stored in ice-cold 0.3 M sucrose solution until all the glands have been dissected (1–2 hours). The glands vary greatly in size, according to the age and sex of the animal, but on average each gland weighs 1.8 g.

About 4–6 medullas at a time are finely chopped on a Perspex board, using a kitchen knife. The chopping must be continued until the tissue is well mashed and has the consistency of a fine mince. The mince is placed in the glass mortar of a Potter-Elvehjem homogenizer (Kontes Glass Company, Vineland, New Jersey), and ice-cold 0.3 M sucrose solution is added until a weight-to-volume ratio of approximately 1:5 is reached. The suspension is then homogenized with a Teflon pestle (clearance from mortar 0.08 mm) that is rotated at 950 rpm by a powerful (675 W) electric motor. The end of the stainless steel shaft of the pestle should be covered with a piece of thick-walled rubber tubing before it is clamped in the chuck of the motor. Homogenization is continued until the pestle has been passed up and down five times.

Differential Centrifugation. (The value of g in this and subsequent sections is calculated using the radius from the middle of the tube to the center of rotation.)

The homogenate is centrifuged at 2° at low speed (380 g) for 20 minutes. The low speed supernatant is carefully sucked off, and 25-ml aliquots are placed in polyallomer tubes and centrifuged in the A 30 rotor of a Spinco ultracentrifuge (Model L or L2-65B) at 2° at 10,000 rpm (8720 g) for 20 minutes. The supernatant (initial medium-speed supernatant) is carefully sucked off, without disturbing the sediment, and kept ice cold. Then 2 ml of 0.3 M sucrose solution are added to each centrifuge tube and the sediment is resuspended by sucking up and down into the pipette. When the sediment is resuspended, 18 ml of 0.3 M sucrose solution are added to each tube with stirring. This suspension is recentrifuged at 10,000 rpm (8720 g) for 20 minutes in the A 30 rotor of the ultracentrifuge, as before. The resulting supernatant is once again gently sucked off with a pipette, until the "fluffy layer" on the surface of the sediment is reached. The fluffy layer is removed by decantation, the tube being inverted for no more than 2 seconds.

The supernatant so obtained is combined with the initial medium-speed supernatant and recentrifuged at 35,000 rpm (80,790 g) for 60 minutes in the A 40 rotor of the ultracentrifuge at 2°. This gives a sediment (the microsomal fraction) and a final supernatant. The microsomal fraction contains small chromaffin granules, Golgi membranes, and rough endoplasmic reticulum.[1] The final supernatant is a convenient, partly-purified source for the preparation of the enzyme phenylethanolamine N-methyl transferase.[2]

Each pellet resulting from the second medium-speed centrifugation is resuspended with 2 ml of 0.3 M sucrose solution as described above, but this time care should be taken to exclude the residual erythrocytes that lie at the bottom of the pellet. This suspension is called the "large granule fraction" and can be used either as a source of impure chromaffin granules for studies in isoosmotic media or as a starting material for the preparation of highly purified chromaffin granules, as described below. The large granule fraction contains 62% of the catecholamines in the low-speed supernatant, 25% of the activity of the succinate tetrazolium reductase, and 45% of the acid deoxyribonuclease, but only 8.5% of the glucose-6-phosphatase activity and 4.8% of the 5-nucleotidase activity. This clearly shows that the large granule fraction is still appreciably contaminated by mitochondria and lysosomes but contains only a relatively small proportion of the total microsomal elements.[3]

Centrifugation through Hyperosmotic Sucrose Solution. In order to isolate chromaffin granules from the large granule fraction, use is made of the observation that chromaffin granules are recovered as a sediment when the large-granule fraction is centrifuged through hyperosmotic sucrose

[1] F. Dubois, I. Benedeczky, S. F. Bartlett, and A. D. Smith, unpublished observations, 1970.
[2] R. J. Connett and N. Kirshner, *J. Biol. Chem.* **245**, 329 (1970); see also this series, Vol. 17B [243a].
[3] F. Dubois, B.Sc. Thesis, University of Oxford, 1970.

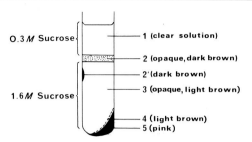

0.3 M Sucrose — 1 (clear solution)

— 2 (opaque, dark brown)

— 2' (dark brown)

1.6 M Sucrose — 3 (opaque, light brown)

— 4 (light brown)
— 5 (pink)

FIG. 1. Diagram representing the appearance of the centrifuge tube after centrifugation of resuspended large granule fraction from adrenal medulla on a 1.6 M solution of sucrose. The large-granule fraction, suspended in 0.3 M sucrose, is layered on a solution of 1.6 M sucrose. After centrifugation at 80,790 g for 1 hour in the A 40 head of the ultracentrifuge, the tube had the appearance shown in the diagram.

solution.[4] Polyallomer tubes and the A 40 rotor of the ultracentrifuge are used for this step. Of the ice-cold 1.6 M sucrose solution, 7.5 ml are placed in each tube and no more than 3 ml of the resuspended large-granule fraction are layered gently onto the dense solution in each tube. After centrifugation at 2° for 60 minutes at 35,000 rpm (80,790 g) several different layers can be distinguished within the tube. These are numbered 1–5 in Fig. 1. The total supernatant is decanted, and the inside of the tube is wiped with paper tissues; 1 ml of 1.6 M sucrose solution is added to the tube, and the loosely adhering layer (4) is resuspended by shaking the tube slightly, the supernatant being decanted and rejected. The washing procedure should be repeated once more. A pink sediment (5) of chromaffin granules is then left at the bottom of the tube. This pink pellet contains 77% of the catecholamine present in the large granule fraction, but only 2% of the fumarase activity, 6.5% of the acid deoxyribonuclease activity, and 11% of the glucose-6-phosphatase activity. The pink sediment is, therefore, a highly purified preparation of chromaffin granules, only very slightly contaminated by mitochondria and lysosomes.

Modifications. The above method gives a pellet containing a mixture of epinephrine-containing and norepinephrine-containing chromaffin granules. The proportions of epinephrine to norepinephrine will vary according to the species. In the pig adrenal medulla, 51% of the catecholamines are norepinephrine and this tissue is a good starting material for the isolation of a fraction enriched in norepinephrine-containing chromaffin granules. This can be achieved by replacing the 1.6 M sucrose solution

[4] H. Blaschko, G. V. R. Born, A. D'Iorio, and N. R. Eade, *J. Physiol.* (*London*) **133**, 548 (1956); A. D. Smith and H. Winkler, *Biochem. J.* **103**, 480 (1967).

in the final step with a 2.05 M sucrose solution: a pellet is obtained in which norepinephrine comprises 88% of the catecholamines.[5]

In order to isolate purified chromaffin granules without exposing them to hyperosmotic solutions, the 1.6 M sucrose solution can be replaced by a solution containing 0.27 M sucrose, 19.5% (w/v) Ficoll in D_2O.[6]

Disassembly of Chromaffin Granules

Exposure of chromaffin granules to hypoosmotic shock, accompanied by slow freezing and thawing, can be used to disrupt the particles, producing soluble and insoluble fractions. These fractions are assumed to correspond to the contents and to the membranes of the chromaffin granules, respectively.

Lysis Procedure. Catecholamines are rapidly oxidized in aqueous solutions above pH 7 and so a buffer of pH 5.9 is used to lyse the chromaffin granules. A suitable buffer can be made by dissolving 6.057 g of Tris and 5.905 g of succinic acid in water, titrating the solution to pH 5.9 with sodium hydroxide solution and making up to 1 liter. A 10-fold dilution of this stock solution gives a buffer which is 5 mM Tris and succinate, and this diluted buffer is used to lyse the chromaffin granules.

Each pellet of highly purified chromaffin granules is suspended with 5 ml of (diluted) buffer, using a pipette. This suspension is frozen in a deep-freeze at $-20°$ and then thawed by standing the centrifuge tubes in water of room temperature. After the suspension has thawed, it is centrifuged at 30,000 rpm (59,360 g) in the A 40 rotor of the ultracentrifuge for 60 minutes. The supernatant (soluble lysate) is decanted and the sediment is resuspended, frozen at $-20°$ for 60 minutes, thawed and centrifuged again, as above. This procedure should be repeated a further three times, in order to remove the bulk of the soluble components of the chromaffin granules.[7]

The insoluble residue obtained in this way is termed "chromaffin granule membranes" and was found to consist of empty sacs and sheets of membrane when examined in the electron microscope. Chemical analysis shows that this membrane residue contains 22% of the total protein, and virtually all the cholesterol and phospholipids of the original chromaffin granules. The soluble lysate contains the bulk of the soluble protein (chromogranins), nucleotides, and catecholamines.

Proteins of the Soluble Lysate. The proteins present in the soluble

[5] H. Winkler, *Naunyn-Schmiedebergs Arch. Exp. Pathol. Pharmakol.* **263**, 340 (1969).

[6] J. M. Trifaró and J. Dworkind, *Anal. Biochem.* **34**, 403 (1970).

[7] H. Winkler, H. Hörtnagl, H. Hörtnagl, and A. D. Smith, *Biochem. J.* **118**, 303 (1970).

lysate are (i) the copper-containing enzyme dopamine β-hydroxylase and (ii) a family of related proteins whose amino acid compositions show a characteristically high ratio of acidic to basic residues.[8] The name "chromogranins" has been used to denote the total soluble proteins from chromaffin granules, including dopamine β-hydroxylase, but it will be used in subsequent sections to mean the proteins of the second category (above) only. While chromogranin A, which makes up 38% of the soluble protein, has been purified and characterized, little is known about the other chromogranins. The effective hydrodynamic radius of chromogranin A increases markedly with decreasing ionic strength, and use is made of this effect in its purification by gel filtration.

Analytical Polyacrylamide Gel Electrophoresis of Soluble Lysate Proteins. The proteins of the soluble lysate are well separated from one another by electrophoresis in 6% (w/v) polyacrylamide gels, using the Tris–glycine buffers described by Clarke,[9] and this technique is used throughout the purification procedures as a means of establishing the identity and homogeneity of individual proteins.

The following solutions are used to make the polyacrylamide gels: (1) 0.28% (v/v) aqueous *N,N,N',N'*-tetramethylethylenediamine (TEMED); (2) 0.14% (w/v) aqueous ammonium persulfate; (3) 24% (w/v) solution of Cyanogum 41 (BDH Chemicals Limited, Poole, Dorset, BH12 4NN, U.K.) in distilled water (Cyanogum 41 is a mixture of acrylamide and *N,N'*-methylenebisacrylamide); (4) Tris–glycine buffer ("gel buffer"), pH 8.6, made by dissolving 29.0 g of glycine and 6.0 g of Tris in distilled water and making up to 980 ml. The first three solutions must be made up fresh, but the buffer may be kept as a stock solution under refrigeration.

The Cyanogum, buffer, and TEMED solutions are mixed in the volume ratios 2:1:1 in a small beaker with a magnetic stirring bar, and 4 volumes of the ammonium persulfate solution are added. Since this initiates the polymerization reaction, all subsequent operations should be carried out as rapidly as possible. The polymerization mixture (1.2 ml per tube) is pipetted into precision-bore glass running tubes (75 mm × 6 mm internal diameter) and very gently overlaid with a layer of distilled water, to give a flat upper surface to the gel. The gel solution becomes slightly opalescent on polymerization, which generally occurs within 20–30 minutes after addition of persulfate to the polymerization mixture.

The newly made gels are stood overnight before use, to allow the polymerization to go to completion. They are then mounted in the electrophoresis apparatus and prerun at a current of 1 mA per gel until the

[8] A. D. Smith and H. Winkler, *Biochem. J.* **103**, 483 (1967).
[9] J. T. Clarke, *Ann. N.Y. Acad. Sci.* **121**, 428 (1964).

voltage stabilizes. (Stock gel buffer diluted 1 in 8 (v/v) with distilled water is used as the electrolyte.) This prerun removes any unreacted persulfate ions remaining after polymerization. After the prerun, the electrolyte is discarded and the outside of each gel tube is wiped dry.

Samples of high ionic strength, such as fractions from ion-exchange chromatography, should be dialyzed against dilute buffer (e.g., 5 mM Tris–sodium succinate buffer—see above) before electrophoresis. All samples are made dense by addition of a small amount of solid sucrose. After drying the inside of the running tube and the upper surface of the gel, a small amount (2–5 μl) of bromophenol blue "tracker dye" solution and the sample are pipetted onto the upper surface of the gel. (The volume of the sample is not critical, although it should not normally exceed 100 μl, the optimum being 20–50 μl. A protein concentration of approximately 1 mg/ml has been found to give satisfactory patterns of stained bands with samples of soluble lysate.) The tube is then remounted in the electrophoresis apparatus, and the sample is carefully overlaid with the electrolyte, made as follows: 29.0 g of glycine and 6.0 g of Tris are dissolved in distilled water, 5 ml of 1 M hydrochloric acid are added, and the solution is made up to 980 ml. A 1 in 10 (v/v) dilution of this stock solution is used as electrolyte.

When all the samples are in place and overlayered with electrolyte, the buffer reservoirs are filled and electrophoresis is begun, at an initial current of 0.5 mA per gel. The tracker dye in the samples soon concentrates to a narrow band and when this band has entered the gel in all of the tubes (usually 5–10 minutes after starting) the current is increased to 1 mA per gel and held constant for the rest of the run. When the dye has migrated to within about 5 mm of the end of the gel, the current is switched off and the gels are removed from their running tubes (this is done by "rimming" with a long needle mounted on a syringe full of water—turning the gel while expelling the water from the syringe) and immersed for 1 to 2 hours in a 0.1% (w/v) solution of Amido Black dye in 7% acetic acid. The stained gels are leached of excess dye, using 7% (v/v) acetic acid, until the protein bands appear against an unstained background. Figure 2 shows the pattern characteristic of the soluble lysate proteins. The diffuse tail on the chromogranin A band is seen only at higher protein loads.

Preparation of Chromogranin A and Soluble Dopamine β-Hydroxylase. Gel filtration on a column (2.2 cm × 90 cm) of Sephadex G-200, equilibrated with 5 mM Tris–sodium succinate buffer pH 5.9 and 0.02% (w/v) in sodium azide, is used to prepare a fraction containing chromogranin A and dopamine β-hydroxylase, free from the other soluble components. The soluble lysate of the chromaffin granules is dialysed against

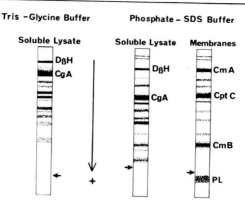

FIG. 2. Polyacrylamide gel electrophoresis of soluble and membrane proteins of bovine adrenal chromaffin granules. The total acrylamide concentration in all gels was 6% (w/v), samples were applied to the top of each gel and migration was toward the anode, which was at the bottom. Abbreviations: DβH, dopamine β-hydroxylase; CgA, chromogranin A; CmA, chromomembrin A; CmB, chromomembrin B; CptC, component C; PL, phospholipid. The arrows mark the migration positions of the bromophenol blue tracker dye bands at the end of the electrophoretic run.

at least three changes of 5 mM Tris–sodium succinate, pH 5.9, containing sodium azide (0.02% w/v), and ultrafiltered to a volume of about 5 ml before application to the Sephadex column. Ultrafiltration must not be allowed to proceed to the point where the protein solution becomes a viscous gel (at a protein concentration of approximately 30 mg/ml). Dialysis, ultrafiltration, and gel filtration should all be carried out at 2°.

The dialyzed, ultrafiltered soluble lysate is applied to the Sephadex G-200 column and eluted by pumping through 5 mM Tris–sodium succinate buffer, pH 5.9, 0.02% (w/v) in sodium azide, at a flow rate of 25 ml/hour. Chromogranin A and dopamine β-hydroxylase are eluted close together as a single UV-absorbing peak, in the region of the excluded volume of the column. Since sodium azide (added to inhibit the growth of bacteria) is a potent inhibitor of dopamine β-hydroxylase, it is necessary to add extra copper, in the form of copper sulfate solution, if it is desired to estimate the enzyme in column fractions. Aliquots of selected fractions are analyzed by polyacrylamide gel electrophoresis, using Tris–glycine buffers, to determine in which fraction contamination by other chromogranins first becomes detectable. All earlier fractions containing UV-absorbing material are then pooled and ion-exchange chromatography on DEAE-Sephadex A25 resin is used to resolve this mixture into its component proteins.

Two to three grams of DEAE-Sephadex A25 ion-exchange resin (Pharmacia Fine Chemicals A.B., Uppsala, Sweden) are swollen in distilled water and converted to the chloride ion form by suspension in 0.1 M hydrochloric acid, followed by washing with distilled water. The resin is then equilibrated with starting buffer, made as follows. A mixture of 10 ml of 5 M HCl with about 900 ml water is titrated to pH 7.5 by addition of a 1 M solution of Tris (55–60 ml) and is then made up to 1 liter with distilled water. After equilibration with this buffer, the resin is packed into a small column (diameter 1 cm) to give a bed height of about 10 cm (volume 8 ml) at a flow rate of 25 ml/hour. The pooled material from the G-200 column is pumped through the column, under which conditions the protein is bound to the resin. The column is then pumped with a further 60 ml of starting buffer. A linear concentration gradient of sodium chloride (limit buffer: 0.8 M sodium chloride, made up in starting buffer) is then applied and fractions of approximately 2 ml are collected. Dopamine β-hydroxylase is eluted as a small UV-absorbing peak at a salt concentration of 50 mM NaCl. The protein in this peak is homogeneous, as judged by electrophoresis in 6% (w/v) polyacrylamide gels, using either Tris–glycine or phosphate–sodium dodecyl sulfate (SDS) buffer systems.

The maximum of the second UV-absorbing peak occurs at a sodium chloride concentration of 0.25 M, but a long "tail" extends from the maximum toward higher salt concentratinos. The material eluting at the peak maximum is chromogranin A, together with a very small amount of dopamine β-hydroxylase—detectable on polyacrylamide gel electrophoresis when high protein loads are used. Chromogranin A is the major protein throughout the second peak, but later fractions contain small quantities of additional components of unknown composition (presumably other chromogranins) which appear as faster-running bands on electrophoresis with either buffer system. In view of the good separation between the two peaks from the DEAE-Sephadex column, it seems likely that elution by a discontinuous step in salt concentration would resolve the dopamine β-hydroxylase from the rest of the chromogranins (principally chromogranin A), without the need to form a linear salt gradient. An alternative procedure for the rapid isolation of dopamine β-hydroxylase from the soluble lysate, by precipitation of chromogranins with the cationic detergent N-cetylpyridinium chloride, has been described by Hörtnagl et al.[10]

Proteins of the Chromaffin Granule Membrane. The protein composition of the chromaffin granule membrane fraction has been described by Hörtnagl et al.[11] These authors distinguished three major proteins by

[10] H. Hörtnagl, H. Winkler, and H. Lochs, *Biochem. J.* **129**, 187 (1972).
[11] H. Hörtnagel, H. Winkler, J. A. L. Schöpf, and W. Hohenwallner, *Biochem. J.* **122**, 299 (1971).

electrophoresis, using either phenol–acetic acid–urea buffer after the method of Takayama *et al.*[12] or a sodium borate–SDS buffer,[7] and named these proteins "chromomembrin A," "component C" and "chromomembrin B" (in order of increasing mobility toward the anode in the SDS-containing system). Treatment with SDS (5 mg/mg protein) results in the solubilization of 90% or more of the total granule protein.

These workers also described a method for the purification of chromomembrin A and chromomembrin B, by gel filtration on a column of Sephadex G-200 in the presence of SDS-containing buffer. They further demonstrated that chromomembrin A is the membrane-bound form of the enzyme dopamine β-hydroxylase[10] and presented strong evidence for identifying component C as being chromogranin A. Different electrophoretic and gel-filtration systems are used in the authors' laboratory, but the results agree broadly with those reported by Winkler and co-workers.

Electrophoresis of SDS-Solubilized Membrane Proteins. Stock (1 M) sodium phosphate buffer pH 6.5[13]: dissolve 81.0 g of disodium hydrogen orthophosphate (anhydrous) and 67.0 g of sodium dihydrogen orthophosphate (dihydrate) in distilled water and make up to 1 liter.

The polyacrylamide gels are made in the same way as for electrophoresis with the Tris–glycine system (see earlier) with the following differences: (1) The Tris–glycine buffer is replaced by the same volume of a solution which is 0.8 M in the above sodium phosphate buffer and 0.8% (w/v) in SDS. (This solution must be warmed slightly before use, as SDS is not sufficiently soluble in phosphate buffer to give the required concentration at room temperature.) (2) The ammonium persulfate solution should be deaerated for a few minutes, prior to its addition to the polymerization mixture.

Samples of chromaffin granule membranes are solubilized by adding a volume of solution [1% (w/v) SDS in 10 mM sodium phosphate buffer, 10% (w/v) in sucrose] sufficient to bring the SDS:protein ratio to greater than 5:1 (w/w) and then heating in a bath of boiling water for 2–5 minutes. For the best results, the resulting orange solution should be centrifuged at room temperature in the A40 rotor of the ultracentrifuge at 30,000 rpm (59,360 g) for 60 minutes and the (clear) supernatant used for electrophoresis. The gels are prerun at a current of 5 mA/gel for at least 60 minutes before use, employing an 0.1% (w/v) solution of SDS in 0.1 M sodium phosphate buffer as electrolyte. After sample loading, the electrophoretic separation is begun, at an initial current of 5 mA/

[12] K. Takayama, D. H. MacLennan, A. Tzagoloff, and C. D. Stoner, *Arch. Biochem. Biophys.* **114**, 223 (1966).

[13] J. V. Maizel, Jr., *in* "Fundamental Techniques in Virology" (K. Habel and N. P. Salzman, eds.), p. 334. Academic Press, New York, 1969.

gel, using this same [0.1% (w/v) SDS, 0.1 M sodium phosphate] buffer as electrolyte. After 20 minutes, the current is increased to 8 mA/gel and the run is continued at this level until the dye has migrated a sufficient distance (usually for a further 2–2.5 hours after increasing the current). The gels are fixed by shaking for 2 hours in a 25% (w/v) solution of trichloroacetic acid, after which the fixative is discarded and the gels are stood overnight in a fresh quantity of fixative.

The following day, the gels are stained by shaking for 1–2 hours in a 25% (w/v) solution of trichloroacetic acid which has been saturated with Amido Black [this is about 0.1% (w/v) in dye] and are then destained by leaching with 7% (v/v) acetic acid, as before. Proteins are stained a dark blue-black by this method, and phospholipids are also detected (see Fig. 2), appearing as a diffuse greenish band, running ahead of the dye marker. The band patterns obtained with SDS-solubilized membranes and soluble lysate proteins (treated with excess SDS) are shown in Fig. 2. The staining of chromomembrin B is somewhat variable, and this rather broad band sometimes appears to have a doublet structure, particularly at righer acrylamide concentrations.

Gel Filtration of SDS-Solubilized Membrane Proteins. The SDS-solubilized membrane proteins may be separated on a preparative scale by gel filtration, at room temperature, on a column (90 cm × 3.5 cm) of Sepharose 6B, eluted with 50 mM Tris·HCl buffer (pH 7.5) 0.1% (w/v) in SDS. The membranes are dissolved by suspending them in 50 mM Tris·HCl buffer (pH 7.5) which is 1.0% (w/v) in SDS. After standing for 2–5 hours, or overnight, the solution is centrifuged at room temperature in the A40 rotor of the ultracentrifuge at 30,000 rpm (59,360 g) for 60 minutes. The clear supernatant is withdrawn and up to 10 ml are applied to the column, which is then eluted at a flow rate of 25 ml/hour, collecting fractions of 2.5–3.0 ml. Eluted peaks are detected by measuring their absorption at 280 nm. Some unsedimented, particulate matter is eluted as a small peak at the exclusion volume of the column (ca. 190 ml). This is followed by a series of incompletely resolved peaks, with K_d values in the range 0.2–0.55, and by a small peak of unidentified UV-absorbing material, eluting at the inclusion volume of the column (ca. 670 ml).

Chromomembrin A is eluted in the first (nonexcluded) peak. Like Hörtnagl and co-workers, we find it necessary to take only those fractions eluted on the rising side of the peak for pooling, but we find that this "purified chromomembrin A" is still appreciably contaminated by two minor components (not described by Hörtnagl *et al.*) with electrophoretic mobilities intermediate between those of chromomembrin A and component C (see Fig. 2). Component C is the next protein to be eluted from the column but, like Hörtnagl *et al.,* we do not find that it is well separated from the other proteins. The next peak is usually the major peak, in terms of UV absorption, and it contains chromomembrin B, reasonably free of other proteins. However, there is usually a shoulder on the descending limb of this peak. The absorption at 408 nm (the wavelength of maximum absorption for the membrane cytochrome 559*b* under these conditions) reaches a maximum in the fractions making up this shoulder, and chemical analysis[14] shows that phospholipids are also eluted at this point. The latter observation is in contrast to the results of Hörtnagl *et al.,* who reported that phospholipids were completely resolved from the proteins after gel filtration on Sephadex G-200.

[14] G. R. Bartlett, *J. Biol. Chem.* **234**, 466 (1959).

[40] Isolation of Melanin Granules

By I. Aradvindakshan Menon and Herbert F. Haberman

Melanin granules (melanosomes) are the end product of melanization occurring within the melanocyte.[1,2] Melanization takes place in different stages. The first step is probably the synthesis of tyrosinase on the rough endoplasmic reticulum. The next step would be the formation of the premelanosomes. The term premelanosome is generally used to refer to all stages in the genesis of melanosomes that precede the fully melanized state. Premelanosomes exhibit various degrees of melanization and have been arbitrarily subdivided into various stages, e.g., early, intermediate, and late.[1,2] The mechanism of formation of the early stages of premelanosomes has not been adequately clarified. Some studies have shown that pigment granules in the early stages (premelanosomes) have hollow vacuoles which are formed from the enlargement and fusion of the Golgi vesicles and which contain tenuous materials in the form of folded fibrils or incomplete lamellae.[3-6] On the other hand, other studies suggest that unpigmented thin fibrillar melanosomes seem to be formed from the rough

[1] T. B. Fitzpatrick, W. C. Quevedo, Jr., A. L. Levene, V. J. McGovern, Y. Mishima, and A. J. Oettle, *Science* **152**, 88 (1966).
[2] T. B. Fitzpatrick, W. C. Quevedo, Jr., A. L. Levene, V. J. McGovern, Y. Mishima, and A. J. Oettle, *in* "Structure and Control of the Melanocyte" (G. D. Porta and O. Muhlock, eds.), p. 1. Springer-Verlag, Berlin and New York, 1966.
[3] S. R. Wellings, and B. V. Siegel, *J. Ultrastruct. Res.* **3**, 147 (1959).
[4] M. Seiji, K. Shimao, and M. S. C. Birbeck, *Ann. N.Y. Acad. Sci.* **100**, 497 (1963).
[5] M. Seiji, *Advan. Biol. Skin* **8**, 189 (1967).
[6] F. H. Moyer, *Ann. N.Y. Acad. Sci.* **100**, 584 (1963).

endoplasmic reticulum.[6,7] It has been proposed that the term melanosome be used to designate the fully pigmented melanin-containing organelle only.[1,2] Melanin synthesis begins with the deposit of granular melanin on both sides of the lamellae and continues to the extent that they obscure these inner structures.[8] It is not definite whether tyrosinase is structurally integrated into the protein matrix or is limited to the outer membrane of the melanosomes.[9] The term melanin granule is used to include all melanin-containing particles that can be observed with the light microscope.[1,2] In our preparation, this would include premelanosomes with varying degrees of melanization as well as fully melanized melanosomes.

It is possible to increase the tyrosinase activity of melanosomal preparations by a variety of physical and chemical treatments.[10-14] The mechanism of this activation phenomenon is not known.

Maintenance of Melanoma

At the present time, the most widely used source of melanin granules is a number of melanotic melanomas, e.g., Harding-Passey Cloudman S91, B16, or Fortner's melanoma. The method described below was developed using melanotic B16 melanoma originally obtained from Jackson Laboratories and subsequently obtained from Dr. A. Kopf. The tumors were maintained by serial transplantation into C57BL/6J mice (Jackson Laboratory, Bar Harbor, Maine) every 2 weeks. The excised melanoma tissue was mixed with normal saline in the ratio of 1:9. The tumor was dispersed by stirring with a glass rod for 2 minutes. Coarse tumor chunks were removed by passage through four layers of gauze; 0.1 ml of tumor suspension was injected subcutaneously into the axillary region of each mouse. The tumors are usually palpable after 1 week, and sufficiently large tumors are obtained 2–3 weeks after transplantation. A recurring problem with mouse melanomas is the spontaneous transformation of the melanotic tumor into amelanotic tumor. It is possible to overcome

---

[7] G. G. Maul, *J. Ultrastruct. Res.* **26,** 163 (1969).

[8] K. Jimbow and A. Kukita, *in* "Biology of Normal and Abnormal Melanocytes" (T. Kawamura, T. B. Fitzpatrick, and M. Seiji, eds.), p. 171. Univ. Park Press, Baltimore, Maryland, 1971.

[9] W. C. Quevedo, Jr., *in* "Biology of Normal and Abnormal Melanocytes" (T. Kawamura, T. B. Fitzpatrick, and M. Seiji, eds.), p. 99. Univ. Park Press, Baltimore, Maryland, 1971.

[10] I. A. Menon and H. F. Haberman, *J. Invest. Dermatol.* **52,** 45 (1969).

[11] I. A. Menon and H. F. Haberman, *Life Sci.* **8,** 207 (1969).

[12] I. A. Menon and H. F. Haberman, *Arch. Biochem. Biophys.* **137,** 231 (1970).

[13] M. H. Van Woert, F. Korb, and K. N. Prasad, *J. Invest. Dermatol.* **56,** 343 (1971).

[14] J. F. McGuire, *Biochem. Biophys. Res. Commun.* **40,** 1084 (1970).

this problem by using melanotic melanoma cells from tissue culture for transplantation to reestablish the pigmented melanoma line.

Preparation of Melanin Granules

Upon disruption of melanocytes and separation of the various sub-cellular organelles by differential centrifugation, melanin granules usually sediment along with mitochondria. Seiji et al.[4] first described a procedure for the separation of melanin granules from mitochondria. They used a discontinuous sucrose gradient system, and after centrifugation at high speed they obtained a fraction rich in tyrosinase activity and melanin separate from the fraction containing high succinic oxidase activity. The fraction rich in tyrosinase was found to contain mostly melanin granules upon examination under electron microscopy. Other workers have subsequently used similar procedures for the separation of melanin granules. The method described below (Fig. 1) is a modification of the original method used by Seiji et al.,[4] and this has been found to be a more convenient rapid method for the preparation of melanin granules in large quantities.

Step 1. Preparation of Mixture of Mitochondria and Melanin Granules. The melanoma tissue excised approximately 2 weeks after transplantation is minced with scissors. Four layers of gauze are folded on top of a 50- or 100-ml beaker containing 0.25 M sucrose. The volume of the sucrose is such that approximately one-fourth of the gauze is dipped in the sucrose solution. The minced melanoma tissue is dropped into the gauze and gently mixed with the pestle of a Teflon homogenizer. Most of the melanoma cells are separated and passed through the gauze into the sucrose solution. Most of the connective tissue remains on top of the gauze. The suspension of the melanoma cells is homogenized using a Potter-Elvehjem homogenizer. The homogenate is centrifuged twice at 600 g for 10 minutes. The supernatant is then centrifuged at 11,000 g for 10 minutes. The sediment is resuspended in the medium and recentrifuged at 11,000 g for 10 minutes. The sediment is then suspended in the same medium. This preparation contains most of the melanin granules and mitochondria.

Step 2. Separation of Melanin Granules from Mixture of Melanin Granules and Mitochondria. The above suspension containing melanin granules and mitochondria is used for obtaining purified preparation of melanin granules. This is best accomplished using discontinuous sucrose gradient centrifugation. A swing-out rotor, e.g., SW 40 or SW 50 of Beckman (Spinco) ultracentrifuge or rotor No. 59430 of MSE superspeed centrifuge, is conveniently used for this purpose. The mixture of melanin granules and mitochondria is layered over 1.7 M sucrose and

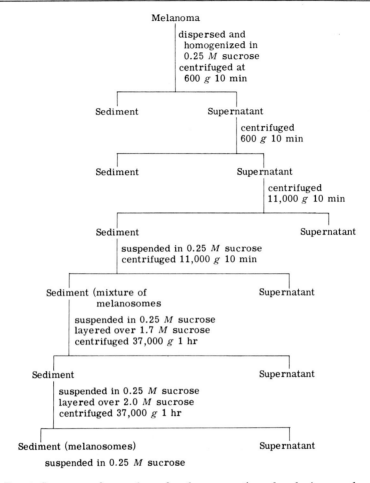

Fig. 1. Summary of procedures for the preparation of melanin granules.

centrifuged at 20,000 rpm (37,000 g) for 1 hour. Most of the melanin granules and a small percentage of mitochondria are sedimented by this centrifugation. The layer on top of the 1.7 M sucrose contains most of the mitochondria and some melanin granules. The sediment is suspended in 0.25 M sucrose and layered over 2.0 M sucrose and centrifuged again at 20,000 rpm for 1 hour. The sediment obtained may be suspended in a convenient medium as required for further experiments, e.g., 0.25 M sucrose.

The mixture of mitochondria and melanin granules obtained on top of the 1.7 M sucrose in the first centrifugation could also be used for obtaining pure mitochondria free from melanin granules. For this purpose the suspension is centrifuged at 100,000 g for 1 hour. The sediment obtained is suspended in 0.25 M sucrose and layered over 1.5 M sucrose and centrifuged at 20,000 rpm for 30 minutes. The small amount of melanin granules present in the mixture is then sedimented and pure mitochondria remain on top of the 1.5 M sucrose. The mitochondria may then be sedimented by high speed centrifugation, e.g., at 100,000 g for 1 hour.

Criteria of Purity

The purity of the preparation of melanin granules could be determined by both biochemical and morphological parameters. Tyrosinase or DOPA oxidase activity has been used as a marker for melanin granules. The tyrosinase activity of the preparation is dependent, however, upon the precise method of isolation.[10-14] From our experience the cytochrome oxidase activity is the best biochemical marker for mitochondria although succinic oxidase could also be used. However, examination by electron microscopy is an essential requirement. Our standard procedure for electron microscopic investigation is as follows. The pellets are fixed with 3% glutaraldehyde in 0.1 M cacodylate buffer pH 7.2 for 90 minutes.

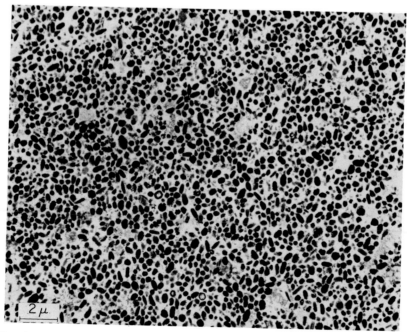

FIG. 2. Electron micrograph of a purified preparation of melanosomes. ×5220.

After 45 minutes the pellets are diced into 1-mm cubes. The specimens are rinsed with cacodylate buffer with 0.2 M sucrose pH 7.2 and post-fixed for 90 minutes in 1% Palade's osmium tetroxide pH 7.2. In block staining is carried out with 0.5% aqueous uranyl acetate. The specimens are dehydrated with graded acetone and embedded in Epon 812. Sections are cut on an ultramicrotome and stained with saturated uranyl acetate at 46° for 15 minutes and Reynolds lead citrate for 15 minutes. An electron micrograph of the melanin granule obtained by the above method is given in Fig. 2.

Comments on the Method of Isolation of the Melanin Granules

The samples of melanin granules obtained by the method described above in detail was found to be the purest preparation of melanin granules obtained by the various methods. Nevertheless, it has the disadvantage that the tyrosinase activity of this preparation could not be activated by further treatments. In the course of attempts to modify this procedure, density gradient centrifugation using Ficoll was investigated. This procedure was found to yield melanin granules having a latent tyrosinase activity that could be activated, as in the case of the mixture of melanin granules and mitochondria. However, the purity of these samples was inferior to those obtained by the above method. Therefore, if it is required to obtain melanin granules containing latent tyrosinase activity, this method is not suitable. For this purpose either a less pure preparation obtained on Ficoll gradient or the mixture of mitochondria and melanin granules could be employed.

Since the separation of melanin granules from mitochondria is not possible if the sucrose gradient contained KCl and Mg^{2+}, it is advisable not to add any type of salts to the sucrose medium.

Acknowledgments

The present procedures were developed in the course of research supported by Ontario Cancer Treatment and Research Foundation. We also wish to acknowledge the assistance of K. Y. Wong and H. Seiden.

[41] The Isolation of Secretory Granules from Mast Cells

By Börje Uvnäs

The most convenient sources of mast cells are the peritoneal and thoracic cavities of the rat. Mast cells can be obtained in pure suspensions (over 95% mast cells) by density gradient centrifugation of the cell suspension obtained by flushing these cavities with salt solution.

After disruption of the cells the basophile granules can be isolated virtually free from other subcellular particles by differential centrifugation. The granules from rat mast cells are rather insoluble in water and dilute salt solutions. This facilitates further purification and also allows repeated experimental procedures to be carried out without annoying losses of material.

Several techniques have been described for the isolation of rat mast cells: the earlier ones were based on sucrose-gelatin or sucrose gradient centrifugation.[1,2] This technique yields cells and granules suitable for some types of morphological and biochemical studies. However, such cells are not suitable for studies of degranulation and histamine release, since their ability to respond to histamine-releasing agents, such as compound 48/80 and antigens, is reduced. The cells also undergo considerable morphological changes.[3] We assumed that this functional and morphological deterioration was due to the exposure of the cells to hypertonic sucrose solutions during the density gradient centrifugation. We therefore developed a technique of our own, involving density gradient centrifugation through Ficoll, a high-molecular-weight neutral polysaccharide, dissolved in isotonic salt solution. Owing to its high molecular weight, the osmotic influence of the Ficoll solution on the mast cell is negligible. Other workers have developed techniques with albumin as the centrifugation medium.[4,5] These methods also yield mast cells with their responsiveness to histamine-releasing agents retained: the cells are probably also well suited for the isolation of granules. Since I do not have any experience with this technique, I will refrain from any comments on it.

[1] I. Padawer and A. S. Gordon, *Proc. Soc. Exp. Biol. Med.* **88**, 29 (1955).
[2] D. Glick, S. L. Bonting, and D. Den Boer, *Proc. Soc. Exp. Biol. Med.* **92**, 357 (1956).
[3] G. Asboe-Hansen and D. Glick, *Proc. Soc. Exp. Biol. Med.* **98**, 458 (1958).
[4] G. T. Archer, *Nature (London)* **182**, 726 (1958).
[5] I. Muta and W. Dias de Silva, *Nature (London)* **186**, 245 (1960).

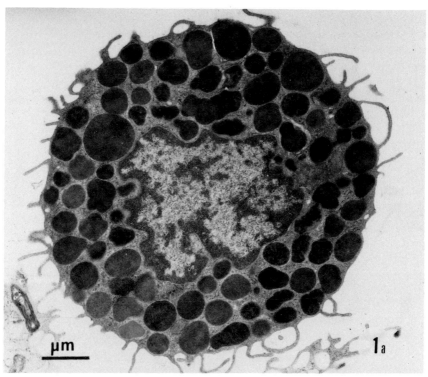

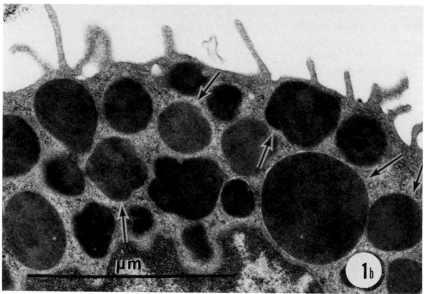

Since mast cells with retained responsiveness to histamine-releasing agents are required for the isolation of granules actively released from mast cells during degranulation, a detailed description of our isolation technique is given.

Over the years our isolation procedure has undergone various minor modifications with the aim of increasing the yield and quality of the isolated material. The technique described below is the one at present used in our laboratory.

Technique for Isolation of Mast Cells

Male Sprague-Dawley rats weighing 350–400 g (old rats give a higher yield of mast cells) anesthetized with ether are decapitated and bled. A small incision is made in the abdominal midline, and 9 ml of buffered NaCl solution pH 6 (1 vol buffer + 9 vol NaCl solution) is pipetted into the abdomen. After carefully massaging for 1.5 min, the fluid in the abdominal cavity is collected with a drop pipette and carefully layered onto a discontinuous 30 to 40% Ficoll gradient in a centrifuge tube. Thoracic mast cells are obtained by flushing the thoracic cavities with the above-mentioned solution; about 4 ml are introduced through a small incision made in the diaphragm. After a 30-second massage, the resulting cell suspension is removed and layered onto the gradient. After 10 minutes of centrifugation at 350 g, a layer containing leukocytes, small mast cells and occasional erythrocytes is found on the upper surface of the gradient. A diffuse band containing mast cells is formed at the interface between the 30 and 40% Ficoll layers. The supernatant and the cell layer on the surface of the Ficoll gradient are discarded. The rest of the Ficoll is transferred with a syringe to a new centrifuge tube and diluted with buffered A solution containing albumin (1 volume of buffer + 9 volumes of diluted A solution + 1 mg of albumin per milliliter). After 10 minutes of centrifugation at 350–450 g, the supernatant is removed, and the cell precipitate is washed in 2 ml of buffered A-albumin solution. Alternatively, the cells are suspended in buffered A-albumin solution and transferred to one large tube for washing. The cells may be washed once more with buffered A-albumin solution or ice-cold isotonic sucrose, depending on what the mast cells are to be used for.

The presence of albumin in the wash fluid is of importance for the reactivity of the mast cells to agents releasing histamine by degranulation; when stored at +4° they retain their responsiveness for 1 or 2 days.

FIG. 1. (a) Isolated rat peritoneal mast cell. Note the electron dense, homogeneous granules. ×12,500. (b) Higher magnification of (a). The granules are surrounded by a unit membrane (arrows). ×56,000.

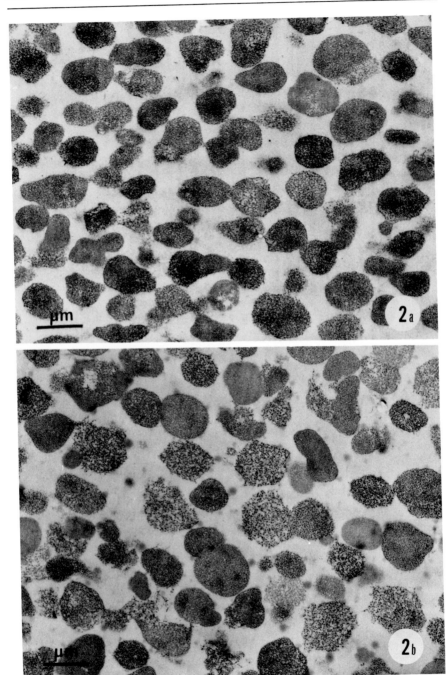

Materials

1. Physiological saline (0.9% NaCl)
2. A solution: NaCl 145 mM, KCl 2.7 mM, CaCl$_2$ 0.9 mM
3. Sörensen phosphate buffer: (a) KH$_2$PO$_4$, 9.078 g/liter; (b) Na$_2$HPO$_4$·2H$_2$O, 11.876 g/liter; pH 6, 7.5 ml a + 2.5 ml b; pH 7, 3.0 ml a + 7.0 ml b
4. Ficoll: (a) 30% in buffered physiological saline pH 7 (1 volume of buffer + 9 volumes of physiological saline). The final solution contains 2 mg of glucose and 1 mg of human serum albumin per milliliter; (b) 40% solution prepared as above

Preparation of the Ficoll solution: The Ficoll is suspended in most of the calculated buffer solution and heated in a water bath at 60°, with occasional stirring until dissolution is complete. The Ficoll solution is cooled, and the rest of the buffer solution containing glucose and albumin is added to give a final concentration of 2 mg of glucose and 1 mg of albumin per milliliter.

Ficoll Gradients. One milliliter of 30% Ficoll is carefully layered onto 1 ml of 40% Ficoll solution in a 10 ml of Pyrex tube and stored frozen until used.

Isolation of Mast Cell Granules

The technique chosen for isolation of mast cell granules depends on the experimental purpose for which the granules are to be used. As seen in Fig. 1a and b, mast cell granules *in situ* are surrounded by a unit membrane. However, granules expelled from mast cells during degranulation have lost their membranes (Fig. 2a). Similarly, granules isolated from mast cells which have been lysed in distilled water are devoid of a surrounding membrane (Fig. 2b). Granules released by degranulating agents or water lysis are thus well suited for studies concerning the chemical composition, biochemical properties, and the characteristics of the storage mechanism, etc., of the water-insoluble granule matrix. However, it should be remembered that changes in the chemical composition and physicochemical and biochemical properties due to loss of the surrounding membrane and of granule material dissolved in the suspension medium have to be taken into account. The absence of a granule membrane

---

Fig. 2. (a) Granules isolated from compound 48/80-treated mast cells. Note the swollen appearance, the decreased electron density, and the loss of perigranular membranes. ×12,000. (b) Granules isolated from mast cells lysed in distilled water. No perigranular membranes can be seen. The same changes in size and electron density as in (a). ×12,000.

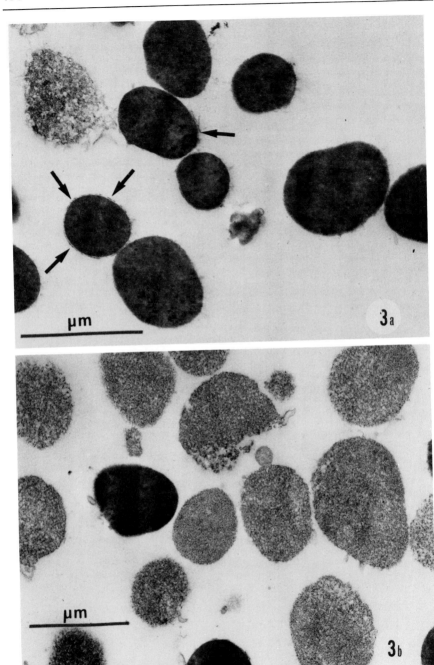

facilitates studies on the interchange between the granules and the surrounding medium; on the other hand, it obviously does not permit any studies to be made on the influence of the granule membrane on the uptake and storage capacity of the granule.

Membrane-bound granules can be isolated from mast cells disrupted by ultrasonic disintegration (sonication). Such a technique was described by Lagunoff et al. in 1964,[6] and we also used it initially. However, the intensity of sonication is very critical, and it is difficult to reproduce the sonication conditions exactly. In our sonication experiments we obtained a mixture of membrane-bound and membrane-free granules. The higher the sonication intensity, the higher the percentage of granules devoid of membranes; and the lower the intensity, the higher the percentage of granules with preserved membranes (Fig. 3a and b). However, at lower sonication intensities the yield of isolated granules is reduced. So far we have been able to obtain up to 80% membrane-bound granules.

Isolation of membrane-bound granules allows studies to be made on the total composition of the granule constituents. It also permits studies to be made on water soluble granules and on the properties of the granule membrane. However, it should be noted that the admixture of granules devoid of membranes complicates the interpretation of the results. For instance, the amine storing capacity of the granules will be markedly influenced by the proportions of membrane-bound and membrane-free granules. Since the granule matrix has the properties of a weak cation exchange material[7] the presence of cations in the granule suspension medium will lead to an immediate depletion of histamine from membrane-free granules. On the other hand, the membrane-bound granules may still retain their histamine even in a cation-containing medium, e.g., physiological saline.

It is probably possible to separate granules with and without membranes by density gradient centrifugation, but we have not yet carried out such experiments.

Isolation of Granules from Mast Cells Lysed in Deionized Water. Isolated mast cells are suspended in deionized water (2 ml per rat), and

[6] D. Lagunoff, M. T. Philips, O. A. Isen, and E. P. Benditt, *Lab. Invest.* **13**, 1331 (1964).
[7] B. Uvnäs, C.-H. Åborg, and A. Bergendorff, *Acta Physiol. Scand.* **78**, Suppl. 336 (1970).

FIG. 3. (a) Granules isolated from sonicated mast cells (MSE 100 W ultrasonic disintegrator, 4 μm, 5 seconds). Note the predominance of electron dense granules surrounded by perigranular membranes (arrows). ×32,500. (b) Granules isolated from sonicated mast cells (MSE 100 W ultrasonic disintegrator, 4 μm, 60 seconds). Note the predominance of swollen, less electron dense granules without perigranular membranes. ×27,000.

the suspension is adjusted to pH 7.1 with 0.1 M NaOH. After 10 minutes of repeated shaking, the lysed cell suspension is centrifuged (350 g, 10 minutes, 4°). This centrifugation precipitates coarse cell debris. The granule-containing supernatant is recentrifuged (3000 g, 25 minutes, 4°), and the resulting pellet is either resuspended and used immediately or dried over P_2O_5. Granules dried in this way retain their capacity to take up and store inorganic cations and biogenic amines.

Isolation of Granules Released from Compound 48/80-Treated Mast Cells. Isolated mast cells are incubated (10 minutes, 37°) in solution A buffered with 10% v/v Sörensen buffer, pH 7.0, containing human serum albumin, 1 mg/ml, and compound 48/80, 2 μg/ml. The suspension is then centrifuged (350 g, 10 minutes 4°), and the resulting supernatant is set aside. The precipitated cells are washed twice with the buffered salt solution mentioned above and centrifuged at 350 g for 10 minutes after each wash. The three granule-containing supernatants are bulked and centrifuged (3000 g, 25 minutes, 4°). The granule material can be used immediately or stored dried over P_2O_5.

Isolation of Granules from Sonicated Mast Cells. Isolated mast cells are suspended in ice-cold 0.34 M sucrose (cells from 5 rats per 10 ml of sucrose); 5-ml aliquots cooled in ice are sonicated with a MSE 100-W ultrasonic disintegrator operating at 4 μm amplitude setting for 5, 10, 30, or 60 seconds. The sonicates are centrifuged (350 g, 10 minutes, 22°), and the granule-containing supernatants are then recentrifuged (3000 g, 20 minutes). The granule-containing sediments were suspended in isotonic salt solutions at pH around 7 to avoid damage to the granule membranes. These granules cannot be dried without damaging the surrounding membrane.

For electron microscopic studies the granule-containing pellets are fixed with 2.5% glutaraldehyde in Millonig phosphate buffer (NaH_2PO_4 + NaOH) pH 7.3 in Beckman polyallomer tubes containing a small amount of prepolymerized Araldite in which a conical hole has been bored. Fixation is continued for 40 minutes at room temperature, and the samples are then centrifuged (3000 g, 20 minutes). The sediments are washed twice with Millonig phosphate buffer pH 7.3 and then postfixed in 1% OsO_4 in the same buffer for 60 minutes at 4°. The suspensions are centrifuged (3000 g, 20 minutes, 4°) and the pellets washed with Millonig phosphate buffer. After dehydration in an ethanol series, the pellets are transferred through propylene oxide to Araldite in which they are prepolymerized overnight at 4°. Polymerization is carried in fresh Araldite for 6 days at 60°.

Thin sections are made on a Reichert ultramicrotome Om U 3 and, stained with uranyl acetate and diluted lead citrate before examination in a Philips EM 300 electron microscope at 80 kV.

[42] The Isolation of Neurosecretory Granules from the Posterior Pituitary

By D. B. HOPE and J. C. PICKUP

Principle

A procedure for the isolation of highly purified neurosecretory granules from bovine posterior pituitary lobes, by differential and density gradient centrifugation of tissue homogenates, was originally developed in this laboratory by Dean and Hope in 1967.[1] The reproducibility of the method is evident in the subsequent work by Poisner and Douglas.[2] Since then we have modified the homogenization and centrifugation conditions[3] so as to substantially increase the yield of granules, while maintaining the almost complete separation from other subcellular organelles. The method has recently been applied to a study of the subcellular fractionation of posterior pituitary lobes from the pig.[3a]

Procedure

Preparation of Tissue. Entire pituitary glands are removed from the animal as soon as possible after death and placed on ice. The posterior lobe is dissected away from the anterior lobe at the cleft between the pars distalis and the pars intermedia. This is particularly easy in the pig, because of the absence of surrounding connective tissue and the prominence of the hypophysial cleft in the young animals.[4] In the ox, however, the pituitary gland is firmly encased in dura mater, particularly on the dorsal surface. Recently we removed the thin covering of pars intermedia from the posterior lobe for work on the ATPase of the neurosecretory granules (S. Nakamura and D. B. Hope, unpublished work). The pars nervosa appears white against the slightly pigmented pars intermedia. Small amounts of pars nervosa tissue recovered are essential for the localization of enzymes in neurosecretory granules free from granules arising from the pars intermedia. The ox has a small protuberance of tissue projecting into the cleft on the ventral surface of the pars nervosa (cone of Wulzen).[4] It is histologically similar to the adenohypophysis and should be removed.

[1] C. R. Dean and D. B. Hope, *Biochem. J.* **104**, 1082 (1967).
[2] A. M. Poisner and W. W. Douglas, *Mol. Pharmacol.* **4**, 531 (1968).
[3] J. C. Pickup and D. B. Hope, *Biochem. J.* **123**, 153 (1971).
[3a] J. C. Pickup, C. I. Johnston, S. Nakamura, L. O. Uttenthol, and D. B. Hope, *Biochem. J.* **132**, 361 (1973).
[4] B. Hanström, *in* "The Pituitary Gland" (G. W. Harris and B. T. Donovan, eds.), Vol. 1, p. 1. Butterworth, London, 1966.

The posterior lobes were weighed (about 300 mg each in the ox and 50 mg in the pig) in cold 0.3 M sucrose and chopped to a fine mince on a Perspex board with a stainless steel knife cooled to 0° before use.

Homogenization. Six bovine glands (1.8 g) are routinely fractionated at one time in this laboratory. The minced tissue is transferred to a cold, smooth-walled, glass homogenizer tube and washed in with 15 ml of 0.3 M sucrose at 0°. Homogenization is carried out with a Teflon pestle (Kontes Glass Co., Vineland, New Jersey) rotating at 950 rpm with a radial clearance of 0.15 mm. Six upward and downward thrusts are given over a period of about 30 seconds, making sure that the pestle is pressed firmly against the bottom of the tube each time.

The motor now used to drive the pestle is a 0.25 horsepower, Hoover A.C. (950 rpm) motor, Type 7344 GBR, Ser. No. 31945 VM. Earlier work was done with a 0.033 horsepower, variable speed (max. 2000 rpm) motor (Fluid Equipment Co. Ltd., 83 Seafield Road, London, N. 11; Ser. No. 1/54/2). Further comparative experiments with these and other motors would be necessary to establish to what extent the quantitative differences in the results reported in our two publications[1,3] are a consequence of the different conditions of homogenization.

Centrifugation. The procedure for the differential centrifugation of homogenates of posterior pituitary lobes is summarized in Fig. 1. Centrifugation of the homogenate for 15 minutes at 1500 rpm (1100 g) at

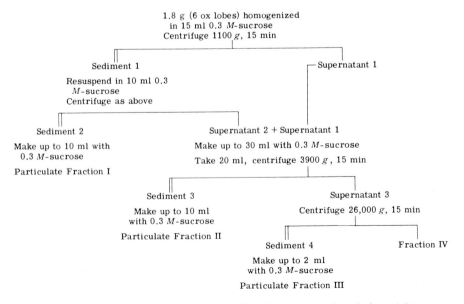

FIG. 1. Summary of the differential centrifugation of posterior pituitary lobes.

$0°$ in suitable refrigerated centrifuge, e.g., M.S.E. Mistral 6L, removes cell debris and nuclei. The supernatant is removed with a Pasteur pipette and retained (supernatant 1); the sediment (sediment 1) is resuspended in 10 ml $0.3 M$ sucrose at $0°$ and centrifuged as before. The resulting sediment (sediment 2) is made up to 10 ml with cold $0.3 M$ sucrose (fraction I) and the supernatant (supernatant 2) is combined with supernatant 1 and made up to 30 ml with cold $0.3 M$ sucrose (supernatant 1 + supernatant 2). Then 20 ml of the cytoplasmic extract (supernatant 1 + supernatant 2) is centrifuged at 8000 rpm for 15 minutes (3900 g) in a refrigerated ultracentrifuge, e.g., Spinco Model L, using the A 40 rotor to give a sediment (sediment 3) and a supernatant. This supernatant is centrifuged in the A 40 rotor at 20,000 rpm (26,000 g) for 15 minutes to give a sediment (sediment 4) and a final supernatant (fraction IV). Sediment 4 is resuspended in cold $0.3 M$ sucrose by sucking up and down several times with a Pasteur pipette, and the volume is made up to 2.0 ml with the sucrose solution to give fraction III.

Typical Results from Differential Centrifugation. Figure 2 shows subcellular-distribution histograms of enzymatic, hormonal activities and of RNA in fractions obtained by the differential centrifugation of bovine posterior pituitary lobes. The results are presented as relative specific activities (percentage of recovered activity:percentage of recovered pro-

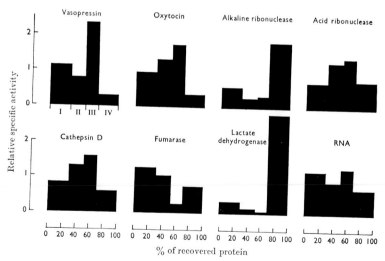

FIG. 2. Subcellular distribution histograms of enzymatic and hormonal activities and RNA in fractions obtained by differential centrifugation of homogenates of bovine posterior pituitary lobes. From J. C. Pickup and D. B. Hope, *Biochem. J.* **123**, 153 (1971).

tein) plotted against the percentage of protein in each fraction.[5] Vasopressin and oxytocin were used as neurosecretory granule markers, fumarase for mitochondria, cathepsin D and acid ribonuclease for lysosomes, and lactate dehydrogenase as a marker for the cytosol. It is clear that fraction III is richest in the neurosecretory granules: as much as 38% of the total glandular vasopressin and 29% of the oxytocin is found in this fraction, which is also relatively free of mitochondria.

Hormonal Content of the Final Supernatant (Fraction IV). The percentage of the polypeptide hormones found in fraction IV is particularly important because of the suggestion that Ca^{2+} ions act in the release mechanism primarily by dissociating an extragranular pool of neurophysin–hormone complex.[6] The hormone in the supernatant may be equivalent to the readily releasable pool which constitutes about 10–20% of the total glandular hormone.[7] This pool of hormone (vasopressin) could be released from the neurohypophysis by the stimulant effect of cold[8] during transit from the slaughterhouse to the laboratory. The experimenter should be aware that the amount of hormone in fraction IV varies slightly from one experiment to another. This may be for a variety of reasons: (1) granules that are more fragile when obtained from glands collected some hours after the death of the animal; (2) the physiological state of the animal—for instance, the animal may be dehydrated as shown by Barer et al.[9]; (3) the shearing force during homogenization: the granules are ruptured by high shearing forces; (4) the ratio of weight of tissue to the volume of sucrose solution used in the homogenization. Reducing the weight of tissue to the volume of homogenate ratio from 1.8 g/15 ml (as described above) to about 15 mg/15 ml[3a] increases the proportion of hormone in fraction IV to 50%.

Density-Gradient Centrifugation. A continuous nonlinear sucrose density gradient is prepared 24 hours before use and kept at 4°: 2.0 ml of 2.0 M sucrose is pipetted into each of three cellulose nitrate centrifuge tubes (1.25 × 5.0 cm) followed by the careful layering on of 1.0 ml of 1.40 M sucrose, and 0.5 ml of 1.35 and 1.30 M sucrose to each tube. The sucrose is allowed to diffuse for a period of 24 hours, 0.5 ml of resuspended fraction III is layered over each of three gradients, and the

[5] C. de Duve, B. C. Pressman, R. Gianetto, R. Wattiaux, and F. Appelmans, *Biochem. J.* **60**, 604 (1955).
[6] M. Ginsburg and M. Ireland, *J. Endocrinol.* **35**, 289 (1966).
[7] H. Sachs, P. Fawcett, Y. Takabatake, and R. Portanova, *Recent Progr. Horm. Res.* **25**, 447 (1969).
[8] W. W. Douglas and A. Ishida, *J. Physiol. (London)* **179**, 185 (1965).
[9] R. Barer, H. Heller, and K. Lederis, *Proc. Roy. Soc. Ser. B* **158**, 388 (1963).

tubes are centrifuged at 35,000 rpm for 60 minutes (101,000 g) in a Spinco model ultracentrifuge (SW 39 or SW 50 swing-out rotor). Subfractions are collected by cutting the tube with a Schuster centrifuge-tube cutter. The density of each subfraction can be measured by weighing a sample in a precooled 0.2-ml constriction pipette calibrated with distilled water. If the fractions are collected in preweighed tubes the volumes can readily be calculated from the density of each sample.

Typical Results. The distribution of enzymatic and hormonal activities and of RNA in subfractions from the nonlinear sucrose density gradient are shown as histograms in Fig. 3. The appearance of the tube after centrifugation is shown diagrammatically. Six subfractions are generally taken, including an intense band near the top (B) and a broader band

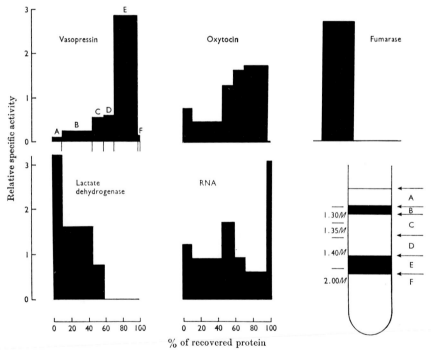

Fig. 3. Distribution histograms for enzymatic and hormonal activities and RNA in subfractions from a nonlinear sucrose density gradient. Fraction III from the differential centrifugation was resuspended in 0.3 M-sucrose, layered over the gradient and centrifuged at 101,000 g for 1 hour. The appearance of the bands after centrifugation is also shown. Arrows indicate the position at which the tube was sliced to obtain the six subfractions A–F. From J. C. Pickup and D. B. Hope, *Biochem. J.* **123**, 153 (1971).

VOLUMES, DENSITIES, AND EQUIVALENT SUCROSE MOLARITIES OF SUBFRACTIONS
OBTAINED FROM THE DENSITY GRADIENT

Subfraction	Volume (ml)	Density, 0°C	Molarity, 20°C
A	1.11	1.077	0.57
B	1.55	1.146	1.09
C	1.35	1.178	1.34
D	1.28	1.201	1.50
E	1.90	1.226	1.70
F	4.80	1.263	2.00

near the bottom (E) of the gradient. Typical volumes (pooled from three gradients), densities, and equivalent sucrose molarities of the subfractions A–F are given in the table.

Band E represents highly purified neurosecretory granules; approximately 75% of the vasopressin and 44% of the oxytocin applied to the centrifuge tube are recovered in this subfraction. Subfractions C and D are also quite rich in oxytocin, indicating that oxytocin and vasopressin may be stored in separate particles.[10]

The top band (B) represents mitochondria (fumarase) and some sedimentable form of cell sap (lactate dehydrogenase). The latter may be pinched-off nerve endings (neurosecretosomes),[11] analogous to the synaptosome, or processes of the neurohypophysial glial cells (pituicytes), particularly as subfraction B contains so much RNA (approximately 33% of that applied to the tube).

Analysis of Fractions

Because of the difficulty in pipetting the original homogenate recoveries are best calculated by taking the sum of fraction I and the cytoplasmic extract (supernatant 1 + supernatant 2) as a measure of activity in the whole tissue.

Protein is determined by the method of Lowry *et al.*[12] with bovine serum albumin as standard. Because of interference by high sucrose concentrations in this method, it is advisable to precipitate the protein with trichloroacetic acid, centrifuge, and decant the supernatant. The protein content of the pellet can then be measured after a preliminary dissolution in approximately 0.5 ml of 1 *M* sodium hydroxide for 30 minutes.

[10] C. R. Dean and D. B. Hope, *Biochem. J.* **106**, 565 (1968).
[11] E. Bindler, F. S. LaBella, and M. Sanwal, *J. Cell Biol.* **34**, 185 (1967).
[12] O. H. Lowry, N. J. Rosebrough, A. L. Farr, and R. J. Randall, *J. Biol. Chem.* **193**, 265 (1951).

Mitochondria are detected, for example, by the fumarase assay of Racker[13] or the succinate dehydrogenase assay of Pennington,[14] as modified by Porteus and Clark.[15]

Lysosomes are detected in this laboratory by estimating cathepsin-D[3] or acid ribonuclease activities.[3]

Neurosecretory Granules. Samples for bioassay are made 0.2% (v/v) with respect to acetic acid and heated in boiling water for 5 minutes. Pressor activity is assayed by the method of Dekanski[16] on male albino rats anesthetized with urethan (1.2 g/kg i.p.) and treated with phenoxybenzamine (1 mg/kg i.v.). The blood pressure of the rat is measured with a Bell and Howell pressure transducer fitted to a Devices R2 single-channel recorder.[1]

Oxytocic activity is assayed on the isolated rat uterus by the method of Holton[17] with the Mg^{2+}-free van Dyke-Hastings solution suggested by Munsick.[18] The uteri are removed from virgin rats that have been injected subcutaneously with 0.07 ml of Oestradiol Benzoate Injection B.P. (1 mg/ml in ethyl oleate 20% and arachis oil 80%) 18 hours before the assay.[19] The tension developed by the uterus is measured by a strain gauge connected to the recorder.[1]

Local standards of synthetic oxytocin and vasopressin are standardized against the IIIrd International Vasopressor, Antidiuretic and Oxytocic Standard,[20] and the results are calculated by using the (1 + 2) method of Gaddum.[21]

The soluble proteins from bovine neurosecretory granules (subfraction E) were originally examined by lysis of a granule pellet in dilute Tris-succinate buffer pH 8.0, followed by starch-gel electrophoresis of the soluble fraction.[1] Two major and one minor protein bands were seen, which are identical in electrophoretic mobility to the native hormone-binding proteins, neurophysins I, II, and -C, which have been isolated from acetone-dried posterior pituitary lobe powder.[22] More recently, we have been able to use a radioimmunoassay specific for porcine neurophysins I and II to

[13] E. Racker, *Biochim. Biophys. Acta* 4, 211 (1950).
[14] R. J. Pennington, *Biochem. J.* 80, 649 (1961).
[15] J. W. Porteus and B. Clark, *Biochem. J.* 96, 159 (1965).
[16] J. Dekanski, *Brit. J. Pharmacol. Chemother.* 7, 567 (1952).
[17] P. Holton, *Brit. J. Pharmacol. Chemother.* 3, 328 (1948).
[18] R. A. Munsick, *Endocrinology* 66, 451 (1960).
[19] R. A. Munsick and S. J. Jeronimus, *Endocrinology* 76, 90 (1965).
[20] D. R. Bangham and M. V. Musset, *Bull. W.H.O.* 19, 325 (1958).
[21] J. H. Gaddum, *in* "Pharmacology," 5th ed., p. 520. Oxford Univ. Press, London and New York, 1959.
[22] M. D. Hollenberg and D. B. Hope, *Biochem. J.* 106, 557 (1968).

study the distribution of these proteins in subcellular fractions from homogenates of porcine posterior pituitary lobes.[3a]

Cytosol may be conveniently assayed by the method of Wróblewski and La Due[23] for lactate dehydrogenase.

RNA can be estimated by the Schmidt and Tannhauser[24] procedure as described by Pickup and Hope.[3]

[23] F. Wróblewski and J. S. La Due, *Proc. Soc. Exp. Biol. Med.* **90**, 210 (1955).
[24] G. Schmidt and S. J. Tannhauser, *J. Biol. Chem.* **161**, 83 (1945).

[43] Methods for Preparation of Secretory Granules from Anterior Pituitary Glands

By W. H. McShan

The fractionation of cellular organelles of the anterior pituitary gland by differential centrifugation began in 1948.[1] Hormones, enzymes, nucleic acids, and granule fractions were studied. In these studies, certain of the hormones were solubilized, presumably as a result of breakage of the granules.[2-4] Emphasis was then placed on the purification of the granules, in which the hormones were assumed to be stored. Methods were developed for the preparation of relatively purified fractions of the small granules from the basophilic and the large granules from the acidophilic pituitary cell types.[5-7] The granules in certain of these fractions were solubilized and used for the study of growth hormone (GH), prolactin (PL),[8] and thyrotropin (TSH).[9] Glands from the rat and the cow have been used mainly for isolation purposes. This review will present in detail both the most effective and the more recent methods for the preparation of highly purified acidophilic and basophilic granules.

[1] H. R. Catchpole, Abstract, *Fed. Proc., Fed. Amer. Soc. Exp. Biol.* **7**, 19 (1948).
[2] W. H. McShan and R. K. Meyer, *Proc. Soc. Exp. Biol. Med.* **71**, 407 (1949).
[3] W. H. McShan, R. Rozich, and R. K. Meyer, *Endocrinology* **52**, 215 (1953).
[4] F. S. LaBella, and J. H. U. Brown, *J. Biophys. Biochem. Cytol.* **4**, 833 (1958).
[5] F. S. LaBella and J. H. U. Brown, *J. Biophys. Biochem. Cytol.* **5**, 17 (1959).
[6] M. W. Hartley, W. H. McShan, and H. Ris, *J. Biophys. Biochem. Cytol.* **7**, 209 (1960).
[7] J. F. Perdue and W. H. McShan, *J. Cell Biol.* **15**, 159 (1962).
[8] H. G. Kwa, E. M. Van der Bent, C. A. Felkamp, P. Rümke, and H. Bloemendal, *Biochim. Biophys. Acta* **111**, 447 (1965).
[9] M. E. Krass, F. S. LaBella, and R. Mailhot, *Endocrinology* **84**, 1257 (1969).

Granules Obtained from Rat Pituitary Glands

Chromatography on Celite[10]

Ten adult male rats (Holtzman Company, Madison) were killed by cervical dislocation. The pituitary glands were removed immediately, separated from the posterior lobes, and placed in an ice-cold moist chamber. The glands were homogenized in a solution consisting of 0.25 M sucrose and 7.3% polyvinylpyrrolidone (PVP) adjusted to pH 7.4 to give a 5% homogenate. Homogenization was done in a ground-glass, sharp-pointed, cold homogenizing tube by mashing rather than grinding with the pestle to avoid undue breaking of the granules and solubilization of the hormones.

The homogenate was centrifuged in the multispeed head of a Model PR-2 International refrigerated centrifuge at 275 g for 10 minutes to remove blood cells, nuclei, and unbroken cells. The supernatant S1 was filtered on a Millipore filter with 5.0 μm diameter pores in a Swinny adapter fitted on a 5-ml syringe. The filtered S1 (FS1) served as the preparation for further fractionation by chromatography on columns of Celite and centrifugation on discontinuous density gradients.

The Celite columns were prepared as follows. A 10% (w/v) suspension of Celite Nos. 545, 535, or 503 (Johns-Manville Company, Chicago) in 0.25 M sucrose was allowed to settle, and the fine suspended particles were decanted. This procedure was repeated four times over a 2-hour period. A 1.5 $\times$ 20-cm column with a 3-cm length of 1 mm diameter capillary tubing at the bottom end of the tube was used. A small disk of silk mesh cloth was placed in the bottom of the column as support for a 1-cm layer of No. 203 glass microbeads (Microbeads, Inc., Jackson, Mississippi). A disk of Reeve Angel No. 202 filter paper was placed in the tube immediately over the glass beads. The Celite slurry was pipetted into the column to a level of 12.8 cm. A reservoir containing 0.25 M sucrose was then attached to the column, which was washed with the sucrose solution for 1 hour. The pH of the eluted sucrose was close to neutrality.

One milliliter of fraction FS1 was carefully placed on top of the Celite column. Four to five minutes were required for the fraction to pass into the column which was developed with 0.25 M sucrose, and 25–30 minutes later, 5 ml of cloudy suspension was collected. This suspension was divided into three equal parts (1.6 ml), and each aliquot was layered on discontinuous density gradients which were prepared previously by successively layering as follows: 0.5 ml 80% (w/v) sucrose, 0.9 ml 65% sucrose,

[10] W. C. Hymer and W. H. McShan, *J. Cell Biol.* **17**, 67 (1963).

0.9 ml 50% sucrose, and 0.9 ml 35% sucrose. The three gradient tubes were placed in the buckets of the SW 39L rotor and centrifuged at 100,000 g for 1 hour in a Spinco Model L centrifuge. A dense white band was readily identified at the 65 to 50% sucrose interface, a more diffuse lighter band was present at the 50 to 35% sucrose interface, and a faint band was present at the layering to 35% sucrose interface. These bands were designated, from top to bottom, as G1, G2, and G3.

The layers were removed from the gradient tubes with a long bent-tipped needle fitted to a 1-ml syringe. The syringe was held in a special holder which made it possible to lower the tip of the needle to the desired layer and withdraw it into the syringe.[6] The lower bands (G3) from the three gradients were recovered, combined, diluted to 5 ml with 0.88 M sucrose, and centrifuged in the swinging-bucket rotor at 40,000 g for 1 hour. A small white pellet designated as G3LSP (low speed pellet) was recovered in the bottom of the centrifuge tube. The supernatant from the G3LSP was decanted into another tube and centrifuged at 100,000 g for 1 hour. A small white pellet was recovered; it was designated G3HSP (high speed pellet).

The G2 layer was similarly obtained from the three gradient tubes, diluted with 0.88 M sucrose, and centrifuged at 100,000 g for 1 hour to obtain a small white pellet designated G2HSP. A pellet was not recovered from the top G1 layer by centrifugation at 100,000 g for 3 hours. This method is outlined in Fig. 1.

The above three pellets were prepared for bioassays to determine the hormones associated with each granule fraction and used for biochemical studies. The purity of the granules is shown in the electron micrographs in Figs. 2 and 3. Granules from pituitary glands of young male chickens have also been prepared by this procedure.[11]

The Celite should be washed well with 0.25 M sucrose before and after addition to the column until the effluent is very close to neutrality. Satisfactory results were not obtained in that granules were not eluted when the Celite was washed and the columns were developed with (1) 0.25 M sucrose which contained 7.3% PVP; (2) same as (1) with the addition of 0.9% saline; and (3) 0.25 M sucrose in 10 mM pH 7.0 phosphate buffer. These treatments may have resulted in changes in the charge or physical properties of the Celite that prevented elution of the granules.

Continuous Density Gradient Centrifugation[12]

The 5% pituitary homogenates were prepared and treated as given above under chromatography on Celite, except that the FS1 fraction was

[11] W. H. McShan, *Mem. Soc. Endocrinol.* 19, 161 (1971).
[12] A. Costoff and W. H. McShan, *J. Cell Biol.* 43, 564 (1969).

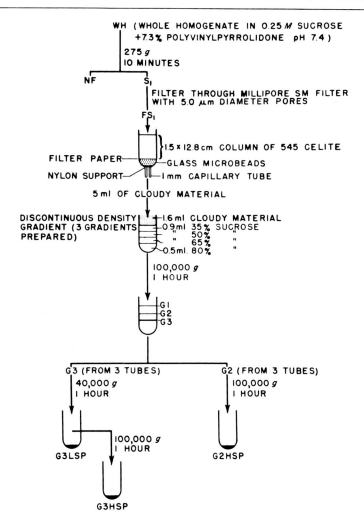

FIG. 1. Method for isolation of pituitary secretory granules. From W. C. Hymer and W. H. McShan, *J. Cell Biol.* **17**, 67 (1963).

obtained by filtration of the S1 through SS Millipore filters with pores 3.0 μm in diameter.

The columns were prepared as follows. Diodrast, the diethanolamine salt of 3,5-diiodo-4-pyridone-*N*-acetic acid, (Winthrop Laboratories, Evanston, Illinois) was obtained as a 35% (w/v) aqueous solution and mixed with sucrose for making the gradients. The solutions used were 6 and 45% (g/100 ml) sucrose each containing 17.5% (g/100 ml) Diodrast and 5×10^{-4} M EDTA. The pH of these solutions was adjusted to

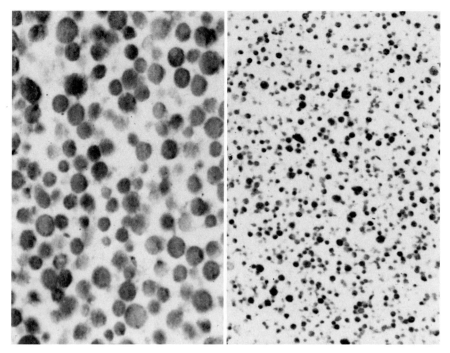

FIG. 2. An electron micrograph of a thin section through the large acidophilic granule pellet (G3LSP). ×25,000. From W. C. Hymer and W. H. McShan, *J. Cell Biol.* **17**, 67 (1963).

FIG. 3. An electron micrograph of a section of the small basophilic granule pellet (G3HSP). ×12,000. From W. C. Hymer and W. H. McShan, *J. Cell Biol.* **17**, 67 (1963).

7.2 with KOH. The machine used for preparing the gradients was calibrated to deliver a total of 4.4 ml into each tube.[6]

The FS1 fraction (1 ml) was layered on the surface of the gradient with a 2-ml syringe. The tubes were placed in a SW 39 swinging-bucket head and centrifuged in a Beckman Spinco Model L centrifuge for 2 hours at 100,000 g. The zones of the gradient were individually removed with a long needle attached to a 2-ml syringe placed in a special holder that enabled careful removal of the desired layer.[6]

The zones were removed from the top of the gradient and were identified as shown in Fig. 4. The SA top layers containing the soluble hormones were removed first. Zones B, C, and D1 were individually removed, diluted with 0.88 M sucrose, and centrifuged at 100,000 g for 1 hour to obtain the respective high speed pellets. Zones D2 and E1 were removed and slowly filtered through moistened Nuclepore filters (General Electric

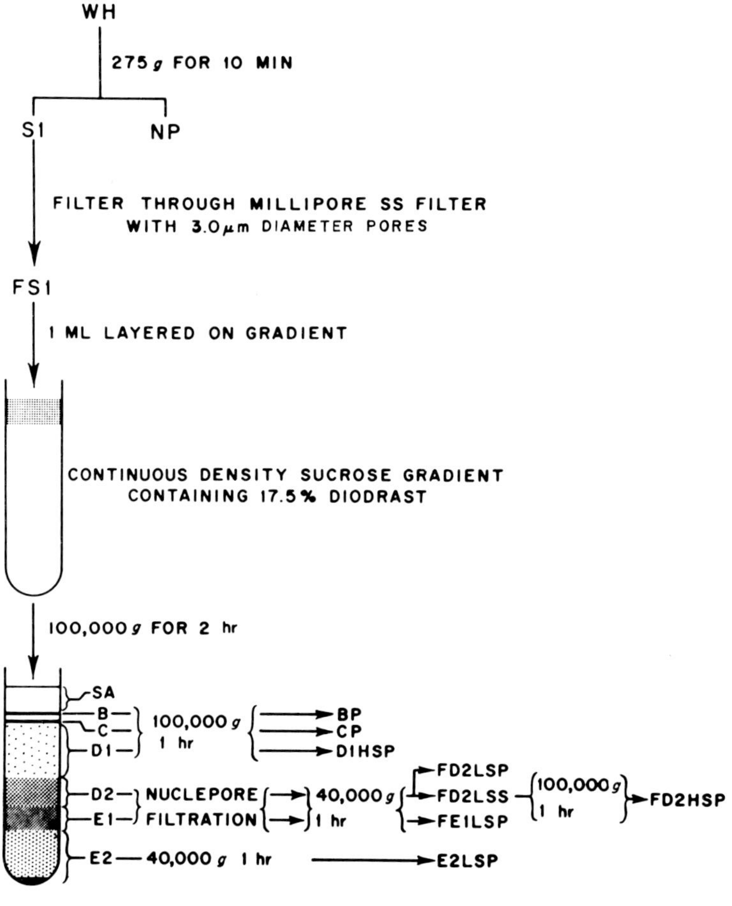

WH

275 g FOR 10 MIN

S1   NP

FILTER THROUGH MILLIPORE SS FILTER
WITH 3.0 μm DIAMETER PORES

FS1

1 ML LAYERED ON GRADIENT

CONTINUOUS DENSITY SUCROSE GRADIENT
CONTAINING 17.5 % DIODRAST

100,000 g FOR 2 hr

SA
B — } 100,000 g {——→ BP
C —   1 hr {——→ CP
D1 —   {——→ D1HSP

                              ┌—→ FD2LSP
D2 —} NUCLEPORE {—→} 40,000 g {—→ FD2LSS —{ 100,000 g } →FD2HSP
E1 —} FILTRATION {—→} 1 hr   {—→ FE1LSP   1 hr

E2 — 40,000 g 1 hr ————————→ E2LSP

Fig. 4. Method for isolation of rat pituitary granules. The designation of the different fractions is indicated. From A. Costoff and W. H. McShan, *J. Cell Biol.* 43, 564 (1969).

Co., Pleasanton, California) with pores 0.5 μm in diameter mounted in a Swinny adapter attached to a 2.0-ml Luer-Lok syringe. These filtered solutions, FD2 and FE1, were diluted and centrifuged at 40,000 g. The resulting filtered low speed pellets were designated FD2LSP and FE1LSP. The FE1LSP pellet is shown in Figure 5. When the supernatant FD2LSS from the FD2LSP pellet was centrifuged for 1 hour at 100,000 g a small pellet FD2HSP was obtained. The lowest zone, E2, which usually included a diffuse pellet at the bottom of the tube, was removed, diluted, and centrifuged at 40,000 g for 1 hour to obtain a pellet of large granules.

This method is repeatable and gives consistent results. The TSH and

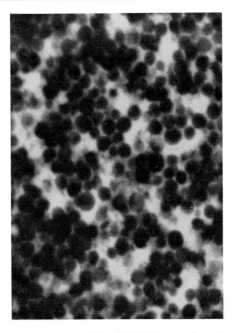

FIG. 5. An electron micrograph of the FE1LSP granule pellet with which growth hormone is associated. From A. Costoff and W. H. McShan, *J. Cell Biol.* **43**, 564 (1969).

ACTH were recovered in a single pellet, as were the FSH and LH. These granules were not separated by this method. Growth hormone (GH) and prolactin (PL) were recovered in separate pellets. The diameters of sections of granules from intact tissue and isolated granules from the corresponding cell types were measured, and they compared favorably. The granules were of high purity as indicated by electron microscopy. The biological activities of the main granule fractions recovered from the continuous density gradients ranged from 51.4% for LH to 89.3% for PL. The results indicate that the large acidophilic granules from the rat pituitary are more stable than the small basophilic granules.

Granules Obtained from Bovine Pituitary Glands

Procedures for Large and Small Granules[13]

The procedures were done at 4°. Bovine glands were homogenized in a solution of 0.33 M sucrose plus 30 mM Tris buffer of pH 7.3. The 5%

[13] J. T. Tesar, H. Koenig, and C. Hughes, *J. Cell Biol.* **40**, 225 (1969).

homogenate was centrifuged at 900 g to remove the nuclear fraction, filtered through Millipore filters having 5, 1.2, and 0.8 μm diameter pores, centrifuged at 4000 g for 20 minutes twice to give the LP1 suspension, which was placed on a 65, 57, 55, and 50% layered sucrose discontinuous gradient and centrifuged at 75,000 g for 90 minutes to give the large-granule fraction LGF. The latter was suspended and centrifuged on a 57 to 55% sucrose gradient to give the final large-granule pellet LGF.

Centrifugation of a 10% homogenate twice at 900 g and the resultant supernatant SS1 at 5000 g was followed by filtration of the supernatant SS2 through Millipore filters of 1.2, 0.65, 0.45, and 0.30 μm pore sizes, and centrifugation of the FSS2 at 14,000 g twice to give the SP2 fraction. The SP2 suspension was layered on a continuous gradient of 10 to 40% sucrose in 17.5% Diodrast and centrifuged at 75,000 g for 3 hours. The

Fractionation Procedure

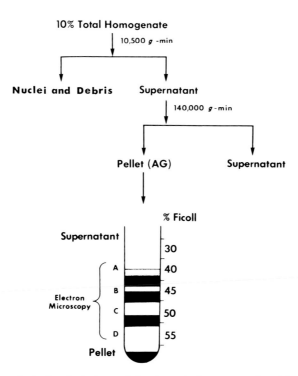

FIG. 6. Method for isolation of bovine pituitary granules. F. LaBella, M. Krass, W. Fritz, S. Vivian, S. Shin, and G. Queen, *Endocrinology* 89, 1094 (1971).

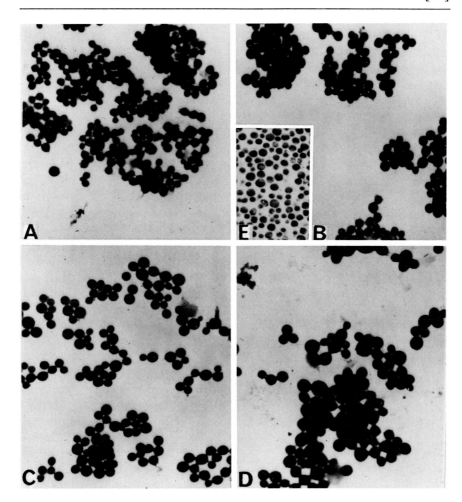

FIG. 7. Electron micrographs of fractions A, B, C, and D. (× approximately 6100). Plates A–D are smears of osmic acid-fixed fractions. Plate E is a section of an osmic acid-fixed pellet (× approximately 4600). From F. LaBella, M. Krass, W. Fritz, S. Vivian, S. Shin, and G. Queen, *Endocrinology* **89**, 1094 (1971).

SGF fraction obtained was diluted with 1 *M* sucrose, layered on a 55 to 50% sucrose gradient, and centrifuged at 105,000 *g* for 80 minutes to give the final small granule pellet SGF.[13]

The small (TSH, FSH, and LH) and large granules (GH and PL) were highly purified, as indicated by electron micrographs, biological activities, and enzyme studies, except that there was a significant amount of alkaline protease associated with the small-granule fraction.

Preparation of Growth Hormone and Prolactin Granules[14]

Bovine pituitary glands were collected and kept on ice until received at the laboratory. The following procedures were performed at 4°. The lobes were freed of connective and other tissues, minced, and homogenized in 8.5% sucrose (w/v) in a motor-driven Teflon-pestle glass vessel to give a 10% homogenate (w/v); 25–30 g were homogenized for a single experiment. The homogenate was centrifuged at 10,500 *g* to remove nuclei and other debris. Heparin was added to the supernatant fraction and to other sucrose media in a final concentration of 0.02%.[5] This supernatant was centrifuged at 140,000 *g* to give the AG fraction which consisted of a white pellet and a fluffy overlying layer. The latter was decanted from the white pellet by use of the sucrose medium. The AG fraction was resuspended twice and centrifuged in a sucrose medium.

A discontinuous density gradient was made by layering 2.5 ml each of 55 and 50%, and 2 ml each of 45, 40, and 30% Ficoll (w/v) in 8.5% sucrose into 1 × 14 cm tubes. The AG fraction (50 mg) was suspended in 1 ml of 8.5% sucrose and layered on the surface of the gradient. This was followed by centrifugation at 400,000 *g* in an International B60 centrifuge with a SB 283 swing-out rotor. The density bands were removed from top to bottom with a Pasteur pipette. This procedure is outlined in Fig. 6. The electron micrographs shown in Fig. 7, and the lack of contamination as shown by enzyme determinations, indicate that these are the most highly purified pituitary granules that have been reported.[14]

Acknowledgments

Special thanks are extended to W. C. Hymer, Allen Costoff, Charles Hughes, and F. S. LaBella for permission to use their material in this review.

[14] F. LaBella, M. Krass, W. Fritz, S. Vivian, S. Shin, and G. Queen, *Endocrinology* **89**, 1094 (1971).

[44] The Isolation of Lung Lamellar Bodies

By CHARLES H. WILLIAMS

The lamellar body of the lung is a highly organized, multilayered, membranous organelle primarily localized within the granular pneumocyte, or type II, cells. The granular pneumocyte cells are distributed in the surface layer of flattened epithelial cells at the air–water interface in the alveolar space of lung tissue.[1] These highly differentiated cells are thus strategically located in lung tissue, and Macklin has suggested that they have an exocrine function.[2] Studies in several laboratories suggested that the exocrine function of the granular pneumocyte was to produce "lung surfactant."[3-6]

Electron microscopic studies appeared to show lamellar bodies being discharged into the alveolar space.[7] Other studies of lung tissue correlated the formation of lamellar bodies in the granular pneumocyte cells with an increase in surfactant activity, or conversely, a decrease in surfactant activity correlated with a disappearance of lamellar bodies.[8-10] It has also been suggested that the granular pneumocyte has phagocytic activity.[10-13]

The isolation and subsequent biochemical study of this unique subcellular organelle will aid in understanding the role of the lamellar body in lung biochemistry and physiology.

Materials and Methods

Lung Tissue. Bovine,[14] alpaca, or guinea pig lung tissue has been used to prepare lamellar body fractions. The lung tissue, obtained at the abattoir, was immediately placed on ice and transported to the laboratory. A gross pathological examination of the lung tissue was conducted, and diseased tissue was rejected. The healthy lung tissue was carefully trimmed to exclude as much as possible of large bronchi and collagenous material. It was then ground twice in a motor-driven meat grinder to a very fine consistency.

Isolation Procedure (Fig. 1). Ground lung tissue (500 g) was washed with three volumes of mannitol preparation medium (MPM) at 0°C. The MPM solution was 0.21 M in mannitol, 0.07 M in sucrose, 10 mM in

[1] E. M. Scarpelli, "The Surfactant System of the Lung," p. 34. Lea & Febiger, Philadelphia, Pennsylvania, 1968.
[2] C. C. Macklin, *Lancet* **266**, 1099 (1954).
[3] C. S. Faulkner, *Arch. Pathol.* **87**, 521 (1969).
[4] E. M. Scarpelli, *Advan. Pediat.* **16**, 177 (1969).
[5] V. E. Goldenberg, S. Buckingham, and S. C. Sommers, *Lab. Invest* **20**, 147 (1969).
[6] Y. Kikkawa, E. K. Motoyama, and L. Gluck, *Resp. Distress* **52**, 177–209 (1968).
[7] K. Bensch, K. E. Schaefer, and M. E. Avery, *Science* **145**, 1318 (1964).
[8] K. E. Schaefer, M. E. Avery, and K. Bensch, *J. Clin. Invest.* **43**, 2080 (1964).
[9] G. W. Brumley, V. Chernick, W. A. Hodson, C. Normand, A. Fenner, and M. E. Avery, *J. Clin. Invest.* **46**, 863 (1967).
[10] K. Hatasa and T. Nakamura, *Z. Zellforsch. Mikrosk. Anat.* **68**, 266 (1965).
[11] A. H. Niden, *Science* **158**, 1323 (1967).
[12] B. Corrin, *Thorax* **24**, 110 (1969).
[13] B. Corrin and A. E. Clark, *Histochemie* **15**, 95 (1968).
[14] C. H. Williams, W. J. Vail, R. A. Harris, D. E. Green, and E. Valdivia, *Prep. Biochem.* **1**, 37–45 (1971).

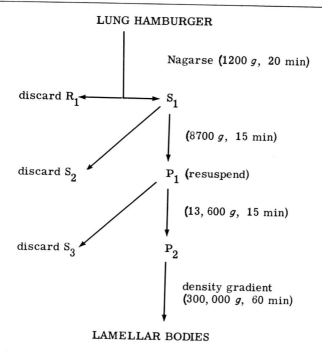

FIG. 1. Schematic outline of the isolation procedure for bovine lamellar bodies (No. 82185-C-1).

maleate neutralized with Tris to pH 6.5 and 1.0 mM in $MgCl_2$. The lung suspension was adjusted to pH 6.5 by the addition of 1.0 M Tris·HCl during the washing. The washed ground lung was collected by filtration through cheesecloth and resuspended in two volumes of MPM. After adjustment to pH 6.5, the suspension was digested for 5 minutes with Nagarse, a proteolytic enzyme, at 0–3° using 2 mg of Nagarse enzyme per gram of lung tissue. The lung suspension was continuously stirred, and the pH was maintained between 6.4 and 6.5 by the addition of 1.0 M Tris during digestion. The digested lung was then briefly homogenized at 2500 rpm with two passes of a glass-Teflon Potter-Elvehjem homogenizer. The clearance for the homogenizer should be approximately 35 μm.

The homogenized lung preparation was poured into 250-ml glass centrifuge bottles previously cooled to 3° and sedimented at 1200 g for 20 minutes in an International Model PR-2 centrifuge at 0°. See Fig. 1 for a schematic flow sheet.

The supernatant (S_1) was decanted and filtered through cheesecloth, adjusted to pH 6.5, and centrifuged at 8700 g for 15 minutes. The pellet contained three distinct layers: a dark amber-colored layer of heavy mitochondria, a light amber-colored layer of swollen and damaged mitochondria, and a nearly white layer which contained the lamellar bodies. The supernatant (S_2) contained the microsomal fraction and soluble cellular components. After the S_2 was removed and discarded, the upper layer of the pellet was resuspended by gently washing the pellet with a few milliliters of MPM media. The red blood cell portion of the pellet and the mitochondria were discarded. The suspension was resedimented

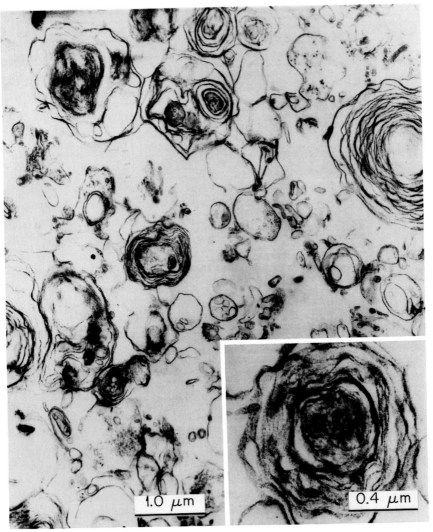

FIG. 2. Electron micrograph (No. 7896-C-1) of the isolated lamellar body fraction before sucrose density gradient centrifugation.

at 13,600 g for 15 minutes. The layers of the pellet were more sharply defined after washing and resedimentation. The wash supernatant (S_3) was discarded and the lamellar body fraction was resuspended by adding 2 ml of MPM to the centrifuge tube and gently rinsing the lamellar body layer off the surface of the mitochondrial pellet with a Pasteur pipette. The lamellar body fraction was easily resuspended without homogenization. The yield of lamellar bodies from 500 g of ground lung tissue was a suspension containing 7–10 mg dry weight of protein.

Density Gradient Centrifugation. Sucrose density gradient fractionation was used to separate contaminating membranes and organelles from the lamellar body fraction. A continuous sucrose gradient from 0.8 M to 1.5 M sucrose was prepared for density gradient separation of the lamellar body preparation. The lamellar body fraction was carefully layered on top of the sucrose gradient and centrifuged at 300,000 g for 1 hour in a Spinco Model L2-65B ultracentrifuge equipped with a SW 65 rotor. The lamellar body fraction separated into four fractions on the sucrose gradient. The top layer was white in our best preparations. This band contained membrane fragments and lamellar bodies. The middle band contained tightly packed lamellar bodies and very few membrane fragments; however, the yield of this fraction was very low. The lower band consisted of microbodies and other extraneous components. The dark-colored pellet at the bottom of the tube consisted mainly of mitochondria which contaminated the lamellar body preparation. The upper band and middle band were used for further studies.

TABLE I

ENZYMATIC ACTIVITY OF UPPER LAMELLAR BODY FRACTION

Enzyme	Activity[a]
Glutamic dehydrogenase	None
Cytochrome oxidase	None
ATPase	114
Acid phosphatase	593
Alkaline phosphatase	185

[a] Measured as nanomoles per minute per milligram of protein. Glutamic dehydrogenase activity was measured by the method of E. Schmidt [*in* "Methods of Enzymatic Analysis" (H. U. Bergmeyer, ed.), p. 752. Academic Press, New York, 1963]. Cytochrome oxidase activity was measured by the method of D. E. Griffiths and D. C. Wharton [*J. Biol. Chem.* **236,** 1850 (1961)]. ATPase activity was measured by the method of O. Lindberg, and L. E. Ernster [*in* "Methods of Biochemical Analysis" (D. Glick, ed.), Vol. 3, p. 1. Wiley (Interscience), New York, 1956. Acid and alkaline phosphatase activities were determined with Sigma Test Kits, Sigma Chemical Co., St. Louis, Missouri.

TABLE II

PHOSPHOLIPID COMPOSITION OF UPPER LAMELLAR BODY FRACTION[a]

Phospholipid	Percent
Phosphatidylcholine	46.6 ± 5
Phosphatidylethanolamine	20.4 ± 3
Sphingomyelin	15.9 ± 3
Phosphatidylserine	6.4 ± 2
Phosphatidylinositol	1.2 ± 0.5
Diphosphatidylglycerol	0.37
Unidentified	0.77
Other	4.26 ± 2

[a] Of the total lipid, 71.6% was phospholipid. The lipid:protein ratio was 1.15:1 (w/w) in the upper lamellar body. Phospholipids were separated on Supelco Inc., Redicoat TLC plates using the two-dimensional system of J. D. Turner and G. Rouser [*Anal. Biochem.* **38**, 426 (1970)]. Phosphorus analysis was performed with the method described by G. R. Bartlett [*J. Biol. Chem.* **234**, 466 (1959)].

Characterization of the Lamellar Body Preparation. Initially, electron microscopic studies of the various cellular fractions were employed in order to evaluate the preparative procedure. Light and heavy mitochondrial fractions, microsomal fractions and lamellar body fractions were studied. The micrograph shown in Fig. 2 confirmed the fact that lamellar bodies could be isolated from lung tissue as relatively intact organelles.

Enzymatic assays were also used to characterize the lamellar body fraction. Table I shows that mitochondrial enzymes are absent from the upper layer lamellar body fraction. However, ATPase, acid phosphatase, and alkaline phosphatase activities were present.

Because of the intense interest in this organelle as the possible site of "lung surfactant" synthesis or assimilation, an analysis of the phospholipid composition of this fraction was completed. Table II presents the data on the phospholipid composition.

Discussion

The isolation procedure, as developed, combines the enzymatic digestion of ground lung tissue with a very gentle homogenization in order to preserve the intact structure of the lamellar body. The clearance of the homogenizer as well as the extent of the homogenization are critical factors in the isolation of the lamellar bodies. A Waring Blendor or a homogenizer with a tight-fitting pestle will produce a smoother homogenate, but the lamellar bodies are no longer recognizable by electron microscopy. Excessive homogenization apparently disintegrates the lamellar bodies and releases other intracellular membranes which completely obscure the differential centrifugation steps in the preparative procedure. Homogenization with a very loose Potter-Elvehjem homogenizer releases part of the mitochondrial fraction, but does not release the lamellar bodies. Because of the porous, elastic nature of lung tissue, it was difficult to achieve a smooth homogenate and homogenization was considered to be satisfactory when approximately one-half of the homogenized lung floats and the remainder sediments in the initial centrifugation step.

The isolation and purification of lamellar bodies from lung tissue provides a good starting point for a series of experiments to determine the physiological and biochemical functional properties of this unique subcellular organelle.

Acknowledgments

This work was supported in part by grants from the Consultation Practice Plan of the University of Wisconsin Medical School and United States Public Health Service Program Project Grant GM-12847.

[45] Procedures for the Isolation of Lipofuscin (Age Pigment) from Normal Heart and Liver

By A. N. SIAKOTOS and BERNARD L. STREHLER

The distinctions among the lipofuscins (age pigments) isolated from various mammalian organs and other autofluorescent lipopigments, such as ceroid (the vitamin E deficiency pigment), have been clearly defined only recently.[1,2] Hendley et al.[3] utilized blender homogenization and sonication to isolate lipofuscin from heart. Björkerud[4] also published an isopycnic centrifugation method in 1963 using a less vigorous homogenization procedure. Although Björkerud[5] later summarized the preceding sixty years of research efforts on the isolation and characteristics of lipofuscin and other autofluorescent lipopigments from heart tissues, considerable speculation remained as to the ideal approach for the isolation of

[1] A. N. Siakotos, I. Watanabe, A. Saito, and S. Fleischer, *Biochem. Med.* **4**, 361 (1970).

[2] A. N. Siakotos *et al., in* "Third International Symposium on Sphingolipids, Sphingolipidoses and Allied Disorders (B. Volk and S. Aronson, eds.), p. 53. New York, 1972.

[3] D. D. Hendley, A. S. Mildvan, M. C. Reporter, and B. L. Strehler, *J. Gerontol.* **18**, 144 (1963).

[4] S. Björkerud, *J. Ultrastruct. Res.*, Suppl. 5 (1963).

[5] S. Björkerud, *Advan. Gerontol. Res.* **1**, 257 (1964).

Trim, discard valves, aorta, and adipose tissue; grind heart (2), blend in solution 1 (3, 4, 5), filter (6), pass through Parr bomb (7), centrifugation as follows:

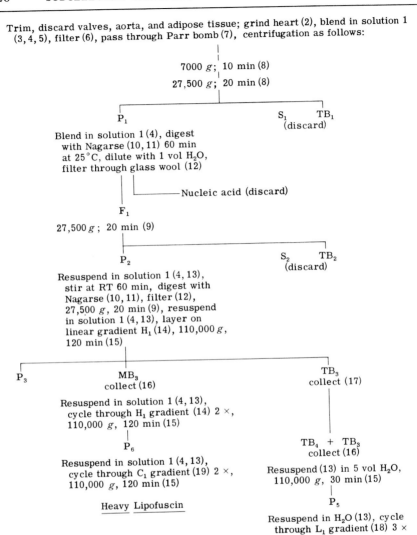

7000 *g*; 10 min (8)

27,500 *g*; 20 min (8)

P₁

S₁   TB₁
(discard)

Blend in solution 1 (4), digest with Nagarse (10, 11) 60 min at 25°C, dilute with 1 vol H₂O, filter through glass wool (12)

Nucleic acid (discard)

F₁

27,500 *g*; 20 min (9)

P₂

S₂   TB₂
(discard)

Resuspend in solution 1 (4, 13), stir at RT 60 min, digest with Nagarse (10, 11), filter (12), 27,500 *g*, 20 min (9), resuspend in solution 1 (4, 13), layer on linear gradient H₁ (14), 110,000 *g*, 120 min (15)

P₃

MB₃
collect (16)

TB₃
collect (17)

Resuspend in solution 1 (4, 13), cycle through H₁ gradient (14) 2 ×, 110,000 *g*, 120 min (15)

P₆

TB₄ + TB₃
collect (16)

Resuspend in solution 1 (4, 13), cycle through C₁ gradient (19) 2 ×, 110,000 *g*, 120 min (15)

Resuspend (13) in 5 vol H₂O, 110,000 *g*, 30 min (15)

Heavy Lipofuscin

P₅

Resuspend in H₂O (13), cycle through L₁ gradient (18) 3 ×

Light Lipofuscin

Fɪɢ. 1. Isolation of lipofuscin from normal human heart. (1) Centrifugations, manipulations, and solutions at 4°. Abbreviations: P, pellet; S, supernatant; TL, top layer; MB, middle band; F, filtrate; R, residue. (2) Meat grinder (607), Sears Roebuck Co., Indianapolis, Indiana. (3) Make to 750 ml per 200 g of heart. (4) 0.4 *M* sucrose, 50 mM Tris, 10 m*M* EDTA, 10 m*M* KCl, 0.01% dextran sulfate 2000, pH 8.0. (5) Blend ground heart 10 seconds in Waring Blendor (Model CB-6). (6) 20-mesh sieve. (7) 900 psi Parr Cell (4635) Disruption Bomb, Parr Instrument Co., Moline, Illinois. (8) Break foam, Sorvall RC-3, HG-4 rotor. (9) Sorvall RC-2B, GSA rotor or I.E.C. B-60, A-54 rotor. (10) 25 mg/100 g heart. (11) Maintain large quantities of pure lipopigments for the careful chemical studies necessary to establish the unique features of lipofuscins from various organs and the experimentally or pathologically induced liver pigment, ceroid.

The methods given below are based on a modification of techniques for brain lipopigments[1]; however, these methods have been modified to confer organ specificity for heart and liver, as well as to avoid vigorous disruptive procedures which may destroy or redistribute lipid-rich lipofuscin components. In addition, specific particle density anomalies observed by earlier investigators have been resolved by controlling salt concentration during the isolation operations. The techniques presented are applicable to gram or kilogram quantities of fresh or frozen heart and liver.

Isolation of Heart Lipofuscin[6]

Reagents and Gradients

Solution 1: 0.4 M sucrose, 50 mM Tris base, 10 mM EDTA, 0.01% dextran sulfate 2000, pH 8.0
Nagarse, Enzyme Development Corp., New York, New York
H_1 gradient: linear, solution 1 to 2.0 M sucrose
L_1 gradient: linear, H_2O to 20% sodium chloride
C_1 gradient: linear, solution 1 to 2.0 M sucrose containing 25% CsCl

Procedure.[6] An outline of the isolation procedure for normal heart lipofuscin is given in Fig. 1. Frozen heart samples were thawed overnight at 4°; the hearts were trimmed free of adipose tissue, aortas, valves, etc., and cut into 1–2-inch cubes. The chopped tissue was ground in a meat grinder[7] and blended for 30 seconds in a 1-gallon Waring Blendor[8] in solution 1 and adjusted to pH 8.0. The homogenate was then filtered through a 20-mesh sieve to remove connective tissue and other debris,

[6] A. N. Siakotos, I. Watanabe, K. Pennington, and M. Whitfield, *Biochem. Med.* 7, 24 (1973).
[7] Model 607, Sears Roebuck & Co., Indianapolis, Indiana.
[8] Model CB-5, Waring Products Co., Winsted, Connecticut.

pH at 7.0–7.5. (12) Filter off nucleic acid and debris through 20-mesh sieve, then through dry Pyrex glass wool (Corning 3950), squeeze dry, repeat once more. (13) Resuspend with Potter-Elvehjem homogenizer, 1200 rpm. (14) Linear gradient H_1, 0.4 to 2.0 M sucrose. (15) I.E.C. B-60, SB-110 rotor. (16) Collect by aspiration, dilute with 1 volume of H_2O and concentrate by centrifugation 27,500 g, 20 minutes (9). (17) Collect by aspiration, add 0.25 volume of 2 M sucrose, and concentrate to TB_4 by centrifugation 110,000 g, 45 minutes (15). (18) Linear gradient L_1, H_2O to 20% NaCl. (19) Linear gradient C_1, solution 1 to 25% CsCl in 2 M sucrose. (20) 250 ml solution 1 per heart.

Grind liver (2), blend in solution 1 (3, 4, 5), filter (6), pass through Parr bomb (7), centrifuge as follows:

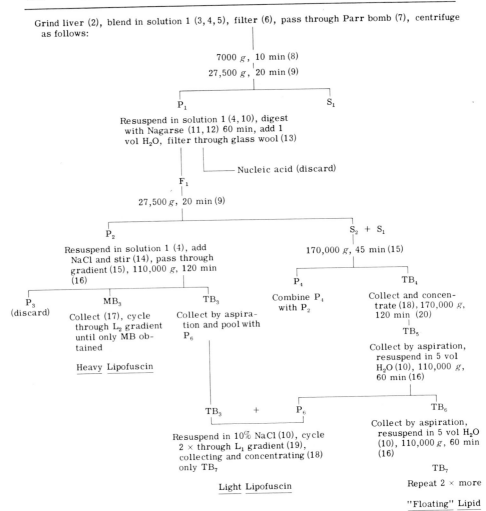

FIG. 2. Isolation of lipofuscin from normal human liver. (1) Centrifugation, manipulations, and solutions at 4°. Abbreviations: P, pellet; S, supernatant; TL, top layer; MB, middle band; F, filtrate; R, residue. (2) Meat grinder (607), Sears Roebuck Co., Indianapolis, Indiana. (3) 30 Seconds in a 1-gallon Waring Blendor (Model CB-6). (4) 0.4 M sucrose, 50 mM Tris, 10 mM EDTA, 10 mM KCl, 0.01% dextran sulfate 2000, pH 8.0. (5) Make to 500 g of liver per 500 ml of solution 1. (6) Filter through 20-mesh sieve. (7) 900 psi, Parr 4635 Cell Disruption Bomb, Parr Instrument Co., Moline, Illinois. (8) Break foam, Sorvall RC-3, HG-4 rotor. (9) Sorvall RC-2B; GSA rotor or I.E.C. B-60, A-54 rotor. (10) Resuspend with Potter-Elvehjem homogenizer, 1200 rpm. (11) 100 mg Nagarse per kilogram of liver. (12) Maintain pH 7.0–7.5. (13) Filter off nucleic acid and debris through 20-mesh sieve, then through dry Pyrex glass wool (Corning 3950); squeeze dry, repeat once more. (14) Add 1 volume of 20% NaCl and stir at room temperature

followed by treatment at 900 psi in a 495 Parr cell disruption bomb.[9] The pressure-treated homogenate was centrifuged at 7000 g for 10 minutes to break the foam and then mixed and centrifuged an additional 20 minutes at 27,500 g in the Sorvall GSA rotor or an I.E.C. A-57 rotor. The floating layer of fat (TB_1) and the supernatant (S_1) were discarded. Pellet (P_1) was resuspended in 500 ml solution 1 plus 26 mg Nagarse per adult heart and digested with stirring for 60 minutes at room temperature, maintaining the pH to 7.5–8.0. The digest was then diluted with 1 volume of distilled water, filtered through a 20-mesh sieve, followed by filtration twice through dry Pyrex glass wool. The filtrate F_1 was then centrifuged at 27,500 g for 20 minutes in the Sorvall GSA rotor to yield a pellet (P_2), supernatant solution (S_2) and a layer of floating lipid (TB_2). Both S_2 and TB_2 were discarded. P_2 was resuspended in solution 1 with 3–4 passes of a Potter-Elvehjem homogenizer at 1200 rpm, layered on a linear gradient H_1 (solution 1 to 2.0 M sucrose), and centrifuged at 110,000 g for 120 minutes in the I.E.C. SB-110 swinging-bucket rotor.

The top band of brown material (TB_3) was collected by aspiration, concentrated by adding 0.25 volume of 2 M sucrose, and centrifuged at 110,000 g for 30 minutes in the I.E.C. SB-110 rotor. The result was a top band (TB_3) noted as a diffuse floating layer of brown pigment (TB_4), which was collected again by aspiration, reconcentrated by adding 0.25 volume of 2 M sucrose, and centrifuged at 110,000 g for 15 minutes in the I.E.C. SB-110 rotor to yield a highly concentrated TB_4. TB_4 was resuspended with a Potter-Elvehjem homogenizer in 5 volumes of H_2O and centrifuged at 110,000 g for 30 minutes in the SB-110 rotor to float off the contaminating "floating lipid." The pellet, P_5, was resuspended in distilled water plus DS 2000 to 0.01% and cycled three times through an L_1 gradient (distilled water to 20% sodium chloride). Each time only the floating layer of brown pigment was collected to yield a final preparation of *light lipofuscin*.

MB_3 or the middle brown band from the H_1 gradient (solution 1 to 2 M sucrose) separation step of the P_2 treatment (Fig. 1) was collected by aspiration, diluted with 1 volume of distilled water and concentrated as a pellet by centrifugation at 27,500 g for 20 minutes in the Sorvall

[9] Parr Instrument Co., Moline, Illinois.

for 60 minutes. (15) Linear gradient L_2, 1.0 to 1.5 M sucrose containing 15% NaCl. (16) I.E.C. B-60 ultracentrifuge, SB-110 rotor. (17) Collect by aspiration, dilute with 1 volume of H_2O, concentrate by centrifugation 27,500 g, 20 minutes (9). (18) Concentrate by adding 0.25 volume of 2 M sucrose, 170,000 g, 30 minutes, I.E.C. A-170 rotor. (19) Linear gradient L_1, H_2O to 20% NaCl. (20) I.E.C. ultracentrifuge B-60, B-29 rotor.

GSA rotor. Only the brown pellet was resuspended with a Potter-Elvehjem homogenizer in solution 1 and cycled through the H_1 gradient for two additional cycles to yield a concentrated pellet (P_6). The pellet (P_6) was resuspended in solution 1 and cycled twice through C_1 gradient (solution 1 to 2 M sucrose containing 25% CsCl). The middle brown band was collected each time and purified to yield a preparation of *heavy lipofuscin*.

Isolation Procedure for Lipofuscin from Normal Liver[6]

An outline of the isolation procedure for normal human liver is given in Fig. 2. Frozen liver samples were thawed overnight at 4° and cut into 1–2-inch cubes. The cubed tissue was passed through a meat grinder[7] and blended for 30 seconds in solution 1 (500 g of liver per liter) in a 1-gallon Waring Blendor.[8] The suspension was filtered through a 20-mesh sieve to remove connective tissue and debris and then passed through the Parr Cell Disruption bomb[9] at 900 psi. The resulting homogenate was centrifuged at 7000 g for 10 minutes in the Sorvall RC-3 (HG-4 rotor) to break the foam and then centrifuged at 27,500 g for 20 minutes in the Sorvall RC-2B (GSA rotor) to yield a pellet, P_1, and supernatant, S_1.

P_1 was collected and resuspended in solution 1 plus 100 mg of Nagarse per kilogram of liver. The suspension was placed on a magnetic stirrer and digested at room temperature for 60 minutes. One volume of cold distilled water was added, and the mixture was filtered first through a 20-mesh sieve and then twice through dry Pyrex glass wool to remove the coagulated nucleic acid. The glass wool filter was squeezed dry each time. The filtrate, F_1, was concentrated by centrifugation: 27,500 g for 20 minutes in the Sorvall RC-2B (GSA rotor). The supernatant solution S_2 was collected and pooled with the supernatant S_1 while pellet P_2 was resuspended in 1 volume of 20% sodium chloride made to 0.01% DS 2000 with a Potter-Elvehjem homogenizer, stirred for 60 minutes at room temperature, and layered on a linear gradient (L_2), solution 1 to 2.0 M sucrose. The gradient tubes and samples, centrifuged in an I.E.C. B-60 ultracentrifuge at 110,000 g for 120 minutes resulted in a pellet (P_3), a middle band of brown pigment (MB_3) and a top band of brown pigment (TB_3). The MB_3 and pellet P_4 were collected and resuspended in 1 volume of 20% sodium chloride containing 0.01% DS 2000 and recycled through the L_2 gradient until no top band was observed. This final middle band was collected, *heavy lipofuscin*. The top band fractions collected from the L_2 gradient steps were collected and pooled with TB_3. The pooled TB_3 and TB_4 fractions were concentrated by adding 0.25 volume of 2 M sucrose and centrifuged to obtain a concentrated top band of pigment, TB_5. TB_5 was resuspended in 5 volumes of H_2O and centrifuged at 110,000 g for 60 minutes in the I.E.C. SB-110 rotor to float off contami-

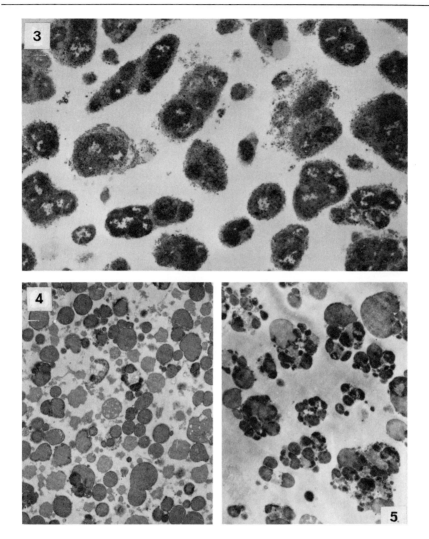

FIG. 3. An isolated preparation of myocardial lipofuscin showing the absence of contaminating structures. ×8000.

FIG. 4. A floating lipid preparation from normal human liver shows many small lipid droplets. Between these spherical droplets, fragmented or minute lipid material is also observed. Most of these lipid bodies are enclosed by membranelike structures. ×6500.

FIG. 5. An isolated liver lipofuscin preparation without interfering nonlipopigment structures. ×7500.

nating "floating" lipid particles (TB$_6$) from the light lipofuscin (P$_6$). P$_6$ was resuspended in distilled water and cycled twice through a linear L$_1$ gradient (distilled water to 20% sodium chloride) at 110,000 g for 60 minutes (SB-110 rotor). Only the top band was collected from the L$_1$ gradient samples. The remaining middle bands were cycled until no further top band material was observed. The final isolate (TB$_7$) was termed *light lipofuscin*. The "floating" lipid layer, TB$_6$, was collected by aspiration, resuspended in 5 volumes of distilled water and purified by centrifugation at 110,000 g for 60 minutes, yielding TB$_8$. This process was repeated for two additional cycles to give a "floating" lipid fraction.

Figures 3–5 show preparations electron micrographs of isolated lipofuscins.

Section III

Subcellular Fractions Derived from Nerve Tissue

[46] Isolation of Myelin from Nerve Tissue

By WILLIAM T. NORTON

Myelin is a sheath that invests most large (>1 μm diameter) nerve fibers of higher organisms. It functions as an insulator, enabling nerve conduction to be considerably more rapid and energetically more efficient. The mammalian myelin sheath is similar in microscopic appearance in both the central nervous system (CNS) and peripheral nervous system (PNS). However, it is generated by different types of cells in these two tissues, and it has a different ultrastructure, somewhat different physical properties, and different chemical composition. There are, as well, compositional differences between species, and, in fact, in any one species spinal cord myelin will be chemically different from brain myelin. It is even possible that myelin from other subregions of the CNS may differ to some extent.

Myelin is an extension of the plasma membrane of the cell which produces it. The generating cell is the oligodendroglial cell in the CNS and the Schwann cell in the PNS. This sheet of membrane wraps around the nerve axon in jelly-roll fashion. In the mature sheath these membranes condense into a compact multilamellated structure, with the protein layers of each unit membrane apparently fusing with the protein layer of the unit membrane apposed to it. In the PNS, which has thicker sheaths than the CNS, there may be as many as 100 layers of double unit membranes (i.e., 200 unit membranes, each consisting of a lipid bilayer with protein coats) surrounding a single large nerve fiber. Reference to diagrams or electron micrographs of such structures will show that the myelin sheath itself consists primarily of this condensed multilayered structure together with varying amounts of cytoplasm of the generating cell.[1,2] These pockets of cytoplasm occur both inside the compacted layers, adjacent to the axon and outside them, as well as within the layers (Schmidt-Lantermann clefts), and at the nodal region, where the sheath generated by one cell ends and that by another cell begins. However, in this chapter, myelin will be defined only as the compact multilayered structure. The methods of isolation have been designed to isolate this structure and exclude cytoplasmic components.

Although myelin has been demonstrated to be continuous with the

[1] A. N. Davison and A. Peters, "Myelination." Thomas, Springfield, Illinois, 1970.
[2] A. Hirano, *in* "The Structure and Function of Nervous Tissue" (G. H. Bourne, ed.), Vol. 5, p. 73. Academic Press, New York, 1972.

plasma membrane of its generating cell, because of its unique properties it is believed to be quite different from the true plasma membrane around the cell perikaryon. These properties are; high lipid content (70–80%); a relatively simple protein composition, and a small number of enzyme activities.

Myelin is present in all parts of the nervous system but is, of course, more concentrated in areas composed mainly of fiber tracts, such as white matter of brain (e.g., corpus callosum, centrum semiovale, and optic nerve) and peripheral nerve trunks, such as sciatic nerve, that contain large motor fibers. In such areas it comprises a very large part of the tissue. It can be calculated that mammalian brain white matter is about 50–60% myelin on a dry weight basis.[3] Even in the whole brain of an adult rat, myelin is approximately 25% of the dry weight, and accounts for 40% or more of the total brain lipid.[3] Spinal cord, which has more white matter than brain, has more myelin than does whole brain.

It is obvious that myelin is a very plentiful substance. This fact, combined with its specific physical properties, enable it to be readily isolated in high yield and high purity by conventional subcellular fractionation procedures. A large number of procedures have been described,[4–13] and many of the results of these investigations have been covered in several recent reviews.[1,3,14–25] All the myelin isolation procedures take advantage

[3] W. T. Norton, in "The Cellular and Molecular Basis of Neurologic Disease" (G. M. Shy, E. S. Goldensohn, and S. H. Appel, eds.). Lea & Febiger, Philadelphia, Pennsylvania, in press.

[4] R. H. Laatsch, M. W. Kies, S. Gordon, and E. C. Alvord, J. Exp. Med. 115, 777 (1962).

[5] L. A. Autilio, W. T. Norton, and R. D. Terry, J. Neurochem. 11, 17 (1964).

[6] M. L. Cuzner, A. N. Davison, and N. A. Gregson, J. Neurochem. 12, 469 (1965).

[7] L. A. Horrocks, J. Neurochem. 15, 483 (1968).

[8] A. N. Siakotos, G. Rouser, and S. Fleischer, Lipids 4, 239 (1969).

[9] R. Shapira, F. Binkley, R. F. Kibler, and I. J. Wundram, Proc. Soc. Exp. Biol. Med. 133, 238 (1970).

[10] M. G. Rumsby, P. J. Riekkinen, and A. V. Arstila, Brain Res. 24, 495 (1970).

[11] T. V. Waehneldt and P. Mandel, Brain Res. 40, 419 (1972).

[12] D. D. Murdock, E. Katona, and M. A. Moscarello, Can. J. Biochem. 47, 818 (1969).

[13] E. D. Day, P. N. McMillan, D. D. Mickey, and S. H. Appel, Anal. Biochem. 39, 29 (1971).

[14] A. N. Davison, in "Fortschritte der Pädologie" (F. Linneweh, ed.), Band II, p. 65. Springer-Verlag, Berlin and New York, 1968. (In English.)

[15] A. N. Davison and J. Dobbing, "Applied Neurochemistry." Davis, Philadelphia, Pennsylvania, 1968.

[16] C. W. M. Adams, "Neurohistochemistry." Elsevier, New York, 1965.

[17] J. S. O'Brien, in "Developmental Neurobiology" (W. A. Himwich, ed.), p. 262. Thomas, Springfield, Illinois, 1970.

of two properties of myelin: large vesicle size and low density. When brain is homogenized in sucrose solutions, the myelin layers swell and peel off the axon, reforming in vesicles. This swelling process is evidently independent of sucrose tonicity but is suppressed by the presence of ions in the medium. If ions are present, then intact myelinated axons can be isolated from the homogenate.[26] The myelin vesicles are of the size range of nuclei and mitochondria and will therefore sediment with these fractions during differential centrifugation. Because of their high lipid content these myelin vesicles have the lowest intrinsic density (d) of any membrane fraction of the nervous system. Myelin is less dense than 0.85 M sucrose ($d = 1.11$) and will band above this density during density gradient centrifugation procedures, whereas microsomes, nuclei, and mitochondria will migrate through this region. Some subfractions of myelin have densities closer to 0.5 M sucrose ($d = 1.065$).

The isolation methods fall into two groups, depending on whether the initial step is differential centrifugation or density gradient centrifugation. The former methods are modifications of fractionation schemes designed to isolate all subcellular fractions.[27,28] Myelin is found predominantly in the crude mitochondrial fraction, although a variable amount is also present in the nuclear fraction. This distribution may depend on both the vigorousness of the homogenization procedure and the age (degree of myelination) of the animal. The crude mitochondrial fraction is centrifuged on a sucrose step gradient to yield crude myelin, synaptosome, and mitochondrial fractions.[6] In the latter methods the total tissue homogenate is

[18] J. S. O'Brien, in "Handbook of Neurology" (B. Vinken and G. Bruyn, eds.), Vol. 7, p. 40. North-Holland Publ., Amsterdam, 1970.

[19] W. T. Norton, in "Chemistry and Brain Development" (R. Paoletti and A. N. Davison, eds.), p. 327. Plenum, New York, 1971.

[20] W. T. Norton, in "Basic Neurochemistry" (W. Albers, G. Siegel, R. Katzman, and B. Agranoff, eds.), p. 365. Little, Brown, Boston, Massachusetts, 1972.

[21] J. Eichberg, G. Hauser, and M. L. Karnowsky, in "The Structure and Function of Nervous Tissue" (G. H. Bourne, ed.), Vol. 3, p. 185. Academic Press, New York, 1969.

[22] M. E. Smith, Advan. Lipid Res. 5, 241 (1967).

[23] L. C. Mokrasch, in "Handbook of Neurochemistry" (A. Lajtha, ed.), Vol. 1, p. 171. Plenum, New York, 1969.

[24] L. C. Mokrasch, in "Methods of Neurochemistry" (R. Fried, ed.), Vol. 1, p. 1. Dekker, New York, 1971.

[25] L. C. Mokrasch, R. S. Bear, and F. O. Schmitt, "Myelin," NRP Bulletin Vol. 9, No. 4, MIT Press, Cambridge, Massachusetts, 1971.

[26] G. H. DeVries, W. T. Norton, and C. S. Raine, Science 172, 1370 (1972).

[27] E. De Robertis, A. Pellegrino de Iraldi, G. Rodriguez de Lores Arnaiz, and L. Salganicoff, J. Neurochem. 9, 23 (1962).

[28] J. Eichberg, V. P. Whittaker, and R. M. C. Dawson, Biochem. J. 92, 91 (1964).

centrifuged on a step gradient to yield a crude myelin layer in the first step. This is accomplished either by overlaying the homogenate in isotonic (0.32 M) sucrose on more dense sucrose and allowing the myelin to migrate down to the interface[5] or by having the homogenate as the dense layer and allowing the myelin to rise to the interface.[4] There are a number of variations on these basic methods. Some workers have used zonal centrifugation instead of swinging-bucket rotors,[9,12,13,29] and mixed Ficoll-sucrose gradients have been used instead of sucrose alone.[13]

The advantages of the methods using differential centrifugation as a first step are that microsomal contamination is removed early, and it is possible to collect other conventional subcellular fractions during the same experiment. The disadvantages are that it is essential to process both the crude nuclear fraction and the crude mitochondrial fraction by density gradient steps in order to collect all the myelin; and if only myelin is desired, some of the differential centrifugation steps are unnecessary.

Our laboratory has concentrated on the procedures employing an initial density gradient step. This has the advantage of giving a crude myelin preparation in the first step, relatively free of nuclei and mitochondria.

The crude myelin layer obtained by any one of these general procedures is of varying purity depending on the tissue from which the myelin is being isolated. Adult brain white matter yields reasonably pure myelin; whereas myelin from the whole brain of a young animal in the early stages of myelination might be quite impure. The major impurities appear to be microsomes and axoplasm, the latter being trapped in the myelin vesicles during the homogenization process. Further purification is generally achieved by osmotically shocking the myelin in water. This opens up the myelin vesicles releasing trapped material. The large, but less dense myelin particles can then be separated from the small membranous material by low speed centrifugation, or by density gradient centrifugation. The method given in detail below employs both steps.

Myelin Isolation Procedure[19,30-32]

This procedure was developed specifically for the isolation of myelin from rat brain, and was designed to be applicable to animals at all stages of development. It affords preparations of high yield and of low and

[29] P. E. Braun and N. S. Radin, *Biochemistry* 8, 4310 (1969).
[30] W. T. Norton, S. E. Poduslo, and K. Suzuki, *Abstr. Meeting Int. Soc. Neurochem. 1st Strasbourg*, p. 161 (1967).
[31] K. Suzuki, S. E. Poduslo, and W. T. Norton, *Biochim. Biophys. Acta* 144, 375 (1967).
[32] W. T. Norton and S. E. Poduslo, *J. Neurochem.* 21 (1973), in press.

equal levels of contamination at all ages. Moreover it is equally satisfactory for the isolation of myelin from bovine or human brain white matter. The steps include a density gradient separation of crude myelin, two water shocks and low speed centrifugation steps to remove microsomes, and a final purification by density gradient centrifugation.

Tissue Source. The whole brains from three rats 30 days of age or older, or 5 g of fresh mammalian brain white matter. Modifications for younger rats are given below.

Reagents. Sucrose, 0.32 M, sucrose, 0.85 M, and distilled water, all chilled to 0°.

Equipment. Dounce homogenizer, 40-ml volume (Kontes Glass Co., Vineland, New Jersey). The procedure has been designed for the Spinco SW-25.2 rotor (3 × 60-ml tubes), although any swinging-bucket rotor can be used provided the *g* forces and times are maintained and the ratios of tissue to media volume are adhered to. The centrifugation times refer to times at stated speed and do not include acceleration and deceleration; *g* forces are measured from the center of the tube. Resuspensions or dispersions of the fractions are always done thoroughly in a Dounce homogenizer, and all steps are carried out at 0–4°.

Step 1. Isolation of Crude Myelin. The rats are killed by decapitation, and the entire brain is removed and weighed. The combined weight will be 4.5–6.0 g. (Alternatively, if larger brains are used, dissect 5.0 g of white matter.) The tissue is divided in two parts, and each part is homogenized in approximately 40 ml of 0.32 M sucrose, using 5 strokes of the loose pestle and 7 strokes of the tight pestle. The homogenates are combined and brought to 100 ml with 0.32 M sucrose. One-third of the homogenate is layered over 25 ml of 0.85 M sucrose in each of three tubes, and the tubes are centrifuged at 75,000 *g* (25,000 rpm) for 30 minutes. The layers of crude myelin which form at the interface of the two sucrose solutions are collected with a Pasteur pipette, everything else is discarded.

Step 2. Sucrose Washout. The combined myelin layers are suspended in water by homogenization and brought to a final volume of 180 ml. This suspension is centrifuged at 75,000 *g* (25,000 rpm) for 15 minutes. The supernatant is discarded.

Step 3. First Osmotic Shock and Removal of Small Membrane Fragments. The three crude myelin pellets are again dispersed in a total volume of 180 ml of water and centrifuged at 12,000 *g* (10,000 rpm) for 10 minutes. The cloudy supernatant is discarded.

Step 4. Second Osmotic Shock. The loosely packed pellets are again dispersed in water and centrifuged as described in step 3.

Step 5. Discontinuous Gradient Centrifugation to Obtain Purified

Myelin. The myelin pellets are combined and suspended in 100 ml of 0.32 *M* sucrose. This suspension is layered over 0.85 *M* sucrose in three tubes and centrifuged exactly as described in step 1. The purified myelin is removed from the interface with a Pasteur pipette.

Modification for 15-Day-Old Rats

Step 1. Four brains are used for 100 ml of homogenate instead of three, to maintain the concentration at approximately 5%. The crude myelin layers are set aside, and this step is then repeated in order to process 8 brains per day.

Step 2. Six myelin layers (instead of 3) are made up to 180 ml, thus the myelin from 8 brains is distributed in 3 tubes.

Step 3 and 4. The myelin pellets from step 2 are combined in 120 ml, the myelin from 8 brains being distributed in 2 tubes.

Step 5. The combined myelin pellets from step 4 are taken up in 33 ml of 0.32 *M* sucrose, thus only 1 tube is used for the final gradient for the myelin from 8 brains.

Modification for 20-Day-Old Rats

Step 1. Do this step twice, using 4 brains each time, as for 15-day-old rats.

Step 2. Combine six myelin layers in 180 ml of water (the same as for 15-day-old rats).

Steps 3 and 4. The myelin pellets from step 2 are combined in 180 ml, thus the myelin for eight brains is distributed in 3 tubes.

Step 5. The combined pellets from step 4 are taken up in 66 ml of 0.32 *M* sucrose, thus 2 tubes are used for purification of the myelin from 8 brains.

Whichever modification is followed, the purified myelin obtained in step 5 is treated in the same way. If electron microscopic examination is desired, a sample can be taken for fixation directly without further processing. If enzyme assays are to be done the myelin layers are combined, suspended in water, and centrifuged at 75,000 *g* (25,000 rpm) for 5 minutes to remove most of the sucrose and to concentrate the sample. For determination of dry weight the water washing step is repeated four more times and the myelin is freeze-dried in a tared container. Chemical analyses can be done directly on the water-washed myelin or the freeze-dried samples can be stored at below $-20°$ in a jar over Drierite or silica gel for analysis when convenient.

This method involves 5 centrifugation steps (exclusive of the final washes) with a combined time of 95 minutes. Allowing more than twice that time for acceleration, deceleration, and other manipulations, purified material can be obtained in about 5 hours from the time the initial discontinuous gradient is made. This leaves sufficient time for enzyme assays the same day. If for some reason a preparation must be interrupted it should be done after either step 4 or step 5. Preparations which have been left overnight at 0° (but not frozen) after these steps, and then continued the next day, show no obvious differences in chemical composition, but no studies have been made of the enzyme activities of such preparations.

The yield of dry myelin from adult bovine brain is 500–600 mg from 5 g (wet weight) of white matter. The yield of myelin from rat brain varies with age and is given in the following tabulation:

Age (days)	15	20	30	60	144	190	425
Myelin (mg/brain)	3.5	13	24	41	56	64	90

These recoveries represent approximately 60–70% of the total myelin present in the starting tissue, although exact calculations cannot be made.[3]

The method as described here is not designed to give the highest purity myelin possible, but rather is a compromise to give good yields of high and fairly constant-purity material in a short time. Material of higher purity can be obtained by introducing two variations in the procedure. The first is to repeat the initial density gradient centrifugation step with the crude myelin layer obtained in step 1, before going on to step 2. The second variation is to use a continuous sucrose gradient in step 5 instead of a discontinuous gradient. We have found a continuous gradient generated from equal amounts of $0.32\,M$ and $1.0\,M$ sucrose to be adequate, although shallower gradients may be useful for specific purposes. The first variation helps remove small amounts of trapped large particles, such as intact axons, nuclei, and mitochondria. The second variation gives a continuum of myelin of different densities and different lipid:protein ratios. Contamination is also reduced since there is less chance of dense material being trapped by a thick myelin layer forming at an interface. The use of EDTA in all steps has been examined, but appears not to affect the myelin purity.

Use of a CsCl Gradient[33,34]

The method described above has been found suitable for most tissues. However considerable myelin is lost in steps 3 and 4, the steps which remove microsomes. Any myelin particles of microsomal dimensions will be discarded in these steps along with contaminating membranes. Since contaminating material is heavier than myelin, it is possible to eliminate these differential centrifugation steps and utilize instead a continuous

[33] W. T. Norton and A. N. Davison, unpublished observations.
[34] S. Greenfield, W. T. Norton, and P. Morell, *J. Neurochem.* **18**, 2119 (1971).

gradient step to separate myelin from smaller but denser material. Either sucrose or CsCl gradients have been used, but the latter gives sharper separations.

The first two steps of this method are the same as those described above.

Step 3. Osmotic Shock. The washed crude myelin pellets are dispersed in 180 ml of water and centrifuged at 75,000 g (25,000 rpm) for 15 minutes. This is a repetition of step 2 and ensures hypotonic conditions.

Step 4. The myelin pellets are each dispersed in 10 ml of 0.3 M CsCl or in water and layered over 45 ml of continuous gradients of CsCl varying from 0.3 to 1.3 M, prepared on a 5-ml cushion of 1.3 M CsCl. These tubes are centrifuged at 75,000 g (25,000 rpm) for 1 hour. Mature myelin forms a granular band centering at about 0.85 M CsCl ($d = 1.11$). [In sucrose gradients, the buoyant density of myelin is 1.08 g/ml (~ 0.65 M sucrose).] In this procedure any heavier material forms a band or a pellet below myelin. As before, the myelin band (or bands) is removed by pipette, washed by centrifugation, and freeze-dried for storage.

This procedure gives a higher yield of myelin than the procedure described above and therefore was valuable in a study of myelin from mutant animals characterized by hypomyelination.[34] The reader should be aware that detailed comparisons of myelin prepared by these two methods from the same source have not been made; although the lipid composition does not seem to be altered by contact with CsCl. However, it is known that the myelin basic proteins are solubilized in high salt concentrations, and it is possible that myelin prepared in CsCl could have altered protein ratios.

PNS Myelin

In spite of the differences among myelins from different parts of the nervous system, these methods described here, as well as other well-designed methods, are suitable not only for brain but for spinal cord and PNS as well. The spinal cord myelin will band at a lower density on the final gradients because it has a higher lipid–protein ratio than brain myelin. The only major modification necessary for the isolation of PNS myelin is the selection of a proper homogenization technique. Unlike brain, peripheral nerve has large amounts of collagen and other connective tissue as well as adipose tissue. It is therefore impossible to homogenize nerve with conventional Dounce or Teflon and glass homogenizers. This problem has been solved in several ways. O'Brien *et al.*[35] homogenize in a Waring Blendor fitted with a special rotor-stator device. Adams *et al.*[36] cut the

[35] J. S. O'Brien, E. L. Sampson, and M. D. Stern, *J. Neurochem.* **14**, 357 (1967).
[36] C. W. M. Adams, Y. Abdulla, D. R. Turner, and O. B. Bayliss, *Nature (London)* **220**, 171 (1968).

nerve into small pieces, immerse it in a glycine buffer, and then cut 25 μm frozen sections before homogenizing. Greenfield et al.[37] pulverize the nerve frozen in liquid nitrogen and then isolate myelin by the CsCl method.

Purity

Criteria of purity have been difficult to set because there is no *a priori* way of knowing what are the intrinsic myelin constituents. The types of criteria for myelin should be the same as for any other subcellular fractions: typical ultrastructure and absence of gross contamination, the absence or minimization of markers characteristic of other particles, and the maximization of markers characteristic of myelin. Most workers have used the obvious criterion of electron microscopic appearance. Isolated myelin retains the typical repeating 5-layered structure and repeat period of ~120 Å seen *in situ*. The difficulty of identifying small contaminating membrane vesicles in a field of myelin membranes and the well known sampling problem inherent in EM work, make this characterizaticn unreliable after a certain purity level has been reached. Useful markers for contamination are succinic dehydrogenase (mitochondria) Na+, K+-stimulated ATPase and 5'-nucleotidase (plasma membranes) glucose-6-phosphatase (endoplasmic reticulum), and nucleic acid (ribosomes, nuclei, adsorbed RNA); all of these are low in purified myelin.[4,6,9,19,38] Lactate dehydrogenase (soluble proteins), β-galactosidase (lysosomes), and acetylcholinesterase (neuronal plasma membranes and endoplasmic reticulum) have also been used.[11]

Markers characteristic of myelin are fewer, and some used previously are now to be regarded with suspicion. Many people have used the solubility or near solubility, of the final product in 2:1 (v/v) $CHCl_3:CH_3OH$ as an indicator of purity. However, one class of myelin proteins is insoluble in this solvent, and these proteins vary considerably during development, thus myelin from young animals may be less soluble than that from adults. And, of course, this criterion falls down if PNS myelin is examined which has a different protein composition and therefore different solubility behavior in this solvent.

It is now well established that the enzyme, 2',3'-cyclic nucleotide-3'-phosphohydrolase, is fairly myelin specific. This enzyme is also present, as expected, in oligodendroglia and fractions derived from these cells, but the bulk of the activity can be recovered in purified myelin fractions. This assay is conveniently carried out using the procedure of Olafson et al.[39] with the

[37] S. Greenfield, S. Brostoff, E. H. Eylar, and P. Morell, *J. Neurochem.* **20**, 1207 (1973).

[38] N. L. Banik and A. N. Davison, *Biochem. J.* **115**, 1051 (1969).

[39] R. W. Olafson, G. I. Drummond, and J. F. Lee, *Can. J. Biochem.* **47**, 961 (1969).

addition of sodium deoxycholate to stimulate the enzymatic activity.[40] Recently another enzyme, a cholesterol ester hydrolase, has also been shown to be myelin specific.[41,42] These two markers should be useful in devising new isolation procedures and in assaying myelin contamination in other fractions.

Composition

The lipid composition of myelin is now well established,[13,15,17-25] and a number of investigators have made considerable progress in defining the protein composition as well.[11,34,43-48] So far the most powerful methods for examining myelin proteins have been by acrylamide-gel electrophoresis, either in detergent solutions or in organic solvent buffer mixtures. A cautionary note is necessary here for the reader generally unfamiliar with current myelin research. In spite of the intensity with which these studies have been pursued in the past 10 or 15 years, there are now being initiated a series of "second generation" investigations that may necessitate a re-evaluation of all past research. There was considerable evidence in the past that myelin was not one single substance. Workers have described light and heavy myelin,[5] large and small myelin,[28] and density differences between adult and immature myelin.[49] Attention is now being focused on these different myelin subfractions of different densities, different protein: lipid composition, and different protein patterns and how they may change during myelination.[11,13,50,51] At this writing it is premature to forecast the outcome of this work. It is not known whether these subfractions represent different types of myelin, or different degrees of purity. It is necessary, however, to realize that the methods given in this chapter enable the recovery of all these myelin fractions, and by conventional methods of assay, these preparations are $>95\%$ pure.

[40] T. Kurihara, J. L. Nussbaum, and P. Mandel, *Brain Res.* 13, 401 (1969).

[41] Y. Eto and K. Suzuki, *Biochim. Biophys. Acta* 239, 293 (1971).

[42] Y. Eto and K. Suzuki, *J. Neurochem.* 19, 117 (1972).

[43] L. F. Eng, F. C. Chao, B. Gerstl, D. Pratt, and M. G. Tavaststjerna, *Biochemistry* 7, 4455 (1968).

[44] F. Wolfgram and K. Kotorii, *J. Neurochem.* 15, 1281 (1968).

[45] E. Mehl and F. Wolfgram, *J. Neurochem.* 16, 1091 (1969).

[46] T. V. Waehneldt and P. Mandel, *FEBS* (*Fed. Eur. Biochem. Soc.*) *Lett.* 9, 209 (1969).

[47] E. Mehl and A. Halaris, *J. Neurochem.* 17, 659 (1970).

[48] F. Gonzalez-Sastre, *J. Neurochem.* 17, 1049 (1970).

[49] M. L. Cuzner and A. N. Davison, *Biochem. J.* 106, 29 (1968).

[50] D. H. Adams and M. E. Fox, *Brain Res.* 14, 647 (1969).

[51] P. Morell, S. Greenfield, E. Costantino-Ceccarini, and H. Wisniewski, *J. Neurochem.* 19, 2545 (1972).

[47] Isolation of Synaptosomal and Synaptic Plasma Membrane Fractions

By CARL W. COTMAN

Brain tissue with the variety of highly specialized cells generates many different subcellular particles following homogenization and has presented a particularly difficult problem for obtaining homogeneous fractions. Procedures have been described for the isolation of fractions containing nuclei,[1] myelin,[2] axonal fragments,[3] mitochondria,[4] polysomes,[5] and synaptic endings (synaptosomes) and subcomponents. The synaptosome, because of its crucial role in interneuronal communication, has received perhaps the most study. These particles were initially purified on a sucrose gradient by Whittaker et al.[6] and De Robertis et al.[7] The fractions are sufficiently pure to permit the study of at least certain properties of various transmitter systems in vitro, but are not completely devoid of contaminating particles nor are they homogeneous in one type of synaptosome. Synaptosomes are readily fractionated into a soluble and particulate fraction but less readily separated into their minimum other components: mitochondria, synaptic vesicles, and plasma membrane. Initial methods for subfractionating synaptosomes[8-11] provided partially purified synaptosomal subfractions. Improvements both in purity and yield have been difficult and slow owing in part to implicit limitations in existing approaches and in part to the lack of definitive markers to assay contamination. Pure fractions of various potential contaminates for comparison have not been available, and it is sometimes difficult to determine which fraction is most contaminated.

[1] B. S. McEwen, L. Plapinger, G. Wallach, and C. Magnus, *J. Neurochem.* **19,** 1159 (1972).

[2] W. T. Norton, this volume [46].

[3] G. H. DeVries, W. T. Norton, and C. S. Raine, *Science* **172,** 1370 (1972).

[4] G. Rodriguez de Lores Arnaiz and E. De Robertis, *J. Neurochem.* **9,** 503 (1962).

[5] M. P. Lerner, F. O. Wettstein, H. R. Herschman, J. G. Stevens, and B. R. Fridlender, *J. Neurochem.* **18,** 1495 (1971).

[6] E. G. Gray and V. P. Whittaker, *J. Anat. (London)* **96,** 79 (1962).

[7] E. De Robertis, A. Pellegrino De Iraldi, G. Rodriguez de Lores Arnaiz, and L. Salganicoff, *J. Neurochem.* **9,** 23 (1962).

[8] V. P. Whittaker, I. A. Michaelson, and J. A. Kirkland, *Biochem. J.* **90,** 293 (1964).

[9] E. De Robertis, G. R. de Lores Arnaiz, L. Salganicoff, A. Pellegrino De Iraldi, and L. M. Zeiher, *J. Neurochem.* **10,** 225 (1963).

[10] R. J. Hosie, *Biochem. J.* **96,** 404 (1965).

[11] G. Rodriguez de Lores Arnaiz, M. Aberici, and E. De Robertis, *J. Neurochem.* **14,** 215 (1967).

For example, we have observed that synaptic vesicle fractions prepared by conventional methods[8,9] have higher specific activities of some "microsomal" enzymes than a microsomal fraction.

The design and testing of procedures for preparing synaptosomes and subsynaptosome components is still undergoing development. Although the fractions prepared by present techniques present a near ideal system for seeking a solution to many problems, they are unsatisfactory for others without further refinement. Cytochemical documentation or uniqueness of a property to the synaptic particle is a reassuring advantage. As the study of synaptosomal fractions becomes more concerned with the components and subsystems serving synaptic function, it is critical to take into account the limited purity of the fractions. The procedures we currently find most satisfactory for isolating synaptosomes and synaptic plasma membranes are described below.

Preparation of Synaptosomal Fractions from Rat Brain

Synaptosomes (synaptic bodies) are defined as pinched off nerve endings or synaptic boutons which, after homogenization, reseal and form a discrete particle.[8] Synaptosomes are made up primarily of a plasma membrane, synaptic vesicles, mitochondria, and assorted soluble components. Besides the presynaptic element, a plasma membrane fragment of postsynaptic origin, which is vesicularized or open, is often attached to the nerve ending by a synaptic cleft. Procedures for isolating synaptosomal fractions are based on the pioneering studies of Whittaker et al.[6] and De Robertis et al.[7] and have been modified by a large number of research workers. The basic procedure is a series of differential centrifugations, followed by further resolution on a sucrose or Ficoll-sucrose density gradient. The synaptosomal fraction isolated by Scheme 1 (Fig. 1) is less homogeneous than that isolated by Scheme 2 (Fig. 2), but the time required for the separation by Scheme 1 is near minimal and the yield is larger. These procedures are based on earlier studies by Autilio et al.,[12] Morgan et al.,[13] and Cotman and Matthews.[14] Characterization and compositional analyses on these types of fractions are relatively extensive.[13-20]

[12] L. A. Autilio, S. H. Appel, P. Pettis, and P. L. Gambetti, *Biochemistry* **7**, 2615 (1968).
[13] I. G. Morgan, L. A. Wolfe, P. Mandel, and G. Gombos, *Biochim. Biophys. Acta* **241**, 737 (1971).
[14] C. W. Cotman and D. A. Matthews, *Biochim. Biophys. Acta* **249**, 380 (1971).
[15] C. W. Cotman, R. E. McCaman, and S. A. Dewhurst, *Biochim. Biophys. Acta* **249**, 395 (1971).
[16] C. W. Cotman, W. Levy, G. Banker, and D. Taylor, *Biochim. Biophys. Acta* **249**, 406 (1971).

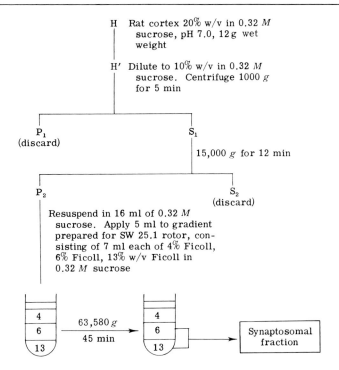

FIG. 1. Scheme 1 for isolation of synaptosomal fraction from rat cortex.

Characteristics of the Synaptosomal Fractions. The yield of a synaptosomal fraction ranges from 12–15 mg of particle protein per gram of brain wet weight (Scheme 1, Fig. 1) to 6–9 mg of protein per gram of brain wet weight (Scheme 2, Fig. 2). As determined by quantitative electron microscopic examination, synaptosomes prepared by Scheme 1 comprise 50% of the total particles (fraction collected between 6 and 13% interface, Scheme 1). Synaptosomes isolated by Scheme 2 constitute approximately

[17] W. C. Breckenridge, G. Gombos, and I. G. Morgan, *Biochim. Biophys. Acta* **266**, 695 (1972).

[18] M. Reith, I. G. Morgan, G. Gombos, W. C. Breckenridge, and G. Vincendon, *Neurobiology* **2**, 169 (1972).

[19] C. W. Cotman, H. Herschman, and D. Taylor, *J. Neurobiol.* **2**, 169 (1971).

[20] J. P. Zanetta, P. Benda, G. Gombos, and I. G. Morgan, *J. Neurochem.* **19**, 881 (1972).

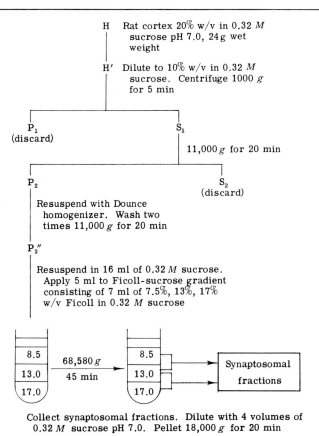

Fig. 2. Scheme 2 for isolation of synaptosomal fraction from rat cortex.

70% of the total particles (8.5–13% or 13–20% fraction, Scheme 2). In general the major contaminants are vesicles, similar in size to synaptosomes which are bounded by a plasma membrane and filled with cytoplasm, but which do not contain synaptic vesicles or have an attached synaptic cleft. Some of the vesicles may be of postsynaptic origin whereas others probably originate as fragments from the soma and processes of neurons and glia. Fractions between 6 and 13% or 8.5 and 13% Ficoll-sucrose have more cytoplasmic vesicles than those sedimenting between 13 and 20%. The 13–20% fraction contains fewer contaminating cytoplasmic vesicles, but is contaminated with free mitochondria. Both the 6–13% and the 8.5–13% Ficoll-sucrose fractions are essentially devoid of free mitochondrial contamination.

Technical Considerations. Once a protocol is adopted, the procedures are extremely reproducible. All sucrose solutions should have a pH between 6.5 and 7.5 and be prepared in glass-distilled water. pH is adjusted with dilute NaOH or HCl; nearly all buffers above 10 mM cause particle aggregration in brain. We carry out all centrifugations in either a Beckman number 30 rotor or SW 25.1 rotor. Homogenization is done in a standard Thomas homogenizer (size C) operated at about 400 rpm with 8 passes up and down. There is a reasonable degree of flexibility in homogenization conditions, and the fractionation procedures for the isolation of synaptosomes are sensitive only to extreme conditions. Underhomogenization results in increased contamination by large membranes, membrane vesicles, and low yields due to loss into P_1, while overhomogenization results in low recoveries due to breakage. All stages in the fractionation should be performed and maintained at ice cold temperatures. Pellets resulting from differential centrifugation should be carefully resuspended by gentle homogenization, and the particle suspensions should be applied to the gradient with a large-holed pipette to minimize damage. Ficoll-sucrose solutions are somewhat unstable and cannot be stored for long periods of time in the refrigerator or freeze-thawed repeatedly. As with any gradient, particularly where the density of the starting zone approaches the buoyant density of a substantial proportion of the particle population, it is important not to overload the gradient, or trapping and impairment of sedimentation results. The quantities of brain used for Schemes 1 and 2 (Figs. 1 and 2) are approximately the limits for the SW 25.1 rotor. The major difference between the procedures in Schemes 1 and 2 is the additional washings of P_2 used in Scheme 2. These washings enrich P_2 in synaptosomes from approximately 27 to 40% of the total number of particles. Small particles are removed due to the more refined classification of particles on the basis of sedimentation coefficients. We·have used Ficoll-sucrose density gradients primarily because they require less centrifugation time and maintain isosmotic conditions throughout the isolation procedure. Although we have not extensively studied the fractionation of synaptosomes from sources other than rat brain, we have noted that synaptosomes from some other species do not fractionate in an identical manner on Ficoll-sucrose gradients.

Isolation of Synaptic Plasma Membrane Fractions

A synaptic plasma membrane (SPM) consists of the plasma membrane from the presynaptic nerve ending, which is often joined to a small segment of postsynaptic plasma membrane via a synaptic cleft. SPM are isolated from synaptosomes by two steps: (1) SPM are released from intact synaptosomes by osmotic shock; and (2) SPM are separated from other components on a sucrose density gradient. Ideally all SPM should be resolved from all other components and 100% of the SPM population recovered. Fractional recoveries introduce the ambiguity that specialized classes of membranes, synaptic or even nonsynaptic, are examined. In practice a complete separation is difficult because SPM, other membranes, and mitochondria overlap considerably in buoyant density on sucrose gradients. In practice this is difficult since SPM and mitochondria overlap considerably in buoyant density on sucrose gradients. As originally suggested by Davis and Bloom,[21] the buoyant density of mitochondria can be increased by enzymatically depositing a dense formazan in the mitochondrial compartment. We have found that the NADH reduction of iodonitroneotetrazolium violet (INT) results in sufficient deposition of formazan in synaptosomal mitochondria to increase their isopycnic density and resolves most SPM from nearly all mitochondria on a sucrose gradient. The general procedure we have used is outlined in Fig. 3.

Characteristics of the Synaptic Plasma Membrane Fraction. As assayed by (sodium + potassium)-activated adenosine triphosphatase (Na-K ATPase), 65–75% of the SPM can be resolved from 97% of the mitochondria as monitored by cytochrome oxidase. Na-K ATPase specific activities are approximately 5 to 12-fold enriched from the total homogenate while cytochrome oxidase specific activities are 1/50 that of a purified mitochondrial fraction. The fractions, SPM-1 or SPM-2 which band at 0.7–1.0 M and 1.0–1.1 M sucrose (see Fig. 3) have the highest Na-K ATPase specific activity with the lowest cytochrome oxidase activity. SPM-2 fractions are consistently approximately 10- to 12-fold enriched in Na-K ATPase activity.

The NADH-INT procedure permits fractions to be collected at sucrose concentrations as high as 1.1 M sucrose without introducing more than approximately 5% mitochondrial contamination in any fraction. Normally, mitochondrial contamination reaches a minimum 10–15% at densities of 1.1 M sucrose. While the NADH-INT procedure is not altogether ideal, it aids in improving the recovery of SPM, which can be obtained relatively free from contaminating mitochondria. The NADH-INT step can be eliminated in the scheme of Fig. 3, but to avoid mitochondrial contamination in excess of 10%, membrane fractions should not be collected at densities heavier than that of 1.0 M sucrose. Alternative methodologies for isolating fractions of SPM from synaptosomal fractions prepared by Scheme 1 (Fig. 1) in the absence of INT have been described elsewhere[14]; the INT-

[21] G. Davis and F. E. Bloom, *J. Cell Biol.* **47**, 46a (1970).

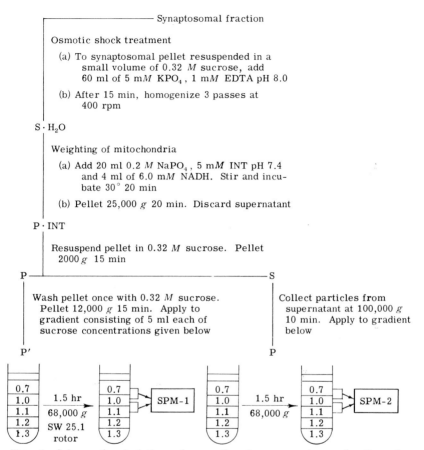

—————————————— Synaptosomal fraction

Osmotic shock treatment

   (a) To synaptosomal pellet resuspended in a
       small volume of 0.32 M sucrose, add
       60 ml of 5 mM KPO$_4$, 1 mM EDTA pH 8.0

   (b) After 15 min, homogenize 3 passes at
       400 rpm

S · H$_2$O

Weighting of mitochondria

   (a) Add 20 ml 0.2 M NaPO$_4$, 5 mM INT pH 7.4
       and 4 ml of 6.0 mM NADH. Stir and incu-
       bate 30° 20 min

   (b) Pellet 25,000 g 20 min. Discard supernatant

P · INT

Resuspend pellet in 0.32 M sucrose. Pellet
2000 g 15 min

P ————————————————————————— S

Wash pellet once with 0.32 M sucrose.
Pellet 12,000 g 15 min. Apply to
gradient consisting of 5 ml each of
sucrose concentrations given below

Collect particles from
supernatant at 100,000 g
10 min. Apply to gradient
below

P'   P

0.7		0.7		0.7		0.7	
1.0	1.5 hr	1.0	SPM-1	1.0	1.5 hr	1.0	SPM-2
1.1	68,000 g	1.1		1.1	68,000 g	1.1	
1.2	SW 25.1	1.2		1.2		1.2	
1.3	rotor	1.3		1.3		1.3	

FIG. 3. Scheme for isolation of synaptic plasma membrane fractions from synaptosomal fraction.

formazan procedure as described above can also be used in conjunction with this procedure. Recently a method to rapidly obtain an SPM fraction directly from the crude mitochondrial fraction, using the INT-formazan procedure, has been reported.[22] In neither case, however, are the specific activities of Na-K ATPase as high as the SPM-2 fraction isolated by the scheme shown in Fig. 3.

Technical Considerations. The extensive homogenization following os-motic shock facilitates the quantitative separation of SPM from mitochon-

[22] C. W. Cotman and D. Taylor, *J. Cell Biol.* **55**, 696 (1972).

dria, but is not absolutely necessary. Milder shear is satisfactory, but results in a greater proportion of the membrane banding at heavier densities. INT at 5 mM solubilizes slowly. Occasionally an excessively large proportion of Na-K ATPase activity is found in the mitochondrial pellet. This appears to be in part caused by the incomplete removal of salts or incomplete resuspension of pellets after hard packing by centrifugation.

[48] The Isolation of Nuclei from Normal Human and Bovine Brain

By A. N. SIAKOTOS

General Procedure

Principle. Generally, preparations of brain nuclei are contaminated with endothelial cells because these two particulates have similar densities. This isolation procedure for nuclei from brain is designed to resolve nuclei from endothelial cells and other cellular and subcellular elements.[1] Other problems associated with the preparation of brain nuclei include their susceptibility to postmortem changes; therefore, preparations of nuclei free of cytoplasmic components cannot be obtained from material unless the tissue is cooled within 1–2 hours of death. These postmortem changes are most evident with human brain because of the differential stability of glial and neuronal nuclei. Preparations of bovine brain nuclei generally are very rich in neuronal nuclei while nuclear preparations isolated from human brain contain a higher percentage of the smaller and more stable glial nuclei. Therefore, the relative ratios of glial to neuronal nuclei obtained in a preparation of brain nuclei are dependent to a large degree on the postmortem state of the brain specimens as well as the rigid maintenance of the +4° temperature during all phases of the isolation and purification procedure.

Reagents and Solutions

Solution 1: 0.4 M sucrose containing 10,000 units of heparin per liter
Solution 2: 0.4 M sucrose
Solution 3: 1 liter of 0.4 M sucrose containing 20,000 units of heparin and 1 ml of 0.1 M ATP, pH 6.1
Solution 4: 0.4 M sucrose containing 0.1 M CaCl$_2$, 1 mM ATP,

---

[1] A. N. Siakotos, G. Rouser, and S. Fleischer, *Lipids* **4**, 234 (1969).

0.5 mM 2-N-morpholinoethanesulfonic acid buffer (MES), pH 6.1

Solution 5: 2.0 M sucrose

Solution 6: 0.83% NH$_4$Cl solution

Special Apparatus

11 × 40 cm plexiglas column with 200-mesh stainless steel support screen

4.5 × 40 cm plexiglas column with 200-mesh stainless steel support screen

Sephadex G-25, coarse beads. Pharmacia Fine Chemicals, Inc., Piscataway, New Jersey

Isolation Method for Brain Nuclei

An outline of the isolation procedure is given in Fig. 1. Fresh brain (1.2–1.4 kg) was cut into ¼ × ½ inch coronal sections and placed into 1 liter of cold solution 1 as soon as possible postmortem. Brain nuclei are very susceptible to postmortem changes, and pure preparations cannot be obtained from brain unless the tissue is sectioned and cooled rapidly within 1–2 hours of death in solution 1 (0.4 M sucrose plus 10,000 units of heparin per liter). All operations, including subsequent filtration steps, are carried out in the cold (+4°). The cooled sections of brain were then chopped in 1 liter of solution 2 (0.4 M sucrose) with a food chopper, drained free of blood on a kitchen strainer, and homogenized in a loose-fitting (0.01-inch clearance) Potter-Elvehjem homogenizer, 5–6 passes at 1200 rpm.

The homogenate was diluted to 3 liters with solution 2 and passed through a 10-mesh kitchen sieve. The filtrate was centrifuged 1000 g for 30 minutes in a Sorvall RC-3 centrifuge (HG-4 rotor). The supernatant S$_1$ was recentrifuged at 1000 g for 30 minutes (Sorvall RC-3, HG-4 rotor) and the resulting supernatant S$_2$ was discarded. The pellets P$_1$ and P$_2$ were pooled, resuspended in 1.3 liters of solution 2 and centrifuged at 1000 g for 30 minutes (Sorvall RC-3, HG-4 rotor). The pellet, P$_3$, was collected and the supernatant S$_3$ was discarded. Pellet P$_3$ was resuspended in 0.5 liter of solution 4 (0.4 M sucrose, 0.1 mM CaCl$_2$, 1 mM ATP, 0.5 mM MES buffer, pH 6.1), then mixed with 1 volume of solution 5 (2.0 M sucrose) and centrifuged at 7000 g for 45 minutes. The supernatant S$_4$ was collected, resuspended with a Potter-Elvehjem homogenizer to disperse the floating layer of myelin and cell debris. The redispersed S$_4$ fraction is recentrifuged at 7000 g for 45 minutes to precipitate additional nuclei and provide a pellet P$_5$. P$_4$ from the first 7000 g centrifugate and P$_5$ are pooled and resuspended in 1 liter of solution 4, then passed through a 11 × 10 cm

Chop fresh brain (1.2-1.4 kg) in 1 liter solution 2 (2), drain, homogenize (3) in solution 3(4), dilute to 3 liters(2), pass through a 10-mesh sieve (5). Centrifugations as follows:

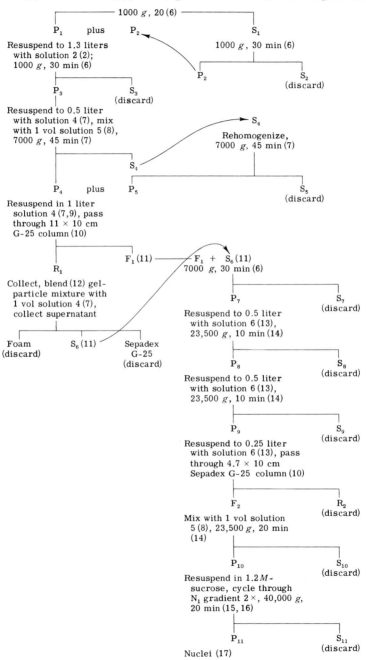

Nuclei (17)

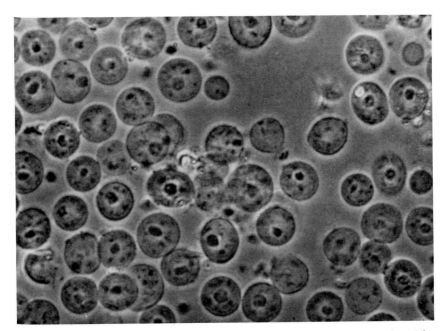

FIG. 2. Phase contrast micrograph of a bovine brain nuclear preparation. The large nuclei with one prominent nucleolus are presumably derived from neurons, and the smaller binucleolar nuclei are probably derived from glial cells. ×530.

Sephadex G-25 column.[2] The Sephadex beads are equilibrated for 6–12 hours in solution 4 and packed in the same solution. As the suspension percolates through the column, the flow rate will be reduced as the filter bed surface retains material, a large squared-off plastic spatula is employed to

---

FIG. 1. Isolation of brain nuclei. (1) All centrifugations, manipulations and solutions at 4°. Abbreviations: P, pellet; S, supernatant; R, visible residue in column bed; F, filtrate. (2) Solution 2: 0.4 M sucrose. (3) Potter-Elvehjem homogenizer, 200-ml capacity, 5–6 passes at 1200 rpm, 0.010 in. clearance. (4) Solution 3: 1 liter 0.4 M sucrose containing 20,000 units of heparin and 1 ml of 0.1 M adenosine triphosphate (ATP). (5) Final filtrate represents approximately 550 g of tissue. (6) Sorvall RC-3, HG-4 rotor. (7) Solution 4: 0.4 M sucrose containing 0.1 mM $CaCl_2$, 1 mM ATP, 0.5 mM 2-N-morpholinoethanesulfonic acid buffer (MES), pH 6.1. (8) Solution 5: 2.0 M sucrose. (9) Plus 1000 units of heparin. (10) Sephadex G-25, coarse, swell gel in solution 4 for 30–60 minutes; apply sample, pass 1 bed volume of solution 4 through, stir residue layer, repeat with two additional bed volumes. (11) F_1, S_6 combined for recovery of nuclei. (12) 1–2 seconds in 500-ml Waring Blendor. (13) Solution 6: 0.83% NH_4Cl solution. (14) Lourdes Model A, VRA rotor. (15) N_1 gradient: linear, 1.2 M to 2.25 M sucrose. (16) Spinco Model L-2, SW 25.2 rotor. (17) Yield 1–2 g.

[2] Refer to Fig. 1 of this series, Vol. 32 [71].

gently stir the filter bed–endothelial cell mixture down to 1–2 cm. This stirring operation is continued deeper into the filter bed with three additional 500-ml increments of solution 4 until the filtrate, F_1, is clear and relatively free of subcellular particles (by phase contrast microscopy). The F_1 filtrate is reserved for subsequent purification. Next increments of the Sephadex filtration media plus filter-residue mixture are poured into a Waring Blendor with a glass jar, 1 volume of solution 4 is added, and the mixture is blended for 1–2 seconds. The blended mixture is allowed to stand for 5–10 minutes, and the red-brown foam floating at the surface is collected with a long-handled teaspoon and discarded. The foaming process (blend 1–2 seconds and allow to stand 5–10 minutes) is repeated approximately three times until significant quantities of foam are no longer obtained, then the combined supernatants (S_6) from the Waring Blendor jar contents are combined with F_1 from the first Sephadex filtration procedure, and the Sephadex is discarded.[2]

The pooled F_1 filtrate and S_6 supernatant are centrifuged at 7000 g for 30 minutes in the RC-3 centrifuge (HG-4 rotor) and the resulting pellet P_6 is resuspended in 0.5 liter of 0.83% aqueous NH_4Cl solution to

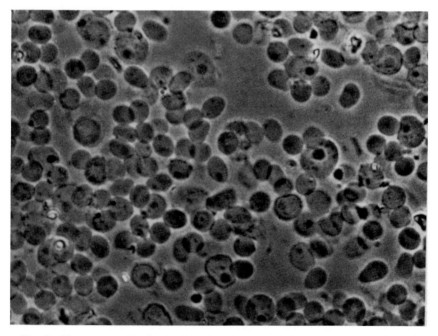

FIG. 3. Phase contrast micrograph of a human brain nuclear preparation. Note the predominance of glial nuclei, and the relative lack of neuronal nuclei as compared with the bovine brain preparation (Fig. 2). ×530.

destroy contaminating red blood cells and centrifuged at 23,500 g (Sorvall RC-2B, VRA rotor) for 10 minutes. The pellet P_7 is resuspended in 0.83% aqueous NH_4Cl and recentrifugated at 23,500 g for 10 minutes (RC-2B, VRA rotor). The NH_4Cl treated pellet is then resuspended in solution 4 and passed through a 4.7 × 10 cm Sephadex G-25, coarse, column prepared in the same manner as the 11 × 10 cm Sephadex column (see above). The filter bed plus residue are stirred and washed with four column volumes of solution 4. The Sephadex filter and residue are discarded.[3] The filtrate (F_2) is diluted with 1 volume of solution 5 and centrifuged at 23,500 g for 20 minutes (Sorvall RC-2B, VRA rotor). The supernatant (S_8) is discarded and the pellet (P_8) resuspended in a mixture of 1:1 solution 4 and solution 5 and cycled twice through a linear gradient (N_1) of 1.2 M sucrose to 2.25 M sucrose in an I.E.C. B-60 ultracentrifuge (SB-110 rotor) at 40,000 g for 20 minutes. Only the pellets are collected and purified. After a final purity check by phase contrast microscopy (Figs. 2 and 3), the gradient purification procedure is interrupted, and the preparation of nuclei is complete. The yield from 600 g of crude gray matter is 1–2 g.

[3] The Sephadex is recovered by extensive washing with tap water, distilled water, acetone, and finally methyl alcohol. The washes are carried out in a large beaker, and the supernatant fluid is decanted with each wash. After the final treatment with methyl alcohol, the Sephadex is air-dried in a sintered-glass Büchner funnel with suction.

[49] The Isolation of Lysosomes from Brain

By HAROLD KOENIG

The methods which are used for the isolation of lysosomes are based upon the well-established procedures of cell fractionation, namely, differential centrifugation and isopycnic density gradient centrifugation. The background, theoretical aspects, and problems related to the isolation of lysosomes have been surveyed in several recent reviews.[1-4] The preparation of pure lysosomal fractions from mammalian tissues is a challenging and difficult undertaking. Nowhere are the difficulties attendant to the isolation

[1] C. de Duve, *J. Theoret. Biol.* **6**, 33 (1964).
[2] C. de Duve, *Harvey Lect.* **59**, 49 (1965).
[3] C. de Duve, *in* "Enzyme Cytology" (D. B. Roodyn, ed.), p. 1. Academic Press, New York, 1967.
[4] H. Beaufay, *in* "Lysosomes in Biology and Pathology" (J. T. Dingle and H. B. Fell, eds.), Vol. 2, p. 515. North-Holland Publ., Amsterdam, 1969.

of lysosomes more in evidence than in brain.[5,6] These difficulties stem from a number of peculiarities of neural tissues which will now be briefly considered.

1. Diversity of cell types. Nerve cells vary widely in their morphology with respect to Nissl bodies, chromatin pattern, and dendritic and axonal processes, according to regions, nuclear groups, functional specialization, and so forth. The glia, which greatly outnumber nerve cells, display an analogous, if less dramatic, polymorphism. It is to be expected that the physicochemical heterogeneity normally manifested by lysosomes will be greatly accentuated by the cellular diversity of neural tissue.

2. Compartmentation of lysosomes. Lysosomes occur in small, but significant, numbers in axons and nerve endings.[7,8] These presently constitute a population of lysosomes which cannot be readily isolated, as the hypoosmotic stress used to rupture axonal and synaptosomal membranes necessarily disrupts the lysosomes.

3. Interference by neural structures. A number of apparently abnormal structures in mitochondrial fractions of brain homogenates, including condensed nerve endings and mitochondria, the latter without mitochondrial marker enzymes, are denser than their normal counterparts and cosediment with lysosomal dense bodies on isopycnic centrifugation.[9] Further difficulty arises from the fact that myelin and nerve endings, which constitute a large portion of the mitochondrial fraction, are substantially less dense than the bulk of mitochondria and lysosomes. The ability of lysosomes to accumulate Triton WR-1339 and dextran, and to be rendered more buoyant thereby, has been exploited for the preparation of a highly purified lysosomal fraction from liver and other parenchymatous tissues.[10,11] Neural lysosomes of immature rat brain behave similarly when Triton WR-1339 is administered intrathecally, a substantial portion of the acid hydrolase activities sedimenting in light subfractions on density gradient centrifugation.[12]

[5] H. Koenig, in "Lysosomes in Biology and Pathology" (J. T. Dingle and H. B. Fell, eds.), Vol. 2, p. 111. North-Holland Publ., Amsterdam, 1969.

[6] H. Koenig, in "Handbook of Neurochemistry" (A. Lajtha, ed.), Vol. 2, p. 255, Plenum Press, New York, 1969.

[7] H. Koenig, Science 164, 310 (1969).

[8] M. K. Gordon, K. G. Bench, G. G. Deanin, and M. W. Gordon, Nature (London) 217, 523 (1968).

[9] H. Koenig, D. Gaines, T. McDonald, R. Gray, and J. Scott, J. Neurochem. 11, 729 (1964).

[10] R. Wattiaux, and P. M. Wibo, Baudhuin, Lysosomes Ciba Found. Symp. 1963, p. 176 (1963).

[11] F. Leighton, B. Poole, H. Beaufay, P. Baudhuin, J. W. Coffey, S. Fowler, and C. de Duve, J. Cell Biol. 37, 482 (1968).

[12] O. Z. Sellinger, L. M. Nordrum, and V. Idoyaga-Vargas, Brain Res. 26, 361 (1971).

Neutral red and other basic dyes concentrate within neural lysosomes *in vivo,* producing ultrastructural changes in these particles and reducing their buoyant density.[13] Unfortunately, neural lysosomes which are altered through the uptake of Triton WR-1339[12] or basic dyes[13] cosediment with fractions enriched in myelin and in nerve endings.

4. Microsomal constituents. A substantial portion of the lysosomal enzymes, 20–35% of the total, is associated with the microsomal fraction.[9] In part, this is due to the fact that very small lysosomes, both free and within diminutive nerve endings and axons, sediment with the microsomal components.[9] In addition, lysosomal hydrolases are associated with endoplasmic reticulum and elements of the Golgi apparatus.[5,6] These enzymes presumably represent newly synthesized molecules which are migrating through the membrane system of the cell en route to their storage site, the lysosome.[14] These extralysosomal hydrolases make an indeterminate contribution to various subcellular fractions. Some Golgi elements evidently sediment in the mitochondrial fraction and contaminate the nerve ending fractions[15] (H. Koenig and C. Hughes, unpublished observations, 1973).

Despite these difficulties, lysosome-enriched fractions can be prepared from brain homogenates by a combination of differential and density gradient centrifugation. The best preparations rarely show more than a 5- to 10-fold purification of lysosomal hydrolase activities over the homogenate. Electron microscopic examination of these enriched fractions divulges that lysosomal dense bodies are but little more than contaminants, representing perhaps 5% of these preparations.[5,6,9] For many purposes, e.g., identification of putative lysosomal constituents, *in vitro* studies of enzyme latencies and lysosomal fragility, etc., these fractions are quite useful. However, meaningful studies of the chemical composition of lysosomes can be carried out only in pure samples of these organelles. The purity of the lysosomal fractions can be greatly improved by starting with neuronal cell body fractions instead of whole brain or cerebral cortex. Idoyaga-Vargas *et al.*[15a] prepared particulate fractions from homogenates of isolated neuronal cell bodies by differential centrifugation and subfractionated these fractions by isopyknic centrifugation in a linear sucrose gradient (35–60%). They obtained a fraction that was enriched up to 15-fold over the starting homogenate (or about 50-fold over the cerebral cortex)

[13] H. Koenig, *in* "Barrier Systems in the Brain" (A. Lajtha and D. H. Ford, eds.), p. 87. Elsevier, Amsterdam, 1968.

[14] C. de Duve and R. Wattiaux, *Annu. Rev. Physiol.* **28**, 435 (1966).

[15] L. Seijo and G. R. de Lores Arnaiz, *Biochim. Biophys. Acta* **211**, 595 (1970).

[15a] V. Idoyaga-Vargas, J. C. Santiago, P. D. Petiet, and O. Z. Sellinger, *J. Neurochem.* **19**, 2533 (1972).

in acid hydrolases and consisted principally of lysosomal dense bodies on electron microscopic examination.

Carrier-free continuous electrophoresis separates particles according to their electrokinetic properties, and has been used with outstanding success to purify lysosomes from rat liver,[16] and nerve endings from brain.[17] Liver lysosomes show the greatest anodic mobility of any organelle and display appreciable electrophoretic and enzymatic heterogeneity.[16] The remarkable anodic mobility of lysosomes is correlated with their highly selective capacity to accumulate cationic dyes *in vivo* and *in vitro*.[5,6,9] These properties indicate that lysosomes possess a large surplus negative charge at physiological pH, due largely to the presence of a soluble, acidic lipoglycoprotein in the lysosomal matrix.[5,6,18] Ryan *et al.*[17] used this method to separate subcellular particles from guinea pig brain, but did not report the distribution of lysosomal enzyme markers in their fractions; however, an electron micrograph (Fig. 5 of Ryan *et al.*[17]) shows structures resembling lysosomal dense bodies in a synaptosomal fraction.

Methods

Homogenization

Lysosomal particles are exceedingly susceptible to disruption by physical and chemical agents. To minimize damage to lysosomes, homogenization of brain tissue must be performed by gentle mechanical means in a suitable suspending medium. Sucrose ($0.25–0.32\ M$) is the most widely used solute to prevent the swelling and rupture of lysosomes. "Physiological" salt solutions do not provide sufficient osmotic protection to cerebral lysosomes, and cannot replace nonionic solutes, such as sucrose. Saline solutions cause the agglutination of organelles in homogenates of liver and alter their sedimentation characteristics.[19] Mg^{2+} ($0.01\ M$) causes cerebral lysosomes to clump and sediment with a heavy particulate fraction.[20] It should be noted that lysosomal particles which have been harvested from hypertonic sucrose media cannot be returned to isotonic sucrose or less hypertonic sucrose media with impunity as these are "hypotonic" with respect to the original medium, and extensive swelling and rupture of lysosomes will result (see Koenig,[5] p. 137).

Homogenization is generally performed in glass homogenizers of the

[16] R. Stahn, K.-P. Maier, and K. Hannig, *J. Cell Biol.* **46**, 576 (1970).
[17] K. J. Ryan, H. Kalant, and E. L. Thomas, *J. Cell Biol.* **49**, 235 (1971).
[18] A. Goldstone, E. Szabo, and H. Koenig, *Life Sci.* **9**, 607 (1970).
[19] C. de Duve and J. Berthet, *Int. Rev. Cytol.* **3**, 225 (1954).
[20] O. Z. Sellinger and L. M. Nordrum, *J. Neurochem.* **16**, 1219 (1969).

Potter-Elvehjem type equipped with a motor-driven Teflon pestle rotating at moderately high speeds. A single pass of rat brain against the pestle usually suffices to disperse the tissue, even without prior mincing. The homogenizer of Emanuel and Chaikoff[21] reportedly gives the highest yield of particle-bound hydrolases in some tissues.[22,23] However, whatever the means used, the homogenization procedure must be carefully standardized if reproducible results are to be obtained.

Differential Centrifugation

When the usual schema for differential centrifugation is applied to brain homogenates, lysosomal marker enzymes are recovered in all the primary subcellular fractions and none shows a significant enrichment of these enzymes.[9] However, if rat brain homogenates are fractionated according to the procedure devised by de Duve *et al.*[24] for rat liver, a lysosome-enriched fraction is obtained.[5,6] This method differs from the classical method in that the large cytoplamic granules are isolated in two successive steps, yielding a "heavy" mitochondrial fraction which contains the bulk of protein and mitochondrial enzyme activity associated with the mitochondrial fraction in the classical procedure, and a "light" mitochondrial fraction containing the remainder of these components with lysosomal hydrolases enriched about 4- to 5-fold over the homogenate. In rat brain homogenates, the "light" mitochondrial fraction is enriched 2- to 3-fold over the homogenate. The "heavy" mitochondrial fraction is deposited by ultracentrifuging the nuclear supernatant ($800 \, g$, 10 minutes) from a 20% homogenate of brain in $0.25 \, M$ sucrose (weight in volume) at $3300 \, g$ for 10 minutes. The resulting supernatant is ultracentrifuged at $16,000 \, g$ for 20 minutes to sediment the "light" mitochondrial fraction. A modification of this method is shown in Fig. 1.

Density Gradient Centrifugation

Isopycnic centrifugation of a crude mitochondrial fraction over a sucrose gradient is more effective than differential centrifugation for concentrating cerebral lysosomes. Both discontinuous[5,6,9] and continuous[25] sucrose gradients have been employed for this purpose. This method depends upon the fact that lysosomes in brain, as in other tissues, tend to be denser than

[21] C. F. Emanuel and I. L. Chaikoff, *Biochim. Biophys. Acta* **24**, 254 (1957).
[22] A. L. Greenbaum, T. F. Slater, and D. V. Wang, *Nature (London)* **188**, 318 (1960).
[23] C. M. Szego, B. J. Seeler, R. A. Steadman, D. F. Hill, A. K. Kimeon, and J. A. Roberts, *Biochem. J.* **123**, 523 (1971).
[24] C. de Duve, B. C. Pressman, R. Gianetto, R. Wattiaux, and F. Appelmans, *Biochem. J.* **60**, 604 (1955).
[25] O. Z. Sellinger and R. A. Hiatt, *Brain Res.* **7**, 191 (1968).

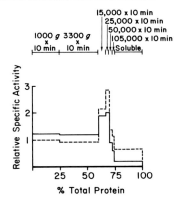

FIG. 1. Subcellular distribution of acid phosphatase (- - -) and β-glucuronidase (—) activities in rat brain homogenate. Conditions for differential centrifugation are given at the top of the figure. Small, but significant, differences are seen in the distribution of the two hydrolases. From H. Koenig, *in* "Lysosomes in Biology and Pathology" (J. T. Dingle and H. B. Fell, eds.), Vol. 2, p. 111. North-Holland Publ., Amsterdam, 1969.

mitochondria and are more likely to sediment through 1.4 *M* sucrose than the latter. This finding also forms the basis for simpler gradient systems. The simplest system involves ultracentrifuging a crude mitochondrial suspension in 0.25 *M* sucrose over 1.4 *M* sucrose to give a lysosome-enriched pellet, but other constituents are not separated by this method. In a system of intermediate complexity, a mitochondrial suspension in 0.25 *M* sucrose is ultracentrifuged on a discontinuous gradient consisting of 0.9 *M* and 1.4 *M* sucrose to yield a myelin fraction floating on 0.9 *M* sucrose, a combined nerve ending-mitochondria fraction floating on 1.4 *M* sucrose, and a lysosomal pellet.[20]

Isopycnic Centrifugation in a Discontinuous Sucrose Gradient. The method herein described has been used in the author's laboratory since 1963.[9] *Tissue fractionation.* Rats, usually 10 per experiment, are decapitated, and the forebrains are rapidly removed, weighed, and placed in 9 volumes of ice cold 0.32 *M* sucrose. All subsequent manipulations are carried out at 0–4°. The tissue is dispersed in a homogenizer of the Potter-Elvehjem type equipped with a Teflon pestle by a single pass against the rapidly rotating pestle. The homogenate is spun at 800 *g* for 10 minutes. The resulting pellet, consisting of nuclei, unbroken cells, erythrocytes, vascular fragments and other tissue debris, is washed twice with 0.32 *M* sucrose and designated the nuclear (N) fraction. Washings and the original supernatant are combined and spun at 15,689 *g* for 20 minutes in the Spinco Model L preparative ultracentrifuge with the Spinco rotor No. 25.1. The resulting pellets are washed once with 0.32 *M* sucrose and suspended in a volume of 0.32 *M* sucrose equal to twice the weight of the original

A — Myelin and membranes

B — Nerve endings

C — Nerve endings, some mitochondria

D — Mitochondria

E — Lysosomes, degenerating mitochondria

FIG. 2. Sucrose density gradient system used for subfractionation of the crude mitochondrial (M) and microsomal (P) fractions by isopycnic centrifugation. From H. Koenig, D. Gaines, T. McDonald, R. Gray, and J. Scott, *J. Neurochem.* **11**, 729 (1964).

fresh tissue to give the primary mitochondrial (M) fraction. The supernatant and washings from the M fraction are combined and centrifuged in the Spinco No. 40 rotor at 81,000 g for 95 minutes. The resulting primary microsomal (P) pellet is suspended without washing in a volume of 0.32 M sucrose equal to twice the free tissue weight. The supernatant from the P fraction is designated the soluble (S) fraction.

The M fraction, and when desired the P fraction, are subfractionated by centrifugation on a discontinuous sucrose density gradient. Gradient tubes are prepared by successive layering with 5 ml each 1.4, 1.2, 1.0, and 0.8 M aqueous sucrose into polyethylene tubes designed for the Spinco rotor No. 25.1 (see Fig. 2). The M or the P fraction (5 ml) suspended in 0.32 M sucrose is carefully introduced on to the top of each tube containing the gradient system previously chilled to 0° and spun in the Spinco Model L ultracentrifuge at 63,500 g for 120 minutes. Separation of the grossly visible fractions is achieved from M, designated the A (0.32 M–0.8 M), B (0.8 M–1.0 M), C (1.0 M–1.2 M), and D (1.2 M–1.4 M) fractions, by sectioning the tube at 1.7, 3.2, and 4.2 cm, respectively, from the surface of the liquid. The remaining pellet is referred to as the E fraction. The P fraction is subdivided into three layers, designated the A_1 (0.32 M–0.8 M), B_1 (0.8 M–1.0 M), and C_1 (1.0 M–1.2 M–1.4 M) fractions, by sectioning the gradient tubes 2 and 3.7 cm from the surface of the liquid, and a pellet D_1.

Results

Biochemical Observations. About 40–60% of the total acid hydrolase activities is recovered in the M fraction, and 20–30% in the P fraction. Considerable enzymatic heterogeneity is evident in the lysosomal populations, as shown by substantial variations in subcellular distribution of 5 acid hydrolases[9] (see Table I).

The distribution of protein and of marker enzymes in the mitochondrial

TABLE I

DISTRIBUTION OF PROTEIN, SUCCINATE DEHYDROGENASE, AND ACID HYDROLASES IN PRIMARY FRACTIONS (N, M, P, AND S) OF HOMOGENATES OF RAT BRAIN[a,b]

Constituent	No. of experiments	Concentration[c]	Percentage distribution				Relative specific activity			
			N	M	P	S	N	M	P	S
Protein	8	127	9.5	43.2	22.1	25.2	—	—	—	—
Succinate dehydrogenase	7	474	3.5	89.7	2.6	4.2	0.37	2.08	0.12	0.17
Acid glycerophosphatase	5	32.4	8.9	48.2	22.9	20.0	0.94	1.12	1.04	0.79
Acid p-nitrophenolphosphatase	4	1259	6.8	37.2	19.3	36.7	0.72	0.86	0.87	1.46
β-Glucuronidase	5	417	16.6	51.0	21.2	11.2	1.75	1.18	0.96	0.44
Acid DNase	4	99.6	3.3	58.3	24.2	14.2	0.35	1.35	1.10	0.56
Acid RNase	2	67.2	5.8	42.4	21.8	30.0	0.61	0.98	0.99	1.19
Cathepsin D	2	883	13.0	65.3	13.6	8.1	1.36	1.51	0.62	0.32

[a] Data from H. Koenig, D. Gaines, T. McDonald, R. Gray, and J. Scott, *J. Neurochem.* **11**, 729 (1964).

[b] Homogenates, 0.32 M sucrose, of rat brain were fractionated by differential centrifugation into the following fractions: nuclear (N), mitochondrial (M), microsomal (P), and soluble (S). Relative specific activity = (% of total constituent in fraction)/(% of total protein in fraction).

[c] Concentration = N + M + P + S. Protein in milligrams per gram of fresh tissue; enzymes in units of activity per gram of fresh tissue.

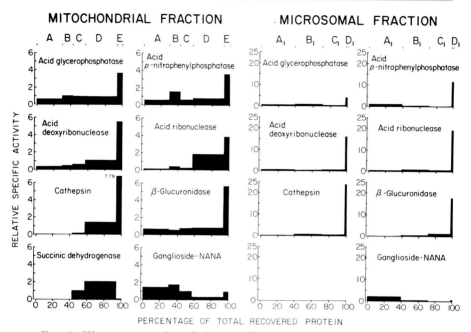

FIG. 3. Histograms of the relative specific activities of acid hydrolases in the mitochondrial and microsomal subfractions of brain prepared by isopycnic centrifugation over the density gradient shown in Fig. 2. The high relative specific activities of the various acid hydrolases and the low relative specific activity of the mitochondrial reference enzyme, succinate dehydrogenase, in fractions E and D, are noteworthy. From H. Koenig, D. Gaines, T. McDonald, R. Gray, and J. Scott, *J. Neurochem.* **11,** 729 (1964).

(M) subfractions from the discontinuous gradient is shown in Fig. 3 and Table II. The bulk of acid hydrolases, about 60–70% of the total in the M fraction, are recovered in the densest fractions D and E. The E fraction, denser than 1.4 M sucrose, generally contains 20–40% of the hydrolase activities together with about 3.5–5.0% of the protein, the relative specific activities ($M = 1$) ranging from 4 to 6.

The results from the P fraction are shown in Fig. 3 and Table III. Acid hydrolase activities are distributed throughout the gradient. About 20–45% of the total acid hydrolase activities is recovered with only 1.5% of protein in the pellet from 1.4 M sucrose. The relative specific activities of the lysosomal enzymes in this fraction (D) are thus very high, 12–24 ($P = 1$).

Activation and Solubilization of Lysosomal Hydrolases. The hydrolase activities in fresh intact brain lysosomes are largely latent, physical or chemical disruptive treatments being required to fully activate these enzymes.[5,6,9] The mechanism for this structural latency has been fully dis-

TABLE II

DISTRIBUTION OF PROTEIN, SUCCINATE DEHYDROGENASE, AND ACID HYDROLASES IN SUBFRACTIONS OF M (A, B, C, D, AND E)[a,b]

Constituent	No. of experiments	Percentage recovery	Percentage distribution					Relative specific activity				
			A	B	C	D	E	A	B	C	D	E
Protein	4	120	30.0	13.4	14.2	37.1	5.3	—	—	—	—	—
Succinate dehydrogenase	4	40	0.8	3.8	14.9	80.0	0.5	0.03	0.28	1.05	2.16	0.09
Acid glycerophosphatase	3	80.7	19.3	14.5	12.9	33.5	19.8	0.64	1.08	0.91	0.90	3.73
Acid p-nitrophenolphosphatase	3	102.0	20.0	21.5	9.8	29.3	19.4	0.67	1.60	0.69	0.79	3.66
β-Glucuronidase	4	92.4	21.8	8.2	10.9	28.9	30.2	0.73	0.61	0.77	0.78	5.70
Acid DNase	3	52.4	11.8	6.6	9.5	42.5	29.6	0.39	0.49	0.67	1.15	5.58
Acid RNase	2	149.0	4.7	6.0	3.6	66.6	19.1	0.16	0.45	0.25	1.80	3.60
Cathepsin D	2	85.2	0.8	1.0	2.9	54.0	41.3	0.03	0.07	0.20	1.46	7.79

[a] Data from H. Koenig, D. Gaines, T. McDonald, R. Gray, and J. Scott, *J. Neurochem.* **11**, 729 (1964).

[b] A 0.32 M sucrose suspension of the M fraction was carefully introduced over a density gradient system containing 5 ml each of 0.8, 1.0, 1.2, and 1.4 M aqueous sucrose in tubes and spun in the Spinco rotor 25.1 at 53,500 g for 2 hours. Four fractions A–D, and a pellet from 1.4 M sucrose, fraction E, were obtained, sedimented, resuspended in 0.25 M sucrose, and assayed for the various constituents. Relative specific activity of the M fraction = 1.0. Relative specific activity = (% of total constituent in fraction)/(% of total protein in fraction).

TABLE III

DISTRIBUTION OF PROTEIN, SUCCINATE DEHYDROGENASE, AND ACID HYDROLASES IN SUBFRACTIONS OF P (A_1, B_1, C_1 AND D_1)[a,b]

Constituents	No. of experiments	Percentage recovery	Percentage distribution				Relative specific activity			
			A	B	C	D	A	B	C	D
Protein	2	111	39.1	32.7	26.3	1.9	—	—	—	—
Acid glycerophosphatase	2	39.7	37.1	34.4	20.4	8.1	0.95	1.05	0.78	4.26
Acid p-nitrophenolphosphatase	1	29.7	47.9	18.4	11.0	22.7	1.23	0.56	0.42	11.9
β-Glucuronidase	1	42.0	11.1	17.5	37.5	33.9	0.28	0.54	1.43	17.8
Acid DNase	2	44.6	34.2	13.9	21.3	30.6	0.87	0.43	0.81	16.1
Acid RNase	2	45.2	17.2	28.0	18.3	36.5	0.44	0.86	0.70	19.2
Cathepsin D	2	38.2	12.3	24.4	17.7	45.6	0.31	0.75	0.67	24.0

[a] Data from H. Koenig, D. Gaines, T. McDonald, R. Gray, and J. Scott, *J. Neurochem.* **11,** 729 (1964).

[b] A 0.32 *M*-sucrose suspension of the P fraction was subfractionated by centrifugation on a density gradient system as described in Table II. Three fractions (A_1–C_1) and a pellet (fraction D_1) were obtained, and the various constituents were measured. Relative specific activity of the P fraction = 1.0. Relative specific activity = (% of total constituent in fraction)/(% of total protein in fraction).

TABLE IV
ACTIVATION AND SOLUBILIZATION OF ACID PHOSPHATASE IN A BRAIN
MITOCHONDRIAL FRACTION[a,b]

| Treatment | Acid phosphatase activity, percentage of total | | |
| | Supernatant | Pellet | |
		Free	Latent
Control	24	13	63
Grinding (3 minutes, 4°)	49	44	7
H_2O (1.5 hours, 20°)	51	44	5
Autolysis (3 hours, 37°, pH 5)	44	31	25
Freeze-thawing (10 cycles)	40	29	31
Triton X-100 (0.5 hour, 20°, 0.1%)	51	49	0

[a] Data from H. Koenig, D. Gaines, T. McDonald, R. Gray, and J. Scott, *J. Neurochem.* **11**, 729 (1964).

[b] Particulate fractions of brain were obtained by spinning the nuclear supernatant at 25,000 g for 10 min. The pellets were resuspended in 0.25 M sucrose (1 ml = 0.1 g fresh tissue) and treated as described. After releasing treatments, suspensions were spun at 150,000 g for 90 min. The resulting supernatant fractions and pellets were analyzed for acid phosphatase activity. Total activity in the pellets was determined with the detergent, Triton X-100 (0.1%).

cussed elsewhere.[5,6] The accessibility of these enzymes is usually measured by incubating particulate suspensions in 0.25 M sucrose with substrate in the absence and presence of 0.2% Triton X-100 to give the "free" and "total" activities, respectively. Enzyme activation is accompanied by solubilization of enzymes (see Table IV). However, the extent of solubilization varies for the different hydrolases and is affected by the mode of treatment, particularly the pH of the medium (H. Koenig, A. Goldstone, and A. Patel, unpublished observations). The solubility of the various acid hydrolases in kidney[26] and brain (A. Patel and H. Koenig, unpublished observations), i.e., percent of enzyme activity released into solution after physical disruption of lysosomal particles, is mainly a function of their electrical charge. It has been shown that the acidic (anodic) isoenzymes of a number of lysosomal hydrolases are rendered soluble, while the corresponding basic (cathodic) isoenzymes remain bound to sedimentible material after freeze-thawing or sonication, 0.2% Triton X-100[26] or high pH (unpublished observations) being required to solubilize the bound enzymes. The acid hydrolases that are associated with the nerve-ending fraction are more re-

[26] A. Goldstone, P. Konecny, and H. Koenig, *FEBS (Fed. Eur. Biochem. Soc.) Lett.* **13**, 68 (1971).

TABLE V

SOLUBILITY OF ACID HYDROLASES IN NERVE-ENDING AND MITOCHONDRIAL-LYSOSOMAL
FRACTIONS FROM RAT BRAIN[a,b]

Constituent	Nerve-ending fraction		Mitochondrial-lysosomal fraction	
	% Soluble	% Sedimentible	% Soluble	% Sedimentible
Acid phosphatase	21.7	78.3	54.0	46.0
Aryl sulfatase	26.0	74.0	72.9	27.1
Acid esterase	4.1	95.9	20.2	79.8
β-N-Acetylhexosaminidase	31.1	68.9	51.0	49.0
β-Galactosidase	30.8	69.2	52.0	48.0
Cathepsin D	17.6	82.4	41.8	58.2
β-Glucuronidase	37.3	62.7	70.0	30.0

[a] A. Patel and H. Koenig, unpublished observations, 1973.

[b] The crude mitochondrial fraction in 0.32 M sucrose was ultracentrifuged for 2 hours at 63,500 g over a gradient consisting of 0.7 M and 1.2 M sucrose. The nerve-ending fraction was recovered at the interface between 0.7 and 1.2 M sucrose, and the mitochondrial-lysosomal fraction as a pellet from 1.2 M sucrose. Fractions were suspended in 0.1 M acetate buffer, pH 5.0, and sonicated for 2 minutes in 8 bursts of 15 seconds each, to release soluble enzymes into solution. The bound sedimentible enzymes were deposited by ultracentrifuging sonicates at 100,000 g for 60 minutes. All operations were at 4°. Results are expressed as percent of total enzyme activities recovered in the soluble fraction and the pellet.

sistant to solubilization than those associated with the M-L fraction[15] (Table V). The effects of several releasing treatments on the latency and solubility of acid phosphatase activity are shown in Table IV. The effects of pH and Triton X-100 concentration on the solubilization of several brain lysosomal hydrolases are shown in Tables VI and VII.

Cytochemical and Ultrastructural Observations. Lysosomal particles are readily identified in subcellular fractions by cytochemical staining for acid hydrolase activities, by their natural fluorescence, and by metachromatic staining with the cationic fluorochrome, acridine orange.[5,6,9] Acridine orange staining is strongly recommended for the routine monitoring of lysosomes in subcellular fractions. For enzyme cytochemistry, tissue fractions suspended in 0.25 M sucrose are fixed for 1–2 hours at 4° with 5% neutral formalin (2% formaldehyde), centrifuged, washed with cold water, and recentrifuged. The pellets are resuspended in water to give a constant concentration of tissue protein. Standard drops are placed on 1 × 3-inch glass slides and dried at 4°. These preparations are stained for one or more of the following lysosomal reference enzymes: acid phosphatase activity,[27]

[27] G. Gomori, *Stain Technol.* **25**, 81 (1950).

TABLE VI

Effects of pH and Triton X-100 on Solubilization of Protein, Phospholipid, and Acid Hydrolases in a Mitochondrial-Lysosomal Fraction from Rat Brain[a,b]

| | % Soluble | | | | | | | |
| | No Triton X-100 | | | | 2.0% Triton X-100 | | | |
Constituent	pH 5	pH 7	pH 10	Residue	pH 5	pH 7	pH 10	Residue
Protein	21.0	12.4	10.0	55.6	50.2	18.8	11.5	19.5
Phospholipid	6.6	9.4	11.4	72.6	64.0	16.0	10.1	9.7
Acid phosphatase	54.0	8.4	3.3	34.3	79.9	9.7	7.5	2.9
Aryl sulfatase	72.9	21.4	3.4	2.3	89.0	9.5	1.0	0.5
Acid esterase	20.2	13.4	11.2	55.4	83.1	9.8	4.8	2.3
β-N-Acetyl-hexosaminidase	51.0	32.8	9.6	0	75.8	20.1	2.3	1.8
β-Glucuronidase	47.2	41.2	3.2	8.4	75.7	19.4	3.1	1.8
Cathepsin D	41.8	24.6	17.5	16.1	63.2	12.6	12.6	11.6

[a] A. Patel and H. Koenig, unpublished observations, 1973.

[b] The mitochondrial-lysosomal fraction was isolated as described in legend to Table V. The pellet was suspended in 0.1 M acetate buffer, pH 5, sonicated for 2 min and centrifuged at $100,000 \times g$ for 30 min. The residue was taken up in 0.1 M Tris buffer, pH 7, resonicated for 2 min, and centrifuged. The residue was resuspended in 0.1 M NaOH-glycine buffer, pH 10, sonicated and centrifuged. In another experiment, the same sequence was repeated with the addition of 2% Triton X-100 to all buffers. Results are expressed as % of the total protein, phospholipid, and enzyme activities extracted in buffers, pH 5, 7, and 10, and in the pellet.

acid DNase,[28] acid RNase,[28] and aryl sulfatase.[29] Fresh suspensions of fractions in 0.25 M sucrose can be examined in the fluorescence microscope without prior staining and also after staining with the basic fluorochrome, acridine orange (0.0005–0.005%).

For ultrastructure, fractions are pelleted and fixed overnight in cold 2% glutaraldehyde buffered with 0.1 M cacodylate, pH 7.4, followed by fixation in 1–2% osmium tetroxide for 1–2 hours, all at 4°. Pellets are dehydrated and embedded in Araldite or Epon, ultrathin sections stained with heavy metals and examined. For electron cytochemical staining of acid hydrolase activities, aldehyde-fixed pellets are used. Small fragments of pellets are incubated in Gomori's medium[27] for acid phosphatase activity or in a nitrocatechol sulfate medium[29] for aryl sulfatase activity, and then osmicated and processed for electron microscopy.

[28] A. Vorbrodt, J. Histochem. Cytochem. 9, 647 (1961).
[29] S. Goldfischer, J. Histochem. Cytochem. 13, 520 (1965).

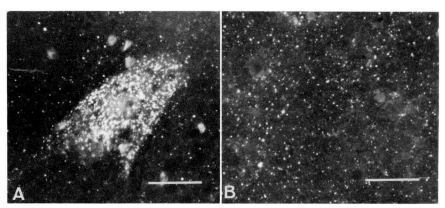

FIG. 5. Fluorescence photomicrographs. ×424. Scale lines = 20 μm. (A) Motor neuron from rat spinal cord stained intravitally by acridine orange (25 mg/100 g) injected intraperitoneally 1 hour before death. Squash preparation. From H. Koenig, *in* "Lysosomes in Biology and Pathology" (J. T. Dingle and H. B. Fell, eds.), Vol. 2, p. 111. North-Holland Publ., Amsterdam, 1969. (B) Lysosome-enriched fraction E prepared from rat brain homogenates and stained with acridine orange (0.001%). From H. Koenig, D. Gaines, T. McDonald, R. Gray, and J. Scott, *J. Neurochem.* **11**, 729 (1964).

pensions, but is abolished by aging particulate suspensions at 4° for 8–20 hours. Fraction E contains the greatest concentration of particles displaying acridine orange metachromasia (see Fig. 5).

Lysosomes in brain and other tissues also exhibit metachromatic staining with acridine orange and other basic dyes *in vivo*. This metachromatic staining *in vitro* is blocked by competing cations and low pH, and is augmented by high pH, indicating that basic dyes are bound by electrostatic interaction forces to lysosomal particles. An acidic lipoglycoprotein, representing about 20–35% of the soluble protein in rat liver and kidney lysosomes, is responsible for the basic dye binding of lysosomal particles.[5,6,18] A similar constituent has been identified in Triton X-100 extracts from rat brain fractions (R. Mylroie and H. Koenig, unpublished observations). This component has the greatest anodic mobility on electrophoresis in acrylamide gels and gives staining reactions characteristic of lysosomal acidic lipoglycoprotein,[18] including reactions for lipid (Sudan Black B), carbohydrate (periodic acid–Schiff reaction), and protein (Amido Schwarz). Moreover, it stains metachromatically with acridine orange. The concentration of this lipoglycoprotein roughly parallels the specific acid hydrolase activities in mitochondrial subfractions from rat brain, indicating that it is a lysosomal constituent in neural tissue. This acidic lipoglycoprotein is thought to serve as an inhibitor of the bound

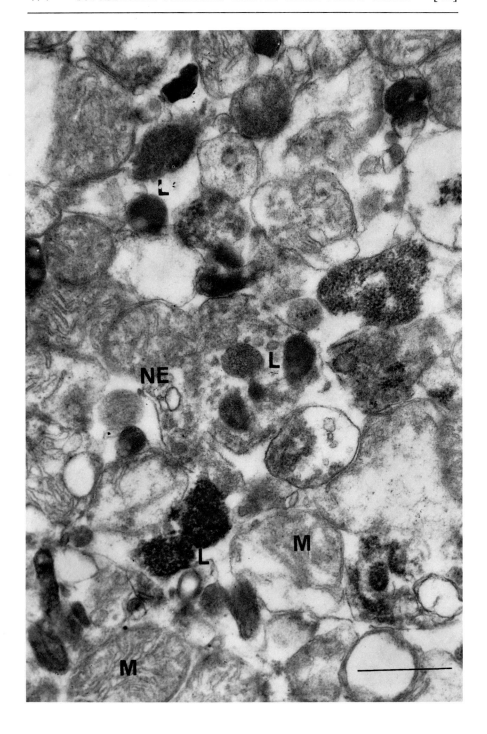

lysosomal enzymes,[5,6,18] and as an acidic buffering medium to maintain an acid pH in phagocytic and autophagic vacuoles.[32]

Ultrastructural Observations. Fraction A consists principally of fragments of myelin sheaths and axons varying in size and state of preservation. Fraction B is made up of relatively compact pinched-off nerve endings. In addition, a heterogeneous array of vesicular and membranous components and occasional osmiophilic dense bodies are present. Fraction C resembles fraction B, but also contains a moderate number of mitochondria. Fraction D consists mainly of free mitochondria, although occasional nerve endings, osmiophilic dense bodies, and assorted membranous and granular structures also occur. Fraction E contains chiefly mitochondria and osmiophilic dense bodies. Vesicular structures, ribosomes, and ill-defined structures are present in lesser degree (see Fig. 6). The majority of mitochondria differ sharply from those present in fraction D in that they are smaller and more compact and appear to be degenerating. Although their numbers vary somewhat in different portions of the pellets, dense bodies are most numerous in fraction E. Dense bodies are usually bounded by a unit membrane and contain compact, finely granular, osmiophilic matrix, which occasionally shows one or more electron-lucent vacuoles.

Fraction A_1 of the microsomal fraction consists mainly of small vesicular components, many of which seemed to be myelin fragments. Some membranous structures probably could represent fragments of smooth endoplasmic reticulum and Golgi lamellae. Occasional compact spheroids resembling minute dense bodies are seen. Fraction B_1 is heterogeneous in appearance. It contains vesicular components of various sizes, membranous fragments, small mitochondria, rough endoplasmic reticulum, some free ribosomes, small pinched off nerve endings, and a few dense bodies. Fraction C_1 resembles B_1 but contains more small mitochondria and fewer nerve endings. Fraction D_1 contains small vesicles, fragments of membranes and free ribosomes. Some portions of these pellets closely resemble fraction E in the presence of numerous osmiophilic dense bodies and compact or shrunken mitochondria; however, these are generally much smaller than the corresponding structures in E (see Figs. 7 and 8).

[32] J. L. Mego, *Biochem. J.* **122**, 445 (1971).

FIG. 6. Electron micrograph of fraction E, composed mainly of mitochondria (M), lysosomes (L), occasional abnormal nerve endings (NE) containing lysosomes, and other, unidentified structures. Lysosomes are dense bodies with a finely granular, osmiophilic matrix and a limiting membrane. ×24,960. Scale line = 1 μm. From H. Koenig, D. Gaines, T. McDonald, R. Gray, and J. Scott, *J. Neurochem.* **11**, 729 (1964).

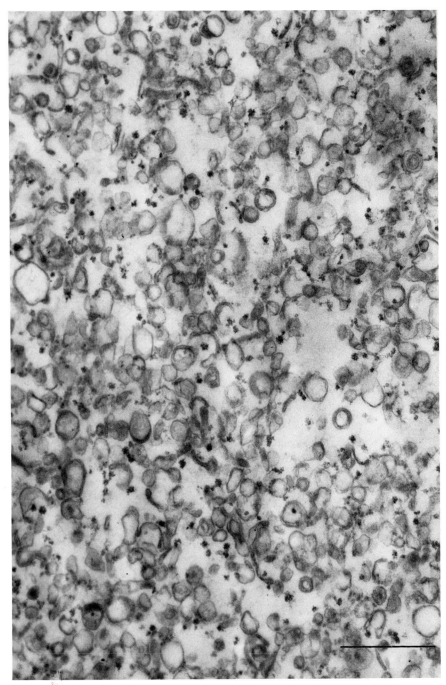

FIG. 7. Electron micrograph of fraction D_1. This portion of the pellet consists of smooth and rough-surfaced endoplasmic reticulum, clusters of free ribosomes, and membrane fragments. ×24,960. Scale line = 1 μm. From H. Koenig, D. Gaines, T. McDonald, R. Gray, and J. Scott, *J. Neurochem.* **11,** 729 (1964).

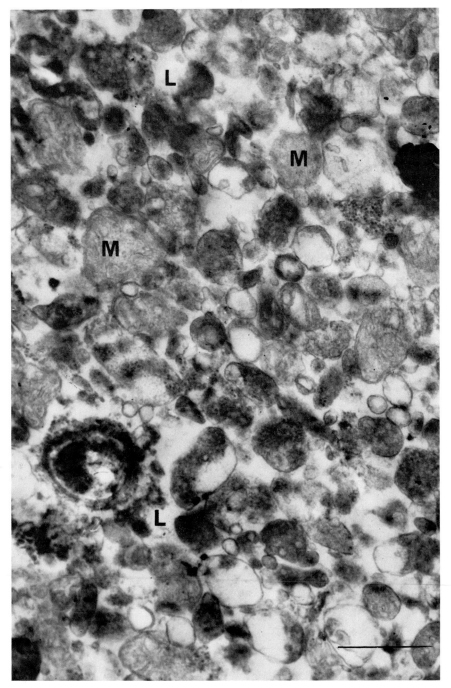

Fig. 8. Electron micrograph of fraction D_1 showing a portion of the pellet which contains numerous small compact mitochondria (M) and lysosomes (L) as well as ribosomes and vesicular components. ×24,960. From H. Koenig, D. Gaines, T. McDonald, G. Gray, and J. Scott, *J. Neurochem.* **11**, 729 (1964).

[50] Procedures for the Isolation of Brain Lipopigments: Ceroid and Lipofuscin

By A. N. SIAKOTOS

Considerable speculation exists in the literature on the role of lysosomes and residual bodies (lipopigments) in the molecular events of the normal and diseased brain cell. This controversy has been compounded by the absence of a method for isolating pure brain lysosomes, the most important subcellular organelle in the dynamics of recycling of cellular constituents as proposed by de Duve.[1] The availability of our previously published method for two brain lipopigments,[2] lipofuscin and ceroid, established the isolation methodology for isolating most of the subcellular participants in the sequential catabolic role of events initiated by brain lysosomes that is, the residual or undegradable products of catabolism (the autofluorescent lipopigments or residual bodies). In the past the identification of autofluorescent lipopigments found in normal and pathological tissues has been very uncertain. Attempts to distinguish these two lipopigments on the basis of different histochemical reactions, solubility, ultrastructure, and the method of induction have led to considerable confusion. The recommended isolation procedures described below were developed in order to establish clear and distinct differences between these two lipopigments,[2] lipofuscin and ceroid. *Lipofuscin* is considered to be the "age pigment" accumulating in normal and abnormal cells in relatively higher concentrations with increasing chronological age. The term *ceroid* is applied to a second autofluorescent lipopigment which was first isolated from a postmortem human brain of a patient affected with a neuropathological condition, classified as "ceroid storage disease."[3] Ceroid has always been associated with some form of pathology and has never been isolated from human organs from normal patients.

Therefore, the methods described establish an operational definition for lipofuscin (age pigment) and ceroid by physical-chemical separation of the two pigments. The isolation procedures are specific for each lipopigment and can be applied to small and large brain samples.

[1] C. de Duve, and R. Wattiaux, *Annu. Rev. Physiol.* **28**, 435 (1966).
[2] A. N. Siakotos, I. Watanabe, A. Saito, and S. Fleischer, *Biochem. Med.* **4**, 361 (1970).
[3] A. S. Levine *et al., Pediatrics* **42**, 483 (1968).

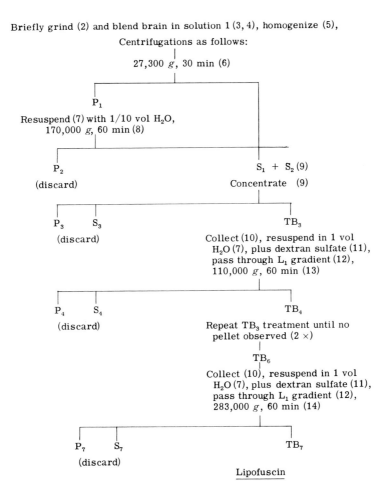

Briefly grind (2) and blend brain in solution 1 (3, 4), homogenize (5),
Centrifugations as follows:

27,300 g, 30 min (6)

P_1
Resuspend (7) with 1/10 vol H_2O,
170,000 g, 60 min (8)

P_2
(discard)

$S_1 + S_2$ (9)
Concentrate (9)

P_3 S_3
(discard)

TB_3
Collect (10), resuspend in 1 vol
H_2O (7), plus dextran sulfate (11),
pass through L_1 gradient (12),
110,000 g, 60 min (13)

P_4 S_4
(discard)

TB_4
Repeat TB_3 treatment until no
pellet observed (2 ×)

TB_6
Collect (10), resuspend in 1 vol
H_2O (7), plus dextran sulfate (11),
pass through L_1 gradient (12),
283,000 g, 60 min (14)

P_7 S_7
(discard)

TB_7
Lipofuscin

FIG. 1. Isolation of lipofuscin. Key: (1) Centrifugations, manipulations, and solutions at 4°. Abbreviations: P, pellet; S, supernatant; TL, top layer; MB, middle band. (2) Solution 1: 0.4 M sucrose, 50 mM Tris base, 10 mM EDTA, pH 8.6. (3) 0.5 kg brain/liter. (4) Pass brain through meat grinder, blend ground brain and solution 1 for 10 seconds in a Waring Blendor. (5) Pass through Parr cell disruption bomb at 900 psi. (6) Sorvall RC-2B, GSA rotor. (7) 5–6 passes at 1200 rpm in a 200-ml Potter-Elvehjem homogenizer. (8) I.E.C. B-60, A-170 rotor. (9) Collect by aspiration, concentrate by repeated centrifugation, 170,000 g, 60 minutes, collecting only TL (brown pigment). (10) Collect by aspiration. (11) Final concentration 0.01% dextran sulfate, pH 7.5. (12) L_1 gradient, linear H_2O to 20% sodium chloride. (13) I.E.C. B-60, SB-110 rotor. (14) I.E.C. B-60, SB-285 rotor.

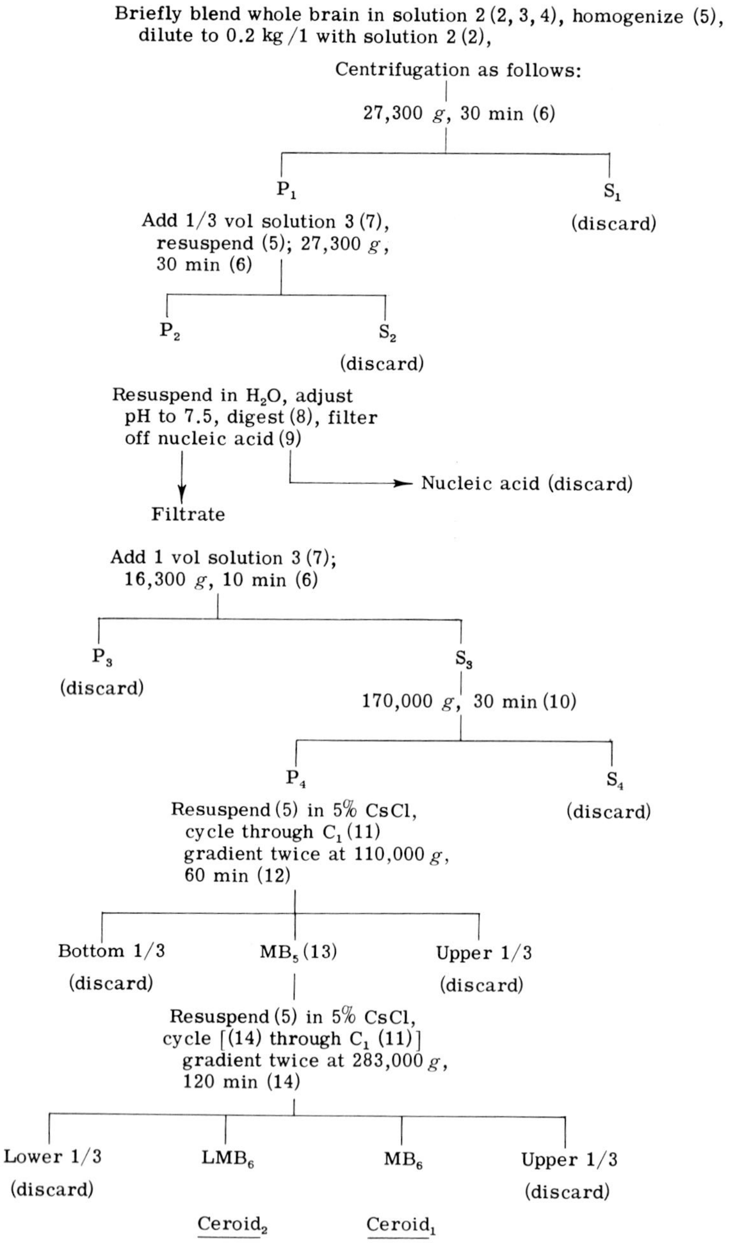

Briefly blend whole brain in solution 2 (2, 3, 4), homogenize (5),
dilute to 0.2 kg /1 with solution 2 (2),

Centrifugation as follows:

27,300 g, 30 min (6)

P₁   S₁

Add 1/3 vol solution 3 (7),   (discard)
resuspend (5); 27,300 g,
30 min (6)

P₂   S₂

(discard)

Resuspend in H₂O, adjust
pH to 7.5, digest (8), filter
off nucleic acid (9)

⟶ Nucleic acid (discard)

Filtrate

Add 1 vol solution 3 (7);
16,300 g, 10 min (6)

P₃   S₃

(discard)

170,000 g, 30 min (10)

P₄   S₄

Resuspend (5) in 5% CsCl,   (discard)
cycle through C₁ (11)
gradient twice at 110,000 g,
60 min (12)

Bottom 1/3   MB₅ (13)   Upper 1/3

(discard)   (discard)

Resuspend (5) in 5% CsCl,
cycle [(14) through C₁ (11)]
gradient twice at 283,000 g,
120 min (14)

Lower 1/3   LMB₆   MB₆   Upper 1/3

(discard)   (discard)

Ceroid₂   Ceroid₁

FIG. 2. Isolation of ceroid. Key: (1) Centrifugations, manipulations, and solutions at 4°. Abbreviations: P, pellet; S, supernatant; TL, top layer; MB, middle band; LMB, lower middle band. (2) Solution 2: 0.8 M sucrose, 50 mM Tris base, 10 mM KCl, pH adjusted to 7.5–7.8. (3) 0.5 kg brain/liter. (4) Blend 10 seconds in Waring

Isolation of Brain Lipofuscin[2]

Reagents and Gradients

Solution 1: 0.4 M sucrose, 50 mM Tris base; 10 mM EDTA, 0.01%
dextran sulfate 2000, pH 8.0
Aqueous sodium chloride, 20%
L_1 gradient: linear H_2O to 20% sodium chloride

Procedure. All postmortem human brain samples were examined as
fresh coronal sections by neuropathologists, with appropriate sections
taken for formalin fixation, and stored frozen at $-76°$. An outline of the
isolation procedure is given in Fig. 1. The tissue mass was cut into sections, approximately 1- to 2-inch cubes, ground,[4] and blended[5] in solution
1 for 20–30 seconds in a 1-gallon Waring Blendor at low speed (14,000
rpm). The proportion of brain to fluid was maintained at 0.5 kg of brain
per liter of solution 1. The suspended brain tissue was passed through a
Parr cell disruption bomb[6] at 900 psi. At this stage the pH was approximately 7.0–7.5. The preparation was then centrifuged in a Sorvall RC-2B
at 27,300 g for 20 minutes, yielding a pellet (P_1) and supernatant fraction
(S_1). The supernatant fluid and brown pigment on the upper inside surface of the centrifuge bottle were pooled. The pellet (P_1) was collected
without dilution, and the bottles were rinsed with approximately one-tenth
the total P_1 volume with distilled water; the mixture of P_1 and rinsings were
then resuspended by homogenization, five to six passes at 1200 rpm, in
a loose-fitting Potter-Elvehjem homogenizer (200-ml capacity) and centrifuged at 170,000 g in an A-170 I.E.C. rotor. The small quantity of
pigment and supernatant fluid (S_2) was collected and pooled with S_1. The
residue or pellet (P_2) was discarded. The combined S_1 and S_2 were concentrated by repeated centrifugation in the A-170 rotor (170,000 g for
60 minutes), and only the floating brown pigment (TL) was collected by
aspiration. The aspirated material was resuspended in solution 1 with a
Potter-Elvehjem homogenizer and recentrifuged in the A-170 rotor. This

[4] Model 607, Sears, Roebuck and Co., Indianapolis, Indiana.
[5] Waring Blendor, Model CB-5, Waring Products Co., Winsted, Connecticut.
[6] Parr Cell Disruption Bomb, Model 598, Parr Instrument Co., Moline, Illinois.

Blendor. (5) 5–6 passes with loose-fitting Potter-Elvehjem homogenizer at 1200 rpm.
(6) Sorvall RC-2B, GSA rotor. (7) Solution 3: 2.0 M sucrose. (8) Digest with
stirring 30 minutes at room temperature with 1 mg Nagarse per milliliter. (9) Filter
through dry glass wool and squeeze dry, repeat once more. (10) I.E.C. B-60, A-170
rotor. (11) C_1 gradient: linear H_2O to 40% CsCl. (12) I.E.C. B-60, SB-110 rotor.
(13) Collect MB (yellow) by aspiration. (14) I.E.C. B-60, SB-283 rotor.

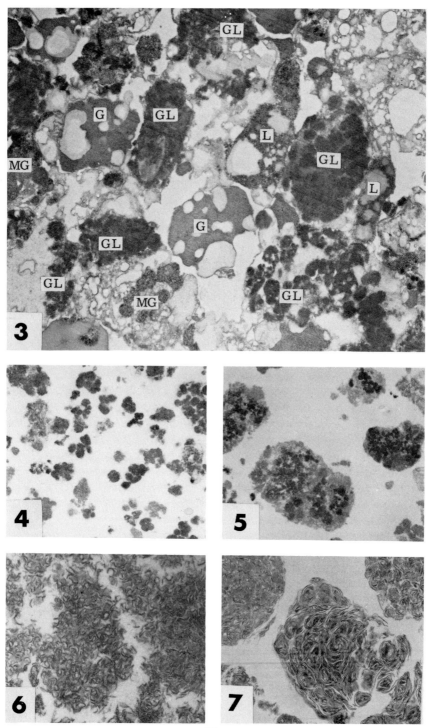

process was continued until the total volume of S_1 and S_2 were reduced to approximately 300–600 ml. The pellet (P_3) and supernatant fluid (S_3) were discarded. TB_3 was resuspended in 1 volume of distilled water containing dextran sulfate (pH 7.5) at a final concentration of 0.01%, layered on a linear gradient (L_1) of distilled water to 20% sodium chloride, and centrifuged at 110,000 g for 60 minutes in a SB-110 rotor. This gradient purification step was repeated twice. Each time the top brown layer (TB) was collected by aspiration, resuspended in 1 volume of distilled water with dextran sulfate to give a final concentration of 0.01%, layered on the L_1 gradient. Finally the top band (TB) was collected and centrifuged at 285,000 g for 60 minutes in a SB-285 rotor. The top brown band (TB_7) was collected. The lipofuscin preparation was examined by phase and fluorescence microscopy and judged to be pure when free of contaminating structures (nonfluorescent material).

With some pathological brain specimens, such as those from some of the leukodystrophies of brain that had been stored at temperatures above $-75°$, the presence of floating lipid, generally white, cannot be resolved by this method. With these samples the preparation is purified by resuspension in 5 volumes of water, and centrifugation at 110,000 g for 60 minutes in the SB-110 rotor. Under these conditions the floating lipid appears as a white floating band at the top of the gradient tube and is discarded, and the lipofuscin is isolated as a brown pellet.

Isolation of Brain Ceroid[2]

Reagents and Gradients

Solution 2: 0.8 M sucrose, 50 mM Tris base, 10 mM KCl, 0.01% dextran sulfate 2000, pH 8.0
Solution 3: 2.0 M sucrose

FIG. 3. Lipofuscin from normal human brain. Classic granular lipofuscin (G) with a granular matrix and light areas of lipid. The remaining types of lipofuscin are associated with large amounts of vacuolated lipid material, presumably neutral lipid. The granulolinear bodies (GL) show areas of deficient granular and lamellar areas. The macrogranular bodies (MG) contain coarse granular material. ×10,400.

FIG. 4. The C_1 or light ceroid from a case (I) of neuronal ceroid lipofuscinosis. This pigment consists of clusters of granular electron dense material. ×10,400.

FIG. 5. C_2 or heavy ceroid bodies from case I showing numerous spherical areas of high density in the matrix. ×8800.

FIG. 6. The curvilinear ceroid from case II of neuronal ceroid lipofuscinosis. ×24,000.

FIG. 7. Ceroid bodies in a dog model of neuronal ceroid lipofuscinosis (Koppang). ×10,400.

Crystalline bacterial proteinase (Nagarse)[7]
C_1 gradient: linear solution 2–2.0 M sucrose plus 25% CsCl

Procedure. Ceroid was isolated from postmortem brain samples of two patients (cases I and II) with neuronal ceroid-lipofuscinosis (Batten's disease).[2] Tissue was fractionated either after being stored frozen at $-75°$ or in the fresh state immediately after autopsy. An outline of the ceroid isolation procedure is given in Fig. 2. The tissue was thawed at $4°$ and briefly blended (20–30 seconds) in a 500-ml Waring Blendor at low speed. The proportion of brain of fluid was maintained at 0.5 kg of brain per liter of solution 2. The blended tissue was homogenized with a loose-fitting Potter-Elvehjem homogenizer, five to six passes at 1200 rpm, and then diluted to 0.2 kg/liter with solution 2. At this stage the pH of the homogenate was 7.0–7.5. The preparation was centrifuged at 27,300 g for 20 minutes in a Sorvall RC-2B (GSA rotor). The resulting supernatant fluid (S_1) was discarded, but the pellet (P_1) was collected and a volume of solution 3 (2.0 M sucrose) equal to one-third the P_1 pellet volume was added; the mixture was resuspended with a Potter-Elvehjem homogenizer. This mixture was centrifuged at 27,300 g for 30 minutes in the GSA rotor. Again the supernatant fraction (S_2) was discarded, but the pellet (P_2) was resuspended in distilled water, the pH was adjusted to 7.0 with Tris base, 1 mg of Nagarse (dry powder) was added per milliliter of suspension, and the mixture was agitated with a magnetic stirrer for 30 minutes at room temperature. The digested mixture was diluted to 2 volumes with water and carefully filtered through a pad of dry glass wool (Corning 3950). The wet glass wool filter was squeezed free of excess fluid, and the pad was discarded. This process of filtration through dry glass wool was repeated once more. Strands of interfering nucleic acid were irreversibly adsorbed to the glass wool and discarded. One volume of solution 3 was added to the filtrate and the suspension was centrifuged 16,300 g for 10 minutes in the GSA rotor. The pellet (P_3) was discarded, and the supernatant fraction (S_3) was centrifuged at 170,000 g for 30 minutes in an A-170 rotor. The resulting fraction (S_4) was discarded, but P_4 was collected, resuspended in solution 2, layered on a linear gradient (C_1), and centrifuged at 110,000 g for 60 minutes in the SB-110 rotor. A yellow middle band was recovered by aspiration and was recycled once more through the C_1 gradient under the same conditions. The middle band from this procedure was collected and cycled through the C_1 gradient twice, with the SB-285 rotor at 285,000 g for 120 minutes. Under these conditions occasionally a second minor band (LMB_6) was observed just below the major yellow middle band (MB_6).

[7] Nagarse obtained from Enzyme Development Corp., New York, New York.

These fractions were designated as C_2, a heavy ceroid, and C_1, or light ceroid, respectively. Examination by phase and fluorescence microscopy was employed to determine the purity of a preparation by the absence of contaminating material (Figs. 3–7).

Occasionally MB_5 was highly contaminated with materials other than ceroid, particularly with brain specimens containing only trace amounts of ceroid. These preparations were treated again with Nagarse as with the pellet P_2, and the glass wool filtrate from this digestion step was applied to the C_1 gradient and the purification continued as above.

Section IV

Subcellular Fractions Derived from Plant Tissue

A. GENERAL ASPECTS
Articles 51 through 56

B. SPECIFIC ORGANELLES AND DERIVED COMPONENTS
Articles 57 through 63

[51] The Diversity of Plant Organelles with Special Reference to the Electron Cytochemical Localization of Catalase in Plant Microbodies[1]

By E. H. NEWCOMB and W. M. BECKER

The Organization of Higher Plant Cells

Our objective in this chapter is first, to provide a brief introduction to the organization of the higher plant cell as a guide to the more detailed chapters on methodology which follow, and second, to describe the cytochemical method for localizing catalase in plant microbodies. In comparing and contrasting plant and animal cells, we will point out the similarities in organization of those subcellular components involved in processes common to both kingdoms and underscore the extent to which the distinctive structural features of plant cells can be correlated with metabolic activities or aspects of anatomical design unique to plants.

The number of generally distributed, morphologically distinct types of cytoplasmic systems or organelles observable in plant cells with the electron microscope is not great. The list includes at present the granular (rough) and agranular (smooth) endoplasmic reticulum (ER), ribosomes, dictyosomes (Golgi apparatus) and associated vesicles, mitochondria, plastids, microbodies, and microtubules. Other prominent structural components of plant cells include, besides the nucleus, the cell wall, vacuoles, and the plasmodesmata or cytoplasmic connections which run through the walls between adjacent cells. Other cytoplasmic components, such as spherosomes, are found in some cells, but they are not of general distribution or importance.

Most of the above components are common to both plant and animal cells, and their general distribution reflects the strikingly similar manner in which all eukaryotic cells organize and effect such fundamental functions as information storage, energy metabolism, protein synthesis, and intracellular movement. Some features, such as cell walls and plastids, are unique to plants and therefore of greater interest in the present context, but in general an ultrastructural comparison of plant and animal cells reveals more similarities than differences in basic cellular design.

However, even among those organelles found in both kingdoms, there

[1] The preparation of this chapter was supported by Grant GB 15246 from the National Science Foundation.

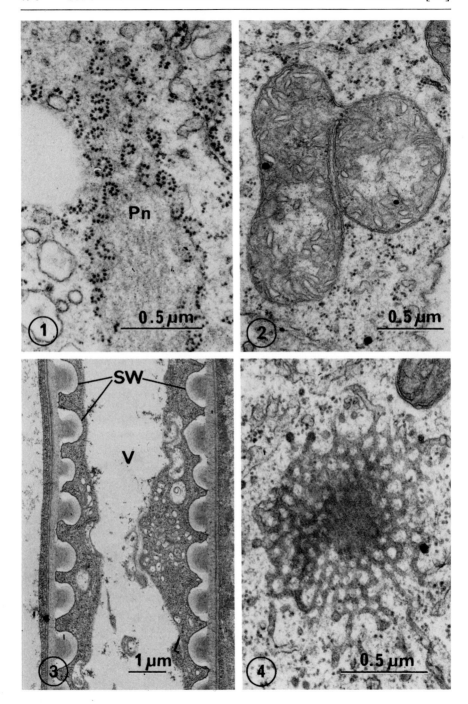

are often characteristic differences in numbers, structure, and time and place of occurrence that are related to fundamental differences between plant and animal cells. In general, the organs and tissues of plants lack the diversity and degree of specialization common to those of most animals. This is in turn reflected in cellular fine structure, with animal cells and their organelles showing a far greater range of variation in form and function than those of plants. Consider, for example, the endoplasmic reticulum ("ER"). In general, meristematic plant cells have a paucity of ER, and carry out protein synthesis on free polyribosomes. Differentiating plants cells usually have more rough ER and have both free and bound polyribosomes (Fig. 1), although they seldom if ever approach the degree of ER proliferation seen in secretory animal cells.

Similarly, whereas in animals mitochondria differ considerably in structure and numbers in cells from different tissues, in plants mitochondria (Fig. 2) appear quite similar regardless of source. This is in part a consequence of the general lack in plant tissues of the particularly intense respiratory activity associated with many animal tissues, e.g., flight muscle, which necessitates large numbers of mitochondria and numerous, closely packed cristae.

Several unique features in the design of plant cells have profound consequences for cellular activities in general and organelle characteristics in particular. One of the most obvious features of plant cells which distinguishes them from animal cells is their enclosure in a rigid, nonliving wall (Fig. 3) composed primarily of cellulose and other polysaccharides. The cell wall can be seen as an evolutionary consequence of autotrophy: only by encasing each cell in a wall and cementing adjacent cells together were plants able to achieve the rigidity and strength necessary for the successful evolution of a rooted, sessile organism obliged to compete literally for its place in the sun.

As an organism, the plant is greatly influenced by the walls which surround its cells: its mode of growth, means of response to stimuli, water relations, and mechanisms of reproduction are only a few of the special

---

FIG. 1. Polyribosomes on the surface of a dilated cisterna of endoplasmic reticulum containing an accumulation of protein (Pn). In an epidermal cell of a root tip of radish (*Raphanus sativus*). Micrograph by H. T. Bonnett, Jr. ×43,000.

FIG. 2. Mitochondrion presumed to be dividing in a root tip cell of bean (*Phaseolus vulgaris*). Micrograph by W. P. Wergin. ×35,000.

FIG. 3. Portion of a differentiating xylem vessel near the tip of a root of lentil (*Lens culinaris*). The prominent bulges (SW) represent secondary wall being laid down on the primary wall. V, vacuole. Micrograph by B. A. Palevitz. ×10,000.

FIG. 4. Tangential section through a dictyosome in a root tip of bean (*Phaseolus vulgaris*). ×46,000.

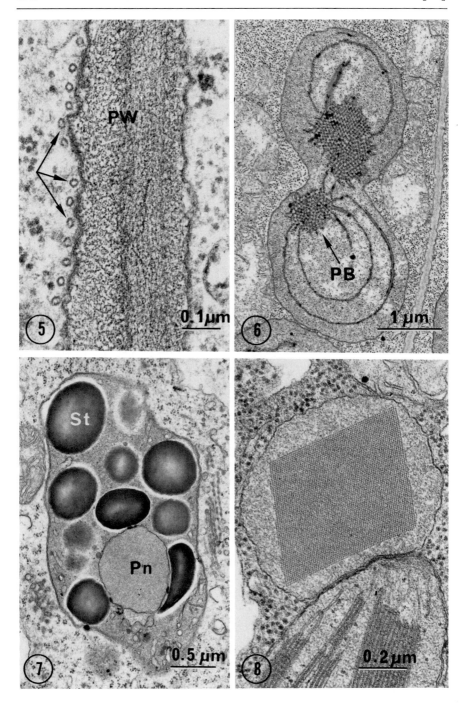

features that can be viewed as indirect consequences of a rigid cell wall. At the cellular level, the impact of the cell wall is equally profound. The biosynthesis, transport, and ordered deposition of the variety of polysaccharides of which the wall consists must surely dominate the biochemical activities of a number of systems within the plant cell during the period of rapid wall growth.

In time of development, location, and properties, two types of walls are distinguishable: primary and secondary. The primary wall is deposited when the daughter cell is first formed, and in some cell types it is the only wall produced. It grows in surface area during growth of the cell; that is, it extends plastically in response to the turgor pressure of the cell and simultaneously incorporates new wall components. In its general organization, the primary wall consists of a framework of cellulose microfibrils embedded in an amorphous matrix.

In some cell types, a secondary wall is deposited upon the primary wall as the cell differentiates (Fig. 3). Thus the secondary wall merges with the primary wall on the outside and is in contact with the plasmalemma on the inside. This type of wall becomes thicker through the deposition of additional materials onto its inner surface. Since it is compact and rigid and is not capable of growth by plastic extension, its deposition effectively stops further growth of the cell.

Both dictyosomes and microtubules are believed to be intimately involved in wall formation. Dictyosomes (Fig. 4) appear to play the central role in the synthesis of polysaccharides and their packaging in vesicles for transport to the growing wall. Typically in an enlarging cell dictyosomes produce vesicles of two types: large ones bounded simply by a membrane, and small ones bearing a reticulate or honeycomblike layer external to the membrane. The large vesicles frequently contain a fibrillar material closely resembling the visible structure in the wall. Much evidence indicates that these vesicles travel from the dictyosomes to the plas-

---

Fig. 5. Transverse section of microtubules (arrows) lying in the cytoplasm adjacent to the growing primary side wall (PW) of an enlarging cell in a root of bean. Micrograph by B. A. Palevitz. ×100,000.

Fig. 6. Etioplast in a palisade cell of an etiolated leaf of an 11-day-old seedling of bean. The etioplast, which appears to be dividing, has two conspicuous prolamellar bodies (PB). Micrograph by P. J. Gruber. ×17,000.

Fig. 7. Amyloplast containing both starch bodies (St) and protein (Pn) in a root tip cell of bean. ×24,000.

Fig. 8. Microbody with a large crystalline inclusion in a mesophyll cell of a leaf of tobacco (*Nicotiana tabacum* cv. Wisconsin 38). At lower right the microbody is in contact with a chloroplast and conforms to its contour. Micrograph by Sue Frederick Gruber. ×64,000.

malemma, where they fuse and add their contents to the growing wall. Autoradiography at the electron microscopic level has shown that when radioactively labeled sugars are supplied to growing cells, the radioactivity appears first in polysaccharides in the dictyosomes, and later in the cell wall. The small "coated" vesicles also appear to fuse with the plasmalemma, but whether they contribute to the wall is unknown.

Microtubules (Fig. 5) appear to play a prominent role in the growth of the primary wall and deposition of the secondary wall. They are characteristically localized in the cytoplasm near the plasmalemma when the wall is growing, and appear to control the direction of microfibril deposition and hence the direction of cell enlargement. It has been observed repeatedly that the newly deposited microfibrils are always aligned parallel to the direction taken by the microtubules. Although it is not yet understood how the latter may exert control over the orientation of the wall microfibrils, there is evidence that the microtubules are attached to the plasmalemma by bridges.

A second distinctive feature of plant cells is observed in the mechanism of cell enlargement. In animals, this is achieved almost entirely by increases in the amount of cytoplasm, but in plants the relatively enormous enlargement that many cells undergo involves a great increase in vacuolar size. Vacuoles (Fig. 3) are membrane-bound regions within the cell which are filled with dilute liquid (vacuolar sap) instead of cytoplasm, and typically contain salts, sugars, minerals, and some dissolved proteins. They apparently also contain a number of the lytic enzymes which in animal cells are localized in separate, single membrane-bounded organelles called lysosomes. The immature plant cell contains large numbers of small vacuoles, which increase in size and coalesce as the cell enlarges. In the mature cell, up to 90% of the internal volume may be occupied by a single large vacuole, with the cytoplasm confined to a thin peripheral layer sandwiched between the vacuole and the plasmalemma. Obviously, cell growth can occur only if the wall yields to the stress generated by vacuolar coalescence and expansion.

A third distinguishing characteristic of plant cells is the presence of plastids, since these are the only major cytoplasmic organelles unique to plants. Plastids range in diameter from about 0.5 to 5 or 6 μm and contain a moderately electron-opaque matrix or stroma surrounded by a double membrane. The cisternae, lamellae, and microtubule-like structures frequently observed in the stroma all appear to originate from the inner membrane of the envelope.

Plastids are of several types and perform a variety of functions, prominent among which is food storage. Meristematic cells contain small, immature proplastids which reproduce by division and develop into one of several different types of mature plastids. In young leaves prior to their exposure to light, the proplastids develop into precursors of chloroplasts termed etioplasts (Fig. 6). These etioplasts commonly contain a para-crystalline membranous structure, the prolamellar body, from which the lamellar system of the chloroplast arises upon illumination. Chloroplasts, of course, contain chlorophyll and other light-trapping pigments, as well as all the enzymes necessary to accomplish the photolysis of water and the reductive fixation of carbon dioxide into carbohydrate. These carbohy-drates, even if stored temporarily as starch, are soon moved out of the plastid and are either utilized locally or translocated for storage or use elsewhere in the plant.

Chromoplasts occur in a number of flowers, in many fruits, and in some roots. They contain varying proportions of the water-insoluble carotenoid pigments, carotenes and xanthophylls, and are thereby colored various shades of yellow, orange, or red. Chromoplasts either develop directly from colorless proplastids, or from chloroplasts by alteration of their structure and composition, as in the ripening of green tomatoes.

Colorless plastids are sometimes referred to collectively as leucoplasts. Included are the starch-storing amyloplasts, the protein-storing proteo-plasts (or proteinoplasts) and the oil-storing elaioplasts. Some leucoplasts store both starch and protein (Fig. 7).

It is now well established that plastids, like mitochondria, contain DNA and RNA and are to some extent genetically independent of the nucleus. They contain ribosomes which are distinguishable from cytoplasmic ribo-somes, but are active in protein synthesis upon isolation. Both messenger RNA and transfer RNA are also present in small amounts in plastids. Thus plastids contain both the information and machinery necessary to make at least some of their own enzymes and structural proteins, although it is clear that many of their genetic traits are in fact under nuclear control. The nature and extent of plastid autonomy are research subjects of great current interest.

A discussion of the distinctive structural features of plant cells should also include reference to the microbodies, which, although not unique to plant cells, are far more ubiquitous than in animal cells and are known to participate in several major metabolic pathways found exclusively in plants.

Microbodies (Fig. 8) are characterized by a single bounding mem-brane, a granular to fibrillar matrix and, at least in early stages of develop-ment, a close association with cisternae of the endoplasmic reticulum, from which they apparently arise. Microbodies exhibit spherical, elongate, or irregularly shaped profiles in section and are usually less than 1 μm in greatest diameter. Dense, amorphous nucleoids and crystalline inclusions are frequently present in the matrix.

Biochemical studies have established that microbodies invariably contain one or more oxidative enzymes that generate hydrogen peroxide, and catalase, an enzyme that decomposes hydrogen peroxide. Microbodies in which these enzymes have been demonstrated are termed peroxisomes. Peroxisomes isolated from green leaf tissues have been shown to possess enzymes involved in the metabolism of glycolate, a two-carbon product of photosynthesis. Similar particles have been obtained from the endosperm and cotyledons of various fat-storing seedlings. These particles, however, contain the enzymes of the glyoxylate cycle required for the conversion of fat to carbohydrate during seed germination and have therefore been termed glyoxysomes.

Organelles resembling typical microbodies are known from electron microscopy to be particularly abundant both in the chlorophyllous cells of green leaves from which peroxisomes have been isolated and in the cells of fat-storing endosperm of cotyledons from which glyoxysomes are obtained. In the storage cells of the seedlings the microbodies are appressed to the lipid deposits, while in leaf cells they are usually associated with the chloroplasts (Figs. 8 and 9). Clearly the observation that there are microbodies within the cells of these tissues does not prove that they are identical with the particles isolated in biochemical studies. Additional, more definitive evidence comes, however, from electron microscopic examination of isolated particles and from cytochemical enzyme localization at the level of the electron microscope. Such evidence is not only relevant to the question of microbody function, but is also of more general interest, since the rigorous establishment of an identity between structures seen by the electron microscopist and particles isolated by the biochemist is important in many correlative studies in cell biology.

In the case of the microbody, the particles isolated from green leaves, from endosperm, and from cotyledons have been shown by electron microscopy to bear a striking resemblance to the organelles seen *in situ*. In addition, it has been demonstrated by electron cytochemical assay, using the methodology pioneered by cell biologists studying animal microbodies, that microbodies are the major if not exclusive sites of catalase activity in these plant tissues, and are thus identical to isolated peroxisomes and glyoxysomes, for which catalase is a common marker enzyme. Since the reaction for catalase is one of the few cytochemical enzyme assays which have thus far been successfully adapted to the level of the electron microscope in animal[2-4] and plant[5] cells, it is described in detail below.

[2] M. E. Beard and A. B. Novikoff, *J. Cell Biol.* **42**, 501 (1969).
[3] H. D. Fahimi, *J. Histochem. Cytochem.* **16**, 547 (1968).
[4] A. B. Novikoff and S. Goldfischer, *J. Histochem. Cytochem.* **16**, 507 (1968).
[5] S. E. Frederick and E. H. Newcomb, *J. Cell Biol.* **43**, 343 (1969).

The Electron Cytochemical Localization of Catalase in
Plant Microbodies

Preparation and Fixation of Tissues. Leaves, roots, or other plant parts
can be used successfully, although younger tissues with active cytoplasm
and small vacuoles give better results than older tissues, whose cells have
harder walls and large central vacuoles. When leaves are used, the lower
epidermis should if possible be stripped from regions between major veins
to allow better penetration of fixative and incubation medium. The stripped
pieces of leaf tissue are then segmented in the fixing solution into squares
approximately 1 mm on a side. Tissues other than leaves are similarly
diced or segmented. The fixing solution consists of 3% glutaraldehyde
in 50 mM potassium phosphate buffer at pH 6.8. Fixation is at room temperature for 1.5–2.0 hours, after which the segments of tissue are rinsed
for 15–20 minutes in at least three changes of phosphate buffer of the
same composition as that of the fixative.

Tissue Incubation. After the tissue segments have been fixed in glutaraldehyde and rinsed, they are incubated in a solution of DAB (3,3'-
diaminobenzidine, Sigma Chemical Co., St. Louis, Missouri). Incubation
can be carried out conveniently in small, covered, glass petri dishes or in
corked vials for 50–60 minutes at 37°. The containers should be agitated
periodically to insure that the tissue segments remain immersed.

The standard incubation medium is that of Novikoff and Goldfischer,[4]
as described by Beard and Novikoff.[2] It contains 10 mg DAB, 5 ml of
50 mM propanediol buffer (2-amino-2-methyl-1,3-propanediol, Sigma
Chemical Co.) at pH 10.0, and 0.1 ml of 3% H_2O_2. This medium should
be prepared immediately before use, and the pH adjusted to 9.0 prior to
addition of the tissue.

Several control procedures can be applied to determine whether any
observed activity is ascribable to catalase. The most definitive is preincubation for 20–30 minutes with propanediol buffer containing 20 mM aminotriazole (3-amino-1,2,4-triazole, Aldrich Chemical Co., Milwaukee, Wisconsin), followed by incubation in the standard DAB medium which also
contains 20 mM aminotriazole. Another control procedure of value consists of preincubating the tissue for 20 minutes in propanediol buffer containing 10 mM KCN, followed by incubation in standard DAB incubation mixture containing 10 mM KCN. The value of these two control
procedures is discussed below.

After the incubations the tissues are rinsed for 15–20 minutes with
several changes of 50 mM phosphate buffer.

*Postfixation and Preparation of Tissues for Examination in the Elec-
tron Microscope.* The tissues are postfixed in 2% OsO_4 in 50 mM phos-

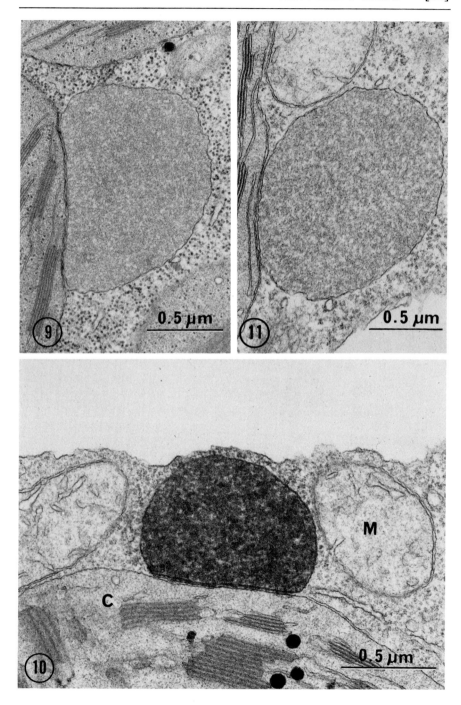

phate buffer at pH 6.8 for 2 hours. They are then dehydrated in a series of increasing concentrations of acetone or ethyl alcohol followed by propylene oxide, and then embedded in Araldite-Epon. Silver gray sections are cut on the ultramicrotome, mounted on copper grids, and stained in 2% aqueous uranyl acetate for 10 minutes followed by lead citrate for 5 minutes before viewing in the electron microscope.

Observations. The fixation of suitable plant tissues in glutaraldehyde followed by their incubation in a medium containing DAB and hydrogen peroxide results in a remarkably sharp localization of electron-opaque material within structures clearly identifiable as microbodies, as illustrated in Fig. 10. The matrix of the microbodies is completely permeated by the highly electron-opaque, coarsely granular deposit. The microbody membrane may appear thicker and more opaque than in unstained microbodies, and electron-opaque material may accumulate to a small extent exterior to the membrane.

In the experience of our laboratory with leaf tissues of a variety of monocots and dicots,[5,6] under the above conditions the reaction product is always sharply localized in the microbodies or in the microbodies and the immediately adjacent cytoplasm. There is little evidence that either the general cytoplasmic matrix or other organelles, including the mitochondria and chloroplasts, are stained. However, in some sections there is a slight deposition of dense material in the cell walls and intercellular spaces.

Certain modifications in the routine experimental procedure result in substantial variation in the intensity of the catalase reaction. Thus differences in opacity of the reaction product can be correlated with the duration of incubation and the temperature. Incubation with the DAB mixture for 30 minutes at 37° gives markedly less accumulation of reaction

[6] S. E. Frederick and E. H. Newcomb, *Planta* 96, 152 (1971).

FIG. 9. Microbody appressed to a chloroplast in a mesophyll cell of a mature tobacco leaf. Micrograph by Sue Frederick Gruber. ×40,000.

FIG. 10. Microbody in a mesophyll cell from a segment of tobacco leaf incubated in medium containing diaminobenzidine (DAB). Dense reaction product (osmium black), attributable to the action of catalase, is confined to the microbody and is not observed in the mitochondria (M) or the chloroplast (C). Micrograph by Sue Frederick Gruber. ×50,000.

FIG. 11. Appearance of a microbody in a tobacco leaf cell when incubation in the DAB medium is carried out in the presence of 20 mM aminotriazole. The general appearance of the microbody matrix and its density relative to the surrounding cytoplasm closely resemble those of cells not exposed to DAB (Figs. 8 and 9). Aminotriazole has inhibited any observable activity of catalase. Micrograph by Sue Frederick Gruber. ×39,000.

product than incubation for 50–60 minutes at 37°. Incubation for these time periods at room temperature results in little or no dense material in the microbodies.

The use of diaminobenzidine to localize catalase with the electron microscope depends upon oxidation of the DAB to a polymeric form that can interact with osmium tetroxide to yield electron-opaque osmium black.[7] However, the oxidation of DAB can be catalyzed by several hemoproteins, including peroxidase, cytochrome oxidase, and myoglobin. Among these, peroxidase is a particularly likely candidate, especially since there is a large body of evidence for the occurrence of both soluble and wall-localized peroxidases in plant tissues, and since catalase and peroxidases are so similar in their peroxidatic activity that distinction between them by cytochemical means is difficult.

Evidence that the accumulation of electron opaque material in plant microbodies is due to catalase rather than peroxidase is provided by experiments employing 3-amino-1,2,4-triazole and is reinforced by the results obtained with potassium cyanide. There is some evidence to indicate that aminotriazole is a much more potent inhibitor of catalase than of peroxidase,[8,9] whereas cyanide is an inhibitor of both catalase and peroxidase. When aminotriazole is included in the incubation medium at a final concentration of 20 mM, it completely eliminates staining of the microbodies, but allows the slight accumulation of dense product to take place in wall structures. A microbody and other organelles in tissue incubated in DAB in the presence of aminotriazole are shown in Fig. 11.

Cyanide, on the other hand, not only causes partial inhibition of microbody staining, but also prevents the rapid browning or blackening of leaf segments that occurs during their exposure to the DAB medium. It seems likely that the browning of leaf segments observed during incubation and the slight deposition of reaction product within the cell walls are due to the oxidation of DAB by peroxidases to a dark product that accumulates diffusely throughout the cytoplasm. This interpretation is supported by the observation that aminotriazole does not reduce the brown color while cyanide eliminates it.

[7] A. M. Seligman, M. J. Karnovsky, H. L. Wasserkrug, and J. S. Hanker, *J. Cell Biol.* 38, 1 (1968).
[8] W. G. Heim, D. Appleman, and H. T. Pyfrom, *Amer. J. Physiol.* 186, 19 (1956).
[9] E. Margoliash and A. Novogrodsky, *Biochem. J.* 68, 468 (1958).

[52] Plant Cell Fractionation

By C. A. PRICE

Purity and structural integrity should be prerequisites for studies on the metabolism or composition of subcellular particles, but this ideal has rarely been approached in practice. For example, chloroplasts may be obtained intact and capable of reducing CO_2 to carbohydrates at rates comparable to those of intact leaves,[1] but the preparations are usually crude pellets, so there is uncertainty over which of the chemical components and enzyme activities observed in these preparations are actually in the chloroplasts. After investigators learned to purify chloroplasts by density gradient centrifugation, it was found that enzymes, such as catalase and glycolate oxidase that were long thought to be in chloroplasts, were actually in peroxisome instead (cf. Tolbert, this volume [74]). Similarly, the peroxidase activity that had been attributed to plant mitochondria was found to be in another fraction, presumably peroxisomes, when the particles were separated on a sucrose gradient.[2]

Excepting those organelles which are *peculiar* to plants, the properties of organelles from plant tissues are not fundamentally different from those of animals. This generalization means that plant nuclei, mitochondria, ribosomes, etc., normally sediment at the same speeds, band at the same densities, and show the same electrophoretic mobilities as the corresponding particles from animals.

What is fundamentally different in the case of plants, and often peculiar to individual plant tissues, are the difficulties met in separating intact plant organelles from the rest of the plant cell. The cellulose walls of plant tissues are so tough that shear forces sufficient to rupture the cell walls may also be sufficient to rupture fragile organelles as well. The presence of large vacuoles containing tannins, acids, and possibly hydrolases[3] is an even more serious impediment to the recovery of intact organelles.

Thus, the first requirement of a successful fractionation of plant tissue is that the cells be broken under mild conditions. Table I contains a list of the principal techniques. The choice depends on the character of the

---

[1] R. G. Jensen and J. A. Bassham, *Proc. Nat. Acad. Sci. U.S.* 46, 1095 (1966).

[2] M. Plesnica, W. D. Bonner, Jr., and B. T. Storey, *Plant Physiol.* 42, 366 (1967).

[3] Mature plant tissues are typically rich in hydrolases, which are unlikely to be floating freely in the cytosol. The subcellular localization of these enzymes has been only rarely documented, but Matile[4] has presented convincing evidence that the vacuoles of meristematic tissues do contain hydrolases and are analogous to the lysosomes of animals.

[4] P. Matile, *Z. Naturforsch. B* 21, 871 (1966).

TABLE I

TECHNIQUES FOR BREAKING PLANT CELLS AND TISSUES

Technique	Apparatus	Comments	Manufacturer or reference
Grinding	Mortar and pestle	Often used with sand, alumina, or other abrasive	Various supply houses
	Potter-Elvehjem homogenizer	Shear results principally from vertical movements	cf. Anderson[a]
Extrusion	French Press	To 1400 kg · cm^{-2}	American Instrument Co.
	Ribi Press (RF-1)	To 4200 kg · cm^{-2}	Ivan Sorvall, Inc.; Ribi and Milner[b]
	"Pea Popper"	Tissue is squeezed between rollers	Rho and Chipchase[c]
	Parr nitrous oxide bomb	Disruption occurs upon release of pressure	Avis[d]
Cutting and slicing	Waring Blendor, polytron mixer	Shear results from rapid motion of blades	Various supply houses

Method			Reference
"Mixer 66"			Moulinex Co., 61 Alencon, France; cf. Bonner[e]
Special cutter	Less shear than in most blades		Morré[f]
Sonication	Probably involves shear produced by cavitation		Hughes[g]
	Various		
Enzymatic disruption	Cell wall hydrolases	Purity of hydrolases remains a serious hazard	D'Allessio and Trim[h]; Takebe et al.[i]

[a] N. G. Anderson, in "Physical Techniques in Biological Research" (G. Oster and A. W. Pollister, eds.), Vol. III, p. 299. Academic Press, New York, 1956.

[b] E. Ribi and K. C. Milner, in "Methods in Immunology and Immunochemistry" (C. A. Williams and M. W. Chase, eds.), Vol. 13. Academic Press, New York, 1967.

[c] J. H. Rho and M. I. Chipchase, J. Cell Biol. 14, 183 (1962).

[d] P. J. G. Avis, in "Subcellular Components" (G. D. Birnie, ed.), 2nd ed., p. 1. Univ. Park Press, Baltimore, Maryland, 1972.

[e] W. D. Bonner, Jr. and E. L. Christensen, in "Microsymposium on Particle Separation from Plant Materials" (C. A. Price, ed.), p. 24. Oak Ridge National Laboratories, CONF-700119. National Technical Information Service, Springfield, Virginia, 1970.

[f] D. J. Morré, this series, Vol. 22, p. 130.

[g] D. E. Hughes, J. Biochem. Microbiol. Technol. Eng. 3, 405 (1961).

[h] G. D'Allessio and A. R. Trim, J. Exp. Bot. 19, 831 (1968).

[i] I. Takebe, Y. Otsuki, and S. Aoki, Plant Cell Physiol. 9, 115 (1968).

specific tissue and the exigencies of the organelle sought. The second requirement is that the organelles be protected from the vacuolar contents. The usual strategy is to neutralize the toxic agents of the central vacuole before or during cell breakage (Table II). For example, Watts and Mathias[5] dramatically improved the yield of polyribosomes from plant tissues by adding Bentonite, an inhibitor of RNase, during extraction; Anderson and Key[6] used diethylpyrocarbonate to the same effect. The latter agent has also been used to destroy RNase in sucrose gradients used much later in purification. Other strategies to minimize the effects of vacuoles on plant particles are to keep the ratio of tissue to breakage medium low (1 to 10) and to separate the organelles from soluble elements of the cell brei as rapidly as possible.

Because of large central vacuoles a plant cell brei is certain to be more dilute in protein than preparations from typical animal tissues. Soluble protein (e.g., serum albumin) is frequently added to the breakage medium. Although beneficial effects are often observed, it is not clear whether the proteins act against vacuolar toxins or have more direct effects on the organelles.

Since most membrane-bound organelles require an osmoticum for the maintenance of their structure, inclusion of osmotica in breakage media will minimize the contamination of soluble and small particle fractions with the contents of the membrane-bound organelles. Sucrose is the osmoticum most commonly used,[7] but sorbitol or mannitol may be less damaging to some membranes.

Another requirement of successful separations is the avoidance of aggregation of cell particles. Divalent cations, notably Mg^{2+}, produce severe clotting, which can cause many small and dissimilar particles to be carried along with large ones in a centrifugation. Careful choice of buffer and salts consistent with the integrity of organelles on the one hand and the avoidance of adhesion on the other is absolutely essential. Gorczynski et al.[8] recommended 2-naphthol-6,8-disulfonic acid (dipotassium salt) as a dispersing agent.

Centrifugation remains the principal method for separating plant particles. Although Schneider's[9] protocol for fractionating liver by differential centrifugation has guided a whole generation of biochemists, it was never

[5] R. L. Watts and A. P. Mathias, Biochim. Biophys. Acta 145, 828 (1967).

[6] J. M. Anderson and J. L. Key, Plant Physiol. 48, 801 (1971).

[7] Sam Granick introduced to biology the use of sucrose as an osmoticum in 1938 (Amer. J. Bot. 25, 558). He reported that chloroplasts lost their characteristic appearance when exposed to salts alone.

[8] R. M. Gorczynski, R. G. Miller, and R. A. Phillips, Immunology 19, 817 (1970).

[9] W. C. Schneider, in "Manometric and Biochemical Techniques" (W. W. Umbreit, R. H. Burris, and J. F. Stauffer, eds.), 5th ed., p. 196. Burgess, Minneapolis, Minnesota.

TABLE II
Protective Agents Used in the Preparation of Plant Brei

Agent	Target substance
Cyanide	Heavy metal oxidases
Polyvinylpyrrolidone	Polyphenols; quinones
Ascorbic acid	Quinones
Mercaptoethanol, glutathione, cysteine, Cleland's reagent, other thiols	Quinones, disulfides, peroxidases
Diethylpyrocarbamate, Bentonite, spermidine and other polyamines	Ribonucleases
Serum albumin	Surface denaturation
EDTA	Heavy metals
NH₃ infiltration	H+
Coconut milk	?

claimed to be satisfactory for plant tissues. At best the crude particle suspensions obtained serve as a starting point for authentic purification; at worst, organelles may be aggregated and denatured by pelleting against the walls of a tube.

Since the 1950's, density gradient centrifugation has been recognized as the method of choice for the separation of biological particles[10–12] (see also, this volume [73]), but only a few milligrams of particles can be handled in conventional swinging-bucket rotors. In addition, the problems of "wall effects" and "droplet sedimentation" are extremely severe and strongly limit resolution, especially for relatively large particles, such as mitochondria.

With the development of zonal rotors,[13] the problem of density gradient centrifugation in quantity has been overcome and the other difficulties minimized. It has become possible, at least in principle, to separate particles of specified sedimentation coefficients and equilibrium densities in quantities suitable for most biochemical and physiological investigations.

Fortunately, as Anderson has pointed out, different classes of subcellular particles tend to have unique combinations of sedimentation coefficients and equilibrium densities, so that we may "locate" them in S-ρ space (Fig. 1). In practice one can either separate particles according to

[10] D. H. Moore, in "Physical Techniques in Biological Research" (D. H. Moore, ed.), 2nd ed., Vol. 2, Part B, p. 285. Academic Press, New York, 1969.

[11] C. A. Price, in "Manometric and Biochemical Techniques" (W. W. Umbreit, R. H. Burris, and J. F. Stauffer, eds.), 5th ed., pp. 213–243. Burgess Publ., Minneapolis, Minnesota.

[12] C. A. Price, "Centrifugation in Density Gradients." Academic Press, New York. In preparation.

[13] N. G. Anderson (ed.), Nat. Cancer Inst. Monogr. 21 (1966).

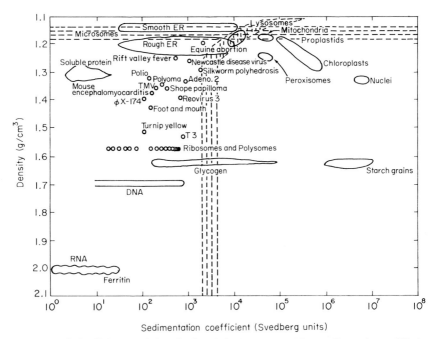

FIG. 1. Subcellular particles displayed in S-ρ space. The ordinate is equilibrium density, and the abscissa is the apparent sedimentation coefficient S^* on a logarithmic scale. Note that most classes of subcellular macro- and multimolecular structures, when considered as particles, occupy unique regions in S-ρ space. Fractions obtained in rate-zonal and isopycnic separations are shown by vertical and horizontal dashed lines. From C. A. Price, "Centrifugation in Density Gradients." Academic Press, New York, in preparation.

their sedimentation coefficients ("rate-zonal," "zonal," or "S-rate" separations) or according to their densities ("equilibrium density" or "isopycnic" separations). These two kinds of separation correspond to vertical and horizontal "cuts" in S-ρ space. Complete purification of a particle often requires the exploitation of both coordinates; such an "S-ρ" separation is carried out by first fractionating the cell brei according to S, and then according to ρ. The two kinds of centrifugation are shown in Fig. 2, and the results are displayed by the vertical and horizontal cuts in Fig. 1. The S-ρ strategem is usually necessary for the purification of membrane-bound organelles from plants, especially from green tissues.[14] The special problem of plant organelles is that three major classes of particles—mitochondria, peroxisomes, and plastids—have similar S-ρ coordinates. Moreover the

[14] V. Rocha and I. P. Ting, *Arch. Biochem. Biophys.* **140**, 398 (1970).

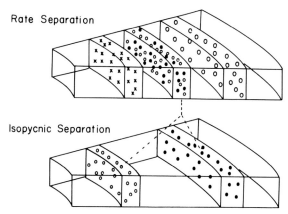

Rate Separation

Isopycnic Separation

FIG. 2. Strategy of S-ρ separations. In the upper figure particles have separated into a gradient according to their sedimentation *rate;* a middle band is pictured as containing two kinds of particles with equal sedimentation rates. In the lower figure this mixture of particles was sedimented into a second gradient, and the particles were allowed to move to their equilibrium density or isopycnic position from C. A. Price, *in* "Microsymposium on Particle Separation from Plant Materials" (C. A. Price, ed.), p. 1. Oak Ridge National Laboratories CONF-700119. National Technical Information Service, Springfield, Virginia, 1970.

ubiquity of chloroplast lamellae[15] in leaf breis adds another particle to this already crowded region.

The following is a description of the techniques of density gradient centrifugation as applied to plant particles.

Choice of Gradient Materials

The vast majority of density gradients are constructed of sucrose, but other substances (Table III) may be advantageous for special purposes. The ideal gradient material would be dense, freely soluble in water, nonviscous, nonosmotic, physiologically inactive, transparent in visible and UV light, and cheap.

Sucrose is a remarkably good candidate, but it is osmotically active, which can be a fatal flaw when separating membrane-bound organelles. It

---

[15] When plant cells are broken, a substantial fraction of chloroplasts lose their outer envelopes and most of their soluble components, but the lamellae hang together in a particle about the same size as intact chloroplasts.[16] Because the lamellae contain most of the pigments of the chloroplast, their appearance in ordinary light microscopy led many workers to mistake these bundles of lamellae for chloroplasts. They have been variously called "class II chloroplasts," "stripped chloroplasts," etc.

[16] A. Kahn and D. von Wettstein, *J. Ultrastruct. Res.* 5, 557 (1961).

TABLE III

PHYSICAL PROPERTIES OF SOME GRADIENT MATERIALS (COMPILED FROM VARIOUS SOURCES)

Substance	Concentration	Stock solution (stable at 4°)			
		Density[a]		Viscosity[b]	Refractive index[c]
CsCl	60% w/w	1.7900	(20°)	—[f]	1.4074
D_2O	100%	1.105	(20°)[e]	—	1.33844
Ficoll	46.5% w/w	1.1629	(4°)	1020	1.3764[g]
Glycerol	100%	1.2609	(20°)	1490 (20°)	1.4729
Silica sols[d]	40.1% w/w	1.295	(25°)	27 (25°)	—
Sorbitol	60% w/w	1.2584	(4°)	102.9 (4°)	1.4402
Sucrose	65% w/w	1.32600	(4°)	56.5 (20°)	1.4532

Various synthetic organics which are highly soluble in water[h]

[a] In grams/cc^{-1} (at temperature in °C).

[b] In centipoise.

[c] At 20°.

[d] Available under the trade name of Ludox (du Pont de Nemours Chemical Co.).

[e] J. B. Ifft, D. H. Voet, and J. Vinograd, *J. Phys. Chem.* **65,** 1138 (1961).

[f] The viscosity of CsCl solutions at 0° as a function of density has been determined by R. Kaempfer and M. Meselson, this series, Vol. 20, p. 521.

[g] A 30% w/v solution.

[h] J. H. Parish, J. R. B. Hastings, and K. S. Kirby, *Biochem. J.* **99:**19P (1966).

also affects many membranes adversely.[17,18] Membranes are less permeable to mannitol, as first noted by the canonized Ursprung and Blum. Sorbitol, which is more soluble than mannitol, is preferred for density gradients.

Since sorbitol is not usually metabolized, it has an advantage over sucrose in the size-fractionation of whole cells in density gradients.[19,20] We found it superior to sucrose for the size-fractionation of yeast mitochondria,[21] and Brown *et al.*[22] preferred it for the continuous-flow harvesting of spinach chloroplasts.

One can construct *isosmotic* gradients of dextrans, which are glucose polymers, or Ficoll, a sucrose polymer. Ficoll was the only gradient material suitable for the separation of *Euglena* chloroplasts among a number

[17] K. J. Tautvydas, *Plant Physiol.* **47,** 499 (1971).

[18] E. D. Day, P. N. McMillan, D. D. Mickey, and S. H. Appel, *Anal. Biochem.* 39, 29 (1970).

[19] H. O. Halvorson, B. L. A. Carter, and P. Tauro, this series, Vol. 21, p. 462.

[20] J. Sebastian, B. L. A. Carter, and H. O. Halvorson, *J. Bacteriol.,* p. 1045 (1971).

[21] C. J. Avers, A. Szabo, and C. A. Price, *J. Bacteriol.* **100,** 1044 (1969).

[22] D. H. Brown, E. E. Barnett, B. W. Harrell, and J. N. Brantley, *in* "Particle Separation from Plant Materials" (C. A. Price, ed.), ORNL-Man Program Symposium, CONF-700119, 1970.

tried.[23] Solutions of these polymers are not very dense, but since membrane organelles do not lose water when exposed to them, equilibrium density occurs in correspondingly less dense solutions. Moreover, they appear to stabilize nuclei and chloroplasts[24-26] and mitochondria.[27] The nature of this stabilization has not been explained.

Silica sols are another source of isosmotic gradients. Mateyko and Kopac[28] claimed they were the solutions of choice for the isopycnic centrifugation of animal cells, and Pertoft[29] has explored them for the separation of lymphocytes and virus particles. They have the advantage of negligible osmotic pressure, high density, and low viscosity, but they are sensitive to salts, especially magnesium, and low pH, in which they form gels. Pertoft recommends that they be fortified with protein or polyvinylpyrrolidone to minimize toxicity. Lyttleton[30] used silica-albumin gradients for the isopycnic separation of chloroplasts, and we have found[31] that marine algae, some of which are very dense, can be banded successfully in silica gradients.[32]

Alkali salts, such as CsCl, are employed when extremely high densities are required, but many particles, e.g., viruses and ribosomes, become unstable at concentrations required for equilibrium density. Solutions of CsCl at 0° are extremely non-Newtonian; i.e., viscosity *decreases* with concentration. Kaempfer and Meselson[33] have exploited this property in the design of *acceleration* gradients (see "Rate Separations" below).

Still another gradient material is Renograffin. It is extremely dense and appears to be physiologically inactive. Matile[4] has used it successfully for the separation of vacuoles from maize roots. However, it has the disadvantage of being opaque in the ultraviolet.

Rate Separations

The separation of particles according to their sedimentation coefficients is primarily a separation according to size. Thus, as we noted above, differ-

[23] A. Vasconcelos, M. Pollack, L. R. Mendiola, H.-P. Hoffmann, D. H. Brown, and C. A. Price, *Plant Physiol.* **47**, 217 (1971).
[24] L. Kuehl, *Z. Naturforsch. B* **19**, 525 (1964).
[25] S. I. Honda, T. Hongladarom, and G. G. Laties, *J. Exp. Bot.* **17**, 460 (1966).
[26] L. R. Mendiola, C. A. Price, and R. R. L. Guillard, *Science* **153**, 1661 (1966).
[27] G. G. Laties and T. Treffry, *Tissue Cell* **1**, 575 (1969).
[28] G. M. Mateyko and M. J. Kopac, *Ann. N.Y. Acad. Sci.* **105**, 219 (1963).
[29] H. Pertoft, *J. Nat. Cancer Inst.* **44**, 1251 (1970).
[30] J. W. Lyttleton, *Anal. Biochem.* **38**, 277 (1970).
[31] C. A. Price, M. Goldstein, L. R. Mendiola-Morgenthaler, and R. R. L. Guillard. In preparation.
[32] We employed "Ludox AM," but found that it must be thoroughly dialyzed before use.
[33] R. Kaempfer and M. Meselson, this series, Vol. 20, p. 521.

ent classes of particles—such as nuclei, chloroplasts, mitochondria, or ribosomes—can be partially or completely resolved by rate separation. In addition, particles within a given class may be fractionated according to size. The resolution in density gradients of 70 S and 80 S ribosomes, of 16 S and 23 S RNA, and of polysomes is an example that is now commonplace. However, there are a number of additional instances in the plant world where particles may be usefully fractionated according to size: Algal, fungal, and bacterial cells of uniform size may be selected out of exponentially growing cultures for initiating synchronous cultures or, since cell size corresponds to developmental state, for studying developmental stages directly.[19,20] The same technique in principle can be done with populations of mitochondria[21] and chloroplasts.[34] Neal et al.[35] showed that the sedimentation rates of yeast mitochondria depended on their state of respiratory repression.

Resolution, defined as zone width divided by the separation between zones, is always a problem in rate separations. Zone width can be minimized by employing a variety of techniques in loading and unloading the gradient[11,36] (cf. this volume [73]). Optimum separation between zones requires control over the *shape* of the gradient. There are some intrinsic advantages in achieving a simple proportion between the sedimentation coefficient of particles and the distance migrated. Martin and Ames[37] described such a gradient for proteins. Noll[38] formalized this relation in *isokinetic gradients,* in which the radial rate of sedimentation dr/dt is made constant for particles of any given density (see [73], this volume).

$$\frac{dr}{dt} = \text{constant} = \frac{r[\rho_p - \rho_m(r)]}{\eta_m(r)} \tag{1}$$

Kaempfer and Meselson[33] showed that even better resolution could be achieved in *acceleration gradients,* in which particle velocities increased with increasing distance from the axis of rotation. Although their recipe was specifically for the separation of ribosomes in gradients of CsCl, the same effect could be achieved by imposing a decreasing gradient of a viscous material, such as polyethylene glycol, on a density gradient of a less viscous material.

Both isokinetic and acceleration gradients are advantageous for gradients in swinging buckets; they are inappropriate in the cylindrical geometry

[34] C. A. Price, J.-J. Morgenthaler, and L. R. Mendiola-Morgenthaler. In preparation.
[35] W. K. Neal, II, H.-P. Hoffmann, and C. A. Price, *Plant Cell Physiol.* **12**, 181 (1971).
[36] H. Noll, *in* "Techniques in Protein Biosynthesis" (P. N. Campbell and J. R. Sargent, eds.), Vol. 2, p. 101. Academic Press, London (1969).
[37] R. G. Martin and B. N. Ames, *J. Biol. Chem.* **236**, 1372 (1961).
[38] H. Noll, *Nature (London)* **215**, 360 (1967).

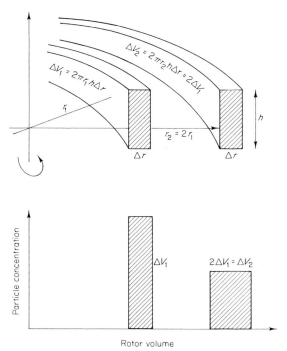

FIG. 3. Sectorial dilution. If a band of particles keeps the same zone width Δr, the zone *volume* ΔV will double as the radius of the particle zone r doubles. From C. A. Price, *in* "Microsymposium on Particle Separation from Plant Materials" (C. A. Price, ed.), p. 1. Oak Ridge National Laboratories CONF-700119. National Technical Information Service, Springfield, Virginia, 1970.

of zonal rotors. The reason is that *sectorial dilution* causes the volume of an annular zone of constant width to increase in proportion to the radius (Fig. 3). Thus while the volumetric distance between two particle zones might increase, the widths of the zones would also increase. Schumaker[39] suggested some time ago that one should be able to *decrease* zone widths if the gradient were sufficiently steep that the leading edge of a particle zone were decelerated with respect to the trailing edge (Fig. 4). Using this principle, Spragg *et al.*[40] constructed a computer program for generating *isometric* gradients in which radial dilution is exactly balanced by gradient-induced zone narrowing. Experiments with serum globulins produced surprising results: resolution was improved, but particle zones did not maintain constant width. *Anomalous zone broadening* was observed whether sedimentation occurred or not.

[39] V. N. Schumaker, *Separ. Sci.* 1, 409 (1966).
[40] S. P. Spragg, R. S. Morrod, and C. T. Rankin, Jr. *Separ. Sci.* 4, 467 (1969).

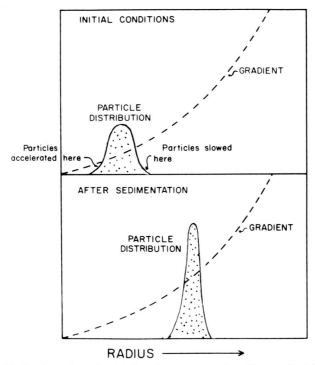

Fig. 4. Mechanism of gradient-induced zone narrowing. If a gradient is sufficiently steep, the particles in the loading edge of a zone will be decelerated with respect to the particles in the trailing edge; as the zone moves into the gradient, its radial width will decrease. The principle of gradient-induced zone narrowing is employed in the construction of *isometric* and *equivolumetric gradients*. From C. A. Price "Centrifugation in Density Gradients." Academic Press, New York, 1973.

An extreme application of gradient-induced narrowing of zones is the step gradient. In this case, a zone may be concentrated to almost infinite thinness as it moves against a steep step in the gradient.[41] But the physics of this process are complex; there is, for example, a real possibility that overloading of the gradient will result. A possible artifact is that zones may be found resting on each of several steps, but the same particles may be spread through several zones.

In another class of gradients, we can ask that the particles move through equal increments of volume at a constant rate. The shape of these *equivolumetric* gradients for a given particle density in zones of negligible width is described by Eq. (2).[42,43]

[41] G. B. Cline and R. B. Ryel, this series Vol. 22, p. 168.
[42] M. S. Pollack and C. A. Price, *Anal. Biochem.* **42**, 38 (1971).
[43] C. A. Price and T.-S. Hsu, *Viert. Naturforsch. Ges. Zürich* **116**, 367 (1971).

$$\frac{r^2}{m(r)}\,[\rho_p - \rho_m(r)] = \text{constant} \qquad (2)$$

We find that such gradients produce zones of constant width with reduced anomalous zone broadening and hence improved resolution. As expected from the model, the volumetric distance migrated by homologous particles is proportional to their sedimentation coefficients. An example of a separation of *Euglena* polysomes in an equivolumetric gradient is shown in Fig. 5. Berns *et al.*[44] have found this gradient shape to be advantageous for the resolution of messenger RNA.

Gradient shape may also be used to improve capacity. Following Britten and Roberts,[45] *wedge-shaped* or *inverse sample gradients* (Fig. 6) are widely employed to minimize *droplet sedimentation* that can quickly spoil a gradient when loaded with a dense suspension of particles. For reasons that are qualitatively but not quantitatively clear,[46] the danger of droplet sedimentation is greater the larger the particle. Whatever the reason, one can easily demonstrate that the resolution of chloroplasts and organelles

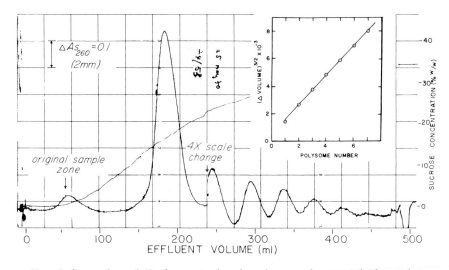

FIG. 5. Separation of *Euglena* cytoplasmic polysomes in an equivolumetric gradient. A crude suspension of *Euglena* ribosomes was layered over an equivolumetric gradient in the B-XXXA rotor (IEC) and centrifuged for 1.0×10^{11} radians² per second. From M. S. Pollack and C. A. Price, *Anal. Biochem.* **42**, 38 (1971).

[44] A. J. M. Berns, R. A. de Abren, M. van Kraaikamp, E. L. Benedetti, and H. Bloemendal, *FEBS Letts.* **18**, 159 (1971).

[45] R. J. Britten and R. B. Roberts, *Science* **131**, 32 (1960).

[46] H. B. Halsall and V. N. Schumaker, *Biochem. Biophys. Res. Commun.* **43**, 601 (1971).

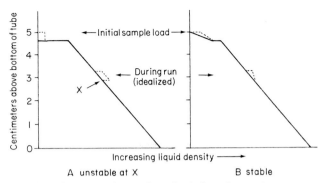

FIG. 6. Stability of rectangular and wedged-shaped sample zones. R. J. Britten and R. B. Roberts [*Science* **131**, 32 (1960)] represented the relative instability of rectangular particle zones (dotted lines) as due to density inversions that would occur when the zone enters the gradient. In fact, instability can occur without any sedimentation, owing to the diffusion of solute across the steep gradient between the sample zone and the underlying gradient.

of similar size is substantially improved by loading the sample as an inverse gradient.

The shape of the underlying gradient may also be manipulated to improve capacity. Berman[47] pointed out that a particle zone that is initially stable may *become* unstable through gradient-induced zone narrowing (Fig. 7). He specifically proposed a class of *hyperbolic* gradients

$$\rho_m(r) = \rho_p - \frac{k}{r^n} \qquad (3)$$

where k and n are arbitrary constants. He showed that if particles are initially loaded to capacity in a wedge-shaped zone, stability during migration should depend on the value of n, which controls the steepness of the gradient. For $0 < n \leq 1$ the gradient will always be stable; for $1 < n \leq 2$ the gradient *may* become unstable; and for $2 < n$ the gradient will always become unstable. Spragg and Rankin[48] found capacities consistent with Berman's predictions when small amounts of viruses were loaded onto very shallow gradients. We then found that Berman gradients will support up to 2 g of ribosome subunits,[49] which is about 60% of theoretical. We have also used Berman gradients for the large-scale isolation of cytoplasmic ribosomes from *Euglena*.[50]

[47] A. S. Berman, *Nat. Cancer Inst. Monogr.* **21**, 41 (1966).
[48] S. P. Spragg and C. T. Rankin, Jr., *Biochim. Biophys. Acta* **141**, 164 (1967).
[49] E. F. Eikenberry, T. A. Bickle, R. R. Traut, and C. A. Price, *Eur. J. Biochem.* **12**, 113 (1970).
[50] M. S. Pollack, E. F. Eikenberry, and C. A. Price, unpublished.

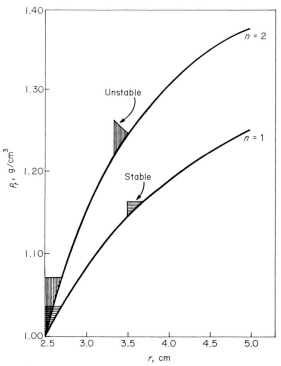

FIG. 7. Density inversion during sedimentation. When $n = 1$, a sample zone that is initially stable will be stable throughout its sedimentation. But in a steeper gradient ($n \geq 2$), the initially stable gradient will become unstable through zone narrowing. From A. S. Berman, *Nat. Cancer Inst. Monogr.* **21**, 41 (1966).

Isopycnic Separations

Separations based on equilibrium density are much easier to perform than rate-zonal separations and often yield exquisite resolution. A good example is the separation of mitochondria and peroxisomes from castor bean endosperm.[51] Moreover, the capacities of gradients in this mode are enormous, in principle limited only by the physical displacement of gradient by particles. Small wonder that isopycnic sedimentation was for many years synonymous with density gradient centrifugation and is still the most common mode.

However, there are limitations. The densities of some subcellular particles are too great for compatible gradient materials. For example, ribosomes will band in gradients of CsCl ($\rho_{eq} \simeq 1.65$), but the high concen-

[51] R. W. Breidenbach and H. Beevers, *Biochem. Biophys. Res. Commun.* **27**, 462 (1967).

trations of the salt extract about half of the ribosomal proteins. The particles that band are no longer ribosomes but so-called *core particles*. Similarly, mitochondria and chloroplasts may be greatly altered by exposure to 30, 40, or 50% w/w sucrose.

A second kind of hazard in isopycnic sedimentation arises from the high centrifugal fields that are often required.[52] We know that chloroplasts may be stripped by the hydrostatic pressures generated by high rotor speeds, and this is true for other large membrane-bound organelles.[53] Another artifact of high speed sedimentation, pressure-induced dissociation of ribosomes, has been described by Infante and Baierlein.[54]

We remarked above that resolution in isopycnic separations could be very good, which is to say that zone widths may be very small.

At first blush one might imagine that at perfect equilibrium a population of homogeneous particles might be distributed in an infinitely thin zone at $\rho_m = \rho_{eq}$, but even relatively large particles have finite diffusion coefficients. An infinitely thin particle zone would imply an infinite rate of diffusion away from that zone of equilibrium. What actually obtains therefore is a zone of finite width in which equilibrium exists between the centrifugal field, which tends to concentrate the particles toward the center, and outward diffusion.

The zone tends to be a Gaussian distribution, and the standard deviation with respect to r has been calculated[55] (Eq. 4).

$$\sigma_r = [RT\rho_0/(M^*(d\rho/dr)_0\omega^2 r_0)]^{1/2} \tag{4}$$

where $R \equiv$ universal gas constant, $T \equiv$ absolute temperature, $M^* \equiv$ particle weight (daltons), $d\rho/dr \equiv$ gradient slope, and subscript $0 \equiv$ denotes the center of the particle zone.

This relation has been used to estimate the heterogeneity of DNA in CsCl gradients. In principle, it can be used to estimate the heterogeneity of any particle population.

Heterogeneity may be more obvious: organelles sometimes resolve into distinct zones of different densities. Sarkissian and McDaniel,[56] for example, reported bands of mitochondria in isopycnic sedimentation of hybrid maize mitochondria. The ρ_{eq} of the bands corresponded approximately to those of the parental types. We have shown that the ρ_{eq} values of mito-

[52] Values of (speed)$^2 \cdot$time need to be about 100 times greater for isopycnic compared to rate-zonal separations.

[53] R. Wattiaux and S. Wattiaux-De Coninck, *Biochem. Biophys. Res. Commun.* **40**, 1185 (1970).

[54] A. A. Infante and R. Baierlein, *Proc. Nat. Acad. Sci. U.S.* **68**, 1780 (1971).

[55] M. S. Meselson, F. W. Stahl, and J. Vinograd, *Proc. Nat. Acad. Sci. U.S.* **43**, 581 (1957).

[56] I. V. Sarkissian and R. G. McDaniel, *Proc. Nat. Acad. Sci. U.S.* **57**, 1262 (1967).

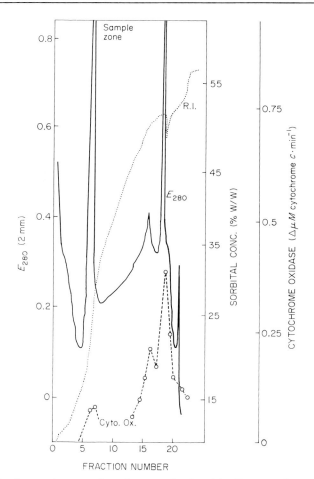

Fig. 8. An isopycnic separation of yeast mitochondria. Derepressing yeast sphero-plasts were disrupted, the clarified brei was layered over a gradient of 40 to 60% w/w sorbitol in 0.1 mM EDTA, and centrifuged for 30 minutes at 30,000 rpm in the B-XIV rotor. The principal mitochondrial band corresponding to repressed mito-chondria appears in a narrow zone $\rho = 1.21$; a smaller band corresponding to de-repressed mitochondria occurs at $\rho = 1.19$. From W. K. Neal, II, H.-P. Hoffmann, C. J. Avers, and C. A. Price, *Biochem. Biophys. Res. Commun.* **38**, 414 (1970).

chondria from repressed yeast differ from those of derepressed cells[35,57] (Fig. 8). Luck[58] employed induced difference in ρ_{eq} values of *Neurospora* mitochondria to test alternate hypotheses of mitochondrial replication. These studies should encourage us to look to both S and ρ coordinates of

[57] W. K. Neal, II, H.-P. Hoffmann, C. J. Avers, and C. A. Price, *Biochem. Biophys. Res. Commun.* **38**, 414 (1970).
[58] D. J. L. Luck, *J. Cell Biol.* **24**, 461 (1965).

mitochondria and other organelles as indices of their compositional and physiological states. Some other examples of isopycnic separation of plant particles follow:

Golgi Apparatus. Morré *et al.*[59] were able to separate the Golgi apparatus (dictyosomes) of plant tissues by isopycnic centrifugation on step gradients of sucrose, but only after the membranes were fixed with glutaraldehyde. In the absence of glutaraldehyde, only single vesicles were recovered. The unstacking of the dictyosomes was partially prevented by the addition of coconut milk, but the generally poor state of the unfixed organelles recovered from gradients means that the proper elixir has not yet been discovered.[60,61,61a]

Bacteroids. A simple isopycnic sedimentation suffices to separate bacteroids from most other cell constituents of soybean root nodules.[62]

Protein Bodies. Mitsuda *et al.*[63] reported the separation "in good yield" of protein bodies from enzymatic digests of rice polish followed by centrifugation in step gradients of sucrose.

Continuous-Flow Harvesting

Biochemists frequently need to harvest particles from relatively large volumes of dilute suspensions. Sharples-type centrifuges are often suitable, but many particles (e.g., viruses) are irreversibly denatured after being pelleted against rotor walls. An alternative is continuous-flow harvesting with isopycnic banding,[41] in which the rotor volume consists of a narrow annulus between the rotor core and the wall. A density gradient is generated by diffusion between a dense solution near the wall and a lighter solution a few millimeters inboard. As a particle suspension entering one channel of the rotor is led along the surface of the density gradient, the particles migrate into the gradient and seek their equilibrium density. The spent fluid passes out through another channel of the rotor. At the end of a run involving 100 liters or more of suspension, the gradient is unloaded and the particles recovered in fractions according to their ρ_{eq}.

Lammers[64] has described the collection of various kinds of plankton

[59] D. J. Morré, H. H. Mollenhauer, and J. E. Chambers, *Exp. Cell Res.* **38**, 672 (1965).

[60] D. J. Morré, this series, Vol. 22, p. 130.

[61] W. J. VanDerWoude, C. A. Lembi, and D. J. Morré, *Biochem. Biophys. Res. Commun.* **46**, 245 (1972).

[61a] P. M. Ray, T. L. Shininger, and M. M. Ray, *Proc. Nat. Acad. Sci. U.S.* **64**, 605 (1969).

[62] J. A. Cutting and H. M. Schulman, *Biochim. Biophys. Acta* **192**, 486 (1969).

[63] H. Mitsuda, K. Murakawi, T. Kusano, and K. Yasumoto, *Arch. Biochem. Biophys.* **130**, 678 (1969).

[64] W. T. Lammers, *in* "Water and Water Pollution Handbook" (L. L. Ciaccio, ed.), Vol. 2, p. 593. Dekker, New York, 1971.

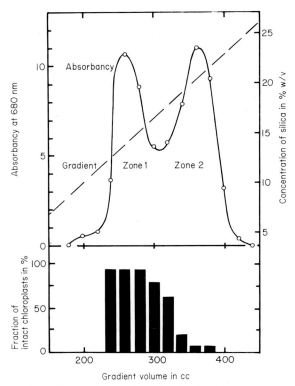

FIG. 9. Isopycnic sedimentation profile of spinach chloroplasts recovered from a silica gradient after continuous-flow harvesting. The upper solid curve represents the approximate distribution of chlorophyll in the gradient; the lower curve figure shows the proportion of intact chloroplasts along the gradient, as estimated by phase-contrast microscopy. The harvest was carried out in a CF-6 rotor containing a gradient of Ludox [C. A. Price, E. N. Breden, and A. C. Vasconcelos, *Anal. Biochem.* **54**, 239 (1973)].

from natural waters with high speed, continuous-flow zonal rotors. Brown *et al.*[22] employed the K-X rotor for the continuous-flow harvesting of spinach chloroplasts. We have used the CF-6, a relatively simple, low-speed rotor for the same purpose (Fig. 9).

A caveat for the separation of membrane-bound organelles or other fragile particles in continuous-flow zonal rotors is that excessive shear may easily develop as particle suspensions are pumped across spinning seals through narrow channels.[65] The channel diameters must be at least 2.5 mm in diameter.

---

[65] This is a problem in batch as well as continuous-flow zonal rotors. DNA requires enlarged channels [H. B. Halsall and V. N. Schumaker, *Nature (London)* **22**, 772 (1969)] and synaptosomes are severely battered in ordinary zonal seals (V. N. Whittaker, personal communication).

[53] Techniques for Overcoming Problems of Lipolytic Enzymes and Lipoxygenases in the Preparation of Plant Organelles

By T. GALLIARD

The study of subcellular organelles ideally requires that the structure, composition, and functional activities of isolated organelles should remain unchanged during isolation procedures. Potentially, it should be possible to isolate from tissues all organelles that exist within the cell as discrete entities, physically separated from other particles and from the cytosol by limiting membranes. In fact many such constituent bodies of plant cells have not been isolated and, of organelles that have been prepared, the structures and properties frequently indicate that marked changes occur during isolation, e.g., mitochondria isolated from plants generally compare poorly with preparations of animal mitochondria.

Most of the problems involved in the isolation of organelles are due to the structure of their limiting membranes. In some cases, the problems are physical; the relatively drastic procedures required to break plant cell walls also rupture the delicate membranes of, for example, the tonoplast surrounding the vacuole. Another major problem lies in the lipoprotein nature of the internal structures and limiting membranes of organelles. Many plant tissues contain active enzymes that attack the lipids of these membranes, and it is these enzymes which are the subject of this article. It should be emphasized that the following discussion will not describe ready-made techniques for avoiding all the problems of enzymatic breakdown of membrane lipids. Unfortunately, the state of the art is not yet sufficiently well developed. At best, the problems can be identified and some remedial procedures proposed.

Problems Involving Lipid-Degrading Enzymes

1. Hydrolytic enzymes can directly attack the lipid components of membrane structures leading to structural breakdown and functional impairment.

2. Free fatty acids produced by the action of lipolytic acyl hydrolases can inhibit the activities of organelles, e.g., mitochondria[1] and chloroplasts.[2]

3. Free fatty acids may also stimulate the enzymatic breakdown of membrane lipids. For example, linoleic acid causes a dramatic increase in the hydrolysis of mitochondrial lipids catalyzed by a lipolytic acyl

---

[1] L. Dalgarno and L. M. Birt, *Biochem. J.* **87**, 586 (1963).
[2] R. E. McCarty and A. T. Jagendorf, *Plant Physiol.* **40**, 725 (1965).

hydrolase found in plants.[3] This effect appears to be due to structural changes in the membrane induced by fatty acids, resulting in exposure to enzymatic attack of the susceptible lipid acyl ester bonds.

4. The predominant fatty acids in membrane lipids of most plant tissues are linoleic and linolenic acids. These two fatty acids are the best natural substrates for lipoxygenase enzymes that catalyze the production of fatty acid hydroperoxides. Although little is known of the action of lipid hydroperoxides in plant tissues, their effects on membrane structures of mitochondria,[4] microsomes,[5] and lysosomes[6] of mammalian tissues are well known.

Endogenous free fatty acids would also give rise to the effects listed above. However, the concentration of free fatty acids in most plant tissues (with the possible exception of senescent tissue[7]) is low in comparison with the amounts that may be liberated by enzymatic action during extraction.

Enzymes Causing Lipid Breakdown in Plant Tissues

Phospholipase D (phosphatidylcholine:phosphatidohydrolase, EC 3.1.4.4). This enzyme catalyzes the hydrolysis of phospholipids to phosphatidic acid. The enzyme occurs widely in the plant kingdom[8] and is particularly active in some storage tissues, e.g., carrot root, pea and marrow seeds, artichoke tubers, and storage organs of cabbage and celery. Both soluble and particulate forms of the enzyme have been described with pH optima in the region pH 5–6.[9] Calcium ions are essential for optimal activity.

Lipolytic Acyl Hydrolases. Enzymes that liberate free fatty acids are potentially the most troublesome in subcellular preparations from plants. The nomenclature of these enzymes is poorly defined, and such enzymes have been described under a variety of names in a wide range of plants. Whether these enzymes are as specific for substrates as their names imply is subject to some doubt.[10] The following categories are therefore somewhat arbitrary.

a. *Lipases* are normally defined as enzymes attacking water-insoluble

[3] T. Galliard, *Eur. J. Biochem.* **21**, 90 (1971).

[4] E. E. Hunter, A. Scott, P. E. Hoffster, J. M. Gebicki, J. Winstein, and A. Schneider, *J. Biol. Chem.* **239**, 614 (1964).

[5] E. D. Wills, *Biochem. J.* **113**, 315 (1969).

[6] C. C. Tsen and H. B. Collier, *Can. J. Biochem. Physiol.* **38**, 957 (1960).

[7] M. S. Baddeley and E. W. Simon, *J. Exp. Bot.* **20**, 94 (1969).

[8] R. H. Quarles and R. M. C. Dawson, *Biochem. J.* **112**, 787 (1969).

[9] T. Galliard, *in* "Form and Function of Phospholipids" (G. B. Ansell, R. M. C. Dawson, and J. N. Hawthorne, eds.). Elsevier, Amsterdam, 1973.

[10] T. Galliard, *Biochem. J.* **121**, 379 (1971).

neutral lipids, e.g., triglycerides. Enzymes of this type are well known in cereal and oil seeds,[11] although frequently assay methods have not permitted distinction from esterases active on water-soluble esters.[12] Although lipases may not directly attack membrane-bound polar lipids, fatty acids released from triglycerides could affect membrane structures as described above.

b. *Phospholipase B* activity that causes liberation of both fatty acids from diacyl phospholipids has been described in some seeds.[9]

c. *Galactolipase,* an enzyme that removes both fatty acids from mono- and digalactodiglycerides (the major lipids of chloroplast membranes) has been demonstrated in leaves[13] and purified from *Phaseolus vulgaris.*[14] The inactivity of chloroplast preparations from certain leaves has been ascribed to the action of galactolipase during isolation procedures.[2]

d. *Sulpholipase,* activated by disruption of leaf cells, hydrolyzes the plant sulfolipid (sulfoquinovosyl diglyceride) present in chloroplasts, liberating free fatty acids.[15]

e. *Lipolytic acyl hydrolase,* the name given to an enzyme that deacylates a wide range of lipids including phospholipids and galactolipids (but not triglycerides). It is present to some extent in many plants, but in those studied is most active in potato tubers.[10] It is possible that an enzyme of this nature is responsible for the phospholipase B and galactolipase activities described above.

With the exception of true lipases (glycerol ester hydrolase; EC 3.1.1.3), for which pH optima from 4 to 9 have been reported in various tissues,[16] most of the lipolytic acyl hydrolases do not have metal ion requirements and do not show sensitivities to inhibitors suitable for use in preparation of functional organelles.

Lipoxygenase (EC 1.99.2.1). The general name refers to a group of enzymes that catalyze the formation of hydroperoxides from unsaturated fatty acids of the type representing the major fatty acids of plant tissues, i.e., linoleic and linolenic acids. Lipoxygenases occur in many plants and the various types of the enzyme differ in several properties including pH optima and the nature of the products formed. The classical lipoxygenase from soybean is active at pH 9.0[17] whereas other lipoxygenases have

[11] E. J. Barron, *in* "Modern Methods of Plant Analysis" (J. F. Linskens, B. D. Sanwal, and M. V. Tracey, eds.), Vol. 7, p. 455. Springer-Verlag, Berlin and New York, 1964.

[12] P. Desnouelle and P. Savary, *J. Lipid Res.* 4, 369 (1963).

[13] M. Kates, *Advan. Lipid Res.* 8, 225 (1970).

[14] P. J. Helmsing, *Biochim. Biophys. Acta* 178, 519 (1969).

[15] A. A. Benson, *Annu. Rev. Plant Physiol.* 15, 1 (1964).

[16] E. D. Wills, *Advan. Lipid Res.* 3, 197 (1963).

[17] A. L. Tappel, this series, Vol. 5, p. 539.

optimal activities in the range pH 5.5–6 and are inactive at high pH.[18] In general, lipoxygenases have no prosthetic groups, no metal requirements, and no active thiol groups,[17] and so inhibition of these enzymes in systems suitable for organelle isolation is difficult. The most satisfactory means of reducing lipoxygenase activity in these systems appears to be the limitation of substrate availability by complexing the free fatty acids with bovine serum albumin.

Conditions to Minimize Lipid Degradation during
    Isolation of Organelles

Isolation Techniques. Most lipolytic enzymes and lipoxygenases are found in the supernatant fraction of cell extracts (possibly owing to liberation from fragile organelles during tissue disruption). Thus a crude homogenate will contain organelles in a medium that also contains the destructive enzymes. Two conditions, which are difficult to reconcile, are important here. Gentle disruption methods will cause less physical breakdown of organelle structure and, possibly, will reduce the liberation of degradative enzymes. On the other hand, a rapid isolation method that removes required organelles from a crude homogenate containing these enzymes would be advantageous. Because of the nature of plant cell walls, methods that physically break these also rupture vacuoles and other organelles. Ideally, the starting material for organelle preparation would be isolated protoplasts obtained by enzymatic digestion of cell walls from plant cells. However, this is practicable for only a few plant tissues at present[19] and cannot yet be considered as a general method. The best solution to these problems to date represents a compromise between disruption methods that give a satisfactory yield of organelles and rapid isolation methods. Such rapid techniques have been shown to have beneficial effects in the preparation from plant tissues of mitochondria[20] and chloroplasts.[21]

pH. Most lipolytic enzymes and lipoxygenases in plant tissues have pH optima in the pH range of homogenates of many plant tissues, commonly pH 5–6, and have little activity above pH 7.5–8. The literature on preparation of functional organelles indicates the almost universal requirement for the extraction medium to be above pH 7. For instance, *Phaseolus* chloroplasts prepared at pH 6 had no Hill activity, whereas chloroplasts prepared at pH 8.5 were active.[2] Thus a high pH reduces the

[18] T. Galliard and D. R. Phillips, *Biochem. J.* **124**, 431 (1971).
[19] A. W. Ruesink, this series, Vol. 23, p. 197.
[20] J. M. Palmer, *Nature (London)* **216**, 1208 (1967).
[21] D. A. Walker, this series, Vol. 23, p. 211.

activity of most lipid-degrading enzymes although excessively high pH values (above pH 8.5) may be undesirable for other reasons.[22]

Chelating Agents. Addition of EDTA, etc., to extraction media inhibits the activity of Ca^{2+}-dependent phospholipase D and may reduce the activity of some lipases. However, other lipolytic enzymes and lipoxygenases are not affected by chelating agents, which, therefore, must be considered as of only limited value in reducing the enzymatic degradation of lipids during organelle preparation.

Antioxidants. Phenolic oxidation reactions in plant extracts present serious problems in the isolation of subcellular particles. The choice of inhibitors to prevent such processes depends on the tissue being used.[23] Oxidative breakdown of fatty acid hydroperoxides is inhibited by antioxidants, but lipoxygenase activity itself is not prevented by antioxidants at concentrations suitable for use in organelle preparations.

Sulfhydryl Compounds. These are commonly used as reducing agents in the preparation of plant extracts. In general they have little effect on lipolytic enzymes or lipoxygenases, although a partial inhibition of galactolipid hydrolysis by cysteine has been reported.[14]

Specific Enzyme Inhibitors. It is not possible to recommend inhibitors of lipid degradation that would be suitable for use in the preparation of functional organelles. However, some crude homogenates from plant tissues appear to contain endogenous nondialyzable inhibitors of phospholipase D activity.[24,25]

Bovine Serum Albumin. BSA is an essential ingredient of media used in the isolation of good chloroplasts[26] and mitochondria.[27,28] The beneficial effect of BSA is probably due to its ability to bind free fatty acids of endogenous origin or fatty acids formed by the action of acyl hydrolase enzymes during extractions. The deleterious effects of free fatty acids were mentioned earlier. Relatively high concentrations of BSA (0.1–1%) are required for optimal activities of chloroplasts[26] and mitochondria.[28] Although it has been assumed hitherto that BSA does not directly inhibit the production of free fatty acids by lipolytic enzymes but serves to prevent further damage to membrane structures by the products of lipid deacylation, recent (unpublished) work from this laboratory has indicated that BSA does inhibit the lipolytic acyl hydrolase from potato tubers.

[22] T. Punnet, *Plant Physiol.* 34, 283 (1959).
[23] W. D. Loomis and J. Battaile, *Phytochemistry* 5, 423 (1966).
[24] H. L. Tookey and A. K. Balls, *J. Biol. Chem.* 218, 213 (1956).
[25] M. Kates, *Can. J. Biochem. Physiol.* 32, 571 (1954).
[26] M. Friedlander and J. Neumann, *Plant Physiol.* 43, 1249 (1968).
[27] W. D. Bonner, this series, Vol. 10, p. 126.
[28] H. Ikuma, *Plant Physiol.* 45, 773 (1970).

Consideration of the above factors indicates that a reduction of the effects of lipid-degrading enzymes on cell organelles during their isolation can be expected with most plant tissues when the following conditions are used: (a) generally accepted criteria for the isolation of good subcellular preparations must be satisfied, i.e., suitable disruption procedures, low temperatures, acceptable osmotica and buffers; (b) fractionation and isolation procedures must be rapid; (c) extraction media should normally be buffered at the highest pH that does not cause other deleterious effects (pH 8.0–8.5 is satisfactory in many cases); (d) fatty acid-free bovine serum albumin (0.1–1%) must be present in the extraction medium; (e) chelating agents may be beneficial in some cases.

Choice of Plant Material

It should be emphasized that, in the author's experience, even when the above conditions are fulfilled, it is not possible to eliminate all lipolytic activity in extracts from tissues that contain very active enzymes, e.g., cabbage leaves and carrot root (high in phospholipase D activity) and potato tuber (high in lipolytic acyl hydrolase). It appears that some breakdown occurs instantaneously upon cell rupture. On the other hand, extracts of spinach leaf tissue, which is low in lipolytic enzyme activity and is a tissue of choice for the preparation of chloroplasts, show little evidence of lipid degradation over short periods.

Although the conditions outlined above will serve in most cases to minimize the effects of lipolytic enzymes and lipoxygenases, the possible contribution of these enzymes to the problems of organelle preparation for a given plant material can be determined only by direct experiment with that tissue. The method described below will detect the products of lipid breakdown under the conditions chosen for the isolation of organelles from a given plant material.

Reference to the literature for data on the activities of these enzymes for a particular plant tissue is not reliable; these enzyme activities are known to vary with the stage of development of the tissue. Moreover, very large differences have been observed between the activities of a given lipolytic enzyme in different varieties of the same species of plant material.[29]

A Method for the Detection of Lipid Breakdown in Plant Extracts

To determine the extent of lipid breakdown that takes place during isolation procedures, the most direct method involves a comparison of

---

[29] T. Galliard and J. A. Matthew, *J. Sci. Fd. Agric.* **24**, 623 (1973).

the lipid composition of the intact tissue with that of a total homogenate prepared as for the isolation of subcellular fractions and treated under the same conditions of temperature, time, etc., as would be used in an extraction. Similar comparisons can be made between the lipid composition of an organelle preparation when first isolated and after storage for a required period.

Lipid Extractions[30]

Homogenate. An aliquot (1 volume) of a crude (unfiltered) homogenate containing the equivalent of 1 g fresh weight of tissue is added to 5 volumes of refluxing methanol–water (4:1 v/v). After further refluxing for 5 minutes, the mixture is cooled, then methanol (1 volume) and chloroform (2.5 volumes) are added. The mixture is allowed to stand 30 minutes, with occasional shaking, at room temperature. Chloroform (2.5 volumes) and 0.2 M sodium acetate buffer, pH 4.0 (2.5 volumes) are added. The mixture is shaken well, then the two-phase system is allowed to separate. The lower, chloroform, phase is removed and concentrated under a stream of N_2.

Intact Tissue. The plant material is chopped or diced, and a representative sample is added to refluxing aqueous methanol as above (5 ml/g fresh weight). After the addition of methanol and chloroform as above, the mixture is homogenized with a small disintegrator, allowed to stand 30 minutes, and then extracted as above.

Thin-Layer Chromatography (TLC) Analysis of Chloroform-Soluble Fraction[30]

Suitable aliquots (equivalent to 10–100 mg fresh weight of tissue) of the concentrated chloroform solutions are applied to thin layers of activated silica gel G together with reference lipids. Polar lipids are separated by development in chloroform–methanol–acetic acid–water (170:30:20:6, v/v), nonpolar lipids are separated on a second thin-layer plate developed in petroleum ether (b.p. 60–80°)–diethyl ether–acetic acid (70:30:1). Lipids are detected with I_2 vapor and partial charring with 50% aqueous H_2SO_4.

Interpretation of Lipid Analyses

Lower amounts of acyl lipids in homogenates relative to intact tissue indicate lipolytic activity. Appearance of phosphatidic acid, free fatty acids, and fatty acid peroxidation products (all of which do not occur in more

[30] T. Galliard, *Phytochemistry* 9, 1725 (1970).

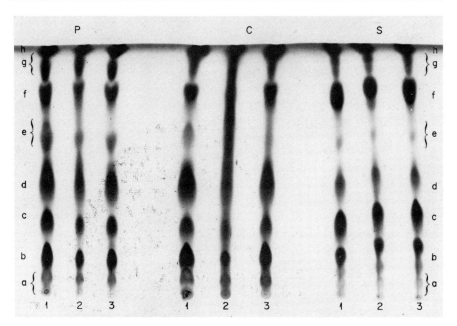

Fig. 1. TLC separations of polar lipids in extracts of plant tissues. Tissues used were: potato tuber (P), cabbage leaves (C) and spinach leaves (S). In each case, illustrations are given of the polar lipid composition of intact tissue (1), of a water homogenate (2) and of a homogenate in a medium containing 0.1 M Tris HCl (pH 8.5), 0.3 M sucrose, 0.2% BSA and 10 mM EDTA (3). Details are given in the text; homogenates were prepared at 0° and held at 0° for 15 minutes before extraction. Silica gel G (Merck) layers (0.25 mm) on glass plates were used. The developing solvent was chloroform–methanol–acetic acid–water (170:30:20:6, v/v). Lipids were detected with I_2 vapor. The identities of the spots lettered in Fig. 1 are as follows: (a) phosphatidylinositol + trigalactosyl diglyceride; (b) phosphatidylcholine + sulfolipid; (c) digalactosyl diglyceride + phosphatidylglycerol; (d) phosphatidylethanolamine; (e) cerebrosides + sterol glycosides; (f) monogalactosyl diglyceride; (g) esterified sterol glycosides, free fatty acids, and fatty acid oxidation products; (h) neutral lipids (triglycerides, pigments, hydrocarbons, waxes, etc.).

than trace amounts in most plant tissues) are evidence of phospholipase D, lipolytic acyl hydrolase, and lipoxygenase activities, respectively.

Examples are given in Fig. 1 of the effects of homogenization on the lipid composition of tissues high in lipolytic acyl hydrolase activity (potato tuber) and phospholipase D activity (cabbage leaves) and of tissue (spinach leaves) that has relatively little hydrolytic activity toward lipids. The marked losses of phospholipids and galactolipids in water homogenates of potato tuber and of phospholipids in cabbage leaves are clearly illustrated; phosphatidic acid, the product of phospholipase D activity, causes the pronounced streaking of the TLC separation in C2 (Fig. 1).

It is also clear that homogenization in the selected medium significantly inhibits the effects of the lipolytic enzymes. By comparison, very similar lipid patterns were observed in intact tissue and in homogenates of spinach leaves.

[54] Overcoming Problems of Phenolics and Quinones in the Isolation of Plant Enzymes and Organelles

By W. D. LOOMIS

Plant tissues present special problems in the isolation of enzymes and organelles, owing primarily to the presence of a large variety, and frequently large quantities, of "secondary products"[1-5] and of phenol oxidases.[5] Considerable progress has been made in solving these problems, but there are still no generally applicable standard methods for isolating high-quality enzymes and organelles from plant tissues. Indeed, because of the great chemical diversity of plant secondary products, it seems unlikely that there ever will be such standard methods for plant tissues. Therefore, it is essential for the plant biochemist to understand the kinds of problems that are involved and the kinds of techniques that have been used, or that are available, to solve them, and to use this knowledge to devise specific solutions to his own specific problem. We have previously reviewed this area,[5,6] as has Anderson,[7] but new knowledge and new insights have been acquired since then.

Our purpose here will be to describe, in as general a way as possible, the special problems of plant enzymology, and the "bag of tricks" that have been developed to deal with these problems. We will be concerned primarily, but not entirely, with phenols and quinones. We will also try to suggest new techniques that might be of value and that should be tested, and to remind readers of old techniques that may not have been exploited to their full capacity.

In a typical mature plant cell the active protoplasm occupies a

[1] W. Karrer, "Konstitution und Vorkommen der organischen Pflanzenstoffe (exclusive Alkaloide)." Birkhaeuser, Basel, 1958.
[2] T. Robinson, "The Organic Constituents of Higher Plants," 2nd ed. Burgess, Minneapolis, Minnesota, 1967.
[3] V. L. Singleton and F. H. Kratzer, *J. Agr. Food Chem.* **17**, 497 (1969).
[4] R. H. Thomson, "Naturally Occurring Quinones," 2nd ed. Academic Press, New York, 1971.
[5] W. D. Loomis and J. Battaile, *Phytochemistry* **5**, 423 (1966).
[6] W. D. Loomis, this series, Vol. 13, p. 555.
[7] J. W. Anderson, *Phytochemistry* **7**, 1973 (1968).

peripheral shell between a large central vacuole and a rigid cell wall. A typical plant tissue thus consists largely of nonprotoplasmic materials, notably vacuole contents, cell wall constituents and starch. Protoplasm is quantitatively a minor component. For example, spinach leaves contain only 2.2% crude protein, and apple fruit only 0.3% on a fresh weight basis, compared to 19.8% protein in beef liver.[8]

In the living plant cell, the small amount of protoplasm is protected from large amounts of potentially damaging materials by compartmentalization, and by the fact that many of these materials are present in "detoxified" forms, such as glycosides. When a plant tissue is homogenized in order to isolate enzymes or organelles, the compartmentalization is destroyed and everything is mixed, with effects ranging from undesirable to devastating, depending on the tissue, and on the isolation techniques. Isolation techniques may vary depending on the species of plant, its physiological condition, the lability of the system being isolated, and whether one demands a completely "native" system or is willing to accept minor modifications. Thus, the procedures required may range from simple to very involved.

Problems Due to Plant Secondary Products

Reactions of Phenols and Quinones with Proteins

Plant phenolic compounds are very diverse, but there are two main biogenetic groups: the phenylpropanoid compounds (including hydrolyzable tannins), and the flavonoids (including condensed tannins). Their reactions with proteins may be divided into four principal classes:

1. Hydrogen bonding. Isolated phenolic hydroxyl groups form very strong hydrogen bonds with the oxygen atoms of peptide bonds. This is one of the strongest types of hydrogen bonds known and cannot be dissociated by conventional techniques, such as dialysis or gel filtration.[5]

2. Oxidation to quinones, followed by covalent coupling reactions or by oxidation of protein functional groups by the quinone. Quinones are powerful oxidizing agents. They also tend to polymerize, and they condense readily, by 1,4-addition, with reactive groups of protein, notably -SH and -NH$_2$ groups.[5,9,10] This is the principal basis of the browning reactions in plant tissues and extracts. Both o- and p-quinones are formed. o-Quinones are more reactive than p-quinones; in fact they are generally very unstable.

---

[8] Composition of Foods, in "Documenta Geigy, Scientific Tables." 6th ed. (K. Diem, ed.), p. 500. Geigy Pharmaceuticals, Division of Geigy Chemical Corporation, Ardsley, New York, 1962.

[9] W. S. Pierpoint, Biochem. J. 112, 609 (1969).

[10] W. S. Pierpoint, Biochem. J. 112, 619 (1969).

3. Ionic interactions. Phenolic hydroxyl groups in general have pK_a values of 8.45 (pK_1 of phloroglucinol) or higher,[11] and at high pH's they may form salt linkages with the basic amino acid residues of proteins. However, plant phenolic compounds of the phenylpropanoid group typically contain carboxyl groups as well, and may, therefore, be negatively charged even at neutral pH and below.

4. Hydrophobic interactions. The aromatic ring structures of phenolic compounds are essentially hydrophobic, and thus have an affinity for other hydrophobic materials, including hydrophobic regions of proteins.

This multiplicity of protein–phenolic interactions requires a variety of countermeasures to remove materials that may be tightly bound by noncovalent attractions, while preventing the formation of covalent complexes. Isolated phenolic hydroxyl groups, characteristic of, for example, the A ring of flavonoid compounds, are the most active in forming H-bonded complexes. Phenolic hydroxyl groups that are located close to other H-bonding groups generally form H bonds with them, and this may greatly reduce their capacity to form external H bonds. This is true, for example, of 1,2-dihydroxy and 1,2,3-trihydroxy phenolic groups, found especially in many phenylpropanoid phenolic compounds and in the B ring of many flavonoid compounds. However, these groups are in general readily oxidized to o-quinones, with the danger of subsequent covalent condensation.

The general approach in isolating plant enzymes and organelles is thus to remove phenolic compounds and other secondary products as quickly as possible, and at the same time to prevent the formation of covalent complexes. Prevention of damage due to phenolics and quinones is in general best accomplished by adding adsorbents or protective agents that can compete with the plant materials in reacting with the phenolics and quinones, and at the same time preventing oxidation of the phenolics.

Effects of Phenols on Mitochondria

Plant mitochondria have been reported to differ from animal mitochondria in characteristic ways,[12] such as being less tightly coupled. Packer *et al.*[13] concluded that the evidence suggests that these differences are due, in every case, to damage to the plant mitochondria during isolation procedures.

There is no question that plant phenolics, because of their reactions

[11] G. Kortüm, W. Vogel, and K. Andrussow, "Dissociation Constants of Organic Acids in Aqueous Solution." Butterworth, London, 1961. Reprinted from *Pure Applied Chem.* 1, Nos. 2 and 3 (1961).

[12] H. Ikuma, *Annu. Rev. Plant Physiol.* 23, 419 (1972).

[13] L. Packer, S. Murakami, and C. W. Mehard, *Annu. Rev. Plant Physiol.* 21, 271 (1970).

with proteins, interfere generally with the isolation of plant mitochondria. In addition, it appears that they may specifically uncouple oxidative phosphorylation. Oxidative phosphorylation by mitochondria is uncoupled by a variety of phenols, including dicoumarol.[14,15] Stenlid[16] reported that several flavonoids, at concentrations around $10^{-4} M$, inhibit ATP synthesis by plant mitochondria. Their effectiveness is comparable to that of dinitrophenol,[16] and earlier investigations by Stenlid suggested an uncoupling mechanism. The uncoupling of oxidative phosphorylation by phenols is not understood, but Weinbach and Garbus[14,17] found that it could be prevented or reversed by the addition of bovine serum albumin (BSA) as a phenol adsorbent.

Some Other Interfering Plant Products

Other plant products beside phenols and quinones can be expected to interfere in the isolation of plant enzymes and organelles. The polyanionic pectic materials are ubiquitous, and it has been reported[18] that they inhibit photophosphorylation in isolated chloroplasts. Terpenes, resin acids, and other secondary products will interfere in specific tissues. It should be mentioned that 1,4-addition reactions similar to the quinone additions are also characteristic of α,β-unsaturated carbonyls (Michael addition). Recently it has been reported[19] that sesquiterpene lactones with α-methylene groups (found particularly in the Compositae) combine with skin proteins and are allergenic. The heterocyclic (nonphenolic) ring of some classes of flavonoids, as well as the corresponding C_3-chain of isoflavones, have the same α,β-unsaturated carbonyl configuration. So do the coumarins.[2] It would be surprising if these did not react to form covalent bonds with protein in a similar fashion to the quinones. Carotenoid pigments bind tightly to proteins isolated from orange peel (O. Cori, personal communication).

Plant Materials

Selection of Plant Material

If the investigations to be undertaken do not demand a particular species or a particular organ, difficulties can usually be minimized by careful choice of plant material. Cultivated vegetable species in particu-

[14] E. C. Weinbach and J. Garbus, *J. Biol. Chem.* **241**, 169 (1966).

[15] W. D. Bonner, Jr., *in* "Plant Biochemistry" (J. Bonner and J. E. Varner, eds.), p. 89. Academic Press, New York, 1965.

[16] G. Stenlid, *Phytochemistry* 9, 2251 (1970).

[17] E. C. Weinbach and J. Garbus, *J. Biol. Chem.* **241**, 3708 (1966).

[18] W. S. Cohen and A. T. Jagendorf, *Arch. Biochem. Biophys.* **150**, 235 (1972).

[19] G. Dupuis, J. C. Mitchell, and G. H. N. Towers, Phytochemical Society of North America, Abstracts, 11th Meeting, Monterrey, Mexico, 1971.

lar have been selected through the centuries for lack of astringency or bitterness (i.e., lack of "tannins"). If one is not studying chloroplast components, the use of nongreen vegetable tissues or of etiolated seedlings is frequently recommended.[20] Bonner[20] has pointed out that etiolated seedlings are especially useful because they can provide reproducible, pigment-free material at any time of the year. Etiolated tissues are probably, in general, low in phenolic materials as well as chloroplast pigments.

Absence of pigments or astringent materials does not guarantee that a plant tissue will be favorable for enzyme isolation. For example, bananas, which are nearly colorless, and quite bland in flavor, are notoriously difficult subjects for biochemical studies. This correlates with the fact that injured or overripe bananas brown rapidly. In our experience, good cell-free systems from plants are characterized by lack of oxidative browning and by a low absorbance in the 260-nm region (A. J. Burbott and W. D. Loomis, unpublished). Most of the 260 nm absorption in plant extracts is due to phenolic compounds rather than to nucleic acids or proteins.

Storage of Plant Tissues and Cell-Free Preparations

In the living plant cell, compartmentalization protects the protoplasm components from secondary products. Therefore, living tissues can usually be stored for some time if kept moist and cold—but not frozen. Freezing destroys cell compartmentalization, and oxidative changes may occur in ordinary frozen storage. For example, Klucas et al.[21] have found that nitrogenase can be isolated satisfactorily from soybean nodules frozen in liquid nitrogen and stored at $-70°$. In ordinary frozen storage of the nodules, the activity of the enzyme is destroyed (H. J. Evans, personal communication).

Freeze-dried plant preparations are, in our experience, often very unstable, and browning may occur even during the drying process. It is our impression that freeze drying of crude plant preparations or tissues is not generally desirable, as it leaves the protoplasmic components in intimate contact with secondary products, in a powder that has an unusually large surface area in contact with the atmosphere.

Isolation Techniques

Control of pH

It has been traditional in isolating plant enzymes and organelles to use concentrated buffers of high pH in order to neutralize "vacuole acids." However, the principal "acidic" materials in most plant tissues are phenols,

[20] W. D. Bonner, Jr., this series, Vol. 10, p. 126.
[21] R. V. Klucas, B. Koch, S. A. Russell, and H. J. Evans, Plant Physiol. 43, 1906 (1968).

rather than carboxylic acids. These, when in the nonionized acidic form, hydrogen bond tightly to the oxygen of peptide bonds. At high pH's, phenols are ionized to a considerable extent and are not strongly hydrogen bonded to protein. On the other hand, these higher pH's promote phenol oxidation and reduce the effectiveness of phenol adsorbents such as PVP.[5] Newer techniques for plant enzyme and organelle isolation generally use a pH of about 6.7 to 7.2.[5,20,22–27] Verleur[28] reported that an extraction medium buffered at pH 6.5 produced the highest quality mitochondria from potato tubers. Tautvydas[29] found that a medium of even lower pH, 6.0 to 6.1, gave optimal yields of pea nuclei.

Several investigators have stressed the importance of pH control in isolating tightly coupled mitochondria, and ribosomes, from plant tissues and have recommended monitoring the pH of the extract with a glass electrode pH meter and adjusting it by addition of KOH solution[20,24–27,30] rather than depending entirely on buffers. Ikuma and Bonner[20,24] recommended maintaining a pH of 7.2 for isolation of mitochondria from etiolated mung bean seedlings. Romani et al.[25] found that in isolating pear fruit mitochondria the best results were obtained by maintaining the pH between 6.7 and 6.9. Matlib et al.[26] maintained the pH between 6.8 and 7.2 to isolate tightly coupled mitochondria from etiolated Vicia faba seedlings. Ku and Romani[30] maintained a monitored pH of 6.5 in isolating ribosomes from pear fruit, whereas Marei and Romani[27] maintained the pH at 6.8 to isolate fig fruit ribosomes.

The paper of Andersen and Sowers[31] has been quoted on occasion to suggest that very low pH's are needed for effective adsorption of phenols by insoluble PVP. It should be pointed out though that Andersen and Sowers tested the binding of rutin by Polyclar AT at only 3 pH's: 8.5, 6.0, and 3.5. The binding of rutin can be calculated from their data as follows: pH 8.5, 36%; pH 6.0, 75%; pH 3.5, 79%. The difference between binding at pH 6.0 and binding at pH 3.5 is in fact very small, and the large jump between pH 6.0 and 8.5 is consistent with the reports

[22] J. K. Palmer, in "Phenolics in Normal and Diseased Fruits and Vegetables," Symposium of the Plant Phenolics Group of North America (V. C. Runeckles, ed.), p. 7, 1964.

[23] J. T. Wiskich, this series, Vol. 10, p. 122.

[24] H. Ikuma and W. D. Bonner, Jr., Plant Physiol. 42, 67 (1967).

[25] R. J. Romani, I. K. Yu, and L. K. Fisher, Plant Physiol. 44, 311 (1969).

[26] M. A. Matlib, R. C. Kirkwood, and J. E. Smith, J. Exp. Bot. 22, 291 (1971).

[27] N. Marei and R. J. Romani, Biochim. Biophys. Acta 247, 280 (1971).

[28] J. D. Verleur, Plant Physiol. 40, 1003 (1965).

[29] K. J. Tautvydas, Plant Physiol. 47, 499 (1971).

[30] L. L. Ku and R. J. Romani, Plant Physiol. 45, 401 (1970).

[31] R. A. Andersen and J. A. Sowers, Phytochemistry 7, 293 (1968).

cited above, to the effect that pH control is critical in the isolation of plant enzymes and organelles, and that optimum results are usually obtained in the range from approximately pH 6.5 to pH 7.2.

The value of neutral to slightly acidic pH's in extracting plant tissues is probably due largely to the ionization and autoxidation of phenolic compounds at higher pH's. Gregory and Bendall[32] reported that in assays of tea leaf polyphenol oxidase, with pyrogallol and 4-methylcatechol as substrates, nonenzymatic blanks were high at pH's above 6.0 and prohibitively high above pH 6.6. Rather surprisingly, these pH values correspond to only about 0.1% and 0.4% ionization of the first-dissociated phenolic hydroxyl group of the substrate. Wehr[33] found that autoxidation of catechol increased rapidly as the pH was raised to 7.5 and above.

The effect of pH on the ionization of groups other than phenolic hydroxyls should also be kept in mind, e.g., organic phosphates and carboxylic acids. The pK_a (4.76) of acetic acid is representative of carboxylic acids. Nonionized carboxyl groups are active in hydrogen bonding[5]; ionized carboxyl groups can form ionic complexes.

Use of Polymers

Polymers, both natural and artificial, are frequently added to the extracting media used in isolating plant enzymes or organelles. In many cases the use of polymers has been reported to be either essential or highly beneficial. In some cases they serve as phenol adsorbents or quinone scavengers. In other cases there appear to be other effects in addition.

Polyvinylpyrrolidone. We have previously described the use of soluble polyvinylpyrrolidone (PVP) in isolating plant mitochondria and of insoluble cross-linked PVP (Polyclar AT from GAF Corporation) in isolating soluble enzymes from plant tissues.[5,6] PVP is effective in binding those phenolic compounds that form strong H-bonded complexes, i.e., those with isolated hydroxyl groups. Use of soluble PVP in isolating plant mitochondria and ribosomes has been described by several authors.[23,25,27,30] In using PVP it is essential that purified grades be used. Polyclar AT is produced for use in beverage processing, and it may contain traces of metal ions and of the vinylpyrrolidone monomer (L. Blecher, personal communication). It can be purified for enzyme work by boiling for 10 minutes in 10% HCl and then washing with glass-distilled water[5,6] until free of Cl⁻. We have recently found that the washing process can be speeded greatly by neutralizing with

[32] R. P. F. Gregory and D. S. Bendall, *Biochem. J.* **101**, 569 (1966).

[33] H. M. Wehr, "Reactions of Protein with Phenols and Quinones: Evaluation of Amino Acid Modification and Protein Digestibility." Ph.D. Thesis, Oregon State University, Corvallis, Oregon, 1973.

KOH after several changes of glass-distilled water, and then continuing the distilled water washing. It appears that PVP binds hydrogen ions as well as phenols. The purified Polyclar AT can be dried for storage if desired, but it should not be ground to remove lumps, as this produces traces of soluble PVP. Likewise, if Polyclar AT is present during homogenization of tissues, traces of soluble PVP may be produced. If the polymer is dried, it should be hydrated before use by soaking for several hours.

In using soluble PVP for isolation of mitochondria, Hulme et al.[34] emphasized the essentiality of using a pharmaceutical grade, as ordinary grades were not sufficiently pure. Pharmaceutical grade soluble PVP was formerly available only in molecular weight ranges of about 28,000[34] and 40,000. It is now available commercially in three molecular weight ranges from GAF Corporation. These are Plasdone C-15, average molecular weight 10,000; Plasdone C-30, average molecular weight 40,000; and Plasdone C-90, average molecular weight 360,000. This provides the possibility of selecting a soluble PVP that can be separated by gel filtration from any given organelle or soluble enzyme. When PVP is used it is essential that the exact type be specified. Insoluble PVP is sometimes designated as polyvinylpolypyrrolidone (PVPP).

Hydrophobic Synthetic Resins. Recently, highly porous synthetic adsorbents of very high surface area have been produced by fusing very small microspheres of various polymers into beads.[35] Two of these, both consisting of porous polystyrene beads, have been tested and show great promise as agents for removing hydrophobic materials from plant extracts. Amberlite XAD-2, the older of the two, has a surface area of 330 m²/g dry weight. Amberlite XAD-4, a newer product, has a surface area of 750 m²/g dry weight. Both of these, and other related adsorbents of high surface area are available commercially from Rohm and Haas Company, Independence Mall West, Philadelphia, Pennsylvania 19105. Technical specifications can be obtained from the manufacturer's literature.[36,37]

Amberlite XAD-2 has been used to remove detergents used in solubilizing membrane-bound enzymes.[38] We have tested both XAD-2 and XAD-4 as adjuncts in plant protein isolation, and XAD-4 appears especially promising. A combination of XAD-4 and Polyclar AT produces crystal-clear protein-containing extracts from potato tubers and from wal-

[34] A. C. Hulme, J. D. Jones, and L. S. C. Wooltorton, *Phytochemistry* 3, 173 (1964).
[35] J. Paleos, *J. Colloid Interface Sci.* 31, 7 (1969).
[36] Amberlite XAD-2. Technical Bulletin, Ion Exchange Department, Rohm and Haas Company, Philadelphia, Pennsylvania, 1971.
[37] Amberlite XAD-4. Preliminary Technical Notes, Rohm and Haas Company, Philadelphia, Pennsylvania, 1971.
[38] I. Shechter and K. Bloch, *J. Biol. Chem.* 246, 7690 (1971).

nut (*Juglans regia*) hulls (J. Lile and W. D. Loomis, unpublished). In these two test tissues neither polymer alone is capable of removing all the pigments. XAD-4 also removes terpenoid compounds from plant extracts and has made it possible to study terpene metabolism in cell-free systems from peppermint by direct gas-chromatographic analysis, without labeled substrates.[39]

The Amberlite polystyrene beads are supplied in hydrated form, but preservatives and traces of monomers are present. For laboratory use the manufacturer recommends[37] that the beads be washed extensively in a column: first backwashing with water for about 10 minutes, or until complete classification of beads by size is achieved; then washing by downflow with 5 bed-volumes of methanol, followed by 20 bed-volumes of high purity water. If the beads should become dehydrated by exposure to the air, they can be rehydrated by treatment with a polar organic solvent such as acetone or methanol, followed by water. Amberlite XAD-4 is most effective when used in a column. In preliminary tests, Amberlite XAD-7 polyester beads appear very promising, perhaps superior to polystyrene.

Bovine Serum Albumin. One of the most widely used and effective additives in plant organelle isolation is bovine serum albumin (BSA).[20,23-25,40-46]

Dalgarno and Birt[42] reported that the presence of 1% BSA in the isolation medium reduced the dependence of isolated carrot mitochondrial fractions on added cofactors.

Laties and Treffry[44] reported that the presence of a soluble polymer (dextran, Ficoll, polyethyleneglycol, PVP, BSA, or bovine γ-globulin) in the suspending medium was necessary in order to obtain *in vitro* rod-shaped mitochondria resembling *in vivo* mitochondria morphologically. The concentrations of polymer required were reportedly greater than those required for protection against phenolics or other lipids, and it appeared that two separate effects, both beneficial, are involved.

Throneberry[41] found that cotton seedlings contained a toxic principle which inactivated the mitochondria during isolation. The toxic material became tightly bound to the particles during isolation and was not removed by washing with buffered sucrose solutions. Addition of 2% BSA

[39] R. Croteau, A. J. Burbott, and W. D. Loomis, *Biochem. Biophys. Res. Commun.* **50**, 1006 (1973).
[40] C. A. Price and K. V. Thimann, *Plant Physiol.* **29**, 113 (1954).
[41] G. O. Throneberry, *Plant Physiol.* **36**, 302 (1961).
[42] L. Dalgarno and L. M. Birt, *Biochem. J.* **83**, 195 (1962).
[43] B. T. Storey and J. T. Bahr, *Plant Physiol.* **44**, 115 (1969).
[44] G. G. Laties and T. Treffry, *Tissue Cell* **1**, 575 (1969).
[45] J. K. Raison and J. M. Lyons, *Plant Physiol.* **45**, 382 (1970).
[46] G. E. Hobson, *Phytochemistry* **9**, 2257 (1970).

to the isolation and washing media yielded particles with high succinoxidase activity. Other agents tested, including cysteine, EDTA (ethylenediamine tetraacetate), gelatin, egg albumin, and PVP, could not replace BSA. (The grade of PVP used was not specified.) Subsequent studies suggested that mitochondrial phosphorylation was inhibited to a greater degree than was oxidative activity (i.e., uncoupling of oxidative phosphorylation)[47-49] and that the phenolic compound gossypol was the principal culprit.

Weinbach and Garbus[14,17] found that addition of BSA not only prevented uncoupling by phenols, but actually restored oxidative phosphorylation in rat liver mitochondria that had been uncoupled by various substituted phenols, including dicoumarol. The native tertiary structure of the protein appeared to be needed for the phenol binding, as BSA denatured by chemical agents or heat was ineffective.

It seems likely that the unusual effectiveness of BSA in protecting plant organelles during isolation may be due to a high capacity to react with plant phenolic compounds in all of the principal ways that proteins combine with phenols. BSA is known for its capacity to bind lipids by hydrophobic forces, and to bind anions,[50] consistent with its high content of hydrophobic amino acids and lysine.[51] Since it is a protein, it can be expected to bind phenols by hydrogen bonding to the peptide-bond oxygens. BSA appears also to be an effective quinone scavenger,[33] by virtue of its high lysine and cystine content. The covalent condensation of quinones with ε-amino groups of lysine was to be expected,[5] but the reaction of quinones with cystine has, to our knowledge, not been reported. However, it does not seem unlikely that under conditions where catechols or hydroquinones are being oxidized to quinones, there would also be some reduction of disulfides to sulfhydryls. Sulfhydryl groups are known to react avidly with quinones.[5]

Protamine. Pulegone reductase from peppermint is extremely labile in crude extracts,[52] and even after purification under N_2, using Polyclar AT, ascorbate, and gel filtration, it is very labile in the presence of O_2. By adding protamine sulfate to a concentration of 0.1% to 0.2% (titrate with 2% protamine sulfate until a slight precipitate appears), followed by centrifugation, materials were precipitated which darkened rapidly in

[47] G. O. Throneberry, *Plant Physiol.* **37**, 781 (1962).
[48] B. D. Myers and G. O. Throneberry, *Plant Physiol.* **41**, 787 (1966).
[49] D. D. Killion, S. Grooms, and R. E. Frans, *Plant Physiol.* **43**, 1996 (1968).
[50] J. F. Foster, *in* "The Plasma Proteins" (F. W. Putnam, ed.), Vol. 1, p. 179. Academic Press, New York, 1960.
[51] P. F. Spahr and J. T. Edsall, *J. Biol. Chem.* **239**, 850 (1964).
[52] J. Battaile, A. J. Burbott, and W. D. Loomis, *Phytochemistry* **7**, 1159 (1968).

the air. After this treatment the enzyme was very stable and was not inactivated by reagents that modify -SH groups. Guanidinoethyl cellulose was not effective as a substitute for protamine. It appears that there is in peppermint a small amount of phenolic material that is very labile to oxidation and which at neutral pH is bound tightly to protein by ionic forces rather than by hydrogen bonding (A. J. Burbott and W. D. Loomis, unpublished).

Antioxidants and Phenol Oxidase Inhibitors

We previously[5,6] cited some phenol oxidase inhibitors and antioxidants that had been used in plant enzyme isolation, and discussed their use. Kull et al.[53] tested a large number of substrate analogs as possible inhibitors of potato polyphenol oxidase. 4-Chlororesorcinol was far the most effective, inhibition being detectable at 10^{-8} M. Several other antioxidants and phenol oxidase inhibitors have been recommended recently for use in isolation of plant enzymes and mitochondria. Two in particular, mercaptobenzothiazole and metabisulfite, have given excellent results.

Mercaptobenzothiazole. Palmer[22] tested a variety of compounds for their ability to inhibit banana polyphenoloxidase, as assayed by color formation from dopamine. Of the compounds tested, sodium mercaptobenzothiazole was especially effective: at 5×10^{-6} M it completely inhibited color formation for more than 15 minutes. Subsequently[54] he recommended routine use of 10^{-4} M sodium mercaptobenzothiazole in plant enzyme extraction media, stating that concentrations above 10^{-5} M caused prolonged inhibition of both color formation and oxygen uptake. Laties and Treffry[44] modified the method of Honda et al.[55] by substituting 5×10^{-4} M mercaptobenzothiazole for mercaptoethanol in the medium, and by using an electric juice extractor in place of slicing the tissues with razor blades, and isolated potato mitochondria which morphologically resembled in vitro mitochondria. Mercaptobenzothiazole is thought to act both as a copper chelator and as a quinone scavenger.

Sulfites. Anderson and Rowan[56] tested the effects of reducing agents and potassium metabisulfite on the extraction of active peptidases from tobacco leaf disks and reported that extracts prepared in media containing dithionite or metabisulfite had the highest activities. In particular, metabisulfite at 10 mM concentration gave optimal peptidase activity and appeared to inhibit phenol oxidases permanently. Using metabisulfite as an

[53] F. C. Kull, M. R. Grimm, and R. L. Mayer, *Proc. Soc. Exp. Biol. Med.* **86**, 330 (1954).

[54] J. K. Palmer and J. B. Roberts, *Science* **157**, 200 (1967).

[55] S. I. Honda, T. Hongladarom, and G. G. Laties, *J. Exp. Bot.* **17**, 460 (1966).

[56] J. W. Anderson and K. S. Rowan, *Phytochemistry* **6**, 1047 (1967).

antioxidant, Stokes *et al.*[57] isolated coupled mitochondria from potato tubers. Metabisulfite concentration was critical: concentrations of 3 mM or below were ineffectual, 4 mM was optimal, and higher concentrations resulted in somewhat reduced P:O and respiratory control ratios. Anderson[7] has reviewed the effects of antioxidants, and other aspects of plant enzyme and organelle isolation, in detail.

Following the reports of Anderson and co-workers, we tested metabisulfite in place of ascorbate in our medium for isolating pulegone reductase from peppermint leaves (A. J. Burbott and W. D. Loomis, unpublished). Previously this enzyme could only be demonstrated in extracts prepared with Polyclar AT under N_2.[52] Addition of 10 mM potassium metabisulfite made it possible to isolate the enzyme in the air at the same activity as previously, or to double the activity when working under N_2.

Metabisulfite (also called pyrosulfite), bisulfite, sulfite, sulfurous acid, and SO_2 are all interconvertible in aqueous solution, the equilibrium proportions depending largely on temperature, concentration, and pH. Their chemistry is complex, and often it is not clear which species is responsible for a particular reaction.[58] The action of metabisulfite in protecting plant enzymes and mitochondria is probably complex.

The reactivities of the sulfites are such[58] that one might expect reactions with phenolic compounds and quinones, with other secondary products such as aldehydes, especially α,β-unsaturated aldehydes (e.g., coumarins), and probably with proteins. In this respect, phenol oxidases seem to be much more sensitive to sulfites than are other enzymes or cell components. However, it should be borne in mind that there may be side effects. For example, certain disulfide bonds of proteins can be cleaved readily by sulfite, whereas others are highly resistant.[59] Inhibitory effects of bisulfite on photosynthesis and on certain plant enzymes have been reported.[60]

In spite of the cautions just expressed, metabisulfite has proved to be extremely effective in protecting peppermint enzymes and potato mitochondria during isolation. There is probably an added bonus from the antibacterial effects of sulfites.

Borate and Germanate. Borate is known for its ability to form complexes with polyols, including *o*-diphenols,[61,62] and to inhibit diphenol oxi-

[57] D. M. Stokes, J. W. Anderson, and K. S. Rowan, *Phytochemistry* **7**, 1509 (1968).

[58] L. C. Schroeter, "Sulfur Dioxide." Pergamon, Oxford, 1966.

[59] G. E. Means and R. E. Feeney, "Chemical Modification of Proteins," p. 152. Holden-Day, San Francisco, California, 1971.

[60] U. Lüttge, C. B. Osmond, E. Ball, E. Brinckmann, and G. Kinze, *Plant Cell Physiol.* **13**, 505 (1972).

[61] U. Weser, *Hoppe-Seyler's Z. Physiol. Chem.* **349**, 982 (1968).

[62] E. E. King, *Phytochemistry* **10**, 2337 (1971).

dases.[61] Borate has been used occasionally as a protective agent during the isolation of plant enzymes and organelles.[62,63] It was reported[62] that extraction of cotton tissues with borate gave higher yields of soluble protein and greater enzyme activity than extraction in the presence of insoluble PVP, while conventional extraction procedures yielded very little soluble protein and little or no enzyme activity. This result is not surprising, since gossypol, the predominant phenol of cotton, has no isolated hydroxyl groups, but rather has highly interacting hydroxyl and carbonyl groups. One would expect it to form stable complexes with borate, and to be less active in forming hydrogen-bonded complexes.

Weser[61] found that the germanate complexes of o-diphenols (at pH 7.5) are two orders of magnitude more stable than the corresponding borate complexes. Kinetic analysis of the inhibition of mushroom diphenol oxidase indicated that both germanate and borate are competitive inhibitors of this enzyme. At pH 7.5, germanate was a much more potent inhibitor (K_i values near $10^{-5} M$) than borate (K_i values of 2×10^{-3} to $5 \times 10^{-3} M$). Weser's findings suggest that germanate might be a valuable additive in extracting plant enzymes.

Inert Atmospheres

In principle, the simplest way to prevent oxidation of plant phenolic compounds is to eliminate oxygen. Anaerobic isolation procedures (e.g., N_2 in a glove box) have proved to be essential or desirable with several plant enzymes.[21,52,64] In practice, working in an inert atmosphere is usually difficult. However, there are several rather easy methods of reducing O_2 contact during isolation procedures, such as sparging all solutions and solid absorbents with N_2 prior to use, vacuum infiltrating tissues before homogenization, avoiding the use of blenders[17] which whip air into the homogenates, and working quickly.

Gases heavier than air can be used in an open-top box to maintain a relatively inert atmosphere. Argon has been so used. CO_2, being heavier, and less expensive, should also be effective, but one would want to evaluate the effects of a CO_2 atmosphere on the acidity of the extracts.

Rapid Procedures

Rapid Homogenization. Recently several investigators have emphasized the importance of minimizing the time period during which protoplasm components and secondary products are in contact and able to interact. Rapid extraction is an obvious approach. Use of low temperatures is an-

[63] I. Zelitch, this series, Vol. 10, p. 133.
[64] T. C. Hall, B. H. McCown, S. Desborough, R. C. McLeester, and G. E. Beck, *Phytochemistry* 8, 385 (1969).

other. Depending on the plant tissue used, several methods have proven effective in obtaining rapid homogenization and the rapid mixing of the homogenate with the extracting medium.

Rapid homogenization in a chilled mortar in the presence of the extracting solution has been effective with soft tissues such as seedlings. Grinding times recommended have been, for example, 30 seconds[24] or 60–90 seconds.[26] Soft pear fruit tissues have been homogenized satisfactorily by pressing and rubbing them against a stainless steel screen immersed in the extracting solution.[25] Palmer[65] isolated mitochondria from Jerusalem artichoke by grating with a stainless steel grater immersed in the extracting solution.

Bonner[20] has suggested that massive tissues, such as potatoes, can be grated in a commercial salad maker in the presence of the extracting solution.

Grinding of plant tissues in liquid nitrogen and adding the frozen powder to the extracting solution has been successful in isolating soluble enzymes from peppermint leaves,[5] ribosomes from pear[30] and fig[27] fruit, and mitochondria from banana fruit, apple fruit, and potatoes.[66] It should be noted that although the banana mitochondria took up O_2 rapidly, they were not tested for phosphorylating activity.

Honda et al.[55] recommended slicing of tissues with stacked stainless steel razor blades to cut open the cells and liberate organelles with minimum crushing. The method yields high quality organelles but is time consuming and not adapted to processing large amounts of tissue. D. Branton and P. de Silva (unpublished) have adapted this concept to a power-driven tissue slicer, which Beevers and Breidenbach (this volume [58]) have used successfully for isolating glyoxysomes in quantity.

Juice extractors, such as that distributed by the Oster Manufacturing Company, provide a reasonably priced substitute for the Branton-de Silva apparatus and have been very effective in quickly macerating plant tissues, especially relatively solid tissues, and removing most of the solid debris.[44,45] This extractor consists of a basket centrifuge having a circular base plate with sharp teeth. The basket is lined with Miracloth to serve as a filter, and a 2–4 mm layer of Dicalite filter-aid has been found helpful with some species.[45] Fine-mesh nylon screen might be better (see below). In using the juice extractor, homogenizing solution is fed in with, or immediately after, the tissue. Maceration and filtration are accomplished in less than 5 seconds.[44]

Rapid Centrifugation and Filtration. Recently published techniques for

[65] J. M. Palmer, *Nature (London)* **216**, 1208 (1967).
[66] N. F. Haard and H. O. Hultin, *Anal. Biochem.* **24**, 299 (1968).

isolation of plant mitochondria have used higher centrifugation speeds than formerly, thus greatly reducing the overall isolation time and obtaining more-intact mitochondria.[26,65,67,68] Palmer[65] replaced the usual low-speed centrifugation to remove debris, by filtration through nylon screen (50 strands per centimeter) and then precipitated mitochondria by centrifugation at 40,000 g for 1.5 minutes. He succeeded in reducing the overall isolation time from 70 minutes to 7 minutes. Sarkissian and Srivastava,[67] and Matlib et al.[26] used similar techniques to obtain tightly coupled mitochondria from wheat and Vicia faba, respectively.

Filtration through fine nylon screen in place of low-speed centrifugation[26,65] appears to be a valuable method of reducing isolation times. In addition to the instances cited above, Tautvydas[29] achieved a 6-fold increase in the recovery of pea nuclei by fractionating the homogenates on four stacked nylon-mesh filters of diminishing pore size, from 375 μm to 10 μm. The whole filtration, or sieving, procedure required only 5–8 minutes.

If nylon screen can be used for rapid filtration of plant homogenates, it should be possible to achieve similar filtration even faster by using nylon screens or bolting silk of known mesh to line the basket of the juice extractor mentioned above. The juice extractor, with teeth filed off and the bottom plate turned over to expose a smooth surface, is a very satisfactory basket centrifuge. One can even visualize a cascading series of such juice extractors lined with nylon screen of diminishing pore size, analogous to Tautvydas' stacked filters.

Protein Determination

The use of protein as a basis for comparisons of enzymatic activities has achieved almost fetishistic acceptance among biochemists. Unfortunately, plant extracts may yield grossly inaccurate protein values with conventional methods. Plant phenolic compounds in particular interfere seriously[69,70] with several of the most popular methods of protein determination,[71] which were originally designed for animal tissues. For example, the UV method of Warburg and Christian,[71] the biuret method,[72] and the method of Lowry et al.[73] often give plant protein values that are in error

[67] I. V. Sarkissian and H. K. Srivastava, Can. J. Biochem. **48**, 692 (1970).
[68] S. S. Malhotra and M. Spencer, J. Exp. Bot. **22**, 70 (1971).
[69] V. H. Potty, Anal. Biochem. **29**, 535 (1969).
[70] A. J. Burbott, A. Traverso-Cori, O. Cori, and W. D. Loomis, in preparation.
[71] E. Layne, this series, Vol. 3, p. 447.
[72] A. C. Gornall, C. J. Bardawill, and M. M. David, J. Biol. Chem. **177**, 751 (1949).
[73] O. H. Lowry, N. J. Rosebrough, A. L. Farr, and R. J. Randall, J. Biol. Chem. **193**, 265 (1951).

by orders of magnitude.[70] We have compared several of the most popular methods of protein analysis on a crude extract of Jonathan apples which, according to Kjeldahl analysis after precipitation with trichloroacetic acid (TCA), contained 0.086 mg of crude protein per milliliter. The analytical results for the crude extract were as follows (in mg/ml): UV, 4.8; sulfo-salicylic acid turbidity, 0.087; TCA turbidity, 0.030; biuret, red precipitate; Lowry, 2.4; Kjeldahl, 0.73. The only methods that yielded results of even the right order of magnitude with the crude extract were the turbidity methods. With the UV ($E_{280}:E_{260}$) method, using the standard formulas, we have on occasion actually obtained negative values for protein in plant extracts.

The biuret method depends on the blue color formed when Cu^{2+} complexes with peptide bonds under alkaline conditions. However, Cu^{2+} forms colored complexes with phenols as well. These complexes do not have the same absorption spectrum as the protein complexes, but they will give a "protein" value when read in a colorimeter in the usual way. In the instance cited, Cu^{2+} was obviously reduced to Cu^+ by the crude apple extract.

The method of Lowry et al.[73] is undoubtedly the most popular method of protein determination today. This method depends on the ability of copper–protein complexes (biuret complexes) to reduce the Folin phenol reagent.[72-74] The original paper[73] stated clearly that phenols interfere, but this fact is often overlooked.

Potty[69] has recommended a method for plant protein analysis in which the Lowry color values (protein plus phenolics) are corrected for the color due to the phenol reagent without copper (phenolics). Samples containing large amounts of phenolic compounds were pretreated by precipitating and washing with 67% ethanol. The procedure may seem tedious, but the rationale is logical, and the method merits consideration. There is no sense in spending any time at all on protein analyses if the results may be wrong by orders of magnitude.

In our experience, the turbidity methods[71,75] are probably the most reliable quick methods for determining protein in plant extracts. They are less sensitive than some of the other methods and apparently are regarded as unsophisticated. However, they do generally give values of the right order of magnitude.

We routinely compare enzyme activities on a "per gram fresh weight" basis. This appears to be reasonable physiologically, and it is quick, and not subject to gross errors. We find that the UV absorption spectrum is a

[74] S. Chou and A. Goldstein, Biochem. J. **75**, 109 (1960).
[75] M. Kunitz, J. Gen. Physiol. **35**, 423 (1952).

valuable index of "cleanness" in purifying cell-free preparations from plants. Removal of 260 nm-absorbing material, largely plant phenolics, with resulting increase in the $E_{280}:E_{260}$ ratio generally results in more-active and more-stable preparations.

Artifactual Organelles

In view of the fact that plant phenolic compounds may precipitate proteins, and that plant phenolic compounds may also be analyzed as "protein," it is important that isolated "organelles" be identified by something more than the relative centrifugal force at which they were precipitated. This problem is illustrated by tea leaf phenol oxidase. This enzyme had been regarded as a chloroplast component. However, Sanderson[76] demonstrated that conventional methods yielded no soluble protein from tea leaves. When powdered nylon was added as a phenol adsorbent, the solubility of tea leaf phenol oxidase, and of protein in general, was proportional to the amount of powdered nylon added. The "chloroplast" fractions obtained previously consisted largely of tannin–protein complexes. Anderson[7] has cited other similar examples.

Acknowledgments

We wish to acknowledge support in the form of research grants from the National Institute of General Medical Sciences of the U.S. Public Health Service (GM-08818) and from the National Science Foundation (GB-25593). We are grateful to Mr. Louis Blecher of GAF Corporation, 140 West 51st Street, New York, New York, for his very helpful advice as well as for samples of the various grades of PVP.

[76] G. W. Sanderson, *Biochim. Biophys. Acta* **92**, 622 (1964).

[55] Fractionation of Green Tissue

By SHIGERU I. HONDA

While the methods discussed here were used chiefly for chloroplasts, etioplasts, microbodies, and mitochondria from green and etiolated tissues, in principle, they can be used as well for spherosomes and nuclei to be simultaneously fractionated with the other organelles. Special attention is focused upon chloroplasts and etioplasts.

In the biochemical approach, usually attention is directed solely toward a single type of organelle and the fate of other organelles is neglected. However, particularly when nucleic acid metabolism is concerned, this narrow view is completely inadequate because of troublesome contaminations, e.g., chloroplast DNA with nuclear and mitochondrial DNA. Also in studies on mitochondria from green tissues, so-called mitochondrial preparations were highly contaminated with green matter from chloroplasts until the advent of isopycnic fractionations. Even now nongreen parts of chloroplasts—the protuberances[1,2] and stroma—undoubtedly contaminate mitochondria prepared by isopycnic fractionation. While even the reverse case of chloroplasts being contaminated with mitochondria is probably known to all, it is necessary to recall the warning that the contamination is not negligible while small in number or bulk.[3]

In the physiological approach, attention is paid to keep all organelles intact during isolation and fractionation. Then subsequent to purification, breakdown of permeability barriers can be used to express potential biochemical activities or subfractionation used to obtain suborganelle fragments for study. Here the physiological approach is stressed, and some novel but sometimes tedious methods are presented. Details for the established methods of differential centrifugation,[4,5] isopycnic fractionation,[6–12] and rate-zonal centrifugation[13–15] are readily available. Because of the emphasis on recovery of all organelles, some considerations of the isolation medium and tissue extraction as well as fractionations are presented.

Isolation and Purification Media

The simple aqueous media with osmoticum and buffer are inadequate to extract the bulk of the different organelles in intact condition. Further, simple NaCl or sucrose media are actually detrimental to structural integrity

[1] S. G. Wildman, T. Hongladarom, and S. I. Honda, Science 138, 434 (1962).
[2] S. I. Honda, T. Hongladarom, and S. G. Wildman, in "Primitive Motile Systems in Cell Biology" (R. D. Allen and N. Kamiya, eds.), p. 485. Academic Press, New York, 1964.
[3] R. M. Leech, Biochim. Biophys. Acta 71, 253 (1963).
[4] J. T. Wiskich, R. E. Young, and J. B. Biale, Plant Physiol. 39, 312 (1964).
[5] L. Y. Yatsu, T. J. Jacks, and T. P. Hensarling, Plant Physiol. 48, 675 (1971).
[6] J. E. Baker, L.-G. Elfvin, J. B. Biale, and S. I. Honda, Plant Physiol. 43, 2001 (1968).
[7] N. K. Boardman and S. G. Wildman, Biochim. Biophys. Acta 59, 222 (1962).
[8] R. W. Breidenbach and H. Beevers, Biochem. Biophys. Res. Commun. 27, 462 (1967).
[9] R. M. Leech, Biochim. Biophys. Acta 79, 637 (1964).
[10] G. M. Orth and D. G. Cornwall, Biochim. Biophys. Acta 71, 734 (1963).
[11] N. E. Tolbert, A. Oeser, R. K. Yamazaki, R. H. Hageman, and T. Kisaki, Plant Physiol. 44, 135 (1969).
[12] N. E. Tolbert, this series, Vol. 23, p. 665.
[13] C. C. Still and C. A. Price, Biochim. Biophys. Acta 141, 176 (1967).
[14] A. Vasconcelos, M. Pollack, L. R. Mendiola, H.-P. Hoffman, D. H. Brown, and C. A. Price, Plant Physiol. 47, 217 (1971).
[15] A. R. Wellburn and F. A. M. Wellburn, J. Exp. Bot. 22, 972 (1971).

of chloroplasts.[16] Complex media have been derived to control the various factors affecting the integrity of several organelles simultaneously.[17,18] While complex media have been developed for use with chloroplasts with excellent results,[19] i.e., attention focused upon a single type of organelle, according to our experience such media can be inadequate for other types of organelles.

Chloroplasts isolated in complex aqueous media by gentle procedures can now match the protein content of chloroplasts isolated with non-aqueous media. Chloroplasts isolated with nonaqueous media apparently lose very little of their protein, but the adsorption of cytoplasmic constituents occurs.[20] The fate of other organelles is unknown in nonaqueous media.[21]

The following medium—basically the "Honda medium"—is used routinely by us to isolate chloroplasts, mitochondria, nuclei, and spherosomes from various tissues. This medium will permit the retention of structural organelle integrity for over 24 hours after cell extraction such that organelles resemble closely their states *in vivo*, even to the display of dynamic pleomorphic changes of mitochondria and chloroplast protuberances. It is the only medium known that will permit cytoplasmic streaming of a cell to continue after it is cut open in the bathing medium.[18]

The medium is formulated in final concentration as: 250 mM sucrose; 25 mM potassium–Tricine buffer, pH 8.5; 8 mM 2-mercaptoethanol; 5 mM magnesium acetate; 50 mg/ml Dextran-40; 25 mg/ml Ficoll; 1 mg/ml bovine serum albumin. Ficoll (Pharmacia Fine Chemicals) is a form of polymerized sucrose, 400,000 MW. Dextran-40 is a dextran of about 26,000 MW. The effect of dextran is markedly changed by concentration and molecular size. Ficoll and dextran are not completely miscible in water solution and can form separate water phases.[22]

The viscous, incomplete stock solution is made in double-strength and kept in a deep-freeze until use. The medium is completed by addition of bovine serum albumin upon dilution to make the working solution. No bacterial contaminations occur if the solution is kept frozen except when used.

Formulation of the medium and its effects are discussed in detail elsewhere.[18] In brief, the components are added to control or counteract effects

[16] T. Hongladarom and S. I. Honda, *Plant Physiol.* **41**, 1686 (1966).
[17] S. I. Honda, T. Hongladarom, and S. G. Wildman, *Plant Physiol.* **37**, xli (1962).
[18] S. I. Honda, T. Hongladarom, and G. G. Laties, *J. Exp. Bot.* **17**, 460 (1966).
[19] S. M. Ridely and R. M. Leech, *Planta* **83**, 20 (1968).
[20] C. R. Stocking, *Plant Physiol.* **34**, 56 (1959).
[21] C. R. Stocking, this series, Vol. 23, p. 221.
[22] P.-Å. Albertsson, personal communication (1965).

of tonicity, acidity, loss of magnesium, addition of unwanted inorganic ions, harmful polyphenols (tannins), and peroxides formed during lipid (fatty acid) oxidation catalyzed by heavy metals. Osmoticum is used in lower concentration than those used by many workers because isotonicity for organelles is lower than it is commonly thought to be and chloroplast protuberances and stroma are lost during osmotic contraction.[16]

Extraction of Green Tissue

Although two methods for extracting green tissue are commonly used— tissue grinding with mortar and pestle and homogenizing with blender—in terms of gentleness only grinding and cutting with razor blades can be recommended for recovery of intact organelles. It may be that methods that extract a greater bulk of tissue will yield a greater absolute amount of intact organelles, but the problems of contamination from other damaged organelles remain. Even the "intactness" of the organelles can be questioned, as with mitochondria subjected to blending no longer having respiratory control.

Chilled tissue with isolation medium can be ground with a chilled mortar and pestle with the aid of materials facilitating the comminution. Current methods obtain chloroplasts (perhaps other organelles as well) in relatively good yield and intactness with the tissue ground in a nylon bag[23] or the mortar and pestle lined with nylon cloth.[24] The nylon netting also serves to remove debris upon lifting from the mortar. Chilled isolation medium and chilled, deveined tissue cut into small 1×1 cm pieces are ground together in ratios of ca. 1–2 ml of medium to 1 g of tissue. Only a short time of grinding is used (10 seconds to 2 minutes). The short time undoubtedly contributes to the intactness of the extracted organelles. Microscope observations on living cells reveal the disorganizing effect of pressure on organelles. Presumably similar effects result when a much greater force is applied during grinding. However, the use of nylon cloth probably avoids some of this effect if the strong, fine nylon threads of the cloth act as slicers during the grind.

The most gentle extraction method involves the cutting or slicing cells open with sharp razor blades. For microscale isolations, small knives made with shards of razor blades affixed to wooden-applicator sticks with sealing wax or epoxy glue are suitable for cutting leaf tissue stripped of the covering epidermis and moistened with isolation medium.[16] Larger scale isolations (ca. 10 g) can be obtained with 10–50 stacked, thin, stainless steel, double-edged razor blades.[18] The appressed blades are so thin that each

[23] P. S. Nobel, *Plant Physiol.* **42**, 1389 (1967).
[24] C. G. Kannangara, K. W. Henningsen, P. K. Stumpf, L.-A. Appelqvist, and D. von Wettstein, *Plant Physiol.* **48**, 526 (1971).

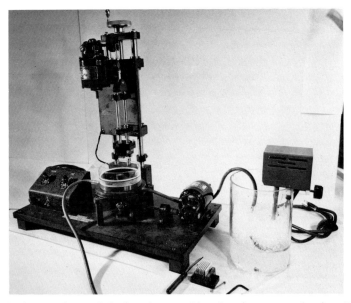

FIG. 1. Automatic, cooled chopping machine for tissue comminution. The cutting rate is adjustable from 50 to 120 strokes/minute while the dish containing the tissue rotates at 3 rpm. (Photograph by C. A. Schroeder.)

cutting edge is separated from each other by about 10^{-3} inch. The closeness and sharpness of the blades result in very efficient cutting of virtually every cell beneath the stacked blade. The tissue is placed upon a plastic or stainless steel block, which is itself on ice. Cutting is done with isolation medium (2:1) covering the tissue. This cutting method is very slow so that it requires 0.5 to 2 hours, depending upon the amount of tissue. However, in combination with the complex medium which can keep organelles intact for 24 hr or more, this gentle method yields organelles in very good condition. An automatic cooled chopper (Fig. 1) removes the tedium from the cutting method.[25]

Fractionation Methods

Evaluations of results from several groups indicate that several fractionation methods have drawbacks, perhaps not generally appreciated.

Any method that involves centrifugal pelleting of a fraction can be suspect. Even a low speed—if it results in packing of the fraction against the centrifuge tube sidewall or bottom—subjects the fraction to high pres-

[25] Plans for this machine can be obtained from S. G. Wildman, University of California, Los Angeles, Department of Botanical Sciences.

sure which can force contents through membranes without necessarily breaking organelle membranes. Subsequent release of pressure does not ensure a return of the lost contents to the organelles. While loss of biochemical activities often have been ascribed to "solubilization loss," it is likely that "squeezing loss" is also involved. Furthermore, aggregation of particles can be promoted so that a low speed preliminary spin can decrease organelle recovery subsequently, e.g., chloroplasts being trapped by nuclear fragments.

While isopycnic fractionation and rate-zonal centrifugation methods can fractionate different organelles of closely similar size, form, and density, the acts of centrifugation cause distortions and physical and physiological injury. The chloroplast envelope can be stripped from chloroplasts during isopycnic centrifugations[9,26] and rate-zonal centrifugation.[14] Vasconcelos et al.[14] suggested that a medium speed of centrifugation instead of a high speed may help preserve the envelopes of chloroplasts from being stripped off. It seems that the medium employed can protect chloroplasts from centrifugation damage if it contains Ficoll and dextran.[14] Etioplasts become distorted in form and can become almost unrecognizable after rate-zonal centrifugation. The etioplasts are damaged to such an extent that they no longer respond to light and cannot carry out morphological transformations for greening.[15] It is not clear whether the motion, hypertonicity, or lack of protective agents are the damaging factors. Just the opposite situation of permitting peroxisomes to become permeable in sucrose allows peroxisomes to be separated from disorganized chloroplasts and mitochondria whereas if Ficoll is used, no separation is effected of intact peroxisomes from intact chloroplasts and mitochondria.[12]

Since descriptions are available for methods commonly used for chloroplasts, etioplasts, mitochondria, and microbodies from green tissue, only a novel, little known and used method of great potential[15,27,28] will be detailed here. Wellburn and Wellburn[15] and Riley et al.[27] published column fractionation methods in which no centrifugation steps are required although a low speed centrifugation step was included by the Wellburns. In addition, the important problem of bacterial contamination can be alleviated since the bacteria also are separated in the fractionation.[15]

Mixtures of plant organelles can be fractionated during their differential rates of percolation through the spaces between beads in a column. The fractionation is effected probably by size discrimination. This column sieving is more efficient than filtration through a thin layer of support medium

[26] P. S. Nobel, *Biochim. Biophys. Acta* **189**, 452 (1969).
[27] V. T. Riley, M. L. Hesselbach, S. Fiala, M. W. Woods, and D. Burk, *Science* **109**, 361 (1949).
[28] S. I. Honda, unpublished experiments, 1964–1971.

TABLE I

ELUTION VOLUMES REQUIRED FOR RECOVERY OF CHLOROPLASTS AND
MITOCHONDRIA WITH DIFFERENT SIZES OF SIEVING BEADS

	Column sieving system Sephadex beads			
	G-10	G-25	G-50	G-50
Size range of beads	40–110	61–80	100–300	100–300
Void volume of system	70 ml	80 ml	50 ml	80 ml
Elution volume:				
Mitochondria	$1.1\ V_0$	$1.0\ V_0$	$1.0\ V_0$	$1.0\ V_0$
Chloroplasts	$1.7\ V_0$	$1.9\ V_0$	$1.5\ V_0$	$1.5\ V_0$

with holes because filters become clogged rapidly. The column sieving system should consist of uniformly sized spherical beads so that the pore spaces are not filled with beads of smaller size or irregular shape with the result of a slowed rate of percolation and fractionation of cell extracts. The sieving is not gel filtration.

In practice, it appears that a mixture of bead sizes is tolerable provided that the smallest bead present is sufficiently large. Discrimination is probably decreased with excessive pore size. For example, with Sephadex beads of different size ranges, the mitochondria percolate through the columns after an effluent volume of $1.0 \times V_0$ ml (Table I) (V_0 is the void volume or interparticulate space of the system). That is, particles much smaller than the pore size can pass through with the liquid of the sample applied to the column without restriction of percolation. On the other hand, chloroplasts being larger are more affected by pore size so that as the system pore size is increased by using larger beads, a smaller effluent volume is required for their elution, and, therefore, a smaller separation from the mitochondria is effected (Table I).

Assuming cubic-closest-packing of the beads, a suitable bead size can be calculated as about 50–100 μm in diameter in consideration of the pore size required to pass the largest particle in the cell extract—the nuclei. Beads can be selected by sieving through stainless steel sieves or by a backwash technique of particle flotation by a constant-rate flow of water at constant temperature.[29]

The composition of the bead material is important with respect to nonpolar and polar groups of the large bead-surface area interacting with the organelles. For example, the column fractionation of organelles reported by Riley et al.[27] undoubtedly was achieved in large part by differential adsorp-

[29] P. B. Hamilton, *Anal. Chem.* 30, 914 (1958).

tion since Celite was the support medium. Fractionation reported here is probably affected only to a small extent by adsorption, except for glass beads, because the first fractions exit just after the first void volume of effluent with a very symmetrical elution pattern without tailing. Wellburn and Wellburn[15] used coarse Sephadex G-50 for fractionation. We have tried glass, polystyrene, and Sephadex G-10 and G-200 beads and, recently, Sephadex G-25 and G-50 beads upon personal communication with the Wellburns.

In our case, beads considerably larger than those calculated as suitable were used because the column pores tend to be clogged with aggregated chloroplasts. When Sephadex was used, partial closing of the pores resulted in part from the beads becoming deformed under pressure produced by a pumping system used for elution. Therefore, downward flow with slight pressure head was used for elution.

Beads were selected for size, packed with Tricine buffer into a column of 25 × 400 mm, and washed with about 4 void volumes each of water, 1 M NaCl, 1 M urea, 50 mM Tricine buffer, pH 8.5, then with cold complex isolation medium. The void volume of ca. 150 ml total bed volume was determined with Blue Dextran (1,000,000 MW, Pharmacia Fine Chemicals) in cold complex isolation medium. If Sephadex G-50, coarse beads, was used, the void volume was near 50 ml. A spinach extract was prepared by the cutting method with complex medium containing Ficoll, Dextran-40, albumin, etc. Five milliliters were introduced at the top of the column with a sample applicator (Pharmacia Fine Chemicals)—its screen serving to filter out debris. No preliminary centrifugation was used. Elution with cell extraction medium commenced at 1 drop per 10 seconds. The effluent was collected in 5-ml fractions as determined by a drop counter. The hydrostatic pressure head for elution was 21 inches of cell extraction medium as maintained with a Mariotte tube. Recognition of fractions is easy by visual inspection, but for quantitative results, the optical density of fractions was monitored at 520 nm (particle refraction) and 260 nm (protein absorption). Light absorption at 260 nm is the more sensitive indicator. Microscope observations showed that the fractions designated as mitochondria or chloroplasts contained these organelles in preponderance, but they were contaminated with small numbers of other organelles including small nuclei and also with a few bacteria. Fractions with fragments of chloroplasts, nuclei, cell walls, etc., appeared far after the fractions of intact chloroplasts. It is likely that aggregations of chloroplasts, etc., were becoming disorganized and fragments were being washed through trapped material at the top of the column.

Although it is not possible to make a close comparison with the results of the Wellburns, a rough one can be made. If the Wellburn first fraction (peroxisomal ?) appeared just after the first void volume of effluent, then the mitochondria appeared in the effluent at ca. $2.3 \times V_0$ ml as contrasted with $1.0 \times V_0$ ml found here (Table I). However, since the Wellburns applied an extract representing about 20–25 times, the amount of cells extracted to a column of about twice the column bed volume used here, it is possible that their system was partially clogged—especially since eluate was pumped through their Sephadex column. They reported a very good recovery of 31% of the etioplasts applied to the column.[15]

If the principle of extracting and fractionating intact organelles is followed, then the centrifugation methods of rate-zonal and isopycnic fractionation will yield fractions of mixed organelles because, in general, the

TABLE II
Buoyant Densities of Organelles in Different Conditions

Organelle	Principal osmoticum	Cell extraction method	Organelle condition[a]	Organelle density (g/ml)	Data source
Glyoxysome	Sucrose	Blending	Intact?	1.25	b
Peroxisome	Sucrose	Blending	Leaky	1.25–1.26	c
	Ficoll	Blending	Nonleaky	1.18–1.23	c
Microbody	Sucrose	Grinding	?	1.20–1.25	d
Etioplast	Sucrose	Blending	Unbroken	1.24	e
	Sucrose	Grinding	Intact	1.22	f
Chloroplast	Sucrose	Blending	Unbroken	1.23	g
	Sucrose	Grinding	Unbroken	1.19	h
	Ficoll	Grinding	Broken	1.18–1.20	h
	Sucrose	Grinding	Broken	1.17	d
	Sucrose	Grinding	Broken	1.16	h
Mitochondria	Ficoll	Blending	Uncontrolled?	1.21–1.23	h
	Sucrose	Blending	Uncontrolled?	1.20	g
	Sucrose	Grinding	?	1.17–1.20	d
	Sucrose	Grinding	?	1.19	b
	Sucrose	Grinding	Controlled	1.16	i

[a] Condition followed by a question mark indicates an inferred condition from characteristics described by the sources upon comparison with other results.
[b] R. W. Breidenbach and H. Beevers, *Biochem. Biophys. Res. Commun.* **27**, 462 (1967).
[c] N. E. Tolbert, A. Oeser, R. K. Yamazaki, R. H. Hageman, and T. Kisaki, *Plant Physiol.* **44**, 135 (1969).
[d] A. H. C. Huang and H. Beevers, *Plant Physiol.* **48**, 637 (1971).
[e] N. K. Boardman and S. G. Wildman, *Biochim. Biophys. Acta* **59**, 222 (1962).
[f] A. R. Wellburn and F. A. M. Wellburn, *J. Exp. Bot.* **22**, 972 (1971).
[g] R. M. Leech, *Biochim. Biophys. Acta* **79**, 637 (1964).
[h] N. E. Tolbert, this series, Vol. 23, p. 665 (1971).
[i] J. E. Baker, L.-G. Elfvin, J. B. Biale, and S. I. Honda, *Plant Physiol.* **43**, 2001 (1968).

already close buoyant densities of the organelles are even closer than when the organelles are disorganized (Table II)—1.22–1.25 g/ml. The peroxisomes upon disorganization float as lipoprotein, and their buoyant density (1.22–1.25 g/ml) is indistinguishable from that of the glyoxysomes and etioplasts, whereas if intact the peroxisomes (1.18–1.23 g/ml) cannot be separated by density difference from mitochondria and chloroplasts (1.16–1.23 g/ml).

In principle, since the gentle column sieving fractionation method easily separates etioplasts from mitochondria and mitochondria from chloroplasts, it should be possible to purify all three from each other from a single extract such as might be obtained from etiolated tissue exposed to light. Whether peroxisomes and glyoxysomes could also be separated in relatively pure state is unknown. If an appropriate medium for upward flow elution of varying density were also used, then this density gradient feature of rate-zonal and isopycnic fractionations possibly could help achieve column sieving fractionation of the microbodies, etioplasts, chloroplasts, and mitochondria. While the sensitivity of etioplasts may be exceptional,[15] a clear case of centrifugation damage has been established for chloroplasts.[9,14] Therefore, since the period of concentration upon purely biochemical properties is passing and emphasis is shifting to physiology, the gentle column sieving fractionation method is presented for consideration. Besides gentleness and ability to fractionate, it has the added advantage of being able to cope with bacterial contaminations—the seriousness of which cannot be overstated for biochemical studies (cf. Jacobson[30]). If Sephadex beads are used, then the molecular sieve or gel filtration effect may also be of advantage in that low molecular weight substances—including noxious agents such as tannins released from vacuoles upon tissue extraction—will immediately enter the bead-gel interior and their influence upon the organelles removed or diminished.

[30] A. B. Jacobson, *J. Cell Biol.* 38, 238 (1968).

[56] Plant Cell Transformation with Bacteria

By CARL V. LUNDEEN

Crown gall is a neoplastic disease of plants which is characterized by the formation of a solid tumor at the site of infection. Such tumors are composed of rapidly dividing cells, are transplantable, and nonself-limiting; i.e., cellular proliferation continues as long as the plant survives. They represent a botanical counterpart of true tumors of animals.

The crown gall system presents an excellent model for the study of abnormal cell physiology and development. Tumors are readily initiated and generally may be brought into tissue culture without difficulty. The study and maintenance of these plant tissues are not complicated by a requirement for complex and largely undefined culture media or by rather sophisticated facilities often needed for animal cell culture.

The pathogen responsible for the transformation of a normal plant cell into a tumor cell is a gram-negative, rod-shaped, motile soil bacterium, *Agrobacterium tumefaciens,* that is closely related to *A. radiobacter.* The mechanism by which this organism brings about the transformation is not clear, but the bacterium is thought to elaborate an as yet uncharacterized tumor-inducing principle (TiP) which causes a heritable cellular change leading to the tumorous state. Once the transformation is accomplished, the presence of the bacterium is not necessary for the continued abnormal proliferation of the tumor cells, either in culture or in a host.[1]

The Bacterium

All the common strains of *A. tumefaciens* grow well on Difco dextrose agar slants. They are aerobic and have a growth optimum at about 28°. Frequency of transfer may be reduced by storage of cultures at 4° after 24 hours of growth at room temperature. Certain strains of the bacteria have remained fully virulent after having been carried in culture continuously for more than 40 years.

A virulent strain[2] of *A. tumefaciens,* e.g., A6 or B6, will produce a rapidly growing, unorganized tumor on many dicotyledonous plant species as well as on certain gymnosperms. Other strains are more limited in their host range and may be virulent for one or a few plant species but innocuous for others. Strain differences also may lead to the formation of tumors with varying degrees of organization. Strain T37, for example, will produce a typical unorganized tumor on the Madagascar periwinkle or on a sunflower plant but produces a teratoma when inoculated on the cut stem surface of tobacco.[3] Different strains inoculated into the same host may also produce tumors of varying size and rate of growth.

Transformation Technique

Several factors are necessary for the experimental production of crown gall tumors: a virulent strain of bacteria, a susceptible dicotyledonous plant suitably injured to initiate the wound-healing response, and a relative

---

[1] P. R. White and A. C. Braun, *Cancer Res.* **2**, 597 (1942).

[2] Virulent strains of *Agrobacterium tumefaciens* are available from the American Type Culture Collection, 12301 Parklawn Drive, Rockville, Maryland 20852.

[3] A. C. Braun, *Bot. Gaz.* (*Chicago*) **114**, 363 (1953).

humidity of 70% or more. The precise method of combining these factors is largely at the discretion of the investigator, since the bacteria are highly efficient in producing tumors. The wounding should not be of such magnitude that it would cause death of or serious injury to the host. The temperature at which the bacterial–host system is maintained during the transformation period should not exceed 27° (see below).

Inoculations may be made directly from a 24–48-hour agar slant culture or from liquid suspensions of the pathogen. Two or three inoculating needle loopfuls of bacteria suspended in 1 ml of sterile tap water or nutrient broth will produce an adequate inoculum. Host plants should be healthy and vigorous. Vegetative plants commonly give rise to larger tumors than do plants in flower.

Inoculation Methods

Stem Puncture. With a virulent strain of pathogen and a suitable host, this technique will produce tumors approaching 100% frequency. An ordinary sterile dissecting needle is used to introduce a drop of suspension or bacteria from an agar slant into the plant. The stem is pierced at one or more internodes and the needle moved in an out two or three times. The needle is then removed, fresh bacteria are picked up, and the process is repeated entering the exit hole of the puncture.

Cut Stem Inoculation. The stem is cut through with a scalpel to produce a flat surface. Bacteria from a slant or suspension are smeared on the surface with a sterile dissecting needle and the surface is covered, e.g., with sterilastic tape, to prevent excessive drying. After about 6 days the tape is removed.

Leaf Inoculation.[4] A suspension of bacteria is rubbed onto the leaf so as to produce abrasion injury. Primary leaves are lightly dusted with No. 400 grit Carborundum, a few drops of the bacterial suspension are placed on them, and the leaves are rubbed gently but firmly with a glass rod. After the leaves have dried, excess Carborundum is washed from them with tap water. Tumors can usually be observed within 1–2 weeks after inoculation.

Carrot Disk Inoculation. Klein and Tenenbaum described a method by which bacterial transformation may be observed *in vitro.*[5] Fresh carrots were obtained at local markets. Three to 4 cm of the top of the root were removed and discarded. Sterilization was effected by removing the periderm of a 2-cm slab with a potato peeler and wrapping the root segment in paper toweling saturated with a commercial preparation of 5.25% sodium hypochlorite (Clorox) for 5–10 minutes. Radial cylinders were removed from

[4] J. A. Lippincott and G. T. Heberlein, *Amer. J. Bot.* **52**, 396 (1965).
[5] R. M. Klein and I. L. Tenenbaum, *Amer. J. Bot.* **42**, 709 (1955).

the segment with a sterile cork borer 13 mm in diameter. Disks cut 2.5 cm or thicker from the secondary phloem adjacent to the cambial layer were immediately placed, with that surface up, on moistened filter paper in sterile petri plates and inoculated with 0.02 ml of bacterial suspension. Macroscopically visible tumor were found after 6–7 days of incubation at 26°.

Stopping Transformation

By interrupting the transformation process, tumors produced on the same host by the same strain of bacteria may be experimentally varied. In the crown gall disease the transformation is a gradual process, extending over a period of 3–4 days. The process is heat-sensitive and may be interrupted by raising the temperature to 32° and holding it there until the wound-healing cycle is complete, after which the cells are no longer susceptible to transformation. The thermal treatment halts only the transformation process; plant cells already transformed are not affected. The growth rate of subsequently developing tumors directly reflects the length of time that transformation is allowed to proceed before interruption. Thus, in the Madagascar periwinkle, transformation interrupted after 34–36 hours gives rise to slowly growing tumors. Interruption after 50 and 60 hours produces moderately fast growing tumors, while cells transformed during a 3–4-day period grow very rapidly. The varying degrees of autonomy of these tissues in culture have been discussed by Braun[6] in terms of biochemical functionalization.

Culture of Crown Gall Tumor Tissue

There are three factors to consider in culturing crown gall tissue from the plant tumor. The tumor must be freed of the pathogen, the surface of the tumor sterilized, and an appropriate culture medium found.

Antibiotic or thermal treatment may be used to destroy the bacterium. In those systems not already examined, the precise method must be determined empirically. Species of plants vary widely in their ability to withstand elevated temperatures and different strains of bacteria vary in their susceptibility to an antibiotic.

In *Vinca rosea* L., the Madagascar periwinkle, a thermal treatment of 46° for 3–5 days will selectively destroy the bacteria without affecting subsequent tumor development.[7] Humidity of 70% or higher should be maintained during the thermal treatment. After sterilization of the tumor surface, tissue excised from the interior may be brought into culture.

[6] A. C. Braun, *Proc. Nat. Acad. Sci. U.S.* **44**, 344 (1958), and references therein.
[7] A. C. Braun, *Amer. J. Bot.* **30**, 674 (1943).

Surface sterilization of stem tumors can be accomplished by cleaning the tumor-bearing stem segment in running tap water, washing it thoroughly with detergent, e.g., a 5% solution of Alconox or 7-X, rinsing it with tap water, and immersing the entire stem segment in 5–10% Clorox for 2–5 minutes. After it has been rinsed several times with sterile water, the stem is dipped into 95% ethanol and placed in a sterile petri plate. The stem is then taken with sterile forceps, flamed briefly to remove excess ethanol, and the tumor is cut off with a sterile scalpel and transferred to another petri plate. With sterile instruments the outer surface of the tumor is removed and pieces of the interior are excised and placed on an appropriate medium.

Manasse and Lipetz[8] reported that the bacterium can be destroyed in 50–75% of *Vinca rosea* tumors by placing 6–7-cm surface-sterilized stem segments bearing one 14-day-old tumor each into 18 × 150-mm tubes containing 15 ml of modified White's 10× agar. The stem is placed apical end up and care is taken that the tumor be above the medium surface. The system is incubated for 7 days at 41° and then brought to 25°. Tumor proliferation was observed in 2–3 weeks. Braun (unpublished results[9]) has found that certain tissue fragments excised from the interior of surface-sterilized tobacco tumors and incubated at 39–40° in liquid White's basic medium for a period of 5–6 days will prove to be bacteria free.

Streptomycin at concentration of 20–80 ppm[10] has been used to free hollyhock tumors from *A. tumefaciens*. The excised tissue from a surface-sterilized tumor is rinsed with sterile streptomycin solution and then cultured on streptomycin-containing medium until it is bacteria free. Susceptibility to streptomycin must be determined for the individual strain of the pathogen.

Culture Media

As a result of the transformation process, the plant tumor cell becomes capable of synthesizing many of the exogenous requirements of normal cells. Crown gall tumor cells can grow profusely on a simple, chemically defined culture medium that does not support growth of normal cells. Exogenous hormones are not required and may inhibit growth. The recipe for White's basic medium, a commonly used medium, is given below.[11]

[8] R. J. Manasse and J. Lipetz, *Can. J. Bot.* 49, 1255 (1971).

[9] A. C. Braun, unpublished results, 1950.

[10] V. N. Gadgil, S. K. Roy, and D. R. Das, *in* "Plant Tissue and Organ Culture—A Symposium" (P. Maheshwari and N. S. Ranga Swamy, eds.), p. 111. Catholic Press, Ranchi, India, 1963.

[11] P. R. White, "A Handbook of Plant Tissue Culture," p. 103. Jacques Cattell Press, Lancaster, Pennsylvania, 1943.

WHITE'S BASIC MEDIUM

Component	Mg/liter
$MgSO_4$	360.0
$Ca(NO_3)_2$	200.0
Na_2SO_4	200.0
KNO_3	80.0
KCl	65.0
$NaH_2PO_4 \cdot H_2O$	16.5
$Fe_2(SO_4)_3$	2.5
$MnSO_4$	4.5
$ZnSO_4$	1.5
H_3BO_3	1.5
KI	0.75
Sucrose	20,000.0
Glycine	3.0
Nicotinic acid	0.5
Pyridoxine	0.1
Thiamine	0.1
Agar	10,000.0

Culture Maintenance

Plant tissues in culture on solid media are maintained by fragmentation. The tissue is transferred to a sterile petri plate, pieces of size 100–300 mg are cut from it and transferred to fresh medium. Areas of tissue which appear to be necrosed are removed and discarded.

[57] Isolation of Plant Nuclei

By JOSEPH P. MASCARENHAS, MYRA BERMAN-KURTZ, and ROBERT R. KULIKOWSKI

The isolation of nuclei from plant tissues poses many more technical problems than are encountered in the isolation of animal nuclei. The major problems are caused by the difficulty of breaking plant cells because of the presence of cell walls. Routine techniques used to break animal cells such as Dounce homogenization, grinding in a Waring Blendor or Sorvall Omni-Mix, sonication, mechanical homogenization, etc., have to be much more vigorous or of longer duration to break open cell walls. Since liberated nuclei are much more fragile than cell walls, a large percentage of them are partially or completely destroyed. The cell walls also trap many nuclei resulting in poor yields. Once the cells have been broken and nuclei released, isolation of plant nuclei is further complicated by the presence of chloroplasts, plastids, and starch grains in the homogenate. These cellular components have sedimentation properties similar to those of nuclei and are thus difficult to eliminate by centrifugation.

Depending on the purpose for the isolation of nuclei there is a choice of several methods of rupturing the cells. If one wishes to study nuclear processes in which transient intermediates are important, such as nuclear RNA synthesis, one would select an isolation method that was rapid even though lower yields of nuclei would be obtained.

There is no entirely satisfactory method yet devised for the isolation of plant cell nuclei, and work on the transient metabolism of the plant nucleus similar to that done with animal nuclei might have to wait until a cell wall-less mutant can be obtained and cultured in tissue culture. It would then be possible to use the techniques developed for animal cells and to isolate nuclei very rapidly and with high yields.

Methods for isolation of nuclei from whole plant tissues, tissue culture cells, plant protoplasts, and germinating pollen grains are described.

Reagents and Their Use

> Isolation buffer. The basic buffer solution contains 10 mM Tris$\cdot$ HCl (pH 7.5), 10 mM NaCl, 1.5 mM MgCl$_2$, 3 mM CaCl$_2$. A divalent cation, either Ca^{2+} or Mg^{2+} or both, is required to maintain the structure of isolated nuclei and to prevent their clumping together during the isolation procedure. The various sucrose solutions are made up as weight/weight solutions in the isolation buffer, e.g., 10% sucrose is 10 g of sucrose in 90 ml of isolation buffer.

> Detergent mixtures. (a) Deoxycholate–Tween 40. A stock solution consists of one part of 10% (w/w) solution of sodium deoxycholate (Mann Research Labs, New York, New York) and 2 parts of 10% (w/w) Tween 40 (Atlas Chemical Industries, Inc., Wilmington, Delaware). It is stored in the freezer at $-20°$. One milliliter of nuclear suspension is treated for 1–3 minutes at 4° with 150 μl of stock solution. (b) Triton X-100 (Rohm and Haas, Philadelphia, Pennsylvania). A 10% (w/w) stock solution stored at $-20°$ is used at 0.2 ml per milliliter of nuclear suspension. The treatment is carried out at 4° for 5–15 minutes, depending on the nuclear preparation.

Both detergent treatments solubilize contaminating cytoplasm, disrupt chloroplasts and mitochondria, and remove the outer nuclear membrane with its contaminating cytoplasm.[1,2] The effectiveness and the time of the detergent treatment is dependent on the concentration of nuclei, organelles, and cytoplasm in the nuclear suspension. It might be necessary to vary both the concentration of detergent and the time of treatment for different tissues to obtain optimal results. On occasion Triton X-100 does not seem to be effective in solubilizing cytoplasm. The exact reason for this is not known, but it appears to be a function of the joint Ca^{2+} and Mg^{2+} concentrations of the nuclear suspension.

STAINING REAGENTS FOR NUCLEI. Most plant nuclei will stain with one or both of the following solutions.

Methyl green[3]: 0.04 g of methyl green, 6 ml of glacial acetic acid, 94 ml of H_2O, and 0.2 ml of a 1 M $CaCl_2$ solution. The methyl green is dissolved in a small quantity of water before adding the acetic acid.

Acetocarmine[4]: 0.5 g of carmine is boiled in 100 ml of 45% acetic acid for 2–4 minutes, cooled, and filtered. Adding a piece of iron such as an old razor blade to this solution for about an hour makes the solution more effective. The solution may be used full strength or diluted 1:2 with 45% acetic acid, depending on the tissue. The preparation has to be gently heated for the stain to react.

CENTRIFUGATION. All low speed centrifugations have been carried out in an IEC clinical centrifuge in a cold room at 4°, and calibrated with a General Radio Strobotac.

Isolation of Nuclei

Isolation from Whole Plant Tissue

Several different procedures have been used in the liberation of nuclei from plant tissue. Some of these include grinding of fresh or frozen tissue with a mortar and pestle, disruption by various types of blendors, French pressure cell, and ultrasound. From our experience we have found that rupturing plant cells using a "pea popper" originally described by Rho and Chipchase[5] gives the most satisfactory results. We use a simplified

[1] S. Penman, I. Smith, E. Holtzman, and H. Greenberg, *Nat. Cancer Inst. Monogr.* 23, 489 (1966).
[2] G. D'Alessio and A. R. Trim, *J. Exp. Bot.* 19, 831 (1968).
[3] L. Kuehl, *Z. Naturforsch. B* 19, 525 (1964).
[4] G. L. Humason, "Animal Tissue Techniques," p. 367. Freeman, San Francisco, California, 1962.
[5] J. H. Rho and M. I. Chipchase, *J. Cell Biol.* 14, 183 (1962).

FIG. 1. Picture of the simplified nuclear "popper" used for rupturing cells. *Inset:* Close up view of the spring loaded lower roller mechanism and screw for adjusting the spacing between the rollers. See text for other details.

manually operated "popper" which can be constructed at low cost. The machine consists of two stainless steel or aluminum counter rotating tissue squeezing rollers (Fig. 1). The rollers are 26 cm long and 2.5 cm in diameter. The lower roller is spring loaded, and the distance between the rollers can be adjusted by two screws built into the housing (see inset to Fig. 1). The normal operating clearance between the rollers is 0.26 mm. The rollers are driven manually by a handle attached to the upper roller. A piece of stainless steel sheet bent down its center is fixed under the rollers to catch the squeezed juice from the tissue and channel it into a beaker. The entire extraction procedure is carried out in a cold room at 4°. Before extraction, the plants, in this case 5-day-old dark grown corn seedlings, are chilled in the cold room for about 30 minutes. The prechilling increases the yield of nuclei and also keeps enzymatic degradation to a minimum. The above-ground shoots (50 g) are collected and chopped into 3–5 mm lengths with a razor blade and spread evenly on a 45-cm length of nylon marquisette cloth (obtainable from most fabric stores). The tissue is spread over the center of the cloth and the sides of the cloth are then folded over the tissue. The cloth with the tissue sandwiched in between is then pulled back and forth through the squeezing rollers several times. Two pairs of hands make this process easier, one to guide the cloth and the other to turn the rollers. The squeezed juice containing the nuclei is collected in a beaker containing a volume equal to that of the expected squeezed juice, of a 10% sucrose solution in isolation buffer. Approximately 20 ml of squeezed juice is obtained from 50 g of corn seedling shoots. A few milliliters of the 10% sucrose buffer are added to the cloth containing the tissue residue to wash out trapped nuclei, and it is then pulled through the rollers once more. The time required for chopping 50 g of tissue and passing it through the rollers is less than 10 minutes. About 40% of the total tissue DNA is recovered in the expressed plant juice. The liberated nuclei in the extract are collected by centrifugation at 150 g for 15 minutes. The pellet at the bottom of the tube consists of starch grains, nuclei, and some pieces of tissue. The pellet is resuspended in about 20 ml of 10% sucrose-buffer, filtered through a Rigimesh stainless steel grade M filter (Pall Trinity Micro Corp., Cortland, New York) to remove the large debris, and centrifuged at 150 g for 10 minutes. The nuclear pellet freed of small starch grains is resuspended in about 0.5 ml of the 10% sucrose-buffer solution, treated with detergent to remove cytoplasmic contamination and to solubilize plastids and mitochondria and layered on 2.5 ml of a 45% sucrose solution and centrifuged at 350 g for 20 minutes. The nuclei are recovered in the pellet. The nuclei are freed of cytoplasm, but the preparation still contains some starch grains. For most practical purposes, however, the starch grains do not interfere with the nuclear processes that will then be studied. The two different detergent treatments described under Reagents, have been used successfully. About 10% of the total DNA in the initial plant material is recovered in the final nuclear pellet.

Procedures have been published in which plant tissue is infiltrated for several hours (12–20) in isolation medium containing gum arabic, prior to high speed blending.[3,6] This treatment results in larger yields, of nuclei (about 30%), but the unknown effects on nuclear metabolism of such long incubations in gum arabic at low temperatures limit the use of the method at the present time. D'Alessio and Trim[2] have described what appears to be a good method for the isolation of nuclei from leaves. Cell-wall digesting enzymes are introduced into the leaves by vacuum infiltration. The walls are partly digested by incubating the mixture at 0° for 30 minutes to 2 hours. Since the walls are weakened by this treatment homogenizing the tissue in an M.S.E. homogenizer results in the release of a large number of undamaged nuclei. Prolonged preincubation of the tissue in gum arabic or enzyme solutions does not seem to affect the ability of the isolated nuclei to synthesize RNA in a cell free system.[2,3,6]

[6] K. J. Tautvydas, *Plant Physiol.* **47**, 499 (1971).

Isolation from Plant Cells in Tissue Culture

Cells of *Centaurea* grown in suspension culture are pelleted (20 g) and the nuclei liberated by passing them between the rollers (adjusted so that there is no space in between) of the "popper," sandwiched between No. 25 Nitex nylon cloth (Tobler, Ernst and Traber, Inc., Elmsford, New York). Cloth with larger mesh openings allows many whole cells to pass through. A narrower piece of nylon cloth than that described for isolation of nuclei from corn seedlings is used. The size of the nylon cloth used depends on the quantity of plant tissue to be extracted. The cell extract enriched in nuclei is collected in the same manner as described for corn seedlings, and large cell debris removed by filtering through a Rigi-mesh stainless steel grade M filter. The nuclei are pelleted at 380 *g* for 7 minutes and resuspended in 0.6 ml of 12% sucrose. The nuclear suspension is treated with either detergent mixture as described and layered on 3 ml of 45% sucrose and centrifuged at 72 *g* for 20 minutes. The nuclei are in the pellet. About 12% of the total initial DNA is recovered in the nuclear pellet. Although this procedure has been worked out for *Centaurea* cells, it should be useful with possibly minor variations for isolating nuclei from other cell cultures, both callus and suspension.

Isolation from Protoplasts

Protoplasts from plant cells can be produced by several different procedures, all of which utilize preparations of cell wall degrading enzymes.[7-9] We have used the method of Bawa and Torrey[9] to prepare protoplasts from *Centaurea* suspension culture cells. The procedure requires an overnight incubation at room temperature in the enzyme mixture and thus is not a suitable procedure for certain types of work with isolated nuclei. These protoplasts and most likely protoplasts from other cells from different plants or plant parts can then be treated essentially as animal cells for isolation of their nuclei.[1] The protoplasts (1 ml packed volume) are resuspended in 5 ml of cold (0 to 4°) hypotonic isolation buffer minus Ca^{2+} and sucrose and transferred to a Dounce homogenizer. The suspension is warmed rapidly to room temperature for 1 minute and then placed in ice and homogenized with two strokes of a tight-fitting pestle (size B). Warming to room temperature for a minute results in better rupturing of the cells. The nuclei are pelleted at 780 *g* for 2 minutes, resuspended in 5 ml hypotonic buffer with a Pasteur pipette, and repelleted. The nuclei are resuspended in buffer, treated with Triton X-100 (10%) at 0.2 ml per

[7] K. N. Kao, W. A. Keller, and R. A. Miller, *Exp. Cell Res.* **62**, 338 (1970).
[8] J. B. Power and E. C. Cocking, *J. Exp. Bot.* **21**, 64 (1970).
[9] S. B. Bawa and J. G. Torrey, *Bot. Gaz.* **132**, 240 (1971).

ml of nuclear suspension for 15 minutes at 4° and repelleted. Yields of 75% and higher have been obtained by this procedure.

Isolation from Pollen Tubes[10,11]

Pollen of *Tradescantia paludosa* is germinated for a few minutes in growth medium.[11] *Tradescantia* pollen tube nuclei are elongated structures and are easily damaged. Intact nuclei can be obtained only if the pelleted pollen tubes are gently ruptured by exposure to a hypotonic buffer. The fragile tips of the tubes burst easily and release the entire contents of the tube. It is thus easier to isolate nuclei from grains that have just begun to germinate than from ungerminated pollen grains. Pollen tubes from 50 mg of pollen are suspended in 12 ml of the isolation buffer (containing 4.5 mM MgCl$_2$ and no Ca^{2+}, and mixed at room temperature with a Pasteur pipette for 1 minute. The suspension is then put in ice and made 10% with sucrose. All subsequent steps are at 0–4°. Pollen grains and tubes are removed by filtering the suspension through a Rigimesh stainless steel grade M filter. The filtrate is then centrifuged at 2960 rpm for 7 minutes. The pellet which contains nuclei and cytoplasm is resuspended in 0.5 ml of 10% sucrose buffer. At this stage one recovers about 50% of the nuclei. If the two types of nuclei, vegetative and generative, are not to be separated from each other, the suspension is treated with Triton X-100 for 15 minutes in ice and layered over 60% sucrose; the nuclei are collected in the pellet after centrifugation in the SB283 rotor of the IEC B-60 centrifuge at 25,000 rpm for 30 minutes. If the vegetative nuclei are to be separated from the generative nuclei, the detergent-treated nuclear suspension is layered on a 22 to 35% linear sucrose gradient made in the buffer in a 4-ml Teflon tube. The gradient is centrifuged at 230 rpm for 8 minutes. Fractions are collected from the gradient either by displacing the gradient from the top with 60% sucrose, or pumping the solution out directly from the bottom of the tube. Aliquots from each fraction are stained with methyl green and observed under the microscope. The cytoplasm remains in the top one-fourth of the gradient. The vegetative nuclei in the next one-fourth and the generative nuclei in the lower half of the gradient. Fractions in between the vegetative and generative nuclei contain a mixture of the two nuclei and are discarded. Sometimes the vegetative nuclei are contaminated with cytoplasmic particles. These can be removed by layering the fraction on 1.5 ml of 60% sucrose in buffer in the SB283 rotor of the IEC B-60 and centrifuging at 25,000 rpm for 30 minutes. The nuclei are recovered in the pellet. Instead of the

[10] K. L. LaFountain and J. P. Mascarenhas, *Exp. Cell Res.* **73**, 233 (1972).
[11] J. P. Mascarenhas and E. Bell, *Biochim. Biophys. Acta* **179**, 199 (1969).

high speed centrifugation, nuclei can also be separated from contaminating cytoplasmic particles by layering them on 45% sucrose and centrifuging them at 350 g for 30 minutes in the clinical centrifuge. The yields of clean vegetative and generative nuclei range between 5 and 10%.

Acknowledgment

Supported in part by U.S. Public Health Service Training Grant GM 02014 and National Science Foundation Grant GU-3859.

[58] Glyoxysomes

By HARRY BEEVERS and R. W. BREIDENBACH

Glyoxysomes is the term introduced to describe organelles uniquely housing the enzymes of the glyoxylate cycle. They were isolated first from the endosperm of the germinating castor bean[1,2] and subsequently from tissues of many different seedlings,[3-5] where the dominant metabolic event is the conversion of fat to sucrose and acetyl-CoA is metabolized almost exclusively through the glyoxylate cycle.[6] From such materials, the glyoxysomes are recovered as a major discrete protein band of buoyant density 1.25 g/cc in sucrose gradients; they are spherical or oblate bodies 0.5–1.0 μm in diameter with finely granular matrix and bounded by a single membrane.[1,2] Most of the malate synthetase and isocitrate lyase from castor bean endosperm are recovered in this band, which is essentially free of mitochondria.[1,2] The other enzymes of the glyoxylate cycle, malate dehydrogenase, citrate synthetase (distinct isoenzymes) and a rather unstable aconitase are also present.[1,2,7] In addition the glyoxysomes are the major intracellular site of a β-oxidation system in which fatty acids are activated[8] and converted to acetyl-CoA.[9,10]

No succinoxidase or cytochromes are present in the glyoxysomes, but they do contain glycolate oxidase,[2] catalase,[2] and uricase.[3] These three

[1] R. W. Breidenbach and H. Beevers, *Biochem. Biophys. Res. Commun.* **27**, 462 (1967).
[2] R. W. Breidenbach, A. Kahn, and H. Beevers, *Plant Physiol.* **43**, 705 (1968).
[3] R. R. Theimer and H. Beevers, *Plant Physiol.* **47**, 246 (1971).
[4] C. P. Longo and G. P. Longo, *Plant Physiol.* **45**, 249 (1970).
[5] T. M. Ching, *Plant Physiol.* **46**, 475 (1970).
[6] D. T. Canvin and H. Beevers, *J. Biol. Chem.* **236**, 988 (1961).
[7] T. G. Cooper and H. Beevers, *J. Biol. Chem.* **244**, 3507 (1969).
[8] T. G. Cooper, *J. Biol. Chem.* **246**, 3451 (1971).
[9] T. G. Cooper and H. Beevers, *J. Biol. Chem.* **244**, 3514 (1969).
[10] D. Hutton and P. K. Stumpf, *Plant Physiol.* **44**, 508 (1969).

enzymes are distinctive components of microbodies from animal tissues which have been intensively studied and named peroxisomes by de Duve and his collaborators.[11-13] It is now clear that organelles with similar structure, buoyant density, and enzyme constituents are widespread in plant tissues, and the general term microbody is used to describe them.[14,15] Within this group several categories can be recognized. From many plant tissues, microbodies with apparently limited enzyme constitution have been isolated.[16] From green leaves an important class of microbodies, the leaf peroxisomes, has been isolated[17-19]; these organelles contain additional enzymes which fit them for a key role in the metabolism of glycolate produced during photosynthesis.[17] The glyoxysomes represent another class of microbodies that are highly specialized and whose metabolic role is the oxidation of fatty acids to acetyl-CoA and its conversion to succinate in the glyoxylate cycle. Distinct organelles with at least some of the enzymes of the glyoxylate cycle have been isolated from microorganisms, including *Tetrahymena*,[20,21] *Euglena*,[22,23] yeast,[24] and *Neurospora*.[25]

Preparation of Glyoxysomes from Castor Bean Endosperm

Glyoxysomes are very susceptible to mechanical damage, osmotic shock, and acid conditions. The general procedure for isolation is (a) gentle disintegration of the tissue in a suitably buffered sucrose solution, (b) recovery of a crude particulate fraction by differential centrifugation, and (c) separation from other organelles by applying the resuspended particulate fraction to a linear sucrose gradient and centrifuging to equilibrium.

[11] C. de Duve and P. Baudhuin, *Physiol. Rev.* **46**, 323 (1966).
[12] C. de Duve, *Ann. N.Y. Acad. Sci.* **168**, 369 (1969).
[13] C. de Duve, *Proc. Roy. Soc. Ser. B* **173**, 71 (1969).
[14] H. H. Mollenhauer, D. J. Morré, and A. G. Kelley, *Protoplasma* **62**, 44 (1966).
[15] H. Beevers, *in* "Photosynthesis and Photorespiration" (M. D. Hatch, C. D. Osmond, and R. O. Slatyer, eds.). Wiley Interscience (1971).
[16] A. H. C. Huang and H. Beevers, *Plant Physiol.* **48**, 637 (1971).
[17] N. E. Tolbert, *Annu. Rev. Plant Physiol.* **22**, 45 (1971).
[18] N. E. Tolbert, this series, Vol. 23, p. 665.
[19] N. E. Tolbert and R. K. Yamazaki, *Ann. N.Y. Acad. Sci.* **168**, 325 (1969).
[20] M. Muller, *Ann. N.Y. Acad. Sci.* **168**, 292 (1969).
[21] J. F. Hogg, *Ann. N.Y. Acad. Sci.* **168**, 281 (1969).
[22] L. B. Graves, L. Hanzely, and R. Trelease, *Protoplasma* **72**, 141 (1971).
[23] L. B. Graves, R. N. Trelease, and W. M. Becker, *Biochem. Biophys. Res. Commun.* **44**, 280 (1971).
[24] A. S. Szabo and C. J. Avers, *Ann. N.Y. Acad. Sci.* **168**, 302 (1969).
[25] M. Kobr, F. Vanderhaeghe, and G. Combepine, *Biochem. Biophys. Res. Commun.* **37**, 640 (1969).

Plant Materials

Castor bean seeds (Baker Castor Oil Company, Plainview, Texas) of several varieties (B-296, Hale, Cimmaron) have been used (depending on availability); there is some variation among seed lots. Maximum yields of glyoxysomes, 0.5–1.0 mg protein per gram fresh weight of tissue, are obtained from the endosperm of seedlings germinated for 4–5 days at 30°. The seeds are first surface-sterilized by immersion in 0.15% hypochlorite for 15 minutes, and then allowed to soak in water for 3–10 hours. They are germinated in moist vermiculite in a dark humidified incubator. After 4–5 days (when the main roots are 6–10 cm long), the endosperm tissue is removed and washed briefly in water.

Grinding

All operations are carried out at 0–4°. Forty grams of tissue is placed in a cold household onion chopper containing 80 ml of ice cold grinding medium of the following composition: 0.5 M sucrose; 0.15 M Tricine buffer pH 7.5; 10 mM KCl; 1 mM MgCl$_2$; 1 mM EDTA; and 10 mM dithiothreitol. The tissue is chopped into tiny pieces (1–2 mm^3), and the slurry is transferred to a prechilled mortar, where it is further ground with a pestle for 1–2 minutes. The brei is filtered through several layers of cheesecloth and squeezed gently. About one-fourth of the original weight of endosperm is not extracted under these conditions and is retained on the cloth. The creamy filtrate is the crude homogenate from 30 g of endosperm tissue. Extracts prepared by mechanical blending show more extensive breakage of organelles.

Isolation of Crude Particulate Fraction

The homogenate is centrifuged for 10 minutes at 270 g, and the pellet and fatty layer are discarded. The remainder, supernatant I, is centrifuged at 10,000 g for 30 minutes to give the crude particulate pellet (approximately 200 mg of protein) and supernatant II, overlain by a fatty layer. Of the total isocitrate lyase (glyoxysome marker) in supernatant I, 50–70% is recovered in the pellet, depending on the degree of breakage of the organelles during grinding. The pellet is then gently resuspended in grinding medium in which the sucrose concentration is increased to approximately 1.0 M. This is done by adding 2–3 ml of medium, breaking up the pellet with a rubber policeman and drawing the mixture repeatedly into a 2-ml pipette and allowing to drain. Portions of the suspension are then layered onto previously prepared sucrose gradients.

Sucrose Gradient Centrifugation

Linear sucrose gradients are constructed, ranging from 30% (w/w) to 60% sucrose in 1 mM EDTA, pH 7.5. Addition of other salts or buffers usually leads to loss of definition of bands. We place a 50-ml gradient in 60-ml cellulose nitrate tubes for centrifuging in a swinging-bucket rotor (Spinco 25.2) and apply a portion of the resuspended crude particulate fraction containing 30 mg protein. The glyoxysomes have reached equilibrium after centrifuging for 3–5 hours at 65,000 g_{av} at 2°. Two major protein bands are visible at this time, the uppermost, at mean density 1.19 g/cc, is a purified mitochondrial fraction, and the lowermost, at mean density 1.25 g/cc, is the glyoxysome fraction. A minor intermediate band at roughly 1.22 g/cc is comprised of proplastids.[26] Fractions from the gradient are collected in the usual way and sucrose concentration determined refractometrically. If only the glyoxysomes are required they can be collected as a whole fraction from a hole pierced in the bottom of the tube. In this fraction 3–4 mg of protein are recovered from the 30 mg of crude particulate protein applied to the gradient. Some 70–80% of the glyoxysome marker enzyme activities (malate synthetase, isocitrate lyase) applied to the gradient are recovered in this fraction; most of the remainder remains in the soluble protein at the top of the gradient. The concentration of sucrose in the glyoxysome fraction is roughly 54%. Even a small dilution at this stage leads to breakage. Further purification can be achieved by increasing the sucrose concentration to 60% and centrifuging in a flotation gradient.[27]

Enzyme Activities

The table lists enzyme activities for which the glyoxysomes are a major intracellular location. References for assays of the individual enzymes are listed. Representative values from the literature for enzyme activities per gram fresh weight of endosperm and per milligram of glyoxysomal protein are included.

For isolation from smaller amounts of tissue (2–5 g) a more tedious but superior method of grinding, using razor blades, may be used[28] and additionally, the crude homogenate may be layered directly onto a linear

---

[26] From some seed lots the proplastids do not separate clearly from glyoxysomes. Inclusion of 10 mM triethanolamine and 10 mM acetate in the gradient improves the separation (E. Theimer and R. R. Theimer, unpublished).

[27] B. Gerhardt and H. Beevers, *J. Cell Biol.* 44, 94 (1970).

[28] T. Kagawa and H. Beevers, personal communication.

ENZYME ACTIVITIES IN EXTRACTS OF CASTOR BEAN ENDOSPERM AND IN
ISOLATED GLYOXYSOMES

Enzyme	Total activity per g fresh wt per minute (enzyme units)	Specific activity in glyoxysomes (μmoles/min/mg protein)	Assay procedure
Isocitrate lyase	6.6	1.0	a
Malate synthetase	8.3	1.2	b
Malate dehydrogenase	6.50	16.6	c
Citrate synthetase	3	1.1	d
Aconitase	15	0.07	e
Catalase^q	4166	5000	f
Uricase	0.3	0.02	g
Glycolate oxidase	0.6	0.07	h
Hydroxypyruvate reductase	0.1	0.1	i
G:O transaminase	3.3	2.3	j
Acetothiokinase	0.1	0.05	k
Fatty acylthiokinase	0.2	0.1	l
Acyl-CoA oxidase	0.5	0.2	m
β-OH-acyl-CoA dehydrogenase	—	3	n
Enoyl-CoA hydratase	—	3	o
Thiolase	2.5	0.2	p

^{a} G. H. Dixon and H. L. Kornberg, *Biochem. J.* **72,** 3P (1959).

^{b} B. Hock and H. Beevers, *Z. Pflanzenphysiol.* **55,** 405 (1966).

^{c} S. Ochoa, this series, Vol. 1, p. 735.

^{d} P. Srere, H. Brazil, and L. Gonen, *Acta Chem. Scand.* **17,** S129 (1963).

^{e} E. Racker, *Biochim. Biophys. Acta* **4,** 211 (1950); and T. G. Cooper and H. Beevers, *J. Biol. Chem.* **244,** 3507 (1969).

^{f} H. Lück, *in* "Methods of Enzymatic Analysis" (H. U. Bergmeyer, ed.), p. 885. Academic Press, New York, 1965.

^{g} M. Müller and K. M. Møller, *Eur. J. Biochem.* **9,** 424 (1969).

^{h} R. W. Breidenbach, A. Kahn, and H. Beevers, *Plant Physiol.* **43,** 705 (1968); J. Feierabend and H. Beevers, *Plant Physiol.* **49,** 28 (1972).

^{i} N. E. Tolbert, R. K. Yamazaki, and A. E. Oeser, *J. Biol. Chem.* **245,** 5129 (1970).

^{j} T. G. Cooper and H. Beevers, *J. Biol. Chem.* **244,** 3507 (1969).

^{k} T. G. Cooper, *J. Biol. Chem.* **246,** 3451 (1971); T. G. Cooper and H. Beevers, *J. Biol. Chem.* **244,** 3514 (1969).

^{l} T. G. Cooper, *J. Biol. Chem.* **246,** 3451 (1971).

^{m} T. G. Cooper and H. Beevers, *J. Biol. Chem.* **244,** 3514 (1969).

^{n} D. Hutton and P. K. Stumpf, *Plant Physiol.* **44,** 508 (1969).

^{o} D. Hutton and P. K. Stumpf, *Plant Physiol.* **44,** 508 (1969).

^{p} T. G. Cooper and H. Beevers, *J. Biol. Chem.* **244,** 3514 (1969); D. Hutton and P. K. Stumpf, *Plant Physiol.* **44,** 508 (1969).

^{q} Measured at pH 7.0, 25° at a substrate concentration of 1.4 mM. Corresponding rate constants are 2.22/second and 2.67/second per milligram of protein.

gradient.[29,30] Under these conditions breakage of organelles is minimized and more than 90% of the total isocitrate lyase is recovered in the glyoxysome fraction.[30] Larger amounts of crude particulate protein—up to 300 mg—can conveniently be separated into sharp bands of constituent organelles on stepped sucrose gradients using the information on equilibrium densities established from linear gradients. For example, on a gradient consisting of 5 ml of 60% sucrose, 10 ml of 57%, 15 ml of 50%, 15 ml of 44%, and 7 ml of 33% the glyoxysomes were recovered in the 50–57% step, the proplastids in that between 44% and 50% and mitochondria between 33% and 44%.[7] Stepped gradients should be used only after isopycnic densities of the organelles are precisely established; even so some sacrifice in recovery is made since some cross contamination of the lighter fraction is likely to occur.

Large-Scale Isolation of Glyoxysomes

Some experimental requirements, for example, purification and biochemical characterization of individual constituent proteins from glyoxysomes, demand larger scale isolation of particles.[31] In general, the requirements of these procedures are the same as those outlined above. However, some special techniques are required to meet these requirements in larger-scale preparations: (1) rapid, gentle, mechanical procedure for tissue homogenization; (2) procedure for rapid concentration of particulates; (3) isopycnic separation on zonal rotors.

In general, three methods have been employed to homogenize tissue for large-scale isolates.

a. Batchwise homogenization by mortar and pestle: 50-g lots of tissue can be homogenized individually and combined. However, unless enough labor and equipment are available to process a number of lots simultaneously, the time required militates against good yields of intact particles.

b. Homogenization by repeated 5-second bursts in a large blender. With this procedure cell breakage is very inefficient, and as the number of bursts is increased to obtain efficient cell rupture the yield of intact particles decreases.

c. Homogenization with the Branton-De Silva tissue slicer: 200-g lots of tissue can be homogenized by two sequential passes through a specially designed slicing machine to be described elsewhere.[32] Lots of 1–2 kg can be processed in approximately 15 minutes with yields of extracted isocitrate lyase that will form glyoxylate at a rate of 0.7–1.0 μmoles per

[29] C. Schnarrenberger, A. Oeser, and N. E. Tolbert, *Plant Physiol.* **48**, 566 (1971).
[30] T. Kagawa, J. M. Lord, and H. Beevers, personal communication.
[31] R. W. Breidenbach, *Ann. N.Y. Acad. Sci.* **168**, 342 (1969).
[32] D. Branton and P. De Silva, personal communication.

minute per gram fresh weight from 5-day-old castor bean endosperm.[33] For comparison, the amount of isocitrate lyase extracted by the mortar and pestle procedure produces 1.0–1.6 μmoles per minute per gram fresh weight and the total extractable enzyme will produce glyoxylate at a rate of 1.8–2.4 μmoles per minute per gram fresh weight. Forty to fifty percent of the enzyme extracted by the tissue-slicer is obtained in the particulate form as compared to 50–70% particulate enzyme obtained with the mortar and pestle procedure.

Isolation of Crude Particulate Fraction

The homogenates are then centrifuged in the Sorvall GSA rotor with 270-ml round-bottom polycarbonate bottles. A low speed pellet (10 minutes at 600 g) is discarded and the high speed pellet (15 minutes at 10,000 g) is resuspended for gradient centrifugation. Approximately 1600 ml can be handled in a single run in the centrifuge. If possible, it is desirable to employ several centrifuges operated simultaneously.

Zonal Centrifugation

The high speed pellets obtained by differential centrifugation are resuspended in 50–100 ml with disposable syringe and cannula and introduced into a Beckman Ti 15 zonal rotor that is already loaded, in order, with the following: 500 ml of a linear sucrose gradient ranging from 30 to 38% w/w, 100 ml of 38% sucrose, 250 ml of 43%, 250 ml of 48%, 250 ml of 55%, and 60% to fill. After the sample is loaded, approximately 100 ml of 20% sucrose is overlayered, the loading assembly is removed, and the rotor is sealed and accelerated to 35,000 rpm.[33]

The gradient is centrifuged for 4 hours and then decelerated to unloading velocity and unloaded by displacement with 65% sucrose. The displacing solution is pumped into the rotor from a filter flask by means of 2–6 pounds of gas pressure from a nitrogen cylinder. The displaced gradient is collected in 20-ml fractions, and fractions containing significant amounts of glyoxysomes are pooled.[33]

[33] R. W. Breidenbach, personal communication.

[59] Vacuoles and Spherosomes

By PHILIPPE MATILE and ANDRES WIEMKEN

Vacuoles

Plant vacuoles cannot be isolated according to a universal scheme of cell fractionation. This circumstance is chiefly due to the wide variation of physical properties of these organelles (size, shape, density, etc.). In addition, the mechanically resistant cell walls require drastic procedures of tissue homogenization which inevitably result in the destruction of large vacuoles. The disintegration of parenchymatous tissues using conventional techniques will always result in bursting and fragmentation of these organelles. The soluble substances contained in these vacuoles are, therefore, mainly present in the soluble fraction of extracts. On the other hand, meristematic cells may contain small vacuoles that are not destroyed upon grinding or blending of the tissue. It appears that the preparation of vacuoles depends largely on the selection of suitable objects. One of these is represented by spheroplasts[1] whose fragility permits a gentle lysis and the release of even large vacuoles.

Isolation of Vacuoles from Spheroplasts

The procedure involves the following steps: (i) preparation of osmotically stabilized spheroplasts by enzymatic digestion of the cell walls, (ii) osmotic lysis of spheroplasts and release of intact vacuoles, and (iii) isolation and purification of vacuoles.[1a] The following method has been successfully used in several laboratories.

Spheroplasts. A growing population of *Saccharomyces cerevisiae* is collected by centrifugation. After washing with distilled H_2O, the cells are suspended in a medium containing buffered osmoticum (0.6 M sorbitol, 10 mM citrate buffer, pH 6.8) supplemented with 0.14 M cysteamine-HCl (approx. 1×10^9 cells/ml). The suspension is incubated for 20 minutes at 27°. The cells are now sedimented and washed once with the above osmoticum (no supplement). The subsequent incubation in the presence of snail gut enzyme (Helicase, lyophilized product of l'Industrie Biologique Française S.A. Gennevilliers, Seine, France; Glusulase, or corresponding preparations) is carried out under the following conditions: 1 ml of Helicase solution (30 mg of enzyme per milliliter of

---

[1] The term spheroplast has now been generally adopted in cases where the question whether the wall has been completely eliminated or not is irrelevant.

[1a] P. Matile and A. Wiemken, *Arch. Mikrobiol.* **56**, 148 (1967).

osmoticum) is used for suspending 1 g of packed cells; incubation at 30°
in an oscillating water bath. Within 45–90 minutes the cells are converted into protoplasts (controls in phase contrast microscope; note the
spherical shape of spheroplasts) which are now cooled down to 0°, the
subsequent working temperature. Separation of spheroplasts from the Helicase solution is achieved by centrifugation in a swing-out rotor (20 minutes,
1000 g) through a gradient ranging from 0.6 M sucrose to 0.6 M sorbitol.
The sedimented protoplasts are washed twice with a solution of 0.7 M
sorbitol. Note that the successful conversion of yeast cells into protoplasts is possible only in growing cells. Resting cells (e.g., commercially
available cakes of bakers' yeast) must be incubated in an appropriate
culture medium for a few hours prior to the preparation of protoplasts.

Liberation of Vacuoles. Osmotic lysis is induced by placing the packed
protoplasts in a medium containing 0.1 M mannitol or sorbitol, and 8%
(w/v) Ficoll (Pharmacia, Uppsala, Sweden). Resuspension is carried out
by gently stirring and swirling with a glass rod. Microscopical controls are
necessary in order to check the completeness of lysis and release of vacuoles
upon gentle agitation of the suspension. It is important that vacuoles are
freed as completely as possible from adhering cytoplasm. This can be
achieved by moving the suspension repeatedly through a Pasteur pipette.

Isolation and Purification of Vacuoles. The product of lysis is placed
in centrifuge tubes and carefully overlayered with about equal volumes of
7.50% and 7.0% (w/v) Ficoll dissolved in 0.1 M mannitol or sorbitol
in 10 mM citrate buffer, pH 6.8. After centrifugation in a swing-out rotor
(15 minutes, 8500 g) the vacuoles have formed a white layer on top of
this system from where they are carefully removed. Turbidity in the Ficoll
layers indicates the presence of small vacuoles and lipid granules which
have not flotated to the top under the conditions used. The sediment contains, among other products of lysis, vacuoles that have not been completely liberated from the protoplasts. If the preparation of flotated
vacuoles is contaminated with lipid granules, a further purification may
be achieved by sedimenting the vacuoles (30 minutes, 2000 g) after having slowly diluted the Ficoll with 1.5 volumes of 0.6 M sorbitol in 10
mM citrate buffer, pH 6.8.

Critique. Osmotic lysis seems to change the permeability properties of
the tonoplast. The above procedure, therefore, is not suitable for investigating the micromolecular contents of vacuoles. To study the low molecular weight contents osmotic lysis can be substituted by metabolic lysis of
yeast protoplasts in the presence of glucose and chelating agents.[2,3] In

[2] K. J. Indge, *J. Gen. Microbiol.* **51**, 433 (1968).
[3] K. J. Indge, *J. Gen. Microbiol.* **51**, 441 (1968).

this case sedimentation of vacuoles or flotation in isotonic gradients of Ficoll, sorbitol, and sucrose having higher densities than those described above must be applied.

Properties of Isolated Yeast Vacuoles. At 0° the isolated yeast vacuoles are stable for many hours. They are osmotically active, but the tonoplasts of vacuoles obtained from osmotically lysed protoplasts seem to be rather permeable with regard to micromolecules. The lysosomal nature of yeast vacuoles is demonstrated by the presence of various digestive enzymes in a concentrated form. The preparations are virtually free of enzymes localized in mitochondria or in the cytoplasmic matrix. Considerable fractions of the total amino acid pools and polyphosphates have been detected in fractions containing vacuoles prepared from metabolically lysed protoplasts.[4]

Suitable Objects. Fungi which have been used for protoplast formation and subsequent liberation of vacuoles comprehend *Saccharomyces cerevisiae*,[1] *Saccharomyces carlsbergensis*,[3] *Candida utilis*,[5] *Fusarum culmorum*.[6] A variety of fungi have been used for preparing protoplasts.[7] Macroconidia of *Neurospora crassa* whose cell walls had been weakened by the action of lytic enzymes can be osmotically lysed and intact vacuoles be released.[8] The observation of free vacuoles in preparations of higher plant protoplasts[9] suggests that the above technique may be successfully modified for the isolation of higher plant vacuoles.

Isolation of Meristematic Vacuoles from Plant Tissues

The subsequently described procedure for the isolation of small vacuoles (up to 2 μm in diameter) has been worked out using maize seedlings.[10]

Tissue Homogenization. Medium: 0.5 M sorbitol, 50 mM Tris·HCl buffer pH 7.6, 1 mM EDTA. Root tips, ca. 5 mm in length, are washed in ice-cold distilled water and blotted on filter paper. Five grams of wet tips are placed in a mortar together with 5 ml of medium and 2.5 g of quartz sand (grain size 0.1–0.8 mm). The grinding is done with a minimum of pressure; reduction of the quartz to small pieces is avoided. The resulting brei is diluted with ca. 20 ml of medium and subjected to an initial cen-

[4] K. J. Indge, *J. Gen. Microbiol.* **51**, 447 (1968).
[5] A. Wiemken, unpublished results.
[6] I. Garcia-Acha, F. Lopez-Belmonte, and J. R. Villanueva, *Can. J. Microbiol.* **13**, 433 (1967).
[7] J. R. Villanueva, *in* "The Fungi" (G. C. Ainsworth and A. S. Sussman, eds.), Vol. 2. Academic Press, New York, 1966.
[8] P. Matile, *Cytobiologie* **3**, 324 (1971).
[9] E. C. Cocking, *Nature* (*London*) **187**, 962 (1960).
[10] P. Matile, *Planta* **79**, 181 (1968).

trifugation (10 minutes, 500 *g*; elimination of sand, starch, cell walls, etc.).

Isolation of Vacuoles. A mitochondrial fraction which contains the vacuoles is sedimented (15 minutes, 20,000 *g*) and carefully resuspended in medium (0.5 *M* sorbitol, 10 m*M* Tris·HCl buffer pH 7.6). The densely packed white layer at the bottom of the sediment which contains mainly starch is not resuspended. After a second sedimentation under the above conditions the washed mitochondrial fraction is resuspended in ca. 4 ml of medium and loaded in portions of 1 ml onto discontinuous gradients of sucrose [e.g., 2 ml layers of 40% and 15% (w/v) sucrose]. After centrifugation in a swing-out rotor (2 hours, 125,000 *g*) mitochondria and peroxisomes are present in the sediment; larger vacuoles are trapped at the surface of the 15% sucrose layer, small vacuoles at the surface of the 40% sucrose layer, respectively. The corresponding bands are isolated with a Pasteur pipette, diluted with medium, and the vacuoles collected by sedimentation (15 minutes, 20,000 *g*). If continuous linear gradients of sucrose (15% to 40%) are used instead of discontinuous gradients, the vacuoles heavier than 15% sucrose are diffusely distributed in the gradient. This behavior is due to the gradual changes of physical properties of vacuoles in the course of vacuolation which takes place in the root meristem.

Properties of Isolated Meristematic Vacuoles. The preparations are practically free of soluble cytoplasmic, mitochondrial, and peroxisomal enzymes. The lysosomal nature of these vacuoles is demonstrated by the presence of a variety of hydrolases in a concentrated form.[10]

Suitable Objects. Similar techniques have been used for isolating lysosomes from root meristem,[11] tomato,[12] and other tissues.[13,14] Of interest for the isolation of vacuoles are the laticifers of certain species whose latex contains numerous small vacuoles. In this case, tissue extraction can be achieved simply by tapping the extended system of communicating laticifers.[15,16]

Isolation of Aleurone Vacuoles

Aleurone grains (protein bodies) represent specialized vacuoles in reserve tissues of seeds. They are packed with reserve proteins and phytate

[11] P. Coulomb, *C. R. Acad. Sci. Ser. D* **267**, 2133 (1968).
[12] E. Heftmann, *Cytobios* 3, 129 (1971).
[13] P. Coulomb, *C. R. Acad. Sci. Ser. D* **269**, 2543 (1969); *J. Microsc. (Paris)* **11**, 299 (1971).
[14] W. Iten and P. Matile, *J. Gen. Microbiol.* **61**, 301 (1970).
[15] S. Pujarniscle, *Physiol. Veg.* 6, 27 (1968).
[16] P. Matile, B. Jans, and R. Rickenbacher, *Biochem. Physiol. Pflanzen* **161**, 447 (1970).

which, in resting seeds, provide the organelle with considerable stability. Although intact aleurone grains are usually obtained upon grinding of reserve tissues, the vacuolar membrane may be injured and, therefore, soluble constituents be lost. Mobilization of reserves taking place in the course of germination results in a considerable swelling of aleurone vacuoles. Extraction of intact organelles from germinating seeds is therefore difficult to achieve. Moreover, these vacuoles are labile after extraction. The physical properties of aleurone grains from ungerminated seeds are similar to those of starch grains and fragments of cell walls. Hence, isolation and purification require centrifugation techniques that take advantage of small differences in density or sedimentation velocity. In the isolation technique[17] described below, some possibilities for overcoming the difficulties mentioned above are evident.

Aleurone Grains from Cotton Seeds. Cotton seeds contain an oleaginous reserve tissue (contaminating starch excluded). Loss of water-soluble constituents from extracted aleurone grains is avoided by employing glycerol as a nonaqueous medium.

The extraction of 10 g of dehulled dry seeds is carried out in a Waring Blendor in the presence of 30 ml of glycerol. The extract is strained through cheesecloth. Large debris is subsequently removed by centrifuging the extract at 1100 g for 5 minutes. Thereafter aleurone grains are sedimented at 41,000 g (20 minutes). The high viscosity of the glycerol allows a satisfactory separation of the large and heavy aleurone grains from other cytoplasmic particles. The sediment of aleurone grains is resuspended in glycerol and sedimented again under the above conditions.

Result: 75% of the total proteins present in the precentrifuged extract is obtained in the fraction of isolated aleurone grains. Acid protease (100%) and acid phosphatase (77%) are concentrated in this fraction whereas a soluble cytoplasmic enzyme, ethanol dehydrogenase, is completely absent.[17]

Suitable Objects. The following seeds have been used for the isolation of intact aleurone grains, or subfractions thereof, using a variety of centrifugation techniques: wheat,[18] barley,[19] rice,[20] pea,[21,22] soybean,[23,24] broad bean,[25] cotton,[26] hemp,[27,28] and squash.[29]

[17] L. Y. Yatsu and T. J. Jacks, *Arch. Biochem. Biophys.* **124**, 466 (1968).
[18] J. S. D. Graham, R. K. Morton, and J. K. Raison, *Aust. J. Biol. Sci.* **16**, 375 (1963).
[19] R. L. Ory and K. W. Henningsen, *Plant Physiol.* **44**, 1488 (1969).
[20] H. Mitsuda, K. Yasumoto, K. Murakami, T. K. Kusano, and H. Kishida, *Agr. Biol. Chem.* **31**, 293 (1967).
[21] J. E. Varner and G. Schidlovsky, *Plant Physiol.* **38**, 139 (1963).
[22] P. Matile, *Z. Pflanzenphysiol.* **58**, 365 (1968).

Spherosomes

Spherosomes represent vacuole-like organelles that are characterized by a high content in triglycerides (lipid droplets). They are ubiquitous in the plant kingdom. The oleaginous reserve cells of certain seeds contain numerous large spherosomes that are extremely rich in lipids. Since the density of these organelles is less than unit, they can easily be flotated in appropriate media. The following procedure has been worked out for tobacco endosperm.[30,31] It closely resembles techniques that have been used for isolating spherosomes from other seeds.

Isolation of Spherosomes from Tobacco Seeds

A layer of resting, soaked, or germinated seeds is placed on a cooled glass plate. The seeds are covered with medium containing 0.4 M mannitol and 10 mM Tris·HCl pH 7.4. They are now gently squeezed under a second glass plate, which is lightly pressed and moved. The soft endosperm can thereby be extracted without damaging the embryos or seedlings. The resulting milky suspension is filtered through cotton wool (removal of seedlings and seed coats). The extract is subsequently placed in centrifuge tubes and carefully overlayered with an equal volume of 0.2 M mannitol. After centrifugation (30 minutes, 700 g) the flotated spherosomes are aspirated with a Pasteur pipette.

Result. At low temperature isolated tobacco endosperm spherosomes are stable for many hours if kept in the mannitol medium or even in distilled water. At room temperature they rapidly burst and coalesce to smaller lipid droplets. Over 90% of the total endosperm lipids are contained in the isolate. The existence of a limiting membrane is indicated by the presence of phospholipids. Spherosomes isolated from resting and germinated seeds contain lipase and other hydrolase activities in a concentration form.

Suitable Objects. Both oleaginous reserve tissues and nonoily tissues have been used for isolating spherosomes: Douglas fir seeds,[32] peanut

---

[23] M. P. Tombs, *Plant Physiol.* **42**, 797 (1967).

[24] K. Saio and T. Watanabe, *Agr. Biol. Chem.* **30**, 1133 (1966).

[25] G. F. I. Morris, D. A. Thurman, and D. Boulter, *Phytochemistry* **9**, 1707 (1970).

[26] N. S. T. Lui and A. M. Altschul, *Arch. Biochem. Biophys.* **121**, 678 (1967).

[27] A. J. St. Angelo, L. Y. Yatsu, and A. M. Altschul, *Arch. Biochem. Biophys.* **124**, 199 (1968).

[28] A. J. St. Angelo, R. L. Ory, and H. J. Hansen, *Phytochemistry* **8**, 1135 (1969).

[29] J. N. A. Lott, P. L. Larsen, and J. J. Darley, *Can. J. Bot.* **49**, 1777 (1971).

[30] P. Matile and J. Spichiger, *Z. Pflanzenphysiol.* **58**, 277 (1968).

[31] J. U. Spichiger, *Planta* **89**, 56 (1969).

[32] T. M. Ching, *Lipids* **3**, 482 (1968).

cotyledons,[33] castor bean endosperm,[34] cotyledons of legume seeds,[35] cotton seeds, onion bulbs, and cabbages.[36]

[33] T. J. Jacks, L. Y. Yatsu, and A. M. Altschul, *Plant Physiol.* **42**, 585 (1967).
[34] R. L. Ory, L. Y. Yatsu, and H. W. Kircher, *Arch. Biochem. Biophys.* **123**, 255 (1968).
[35] H. H. Mollenhauer and C. Totten, *J. Cell Biol.* **48**, 533 (1971).
[36] L. Y. Yatsu, T. J. Jacks, and T. P. Hensarling, *Plant Physiol.* **48**, 675 (1971).

[60] The Isolation of Plant Protoplasts

By Edward C. Cocking

It has recently been emphasized that isolated higher plant protoplasts are naked cells and that being cells they can under suitable conditions be induced to grow and divide.[1] These naked cells are in many respects ideal single-cell cultures. They have considerable potential for the cloning of plant cells generally. They are also well suited for studies on the fusion of plant cells, and the resultant heterokaryons serve as starting cultures for experimental investigation of the possibility of somatic hybridization of plants.[2] One remarkable attribute of these naked cells is that they can re-build a cell wall. This resynthesis has been fully described and discussed previously[2a]; physiologically it results in the development of cells with the growth potential of cells as ordinarily cultured.

Increasingly, plant biochemists and plant physiologists have become interested in such isolated protoplasts; moreover the removal of the cell wall exposes the plasmalemma directly to the influence of the culture medium. The presence of a cell wall is a complicating factor in studies on uptake generally, particularly so if the investigator is interested in the possible uptake of macromolecules or particles by the isolated protoplasts since the cell wall acts as a very efficient ultrafilter.

Studies on the fusion and endocytotic activity of these naked cells have focused attention on the properties of the plasmalemma. Willison *et al.*[3] have postulated a mechanism for the pinocytosis (endocytosis) of latex spheres by tomato fruit protoplasts. Protoplasts were isolated from tomato fruit locule tissue obtained from tomato plants grown under controlled conditions.[4] Watering of plants should not be carried out for several hours

[1] E. C. Cocking, *Annu. Rev. Plant Physiol.* **23**, 29 (1972).
[2] E. C. Cocking, *Scienza Tecnica* **74** (in press).
[2a] E. C. Cocking, *in* "Dynamic Aspects of Plant Ultrastructure" (A. W. Robards, ed.), 1973 (in press).
[3] J. H. M. Willison, B. W. W. Grout, and E. C. Cocking, *J. Bioenerg.* **2**, 371 (1971).
[4] J. W. Davies and E. C. Cocking, *Planta* **67**, 242 (1965).

before the fruit is detached from the plant. The diced tissue was incubated with 0.5% Macerozyme (All Japan Biochemicals Ltd., Shingikancho, Nishinomiya, Japan) and 5% cellulase (Onozuka 1500, All Japan Biochemicals Ltd., Shingikancho, Nishinomiya, Japan) for 3 hours; the protoplasts released were then washed with 20% sucrose, in which they float.[5] This enzymatic isolation of protoplasts makes use of cell wall degrading enzymes. As normally used these cell wall degrading enzymes are grossly impure, containing many enzymes other than those directly involved in the cell wall degradation. Toxic compounds can also be present, and although it is difficult to generalize, it would seem best always to treat the enzymes with Sephadex before use in order to remove certain enzymes, particularly nucleases, and low molecular weight toxic material.[6]

Experience over several years and extending back to the early work of Tribe,[7] who used mechanical rather than enzymatic methods for protoplast isolation, has indicated that isolated protoplasts are far more resistant to such possible harmful effects when they are in the plasmolyzed condition. One reason for this has recently become clear from the work of Withers and Cocking.[8] From studies using thorium dioxide as an electron dense electron microscopic marker, they showed that extensive uptake from the plasmolyticum took place during plasmolysis probably by large-scale membrane invagination. In view of this extensive uptake, it would seem advantageous to carry out preplasmolysis in a suitable plasmolyticum before subjecting the tissue to digestion in a possibly harmful cell wall-degrading enzyme mixture. As a result the procedures recommended for the isolation of tobacco leaf protoplasts for induced fusion studies with sodium nitrate involved preplasmolysis in 25% sucrose.[8] Pieces of tobacco leaf taken from 50- to 60-day-old *Nicotiana tabacum* var. Xanthi plants, from which the lower epidermis had been removed (for practical details see Power and Cocking[9]) were incubated for 2 hours in 25% sucrose. After this preplasmolysis the leaf pieces were incubated in a mixture of 5% w/v cellulase (Onozuka 1500) and 0.5% w/v Macerozyme in 25% sucrose, at 20°, for periods of between 2 and 4 hours. The released protoplasts were washed by flotation in 25% sucrose.

The following two detailed examples may help the reader to appreciate more fully the procedures required for the isolation of protoplasts, first from cereal leaves, and second from suspension culture cells of rose. The

[5] E. Pojnar, J. H. M. Willison, and E. C. Cocking, *Protoplasma* 64, 460 (1967).
[6] P. K. Evans and E. C. Cocking, *in* "Plant Tissue and Cell Culture" (H. E. Street, ed.), pp. 100–120, 1972.
[7] H. T. Tribe, *Ann. Bot.* 19, 351 (1955).
[8] L. A. Withers and E. C. Cocking, *J. Cell Sci.* 11, 59 (1972).
[9] J. B. Power and E. C. Cocking, *J. Exp. Bot.* 21, 64 (1970).

earlier report of Ruesink[10] may also be useful as a further guide to the basic principles involved.

Isolation of Cereal Leaf Protoplasts

Mature leaves of *Secale cereale* (rye) or *Triticum aestivum* (wheat) or *Hordeum vulgare* (barley) or *Avena sativa* (oats) are employed. The cereal plants are between 21 and 28 days old, and the leaves are surface sterilized by treatment with 70% ethanol for 2 minutes followed by 30 minutes in 3% sodium hypochlorite (0.3–0.425% w/v available chlorine). It is necessary to add Teepol (0.5%) as a wetting agent (BDH Chemicals Ltd., Poole, United Kingdom). It is important to remove thoroughly the hypochlorite which has been used as a surface sterilant. The hypochlorite is removed by three successive washes in sterile water.

After this treatment the leaves are cut into narrow longitudinal threads using a scalpel and the leaf pieces immersed in a filter-sterilized enzyme mixture with 1 g fresh weight of pieces to every 7 ml of enzyme solution. The enzyme solution should be freshly prepared. A satisfactory enzyme solution was made up as follows: 4% Meicelase (Meiji Seika Kaisha Ltd., Tokyo) with either Pectinol R10 (Rohm and Haas, Philadelphia, Pennsylvania) or 1% Macerozyme (All Japan Biochemicals Ltd., Nishinomiya, Japan) and 1% potassium dextran sulfate with a sulfur content of 17.3% (Meito Sangyo Co. Ltd., Nagoya, Japan). These enzymes were dissolved in 0.7 M sorbitol or 0.7 M mannitol. Pectinol R10 is standardized with diatomaceous earth and a solution of the enzyme was prepared by suspending 20 g in 100 ml of distilled water for 2 hours. The insoluble material was removed by filtration and after the addition of the materials listed above the volume was made up to 100 ml. As only 6.5% of the Pectinol R10 is water soluble, this represents 1.3% solution of Pectinol R10. The pH was adjusted to 5.8 with normal HCl. The leaf shreds were incubated in the above medium at 25° for 5 hours. After this, the enzymes were removed and the leaf pieces were rinsed briefly with a sterile washing solution. The sterile washing solution was of the following composition: 0.7 M sorbitol, 1 mM KNO$_3$, 0.2 mM KH$_2$PO$_4$, 0.1 mM MgSo$_4$, 1 mM CaCl$_2$, 1 μM KI, and 10 mM CuSO$_4$.[11] After this washing the leaf pieces were placed in a fresh enzyme solution. They were incubated in this fresh enzyme solution for a further 12 hours at 25°, and after this the leaf shreds were teased apart in the washing solution; this resulted in the release of the protoplasts.

The isolated cereal leaf protoplasts can be readily sedimented by cen-

[10] A. W. Ruesink, this series, Vol. 23, p. 197.
[11] R. G. Jensen, R. I. B. Francki, and M. Zaitlin, *Plant Physiol.* **48**, 9 (1971).

trifugation at 100 g. After centrifugation at 100 g for 5 minutes, the isolated protoplasts were resuspended in sterile 0.8 M sucrose. In 0.8 M sucrose the isolated protoplasts do not sediment, but because of their density they rise to the surface of the sucrose when the suspension is centrifuged; as a result further centrifugation at 200 g for 5 minutes caused intact protoplasts to rise to the surface, while cells, cell debris, and free chloroplasts sedimented to the bottom of the centrifuge tube. The layer of protoplasts was resuspended in a large excess of washing solution and centrifuged at 100 g for 5 minutes. The protoplasts sediment in this washing solution. The protoplasts were then suspended in a known volume of washing solution, and a small sample was taken for counting, using a modified Fuchs Rosenthal hemacytometer, 0.2 mm deep (Hawkesley Gelman, Lancing, England). High yields of protoplasts are obtained—of the order of 10^6 protoplasts per gram fresh weight of leaf.

Such isolated cereal leaf protoplasts as these can be readily suspended at known cell densities in suitable culture media and their growth and development studied further. For further details, see Evans et al.[12]

Isolation of Rose Suspension Culture Protoplasts

Protoplasts isolated from "Paul's Scarlet" rose suspension culture cells have been extensively studied.[13] The suspension culture cells themselves grow rapidly in batch culture on a fully defined medium originally devised by Davies.[13a] They are sterile, and therefore preliminary sterilization of the plant material is not required as in the case of cereal leaves.

The fact that the starting material, i.e., the cells, is already sterile removes many of the uncertainties often present when dealing with material fresh from the plant. Indeed, the isolation of leaf protoplasts is beset with the variation that occurs even from variety to variety of the same species in relation to the efficacy of surface sterilization. The suspension culture of rose cells forms a pipettable suspension which is composed of single cells or small groups of cells. The fact that cells are largely separate from one another eliminates the need to use a pectinase enzyme. Cells from all phases of the growth cycle can be used (but see below).

One gram fresh weight of cells filtered from the medium is added to every 10 ml of enzyme digestion mixture. The cell wall digestion mixture contains Meicelase (Meiji Seika Kaisha Ltd.) or Onozuka Cellulase (1500 or 3000). The concentration of enzyme which can be employed lies between 0.5% and 3%. The pH is adjusted to between 5.5 and 6.0 with NaOH or HCl. The incubation conditions are as follows: 20 ml of incuba-

[12] P. K. Evans, A. G. Keates, and E. C. Cocking, *Planta* **104**, 178 (1972).
[13] R. Pearce, Ph.D. Thesis, Univ. of Nottingham, 1972.
[13a] M. E. Davies, *Phytochem.* **10**, 783 (1971).

tion mixture is placed in 100-ml Erlenmeyer flasks (too great a depth of mixture is deleterious to survival) and held at a constant temperature of 25° for up to 24 hours. At the end of the incubation period the mixture is centrifuged for 1 minute at 300 g, and the protoplasts that collect at the surface of the liquid are transferred with a Pasteur pipette to a suitable concentration of sucrose. The protoplasts are then resuspended and centrifuged once more in the same molarity of sucrose. This procedure is then repeated. This centrifugation procedure results in the protoplasts forming a dense layer at the surface of the plasmolyticum while debris sinks to the bottom. This relatively simple procedure results in the isolation of rose protoplasts, largely entirely free of debris.

The yield of protoplasts and the extent to which cells are absent from the protoplast suspension is related to the age of the suspension relative to the time interval between its current growth and the time when it was isolated and grown up from callus. Excellent preparations of isolated rose protoplasts, essentially completely free of debris, and from cells, were obtained when rose suspension cells were treated in this way when the rose suspension cells had been obtained from liquid cultures which had recently been grown up from callus on agar. Frequently almost all the suspension cells were converted to isolated protoplasts. When cells grown from subsequent passages of the suspension culture were used, it became necessary first to raise the incubation temperatures to 33° and later to use 3% purified cellulase (Cellulase 3000) to obtain satisfactory results. It was found, however, that if the suspension cells were returned to agar for several passages, then the original isolation conditions were once more satisfactory.

It would thus seem likely that progressive changes either in wall composition or in the nature of enzyme inhibitors in the cells are responsible for these effects. It would also seem likely that there are progressive changes either in wall composition or in the level of inhibitors as the number of passages of the suspension cultures increases. As earlier mentioned, it may be advantageous with certain cell systems to preplasmolyze the tissues or cells from which it is desired to isolate protoplasts. It is difficult, however, to generalize in this respect; in the case of the isolation of rose protoplasts, preplasmolysis in sucrose for 0.5 or 24 hours reduced release of protoplasts without increasing total survival.

Added Comments

Although the enzymatic isolation of higher plant protoplasts has now become an acceptable laboratory procedure, each tissue must be investigated systematically in relation to the optimum conditions for protoplast release.

It should be noted that protoplasts can also be isolated from tissues of higher plant cells by first separating the cells, under plasmolyzing conditions, with Macerozyme and then isolating protoplasts using cellulase to degrade the walls of the separated cells.[14] This method is useful in that it enables the isolation of specifically palisade layer protoplasts from leaves. It is, however, more laborious than the mixed enzyme (Macerozyme plus Cellulase) procedure described for the isolation of cereal leaf protoplasts. Moreover, as discussed by Evans, Keates, and Cocking,[12] the sequential method of Takebe and his co-workers does not yield protoplasts from many cereal leaves.

It is also possible to isolate protoplasts mechanically by means of the procedure first introduced by Klercker in 1892.[15] This method does not release protoplasts from meristematic cells since they do not plasmolyze sufficiently. The yield of protoplasts is usually low even from more mature cells. Nevertheless, such mechanically isolated protoplasts are very useful for comparisons between the metabolic activity of protoplasts isolated mechanically and enzymatically. Pilet and his co-workers[16] have recently pioneered the use of mechanically isolated protoplasts in this respect.

[14] T. Nagata and I. Takebe, *Planta* **92**, 301 (1970).
[15] J. Klercker, *Ofvers Vetensk. Akad. Forh., Stockholm*, **49**, 463 (1892).
[16] P. E. Pilet, *C. R. Acad. Sci. Ser. D* **273**, 2253 (1971).

[61] Isolation of Microsomes, Ribosomes, and Polysomes from Plant Tissues[1]

By Joe H. Cherry

The first chemical indication that ribonucleoproteins played a direct role in the synthesis of proteins came in the early 1950's when animals were fed radioactive amino acids. When the labeled tissue was excised, homogenized, and then fractionated, the highest concentration of radioactive amino acids was found in the microsomal fraction.[2] This fraction has been shown to contain ribosome (and polyribosome) particles attached to membrane fragments. Careful kinetic studies on the rate of amino acid incorporation into different proteins support the likelihood that the newly synthesized polypeptides attached to ribosomes were precursors to the enzymes found in the cells. Furthermore, it was found that RNase stopped protein synthesis in isolated subcellular fractions. Experiments on whole

[1] Some of the research presented in this paper was supported by a contract (C00-1313-31) from the U.S. Atomic Energy Commission. This is journal paper 4722 of the Purdue Agriculture Experiment Station.
[2] E. B. Keller, P. C. Zamecnik, and R. B. Loftfield, *J. Histochem. Cytochem.* **2**, 378 (1954).

animals and in isolated cells were soon supplemented by *in vitro* studies employing cell-free systems to incorporate amino acids into protein. Subsequently, the major part of these studies on protein synthesis was accomplished with bacterial systems. Therefore, the present basic concepts of ribosome structure and function have come from studies utilizing bacteria and animals.[3]

Since the initiation of studies of plant ribosomes about twenty years ago,[4] a great deal has been discovered about plant ribosome composition in terms of protein and RNA. Presently, the requirements for the factors involved in *in vitro* protein synthesis by plant ribosomal and polyribosomal systems are being studied in a number of laboratories. It is obvious that the isolation of good ribosomal preparations is a prerequisite to any *in vitro* study of protein synthesis. Since the level of polyribosomes (ratio of monoribosomes to polyribosomes) is thought to provide an indication of *in vivo* protein synthesis, good methodology is required to isolate ribosomal preparations. The major objective of such methodology is to obtain a high yield of "active" ribosomes and polyribosomes with as little degradation as possible. For example, eliminating or reducing the activity of RNase is usually of major importance. At the moment a number of methods are being used to achieve these objectives. The methods vary depending on plant tissues (leaves, roots, embryos, etc.) and the design of the experiment (whether the ribosomal preparation will be used only for amino acid incorporation).

This paper will deal with the various methods available for the isolation of ribosomes and polyribosomes for *in vitro* amino acid incorporation studies and for the analysis of polyribosomes on sucrose gradients.

Isolation of Microsomes

Microsomes are found in a subcellular particulate fraction composed of ribosomes (and polyribosomes) bound to membrane fragments. The microsomal fraction contains in part, the rough endoplasmic reticulum seen in the electron microscope.[5] Preparation of microsomes may be achieved by the following simple method.[6] Homogenize 5 g of tissue (usually meristemic tissues, such as root tips, mesocotyl sections, or embryos, are best) in 10 ml of 0.5 *M* sucrose for 3 minutes using an ice-jacketed glass homogenizer with a power-driven Teflon pestle. Clear the homogenate of cellular debris by filtering through Miracloth (Calbiochem). Remove the mito-

[3] P. C. Zamecnik, *Cold Spring Harbor Symp. Quant. Biol.* **34**, 1 (1969).
[4] G. C. Webster, *J. Biol. Chem.* **229**, 535 (1957).
[5] J. L. Key, J. B. Hanson, H. A. Lund, and A. E. Vatter, *Crop Sci.* **1**, 5 (1961).
[6] J. H. Cherry, R. H. Hageman, and J. B. Hanson, *Rad. Res.* **17**, 724 (1962).

chondria and larger particulate fractions by centrifugation at 40,000 g for 20 minutes. The microsomal fraction is pelleted from this supernatant by centrifugation at 100,000 g for 1 hour. Usually a cleaner fraction is obtained by suspending the pellet in 0.5 M sucrose followed by centrifugation at 100,000 g for 1 hour. The washed microsomal pellet may be suspended in the desired buffer and used as needed.

Isolation of Monoribosomes

Monoribosomes may be obtained from polyribosomes by the removal of the tRNA and mRNA. The transfer RNA portion can be removed by incubating polyribosomes with puromycin which forms a puromycin–tRNA complex. Messenger RNA may be removed from the ribosomes by incubating polyribosomes in 0.5 M NH$_4$Cl for a few minutes at 2°–3°. These and other procedures, however, have several disadvantages; primarily, either the ribosomes are not completely free of mRNA or tRNA or the treatment is so drastic that ribosomal subunits are produced. Therefore, the easiest and simplest way to isolate monoribosomes is to use tissue that does not contain polyribosomes. Usually, unfertilized eggs or unimbibed seeds contain few polyribosomes and therefore are an excellent source of monoribosomes. The procedure devised by Marcus and Feeley[7] gives a high yield of monoribosomes from unimbibed seed (wheat, peanut, etc.). The following method[8] should provide monoribosomes suitable for studies of ribosome characterization and messenger RNA-directed protein synthesis.

Remove and chill 2 g of seed tissue. Grind the tissue with a precooled ($-20°$) mortar and pestle in 2 ml of solution A (0.5 M sucrose, 10 mM MgCl$_2$, 50 mM Tris (adjust pH to 7.8 with succinic acid), 20 mM KCl, 5 mM 2-mercaptoethanol, and 1% deoxycholate). Strain the homogenate through cheesecloth and then centrifuge for 15 minutes at 20,000 g. Layer the supernatant over 4 ml of solution B (2 M sucrose, 10 mM MgCl$_2$, 20 mM KCl, 5 mM 2-mercaptoethanol, and 10 mM Tris-succinate, pH 7.8) and centrifuge in a Spinco SW 39 rotor (or equivalent rotor—angle rotors are adequate for this step) at 105,000 g for 3 hours. Suspend the resultant ribosomal pellet in 1 ml of solution C (10 mM Tris, pH 7.8, with succinic acid), 10 mM MgCl$_2$, 20 mM KCl, and 5 mM 2-mercaptoethanol) and clarify by centrifugation for 10 minutes at 20,000 g. Dilute the supernatant so that the final concentration of ribosomes contains 2–4 mg of ribosomal RNA per milliliter. According to Tester and Dure,[9] one OD$_{260}$ unit is equal to 0.5 mg of ribosomal RNA.

[7] A. Marcus and J. Feeley, *Proc. Nat. Acad. Sci. U.S.* **51**, 1075 (1964).
[8] W. J. Jachymczyk and J. H. Cherry, *Biochim. Biophys. Acta* **157**, 368 (1968).
[9] C. F. Tester and L. S. Dure, *Biochem. Biophys. Res. Commun.* **23**, 287 (1966).

The ribosomal preparation as obtained by the above procedure may be used in a number of experiments where 80 S ribosomes are required.

Isolation of Polyribosomes

As mentioned above, the basic objective of isolating polyribosomes is to obtain a high yield with little degradation. A number of workers have devised techniques suitable for a wide range of tissues. Suitable techniques are available for the isolation of polyribosomes from cotyledons,[10] embryos,[11] tuber tissue,[12] fruit tissue,[13] leaves,[14] and root and stem tissues.[15] Each of these techniques varies in procedure in a number of ways. Some require that the tissue be powdered in liquid nitrogen or dry ice while others employ mortar and pestles for grinding or blade-type homogenizers. In most cases it seems that the basic issue involves RNase. The tissue should be ground in such a manner so that RNase activity does not affect the polyribosome profile. A number of workers[11,14,15] have shown that diethyl pyrocarbonate (DEP) is an effective inhibitor of RNase when the chemical is added to the grinding medium. When high levels (0.5%) of DEP are added to the grinding medium and maintained throughout isolation, the ribosomes are nonfunctional in regard to amino acid incorporation. The careful use of DEP (concentrations at 0.125% and less) may be employed to isolate polyribosome with amino acid incorporation activity greater than that of the control.[15] Two methods are presented below for the isolation of polyribosomes, one employs DEP whereas the other does not.

A. Polyribosome Isolation without DEP

Three grams of plant tissue (roots, tubers, cotyledons, etc.) are ground[16] with a mortar and pestle (previously cooled to $-20°$) with 6 ml of a medium containing 0.25 M sucrose, 50 mM Tris-succinate buffer, pH 7.8, 15 mM KCl, 10 mM MgCl$_2$, 5 mM 2-mercaptoethanol, and 0.5% deoxycholate.[12] Filter the resultant soupy homogenate through glass wool

---

[10] R. B. VanHuystee, W. J. Jachymzyk, C. F. Tester, and J. H. Cherry, *J. Biol. Chem.* **243**, 2315 (1968).

[11] D. P. Weeks and A. Marcus, *Plant Physiol.* **44**, 1291 (1969).

[12] C. T. Duda and J. H. Cherry, *Plant Physiol.* **47**, 262 (1971).

[13] L. I. Ku and R. J. Romani, *Plant Physiol.* **45**, 401 (1970).

[14] R. L. Travis, R. C. Huffaker, and J. L. Key, *Plant Physiol.* **46**, 800 (1970); Y. Eilam, R. D. Butler, and E. W. Simon, *Plant Physiol.* **47**, 317 (1971).

[15] J. M. Anderson and J. L. Key, *Plant Physiol.* **48**, 801 (1971).

[16] Often superior polyribosome profiles are obtained if the tissue is frozen on dry ice or in liquid nitrogen followed by pulverization with a mortar and pestle in either of the two substances. Then the frozen powdered tissue may be gently homogenized in a loose-fitting conical glass homogenizer.

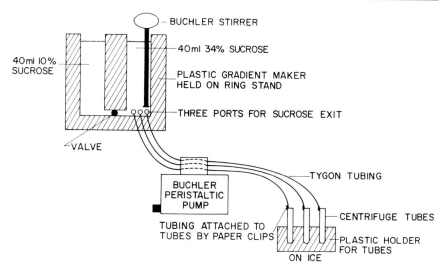

FIG. 1. A system to produce three identical linear gradients from 10 to 34% sucrose.

and centrifuge the filtrate at 20,000 g for 10 minutes. Layer the super-natant over 7 ml of 0.5 M sucrose supported by 2 ml of a 1.6 M sucrose cushion (both solutions contained 10 mM Tris-succinate, pH 7.8, 10 mM $MgCl_2$, 15 mM KCl, and 5 mM 2-mercaptoethanol). Centrifuge this tri-phasic gradient at 150,000 g for 3 hour in a Spinco 50 rotor. Suspend the resultant ribosomal pellet in 1 ml of 10 mM Tris-succinate buffer, pH 7.8, containing 10 mM $MgCl_2$, 15 mM KCl, and 5 mM 2-mercaptoethanol, and then layer the entire suspension on a 25-ml linear sucrose gradient (10 to 34%) containing 10 mM Tris-succinate, pH 7.8, 10 mM $MgCl_2$, and 1 mM spermidine. The sucrose gradients may be produced according to the diagram given in Fig. 1. Other apparatus may be built to give linear sucrose gradients.[17] Centrifuge the gradient in a SW 25 rotor at 23,000 rpm for 3 hours in a Spinco Model L centrifuge. Perform all operations at 0 to 4°. The centrifuged gradient may be monitored at 254 nm by passing the gradient through the photocell of an ISCO spectrophotometer, Model 180 or some other UV analyzer. Alternatively, the tubes may be punctured at the bottom by a special device and the contents of the tube slowly col-lected dropwise. Water (about 1 ml) can be added to the collected sucrose fractions and absorbance determined at 260 nm. Polyribosomal profiles obtained from sugar beet tuber tissue used in these techniques are illus-trated in Fig. 2.

[17] J. H. Cherry, "Molecular Biology of Plants: A Text-Manual." Columbia Univ. Press, New York, 1973.

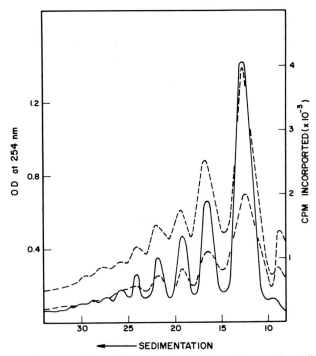

FIG. 2. Polyribosomes isolated from washed (8 hours) sugar beet disks. ————, OD at 254 nm; - - -, incorporation of [³H]uridine into the ribosome materials of control and γ-irradiated tissues.

B. Polyribosome Isolation with DEP

Homogenize 3 g of plant tissue in 15 ml of a Tris buffer (Tris·HCl pH 7.5, 50 mM MgCl₂, 5 mM KCl, 15 mM dithiothreitol, 0.67 mM and sucrose 0.25 M and 0.02 ml of DEP) at high speed for 8 seconds with a Willems Polytron PT20st (a VirTis Model 45 homogenizer may also be used).

The presence of DEP in the homogenization medium changes the pH because carbonic acid is formed.[18] Therefore, an additional amount of Tris should be added to counteract the changing pH caused by DEP. Both the additional Tris and DEP should be added immediately before the tissue is homogenized. Stock solutions of Tris (1.0 M, untitrated) and DEP should be used. Since DEP has a short half-life in water,[11] an aqueous solution should not be made prior to tissue homogenization. According to Anderson and Key,[15] 0.125% DEP in the grinding medium is sufficient to inhibit RNase activity in mung bean embryonic axes. However, Weeks and

[18] I. Fedorcsak and L. Ehrenberg, *Acta Chem. Scand.* **20**, 107 (1966).

Marcus[11] used 1% DEP in the grinding solution to inhibit RNase of wheat embryos.

Centrifuge the homogenate at 20,000 g for 15 minutes and filter through one layer of Miracloth (Calbiochem). Polyribosomes are prepared by centrifuging the supernatants through 3.0 ml of 1.5 M sucrose containing the above Tris-buffer at 150,000 g for 3 hours (Spinco type 50 rotor). The polyribosome pellets are suspended in 1.0 ml of the Tris grinding buffer. Preparations used for sucrose gradient centrifugation are suspended in the same Tris-buffer containing 1.0% Triton X-100 (v/v). The Triton X-100 causes dissociation of some nonribosomal material, which floats on the sucrose gradients. A 0.4-ml sample of ribosomes containing approximately 16 A_{260} units is layered on a 25 ml, 10 to 30% w/v linear sucrose gradient and centrifuged at 23,000 rpm for 3 hours (Spinco SW 25 rotor). Gradients are fractionated as described above (A).

[62] Isolation of Mitochondria from Plant Material

By George G. Laties

It would be both unrealistic and misleading to suggest there is a single ideal or preferable way to isolate mitochondria from plant materials. The very nature of this volume, describing separately as it does techniques for the preparation of membranes from a variety of sources, is indicative of the special, and sometimes unique, aspects which attend the isolation of specific cell components or organelles from different tissues. With respect to the isolation of mitochondria in general, and plant mitochondria in particular, there are both broadly pertinent considerations and considerations specific to the material. The outstanding feature of plant cells in the context of mitochondrial isolation is the ubiquitous central vacuole replete with contents potentially disastrous to mitochondrial well-being. Much of the methodology pertaining to the isolation of mitochondria from plant materials is tacitly or explicitly addressed to the related problems of sparing the mitochondria from the depredations of vacuolar contents, and separating mitochondria from the latter as quickly as possible.

General Considerations

Ideally, isolated mitochondria should display the same characteristics as mitochondria *in vivo*—both morphologically and biochemically. Comparison of isolated mitochondria, with mitochondria in live sections under the phase microscope, permits an effective appraisal of the morphological condition of the isolated organelle.[1] Minimal but useful biochemical criteria of respiratory activity indicative of mitochondrial intactness have been taken to be the lack of need for exogenous cofactors (other than ADP), manifestation of pronounced respiratory control (RC) and P:O ratios approaching the theoretical (see Bonner[2]). The latter criteria have been embraced both because of subjective predilection and the fact that clearly abused mitochondria fail to display the characteristics in question. The full meaning of state 4 respiration remains to be elucidated (see below), and the extent of respiratory control *in vivo* in tissue which shows no respiratory enhancement in response to uncouplers of oxidative phosphorylation must remain a matter of conjecture.

In recent years increasing attention has been focused on cellular microbodies,[3] particularly glyoxysomes[4] and peroxisomes,[5] which sediment with the mitochondria in conventional isolation techniques based on differential centrifugation. There is convincing evidence of biochemical interaction between mitochondria and one or another of these microbodies,[6] and this is particularly true in green leaves characterized by the conventional C-3 photosynthetic pathway with attendant high photorespiration.[6] Thus, where the nature of the study demands it, mitochondrial isolation techniques must provide for separation of mitochondria from other oxidative organelles[7] as well as from chloroplasts.[8] An additional problem is inherent in the changing condition of organelles and endoplasmic reticulum in the maturation of plant cells and tissues. In general membranicity is sharply enhanced with cellular maturation—with respect to both the endoplasmic reticulum and the mitochondrial inner membranes (cristae).[9] Thus membrane contamination of mitochondrial pellets and the density and voluminousness of such pellets may be utterly different in preparations from a given tissue at different stages of development.

Pitfalls in the Preparation of Mitochondria from Plant Tissues

Structural Disorganization and Free Fatty Acids. The most blatant difficulties connected with the preparation of plant mitochondria comprise

---

[1] S. I. Honda, T. Hongladarom, and G. G. Laties. *J. Exp. Bot.* **17**, 460 (1966).

[2] W. D. Bonner, Jr., this series, Vol. 10, p. 126.

[3] E. H. Newcomb and W. M. Becker, this volume [51].

[4] H. Beevers and R. W. Breidenbach, this volume [58].

[5] N. E. Tolbert, this volume [74].

[6] H. Beevers, *in* "Photosynthesis and Photorespiration (M. D. Hatch, C. B. Osmond, and R. Slatyer, eds.), p. 483. Wiley, New York, 1971.

[7] C. A. Price, this volume [52]; [cf. R. Douce, E. L. Christensen, and W. D. Bonner, Jr., *Biochim. Biophys. Acta* **275**, 148 (1972)].

[8] S. I. Honda, this volume [55].

[9] M. Chrispeels, A. E. Vatter, and J. B. Hanson, *J. Roy. Microsc. Soc.* **3**, 85 (1966).

structural disorganization—most often with attendant swelling, and inhibition or perversion of normal mitochondrial function by endogenous free fatty acids (FFA),[10] phenols and/or quinones[11] and by seemingly unnatural ATPase[12] and lipase[13] activity. In extreme cases loss of endogenous cofactors (i.e., NAD, thiamine pyrophosphate, CoA, cytochrome c, Mg^{2+}) may further reflect deterioration. An understanding of the sources of difficulty illuminates to some degree the multitudinous recipes and techniques which have been reported for isolation of plant mitochondria exhibiting pronounced respiratory control, or tight coupling.[14–17] Plant mitochondria swell spontaneously in the presence of a permeant solute (i.e., KCl) under conditions of curtailed electron transport or in the absence of ATP.[17,18] Such swelling is normally reversible, but where degradation of mitochondrial membranes occurs, primarily in response to endogenous lipase activity (i.e., lipase, phospholipase, galactolipase), swelling may proceed beyond all hope.[10,19] Lipases are rife in potato homogenates,[13] and appear to be active in most plant tissues immediately upon homogenization or even in response to rough handling.[13] Ca^{2+} is in many instances an activator of lipase[20] and phospholipase,[21, cf. 13] and the frequent effectiveness of a chelator, usually EDTA, in the extraction medium in improving ultimate mitochondrial behavior[9] may well have to do with Ca^{2+} chelation and consequent suppression of lipase activity.[22] In this connection a word is in order regarding the widespread use of EDTA. EDTA has a broadly similar high affinity for Ca^{2+} and Mg^{2+}, and excessive or prolonged use of EDTA introduces the danger of depleting mitochondrial Mg^{2+}. Parenthetically, the expected utility of the frequent simultaneous use in extraction media of Mg^{2+} and EDTA, the latter frequently at concentrations in excess of added Mg^{2+}, is at best ambiguous. It would seem more appropriate to supplant EDTA with EGTA (ethylene glycol bis(β-aminoethyl ether)-N,N'-tetraacetic acid) which binds Ca^{2+} with an avidity six orders of magnitude greater

[10] M. J. Earnshaw, B. Truelove, and R. D. Butler, *Plant Physiol.* **45**, 318 (1970).
[11] W. D. Loomis, this volume [54].
[12] T. K. Hodges, this series, Vol. 32 [36].
[13] T. Galliard, this volume [53].
[14] H. Ikuma, *Plant Physiol.* **45**, 773 (1970).
[15] I. V. Sarkissian and H. K. Srivastava, *Plant Physiol.* **43**, 1406 (1968).
[16] H. S. Ku, H. K. Pratt, A. R. Spurr, and W. M. Harris, *Plant Physiol.* **43**, 883 (1968).
[17] C. D. Stoner and J. B. Hanson, *Plant Physiol.* **41**, 255 (1966).
[18] J. B. Hanson and T. K. Hodges, *Curr. Top. Bioenerg.* **2**, 65 (1967).
[19] M. J. Earnshaw, and B. Truelove, *Plant Physiol.* **45**, 322 (1970).
[20] E. D. Wills, *Advan. Lipid Res.* **3**, 197 (1965).
[21] R. M. C. Dawson, *Biochem. J.* **88**, 414 (1963).
[22] H. W. Knoche and T. L. Horner, *Plant Physiol.* **46**, 401 (1970).

than that for Mg^{2+}.[23,24] The deleterious consequences of the various types of lipase activity are not confined to the effect on membrane structure. The fatty acids which are released inhibit a variety of respiratory enzymes and uncouple oxidative phosphorylation as well.[10,25] The common use of fat-free bovine serum albumin (BSA) both in extraction and reaction media stems from the effectiveness of BSA in binding free fatty acids,[19,25] and in protecting as well from a swelling-inducing product of phospholipase activity, lysolecithin.[19] Ironically BSA may occasionally stimulate phospholipase by keeping its potential substrate stripped of adhering free fatty acids.[19] In short, it would be better to prevent lipase activity in the first place, and some lipase inhibitors have been reported.[20,26]

Phenolics. A variety of phenolics are to be found in plant cell vacuoles and *both* phenolics and their oxidized products, the quinones, combine with proteins in a variety of ways.[12] Quinone attachment is covalent in nature and in consequence perhaps more insidious. Minimization of the oxidation of phenolics during homogenization is an emphatic desideratum in the isolation of plant mitochondria, and the reduction of quinones formed in spite of all is an ancillary goal. A variety of phenolase inhibitors may be added to the homogenization medium of which metabisulfite[27] and 2-mercaptobenzothiazole[28] (MERCAP) appear to be most effective. MERCAP is not only a chelating agent—being especially effective in binding the copper of polyphenolase[28]—but a sulfhydryl compound as well. In consequence it may do double duty and in certain instances replace the need for EDTA or EGTA *plus* one or another sulfhydryl compound, such as cysteine or mercaptoethanol.[29] Polyvinylpyrrolidone (PVP) both in soluble and solid form (Polyclar AT) binds noxious phenolic compounds,[11] and, while PVP is normally used as the sole protective agent against phenolics, it may prove effective to use PVP and MERCAP together in special cases.

Membrane and Enzyme Contamination. Bonner and Voss (see Bonner[2]) first called attention to contaminating elements sedimenting with both the starch and mitochondrial pellets which lead to a lack of respiratory control when present in mitochondrial suspensions. In consequence a variety of centrifugation and washing techniques have been designed[2,14-17] to separate mitochondria both from heavier sediments (starch and debris)

[23] R. A. Murphy and W. Hasselbach, *J. Biol. Chem.* **243**, 5656 (1968).
[24] G. Schwarzenbach and H. Flaschka, "Complexometric Titrations," pp. 10–11. Methuen, London, 1969.
[25] L. Dalgarno and L. M. Birt, *Biochem. J.* **87**, 586 (1963).
[26] F. H. Mattson, R. A. Volpenhein, and L. Benjamin, *J. Biol. Chem.* **245**, 5335 (1970).
[27] D. M. Stokes, J. W. Anderson, and K. S. Rowan, *Phytochemistry* **7**, 1509 (1968).
[28] J. K. Palmer and J. B. Roberts, *Science* **157**, 200 (1967).
[29] G. G. Laties and T. Treffry, *Tissue Cell* **1**, 575 (1969).

and lighter elements, which may include swollen mitochondria (the fluffy layer) and membranous material.[12] Membranous material is often occluded in the low speed sediment comprising starch and debris, but membrane fragments also frequently contaminate the first relatively high-speed mitochondrial pellet, and resuspension and recentrifugation serve to free the mitochondria of membranous contaminants, which on washing remain in the supernatant fraction. Nonmitochondrial membrane-bound as well as soluble ATPase[12] provides at least one reasonable basis for the lack of respiratory control in impure mitochondrial suspensions. The possibility of membrane-associated lipase provides another.[22,30] Where differential centrifugation fails to yield homogeneous or uncontaminated mitochondria, density gradient centrifugation may prove necessary. While centrifugation through a single concentrated cushion (i.e., 0.6 M sucrose) may effect some purification,[10] multiple discontinuous or continuous density gradient centrifugation is indicated for more meticulous separation.[4,7]

Homogenization or Comminution of Plant Tissues

The extremes of plant tissue disruption range from gentle hand chopping with a razor blade[1] to homogenization at full speed with a power blendor. In general the excessive frothing and shear forces attending full speed blending make the latter a totally unsuitable method. Ku *et al.*[16] have compared razor blade chopping, mortar and pestle grinding, and power blending of tomato fruit flesh and found the methods to yield excellent, fair, and hopeless mitochondria, respectively, with the respiratory control ratio (RCR) as a criterion. Bulky storage tissue or otherwise suitable material may frequently be grated to good advantage, with no further disruption, either with a hand or power-driven vegetable grater. The latter method is gentle, and the yield is frequently diminished no more than 25–30% compared with power blending.[31, see 2] Again, bulky storage tissue can be homogenized in seconds with a vegetable juicer[29] (Oster Mfg. Co., Milwaukee) which consists of an aluminum perforated basket centrifuge with a toothed plate in its base, all contained in a plastic housing. Tissue pieces or cores (25–50 g at a time) are fed through a tubular port which opens just above the toothed plate, which spins together with the basket at approximately 3000 rpm (ca. 750 g). The basket is lined with a strip of Miracloth (Chicopee Mills, 1450 Broadway, New York 10018), a stiff porous "cloth" which effectively filters cell debris. As tissue impinges on the revolving plate, cells are rent in one act, and their contents are filtered virtually instantaneously, the homogenate being received in a chilled beaker

[30] R. L. Ory, L. Y. Yatsu, and H. W. Kircher, *Arch. Biochem. Biophys.* **264**, 255 (1968).

[31] L. L. Poulsen and R. T. Wedding, *J. Biol. Chem.* **245**, 5709 (1970).

in less than 5 seconds. Thus, while disruption is at high speed, mitochondria are not subjected to prolonged shearing, or to extended protein denaturation at liquid–air interfaces as in a blendor. When potato tissue is disrupted as described, starch grains appear to be almost totally unbroken and are readily sedimented for removal.

The Preparation of Mitochondria: Isolation from Potato Tuber Tissue as an Example

The following procedure for the isolation of mitochondria from potato tuber tissue provides ostensibly undegraded mitochondria with good respiratory control and P:O ratios. The procedure should prove equally effective for bulky storage tissues of all kinds, and, except perhaps for the method of comminution, for plant tissues in general. Blocks or cores of potato parenchyma (25–100 g) are cut from within the vascular ring of potato tubers, wrapped in a plastic film, and chilled in ice. A strip of Miracloth (21 × 2¼ inches) is loosely fashioned into a ring and spun into place against the wall of the basket centrifuge of a vegetable juicer (see above) by turning on the power. Five to 10 ml of extraction medium is directed against the spinning Miracloth filter from a pipette. The chilled tissue is then introduced into the port of the running juicer, followed at once by two volumes of ice cold extraction medium (see below) added in one act from a flask or beaker. A single strip of Miracloth will take the debris from several hundred grams of tissue without clogging. For larger quantities of tissue, where speed is desired, an extra standby basket-head may be lined with Miracloth in advance, and switched in a matter of seconds. The filtered homogenate is collected from the juicer spout into a beaker held in ice, and distributed among 50-ml round-bottom plastic centrifuge tubes.

The first centrifugation, in an RC 2-B Sorvall centrifuge with SS-34 angle head, is for 3 minutes at 4000 g. The supernatant fraction is gently poured into fresh cold centrifuge tubes through a Miracloth filter held in a funnel, to remove the small amount of residual froth from the original homogenate. Care is taken to leave the starch pellet undisturbed by sacrificing a milliliter or two of supernatant fluid just above the starch layer. The supernatant fraction is then spun at 39,000 g for 5 minutes, and the second supernatant is discarded. The pellet is resuspended in a small volume of extraction medium by repeatedly gently drawing the pellet into, and ejecting it from, a Pasteur pipette. Further medium is added to each cup to a volume of 40 ml, and the suspension is recentrifuged at 39,000 g for 5 minutes. The supernatant is once again discarded; the pellets are resuspended in small volumes of BSA-free suspending medium (see below), and the latter are combined so that mitochondria from 100 g of tissue are

finally contained in a volume of 1 or 2 ml (approximately 2 mg of mitochondrial N). Considerably more concentrated final suspensions can be
prepared if desired. Aliquots (ca. 0.1 ml) are taken for nitrogen determination following which enough BSA is added at once to the level of
1 mg/ml. Where nitrogen is determined on a total digested Nesslerized
aliquot rather than on precipitated protein, the contribution of the buffer
is readily corrected for. The concentrated suspension is stored in ice until
use. A more conventional preparative technique involves longer centrifugation times at lower relative centrifugal forces. Thus the initial centrifugation
designed to remove starch is carried out at ca. 800 g for 10 minutes, while
subsequent centrifugations are carried out at approximately 14,000 g for
20 minutes. The choice between the two methods is somewhat arbitrary.
The shorter times are favored on the suspicion that protection of mitochondrial integrity by the various addenda to the extraction and suspension
media is not complete, and that the shorter the exposure to potentially degradative influences the better.

Should further purification by density gradient centrifugation prove
desirable, the pellet from the first or second high speed centrifugation can
be resuspended in a small volume of extraction or suspension medium, and

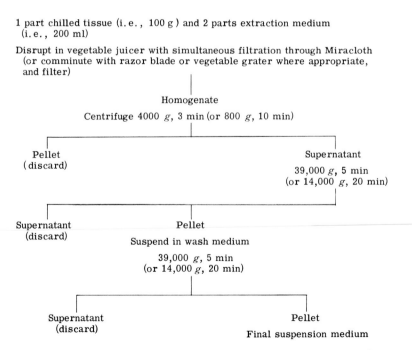

FIG. 1. Flow scheme for the preparation of mitochondria from potato tissue.

the concentrated suspension is then layered on a gradient of choice for further centrifugation.[7]

The procedure is summarized in a flow scheme (Fig. 1).

Reagents and Media

Extraction and Wash Medium
    Mannitol, 0.35 M
    Sucrose, 0.25 M
    N-tris(hydroxymethyl)methyl-2-aminoethane sulfonic acid (TES), 25 mM, pH 7.8
    BSA, 1 mg/ml   ⎫ Added just
    0.1 mg/ml Sodium mercaptobenzothiazole ⎬ before
        (MERCAP)[32]   ⎭ homogenization
    Dextran-40, 5%[33]
Final Suspension Medium
    Mannitol, 0.35 M
    Sucrose, 0.25 M
    TES, 25 mM, pH 7.8
    BSA, 1 mg/ml. Add to suspension immediately after removing aliquots for nitrogen determination.
    Dextran-40, 5%[33]
Reaction Medium
    Mannitol, 0.4 M
    TES, 25 mM, pH 7.4
    Mg^{2+}, 5 mM (SO_4 or Cl)
    Potassium phosphate, 5 mM pH 7.4
    BSA, 1 mg/ml
    Dextran-40, 5%[33]

General Comments

Osmotica. There is considerable latitude in the choice of osmoticum and the concentration thereof. With respect to the osmoticum the extraction medium given above is arbitrary and continues to be used simply because it serves well. Density and viscosity of the medium and membranicity of the homogenate are factors that determine the ultimate choice. The extraction medium given above is more hypertonic, dense, and viscous than

---

[32] The water-soluble sodium salt of 2-mercaptobenzothiazole is obtainable from Hopkin and William Ltd., Chadwell Heath, Essex, England. 2-Mercaptobenzothiazole is generally available, but must be dissolved in a stock alkaline alcoholic solution.
[33] Added at each stage, or occasionally to reaction mixture only, when rodlike mitochondria are desired; greater centrifugal forces may be needed where dextran is used; see text.

is necessary for a variety of plant materials, but it serves to favor the separation of mitochondria from contaminating membranes. Potato juice is essentially equivalent in osmotic pressure to a 0.3 M mannitol solution; 0.3 M or slightly higher mannitol[2,14] makes a perfectly acceptable osmotium, as does sucrose at similar levels.[15,17] One reputed advantage of mannitol over mannitol-sucrose, or sucrose extraction mixtures, is that mitochondria are more readily separated from starch in washing procedures in mannitol solutions.[2,34] Hypertonicity during preparation does not seem to be a problem, and may even be advantageous.[15] The effect of tonicity in the reaction medium may be obscured by the nature of the osmoticum per se. High sucrose concentrations have long been considered inhibitory to respiration of mammalian mitochondria, and high sucrose concentrations in the reaction medium (in excess of 0.125 M) reportedly inhibit oxidations of plant mitochondria.[35] However concentrations of mannitol or glucose up to 1.5 M in the reaction medium do not seem to be troublesome to plant mitochondria.[35]

Chelators and Sulfhydryl Compounds. Where EDTA (ca. 1 mM to 8 mM),[14,16] EGTA (ca. 5 mM)[23,36] and/or cysteine (ca. 4 mM),[16] mercaptoethanol (ca. 4 mM)[1] or other sulfhydryl compounds are included in the extraction medium, the concentrations are frequently diminished or reduced to zero in subsequent washes or in the reaction medium. As pointed out previously, mitochondrial Mg^{2+} may be removed with excessive chelator, and excessive sulfhydryl addenda may have their own deleterious effects, disrupting protein disulfide linkages. The main purpose of both chelators and sulfhydryl compounds is to forestall major damage in the initial homogenate.

Buffers. In general buffers are included in the extraction medium, although to minimize the salt concentration the extract is occasionally kept at the desired pH by dropwise addition of base during homogenization.[2,34] While tris(hydroxymethyl)amino methane(Tris) is a popular buffer, its pitfalls have been described by Good *et al.,*[37] who have developed a series of effective zwitterionic buffers of which TES, included in the recommended extraction medium above, is one. Phosphate in the extraction medium may lead to mitochondria exhibiting unduly high acceptorless respiration[18,39]—i.e., high state 4 rates and consequent low RCR.

Salts. KCl (10 mM) is occasionally included in the extraction[16,38] and

[34] J. T. Wiskich, R. E. Young, and J. B. Biale, *Plant Physiol.* **39,** 312 (1964).

[35] L. C. Campbell, J. K. Raison, and C. J. Brady, in press (see ref. 41).

[36] J. B. Hanson, *Plant Physiol.* **51,** 357 (1973).

[37] N. E. Good, G. D. Winget, W. Winter, T. N. Connolly, S. Izawa, and R. M. M. Singh, *Biochemistry* **5,** 467 (1966).

[38] C. Lance, G. E. Hobson, R. E. Young, and J. B. Biale, *Plant Physiol.* **40,** 1116 (1965).

reaction media[14] on the assertion that it improves the appearance of the pellet and possibly increases RCR values.[38] Mg^{2+} has also been added to the homogenization medium with the avowed purpose of minimizing aggregation.[34] Again one may raise the question of the utility of adding Mg^{2+} together with EDTA at concentrations equivalent to or greater than the Mg^{2+} concentration.[34,35] $EGTA^{[23,24,36]}$ must surely be preferable if Mg^{2+} is desired in the extraction medium.

KCl, at least at high concentrations (ca. 0.2 M), leads to swelling in the absence of active metabolism, simply by dint of permeating the mitochondria.[17,18] It is an open question whether KCl at the levels frequently used (10 mM)[14,38] may be expected to exert the same effect given a longer time. The use of inorganic phosphate in the extraction and wash media poses problems of particular interest. Hanson has described and investigated the basis for mitochondrial respiration characterized by what he has termed "loose coupling," or acceptorless respiration, a condition wherein state 4 respiration is high, and the RCR is low, but where there is little or no outright ATPase activity.[18,39] Loose coupling is dependent on P_i, in particular on intramitochondrial P_i,[39] and is markedly intensified by presenting P_i together with a polycation, such as protamine or polylysine.[39] It is self-evident that loose coupling is a diametric alternative to the exalted and coveted condition of tight coupling. Extraction conditions may influence the preparation of tightly or loosely coupled mitochondria, respectively, without the latter necessarily reflecting a degenerated state. Phosphate in the extraction medium together with relatively long preparation times perhaps favor the production of loosely coupled mitochondria. Thus the fundamental meaning of state 4 respiration, specifically in the absence of outright ATPase, remains a matter of conjecture, together with the question of whether an ideal or theoretical RCR can be enunciated. An ancillary matter relates to the calculation of P:O ratios from recorder traces representing oxygen utilization by coupled mitochondria as measured on the oxygen electrode. More often than not theoretical P:O ratios for plant mitochondrial oxidations are attained only when state 4 rates are subtracted from state 3 rates. However, the original description of the calculation admonishes that no such subtraction be made.[40] Clearly the validity of the calculation process depends on the nature of state 4 and on whether the events of state 4 are or are not preempted by state 3.

Macromolecules. Extraction media recipes have frequently included one or another of a variety of high molecular weight polymers or other macromolecules. Many of these compounds have an uncomprehended if not mystical beneficial effect,[1] while others are added for an explicit pur-

[39] J. B. Hanson, *Plant Physiol.* **49**, 707 (1972); ibid. **50**, 347 (1972).
[40] R. W. Estabrook, this series, Vol. 10, p. 41.

pose. Thus, as has been noted, PVP and Carbowax (polyethylene glycol) protect against phenolics and quinones,[11] while BSA protects against free fatty acids and other lipid products.[10,19,25] In addition to the specific protective functions which are understood, macromolecules at appropriate concentrations dramatically affect the morphology of extracted plant cell organelles, and usually it is only in the presence of given macromolecules at or above a critical concentration that isolated mitochondria morphologically resemble the organelles *in vivo*.[1,29] Mitochondria as prepared by almost all of the methods described or cited herein are spherical—whether swollen or not. The preparation of rodlike, or filiform, mitochondria, such as exist *in vivo* in plant tissues,[1] demands macromolecule addenda either from the outset in the extraction medium or, if no injury is done the mitochondrial membranes during isolation, to the ultimate reaction medium. Effective macromolecules in this connection[1,29] include Ficoll (sucrose-epichlorhydrin polymer), dextran (polyglucose), PVP, Carbowax, BSA, and a variety of other proteins. Effective concentrations range from 2% or greater for Dextran-40 to 8% or greater for Carbowax. The influence of macromolecules is not osmotic, and levels which influence morphology are generally higher than those which exert specific protective action (for example in the cases of PVP and BSA).[29] Macromolecules contribute to a considerable and variable degree to the viscosity and density of the extraction medium. Thus centrifugal forces and times of centrifugation must be varied accordingly. Whether or not isolated mitochondria which morphologically resemble mitochondria *in vivo* differ functionally in any significant way from isolated ostensibly uninjured spherical mitochondria, remains to be determined.

Respiratory Measurements

Measurements of mitochondrial oxidations are normally carried out with the oxygen electrode. The process has been amply described.[2,40] However, with respect to comparative studies one point bears emphasis. In many, if not most, measurements of mitochondrial respiratory activity, where measurements are made so as to demonstrate respiratory control, the first state 3 rate is manifestly lower than subsequent state 3 rates. In fact, the state 3 rate may continue to increase through several consecutive alternations of state 3/state 4 respiration. Raison *et al.*[41] have termed the phenomenon "conditioning," and have convincingly demonstrated that the phenomenon is general, and that the initial low state 3 rates are not restricted to succinate oxidation, where oxaloacetate reputedly inhibits succinoxidase.[42] The meaning of conditioning apart, for comparisons to

[41] J. K. Raison, J. M. Lyons, and L. C. Campbell, *J. Bioenergetics* **4**, 397 (1973).
[42] R. E. Drury, J. P. McCollum, S. H. Garrison, and D. B. Dickinson, *Phytochemistry* **7**, 2071 (1968).

be meaningful, either between substrates or between different mitochondrial preparations, it is important that comparisons be made between fully conditioned mitochondria. Finally, in this connection, the endogenous substrate level of isolated mitochondria may have a marked effect on the state 3 rate in response to a given substrate,[43] as may the concentration of mitochondria in the reaction vessel.[43,44] With plant mitochondria, there is a *very sharp* drop in state 3 oxidative activity as the total mitochondrial protein drops below a critical level in the reaction chamber.[44]

[43] E. M. Tarjan and R. W. Von Korff, *J. Biol. Chem.* **242**, 318 (1967).
[44] J. K. Raison and J. M. Lyons, *Plant Physiol.* **45**, 382 (1970).

[63] Rapid Isolation Techniques for Chloroplasts

By Park S. Nobel

Once out of the plant cell, chloroplasts are rather labile. Consequently, a rapid and gentle procedure has been developed for their isolation that features (a) grinding the leaf material while it is in a special cloth bag, (b) removal of multicellular debris by filtering through cloth with an appropriate pore size, (c) a brief, low-speed centrifugation to obtain the chloroplast pellet, and (d) a total time of only 2 minutes (the procedure is summarized in the table). Pea chloroplasts isolated by this technique are initially 95% Class I or intact, whereas after storage for 30 minutes at 20° the Class I fraction decreases to 56%.[1] (In a phase contrast microscope a Class I chloroplast appears highly refractile, often being surrounded by a halo, and it has a pair of intact limiting membranes; the less refractile Class II chloroplasts are dark green with grana clearly visible and have lost their stromal material through the broken limiting membranes.) The rate of endogenous photophosphorylation decreases 47% when the isolated chloroplasts are stored for only 30 minutes at 20°, because of an outward diffusion of the Mg^{2+} on which it depends.[2] Thus a rapid technique for the isolation of relatively intact chloroplasts has not only the obvious advantage of saving time, but also it minimizes changes in structure[1] and losses of metabolic activities[2,3] during the isolation procedure itself.

The isolation of chloroplasts in aqueous solutions involves the follow-

[1] P. S. Nobel, *Plant Physiol.* **43**, 781 (1968).
[2] D. C. Lin and P. S. Nobel, *Arch. Biochem. Biophys.* **145**, 622 (1971).
[3] P. P. Kalberer, B. B. Buchanan, and D. I. Arnon, *Proc. Nat. Acad. Sci. U.S.* **57**, 1542 (1967).

SUMMARY OF A RAPID TECHNIQUE FOR ISOLATING CHLOROPLASTS[a]

Step	Elapsed time (seconds)
1. Use scissors to cut 10 g of pea leaves into a nylon bag (cloth pore size 60 μm × 80 μm) placed on a top-weighing balance	0–15
2. Transfer the bag to a prechilled mortar containing 10 ml of 0.25 M sucrose, 10 mM TES-NaOH (pH 7.5) or 10 mM Tris·HCl (pH 7.9) at 1° and grind for about 10 seconds with the double-layered bag intervening between the pestle and the mortar	15–30
3. Squeeze the contents of the bag into the mortar and pour the solution from the mortar into a prechilled 50-ml centrifuge tube placed in a bench-top centrifuge (with a balancing tube already in place)	30–40
4. Centrifuge for 60 seconds at an average force of 1000 g	40–100
5. Stop the centrifuge quickly (e.g., using a hand-held cloth), remove the tube, and decant the supernatant fluid	100–110
6. Gently resuspend the chloroplast pellet using a vortex mixer, cover the tube with aluminum foil to keep the chloroplasts in the dark, and put the centrifuge tube on ice	110–120

[a] Modified from P. S. Nobel, *Plant Physiol.* **42**, 1389 (1967); cited by permission.

ing steps: (1) selection of plant material, (2) choice of a suitable osmoticum, buffer, and sometimes other components of the isolation medium, (3) breaking of the leaf cells, (4) removal of unbroken cells and other large debris, and finally (5) separation of chloroplasts from cell wall fragments, mitochondria, nuclei, and other subcellular components. We will discuss these various steps in turn, relying mainly on the techniques developed by Walker[4] and by Nobel[5] for pea chloroplasts.

Plant Material

Chloroplasts are readily isolated from the mesophyll cells of suitable green leaves, which contain from about twenty to a few hundred chloroplasts per cell. Spinach (*Spinacia oleracea* L.) is a convenient plant for chloroplast isolation based on its availability from local markets, the near neutrality of its cell sap, and its low quantities of tannins (which can bind to the plastids during their isolation). Commercial sources can be undependable, but laboratory cultivation is often inconvenient because germination and initial growth of spinach are both relatively slow. Peas (*Pisum sativum* L.) germinate readily and grow rapidly, producing in 2 weeks at 20° about 100 g (fresh weight) of leaves from 100 g (dry weight) of seeds. Other common sources of chloroplasts include various varieties of

[4] D. A. Walker, *in* "Biochemistry of Chloroplasts" (T. W. Goodwin, ed.), Vol. II, p. 53. Academic Press, New York, 1967.
[5] P. S. Nobel, *Plant Physiol.* **42**, 1389 (1967).

beet, various species of tobacco, bush and broad bean, lettuce, pokeweed, and tomato. Since chloroplasts containing large starch grains will generally be broken during centrifugation, the plant material should not be intensely illuminated for long periods prior to chloroplast isolation. (Starch grains are essentially absent from pea chloroplasts in plants maintained at a moderate light intensity of 2000 lux for up to 12 hours.)

Isolation Medium

Perhaps the greatest diversity in the isolation procedures for chloroplasts occurs in the choice of components for the isolation medium. We will briefly discuss these under the headings of osmoticum, buffer, and other additions.

Osmoticum. Ideally, the osmoticum should not enter the chloroplasts and the concentration chosen should lead to the same osmotic pressure as that experienced by the organelles when they are in the plant cell. Obvious candidates like NaCl and KCl have proved to be unsatisfactory as osmotica, because they lead to a deteriorative swelling of the chloroplasts and a loss of those referred to as Class I.[3-6] In fact, nonelectrolytes have proved to be far superior as osmotica. For instance, sucrose, sorbitol, and mannitol are all essentially impermeant for pea chloroplasts.[7] These are the three osmotica that have come into widespread use in chloroplast studies, sucrose being the most common osmoticum employed. If the osmotic pressure of the isolation medium were the same as that of the cell cytoplasm, then the organelles would not be expected to change volume during isolation. For most leafy green vegetables, the osmotic pressure of the cytoplasm corresponds to a sucrose concentration of 0.2–0.4 M. For example, the osmotic pressure of sap expressed from pea leaf mesophyll cells is the same as that of 0.27 M sucrose (the osmotic pressure approximately equals $RT \sum_j c_j$ by the Van't Hoff relation, where c_j is the concentration of species j and the sum is over all solutes present).

Buffer. The selection of a buffer involves a consideration of nearness of the buffer pK to the pH desired, concentration, and whether an anion or a cation added in the buffer titration is less detrimental. Good *et al.*[8] have enumerated various mainly zwitterionic buffers suitable in the pH range from 6.0 to 8.6. The most common ones for chloroplast isolation are N-tris(hydroxymethyl)methyl-2-aminoethanesulfonate (TES, which has a pK of 7.5 at 20° and is generally titrated with a base such as KOH or NaOH) and tris(hydroxymethyl)aminomethane (Tris, which has a pK

[6] D. Spencer and H. Unt, *Aust. J. Biol. Sci.* **18**, 197 (1965).
[7] C.-T. Wang and P. S. Nobel, *Biochim. Biophys. Acta* **241**, 200 (1971).
[8] N. E. Good, G. D. Winget, W. Winter, T. N. Connolly, S. Izawa, and R. M. M. Singh, *Biochemistry* **5**, 467 (1966).

of 8.3 at 20° and is titrated with an acid such as HCl). Tricine [N-tris (hydroxymethyl)methylglycine] has a pK of 8.2 at 20°, but its appreciable binding of divalent cations such as Ca^{2+} and Mg^{2+} must be kept in mind. Another common buffer is phosphate, which has a pK of 7.2 at 20°; however, it is involved in photophosphorylation and other reactions. Buffer concentrations should be kept reasonably low, e.g., 5 to 20 mM, which will minimize the amount of ion added in the titration of the buffer. For rapid isolation techniques, pH values near 7 or slightly above have proved to be satisfactory.

Other Additions. There is rather little systematic experimentation on the effect of other components often found in chloroplast isolation media. The following substances are sometimes present: (a) 1–10 mM $MgCl_2$ (originally added to prevent uncoupling,[9] but not essential for rapid isolation techniques[2]); (b) $MnCl_2$ (can be replaced by $MgCl_2$, which itself is not really necessary); (c) 1–2 mM sodium isoascorbate (added to minimize oxidation of chloroplast components during isolation, particularly in CO_2 fixation studies[10,11]); (d) 2 mM EDTA (may bind Ca^{2+} and other inhibitory divalent cations, but also binds Mg^{2+}—generally not necessary, but see references cited in footnotes 3, 9–11); (e) 0.1–1% bovine serum albumin (binds fatty acids, which apparently inhibit photophosphorylation by bean chloroplasts[12]); (f) 2–5% dextran, Ficoll, or polyvinylpyrrolidone (could adsorb polyphenols released from plant tissues,[13] but generally not necessary for pea or spinach chloroplasts); (g) various compounds containing -SH groups such as dithiothreitol, dithioerythritol, glutathione, or mercaptoethanol (can prevent the oxidation of sulfhydryls in chloroplasts, which is not as crucial for rapid isolation techniques as it is for the more time-consuming conventional ones).

Breaking Cells

We will next present the three different techniques that are commonly employed for breaking open leaf cells. The leaves can be placed in a motor-driven blender (e.g., a Waring Blendor) and macerated for a short period (3–10 seconds) at a moderate speed. (At full speed, cavitation is excessive and the rapidly rotating blades cause much frothing, while long periods of maceration tend to disrupt the chloroplasts.) Generally, about 2 parts of isolation medium to one part of leaf tissue (by weight) are used in the blender. A somewhat less reproducible but gentler technique for breaking cells involves the use of a mortar and pestle. Often

[9] A. T. Jagendorf and M. Smith, *Plant Physiol.* **37**, 135 (1962).
[10] R. G. Jensen and J. A. Bassham, *Proc. Nat. Acad. Sci. U.S.* **56**, 1095 (1966).
[11] W. Cockburn, D. A. Walker, and C. W. Baldry, *Plant Physiol.* **43**, 1415 (1968).
[12] A. O. Gyldenholm and F. R. Whatley, *New Phytol.* **67**, 461 (1968).
[13] S. I. Honda, T. Hongladarom, and G. G. Laties, *J. Exp. Bot.* **17**, 460 (1966).

equal weights of leaves and isolation medium are placed in the mortar and then ground for a short period (10–15 seconds). The use of added abrasives is recommended only if necessary—in fact, the contact between mortar and pestle itself can prove to be too severe, in which case the leaf material may be ground while in a bag (see below). A final category of techniques involves breaking the leaf tissue by chopping or cutting. This varies all the way from using a motor-driven block in which razor blades are inserted[13] to an ordinary kitchen juicer which has a rotating serrated blade that rips the leaf tissue.[14] For any technique, tissues relatively poor in chloroplasts such as large veins (e.g., the midrib of a spinach leaf) and excess stems are generally avoided or removed first.

Removal of Large Debris

The slurry produced upon breaking open plant cells contains subcellular components plus large debris consisting of broken cells, intact cells, and multicellular leaf fragments. Conventional isolation procedures remove the large debris by a brief, low-speed centrifugation (e.g., 4 minutes at 500 g). Rapid techniques for chloroplast isolation achieve this objective by filtering the slurry instead of centrifuging it.[3–6]

In the simpler of the two filtering techniques, the slurry is strained through muslin or cheesecloth. Six to eight layers have proved satisfactory for this purpose, but the optimal number of layers depends somewhat on the manufacture of the cloth (it is of course prudent to examine the effect of all stages of a chloroplast isolation procedure with a phase contrast microscope). Both muslin and cheesecloth are made of cotton and consequently absorb appreciable amounts of the isolation medium. Also, the "pores" are not uniform in size in such cloth and some of the openings may allow cells and multicell fragments to pass through. Other materials with the same general characteristics as cheesecloth and muslin include cotton fiber pads and commercially available filtering cloth (e.g., Miracloth, Chicopee Mills, Inc., 1450 Broadway, New York, New York 10018).

A more sophisticated filtering procedure is to use nonabsorbing material of uniform pore size.[5,6] If the pore opening is chosen to be just smaller than the size of the mesophyll cells, then the filter will retain whole cells and multicellular leaf fragments. For instance, nylon cloth (mesh about 50 per cm) with rectangular holes 60 μm by 80 μm retains pea leaf cells and other large debris, but allows the chloroplasts and other subcellular components to pass through.[5] (It should be pointed out that it is not the mesh but rather the actual opening between the fibers which is critical in determining the filtering action.) Moreover, nylon, dacron, and other syn-

[14] A. B. Tolberg and R. I. Macey, *Biochim. Biophys. Acta* **109**, 424 (1965).

thetic polymers do not absorb water, and so cloth made of such material absorbs very little isolation medium and also it is readily cleaned after each use. Suitable "informal" sources of durable filtering cloth of the proper pore size include the nylon or dacron used to line women's dresses and the cloth used for spinnaker and other sails. A less exotic and more expensive source is Tobler, Ernst, and Traber, Inc. (71 Murray Street, New York, New York 10007), who supply Nitex monofilament nylon bolting cloth of various pore sizes.

To facilitate chloroplast isolation, the filtering cloth can be made into a small bag into which the leaf material is placed. The bag is then set in a mortar containing the isolation medium (approximately the same amount by weight as the leaves to be ground) and the pestle inserted through the top, or open end, of the bag. Subsequent grinding of the leaf material against the inside of the bag has proved to be a very gentle way of breaking the mesophyll cells. In this procedure, the pestle crushes the leaf material against the cloth, the mortar serving as a rigid support behind the bag. (The bag can be made of two layers of cloth for added strength and reliability as the grinding tends to break the weaker fibers.) After grinding, the bag contents are squeezed into the mortar and the solution containing the chloroplasts is poured from the mortar into a centrifuge tube.[5]

Chloroplast Pellet

Rapid isolation techniques use a shorter centrifugation period to form the chloroplast-containing pellet than do the conventional procedures. In fact, the longer times and the higher centrifugal forces commonly used cause the sedimentation of many mitochondria, chloroplast fragments, and other small particles with the chloroplasts. A suitable chloroplast pellet can be obtained by centrifuging for 60 seconds at an average force of 1000 g. (Such a centrifugal force is within the capability of an ordinary bench-top centrifuge, e.g., an International Clinical Centrifuge.) The centrifugation step can be shortened by using a higher centrifugal force (which tends to break some of the chloroplasts) or by using the same force for a shorter time (which lowers the yield of chloroplasts). Centrifugal force varies linearly with radius and thus is lower at the top of the centrifuge tube. Also, the chloroplasts at the top of the suspension have a greater distance to travel along the tube in order to reach the pellet at the bottom. Consequently, it may be advantageous to use an oversize centrifuge tube, e.g., if 10 ml of solution is to be centrifuged, it could be placed in a 50-ml tube to make the centrifugal force more uniform throughout the solution and to reduce the overall path length compared with placing the same amount of solution in a (narrower) 12- or 15-ml tube. After decanting the supernatant fluid following the centrifugation step, the rather loose pellet can be resuspended by placing the centrifuge tube on a vortex mixer for 3–5 seconds (preferably a variable-speed mixer run at a moderate speed). The pellet can also be resuspended using a pipette or a glass rod (the latter to avoid bubbles). Although not necessary, isolation medium can be added to the centrifuge tube during the resuspension of the pellet.

The resuspended chloroplasts will inevitably be contaminated with a certain amount of nuclei, mitochondria, other organelles, chloroplast fragments, pieces of cell wall, whole cells, and perhaps multicellular fragments. Upon examination with a phase contrast (or even an electron) microscope the contamination may be deemed insignificant for the purpose in mind. If the contamination with mitochondria and other small debris is judged to be excessive, isolation medium can be added to the original pellet, the tube recentrifuged (e.g., 60 seconds at 1000 g), the supernatant fluid decanted, and the "washed" chloroplast pellet then resuspended. On the other hand, if the contamination of the chloroplasts is mainly with whole cells and other large debris, a smaller pore size for the filter should be chosen.

Many chloroplast properties are conveniently presented per unit weight of chlorophyll. After extraction into an organic solvent, chlorophyll can be measured spectrophotometrically. For instance, the amount of chlorophyll can be calculated using an equation derived by Arnon[15] appropriate for absorption coefficients determined by MacKinney[16] at 645 and 663 nm for 80% acetone in water (v/v): $(0.802 \text{ absorbance}_{663} + 2.02 \text{ absorbance}_{645}) \times$ (dilution factor/100) gives the chlorophyll concentration in milligrams per milliliter.

Example

The table summarizes a rapid isolation technique used for isolating chloroplasts from many leafy green vegetables, such as pea and spinach. The procedure is carried out with the mortar and pestle, isolation medium, centrifuge tube, and tube adapter for the bench-top centrifuge all prechilled to near 1°. However, when the isolation is performed with the apparatus at 22°, the rate of endogenous photophosphorylation and the fraction of Class I chloroplasts are only slightly less.[5] The lower temperature is preferable, but chloroplast isolation at room temperature without a great loss of activity or deterioration of chloroplast structure is possible with a rapid isolation technique. Moreover, the entire procedure for chloroplast isolation as outlined in the table requires only 2 minutes.

[15] D. I. Arnon, *Plant Physiol.* 24, 1 (1949).
[16] G. MacKinney, *J. Biol. Chem.* 140, 315 (1941).

Section V

Preparations Derived from Unicellular Organisms

[64] Isolation of Spheroplast and Membrane Vesicles from Yeast and Filamentous Fungi[1]

By WILLIAM R. WILEY

Protoplasts and/or spheroplasts formed by the enzymatic removal of the cell walls of bacteria have been very useful sources of cytoplasmic membranes for the investigation of cell membrane structure and function in bacteria. Kaback[2] and his collaborators have used a technique of mild osmotic shock and EDTA-RNase-DNase treatment of spheroplast and protoplast to prepare cytoplasmic membrane vesicles from gram-positive as well as gram-negative bacteria. Studies using these membrane vesicles have contributed greatly to our understanding of the mechanism of energy-coupling and nutrilite translocation in bacteria.

Cytoplasmic membrane vesicles have recently been prepared from yeast[3] and *Neurospora crassa* spheroplast.[4] These structures, which in many aspects are homologous to those of bacteria, may provide an excellent tool for the study of membrane structure and function in eukaryotic cells. In addition, intact spheroplasts of yeast and fungi may prove valuable in the study of cell wall synthesis, enzyme induction, nutrilite transport and for the isolation of particulate enzyme systems from eukaryotic cells.

The principles involved in the preparation of spheroplast from yeast and fungal cells are very similar to those for bacteria; i.e., the cells are treated with agents which degrade or prevent cell wall synthesis. The spheroplast which are released from the cell walls are stabilized in media containing hypertonic sucrose, mannitol, or inorganic salts. In the case of some bacteria a single enzyme exists (e.g., lysozyme) which is capable of complete degradation of the rigid layers of the cell wall. No single enzyme has been isolated which is capable of accomplishing this feat for yeast and fungal cell walls.

There are naturally occurring cell wall-less mutants of some fungi which are much like L forms of bacteria. Emerson and Emerson[5] isolated a spherical, osmotically fragile mutant of *N. crassa* (slime) which is genetically blocked at one or more sites in cell wall synthesis. The cells

---

[1] This work was supported by Battelle Institute.
[2] E. M. Barnes, Jr. and H. R. Kaback, *J. Biol. Chem.* **246**, 5518 (1971).
[3] A. A. Boulton, *Exp. Cell Res.* **37**, 343 (1965).
[4] W. R. Wiley, unpublished results.
[5] S. Emerson and M. R. Emerson, *Proc. Nat. Acad. Sci. U.S.* **44**, 668 (1958).

grow and divide as spheroplastlike structures in the presence of appropriate growth factors containing an osmotic stabilizer.

Use of the terms spheroplast and protoplast in referring to the intact spherical fungal or yeast protoplasm released by digestion of the cell wall is not uniform among investigators working with these organisms. Many of the workers use the term protoplast and spheroplast interchangeably. Since, in many cases, no definitive evidence showing the absence of cell wall material (mannans and glucans) has been demonstrated,[6,7] we chose the term spheroplast to describe the intact protoplasmic structures released following enzymatic hydrolysis of the cell wall.

It is beyond the scope of this review to describe all the methods reported in the literature for the preparation of spheroplast and membrane vesicles from yeast and fungi. Instead we have attempted to include those enzymatic methods which are suitable for the isolation of large quantities of spheroplast and consequently cytoplasmic membrane vesicles and other subcellular structures for biochemical studies.

The table provides a list of sources for enzymes which hydrolyze yeast

SOURCES OF YEAST AND FUNGAL CELL WALL HYDROLYZING ENZYMES

Sources of cell wall hydrolyzing enzymes	Enzymes postulated to be active in cell wall hydrolysis	Cell walls hydrolyzed (genus)	References
Gut juice, *Helix pomatia*	Endo-β-(1 → 6)- and β-(1 → 3)-gluconases	*Saccharomyces, Candida, Neurospora*	a–d
Culture filtrates of *Penicillium melinii*	Endo-α-D-(1 → 3)-gluconase, α-(1 → 3)-glucosidases, and endo-α-D-(1 → 4)-gluconase	*Aspergillus*	e
Culture filtrates of *Streptomyces* sp. (strepzymes)	β-Gluconases	*Phytophthora, Candida, Torula, Torulopsis*	f, g
Bacillus circulans	Endo-β-(1 → 3)- and endo-β-(1 → 6)-gluconases	*Saccharomyces*	h

[a] B. J. Bachmann and D. M. Bonner, *J. Bacteriol.* **78**, 550 (1959).
[b] R. P. Langley, A. H. Rose, and B. A. Knights, *Biochem. J.* **108**, 401 (1968).
[c] F. B. Anderson and J. W. Millbank, *Biochem. J.* **99**, 682 (1966).
[d] A. A. Eddy and D. H. Williamson, *Nature (London)* **179**, 1252 (1957).
[e] K. K. Tung, A. Rosenthal, and J. H. Nordin, *J. Biol. Chem.* **246**, 6722 (1971).
[f] A. A. Boulton, *Exp. Cell Res.* **37**, 343 (1965).
[g] C. Garcia Mendoza and J. R. Villanueva, *Nature (London)* **202**, 1241 (1964).
[h] H. Tanaka and H. J. Phaff, *J. Bacteriol.* **89**, 1570 (1965).

[6] S. Bartnicki-Garcia and E. Lippman, *J. Gen. Microbiol.* **42**, 411 (1966).
[7] B. J. Bachmann and D. M. Bonner, *J. Bacteriol.* **78**, 550 (1959).

and fungal cell walls. The snail gut enzyme preparation has been used more extensively than the enzymes from other sources.

The cell walls of yeast and many fungi are comprised of the polysaccharides, principally glucans and mannans. Most of the mannans have a branched structure with disulfide bridges between the mannans and cell wall proteins. As noted in the table the β-gluconases are present in all enzyme preparations which are capable of hydrolyzing fungal cell walls.

Method for Spheroplast Formation from Yeast Using the Digestive Juices of the Snail *Helix pomatia*

Modification of the Kornblatt and Rudney Method for Yeast[8]

One of the most readily available sources of enzymes capable of degrading yeast cell walls is the gut juice from the snail *Helix pomatia*. While some fractionation of the commercially available preparations is desirable, it is not necessary for the preparation of stable yeast spheroplasts. Anderson and Millbank[9] fractionated whole gut juice preparations into three fractions by gel filtration using Sephadex G-100. Of the three fractions examined (fractions A, B, C) only fraction C was capable of spheroplast formation. Fraction C was devoid of proteases, phosphatases, and lipases, all of which are present in whole snail juice. The authors concluded that at least two β-gluconases, which are present in fraction C, are required for hydrolysis of yeast cell walls (endo-β-1 $\rightarrow$ 3- and endo-β-1 $\rightarrow$ 6-gluconase) and that exoglucosidase present in fractions A and B are responsible for the hydrolysis of β-linked glucose disaccharides remaining after hydrolysis by the endo-β-gluconases.

Growth and Preparation of Cells for Spheroplast Formation

Cells of baker's yeast are grown with vigorous agitation in medium containing mineral salts, 0.5% yeast extract, 1% tryptone, and 2% sucrose. The cells are harvested in mid-log phase by centrifugation at 9500 rpm in a Servall RC-2 refrigerated centrifuge (GSA rotor) and washed two times in 20 mM-Tris containing 10 mM MgSO$_4$. The 9500 rpm cell pellet (15 g wet weight packed cells) is resuspended in a volume of medium (20 mM-Tris, pH 8.0, 25 mM EDTA, and 0.2 M 2-mercaptoethanol), equivalent to approximately three times the wet weight of the cell pellet. The cells are incubated in the 2-mercaptoethanol medium 30 minutes at 30° and harvested by centrifugation. Pretreatment of the yeast with 2-mercaptoethanol is thought to function by reducing disulfide bridges

[8] J. A. Kornblatt and H. Rudney, *J. Biol. Chem.* **246**, 4424 (1971).
[9] F. B. Anderson and J. W. Millbank, *Biochem. J.* **99**, 682 (1966).

between mannans and protein complexes in the cell wall.[10,11] Reduction of the disulfide bridges is thought to enhance degradation of the cell wall mannans by mannanases present in the snail juice. Other disulfide reducing agents will function to reduce the disulfide linkages, however 2-mercaptoethanol is the most suitable agent for this purpose. Following the 20 minutes' incubation in 2-mercaptoethanol the cell pellet is washed and immediately resuspended in 30 ml of either 20 mM phosphate, acetate, or citrate buffer (pH 5.8–6.2) containing 10 mM MgSO$_4$ and either 0.6 M mannitol, 0.8 M MgSO$_4$, 0.5 M sucrose, 1 M NaCl, or 0.8 M LiCl as an osmotic stabilizer.

Preparation of Commercial Snail Juice Enzymes

The snail juice preparation obtained from Endo Laboratories (USA), Boehringer Corporation (USA), or Industrie Biologique Française (Gennevilliers, Seine, France) is diluted 1:1 with 0.5 M NaCl and centrifuged at 15,000 rpm for 10 minutes at 0°. The pellet from the 15,000 rpm spin is discarded and approximately 0.1 volume of a freshly prepared solution of 1% cysteine hydrochloride added to the supernatant. Sutton and Lampen[12] suggest that treatment of the snail juices with cysteine-HCl serves to inactivate preservatives present in the preparation. The snail gut preparations contain between 80 and 200 mg of protein per milliliter and from 173 × 10^3 to 186 × 10^3 units of β-glucuronidase.

Formation and Isolation of Yeast Spheroplast

Yeast cells suspended in 25 ml of buffer containing the appropriate osmotic stabilizer are placed in a water-bath shaker, and the suspension was allowed to equilibrate at 30° before the addition of snail juice enzymes. After equilibration of the cell suspension at 30°, a sufficient volume of snail gut preparation is added to give a final concentration of about 10 mg of snail gut juice protein per milliliter of reaction mixture (as measured by the Lowry *et al.* protein assay). Spheroplast formation begins immediately, as judged by the increased osmotic fragility of the cell suspension. After 30 minutes, approximately 90–95% of the cells are converted to spheroplast. The proportion of whole cells to spheroplast can be enumerated microscopically by counting the whole cells and spheroplast in a Petroff-Hauser counter. Spheroplast are readily differentiated from whole cells using either phase contrast, Nomarski or interferometric microscopy. The relative concentration of whole cells can be determined spectrophoto-

[10] R. Davies and P. A. Elvin, *Biochem. J.* 93, 8p (1964).
[11] J. S. D. Bacon, B. D. Milne, I. F. Taylor, and D. M. Webley, *Biochem. J.* 95, 28c (1965).
[12] D. D. Sutton and J. O. Lampen, *Biochim. Biophys. Acta* 56, 303 (1962).

metrically by diluting the osmotic tonicity of the medium and measuring changes in turbidity of the cell suspension at 600–700 nm.

Spheroplasts are harvested from the snail juice mixture by diluting the suspension with 2 volumes of the hypertonic medium,[8] followed by centrifugation at 1000 g for 10 minutes. The spheroplast pellet from the 1000 g centrifugation is then washed 3 times in 200 ml of hypertonic medium containing 50 mM MgSO₄. Yeast spheroplast prepared in this manner are stable for 24 hours at 4°, as judged by the rates of incorporation of radioactively labeled amino acids into protein and by microscopic examination.[4]

These methods, or minor variations of them, have been used to release spheroplast from *Saccharomyces carlsbergensis*,[9] *Saccharomyces fragilis*,[10] and *Candida utilis*.[13] It should be emphasized, however, that the efficiency of conversion of cells to spheroplast (i.e., time required for complete conversion) using snail juice enzymes varies with the physiological state of the cells and the strain of organism.[9] Therefore it may be necessary to modify the basic methods described here to accommodate the particular organism, the physiological state of the cell, and the parameters which the investigator wishes to examine.

The Strepzyme Method for Yeast Spheroplast Formation[13-16]

Method of Villanueva et al. for Yeast[13,15]

Strepzyme is a lytic enzyme preparation that can be isolated from the culture medium after growth of *Streptomyces* sp. or from *Micromonospora*. The principal value of the strepzyme method for spheroplast formation is that it provides an alternative source of enzymes for yeast whose cell walls may be resistant to the enzymes present in snail gut juices.

The strepzymes are obtained by growing *Micromonospora* AS or *Streptomyces* sp. 6 days with vigorous agitation.[13] The culture fluids are separated from the cells by centrifugation. Solid ammonium sulfate (NH₄)₂SO₄ is added to culture filtrate (50% saturation) and allowed to stand for 6 hours at 5°, after which the (NH₄)₂SO₄ concentration is increased to 80% saturation. The protein precipitate is collected by centrifugation, resuspended in one-half the original volume of distilled water, and dialyzed against water. The dialyzed preparation is then used for yeast spheroplast formation.[13]

[13] S. Gascon and J. R. Villanueva, *Can. J. Microbiol.* **10**, 301 (1964).
[14] C. Garcia Mendoza and J. R. Villanueva, *Nature (London)* **202**, 1241 (1964).
[15] C. Garcia Mendoza and J. R. Villanueva, *Nature (London)* **195**, 1326 (1962).
[16] S. Gascon, A. G. Ochoa, and J. R. Villanueva, *Can. J. Microbiol.* **11**, 573 (1965).

Preparation and Isolation of Spheroplast

Washed yeast cultures grown in the manner described above are suspended in 20 mM K:PO$_4$ buffer, pH 6.8, containing 0.8 M MgSO$_4$. Strepzymes are added to a volume equivalent to one-fourth that of the yeast suspension.[13] The cell suspension is then incubated at 30°. Spheroplast formation from *Candida utilis* is complete after 6–7 hours of incubation at 30°.[13] The spheroplast may be isolated by the procedures described above. This method of spheroplast formation requires a fairly long incubation period which may make this method undesirable for biochemical studies. Spheroplasts prepared by the strepzyme method are stable for several days when stored at 5° under hypertonic conditions.[15]

Method for Isolation of Spheroplast from Filamentous Fungi Using *Helix pomatia* Gut Juices

Modifications of the Method of Bachmann et al.[7] Neurospora crassa

Some filamentous fungi, such as *Neurospora,* are coenocytic, i.e., the cytoplasm is continuous throughout the filament. Therefore the spheroplast structures released by removal of the cell wall may not be analogous to those released from the cell walls of yeast and bacteria. Figure 1 shows an electron micrograph of the "pseudosepta" between two compartments in *Neurospora.* Note that the plasma membrane (PM) and cell wall appear to be continuous from one cell compartment to the next. The dark connecting structure (A) between the two compartments is frequently observed along the mycelia of *Neurospora.* The location of these dark structures suggest that they may actually block cytoplasmic continuity in this coenocytic organism. The question is whether the spheroplast released after removal of the cell wall are derived from a single compartment of the mycelium or from 2, 3, or more compartments. The answer to this question is probably that both types of spheroplast are released and that the continuous membrane surrounding the spheroplast is formed by fracture (near the septa) and reclosure of the plasma membrane. The variations in the diameter of spheroplast released from *Neurospora* mycelia are consistent with this view, however definitive evidence to support this contention is not presently available.

The procedures described for spheroplast formation in this report are applicable to 12–16-hour liquid cultures of *Neurospora.* These cultures are usually referred to as log cultures or shake cultures. Since the physiological state of the cells is an important consideration in the preparation of spheroplast, the procedures for growing and harvesting the cells will be described in some detail.

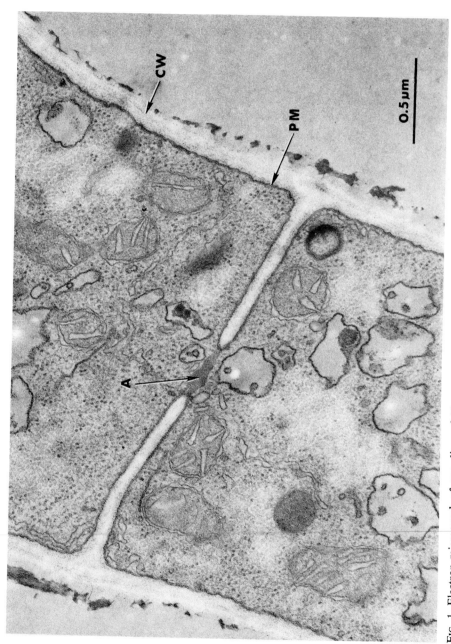

FIG. 1. Electron micrograph of mycelium of *Neurospora crassa*. CW, cell wall; PM, plasma membrane; A, break in continuity of cytoplasm between compartments. This micrograph was taken by Mr. R. R. Adee, Battelle Northwest Laboratories.

Procedure for Growth of Neurospora. Wild-type cultures of *Neurospora* are maintained in screw-cap tubes on agar butts containing Vogel's mineral salts medium and 2% sucrose. Large quantities of conidia for liquid cultures are obtained by inoculating 250-ml Ehrlenmeyer flasks containing 50 ml of Vogel's medium and 2% agar with about 10^{10} conidia. The conidia are germinated and grown for 48 hours at 30°. After 48 hours the culture flasks are removed from the 30° incubator and kept at room temperature under continuous fluorescent irradiation for 3 days. Once conidiation is complete the cultures may be stored at 5° for 2 weeks without loss of viability.

To prepare log phase cultures, conidia from the 250-ml Ehrlenmeyer flasks are resuspended in 30 ml of mineral salts medium and the conidia filtered aseptically through cheesecloth to remove mycelia. The conidia in the filtrate are inoculated into Fernbach flasks containing 1 liter of Vogel's mineral salts medium and 2% sucrose (final concentration conidia, about 5×10^8/ml). The liquid culture is germinated and grown with vigorous agitation on a rotary shaker for 12–16 hours at 30°.

The 12–16 hour log-phase cultures are harvested by vacuum filtration through porous filter paper (hand towels, cut to fit a 135-mm diameter Büchner funnel are excellent for this purpose) and washed twice in 0° distilled water by resuspension and filtration. One liter of a 16-hour culture provides about 4 g dry weight of germinated conidia per liter.

Preparation of Cells for Spheroplast Formation. Preparation of the cells for spheroplast formation involves pretreatment with 2-mercaptoethanol. Cells (40 g wet weight) are resuspended in 120 ml of 20 mM K:PO$_4$ buffer, pH 6.2, containing 50 mM MgSO$_4$ and 10 mM 2-mercaptoethanol and incubated at 37° for 20 minutes. After the 20-minute incubation in 2-mercaptoethanol, the cells are harvested by vacuum filtration, washed with 200 ml of 20 mM K:PO$_4$ buffer, pH 6.2, and resuspended in the following hypertonic medium (volume of medium equivalent to three times the wet weight of cells): 20 mM K:PO$_4$, pH 6.2; 50 mM MgSO$_4$; 10 mM dithiothreitol; 1.0 M NaCl (0.8 M mannitol, 0.8 M MgSO$_4$ or 0.5 M LiCl may be substituted for NaCl); pretreated snail juice enzymes (8–10 mg of snail juice protein per milliliter reaction mixture).

Preparation of Spheroplast. Cells suspended in the above medium are incubated at 37° for 30 minutes with slow agitation (approximately 90 rpm). At the end of the 30 minutes' incubation, the suspension is diluted with 2 volumes of a solution containing 10 mM MgSO$_4$ and 1 mM CaCl$_2$ (pH 6.2) and incubation continued for an additional 5 minutes at 37°. Approximately 95% of the cells are converted to spheroplast by this procedure. Diluting the hypertonic medium is thought to cause osmotic swelling of the spheroplast, which results in increased pressure on the weakened but incompletely hydrolyzed cell wall. The spheroplast are then extruded through one or more pores along the mycelial wall. The sequence of events illustrated by the photographs in Fig. 1 is consistent with this proposal. The light microscope photographs (Fig. 2A, B, and C) were taken with

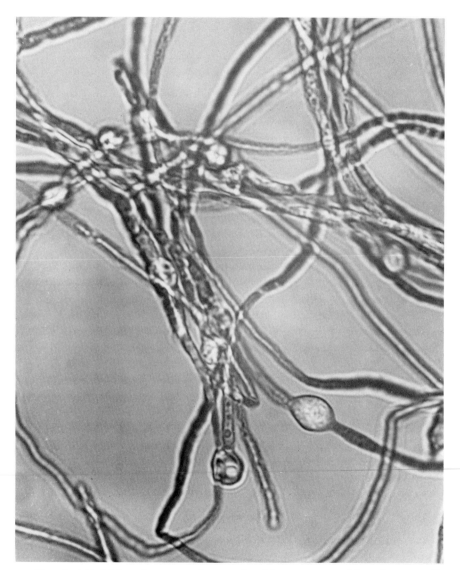

FIG. 2. (A) Light microscope photograph of intact untreated log-phase mycelia of *Neurospora crassa* in 1.0 *M* NaCl. (B) After treatment for 30 minutes with snail juice enzymes. (C) After dilution of osmotic stabilizer (1.0 *M* NaCl) to 0.33 *M*.

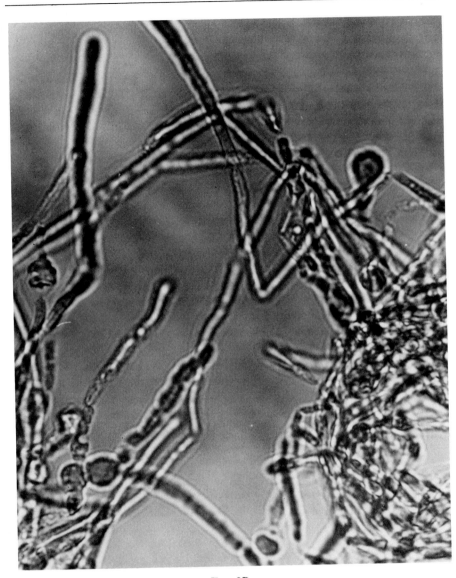

Fig. 2B.

a Zeiss microscope using an interference objective. Figure 2A shows the morphology of the mycelia in hypertonic medium previous to the addition of snail juice enzymes. There are a few bulges along the mycelia, however, most of the mycelia were typical threadlike structures showing similar interference patterns throughout the length of the mycelium. Figure 2B shows the appearance of the mycelia after 30 minutes' incubation in

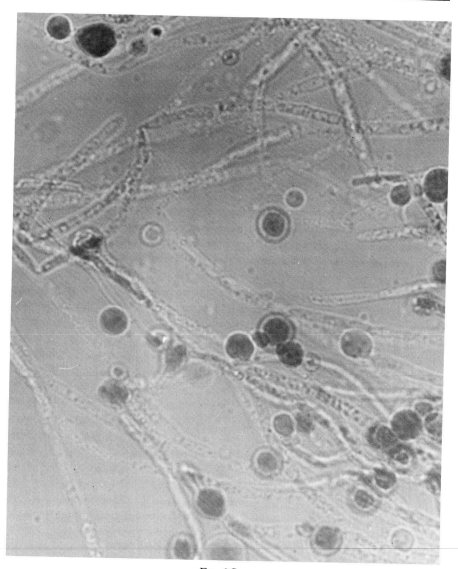

Fig. 2C.

medium containing snail juice. Note that few spheroplast are present at this point and the interference pattern within the mycelial wall in Fig. 2A and B are almost the same. However, when the cells shown in Fig. 2B were diluted with distilled water containing $MgSO_4$ and $CaCl_2$ spheroplast were released from the mycelia (Fig. 2C) and the dark interference pattern observed inside the cell walls in Fig. 2A disappears, although much of the cell wall morphology is maintained. The implication is that the dark interference pattern was, in part, a function of the cytoplasm which after dilution becomes a component of the spheroplast.

Isolation of Spheroplast. The fact that the cell walls are not completely degraded by the snail juice enzymes poses a problem for the isolation of the spheroplast. Spheroplast of *Neurospora,* for example, have a tendency to adhere rather tenaciously to the cell wall debris, which makes it difficult to separate the spheroplast from intact cells and cell wall debris by differential centrifugation.[4] To circumvent this problem the spheroplast preparations are diluted 5-fold with osmotic stabilizing medium and the cell suspension filtered through a 45 × 2.5 cm column of loosely packed (washed) glass wool.[4] Intact cells and large pieces of cell wall material will adhere to the glass wool, while the spheroplast will pass through in the solvent front. The spheroplast in the column eluate are then harvested by centrifugation at 1000 *g* for 5 minutes in a Servall refrigerated centrifuge (GSA rotor). *Neurospora* spheroplast prepared in this manner are stable (as judged by amino acid transport rates and protein synthesis) for about 24 hours at 5° in phosphate buffer containing 50 mM MgSO$_4$, and either 0.6 M NaCl or 0.25 M LiCl. Figure 3 is a light microscope photograph of a spheroplast preparation prepared in this manner. As illustrated by Fig. 2C, the preparations are free of intact cell walls. The diameter of the spheroplast varies from about 4 to 20 μm. It is not possible to determine quantitatively the recovery of spheroplast from the glass wool column; however, we estimate from protein measurement that approximately 70–80% of the spheroplast added to the column are recovered.

Except for differences in osmotic fragility, the biochemical properties of the spheroplast appear to be the same as those for whole cells. The size of mycelia in 0.33 M NaCl range from a diameter of 4–7 μm, while spheroplasts in 0.33 M NaCl have a diameter of approximately 4–20 μm. When the spheroplast are placed in a suitable growth environment at 30° cell walls are regenerated and normal mycelial growth is resumed.[7]

Preparation of Cytoplasmic Membrane Vesicles from Yeast and Filamentous Fungi. The principles involved in the preparation of plasma membrane vesicles from spheroplast of yeast and those of fungi are essentially identical. The procedure involves controlled lysis of the spheroplast by diluting the osmotic stabilizer in the presence of ethylenediaminetetraacetic acid (EDTA), RNase, and DNase.[17] The plasma membranes are disrupted, and the submembrane particles reanneal to form bounded membrane sacs (vesicles). The mechanism of this reaction is unknown.

Homogenization and Lysis of Spheroplast. Spheroplasts prepared in the manner described above are harvested by centrifugation at 1000 *g*,

[17] H. R. Kaback, this series, Vol. 22 [72].

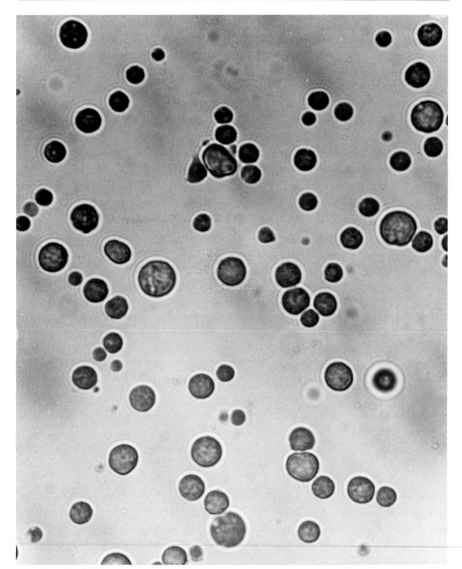

FIG. 3. Light microscope photograph of isolated spheroplast of *Neurospora crassa*.

resuspended in a small volume of hypertonic stabilizing medium (30 mg of spheroplast protein per milliliter). The cell paste is then homogenized (6 strokes) in a glass homogenizing vessel (using a tightly fitting Teflon plunger). Add 500 μg/ml each of RNase and DNase[17] (protease free) to the cell paste and homogenize an additional 4–6 strokes, or until the suspension is uniform. The cell paste is transferred to 50 volumes of 20 mM K:PO$_4$, pH 6.2, containing 0.33 M NaCl and 20 mM EDTA, and the suspension is incubated for 15 minutes at 30°. A sufficient volume of a concentrated solution of MgSO$_4$, adjusted to pH 6.2, is then added to the cell suspension to give a final concentration of 50 mM Mg, and the incubation is continued for an additional 10 minutes.

Removal of Whole Cells. Any whole cells remaining after this treatment are removed by centrifugation at 600 g for 5 minutes.

Isolation and Purification of Cytoplasmic Membranes. The supernatant from the 600 g centrifugation contains mitochondria, cytoplasmic membrane vesicles, soluble proteins and a few membrane "ghosts." The particulate components in the 600 g supernatant are then centrifuged at 39,000 g for 30 minutes. The pellets from the 39,000 g spin are resuspended and homogenized (10 strokes) in 6–15 ml of medium containing 5 mM K:PO$_4$ buffer, pH 6.2, 3 mM dithiothreitol, 20 mM EDTA and 0.33 M NaCl. After homogenization (6–15 strokes) sufficient MgSO$_4$ is added to give a final concentration of 50 mM. The volume of the suspending medium is then increased 20-fold, and the suspension is centrifuged at 39,000 g for 30 minutes. By repeating this procedure 3–5 times, most of the soluble protein contaminants are removed from the membrane preparation. Further purification of the cytoplasmic membranes can be achieved by discontinuous sucrose gradient centrifugation. Washed membrane pellets from the 39,000 g centrifugation are suspended in 20 mM K:PO$_4$ buffer containing 0.33 M NaCl and 10 mM EDTA (3.7 mg protein/ml) layered on the top of discontinuous sucrose gradients comprised of equal volumes of 30, 40, 45, 55, and 65% sucrose in 20 mM K:PO$_4$ buffer containing 10 mM MgSO$_4$. The gradients were centrifuged at 75,500 g for 4 hours at 4°. Membrane vesicles equilibrated at densities on the gradient corresponding to 1.18–1.22 g/cm^3. The vesicles were removed from the sucrose gradient, diluted to 0.25 M sucrose with K:PO$_4$ buffer, harvested by centrifugation at 39,000 g for 30 minutes and resuspended (10 mg of protein per milliliter) in 20 mM K:PO$_4$ buffer containing 20 mM MgSO$_4$ and 0.33 M NaCl. The vesicles were frozen in small aliquots (0.5–1.0 ml) in liquid nitrogen and stored at −80°. The membrane vesicles are stable (as judged from electron microscopy) for at least 6 weeks when stored in this fashion.

Characterization of Yeast and Fungal Cytoplasmic Membrane Vesicles

One of the major criteria for the identification of membrane vesicles from yeast and fungi is their morphological appearance in the electron microscope. Figures 4 and 5 are typical electron micrographs of cyto-

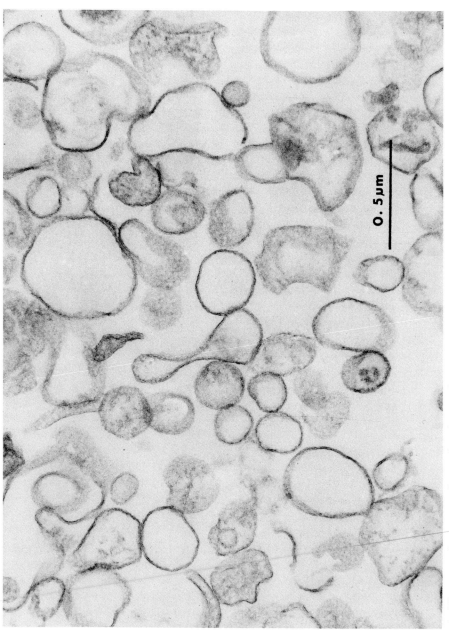

FIG. 4. Electron micrograph of cytoplasmic membrane vesicles prepared from spheroplast of baker's yeast. Electron micrographs (Figs. 4–6) were taken by Mr. R. R. Adee, Battelle Northwest Laboratories.

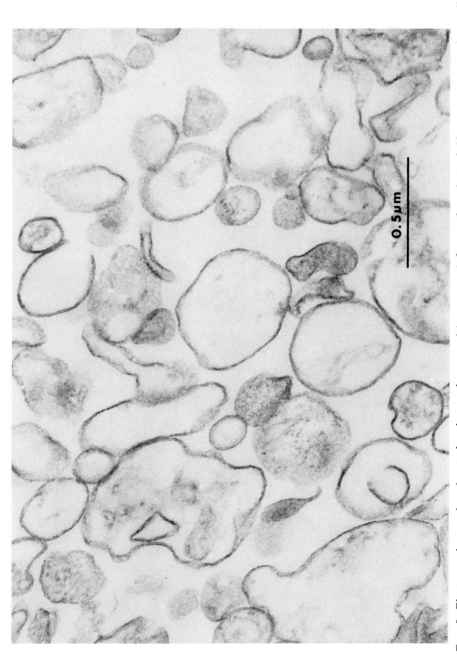

FIG. 5. Electron micrographs of cytoplasmic membrane vesicles prepared from spheroplast of *Neurospora crassa* by snail juice enzyme method.

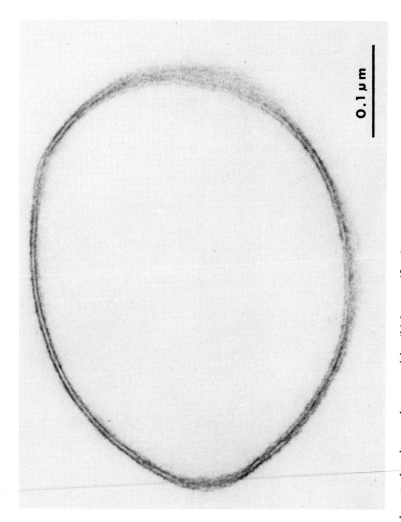

FIG. 6. Single cytoplasmic membrane vesicle (high magnification) showing the single trilaminar structure of the vesicles. Vesicles prepared from *Neurospora crassa* spheroplast membranes.

plasmic membrane vesicles prepared from spheroplast of *Saccharomyces cerevisiae* and *Neurospora crassa,* respectively.[4] As shown, the predominant structures are intact, closed, membranous vesicles which are, in most cases, devoid of internal structure. The boundary of the majority of the vesicles appears to be a single trilaminar membrane (see Fig. 6), which is approximately 80–90 Å thick. Similar thicknesses are observed for the cytoplasmic membranes in electron micrographs of intact cells (Fig. 1). The morphological appearance of vesicles from *S. cerevisiae* and *N. crassa* are almost indistinguishable from one another. Both preparations (Figs. 4 and 5) show a great deal of heterogeneity in size (range, from 0.05 μm to 1.0 μm in diameter). The diameter of the vesicles appears to be directly proportional to the amount of homogenization; the greater the number of homogenizations, the smaller the diameter of the vesicles.

Saccharomyces cerevisiae cytoplasmic membranes are comprised of 49% protein, 39.1% lipid, 7.0 $\pm$ 0.6% RNA, and 4.0–6.0% total carbohydrate.[3,18] Since the membranes shown in Fig. 3 were treated with RNase and DNase during preparation, the content of these constituents in the vesicles was considerably less (0.5% and 1.5%, respectively).

Cytoplasmic membrane vesicles from *N. crassa* prepared in the manner described here contained no detectable RNA or DNA. Approximately 25–32% of the total protein in *Neurospora* spheroplast is present in the vesicle preparations. Less than 1% of the cellular activities of tryptophan synthetase and succinic dehydrogenase was detectable in vesicle preparations from *Neurospora*. Finally, proteins thought to be loosely associated with the cell membrane of *Neurospora,* e.g., the tryptophan-binding protein released by osmotic shock from whole cells,[19] are not associated with the membrane vesicles. This is interpreted to mean that the vesicle preparations were derived from the plasma membrane and that they contain only those proteins that are intrinsic components of the cell membrane. The possibility exists, however, that the preparations contain a heterogeneous population of membranes derived from several cellular structures, e.g., vacuoles, nuclear, and mitochondrial membranes, etc. A more extensive characterization of the enzymatic constitution of the vesicle preparations should resolve this question.

[18] R. P. Langley, A. H. Rose, and B. A. Knights, *Biochem. J.* **108**, 401 (1968).
[19] W. R. Wiley, *J. Bacteriol.* **103**, 656 (1970).

[65] Isolation of Promitochondria from Anaerobically Grown *Saccharomyces cerevisiae*

By Gottfried Schatz and Ladislav Kováč

If *Saccharomyces cerevisiae* cells grow anaerobically on a fermentable carbon source, they cease to synthesize respiring mitochondria.[1] Instead, they form respiration-deficient mitochondrial precursors termed promitochondria.[2-7] These organelles differ from normal yeast mitochondria in their complete lack of cytochromes aa_3, b, c, and c as well as of ubiquinone. They also possess very low levels of ergosterol and an unusual fatty acid composition.[4] When the anaerobically grown cells are exposed to oxygen, promitochondria are adaptively transformed into respiring mitochondria.[8]

Reversible dedifferentiation of yeast mitochondria can also be induced by several other environmental conditions, such as starvation of essential nutrients[9-11] or high external concentrations of glucose.[12] However, the mitochondrial variants accumulating under these conditions still retain significant capacity for respiration and oxidative phosphorylation. Since promitochondria completely lack these functions, they represent the most dedifferentiated mitochondria described so far. A detailed study of promitochondria may thus provide insight into the mechanism of mitochondrial assembly.[13] It may also tell us which mitochondrial functions can occur in the absence of an electron transfer chain.[e.g., 14,15]

[1] P. P. Slonimski, "La formation des enzymes respiratoires chez la levure." Masson, Paris, 1953.

[2] G. Schatz, *Biochim. Biophys. Acta* 96, 342 (1965).

[3] R. S. Criddle and G. Schatz, *Biochemistry* 8, 322 (1969).

[4] F. Paltauf and G. Schatz, *Biochemistry* 8, 335 (1969).

[5] H. Plattner and G. Schatz, *Biochemistry* 8, 339 (1969).

[6] C. H. Damsky, W. M. Nelson, and A. Claude, *J. Cell Biol.* 43, 174 (1969).

[7] K. Watson, J. M. Haslam, and A. W. Linnane, *J. Cell Biol.* 46, 88 (1970).

[8] H. Plattner, M. M. Salpeter, J. Saltzgaber, and G. Schatz, *Proc. Nat. Acad. Sci. U.S.* 66, 1252 (1970).

[9] P. A. Light, R. A. Clegg, C. I. Ragan, and P. B. Garland, *FEBS* (*Fed. Eur. Biochem. Soc.*) *Lett.* 1, 4 (1968).

[10] T. Ohnishi and B. Chance, *in* "Flavins and Flavoproteins" (H. Kamin, ed.), p. 681. Univ. Park Press, Durham, North Carolina, 1971.

[11] J. W. Proudlock, J. M. Haslam, and A. W. Linnane, *Biochem. Biophys. Res. Commun.* 37, 847 (1969).

[12] B. Ephrussi, P. P. Slonimski, Y. Yotsuyanagi, and J. Tavlitzki, *C. R. Lab. Carlsberg., Ser. Physiol.* 26, 87 (1956).

[13] G. Schatz, *in* "Membranes of Mitochondria and Chloroplasts" (E. Racker, ed.),

Although anaerobic growth of yeast invariably causes the formation of nonrespiring promitochondria, the chemical and enzymatic composition of these organelles is profoundly influenced by the composition of the growth medium.[4,6,7] For example, high concentrations of glucose in the culture medium decrease the intracellular concentration of promitochondria[3] and alter their enzyme complement.[16] Similarly, anaerobic growth in the absence of added oleic acid and ergosterol (neither of which can be synthesized in the absence of oxygen) causes the formation of extremely labile promitochondria which differ in many respects from those formed in a medium supplemented with these lipids.[4] It has not yet been possible to isolate these lipid-deficient promitochondria in a reasonably intact state.

In this article, we outline the isolation of promitochondria from yeast cells grown anaerobically in the presence of oleic acid (in the form of Tween 80), ergosterol, and low concentrations of glucose. These conditions permit the indefinite propagation of yeast cells under anaerobic conditions and minimize catabolite repression of promitochondrial formation.

Preparation of Cells

Growth Medium. The growth medium contains per liter: Difco yeast extract, 3 g; Tween 80 (polyoxyethylene sorbitan monooleate), 2.6 g; ergosterol, 12 mg; $CaCl_2$, 0.4 g; NaCl, 0.5 g; $MgSO_4 \cdot 7H_2O$, 0.7 g; KH_2PO_4, 1.0 g; $(NH_4)_2SO_4$, 1.2 g; $FeCl_3$, 5 mg; anhydrous glucose, 3 g; and a few drops of a silicone antifoam emulsion. The medium is sterilized in 12-liter batches (cf. below) for 40 minutes at 120° and quickly cooled to room temperature under a stream of nitrogen.

Design of the Fermenter. As respiratory adaptation of yeast occurs already at very low oxygen tension, great care must be taken to choose a fermenter that is free of leaks. All gas connections should be as short as possible and either of thick-walled rubber or, preferably, stainless steel.[17] Our laboratory uses a Microferm fermenter (New Brunswick Scientific Inc., New Brunswick, New Jersey) equipped with 14 liter fermenter assemblies (working volume 12 liters). To minimize leaks, the exhaust condenser was permanently welded to the top plate, all unnecessary parts were welded shut, and the seal of the impeller shaft is frequently replaced. All connections and valves are of stainless steel. Nitrogen containing less

American Chemical Society Monograph No. 165, p. 251. Van Nostrand-Reinhold, New York, 1970.

[14] G. Schatz and J. Saltzgaber, *Biochem. Biophys. Res. Commun.* **37**, 996 (1969).

[15] G. S. P. Groot, L. Kováč, and G. Schatz, *Proc. Nat. Acad. Sci. U.S.* **68**, 308 (1971).

[16] G. Schatz, unpublished observations.

[17] R. T. Payne, "The Purification of Inert Gases to High Purity," Technical Bulletin, R. D. Mathis Co., Long Beach, California, 1969.

FIG. 1. A New Brunswick Microferm Fermentor modified for anaerobic culture of yeast. A, Catalyst for purifying nitrogen (Deoxo gas purifier); B, unidirectional stainless steel valve; C, stainless steel regulator valve; D, flexible stainless steel tubing (vacuum tested); E, gas filter; F, harvesting port; G, inoculation port with rubber seal; H, unidirectional stainless steel valve separating gas outlet port from mercury trap; I, mercury trap; and K, precision flow meter.

than 1 ppm oxygen is readily obtained by passing "Prepurified Nitrogen" containing 1% hydrogen (Matheson Gas Products, Inc., East Rutherford, New Jersey) through a catalyst (Deoxo Gas Purifier D 50-100, Engelhard Industries, Inc., East Newark, New Jersey). This catalyst operates at room temperature and requires no regeneration. It is protected against backflow from the fermenter vessel by a unidirectional valve of stainless steel. The gas outlet of the fermenter is connected to a mercury trap. The salient features of this modified fermenter are depicted in Fig. 1.

Anaerobic Growth. The stirred culture medium (total volume 12 liters) is inoculated with approximately 2×10^8 aerobically grown yeast cells and flushed with a slow stream of nitrogen (0.5 liter/minute) for 5 hours. The gas inlet valve is then closed, and the cells are allowed to grow for approximately 20 hours at 28°. The fermenter vessels are covered with aluminum foil to prevent photoinhibition of the anaerobic cells.[18] Cell growth is followed turbidimetrically on samples collected through a

---

[18] E. Sulkowski, B. Guerin, J. Defaye, and P. P. Slonimski, *Nature (London)* **202**, 36 (1964).

harvesting port. The back-pressure generated by the mercury trap is sufficient to extrude the culture.

Harvesting the Cells. When the culture has almost reached the stationary phase of growth (see step 5 of the procedure given below), the gas outlet port is closed and the culture is chilled to $0°$. If a New Brunswick Microferm is used, methanol chilled with solid CO_2 can be pumped through the hollow baffle heat exchangers. A concentrated solution of cycloheximide (about 10 mg/ml) is freed from dissolved oxygen by the addition of a few grains of $Na_2S_2O_4$, quickly drawn into a large syringe, and injected into the stirred culture through a neoprene seal to a final concentration of 25 μg/ml. The combined application of chilling and poisoning minimizes respiratory adaptation of the cells during subsequent handling. After 5–10 minutes the carboy is opened and the cells are collected by centrifugation in the cold for 5 minutes at 2000 g. They are washed twice by suspending them in ice-cold 40 mM KP$_i$-buffer pH 7.4 containing 25 μg cycloheximide per milliliter and 0.05% bovine serum albumin. (The albumin facilitates the removal of residual Tween 80 and ergosterol.) The washed cells (final yield about 24 g wet weight) are immediately processed as outlined below.

Isolation of Promitochondria[19]

The following procedure refers to 24 g (wet weight) of cells as a starting material. It can be readily scaled down by at least a factor of twenty. All centrifugation steps refer to the Sorvall SS-34 rotor (average rotating radius 7 cm).

1. The washed cells (24 g wet weight) are suspended with 120 ml of 0.1 M Tris SO$_4$ pH 9.3 containing 25 μg of cycloheximide per milliliter.

2. Mercaptoethanol (4.4 ml) is added and the mixture is incubated for 5 minutes at $30°$.

3. The cells are collected by centrifugation for 5 minutes at 6000 rpm (3000 g_{av}), resuspended in 60 ml of ice-cold 1.5 M sorbitol containing 25 μg of cycloheximide per milliliter and recentrifuged as before. (Care should be taken to adjust all sorbitol solutions to pH 6.5–7.4 with 0.1 N NaOH.)

4. The sedimented cells are suspended in 60 ml of 1.5 M sorbitol– 1 mM EDTA pH 7.4–25 μg of cycloheximide per milliliter and mixed with 5 ml of glusulase (a partially purified gut juice from the snail *Helix pomatia* manufactured by Endo Laboratories, Garden City, New York). Alternately, 1.2 g of lyophilized snail gut juice (Industrie Biologique

---

[19] Exactly the same procedure is used by us to isolate mitochondria from aerobically grown yeast cells except that the addition of cycloheximide to the various solutions is omitted.

Francaise, Gennevilliers, France), dissolved in a minimal volume of 1.5 M sorbitol, may be added.

5. The mixture is gently shaken in a 30° water bath. The formation of spheroplasts is monitored by diluting small aliquots 100-fold with water and with 1.5 M sorbitol, followed by differential cell count. The formation of spheroplasts is the critical step of the entire procedure. With most *S. cerevisiae* strains, spheroplast formation is essentially complete after 30–90 minutes. Occasionally, however, a batch of cells proves to be refractory to the snail enzymes and has to be discarded. A detailed discussion of the various parameters influencing spheroplast formation is presented by Hutchison and Hartwell.[20] In general, cells harvested in the logarithmic growth phase are more susceptible to the snail enzymes than stationary cells. On the other hand, maximal yields of cells and promitochondria are obtained in the early stationary phase. It is usually best to settle for a compromise and to harvest the cells just before the onset of the stationary phase. Under our conditions, this point corresponds to a density of 2.8×10^7 cells per milliliter of culture.

6. When more than 90% of the cells have been converted to spheroplasts, the suspension is centrifuged for 5 minutes at 6000 rpm (3000 g_{av}) at 4°. From this point on, all steps are performed at 0–4°. The supernatant is decanted and saved for reclaiming the snail enzymes (cf. below).

7. The sedimented spheroplasts are washed twice by gentle resuspension in 1.5 M sorbitol–1 mM EDTA–0.1% bovine serum albumin–25 μg of cycloheximide per milliliter and centrifugation for 5 minutes at 6000 rpm (3000 g_{av}).

8. The washed spheroplasts are suspended in 120 ml of 0.6 M mannitol–2 mM EDTA pH 7.4–0.1% bovine serum albumin. The suspension is immediately homogenized in 40-ml aliquots in a Lourdes overhead blendor (Lourdes Instrument Corp., Brooklyn, New York) for 10 seconds at full speed. The type of blendor does not seem to be overly important as other models have been used with good success.

9. The homogenized aliquots are pooled and centrifuged for 10 minutes at 4000 rpm (1200 g_{av}). The supernatant is carefully decanted without disturbing any fluffy layer and centrifuged for 10 minutes at 10,000 rpm (8000 g_{av}).

10. The supernatant is discarded and the inside of the centrifuge tube is wiped with cotton to remove adhering lipid.

11. The pellet is homogenized with a small volume of 0.6 M mannitol–2 mM EDTA pH 7.4 in a glass–Teflon homogenizer, diluted to 20–

[20] H. T. Hutchison and L. H. Hartwell, *J. Bacteriol.* **94**, 1697 (1967).

30 ml with the same solution and centrifuged for 5 minutes at 3700 rpm ($1100\ g_{av}$) to remove residual cells and debris.

12. The supernatant from step 11 is centrifuged for 10 minutes at 15,000 rpm ($17,800\ g_{av}$). The supernatant is discarded and any fluffy layer on top of the promitochondrial pellet is removed by a careful rinse with 0.6 M mannitol–1 mM EDTA pH 7.4. The yield is 40–60 mg promitochondrial protein.[21]

Recycling of the Snail Enzymes[22]

The snail enzyme preparation can be reused up to six times as follows: the supernatant from step 6 is clarified by centrifugation (20 minutes at 105,000 g) and brought to 65% saturation with solid ammonium sulfate (39.8 g/100 ml) and stored at 5°. Immediately before reuse, the precipitated enzymes are collected by centrifugation (5 minutes at 10,000 rpm) and dissolved in 1.5 M sorbitol–1 mM EDTA–25 μg cycloheximide per milliliter. For unknown reasons, the effectiveness of the snail preparation may actually increase during the first few recyclings.

Properties of Promitochondria

For a detailed account of the chemical, morphological, and enzymatic features of isolated promitochondria, the reader is referred to earlier publications.[3-7,8,14,15] In the present context it is important to note that isolated promitochondria are much more fragile than respiring yeast mitochondria. This fragility may in part result from their low content of ergosterol[4] and/or the absence of respiratory catalysts. It is well known that the stability of isolated mitochondria is enhanced by an active endogenous respiration. Because promitochondria are quite fragile, further purification by sucrose gradient centrifugation leads to considerable damage. Although promitochondria obtained by the present method are sufficiently pure for many biochemical studies, they are still contaminated by some other membranes[23] as well as by cytoplasmic ribosomes.

---

[21] If aerobically grown cells are processed in this manner, the yield of mitochondria is 60–100 mg of protein.

[22] This procedure was communicated to us by Dr. R. S. Criddle, whom we wish to thank at this point.

[23] G. Schatz, *J. Biol. Chem.* **243**, 2192 (1968).

[66] The Isolation of Intracellular Membranes of *Escherichia coli* 0111a

By John W. Greenawalt

Although intracellular membranes generally have not been considered characteristic of gram-negative bacteria, an increasing number of reports describe the presence of internal membranes in various strains of *Escherichia coli*. It has been suggested recently that mesosomes may be common to *E. coli,* but that special conditions of fixation and embedding are required in order for them to be observed.[1] *E. coli* 0111a is a thermosensitive strain which, when grown at 40°, accumulates large quantities of internal membranes as vesicles, tubules, and large membrane whorls. When *E. coli* 0111a is grown at 30°, intracytoplasmic membranes are not formed.[2,3] In this laboratory these membranes in *E. coli* 0111a are referred to as "extra membranes" since little is yet known about their biochemical functions. However, other workers have called them mesosomes.[4]

As a foundation for investigations into questions concerning the biosynthesis and assembly of membranes as well as the integration of these processes with other cellular activities, a method for the isolation of extra membrane whorls was developed and is described here. The techniques used to isolate the whorls from *E. coli* 0111a may have general applicability in studies directed toward isolating internal membranes from bacteria cells. It should be noted, however, that the size, complexity, and compactness of the whorls and the amounts of membrane accumulated in *E. coli* 0111a at 40° are unusual and constitute features unique to this strain (Fig. 1).

General Principles. In the method outlined here a two-step procedure is used to obtain a fraction highly enriched in extra membrane whorls; this includes (1) separation of a crude membrane fraction containing whorls which are found on a linear sucrose gradient (18 to 56%, w/w) in a band unique to *E. coli* 0111a grown at 40° and (2) further enrichment of the whorls from this band by rate (differential) centrifugation through a layer of 27% sucrose. These techniques, initially developed on a small scale, have been applied successfully to large-scale preparations utilizing large batches of fermenter-grown cells and zonal centrifugation on a sucrose gradient using a Ti-14 rotor.

[1] R. D. Pontefract, G. Bergeron, and F. S. Thatcher, *J. Bacteriol.* **97**, 97 (1969).
[2] C. Schnaitman and J. W. Greenawalt, *J. Bacteriol.* **92**, 780 (1966).
[3] R. A. Weigand, J. M. Shively, and J. W. Greenawalt, *J. Bacteriol.* **102**, 240 (1970).
[4] B. C. Altenburg and S. C. Suit, *J. Bacteriol.* **103**, 227 (1970).

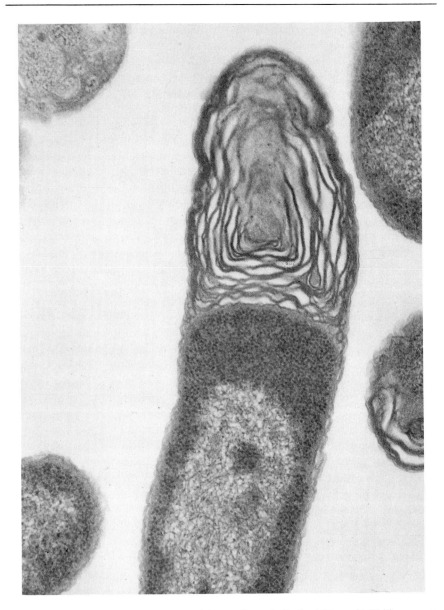

FIG. 1. Electron micrograph of thin section of *Escherichia coli* 0111a grown at 40°. A terminal extra membrane whorl is readily seen. All samples for electron microscopy were fixed with glutaraldehyde followed by OsO₄, embedded in Epon 812 and sections were stained with uranyl acetate and lead citrate as previously described. R. A. Weigand, Ph.D. Thesis, Johns Hopkins School of Medicine. ×50,000.

Two factors are of great importance in facilitating the successful isolation of extra membranes from *E. coli* 0111a. First, careful regulation of the growth temperature to 40 ± 0.5° is essential in obtaining a high proportion of cells which contain large amounts of extra membranes. The amount of extra membrane accumulated in each cell as large whorls varies in response to variations in the temperature.

The second factor is the use of a mild method of cell disruption. Because no enzymatic or chemical markers are known to be specific for the extra membranes, the present multistep isolation procedure utilizes the unique size and complex ultrastructural of the membrane whorls as a means of following the isolation of these extra membranes. Thus, the successful separation of extra membrane whorls from *E. coli* 0111a depends upon being able to break adequately the bacterial cells without destroying the whorls. This can be done in a French pressure cell operated at low pressures; pressures used generally for the extraction and purification of enzymes, for example, completely destroys the distinguishing ultrastructural characteristics of the extra membrane whorls.

In the application of the present technique to the isolation of intracellular membranes from other cells, the effectiveness of the lysing procedure in preserving the structures of interest as well as in disrupting the cells should be investigated. Electron microscopy is a valuable means of monitoring each of the cell fractions obtained in order to determine the preservation of the intracellular membranes of interest.

At the present time two means of monitoring the formation of extra membranes in *E. coli* 0111a are effective: (1) The maximal growth (maximal optical density) achieved at 40° by *E. coli* 0111a is significantly less than that of closely related strains. (2) The extra membranes can be detected in cells observed in the electron microscope. The latter provides the more reliable means of establishing with certainty the presence of extra membranes. This process has been speeded greatly by the finding that extra membranes can be seen in negatively stained whole cells of *E. coli* 0111a.

Materials

Cells. *Escherichia coli* 0111a 11b:B_4:NM Stoke W is designated throughout as *E. coli* 0111a. This strain was obtained originally by Dr. E. C. Heath from the Communicable Disease Center, Atlanta, Georgia, and was obtained from him for these studies.

Growth Medium. The cells are grown routinely in medium containing (in grams per liter of distilled water): casitone (Difco), 17.0; peptone (Difco), 3.0; NaCl, 5.0; K_2HPO_4, 2.5. The original strain of 0111a will not grow at 40° in glucose–mineral salts medium unless this medium is supplemented with the amino acids isoleucine and valine. However, we have subcultured a clone derived from the original strain which does not require the amino acid supplementations for growth at 40° and still accumulates extra membranes. This indicates that extra membrane formation is not linked directly to these nutritional requirements.[5]

Solutions and Reagents

NaCl, Reagent Grade, (0.9%)
Pancreatic DNase (EC 3.1.45) B grade, Calbiochem
Sucrose, Reagent Grade (18.5–58%, w/w)

French Pressure Cell. Model 4-3398 (American Instrument Co., Silver Spring, Maryland) was used, to which was adapted a hydraulic press Model 12-10 S (Wabash Metal Products, Wabash, Indiana) with a variable pressure hydraulic pump, Model P-182 (Blackhawk Industrial Products, Butler, Wisconsin). This unit was assembled as a prototype but may now be purchased through American Instrument Co.

Small-Scale Preparation

Growth of Cells

Cultures are grown at 40° in 500-ml Erlenmeyer flasks containing 200 ml of growth medium. Temperature is regulated at $40 \pm 0.5°$ by using a gyratory water bath shaker rotating at 200 rpm (Model G 76, New Brunswick Scientific Company, New Brunswick, New Jersey). Growth is initiated by addition of a 2.5% (v/v) inoculum of an overnight culture grown at 30°. Growth can be measured by following changes in absorbance of the culture (diluted 1:3) with time at 660 nm. Linear relationships between absorbance, protein and dry weight have been shown.[3] When growth has reached early stationary phase (ca. 4.5 hours), cells are harvested by centrifugation in the cold for 15 minutes at 10,000 g (Model SL centrifuge, Lourdes Instrument Corp., Brooklyn, New York). Cells are washed and resuspended to the concentration desired for French pressing with cold 0.9% sodium chloride solution. Cells grown at 30° into early stationary phase (ca. 7–8 hours) serve as a control.

Cell Breakage

All steps in the fractionation procedure are carried out at 0° to 4°. Washed suspensions of cells grown at 40° or 30° containing approximately 10 mg of protein per milliliter are broken in the French pressure cell at 2000 psi of applied pressure (8000 psi internal cell pressure) and 4000 psi

[5] R. A. Weigand, Ph.D. Thesis, Johns Hopkins School of Medicine, 1972.

applied pressure (16,000 psi internal pressure), respectively. This difference in applied vs. internal pressure of the French cell should be noted since the use of reduced pressure with 40°-grown cells is critical for the preservation of the integrity of the membrane whorls. The broken cell suspensions are treated with pancreatic DNase 1 (100 μg/ml) at 40° for 1 hour to reduce the viscosity of the lysate and to prevent aggregation of cell components.

Cell Fractionation

Step 1. Density Gradient Centrifugation. Aliquots (3.5-ml sample) of the DNase-treated suspension are layered on linear sucrose gradients (27 ml, 18.5 to 55.5%, w/w); the tubes are centrifuged for 12 hours at 25,000 rpm in a Spinco SW 25.1 rotor. Fractions are collected from the bottom of the tubes by a gradient fractionator.

As illustrated in Fig. 2 a fraction (band III) unique to cells grown at 40° bands broadly with a peak at about 42% sucrose (density of about 1.194 g/cc) on the linear sucrose gradient. Examination of all other fractions in the electron microscope from both gradients shown that *only* this fraction contains ultrastructurally intact membrane whorls; the corresponding fractions from 30°- and 40°-grown cells are indistinguishable ultrastructurally. Bands I and II (Figs. 3A and 3B) are largely devoid of membranes, although Band II contains very small membrane fragments. Band IV (Fig. 3C) is an envelope fraction containing both cell wall and membrane components. Cell wall-containing fragments can be distinguished readily from membrane components by their characteristic curled or "C-shaped" appearance (cf. Fig. 3C, 3D).

Analysis of the fractions removed from gradients showed that Band III is characterized by an extremely high phospholipid phosphate (PLP) to

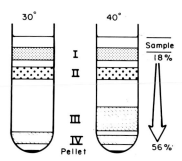

FIG. 2. Separation of extra membranes of *Escherichia coli* 0111a on linear sucrose gradients. Band III is unique to lysates of cells containing extra membranes (grown at 40°). Membrane whorls are present only in this fraction as determined by electron microscopic examination (Fig. 3). See text for details.

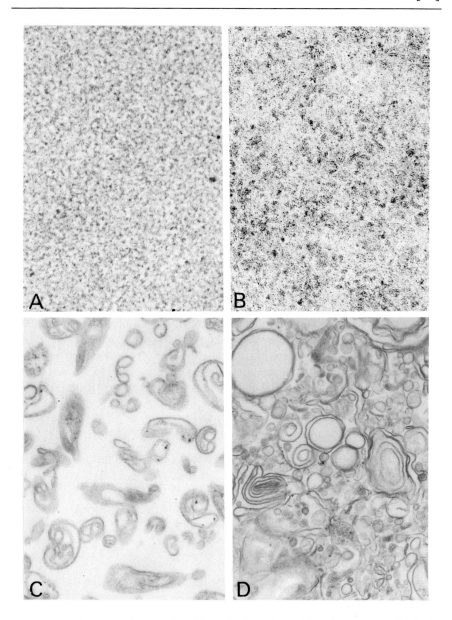

FIG. 3. Electron micrographs of bands formed on linear sucrose gradients by lysates of *Escherichia coli* 0111a grown at 40° (cf. Fig. 2). A. Band I. B. Band II. C. Band IV. D. Band III. ×19,000.

protein ratio with a mean value of 1.55 μmoles of PLP per milligram of protein for the pooled fractions comprising band III.

Further enrichment of the membrane whorls in band III from the cells grown at 40° has been achieved by step 2 in this procedure, i.e., by rate centrifugation of this fraction through a layer of sucrose onto a pad of denser sucrose.

Step 2. Rate Centrifugation. Fractions from the sucrose density gradients containing the extra membrane whorls (band III) are diluted with an equal volume of 0.9% sodium chloride solution. Aliquots (0.5 ml) of each diluted fraction are layered on centrifuge tubes containing a pad (0.6 ml) of 55.5% sucrose and a layer (4.0 ml) of 27% sucrose. The separation is carried out by centrifugation at 15,000 rpm for 25 minutes at approximately 2° using a Spinco SW 50L rotor. The rationale for the separation of the whorls from the other membrane components present in band II is based on the following relationships:

$$ t = \frac{9}{2} \frac{\eta}{\omega^2 r_p^2 (\rho b - \rho)} \ln \frac{R_t}{R_b} \tag{1} $$

where t = time, η = viscosity of medium, ω = angular velocity, r_p = radius of particle, ρb = density of particle, ρ = density of medium, R_t, R_b = radius of rotation at top and bottom of tube, respectively.[6]

Using this equation the length of time (t) required to sediment a spherical particle of known size (r_p) and density (ρb) through a layer of medium of selected density (ρ) and viscosity (η) can be calculated. The separation achieved by the density gradient run (step 1) provides information essential to the application of this equation. That is, the average size of the extra membrane whorls (r_p) can be estimated from electron microscope measurements and the approximate density (ρb) can be determined from the density of sucrose at the position on the linear gradient to which the particles sedimented. It was on the basis of these considerations that it was calculated that the extra membrane whorls in *E. coli* 0111a which had an approximate diameter of 440 nm would be sedimented onto the pad in 25 minutes. Electron microscopic examination showed that, in fact, the fraction obtained was highly enriched in extra membrane whorls as indicated by the electron micrograph provided (Fig. 4).

Large-Scale Isolation of Extra Membranes from
    Fermenter-Grown *E. coli* 0111a Cells

The procedures described above are inadequate for the isolation of large enough quantities of extra membranes to do detailed biochemical and

[6] H. R. Mahler and E. H. Cordes, "Biological Chemistry," pp. 390–393. Harper & Row, New York, 1966.

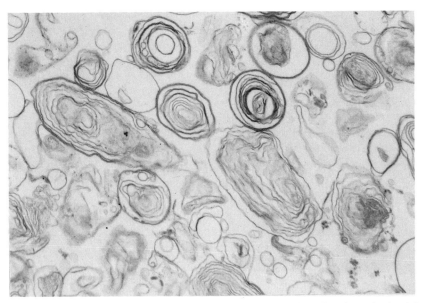

Fig. 4. Electron micrograph of extra membrane-enriched fraction obtained by rate centrifugation of band III from sucrose gradient. Enrichment of extra membrane whorls can be seen. ×19,000.

chemical analyses. However, the principles are applicable to larger scale production and isolation of extra membranes.

Cells

Large quantities of cells containing extra membranes can be obtained by growing *E. coli* 0111a at 40° in a 14-liter fermenter (New Brunswick model MF-114) containing 9 liters of growth medium. The medium is agitated with stirring at 200 rpm and aerated with 1.5 liters of air per minute at 3 psi. The cultures are started using a 2.5% inoculum which is grown at 30° into early stationary phase. The cells are harvested after 4.5 hours of growth when extra membrane production is near maximal levels. The culture is chilled, and the cells are centrifuged at 0° for 30 minutes at 2000 rpm using a Lourdes Model 30R Clinifuge, TR6L head. The cells are washed once with cold 0.9% sodium chloride.

Cell Fractionation

Step 1. Zonal Centrifugation. Extra membrane isolation is carried out by density gradient centrifugation much as described before except that a zonal rotor is used to separate the large volume of cell lysate. Washed, 40°-grown cells are resuspended in 0.9% sodium chloride and disrupted at

2000 psi (applied pressure) in the French pressure cell. A Spinco Ti 14 zonal rotor (650 ml capacity) spinning at 3000 rpm is filled using a Spinco gradient pump (Model 141) with 550 ml sucrose density gradient (linear with volume, 27–56%, w/w) and 100 ml of 56% sucrose as a pad. Part of the pad is displaced by 40 ml of the DNase-treated lysate and 40 ml of 0.9% sodium chloride overlay solution. The rotor is capped and centrifuged for 12 hours at 40,000 rpm. After centrifugation fractions are collected from the center of the rotor by pumping 58% sucrose into the outer portion of the rotor while it is spinning at 3000 rpm.

Analyses of fractions show that a high ratio of PLP per milligram of protein, which corresponds to that of band III, above, is found also here in the region of 39–44% sucrose. Furthermore, electron microscopic examination of all fractions showed that extra membrane whorls contaminated with other membranes are found only in this region of the zonal gradient. Fractions containing extra membrane whorls are diluted with about equal volumes of 0.9% sodium chloride for rate centrifugation. It is necessary to calculate the dilution factor required so that the sucrose in the samples taken from the gradient are slightly less concentrated than that of the sucrose through which the extra membranes are to be centrifuged.

Step 2. Rate Centrifugation. By applying the same calculations as those used for rate centrifugation in the small-scale isolation (Eq. 1, above), it is possible to determine the conditions needed to separate the extra membrane whorls from the other membranes using a Spinco SW 25.1 rotor to handle the larger volumes. It was calculated that the extra membranes can be centrifuged through a 23-ml layer of 27% sucrose onto a pad of 56% sucrose (3.5 ml) by centrifugation at 10,000 rpm for 55 minutes. Fractions are collected by using a gradient fractionator; a fraction highly enriched in membrane whorls are collected at the interface of the 27% sucrose and the 56% sucrose pad.

Properties of the Extra Membrane-Enriched Fraction

Electron microscopic examination showed that the fractions obtained by the small-scale isolation and the large-scale procedures are essentially identical in ultrastructural appearance and degree of enrichment. Thus, Fig. 4 is representative of the fraction highly enriched in extra membranes whether it is obtained by small- or large-scale procedures.

The extra membrane-rich fraction is characterized by a ratio of about 1.25 μmoles of PLP per milligram of protein, which is about 1.5 times the ratio generally accepted for the cytoplasmic membrane of *E. coli*. Qualitatively the phospholipid composition is similar to the cytoplasmic membrane and contains phosphatidylethanolamine and phosphatidylglycerol as major components. Phosphatidylserine is present in small amounts and diphosphatidylglycerol (cardiolipin) appears to be present in even smaller amounts. Lipopolysaccharide components of the cell wall of *E. coli*, as estimated by the presence of 2-keto-3-deoxyoctonate (KDO), are largely absent from the extra membrane fraction. An envelope fraction containing both wall and membrane has a KDO:protein ratio 3.1 times that found in the extra membrane-enriched fraction. Malate dehydrogenase activity is not detectable in the extra membrane-enriched fraction, but NADH oxidase specific activity is about twice that of a cytoplasmic membrane fraction obtained from cells grown at 30° and 3.5 times that of an envelope fraction from 30°-grown cells.

[67] Separation of the Inner (Cytoplasmic) and Outer Membranes of Gram-Negative Bacteria

By M. J. Osborn and R. Munson

The cell envelope of *Salmonella, Escherichia coli*, and related gram-negative bacteria contains three morphologically distinguishable layers: an inner membrane bounding the cytoplasm, a murein (peptidoglycan) layer external to the cytoplasmic membrane, and an additional membranous structure, the outer membrane at the external surface of the cell.[1,2] The two membranes differ both in composition and in function. The lipopolysaccharide of the cell envelope is localized almost exclusively in the outer membrane, together with substantial amounts of phospholipid and a characteristic spectrum of major protein components.[3,4] The inner membrane appears to correspond to the cytoplasmic membrane of gram-positive bacteria, and contains enzyme systems related to electron transfer,[3,4] active transport of solutes,[5,6] and biosynthesis of phospholipids,[7,8] and lipopolysaccharide.[5]

Techniques for separation of inner and outer membrane have been

[1] R. G. E. Murray, P. Steed, and H. E. Elson, *Can. J. Microbiol.* **11**, 547 (1965).

[2] S. DePetris, *J. Ultrastruct. Res.* **18**, 45 (1967).

[3] C. A. Schnaitman, *J. Bacteriol.* **104**, 882, 890 (1970).

[4] M. J. Osborn, J. E. Gander, E. Parisi, and J. Carson, *J. Biol. Chem.* **247**, 3962 (1972).

[5] M. J. Osborn, J. E. Gander, and E. Parisi, *J. Biol. Chem.* **247**, 3973 (1972).

[6] C. F. Fox, J. H. Law, N. Tsukagoshi, and G. Wilson, *Proc. Nat. Acad. Sci. U.S.* **67**, 598 (1970).

[7] R. M. Bell, R. D. Mavis, M. J. Osborn, and P. R. Vagelos, *Biochim. Biophys. Acta* **249**, 628 (1971).

[8] D. A. White, F. R. Albright, W. J. Lennarz, and C. A. Schnaitman, *Biochim. Biophys. Acta* **249**, 636 (1971).

based on isopycnic sucrose gradient centrifugation and/or free-flow electrophoresis of envelope or total membrane fractions obtained either by rupture of bacteria in a French pressure cell or by lysis of spheroplasts. Partial separation of the inner and outer membranes of *E. coli* was first reported by Miura and Mizushima[9]; the method employed isopycnic centrifugation of membranes derived from conventional lysozyme-EDTA spheroplasts. Similar results were obtained by Schnaitman[3] following disruption of *E. coli* in a pressure cell. In this procedure, which avoids use of lysozyme, the murein layer which underlies the outer membrane is not degraded and remains associated with the isolated outer membrane fraction. Improved separation could be obtained by a combination of preparative particle electrophoresis and isopycnic centrifugation.[10] However, the electrophoretic fractionation required the presence of O-antigen chains in the outer membrane lipopolysaccharide.

The above procedures were developed for use with *E. coli,* and neither is applicable to most strains of *Salmonella.* The procedure[4] described here is a modification of the technique of Miura and Mizushima which appears to be generally applicable to gram-negative enteric bacteria, and permits reproducible isolation of inner and outer membrane fractions with improved yield and purity. Spheroplasts are prepared by a slight modification of the method of Birdsell and Cota-Robles.[11] This procedure gives rapid, efficient spheroplasting of bacteria, such as *S. typhimurium,* which are relatively resistant to other methods, and has the further advantages that the outer membrane tends to peel away during spheroplast formation, leaving naked cytoplasmic membrane exposed over most of the surface of the spheroplast. After lysis of the spheroplasts, the total membrane fraction is collected by centrifugation and inner and outer membranes are separated either by sucrose gradient centrifugation or electrophoresis. The latter procedure is limited to organisms in which the lipopolysaccharide contains O-antigen chains but permits more rapid separation of the membranes, and the resulting inner membrane fraction is significantly less contaminated with outer membrane material.

Growth of Bacteria and Preparation of Spheroplasts

Reagents

PPBE Medium: 1% (w/v) Proteose Peptone No. 3 (Difco), 0.1% (w/v) beef extract and 0.5% (w/v) NaCl in distilled or deionized H_2O. Sterilize by autoclave

Minimal Medium M9: M9 salts (10× concentrate): 60 g of

[9] T. Miura and S. Mizushima, *Biochim. Biophys. Acta* **150**, 159 (1968).
[10] D. A. White, W. J. Lennarz, and C. A. Schnaitman, *J. Bacteriol.* **109**, 686 (1972).
[11] D. C. Birdsell and E. H. Cota-Robles, *J. Bacteriol.* **93**, 427 (1967).

Na$_2$HPO$_4$ anhydrous, 30 g of KH$_2$PO$_4$ anhydrous, 5 g of NaCl, 10 g of NH$_4$Cl per liter solution. Sterilize by autoclave. For 1 liter of complete medium, add 100 ml of M9 salts (10×), 10 ml of sterile 0.1 M MgSO$_4$, 10 ml of sterile 10 mM CaCl$_2$, and 20 ml of sterile 20% (w/v) glycerol or sodium lactate (pH 7) to 860 ml of sterile H$_2$O

Sucrose, 0.75 M, in 10 mM Tris·acetate buffer, pH 7.8

Egg white lysozyme, 2 mg/ml in H$_2$O. Prepare fresh daily

1.5 mM EDTA, pH 7.5

Procedure. Bacteria may be grown either in PPBE medium or in minimal medium M9 supplemented with a small amount (0.05 volume) of PPBE. Cultures (50–2000 ml) are grown at 37° with vigorous aeration to a density of approximately 5 × 10^8 bacteria/ml (A_{600} = 0.6 to 0.8), and immediately harvested by centrifugation at 12,000 rpm for 5 minutes at 0–4°. The insides of the centrifuge bottles are carefully wiped to remove residual medium, and the cell pellets are rapidly resuspended in cold 0.75 M sucrose–10 mM Tris, pH 7.8, to a final density of 7 × 10^9 bacteria per milliliter (0.7 ml sucrose solution per 10 A_{600} units of original culture). The suspension is transferred to an Erlenmeyer flask, lysozyme (0.05 ml per milliliter of cell suspension) is added, and the mixture is incubated in ice for 2 minutes. At this point the cells are osmotically fragile but retain their rod shape. Conversion to the spheroplast form is accomplished by slowly diluting the suspension with 2 volumes of cold 1.5 mM EDTA. The EDTA is delivered at a constant rate over a period of 8–10 minutes either by means of a peristaltic pump or (for large volumes) a separatory funnel adjusted to deliver at the desired rate. The cell suspension is constantly swirled in an ice–H$_2$O bath and the diluent is added under the surface of the liquid in order to avoid local hypotonicity. Vigorous agitation and bubbling of air through the suspension is avoided in order to minimize premature lysis. The efficiency of spheroplast formation is monitored by phase contrast microscopy and should be greater than 95%.

Factors Affecting Spheroplast Formation. Good spheroplasting is essential for successful separation of the membranes and is critically dependent on several factors. Cells grown in minimal media are more resistant to spheroplasting by this and other techniques than are cells taken from rich media, such as PPBE, and spheroplast formation and membrane separation tends to be irreproducible and incomplete. This difficulty is largely overcome by supplementation of minimal growth media with a small amount (0.05 volume) of PPBE. The reasons are not known. Efficient plasmolysis of the cells in the hypertonic 0.75 M sucrose–Tris solution is also extremely important, and is adversely affected by a variety of treatments which appear to damage the permeability barrier to sucrose. These include prior exposure of the cells to Tris, washing after harvesting, permitting the culture or cell pellet to stand at 0° for 30 minutes before suspension in hypertonic sucrose, and exposure to the hypertonic sucrose-Tris solution for more than 10 minutes. The resulting spheroplast preparations contain large numbers of prematurely lysed rod-shaped ghosts as well as intact rods, and subsequent separation of the membranes is extremely poor. The efficiency of spheroplasting is also dependent on the concentration of lysozyme. With *S. typhimurium* at the cell density employed, concentrations below 50 μg/ml give incomplete spheroplasting while concentrations above 100 μg/ml result in macroscopic clumping of cells and spheroplasts of poor quality. The optimal concentration of lysozyme may, however, vary somewhat with organism and growth condition. Increasing the time of exposure to lysozyme beyond 2 to 3 minutes or raising the temperature to 25° does not increase the yield of spheroplasts. The rate of addition of EDTA does not appear to be critical, except that dilution of the hypertonic medium must be sufficiently slow to avoid localized hypotonic shock and premature lysis. Some leeway in the final sucrose concentration is also possible. This concentration is 0.25 M in the standard procedure, and the spheroplasts are generally reasonably stable to spontaneous lysis. However, if significant numbers of spheroplast ghosts are visible in phase contrast microscopy, dilution may be carried out as described above with 1.25 volumes of 1.8 mM EDTA (final concentrations, 0.33 M sucrose, 1 mM EDTA).

The nature of the lipopolysaccharide appears to have little effect on the efficiency of spheroplast formation in cells grown in rich media (but see below). However, incomplete and irreproducible spheroplasting has been observed with minimal-grown cells when the lipopolysaccharide has the complete wild-type structure, i.e., when O-antigen chains are present in normal amount. The reasons are not clear.

Lysis of Spheroplasts and Isolation of Total Membrane Fraction

Reagents

0.25 M Sucrose–3.3 mM Tris, pH 7.8–1 mM EDTA. Prepare by mixing 2 volumes of 1.5 mM EDTA, pH 7.5, with 1 volume of 0.75 M sucrose–10 mM Tris·acetate, pH 7.8

25% sucrose (w/w)–5 mM EDTA, pH 7.5: 27.6 g of sucrose and 5 ml of 0.1 M EDTA, pH 7.5 in a final volume of 100 ml

25% sucrose (w/w)–5 mM Tris–5 mM EDTA: 27.6 g of sucrose,

5 ml of 0.1 M EDTA, pH 7.5, and 5 ml of 0.1 Tris·acetate pH 7.8 in a final volume of 100 ml

Procedure. Spheroplasts may be lysed either by osmotic shock or by brief sonication. Sonication is the more convenient and is satisfactory for most purposes. The spheroplast suspension is sonicated in 10–25 ml aliquots in a 2.5 × 8 cm cellulose nitrate tube; a cut-down 1 × 3.5 inch Spinco tube (No. 302237) is convenient. The tube is immersed in an ice–salt bath and the suspension is sonicated for 3 15-second periods with a 20 KC Branson Sonifier or equivalent instrument. The suspension is cooled for 1 minute between bursts in order to maintain a temperature below 10°. If desired, dithiothreitol or 2-mercaptoethanol may be added before lysis and may be included in all subsequent solutions. Exposure to sonic oscillation should not be continued beyond the minimum time required for complete lysis, since prolonged sonication can lead to artificial hybrids between the inner and outer membrane. Such hybrids appear as unseparated envelope fragments in the subsequent separation step.

Osmotic shock provides a more gentle means of lysis and is the method of choice when the component of interest is solubilized or inactivated by sonication. This method is also advantageous when the spheroplast preparation contains significant numbers of intact rods; unlysed cells remaining after osmotic lysis of spheroplasts may be removed by centrifuging the lysate at 1200 g for 15 minutes at 2–4°. The spheroplast suspension is slowly poured into 4 volumes of cold H_2O with magnetic stirring and is stirred for 10 minutes in the cold.

The total membrane fraction is recovered from the sonicate or osmotic lysate by centrifugation. The lysate is first centrifuged at 1200 g for 15 minutes at 2–4° to remove any unlysed cells, and the supernatant fraction is then centrifuged for 2 hours at 60,000–65,000 rpm ($R_{max} = 360,000\ g$) at 2–4°. The membrane pellet is resuspended in a small volume of cold 0.25 M sucrose–3.3 mM Tris–1 mM EDTA, pH 7.8, with the aid of a No. 23 needle and syringe, diluted with the same buffer to a volume approximately equal to that of the original spheroplast suspension, and centrifuged for 2 hours at 360,000 g as before. The washed membrane pellet is carefully suspended as above in 0.8 to 6 ml of cold 25% sucrose–5 mM EDTA, pH 7.5, for separation of the membranes by density gradient centrifugation.

If the electrophoretic technique is to be used for separation of the membranes, the time required for sedimentation of the total membrane fraction can be shortened considerably by addition of Mg^{2+} to the lysate. Addition of Mg^{2+} completely prevents subsequent separation of inner and outer membranes by isopycnic centrifugation, but does not interfere with the electrophoretic separation. Mg^{2+} is added to the 1200 g supernatant fraction to a final concentration of 5 mM (50 μl of 0.1 M $MgCl_2$ per milliliter), and the membranes are sedimented by centrifugation at 50,000 rpm (R_{max} = 225,000 g) for 30 minutes. The pellet is suspended in sucrose–Tris–EDTA as above, Mg^{2+} is then added to a concentration of 5 mM, and the centrifugation is repeated. The washed membrane fraction is suspended in 0.5 to 2 ml of cold 5% sucrose–5 mM Tris–5 mM EDTA, pH 7.7, for electrophoresis.

Recovery of membrane material in the final total membrane fraction is reproducibly greater than 90% as measured by the distribution of radioactive phospholipid and lipopolysaccharide after growth of bacteria with [2-^{3}H]glycerol and [^{14}C]mannose or galactose. Although substantial amounts of lipopolysaccharide can be released from cells by treatment with 10 mM Tris–10 mM EDTA at 37°, no significant release of lipopolysaccharide has been observed under the conditions employed here for spheroplast formation and isolation of membranes.

Separation of Membranes by Isopycnic Centrifugation

Reagents

Sucrose solutions. The indicated amounts of sucrose are dissolved in 120–150 ml of H_2O; 10 ml of 0.1 M EDTA, pH 7.5 are added and the solution is made up to a final volume of 200 ml. Sucrose concentrations are w/w. 55% sucrose: 138.3 g, 50% sucrose: 123.0 g, 45% sucrose: 108.2 g, 40% sucrose: 94.1 g, 35% sucrose: 80.6 g, 30% sucrose: 67.6 g

Procedure. The total membrane fraction is layered on a 30 to 50% sucrose gradient and centrifuged to equilibrium. For small-scale preparations the SW 41 rotor is employed. Step gradients are prepared by layering 2.1 ml each of 50%, 45%, 40%, 35%, and 30% sucrose over a cushion (0.5 ml) of 55% sucrose. All sucrose solutions contain 5 mM EDTA, pH 7.5. The membrane samples (0.8–1 ml) are layered on top of the gradients and centrifugation is carried out at 35,000–38,000 rpm for 12–16 hours at 2–6°. Apparent equilibrium is reached within 12 hours. Up to 1.5 mg of membrane protein (obtained from approximately 200 ml of culture) may be applied to each SW 41 gradient tube. For larger-scale preparations (1–2 liters of culture) the SW 27 rotor is used. Gradients are prepared in the same manner by layering 6.3 ml of each sucrose solution over a 1.5-ml cushion of 55% sucrose. As much as 5 mg of membrane protein (in 2–3 ml) can be accommodated per tube. Centrifugation is continued for 18–40 hours at 25,000 rpm; equilibrium is reached only after 30–36 hours, but adequate separations are achieved by 18 hours.

Gradients are fractionated either by puncturing the bottom of the tube with a piercing device and collecting drops, or by removal of material from the top of the gradient with a coarse needle and peristaltic pump. The latter technique sacrifices resolution but is convenient for collection of visible bands in preparative gradients. Buoyant densities are determined from measurements of refractive index. Membranes are recovered from the gradient fractions by centrifugation at 65,000 rpm for 2 hours at 2–4° after dilution with 1 mM EDTA, pH 7.5 to a final sucrose concentration of 10% or less.

Analysis of the Membrane Fractions. Four discrete membrane bands are obtained by this technique (Fig. 1). Bands L_1 ($\rho = 1.14$ g/cc) and L_2 ($\rho = 1.16$ g/cc) have been characterized as inner (cytoplasmic) membrane, H ($\rho = 1.22$ g/cc) as outer membrane, and the minor band, M, at intermediate density ($\rho = 1.19$ g/cc) as unseparated envelope fragments. Chemical and enzymatic composition of the fractions are summarized in Tables I and II. The separation of cytoplasmic membrane fragments into 2 bands appears to reflect merely a heterogeneity in the degree of contamination by outer membrane material[4]; no qualitative

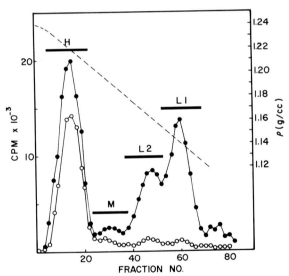

Fig. 1. Isopycnic sucrose gradient centrifugation. Membranes were prepared from *Salmonella typhimurium* G30 (UDP-galactose epimerase negative) after growth for 4 generations in PPBE medium containing [2-³H]glycerol and [¹⁴C]galactose as specific precursors of phospholipid and lipopolysaccharide, respectively. ●—●, [³H]-glycerol; ○—○, [¹⁴C]galactose; - - -, buoyant density (ρ).

TABLE I
CHEMICAL COMPOSITION OF MEMBRANE FRACTIONS[a]

Fraction	Phospholipid (mg/mg protein)	Lipopolysaccharide[b] (mg/mg protein)	Cytochrome (mg/mg protein)	RNA (%)[c]	DNA (%)[c]	Peptide glycan (%)[c]
Total membranes	0.32	0.66	0.025	1.4	0.8	18.7
Inner membrane: L1	0.61	0.09	0.092			2.0
L2	0.53	0.14	0.086			2.6
Outer membrane	0.30	0.98	0.003			13.0

[a] M. J. Osborn, J. E. Gander, E. Parisi, and J. Carson, *J. Biol. Chem.* **247**, 3962 (1972).
[b] Wild-type polymer, containing O-antigen chains.
[c] Percentage of total in cells.

differences in chemical composition or enzyme activities have been detected.

Factors Affecting Separation. As indicated above, addition of Mg^{2+} or other divalent cation to the membranes at any point in the procedure results in extremely poor separation, and virtually all the membrane material is recovered in the position of the **M** band. In addition, this procedure requires maintenance of a low ionic strength in both the sucrose gradient and the isolation of the total membrane fraction. Addition of

TABLE II
LOCALIZATION OF ENZYME ACTIVITIES IN MEMBRANE FRACTIONS[a-d]

Inner membrane	Outer membrane
NADH oxidase	Phospholipase A_1
Succinate dehydrogenase	Lysophospholipase
D-Lactate dehydrogenase	Lysophosphatidic acid phosphatase
L-α-Glycerophosphate dehydrogenase	UDP-glucose hydrolase
α-Methylglucoside phosphotransferase fraction II	
Enzymes of phospholipid biosynthesis	
Enzymes of O-antigen synthesis	
UDP-glucose hydrolase	

[a] M. J. Osborn, J. E. Gander, and E. Parisi, *J. Biol. Chem.* **247**, 3973 (1972).
[b] C. F. Fox, J. H. Law, N. Tsukagoshi, and G. Wilson, *Proc. Nat. Acad. Sci. U.S.* **67**, 598 (1970).
[c] R. M. Bell, R. D. Mavis, M. J. Osborn, and P. R. Vogelos, *Biochim. Biophys. Acta* **249**, 628 (1971).
[d] D. A. White, F. R. Albright, W. J. Lennarz, and C. A. Schnaitman, *Biochim. Biophys. Acta* **249**, 636 (1971).

0.2 M KCl to the wash medium results in a sharp increase in the amount of the L_2 and M bands at the expense of L_1 and H, and inclusion of KCl in the sucrose gradient entirely prevents separation.

Separation is also dependent on the nature of the outer membrane lipopolysaccharide, although the isopycnic centrifugation technique is less sensitive to lipopolysaccharide structure than is the electrophoretic method (see below). The separation is not affected by the presence or the absence of O-antigen chains (although the efficiency of spheroplasting may be sensitive to this parameter; see above), and excellent separations have been obtained with rough mutants of S. *typhimurium* containing incomplete lipopolysaccharides of the types R_a (complete core), R_b (lacking N-acetylglucosamine), and R_c (lacking galactose). With heptoseless mutants (R_e), however, separation into H and L fractions is incomplete. The cytoplasmic membrane bands (L_1 and L_2) are recovered in decreased yield, and the usual outer membrane band H ($\rho = 1.22$) is much reduced or absent. Most of the outer membrane material is recovered in a band of intermediate density ($\rho = 1.18$–1.19), which is heavily contaminated by cytoplasmic membrane as measured by DPNH oxidase activity. Similar difficulties have also been encountered with R_d mutants which contain the heptose region but lack glucose.

Separation of Membranes by Sucrose Gradient Electrophoresis

Reagents

5 mM Tris–5 mM EDTA–10% sucrose (w/w): 20.8 g of sucrose is dissolved in approximately 150 ml of H_2O; 10 ml of 0.1 M Tris·acetate, pH 7.8, and 10 ml of 0.1 M EDTA, pH 7.5, are added, and the volume is adjusted to 200 ml

5 mM Tris–5 mM EDTA–30% sucrose (w/w): 338.1 g of sucrose is dissolved in 700 ml of H_2O; 50 ml each of 0.1 M Tris·acetate, pH 7.8, and 0.1 M EDTA, pH 7.5, are added, and the volume is adjusted to 1 liter

3% agarose: 1.5 g agarose is dissolved in 50 ml of 5 mM Tris–5 mM EDTA–30% sucrose in a boiling H_2O bath

Cathode buffer: 5 mM Tris–5 mM EDTA: 50 ml each of 0.1 M Tris·acetate, pH 7.8, and 0.1 M EDTA, pH 7.5 are mixed and diluted to 1 liter

Anode buffer: 5 mM Tris–5 mM EDTA–30% sucrose. See above.

5 mM Tris–5 mM EDTA–5% sucrose (w/w): 5.1 g of sucrose is dissolved in approximately 70 ml of H_2O. Then 5 ml each of 0.1 M Tris·acetate, pH 7.8, and 0.1 M EDTA, pH 7.5, are added, and the volume is adjusted to 100 ml

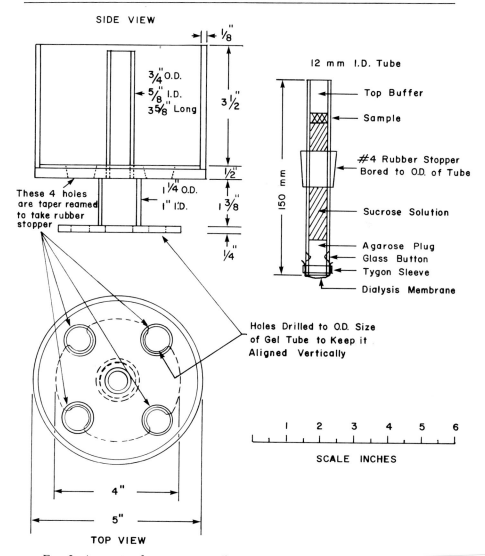

SIDE VIEW

$\frac{1}{8}''$

$\frac{3}{4}''$ O.D.
$\frac{5}{8}''$ I.D.
$3\frac{5}{8}''$ Long

$3\frac{1}{2}''$

$\frac{1}{2}''$

These 4 holes are taper reamed to take rubber stopper

$1\frac{1}{4}''$ O.D.
$1''$ I.D.

$1\frac{3}{8}''$

$\frac{1}{4}''$

Holes Drilled to O.D. Size of Gel Tube to Keep it Aligned Vertically

12 mm I.D. Tube

Top Buffer

Sample

#4 Rubber Stopper Bored to O.D. of Tube

Sucrose Solution

Agarose Plug
Glass Button
Tygon Sleeve
Dialysis Membrane

150 mm

| | 1 | 2 | 3 | 4 | 5 | 6 |

SCALE INCHES

4"

5"

TOP VIEW

Fig. 2. Apparatus for sucrose gradient electrophoresis, see text for details.

Procedure. Spheroplasts are prepared and lysed by the usual technique. $MgCl_2$ is added to the lysate to a final concentration of 5 mM, and the membranes are collected by centrifugation and washed as described above. The final membrane pellet is suspended in 5% sucrose–5 mM Tris–5 mM EDTA, pH 7.7 (0.5–1 ml per 100 A_{600} units of original culture).

Electrophoresis is carried out in a 12 mm i.d. × 15 cm (approximately) glass tube. The bottom of the tube is covered with a section of

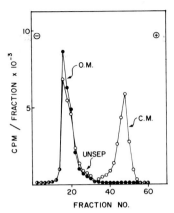

FIG. 3. Sucrose gradient electrophoresis of total membrane fraction. Membranes were prepared as in Fig. 1. ●—●, [^{14}C]galactose; ○—○ [2-^{3}H]glycerol. O.M., outer membrane; C.M., cytoplasmic membrane.

dialysis tubing held in place with a Tygon tubing sleeve (Fig. 2), and the tube is plugged with about 3 ml of 3% agarose in 5 mM Tris–5 mM EDTA–30% sucrose (pH 7.7). A 14-ml linear 10 to 30% gradient of sucrose in 5 mM Tris–5 mM EDTA, pH 7.7 is formed in the tube (the top of the gradient should come 3–4 cm from the top of the tube). The bottom half of the tube must be kept immersed in 5 mM Tris–5 mM EDTA 30% sucrose to prevent leakage of the gradient through the agar plug and dialysis tubing.

Electrophoresis is carried out in a disc gel electrophoresis apparatus, the upper chamber of which is modified to accept the large diameter tubes (Fig. 2). The lower chamber of the apparatus should be jacketed. The lower chamber (anode) is filled with anode buffer and the gradient tubes are placed in the electrophoresis chamber. Cathode buffer is carefully layered over the gradients to fill the tubes, and the upper chamber is then carefully filled with cathode buffer. The samples (up to 1.5–2 mg protein in 0.3–0.5 ml) are layered at the interface between the sucrose gradient and the cathode buffer as for disc gel electrophoresis. Electrophoresis is carried out for 75–80 minutes at 15 mA per tube. The run is carried out in the cold room, and ice water is circulated around the lower chamber.

Fractions are collected from the top of the tube. A device such as the Auto Densiflow (Buchler Instruments) is convenient. Three membrane bands are obtained (Fig. 3). Under the conditions employed, outer membrane shows a low mobility and is found near the top of the gradient, while inner membrane is recovered as a single rapidly migrating anionic band near the middle of the tube. Unseparated material migrates as a minor band just below outer membrane. As indicated above, the separation requires the presence of O-antigen chains in the lipopolysaccharide; in their absence outer membrane comigrates with inner membrane.

[68] Isolation of Cell Walls from Gram-Positive Bacteria

By MILTON R. J. SALTON

It had long been known that bacteria possessed a fairly resistant, rigid wall or envelope structure, but its precise chemical constitution remained quite uncertain until bacterial cell wall chemistry studies were launched some 20 years ago after the development of suitable methods for their isolation.[1] The recognition that the wall structures of bacteria were sufficiently robust to survive cell disruption procedures, such as sonication,[2] treatment in ball mills used for enzyme extraction,[3] and vigorous shaking with minute glass beads,[4] indicated the feasibility of wall isolation for chemical and biochemical studies. The detection of the wall structures as empty sacs or shells in preparations examined in the electron microscope[2,4] demonstrated its great value in "visualizing" these structures and has led to the universal acceptance of the electron microscope as the instrument of choice in monitoring the structural homogeneity of wall preparations.[1,5,6]

After Dawson's[4] demonstration that *Staphylococcus aureus* could be disrupted to give empty walls, cytoplasmic debris, and particles, Salton and Horne[1] developed procedures for the isolation of "clean" wall or envelope fractions. The heat-treatment rupture method was really successful only with gram-negative bacteria, but disruption with glass beads in the Mickle Shaker worked equally well with gram-positive and gram-negative species.[1] The subsequent removal of contaminating cytoplasmic debris by washing procedures proved to be easier with the majority of gram-positive bacteria.[5] The basic principles utilized in this early method[1] still apply, but as indicated below a great variety of cell disruption pro-

[1] M. R. J. Salton and R. W. Horne, *Biochim. Biophys. Acta* **7**, 177 (1951).
[2] S. Mudd, K. Polevitsky, T. F. Anderson, and L. A. Chambers, *J. Bacteriol.* **42**, 251 (1941).
[3] M. Stephenson, "Bacterial Metabolism," p. 16. Longmans, Green, London, 1949.
[4] I. M. Dawson, *Symp. Soc. Gen. Microbiol.* **1**, 119 (1949).
[5] M. R. J. Salton, "The Bacterial Cell Wall," p. 1. Elsevier, Amsterdam, 1964.
[6] H. J. Rogers and H. R. Perkins, "Cell Walls and Membranes," p. 1. Spon, London, 1968.

cedures, equipment, and steps in the subsequent "processing" of the walls of gram-positive bacteria have been introduced.

Principle of Isolation Methods

The isolation of the cell walls from gram-positive bacteria which possess walls varying in thickness from about 15 to 80 nm is based essentially on: (1) mechanical disruption of cells in suspension or paste form; (2) separation of the wall structures from soluble and particulate components of the cell, either by differential or gradient centrifugation procedures and particle separation techniques in polymer mixtures[7]; (3) wall-processing procedures designed to "clean up" the wall fractions to rid them of contaminating particles, denatured protein, membrane remnants, etc. The latter steps may vary from simple washing in buffers[1,5,6] to enzymatic digestions, detergent treatment, and chemical or organic solvent extractions to remove persistent, coseparating components. Although it may not be absolutely essential, it is often a good idea to examine in the electron microscope metal-shadowed and/or negatively stained specimens of cells and wall fractions after the various steps, especially when a new organism is being used or where wall isolation is attempted for the first time.

Cell Growth Conditions and Preparation of Suspension and
    Cell Pastes

The selection of suitable growth media is obviously dictated by the species under investigation but where choices are permissible the task of wall isolation can be simplified by selecting appropriate media. Thus, if possible, media containing suspended matter which may be difficult to separate from the cells during harvesting and washing should be avoided. Contamination of wall fractions with agar can be troublesome and may also have to be avoided for some studies. Media that can be readily washed from the cells with distilled water, saline, or buffers are naturally advantageous.

Intracellular storage granules (e.g., poly-β-hydroxybutyrate (PHB), polymetaphosphate) can be troublesome during wall isolation as reported for *Bacillus megaterium* and other species.[5,8] Growth media and conditions can often dramatically affect the production of such storage granules,[9] and the selection of a medium minimizing granule production will thus simplify wall isolation. Medium composition also affects autolysins, and cells grown in certain media may be prone to lysis; this may complicate the harvest-

[7] P. -Å. Albertsson, *Methods Biochem. Anal.* **10**, 229 (1962).

[8] J. M. Vincent, B. Humphrey, and R. J. North, *J. Gen. Microbiol.* **29**, 551 (1962).

[9] J. F. Wilkinson and J. P. Duguid, *Int. Rev. Cytol.* **9**, 1 (1960).

ing and washing steps, in addition to indicating involvement of wall-degrading enzymes and the need for limitation of their activities during isolation steps.[5,6]

The choice of suspending fluid for the cells to be subjected to the disruption procedure will usually be determined by local factors and experience. Distilled water, 0.9% sodium chloride solutions, various buffers have all proved satisfactory with a variety of gram-positive bacteria. Where problems of dispersion of cells exist (as for example with *Mycobacterium*[10]), suitable surface-active agents may be incorporated in the wash and suspension buffers. Similar general considerations also apply to the preparation of cell pastes and in some instances abrasives are also added to the pastes for more efficient cell disruption.

The final density of the cell suspension will depend largely on the choice of disruption procedure and equipment and the scale of preparation. It has varied from 10 to 20 mg dry weight of cells per milliliter in the Mickler procedure of Salton and Horne,[1] to 15 g of dry cells added to 250 ml of distilled water and 250 ml of glass beads in the larger-scale preparation of cell walls of *Micrococcus lysodeikticus* by Sharon and Jeanloz.[11]

Mechanical Disruption of Bacteria

Because of their thicker cell walls, it has long been known that gram-positive bacteria are more resistant to mechanical disruption.[12] Moreover, some of the other cell disruption techniques available for the preparation of envelopes of gram-negative bacteria (e.g., osmotic lysis of halophiles,[13] glycerol lysis (osmotic shock) as used with *Azotobacter agilis* and several other gram-negative organisms[14]) do not appear to be applicable to gram-positive bacteria. Thus for the preparation of cell walls of gram-positive bacteria disruption of the cells can be best achieved by a variety of mechanical devices which violently agitate bacterial cell suspensions with beads or abrasive additives. The force of the impact of the wall and beads or abrasive is thus sufficient to tear or shear the cell wall structure, fragment the membrane and allow the cell sap to spill out. The sheared ends and what appear to be surface abrasion marks on the cell wall of *Bacillus cereus* obtained by disruption with glass beads are illustrated in Fig. 1 and disruption points in *Staphylococcus aureus* walls are clearly visible

[10] E. Ribi, C. L. Larson, R. List, and W. Wicht, *Proc. Soc. Exp. Biol. Med.* **98**, 263 (1958).

[11] N. Sharon and R. W. Jeanloz, *Experientia* **20**, 253 (1964).

[12] P. K. Stumpf, D. E. Green, and F. W. Smith, *J. Bacteriol.* **51**, 487 (1946).

[13] A. D. Brown, C. D. Shorey, and H. P. Turner, *J. Gen. Microbiol.* **41**, 225 (1965).

[14] S. A. Robrish and A. G. Marr, *Bacteriol. Proc.* **70**, 130 (1957).

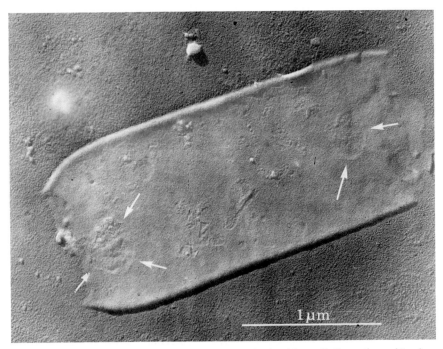

FIG. 1. Isolated cell wall of *Bacillus cereus* obtained by cell disruption with glass beads in a mechanical shaker. Note the sheared ends of the flattened rod-shaped wall. What appear to be surface abrasion marks are indicated by arrows. Preparation was chromium shadowed.

in Fig. 2 (Salton, unpublished electron micrographs). Cell disruption and shearing of the wall can also be achieved by the forces subjected by the French pressure (20,000 psi) apparatus, by cavitational forces exerted by exposure to sound and ultrasound (sonicators) and by subjecting frozen cell pastes mixed with abrasives to high pressure or the force exerted a fly press as in the Hughes press.[15] The relative merits and disadvantages of the array of equipment available for cell disruption will accordingly be reviewed briefly as it may help investigators in selecting the apparatus most suited to their particular needs.

Choice of Cell Disruption Equipment. Two of the earlier pieces of equipment useful for medium to small-scale preparation of cell walls (1–2 g or less) include the Mickle[16] apparatus as used by Salton and Horne[1] and the wrist-action shaker as used by Dawson.[4] In the Mickle apparatus, two plain-walled glass cups each containing 10 ml of bacterial

[15] D. E. Hughes, *Brit. J. Exp. Pathol.* **32**, 97 (1951).
[16] H. Mickle, *J. Roy. Microsc. Soc.* **68**, 10 (1948).

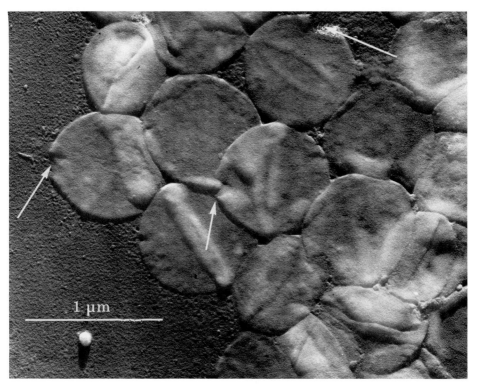

Fig. 2. Cell walls of *Staphylococcus aureus* obtained by mechanical disintegration with glass beads. Arrows indicate rupture points in the walls. Chromium shadowed.

suspension (10–20 mg dry weight of cells per milliliter) and 10 ml of beads (specifications of suitable beads will be discussed below) are held at the end of bars which are subjected to electromagnetic vibration. To achieve maximal disruption, the amplitude of the vibrating bars has to be "tuned" to a maximum by mechanical adjustment. The need for adjustment is a disadvantage and may be necessary in the middle of a run. Under optimal conditions of shaking in the Mickle, cell disruption of more than 99.99% (based on gram-stained smears) can be achieved in 10 minutes, and up to 2 g dry weight of bacterial cells (in aqueous suspensions) can be processed in about 1–2 hours. The apparatus has no cooling device and will generate considerable heat in the suspension and therefore has to be used in a cold room or at ambient temperature with frequent cooling of cups, suspension, and beads. The wrist-action shakers are less vigorous and require a longer period of shaking, the flasks have to be cooled or the whole operation be performed in a cold room, and the processing capacity is generally no better than that of the Mickle apparatus.

Other mechanical cell disintegrators which have been used successfully for wall preparation include the Nossal[17] disintegrator and the Braun shaker[18] and both can be used at room temperature with the aid of suitable cooling devices. The Braun shaker is very efficient and uses bottles with 100-ml capacities; it is thus suitable for preparing larger batches of cell walls. Its main disadvantages are the CO_2 cooling device and problems arising from the design of the bottle holder. An apparatus based on the principle used in the Nossal disintegrator has been described by Ross[19] and has the great advantage of being designed for continuous processing of cell suspensions and temperature control by expanding liquid CO_2 on the oscillating chamber. With cooling, the exit temperature of the suspension was reduced from 30–35° (for a 35-second hold-up time) to 18°–21°. High breaking efficiency was claimed, and the apparatus has an excellent volume capacity of 1–2 liters per hour depending on the "hold-up time" required for complete rupture of the cells.[19] Ross[19] introduced the use of styrene–divinylbenzene copolymer beads (20–50 mesh) for cell disruption in this equipment, and although it was designed for the preparation of subcellular fractions retaining high levels of enzymatic activity, it could clearly be used for wall preparation.

For moderately large-scale isolation of walls of *Streptococcus faecalis,* Shockman *et al.*[20] designed a high-speed shaker head which could be fitted in the refrigerated International centrifuge for mechanical disruption of the cells. This apparatus had the 2-fold advantages of high capacity (6 g dry weight of bacteria could be disrupted in a single run) and operational temperatures near 0°. The main disadvantages of this procedure stemmed from the need for careful adjustment of the head and centrifuge speed, and leveling of the centrifuge to avoid vibration and mechanical damage to the centrifuge.

The use of high-speed blending of bacterial suspension and glass beads (0.2 mm diameter) was introduced by Lammana and Mallette,[21] and this type of procedure was later adapted by Sharon and Jeanloz[11] for large-scale preparations of *M. lysodeikticus* cell walls. In a typical protocol for this procedure, Sharon and Jeanloz[11] placed 15 g dried cells (*M. lysodeikticus* from Miles Chemical Company, Indiana), 250 g of Super-

[17] P. M. Nossal, *Aust. J. Exp. Biol. Med. Sci.* 31, 583 (1953).
[18] M. Merkenschlager, K. Schlossmann, and W. Kurz, *Biochem. Z.* 329, 332 (1957).
[19] J. W. Ross, *Appl. Microbiol.* 11, 33 (1963).
[20] G. D. Shockman, J. J. Kolb, and G. Toennies, *Biochim. Biophys. Acta* 24, 203 (1957).
[21] C. Lamanna and M. F. Mallette, *J. Bacteriol.* 67, 503 (1954).

brite glass beads, 0.1 mm diameter (Minnesota Mining and Manufacturing Company), and 250 ml of distilled water (precooled in ice) in a 400-ml stainless steel chamber (cooled to 0°) which was held at about −5° in a freezing mixture. The vessel was attached to an Omni-Mixer and the bacterial suspension and beads homogenized at top speed for 50 minutes, at which time virtually complete breakage of the cells had been achieved. The temperature inside the mixing chamber was 3°–4° throughout the operation. Cell walls were separated from unbroken cells, membrane fragments, debris, and metal particles by differential centrifugation. The advantages of this procedure are the high capacity and excellent temperature control but the principal disadvantage is contamination of the preparation by metal particles resulting from the abrasion of the chamber and stirrer.

Pressure cell disintegrators have been used for cell and organelle disruption in enzyme isolation[22] and Ribi et al.[23] extended and refined this method in devising an apparatus for wall preparation. For this, a new type of pressure cell was designed so that the steel cylinder containing the bacterial suspension could be subjected to a total load of 35,000–40,000 psi. Disruption in the Ribi Cell Fractionator could thus be achieved by bleeding the suspension through a needle valve cooled to 2° with a temperature control of the "bled" products to 15°. This method has been widely used and has advantages of an excellent capacity (100 ml of suspension, 3 mg bacteria per milliliter per run) and good temperature control. This method gave superior wall preparations of mycobacteria and avoided the problem of aggregate formation encountered with other procedures.[23] The Ribi apparatus also has the distinct advantage of being adaptable to complete enclosure during disruption of pathogens, thereby minimizing aerosol dispersion of potentially dangerous organisms.

Sonic and ultrasonic disintegrators have been used widely for cell disruption in both cell wall and enzyme preparation. The efficiency of cell disruption varies tremendously but the Raytheon 10 kc sonic oscillator has been successfully employed in many laboratories for wall isolation. The advantages of this method of cell disruption are good capacity, excellent temperature control and good retention of many labile biological activities. One of the major disadvantages of this method is that many of the gram-positive cocci (e.g., staphylococci, micrococci, Sarcina spp.) are resistant to sonication, and many require prolonged exposure to sonic or ultrasonic vibration. Another disadvantage arising from the need for prolonged sonication is that fragmentation of the wall occurs, thereby reduc-

[22] H. W. Milner, N. S. Lawrence, and C. S. French, Science 111, 633 (1950).
[23] E. Ribi, T. Perrine, R. List, W. Brown, and G. Goode, Proc. Soc. Exp. Biol. Med. 100, 647 (1959).

ing yields and the recovery of wall structures under the usual conditions of centrifugation. However, these disadvantages can be overcome by establishing the minimum time for a useful degree of cell disruption. The author has also found that the addition of glass beads (0.1–0.2 mm diameter) improves the efficiency of cell disruption of refractory bacteria, but the long-term use of beads may lead to some abrasion of stainless steel suspension containers.

Another useful mechanical procedure for crushing bacterial cells was devised by Hughes[15] and embodied in the apparatus known as the Hughes press. This technique essentially involves the use of a cooled stainless steel block with a hollowed cylinder into which the bacterial paste or suspension and appropriate abrasive are placed. The bacteria are subjected to a short period of mechanical treatment (0.5–5 seconds) by means of fly press exerting a force of 12–15 tons psi through a carefully machined piston fitting into the hollowed receptacle of the block. The cells are forced under this pressure through a reservoir channel in the block. Abrasives used in this procedure included powdered Pyrex glass, silica, and powdered aluminum. If temperatures of $-20°$ to $-35°$ were used the abrasives could be dispensed with. Under optimum conditions of abrasive:cell ratios and temperature, virtually complete disintegration of many gram-positive bacteria could be achieved. Although this procedure is carried out on a batch basis, the overall capacity is good because of the short period of treatment, and temperature control is excellent. The principal disadvantage of the method arises from the use of abrasives, which are frequently needed for a high efficiency of disruption. The use of low temperatures can eliminate the need for abrasives, but this is counterbalanced by incomplete disintegration of bacteria. An essentially similar method using an Edebo or X-press has been developed by Yoshida et al.[24] for the isolation of walls from gram-positive and gram-negative bacteria.

Ball mills for wet crushing of bacteria have long been used for enzyme extraction,[25] and McCarty[26] used such a procedure for cell rupture in preparing group A streptococcal walls. The capacity of these mills is high, but prolonged treatment for a good level of disruption is frequently needed.[24]

Further aspects of cell disruption techniques are discussed critically by Hugo.[25] A summary of a selection of cell disruption techniques and features of these methods is given in the table.

The rapid and efficient disintegration of bacterial cells in mechanical

[24] A. Yoshida, C. G. Hedén, B. Cedergren, and L. Edebo, J. Biochem. Microbiol. Technol. Eng. 3, 151 (1961).
[25] W. B. Hugo, Bacteriol. Rev. 18, 87 (1954).
[26] M. McCarty, J. Exp. Med. 96, 569 (1952).

shaking devices and high-speed blendors requires the presence of glass beads or other suitable beads or abrasive particles. Although not absolutely essential, beads have also been used in sonic vibrators for disruption of streptococci[27,28] and other bacteria.[5] The use of small smooth glass beads was introduced by King and Alexander,[29] and they defined the optimal size for most effective killing (0.13–0.26 mm). Other parameters for cell disruption (e.g., frequencies and duration of shaking, cell density, and suspending media) were determined for the optimal bead size of 0.1–0.2 mm diameter (Ballotini grade 12). Glass beads of diameters outside the 0.1–0.2 mm range are thus less efficient in cell rupture and would consequently result in lowered wall yields. For wall preparation, glass beads have been the particle of choice since they can be continually "regenerated" by washing with acid for further use. The principal disadvantage in their use is the generation of alkali,[30] which could result in loss of alkali-labile components of the walls of gram-positive bacteria (e.g., O-acetyl and ester-linked D-alanyl groups) and the generation of heat during vigorous shaking. Both deleterious effects were avoided by the introduction of styrene–divinylbenzene copolymer beads (Dow Chemical Co.) by Ross.[19] Because of their low density, less energy was required and less heat was developed during mechanical shaking with the plastic beads.[19] These beads have been used for streptococcal wall isolation,[31] but since they are used and discarded and are less suitable for recycling, their widespread adoption will be dependent upon their availability at a modest cost. Other abrasive particles of wide size distribution are unsuitable for the subsequent steps of wall isolation involving filtration or sedimentation of particles and wall recovery by centrifugation.

Isolation of Walls from Disrupted Cell Suspensions

The processing of cell walls should proceed as rapidly as possible at 0° after cell disruption, to minimize enzymatic modification by wall-degrading autolytic enzymes (see Higgins and Shockman[32] for recent discussion of autolysins). Indeed, where these are particularly active, as in *Bacillus* species,[33] it may be necessary to inactivate the enzymes by rapid heating for 5–10 minutes at 100°C prior to further processing or at least heat the crude wall deposits after the first centrifugation. In general,

[27] E. L. Hess and H. D. Slade, *Biochim. Biophys. Acta* 16, 346 (1955).
[28] B. S. Roberson and J. H. Schwab, *Biochim. Biophys. Acta* 44, 436 (1960).
[29] H. K. King and H. Alexander, *J. Gen. Microbiol.* 2, 315 (1948).
[30] J. J. Kolb, *Biochim. Biophys. Acta* 38, 373 (1960).
[31] G. D. Shockman, J. S. Thompson, and M. J. Conover, *Biochemistry* 6, 1054 (1967).
[32] M. L. Higgins and G. D. Shockman, *CRC Critical Rev. in Microbiol.* 1, 29 (1971).
[33] C. Forsberg and H. J. Rogers, *Nature* (*London*) 229, 272 (1971).

COMPARATIVE FEATURES OF SEVERAL TYPES OF CELL DISINTEGRATION EQUIPMENT SUITABLE FOR WALL PURIFICATION

Type and description	Capacity	Abrasive particles	Duration of treatment for complete disruption	Temperature control	References
Mechanical shakers					
Mickle tissue disintegrator	20 ml/run, 1–2 g dry weight bacteria/hour	Glass beads	10 Minutes	None provided, rise from 0 to 20°	a–c
Nossal cell disintegrator	0.5 g wet weight bacteria	Glass beads	2 Minutes	Operating at −1°, rise to 20° in 90 seconds	d
Ross continuous-flow cell disintegrator	3–12 mg cell N/ml 1,000–2,000 ml	Plastic beads	45–96 seconds hold-up time	CO_2 coolant, rise from 6° to 18–21°	
Braun shaker	36 ml (50 mg dry weight/ml)	Glass beads	5/15–20-Minute period	CO_2 coolant, rise to 5°	e, f
Shockman shaker	6 g dry weight/run	Glass beads	15 Minutes	0°	g
Blender					
Omni-Mixer	15 g dry cells	Glass beads	50 Minutes (top speed)	3–4°	h
Sonic oscillator					
Raytheon 9 KC	10 ml suspension (group A streptococci)	Glass beads	25 Minutes	4°	i

Raytheon 9 KC	10 ml (0.2 g dry weight/ml) (*Bacillus megaterium*)	None	10 Minutes	10°	j
Pressure cell and crushing devices				
Ribi pressure cell	100 ml (3 mg/ml)	None	Not given	Rise from 2° to 15°	k
Hughes press	2–8 g wet weight cells	Powdered glass	6 Blows with fly-press (seconds/blow)	Crushed cells frozen	l
Hughes press	2–8 g wet weight cells	None	6 Blows with fly-press (seconds/blow)	−35°	l

[a] M. R. J. Salton and R. W. Horne, *Biochim. Biophys. Acta* **7**, 177 (1951).

[b] M. R. J. Salton, "The Bacterial Cell Wall," p. 1. Elsevier, Amsterdam, 1964.

[c] H. Mickle, *J. Roy. Microsc. Soc.* **68**, 10 (1948).

[d] P. M. Nossal, *Aust. J. Exp. Biol. Med. Sci.* **31**, 583 (1953).

[e] M. Merkenschlager, K. Schlossmann, and W. Kurz, *Biochem. Z.* **329**, 332 (1957).

[f] E. Huff, H. Oxley, and C. S. Silverman, *J. Bacteriol.* **88**, 1155 (1964).

[g] G. D. Shockman, J. J. Kolb, and G. Toennies, *Biochim. Biophys. Acta* **24**, 203 (1957).

[h] N. Sharon and R. W. Jeanloz, *Experientia* **30**, 253 (1964).

[i] B. S. Roberson and J. H. Schwab, *Biochim. Biophys. Acta* **44**, 436 (1960).

[j] M. R. J. Salton, *J. Gen. Microbiol.* **9**, 512 (1953).

[k] E. Ribi, T. Perrine, R. List, W. Brown, and G. Goode, *Proc. Soc. Exp. Biol. Med.* **100**, 647 (1959).

[l] D. E. Hughes, *Brit. J. Exp. Pathol.* **32**, 97 (1951).

wall isolation following cell disintegration by one of the methods chosen from those outlined above will involve the basic steps detailed below.

Step 1. (Optional depending on cell disruption method.) Glass or plastic beads are removed, preferably by filtration on a medium porosity-fritted glass filter, and the beads are washed with a minimum volume of water. Alternatively the beads are removed by decanting off the suspension of ruptured cells and washing the beads with ice-cold water or buffer. If sonication without beads or abrasives has been used, this step can be omitted although filtration is useful in removing coarse aggregates and debris. Filtrates, suspension supernatants, and washings are pooled and subjected to centrifugation.

Step 2. If disintegration is incomplete, a low-speed centrifugation of the suspensions (3000 g, 5–10 minutes) will help to remove intact cells and debris. Cell walls can then be deposited and separated from the "cytoplasm" by centrifugation at 9000–10,000 g for 20–30 minutes. Discard the supernatant fluid.

Step 3. Resuspend the wall deposit in distilled water and wash with distilled water several times on the centrifuge to remove soluble proteins. Additional washing with 1 M NaCl and buffers[1,5] have been useful in more efficient removal of cytoplasmic contaminants.

Step 4. (Optional.) A variety of enzymatic treatments of the wall fractions have been used to ensure removal of nonwall components. These have included digestions with trypsin, pepsin, ribonuclease, and deoxyribonuclease.[26,34] *Caution:* Some of the external protein structures of "native" walls[35,36] may be removed: care should be taken to make sure that the enzyme preparations are devoid of muramidase activities; if a marked drop in wall turbidity is observed the digest supernatant should be checked for solubilized peptidoglycan components.

Step 5. Where enzymatic digests have been carried out, several washes in buffer and distilled water are essential for the removal of enzymes and digests products. After such final washes, the walls should be resuspended to give a uniform suspension, which can then be given a final low-speed centrifugation to remove aggregates and any particulate matter concentrated in the preparation during centrifugations and manipulations. Alternatively, the wall pellet can be mechanically removed by means of a spatula from the layer of foreign matter at the bottom of the centrifuge tube. The purified walls of gram-positive bacteria have a milky white appearance and should be completely devoid of any cellular pigments.

[34] C. S. Cummins and H. Harris, *J. Gen. Microbiol.* 18, 173 (1958).
[35] R. G. E. Murray, *in* "Microbial Protoplasts, Spheroplasts and L-Forms" (L. B. Guze, ed.), p. 1. Williams & Wilkins, Baltimore, Maryland, 1968.
[36] M. V. Nermut and R. G. E. Murray, *Z. Mikrobiol.* 8, 195 (1968).

The above basic steps in wall isolation represent the simplest procedure where conditions for cell disruption are optimized and where there are no complications from the presence of additional cellular structures. Modifications of these procedures are necessary if there is a reasonably high proportion of unbroken cells and/or intracellular granules (e.g., polyhydroxy butyrate, metachromatic, glycogenlike storage granules) which coseparate or cause extensive contamination of the lower layers of centrifuged wall fractions. Under these circumstances two-phase polymer systems, density gradient centrifugation procedures, and flotation in CsCl gradients can be utilized in the separation of the walls and for the examination of the homogeneity of the preparations.[5-7,28,37] Standard centrifugation procedures can then be used for recovering the purified walls from polymer or gradient media.

Walls may be stored in suspension preferably at -10 to $-70°$ or lyophilized after removal of salts and buffers by centrifugation, washing, and/or dialysis. The yields of walls from gram-positive bacteria are generally of the order of 10–20% of the dry weight of the bacteria used,[5] thus representing a recovery of about 50–80% of the wall material. Under optimal conditions recoveries can be even higher, approaching 90–95%. As the thickness of the cell walls of gram-positive organisms varies widely with the cultural conditions, the contribution of the wall to the dry weight of the cells will vary, and hence marked differences in yields can be expected with the same species. The latter will thus be superimposed on differences in the efficiency of wall isolation methods. With *M. lysodeikticus,* the large-scale wall preparation from commercially grown cells gave a yield of 2.0 g of wall from 15 g of cells[11] (a yield of 13%) whereas the same organism grown on tryptic-digest of casein agar medium and walls isolated by the Mickle procedure[1] yielded 0.8 g of wall from 2.3 g dry weight of cells (a yield of 35%).

Criteria for Homogeneity of Isolated Wall Preparations

Although regular staining procedures, such as the gram stain, are useful in following the efficiency of cell disruption, they cannot be relied upon for assessing the morphological homogeneity of wall fractions. Examination of preparations in the electron microscope is the method of choice for the monitoring of the structural homogeneity of walls. The presence of electron-dense cytoplasmic granules, particles, and debris is readily revealed in metal-shadowed preparations (e.g., Figs. 1 and 2). For the rapid examination of wall fractions, negative staining is particularly useful in giving an overall impression of the cleanliness and homogeneity

[37] E. Huff, H. Oxley, and C. S. Silverman, *J. Bacteriol.* 88, 1155 (1964).

Fig. 3. A preparation of group A streptococcal walls from cells disrupted mechanically with glass beads. Negative staining with 2% phosphotungstate. Unpublished electron micrograph by J. H. Freer, M. S. Nachbar, and M. R. J. Salton.

of the wall preparations. This is especially apparent in the illustration of the walls of group A streptococci, negatively stained with phosphotungstic acid presented in Fig. 3. A comparison of negatively stained (or metal-shadowed) whole cells and isolated walls will also give an indication of the loss of external fine-structured layers[35,36] during the isolation procedures. Electron microscopy thus provides one criterion for assessing contamination with particulate material, but nonparticulate, firmly absorbed contaminants would not be revealed.

Other criteria for examining the homogeneity of wall preparations can be established by using physicochemical, chemical, biochemical, and immunological methods. Ultracentrifugation, behavior on density gradient centrifugation, and electrophoretic analysis can be utilized in defining the homogeneity of wall fractions.[5,6,28,37] Moving-boundary electrophoresis was used by Roberson and Schwab[28] in their study of isolation and immunological properties of group A streptococcal walls, and the patterns showed remarkable homogeneity despite the complexity of the crude cell extract. Although "native" isolated walls are usually antigenically complex, immunological tests can be of some practical use in defining the fractions, especially where antisera to surface capsular substances, plasma membrane, and cytoplasmic components are available. Walls of gram-positive bacteria are generally believed to be devoid of enzymes, except perhaps for those in firmly associated membrane fragments. Thus biochemical assays of enzymes known to occur specifically in the membrane or cell cytoplasm can be used as an index of contamination in wall fractions. Similarly, any chemical markers known to be distributed in other cellular fractions can form the basis of analytical tests for homogeneity. Finally where the wall structure is susceptible to complete dissolution by a muramidase, an estimate of the nonwall, particulate contaminant fraction can be made, and this has generally accounted for something less than 1% (dry weight) of the purified wall fractions.[5]

[69] Isolation of the Cell Membrane of *Halobacterium halobium* and Its Fractionation into Red and Purple Membrane

By Dieter Oesterhelt and Walther Stoeckenius

Extremely halophilic bacteria require high NaCl concentrations for growth. Their physiology and biochemistry have been reviewed by Larsen.[1] Five strains of *Halobacterium* have been described in some detail[2]: *H. halobium, H. cutirubrum, H. salinarium, H. trapanicum,* and *H. marismortui.* They show a number of unusual properties, which have been studied mainly in *H. halobium, H. cutirubrum,* and *H. salinarium:* (1) The cell wall of these bacteria consists mainly of protein. Carbohydrates are present in small amounts only. No lipids, peptidoglycans, or teichoic acids have been found.[3–6] (2) Under the usual culture conditions, halobacteria do not synthesize fatty acids. Their lipids contained in the cell membrane carry exclusively dihydrophytol chains in ether linkage to glycerol.[7] (3) Most

[1] H. Larsen, *Advan. Microbiol.* 1, 97 (1967).
[2] "Bergey's Manual of Determinative Bacteriology" (R. S. Breed, E. G. D. Murray, and N. R. Smith, eds.), 7th ed., p. 208. Williams & Wilkins, Baltimore, Maryland, 1957.
[3] D. J. Kushner, S. T. Bayley, J. Boring, M. Kates, and N. E. Gibbons, *Can. J. Microbiol.* 10, 483 (1964).
[4] W. R. Smithies, N. E. Gibbons, and S. T. Bayley, *Can. J. Microbiol.* 1, 605 (1955).
[5] W. Stoeckenius and R. Rowen, *J. Cell Biol.* 34, 365 (1967).
[6] W. Stoeckenius and W. H. Kunau, *J. Cell Biol.* 38, 337 (1968).
[7] M. Kates, L. S. Yengoyan, and P. S. Sastry, *Biochim. Biophys. Acta* 98, 252 (1965).

cellular enzymes studied require salt concentrations of 2 M or higher for optimal activity and are irreversibly denatured by exposure to low salt concentrations or distilled water. (4) The cells lyse when exposed to low salt concentrations or distilled water. Many cell proteins are then released into the medium in soluble form, the ribosomes dissociate, and the cell membrane disintegrates into pieces of widely varying size, which can be separated by centrifugation.

Because it is possible to selectively remove the cell wall from isolated cell envelopes (cell wall + membrane) of these bacteria and because no intracellular membranes are present (with the exception of gas vacuole membranes in some strains; see Stoeckenius and Kunau[6]), clean cell membrane preparations can easily be obtained. After disaggregation of the isolated membrane in water, further fractionation yields membrane fragments which differ in structure and composition. One of these fragments, the purple membrane, has attracted attention, because it exhibits an unusual and very regular structure in the plane of the membrane[8,9] and because it contains a chromoprotein, bacteriorhodopsin, which in composition and structure appears to be closely related to the visual pigments of higher animals. It is responsible for the deep purple color of the purified membrane fraction. The remainder of the cell membrane contains high concentrations of carotenoids[10] and appears orange red.

Culture Conditions

Halobacterium halobium NRL (obtained from the National Research Council, Ottawa, Canada) or a mutant R_1 derived from it have mainly been used, but other *H. halobium* strains have given similar results. R_1 is preferred for isolation of the red and purple membrane because it does not form gas vacuoles, which are present in the NRL and other *Halobium* strains and are difficult to remove completely from the red and purple membrane fractions.

Cells are grown in 10-liter cultures in a MicroFerm fermentor (New Brunswick Scientific, New Brunswick, New Jersey) with illumination by the bank of daylight fluorescent tubes of the MicroFerm. The medium contains in 1000 ml: NaCl, 250.0 g; $MgSO_4 \cdot 7 H_2O$, 20.0 g; trisodium citrate $\cdot$ 2 H_2O, 3.0 g; KCl, 2.0 g; Oxoid Bacteriological Peptone L 37[11] (Colab

---

[8] D. Oesterhelt and W. Stoeckenius, *Nature (London) New Biol.* **233**, 149 (1971).
[9] A. E. Blaurock and W. Stoeckenius, *Nature (London) New Biol.* **233**, 152 (1971).
[10] M. Kelly, S. Norgard, and S. Liaaen-Jensen, *Acta Chem. Scand.* **24**, 2169 (1970).
[11] Bacto-Peptone, Bacteriological Technical (Difco 0885-02; Difco Laboratories, Detroit, Michigan) may be used instead of Oxoid Bacteriological Peptone. Bacto-Tryptone (Difco 0123-01) or Bacto-Peptone (Difco 0118-01), however, are not suitable because they contain a substance which lyses the cells.

Laboratories, Glenwood, Illinois), 10 g. After prefiltration through soft filter paper, the medium is filtered through a sterile Millipore filter of 0.45 μm pore size into the autoclaved culture vessel. Working under sterile conditions is not absolutely necessary, because only extreme halophiles will grow in this medium and washing the fermentor with water should be sufficient. However, if different strains of halobacteria are used in the same laboratory, cross-contamination of the strains becomes a problem. This is probably caused by viable bacteria enclosed in or adhering to microcrystals of salt. Since such a cross-contamination in many cases is not recognized easily, cleanliness and strict maintenance of sterile conditions are advisable.

The strains are maintained on slants of the same medium containing 1.8% agar and transferred every 3 months. After incubation at 37° or 40° until good growth is obtained, the slants are kept at 4°. From these, a 150-ml shake culture in a 500-ml Erlenmeyer flask is grown and used as an inoculum for the fermentor. For the following two or three fermentors, 500 ml stationary phase cells from the preceding fermentor culture are saved each time for inoculation.

Cells will grow at temperatures as high as 57°. However, evaporation of water becomes a problem, and 39° is usually used. The culture is stirred vigorously at 200 rpm and aerated with 240 liters of air per hour. Dow Corning Antifoam A is added as necessary. Under these conditions generation time is 8 hours for *H. halobium* R_1.

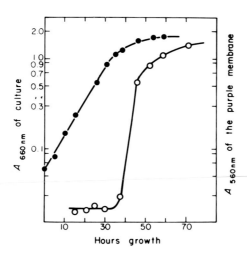

FIG. 1. Growth of *Halobacterium halobium* R_1 and synthesis of its purple membrane. Absorbance values are given for the purple membrane fractions isolated from 50 ml of culture and resuspended in equal volumes of distilled water. ●, Whole cells; ○, purple membrane.

Figure 1 shows the growth curve of *H. halobium* R_1 and synthesis of the purple membrane. Growth is followed by measuring extinction at 660 nm or by counting of cells in a Petroff-Hauser chamber. Synthesis of purple membrane is measured by absorbance of the purple membrane fraction at 560 nm. A sharp increase in the synthesis of the purple membrane occurs at the end of the exponential growth phase. For optimal yield of purple membrane, the culture is usually harvested after approximately 90 hours. Carotenoid formation and the yield of purple membrane are somewhat variable. If the procedure outlined above is strictly followed, 200–500 mg of purified purple membrane are obtained from a 10-liter culture. However, small differences in culture conditions may result in almost complete loss of purple membrane. Among other parameters, light and the O_2 tension in the culture medium are crucial.[12] Complete quantitative data are not available at this time.

Isolation of the Purple Membrane

All operations are carried out at room temperature, the centrifuges are kept at 4°. Cells from a 10-liter culture are harvested by centrifugation for 15 minutes at 13,000 g and resuspended in 250 ml of basal salt (medium without peptone); 5 mg DNase (Rinderpankreas, Roth OHG Karlsruhe, West Germany; DML Worthington 1200 u/mg) is added and the cells are dialyzed overnight against 2 liters of 0.1 M NaCl. This lyses the cells and prevents the development of excessive viscosity from liberated DNA.

The clear red lysate is centrifuged at 40,000 g for 40 minutes and the red supernatant decanted. The reddish purple sediment is resuspended in 300 ml of 0.1 M NaCl using a tight-fitting Teflon pestle (Thomas tissue grinder, A. H. Thomas, Philadelphia) and again centrifuged under the same conditions. This is repeated until the supernatant is almost colorless, which usually takes 2–3 additional runs. The sediment is then resuspended in deionized water and washed again 2 or 3 times until the supernatant appears colorless or faintly purple. The final sediment is taken up in 6–10 ml of water and layered over a linear 30 to 50% sucrose density gradient with a 2 ml of 60% sucrose bottom cushion using 6 tubes of the Spinco L2-65 SW 27 rotor. Typically 17 hours centrifugation at 100,000 g and 15° are necessary for the purple membrane to reach equilibrium at a buoyant density of 1.18 g/cm³.

Figure 2 shows a typical gradient tube with the large purple band and above it a smaller red band. The latter consists of the remaining pieces of red cell membrane which have not been removed by the differential cen-

---

[12] D. Oesterhelt and W. Stoeckenius, *Proc. Nat. Acad. Sci. U.S.*, in press (1973).

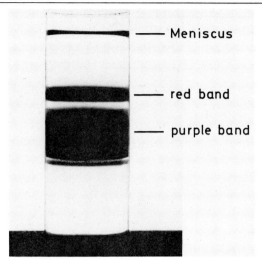

Meniscus

red band

purple band

FIG. 2. Sucrose gradient for purification of purple membrane fraction. The gradient is overloaded to obtain high yields. Smaller amounts centrifuged to obtain equilibrium yield much sharper bands.

trifugation. A small purple band of slightly higher density is sometimes observed. It is probably an artifact due to absorption of a contaminant present in the sucrose. No chemical, spectral, or morphological differences have been detected between the two purple bands. The denser band is absent when ultrapure sucrose is used for the gradient.[13]

The purple bands are collected and sucrose is removed by dilution to 300 ml and repeated centrifugation at 50,000 g for 30 minutes. Losses may occur due to adsorption of membrane material to the wall of the centrifuge tube because of the high concentration of membrane in the bands.

Properties of the Purple Membrane

The purple membrane is a differentiated area of the cell membrane, recognizable in freeze-fractured preparations of intact bacteria.[9] When isolated it appears as oval or round sheets with average diameter 0.5 μm (Fig. 3). The two sides of the membrane have a different surface structure; 25% of the dry weight of the membrane is lipid, 75% protein. Only one protein has been found in SDS-acrylamide gel disc electrophoresis, with an apparent molecular weight of 26,000. The protein forms a regular hexagonal lattice in the plane of the membrane.[9] It contains 1 mole of retinal per 26,000 g of protein, apparently bound as a Schiff base to a lysine residue of the protein. In the native membrane the retinal protein

[13] V. K. Miyamoto, unpublished results.

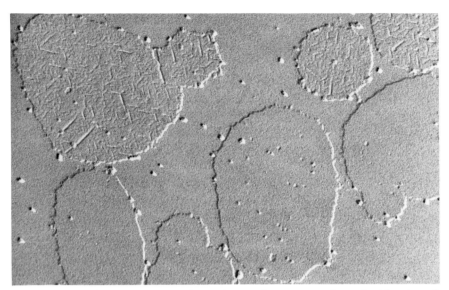

FIG. 3. Electron micrograph of purple membrane sprayed on freshly cleaved mica and shadowed. Note the different appearance of the two surfaces of the membrane. The small particles are ferritin molecules added as a marker. ×43,000.

complex has an absorption maximum at 560 nm (Fig. 4). The ratio of the 280/560 absorption varies slightly from one preparation to another from 2.0 to 2.1. A molar extinction coefficient of 54,000 (1/mole/cm) at 560 nm has been determined. The absorption maximum at 560 nm shifts only slightly upon illumination. The purple complex is stable against reducing agents and hydroxylamine as long as the membrane is intact. Detergents or organic solvents render it labile.

The purple membrane functions as a light-driven proton pump, permitting the cell to use light as an alternative source of energy.[12]

Isolation of the Cell Membrane

The red cell membrane may be obtained from the lysate of whole cells that has been described for the purple membrane. However, we prefer to first isolate cell envelopes and further dissociate these by steps, releasing the wall proteins before disaggregating the membrane. Again, *H. halobium* R$_1$ is preferred over the wild type because gas vacuole membranes are trapped in the envelope vesicles and are difficult to remove from the membrane fractions. The technique for the isolation of membrane vesicles is based on a procedure first described by A. D. Brown *et al.*[14] All centrifuga-

[14] A. D. Brown, C. D. Shorey, and H. P. Turner, *J. Gen. Microbiol.* **41**, 225 (1965).

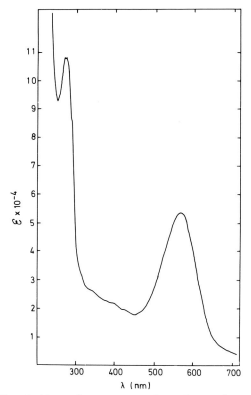

Fig. 4. Absorption spectrum of purple membrane.

tion times and/or speeds may have to be varied to some extent to give optimal yields with cells from different cultures. This is apparently due to small changes in culture conditions which cannot be rigorously controlled.

Cells from a 10-liter culture are resuspended with 300 ml of basal salt solution in a plastic container. They are frozen by placing the container inside a larger vessel filled with liquid nitrogen to the level of the cell suspension. After the liquid nitrogen has evaporated and the cell suspension has reached room temperature, 5 mg of DNase are added and the gelatinous preparation is stirred for 1 hour on a magnetic stirrer. Alternatively, viscosity may be reduced using a Waring Blendor for 1–2 minutes at top speed. The suspension is then centrifuged at 20,000 *g* for 2 hours and the supernatant, which has faint reddish yellow color, is discarded. After resuspension in 300 ml of basal salt, the dark red sediment is centrifuged at 100,000 *g* for 30 minutes. This step is repeated, usually three times, until the supernatant contains only a small and constant amount of protein. For resuspension of all envelope and membrane sediments a Thomas tissue grinder (A. H. Thomas, Philadelphia) fitted with a Teflon pestle is used. The final sediment obtained constitutes the cell envelope preparation (fraction E-BS). It consists mainly of closed vesicles and can be further fractionated to separate cell wall and membrane.

For this purpose fraction E-BS is dialyzed for 3 hours at room temperature against a solution containing 1.0 M NaCl, 20.0 mM MgSO$_4$, 10.0 mM Tris·HCl buffer. This solubilizes the cell wall proteins. The membrane, still in the form of closed vesicles, is then sedimented at 100,000 g for 30 minutes. The supernatant should be nearly colorless, and the centrifugation may be repeated one or two times. The sediment is then dialyzed

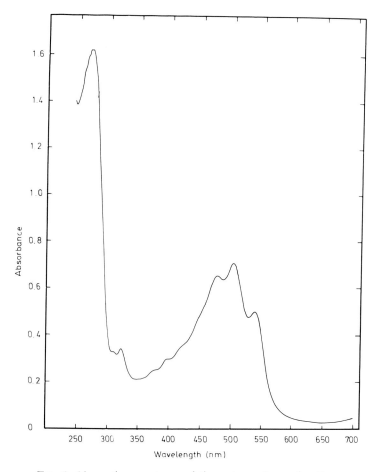

FIG. 5. Absorption spectrum of the red membrane fraction.

against deionized water for 4 hours at room temperature, and the purple membrane is spun down as already described for lysates of whole cells (see page 670).

The red supernatant contains the cell membrane fragments which have a lower density and smaller size than the purple membrane. It is centrifuged for 27 hours at 100,000 *g*, yielding a red sediment and a reddish yellow supernatant. The sediment can be resuspended in water and centrifuged in a sucrose density gradient under the same conditions which are used for the purple membrane (see page 670). However, the small membrane fragments will not reach equilibrium even after centrifugation at 300,000 *g* for 45 hours in a 30 to 50% sucrose gradient in water. The resulting broad red zone shows incomplete separation into several bands. If shallower gradients are used, e.g., 25 to 40% sucrose, fractions can be obtained which differ in their pigment and cytochrome content. This becomes obvious when

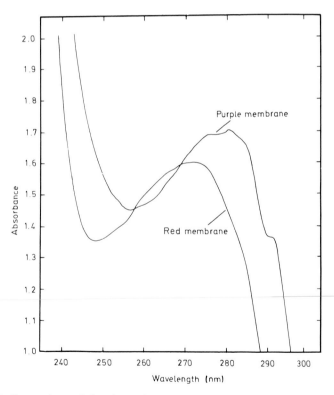

Fig. 6. Comparison of the absorption spectra of the red and purple membrane in the region of protein absorption.

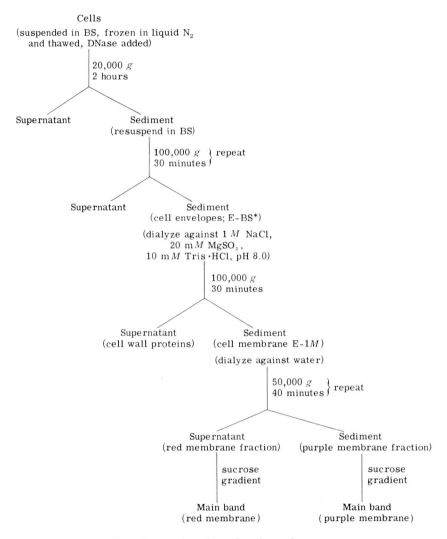

Fig. 7. Fractionation of cell envelopes.

a mutant is used which cannot synthesize the main carotenoid, bacterio-ruberin.[15]

The red membrane fraction can also be obtained by means of a modified procedure. E-BS is first dialyzed for 3 hours at room temperature

---

[15] D. Oesterhelt, *Abstr. Comm. Meet. Fed. Eur. Biochem. Soc.* 8, No. 125 (1972) and D. Oesterhelt, M. Meentzen, and M. Milanytch, unpublished results.

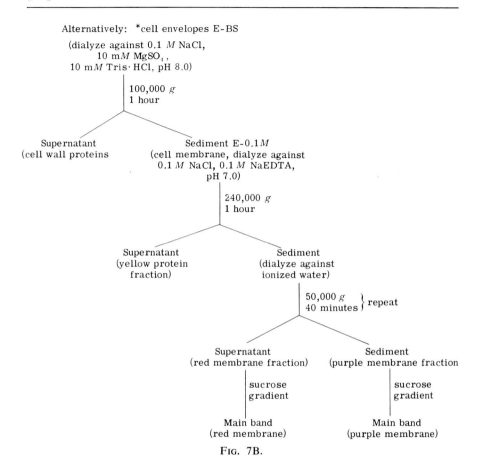

Alternatively: *cell envelopes E-BS

(dialyze against 0.1 *M* NaCl,
   10 m*M* MgSO$_4$,
10 m*M* Tris·HCl, pH 8.0)

100,000 *g*
1 hour

Supernatant
(cell wall proteins

Sediment E-0.1 *M*
(cell membrane, dialyze against
0.1 *M* NaCl, 0.1 *M* NaEDTA,
pH 7.0)

240,000 *g*
1 hour

Supernatant
(yellow protein
fraction)

Sediment
(dialyze against
ionized water)

50,000 *g*
40 minutes } repeat

Supernatant
(red membrane fraction)

Sediment
(purple membrane fraction

sucrose
gradient

sucrose
gradient

Main band
(red membrane)

Main band
(purple membrane)

Fig. 7B.

against 0.1 *M* NaCl containing 20.0 m*M* MgSO$_4$, followed by centrifugation at 100,000 *g* for 1 hour. The sediment (fraction E-0.1) is resuspended
in 0.1 *M* NaCl containing 0.1 *M* sodium EDTA pH 7.0 and dialyzed
against the same solution for 24 hours. Sedimentation at 240,000 *g* for
1 hour yields a red sediment clearly separated from a yellow supernatant.
The sediment, when resuspended in deionized water, dissociates into the
red and purple membrane, which may be separated as already described.

The color of the red membrane fraction is mainly due to the C$_{50}$-carot-
enoid bacterioruberin[10] (Fig. 5). The red and purple membrane fraction
also show characteristic differences in their UV-absorption spectra (Fig.
6). The red membrane has a broad absorption maximum around 270 nm,
while the purple membrane shows a typical tryptophan absorption with a
maximum at 280 nm and a shoulder at 290 nm. The total E-0.1 fraction
has an absorption ratio *A* 270:*A* 503 = 2.3.

The supernatant of the EDTA-treated E-0.1 fraction is designated the yellow protein fraction. Its buoyant density in a CsCl gradient is 1.33 g/cm^3, typical for proteins. The UV-absorption maximum is at 270 nm. In the analytical ultracentrifuge in 0.1 M NaCl, 0.01 M sodium EDTA pH 7.3, a symmetrical peak is observed corresponding to a sedimentation constant $s_{20,w}^0 = 5.9 \times 10^{-13}$ (sec) and a diffusion coefficient $D_{20,w}^0 = 2.2 \times 10^7$ (cm^2/second). In acrylamide disc gel electrophoresis, a main band is seen after amido black staining which is not identical with a smaller yellow band. The yellow color is apparently due to the flavin and cytochrome pigments in the yellow band. The yield of material in the yellow band may vary considerably from one preparation to the next. The yellow protein fraction may therefore appear colorless.

The fractionation of cell envelopes is summarized in Fig. 7.

Note Added in Proof:

Since the article has been written a function of the purple membrane as a light energy transducing system has been suggested. The reader is referred to the following articles:

D. Oesterhelt and B. Hess. Reversible photolysis of the purple complex in the purple membrane of *Halobacterium halobium*. *Eur. J. Biochem.* **37**, 316–326 (1973).

D. Oesterhelt and G. Krippahl. Light inhibition of respiration in *Halobacterium halobium*. *FEBS Letts.* **36**, 72–76 (1973).

D. Oesterhelt and W. Stoeckenius. Functions of a new photoreceptor membrane. *Proc. Nat. Acad. Sci.* **70**, 2853–2857 (1973).

E. Racker and W. Stoeckenius. Reconstitution of purple membrane vesicles catalyzing light-driven proton uptake and adenosine triphosphatase formation. *J. Biol. Chem.* **249**, 662 (1974).

[70] The Isolation of Gas Vesicles from Blue-Green Algae

By A. E. WALSBY

Gas vesicles are hollow, cylindrical, or spindle-shaped structures which, stacked in a regular hexagonal fashion or in loose clusters, comprise the gas vacuoles found only in blue-green algae and bacteria.[1,2] Each gas vesicle consists simply of a rigid wall of protein,[3,4] about 2 nm thick, which entirely surrounds a space devoid of liquid or solid contents. Gases diffuse

[1] G. Cohen-Bazire, R. Kunisawa, and N. Pfennig, *J. Bacteriol.* **100**, 1049 (1969).
[2] A. E. Walsby, *Bacteriol. Rev.* **36**, 1 (1972).
[3] D. D. Jones and M. Jost, *Arch. Mikrobiol.* **70**, 43 (1970).
[4] D. D. Jones and M. Jost, *Planta* **100**, 277 (1971).

freely to and fro across the wall between the space and the surrounding solution.[5]

When subjected to a sufficient pressure, known as the critical pressure [which may vary from less than 200 to more than 600 kN m^{-2} (2–6 atm)] the gas vesicles collapse irreversibly to flattened, envelopelike structures, the gas they contained diffusing away into the surrounding solution as this happens.[6] It is feasible to recover collapsed vesicles by standard, density-gradient centrifugation techniques,[7] but the preparations so obtained are not easily purified[3] and are of limited usefulness. Using special precautions to avoid exposing gas vesicles to pressures in excess of their critical pressure they may be isolated and purified in an intact state.[8] Such precautions are not required in the preparation of any other subcellular structures and particular emphasis is given to this aspect here.

Determination of Critical Pressures

Since several of the stages in the isolation procedure may involve the generation of pressure, it is first essential to determine the pressure that the gas vesicles will withstand. Their collapse in aqueous suspension is accompanied by a marked decrease in turbidity, and nephelometry or colorimetry may be used to assess the degree of collapse ensuing application of any given pressure.[6]

A test tube of the algal or gas vesicle suspension is placed in a simple, air-tight chamber capable of withstanding about 2 MN m^{-2} (20 atm) pressure.[6] The chamber is connected, via high-pressure hose, to a cylinder of compressed air or nitrogen which will deliver gas to 1.4 MN m^{-2} (about 200 psi). The suspension is subjected to a pressure of 50 kN m^{-2} for 30 seconds, and the turbidity (or absorbance) is measured. This process is repeated at steps of 50 or 100 kN m^{-2} up to 1 MN m^{-2} (or until there is no further turbidity change). The percentage gas vesicle collapse at any pressure is then equal to

$$(x - y)/(x - z) \times 100$$

where x is the turbidity of the untreated suspension (with gas vesicles), y of the treated suspension, and z of the suspension in which all the gas vesicles have been collapsed.

Pressure-collapse curves are plotted as shown in Fig. 1. As seen in the figure, the mean critical pressure of the isolated vesicles appears greater than that of the vesicles inside algal cells from the same sample. This is

[5] A. E. Walsby, Proc. Roy. Soc. Ser. B 173, 235 (1969).
[6] A. E. Walsby, Proc. Roy. Soc. Ser. B 178, 301 (1971).
[7] W. Stoeckenius and W. H. Kunau, J. Cell Biol. 38, 336 (1968).
[8] A. E. Walsby and B. Buckland, Nature (London) 224, 716 (1969).

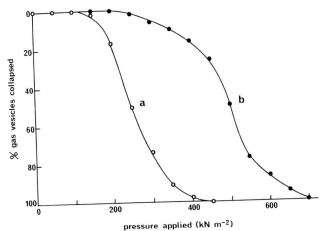

Fig. 1. Collapse of gas vesicles with pressure. Curve a, vesicles inside turgid cell; curve b, vesicles isolated from the same cells. The mean distance between the two curves is equivalent to the cell turgor pressure. Figure modified from A. E. Walsby, *Proc. Roy. Soc. Ser. B* **178**, 301 (1971).

because those vesicles inside the cells are already subjected to cell turgor pressure (equal to the mean separation between the two curves, a and b).[6] The curves shown are fairly typical for blue-green algae, and in the absence of equipment for determining critical collapse pressures, could be used as a rough guide for deciding the acceptable pressure limits in subsequent stages of the procedure.

Concentrating Cell Suspensions

Concentrated cell suspensions, required for quantitative gas vesicle preparations, may be obtained in the following ways.

Flotation. Gas vesicles render algal cells buoyant if they occupy a sufficient proportion of the cell volume. If cultures or samples gathered from water blooms are allowed to stand, the alga forms a cream at the surface which may be drawn off with a fine syringe needle, attached to an evacuated vessel, held in contact with the meniscus. This method may be used to select for cells of high buoyancy, usually associated with a high degree of gas vacuolation.[8]

Accelerated Flotation. The flotation process may be accelerated by centrifugation as long as the pressure generated (see p. 682) does not exceed the apparent critical pressure of vesicles in the turgid cells (Fig. 1, curve a). Gas-vacuolate cells heavier than water may also be recovered, as a pellet, in this way without further loss of gas vesicles.

Filtration. Algal slurries may be further concentrated by filtration, collecting the cells or filaments on a sintered-glass filter of wide diameter (100 mm) and small pore size (5–15 μm). Continuous scraping with a rubber spatula helps to prevent the filter from becoming clogged.[3]

Lysing the Algal Cells

Techniques often employed in breaking open algal cells, such as ultra-sonication and use of a French pressure cell cannot be used because they develop very high pressures on the vesicle walls. Osmotic shock can be used only after weakening the cells to the point that they burst below critical pressure of the gas vesicles. The following methods are recommended for different algae.

Osmotic Shrinkage. When placed in strongly hypertonic sucrose solutions, blue-green algal cells lose water and shrink, usually without plasmolyzing. The cell wall, which seems to be firmly attached to the plasmolemma, is placed under tension and tends to rupture. Osmotic shrinkage is particularly effective in lysing algae of the *Anabaena* and *Nostoc* type, in which the filaments show a weakness where they are constricted at the septa; it is somewhat less effective with algae of the *Oscillatoria* type, with filaments of uniform diameter, and only partially effective with unicellular forms, such as *Microcystis* and *Coelosphaerium.*

The algal slurry should be mixed rapidly with a concentrated solution of sucrose, giving a final concentration of 0.7 M. Rapid mixing is not possible with crystalline sucrose. Equal volumes of alga and 1.4 M sucrose are recommended. Most of the lysis occurs in the first 10–15 minutes, but it may be necessary to leave the mixture stirring for 1 or 2 hours at room temperature to achieve maximum lysis.

Osmotic Shock of Penicillin-Weakened Cells. Penicillin will weaken the walls of algal cells that are actively growing and dividing, by inhibiting the incorporation of the mucopeptides (responsible for tensile strength) into the growing walls.[9] The cells can then be broken by osmotic shock after infiltrating with glycerol, before large turgor pressures are generated, collapsing the gas vesicles.

Jones and Jost[3] recommended incubating cells (5×10^7 per milliliter) in the presence of benzylpenicillin (K salt; Sigma, 200–250 units liters^{-1}) and Mg^{2+} (1 mM) for 5 hours, concentrating by filtration, and resuspending in glycerol (1 M) for 15 minutes. The infiltrated cells are then mixed rapidly with 3 volumes of a Tris·HCl buffer (0.02 M), pH 7.7, and stored for 2 hours at 4 C.

This method is more suitable for *Microcystis* and other unicells than that of osmotic shrinkage, but the sensitivity of cells to penicillin is greatly

[9] R. Y. Stanier and C. B. van Niel, *Arch. Mikrobiol.* **42**, 17 (1962).

diminished with alga in the stationary phase of growth, when gas vacuolation is often highest.[10,11]

Lysozyme Treatment. Lysozyme (muramidase) attacks the mucopeptide components of prokaryotic cell walls. In hypotonic solutions lysis follows as a consequence of osmotic swelling.[9]

Biggins[12] suggested using a final enzyme concentration of 0.05% (w/v) of muramidase (Worthington Biochemical Corporation, Freehold, New Jersey) and incubating at 35 C, (although higher concentrations should be used with dense algal slurries). Incubation times in excess of 3 hours give extensive cell rupture. The method can be combined with subsequent glycerol infiltration and osmotic shock (see above), and is generally effective with cells at all stages of growth. Gas vesicles, whose walls are made only of protein,[3] are not attacked by lysozyme.

The gas vesicles released by cells broken in these ways retain their characteristic shape, fine structure,[13] and dimensions.[14]

Centrifugally Accelerated Flotation of Gas Vesicles

Intact gas vesicles have a density (of about 100 kg m^{-3})[2] much lower than that of any other cell components, and despite their small size float up more rapidly. The process can be accelerated by centrifugation and, by Stokes' law, the rate of rise will be proportional to a, the acceleration.

Pressure Considerations. The acceleration will generate a pressure, p, equal to $h\rho a$, where h is the depth below the meniscus of the liquid and ρ its density. If maximum recovery is required the pressure at the bottom of the centrifuge cup (where it is greatest) should not exceed the critical pressure of the weakest gas vesicles[8] (determined from curve b in Fig. 1). For this reason it is advantageous to employ small values of h; this will, at the same time, reduce the time taken for gas vesicles to rise from the bottom of the cup to the meniscus. On the other hand, it decreases the volume of suspension that can be treated. To counteract this, centrifuge vessels of the widest available cross-sectional area should be used. The ideal solution is a zonal rotor head of wide diameter and large cylinder depth.

The maximum acceleration permitted with vesicles of different minimum critical pressure and different values of h, may be computed from the data in Fig. 2. Alternatively, the value may be calculated as follows.

Sample Calculation. Aliquots of algal lysate in 0.7 M sucrose, and showing the pressure-collapse characteristics of curve b in Fig. 1, formed

[10] R. V. Smith and A. Peat, *Arch. Mikrobiol.* **58**, 117 (1967).
[11] H. Lehmann and M. Jost, *Arch. Mikrobiol.* **79**, 59 (1971).
[12] J. Biggins, *Plant Physiol.* **42**, 1442 (1967).
[13] A. E. Walsby and H. H. Eichelberger, *Arch. Mikrobiol.* **60**, 76 (1968).
[14] M. Jost and D. D. Jones, *Can. J. Microbiol.* **16**, 159 (1970).

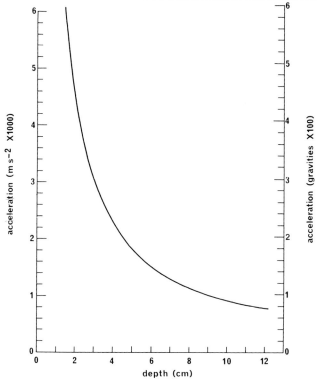

FIG. 2. The centrifugal acceleration a (given in m sec^{-2} on the left, and gravities on the right) generating a pressure of 100 kN m^{-2} (approximately 1 atm) at different depths in a 0.7 M sucrose solution, having a density of 1090 kg m^{-3}. The pressure generated is directly proportional to a. Hence for other values of pressure, p in kN m^{-2}, multiply the acceleration given by $p/100$. Figure modified from B. A. Buckland, Ph.D. Thesis, University of London (1971).

layers 50 mm deep in the centrifuge bottles. The highest permissible g is given by

$$a = p/h\rho$$

where $h = 50$ mm $= 50 \times 10^{-3}$ m; $\rho = 1.09$ g cm^{-3} $= 1090$ kg m^{-3} (density of 0.7 M sucrose); $p = 200 \times 10^{3}$ N m^{-2} at base of cup.

$$a = \frac{200 \times 10^{3} \text{ N m}^{-2}}{50 \times 10^{-3} \text{ m} \times 1090 \text{ kg m}^{-3}}$$

$$= 3670 \text{ m sec}^{-2} \ (390 \text{ gravities})$$

Since the acceleration produced varies linearly with distance from the rotor spindle, this value of a should be applied at the position r' (equal to $r - h/2$, where r is the distance from the rotor spindle to the base of the centrifuge bottle). For example, given that r' is 150 mm, the rotor speed, n, generating the required acceleration is given by

$$n = \sqrt{\frac{a}{(2\pi)^2 r'}}$$

$$= \sqrt{\frac{3670 \text{ m sec}^{-2}}{(2\pi)^2 \times 1.15 \text{ m}}}$$

$$= 25 \text{ sec}^{-1} \text{ (1500 rpm)}$$

Ancillary Considerations

1. From curve b of Fig. 1, it is seen that only 20% of the gas vesicles collapse at 400 kN m^{-2}, twice the minimum critical pressure. This pressure applied at the base of the cup will permit a doubling of a, and the resultant loss of vesicles will be considerably less than 20% (in fact only about 3.2%) because the pressure generated decreases with both h and g decreasing toward the surface.

2. Despite the fact that it increases h, it is highly advantageous to layer water (to a depth of 5–10 mm) over the sucrose-containing lysate.[3] This layer serves to rinse the vesicles free of soluble contaminants as they rise to the surface and allows them to separate from particles rising more slowly from the surface of the lysate.[15]

3. Both the water layer and the lysate should be buffered if their pH falls outside the range of 6 to 9.5, as this is the region of the vesicles' maximum stability.[15]

4. As the gas vesicles rise toward the surface, the effective value of h decreases, allowing a to be increased accordingly. The rate of rise depends on the viscosity of the medium and has been measured as 0.85 and 2.04 m sec^{-1} for a 1.0 m sec^{-2} field of acceleration in 0.7 M sucrose and water, respectively.[15] From these figures the rate at which the acceleration can be increased may be calculated (see example given by Buckland and Walsby[15]).

5. The preparation should not be centrifuged for a time greater than that required to bring all gas vesicles to the surface.[15] This is mainly because with time other particles which are lighter than water may float up and contaminate the vesicle layer. Also gas vesicles left in an unpurified state tend to become weaker on standing, and may collapse.

6. The weakening of gas vesicles, which is apparently due to enzymatic attack, is greatly diminished in the cold (5 C), at which temperature the lysate should be kept in all stages following lysis.

Subsequent Centrifugation Steps. The surface cream of gas vesicles is removed, after completing the first centrifugation, by using a 20 ml-capacity syringe fitted with a fine, square-ended needle held in contact with the

---

[15] B. Buckland and A. E. Walsby, *Arch. Mikrobiol.* **79**, 327 (1971).

meniscus. The vesicles are drawn off in the smallest practicable volume. An air space is kept in the syringe and the vesicle suspension is expressed from it with only slight pressure, to avoid any risk of collapse.

This supernatant fraction is diluted with water (or buffer as necessary), preferably at least 10-fold, the volume being determined by that which can reasonably be centrifuged at the next stage. The same considerations determine the acceleration used, but less time will be required as the viscosity of the preparation is decreased.

The centrifugation procedure is repeated once or twice more, until a fairly white preparation is obtained.

Filtration

The two principal contaminants left by centrifugation are intact gas-vacuolate cells or protoplasts which float up with the vesicles, and soluble substances (mainly sucrose) which diffuse into the surface layers. These contaminants are removed in a two-stage filtration.[15]

1. The suspension is put through a membrane filter (Sartorius, or other makes) of 50 mm diameter, having pores of 1.2 μm diameter. Cells and protoplasts are held back; the gas vesicles pass through. Just before the filter dries, wash with about 5 ml of water, adding drop-by-drop.

2. Transfer the gas vesicle-containing filtrate to a membrane filter of similar diameter, but of 0.05 μm pore size. Use only a moderate vacuum (-30 kN m^{-2}) to draw the suspending solution through, leaving the gas vesicles on the filter. They may be rinsed several times with water or buffer, but *do not allow the filter to dry:* the pressures set up by surface tension at the surface of drying films can easily collapse gas vesicles.[6]

Final Centrifugation

The preparation is improved by a final centrifugation which serves to separate the vesicles from other particulate matter which might have been released from cells breaking at the first filtration. The standard method described above (after resuspending in a large volume of water) is satisfactory. Alternatively, Jones and Jost[3] recommend placing the concentrated vesicles in a narrow layer at the surface over concentrated sucrose and centrifuging at very high accelerations (200,000 g). If this is done, gas vesicles having a critical pressure of 500 kN m^{-2} will survive only in the top 0.25 mm. Accordingly, the rotor should first be spun for sufficient time at a low acceleration to bring the vesicles very close to the meniscus.

Efficiency of the Above Method

Using the method described above, Buckland and Walsby[15] record recoveries of between 50 and 60% (from turbidity measurements) and a percentage purity of 97.6 for the final preparation (according to the results of a radioactive labeling experiment).

Storing the Gas Vesicles

Bacterial growth in crude lysates results in substantial losses over a period of days, apparently due to enzymatic attack on the gas vesicle protein. However, in a highly purified state, aqueous suspensions of intact vesicles may be stored at 5 C for a year or more without appreciable loss.

Vesicles can also be frozen, freeze-dried and then kept in a dried state indefinitely. On simply adding water a suspension of intact vesicles may be reconstituted.[15] Unfortunately, although drying itself has no adverse effects on the vesicles, pressures developed during the freezing process result in substantial weakening and collapse (50% or more). It is known from freeze-etching studies[16,17] that gas vesicles will survive intact when frozen rapidly in liquid Freon held at near liquid nitrogen temperatures. It may be possible to scale up such procedures for the quantitative preparation of frozen and freeze-dried intact vesicles.

[16] M. Jost, *Arch. Mikrobiol.* **50**, 211 (1965).
[17] J. R. Waaland and D. Branton, *Science* **163**, 1339 (1969).

[71] The Isolation of the Amoeba Plasma Membrane and the Use of Latex Beads for the Isolation of Phagocytic Vacuole (Phagosome) Membranes from Amoebae Including the Culture Techniques for Amoebae

By Edward D. Korn

Acanthamoeba castellanii[1] is one of a group of small soil amoebae that can be grown axenically on soluble medium. Because the amoebae are nutritionally dependent on the processes of phagocytosis and pinocytosis, they provide an unusual opportunity for studying the mechanisms of membrane fission and fusion, the relationship between the plasma membrane and the membrane of the phagocytic vacuole (phagosome), and the dynamics and biochemistry of membrane movements.

Culture Techniques

A defined culture medium which will support growth has been described,[2] but for most purposes the soluble culture medium[3] shown in

[1] R. J. Neff, *J. Protozool.* **4**, 176 (1957).

TABLE I
Axenic Growth Medium for *Acanthamoeba castellanii*[a]

	Stock solution	Use for 1 liter of medium	Final concentration of medium
Glucose	—	15 g	1.5%
Proteose-peptone (Difco Labs.)	—	15 g	1.5%
L-Methionine	50 mM	2 ml	0.1 mM
KH$_2$PO$_4$	1.5 M	2 ml	3 mM
MgSO$_4$	1.0 M	0.1 ml	0.1 mM
CaCl$_2$	0.1 M	0.1 ml	10 μM
FeCl$_3$	10 mM	0.1 ml	1 μM
Thiamine HCl	10 mg/ml	0.1 ml	1 mg/liter
Biotin	2 mg/ml	0.1 ml	200 μg/liter
Vitamin B$_{12}$	0.1 mg/ml	0.01 ml	1 μg/liter

[a] To prepare the stock solution of thiamine HCl it is necessary to add NaOH. The growth medium is dispensed into flasks and autoclaved for 20 minutes. When the medium is prepared in 15-liter carboys, autoclaving is extended to 1 hour.

Table I is convenient and sufficient. When prepared in this way the medium is pH 6.3 before autoclaving and pH 5.8 after autoclaving. Some investigators[4] recommend adjusting the medium to pH 7 before autoclaving, but in this laboratory the higher pH has had no effect on the rate of growth of the amoebae.

In this laboratory the cultures are carried in 50-ml Erlenmeyer flasks containing 10 ml of medium, transfers of 0.2 ml being made every 2 weeks with normal sterile technique. The cultures are incubated at 28°–30° without shaking. For larger stationary cultures we use 500 ml of medium and inoculate with 1 ml of a 2-week-old culture. A quantity of cells sufficient for most biochemical studies can be conveniently grown in one or more 3-liter low-form culture flasks containing 1 liter of medium and inoculated with one 2-week-old 50-ml culture. To obtain adequate growth it is necessary to rotate the flasks slowly (about 80 rpm) on a rotating platform shaker at 30°. Under these conditions the amoebae reach a concentration of approximately 1×10^6/ml (about 400 μg of amoeba protein per milliliter) in about 6 days. Growth is logarithmic for the last several days with a generation time of about 24 hours after an initial lag period which may vary between 1 and 3 days. When the cells reach a density greater than about 1×10^6/ml, or when the cultures are main-

[2] K. M. G. Adam, *J. Gen. Michobiol.* **21**, 519 (1959).
[3] R. J. Neff, R. H. Neff, and R. E. Taylor, *Physiol. Zool.* **31**, 73 (1958).
[4] R. N. Band, *J. Gen. Microbiol.* **21**, 80 (1959).

tained for long periods, the amoebae encyst.[5-7] The rate and extent of growth are clearly dependent on a number of factors many of which have not been investigated but which may include the size and shape of the culture flask and the rate of aeration.

Larger quantities of amoebae can be grown in 15 liters of medium in a 5-gallon carboy fitted with a rubber stopper with three glass tubes connected to rubber tubing of appropriate lengths.[8] One glass tube extends to the bottom of the carboy and terminates in a fritted-glass bulb through which O_2 will be bubbled at a rate of 85 ml/minute. A second glass tube extends only to the air space above the medium and serves as an outlet for the O_2. A third tube extends to the bottom of the carboy and is used to sample the culture during growth. A large magnetic stirring bar is placed in the carboy. This is of help in dissolving the glucose and proteose peptone, and in stirring the culture after inoculation. The ends of the rubber tubing attached to the O_2 inlet and outlet tubes are connected by short glass tubes to rubber stoppers that are covered with gauze pads and wrapped in paper to maintain sterility. The carboy is autoclaved for 60 minutes with the stopper clamped in place and the O_2 inlet and sampling tubes tightly closed. The O_2 outlet tube is left open to permit volume changes during heating and cooling. Two glass tubes (approximately 3×20 cm) are filled with nonabsorbent cotton and separately autoclaved and dried. These are fitted to the rubber stoppers attached to the O_2 inlet and outlet tubes after the carboy has cooled to at least 30° (this takes overnight at room temperature) and has been inoculated with one or two 1-liter cultures. The carboy is set on top of a large magnetic stirrer so that the culture can be slowly stirred in addition to the agitation provided by the aeration. Growth is continued for 4–6 days until the cells reach a density of 0.5 to 1×10^6/ml. Samples can be obtained during growth by closing the O_2 outlet tube and opening the sampling tube. The sampling tube should be closed before the O_2 outlet tube is reopened so that sterility is maintained by the positive pressure. If still larger quantities of amoebae are desired these can readily be grown in large fermenters with no difficulty.[8]

Alternative Culture Technique

At the suggestion of Dr. R. J. Neff, Vanderbilt University, we have transferred stationary cultures of *Acanthamoeba* to tubes containing 20

[5] R. N. Band, *J. Protozool.* **10**, 101 (1963).
[6] R. J. Neff, S. A. Ray, W. T. Benton, and M. Wilborn, *Methods Cell Physiol.* **1**, 55 (1964).
[7] B. Bowers and E. D. Korn, *J. Cell Biol.* **41**, 786 (1969).
[8] R. Weihing and E. D. Korn, *Biochemistry* **10**, 590 (1971).

ml of medium continually aerated with compressed air at the maximum practical rate. After several weeks of transferring cells at intervals of 3–4 days the amoebae attained a generation time of about 14 hours and reached a density of about 2.5 to 3×10^6 cells per milliliter without encysting. Rapidly growing cells maintain their growth rate and reach the same high density when transferred to carboys containing 15 liters of medium as described above. This culture technique has the important advantages of providing more cells from cultures which are essentially in continual log phase, but requires more frequent transfers to maintain the cultures in the rapidly growing state.

Isolation of Plasma Membranes

Solutions

Tris·HCl buffer, 10 mM, pH 7.5
10%, 25%, 30%, 35%, 40%, 45%, 50%, and 60% sucrose in buffer (concentrations in g/100 ml of a solution)

Method. A procedure suitable for the isolation of highly purified plasma membranes[9] from 1×10^9 to 1×10^{10} amoebae is outlined in Fig. 1. Other procedures have been described[10,11] and although the membranes prepared by those methods have not been fully characterized there is no reason to think they are not as pure as the membranes prepared by the method described here.

The amoebae are harvested from their culture medium by centrifugation at room temperature at 500 g for 5 minutes, washed once with ice-cold 10 mM Tris·HCl buffer, pH 7.5, and suspended in cold buffer at a concentration of 2.4×10^7 cells/ml. All subsequent procedures are carried out at 0–4° and all solutions are made up in 10 mM Tris·HCl, pH 7.5. Sucrose concentrations are in grams per 100 ml of solution. To determine cell density it is convenient to prepare a standard curve relating absorbance at 660 nm to cell concentration.

The amoebae are left for 15 minutes with occasional gentle shaking to allow them to swell, and then they are homogenized by four gentle down and up cycles of a large, tight-fitting Dounce homogenizer. Approximately 95% of the amoebae are broken with the plasma membranes forming large fragments but with few intact ghosts. The down and up strokes should each take about 8 seconds since it is necessary to ensure

[9] A. G. Ulsamer, P. L. Wright, M. G. Wetzel, and E. D. Korn, *J. Cell Biol.* **51**, 193 (1971).
[10] T. M. G. Schultz and J. E. Thompson, *Biochim. Biophys. Acta* **193**, 203 (1969).
[11] F. J. Chaplowski and R. N. Band, *J. Cell Biol.* **50**, 634 (1971).

that the plasma membrane is not broken into very small pieces by too strenuous homogenization. As each batch is homogenized it is immediately added to a volume of 60% sucrose calculated to give a final concentration of 10% sucrose when all the cells have been homogenized; i.e., the volume of 60% sucrose should be 20% the volume of the cell suspension before homogenization.

The homogenate is then centrifuged at 500 g (maximum) for 20 minutes in the SS 34 rotor of a Sorvall RC2B centrifuge. The pellet is suspended in half of the original volume of 10% sucrose, and the suspension is again centrifuged at 500 g for 20 minutes. This pellet is suspended in the same volume of 10% sucrose and centrifuged at 750 g for 20 minutes. By these low speed centrifugations the large fragments of plasma membrane are effectively separated from soluble molecules, micro-

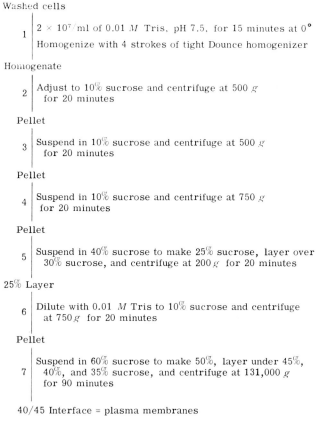

Fig. 1. Isolation of plasma membranes from *Acanthamoeba castellanii*.

somal and ribosomal material, and much of the mitochondrial fraction. It is important to be careful when resuspending the membrane pellet in order not to fragment the plasma membranes further.

The pellet from the centrifugation at 750 g is suspended in an equal volume of 40% sucrose (to give a concentration of sucrose of 25%) and sufficient 25% sucrose is added to attain a total volume equal to half the volume of the original homogenate. As much as 100 ml of this suspension is carefully layered over 40 ml of 30% sucrose in 200-ml glass bottles, which are then centrifuged at 200 g for 20 minutes in a No. 284 rotor of an International PR-2 centrifuge. The centrifuge should be brought to speed slowly to maintain the integrity of the gradient. The 25% sucrose layer is collected being careful not to include any of the material at the interface. By this step the plasma membranes are separated from nuclei, intact cells or cysts, and partial ghosts. The suspension is diluted with 1.5 volumes of buffer to 10% sucrose and centrifuged at 750 g for 20 minutes to recover the plasma membranes as a pellet. Little purification is obtained by this step, but it serves to concentrate the partially purified membranes.

This pellet is suspended in enough 60% sucrose to give a concentration of at least 50% sucrose, and, if necessary, 50% sucrose is added to give a volume of 1 ml for every 10^8 amoebae with which the procedure was begun. This suspension (16 ml) is placed at the bottom of a 3.5 × 1 inch cellulose nitrate tube, overlayered with 8 ml of 45% sucrose, 7 ml of 40% sucrose, and 6 ml of 35% sucrose and centrifuged at 131,000 g for 90 minutes in a Beckman SW 27 rotor. Plasma membranes accumulate at the interface between 40% and 45% sucrose from which they can be collected with ease after first removing the superficial layers. The entire procedure takes about 6 hours from the time of homogenization. The procedure has been highly reproducible in more than 50 isolations by several different investigators.

Characterization of the Plasma Membrane

The purified plasma membrane is free of mitochondria, endoplasmic reticulum, and other contaminants recognizable by electron microscopy (Fig. 2) or detectable by assaying for enzymes usually considered to be markers of other cellular membranes (Table II). The amoeba plasma membranes have a high level of alkaline phosphatase activity and of 5'-nucleotidase activity (which may be the same enzyme, but see Table II, footnote b) but no detectable (Na,K)-ATPase, an enzyme that is usually found in mammalian cell plasma membranes. The low specific activity of acid phosphatase may represent very slight contamination by intact digestive vacuoles which have a specific activity several hundredfold higher.

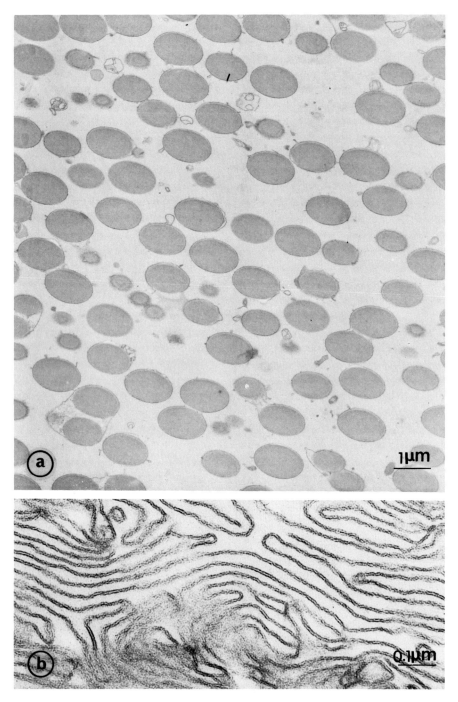

FIG. 2. Isolated plasma membranes of *Acanthamoeba castellanii*. a, ×10,000; b, ×100,000.

The gross chemical composition of the plasma membrane (Table II), in particular the high ratio of sterol to phospholipid, is also indicative of its purity. The phospholipid composition of the plasma membrane is given in greater detail in Table III. The fatty acid compositions of the phospholipids have also been determined.[9] The sterols are 60% ergosterol and 40% dehydroporiferasterol.[9]

The purified plasma membranes contain about 10% of the total cell sterol. Thus the minimal yield of plasma membrane is 10%, assuming that all of the amoeba sterols were in the plasma membranes. Since 0.6% of the cell protein and 2.4% of the amoeba phospholipids are recovered in the plasma membrane fraction, the plasma membrane could contain a maximum of 6% of the total cell protein and 24% of the total cell phospholipids. It is more probable[9] that something like 50% of the cell sterol is contained in the plasma membrane so that the plasma membrane might account for 3% of the cell protein and 12% of the cell phospholipid.

TABLE II

Enzymatic and Chemical Characterization of the *Acanthamoeba* Plasma Membrane[a]

Component	Plasma membrane	Homogenate
	units/mg protein	
Succinic dehydrogenase	0	0.009
NADH-cytochrome *c* reductase	0	0.004
NADPH-cytochrome *c* reductase	0	0.008
Acid phosphatase	0.02	0.35
Glucose-6-phosphatase	0.003	0.017
Alkaline phosphatase[b]	3.3	0.25
5'-Nucleotidase[b]	0.35	0.028
	mg/mg protein	
RNA	0.005	0.09
DNA	0.001	0.005
Total lipid	0.78	0.23
Glycerides	0.08	0.10
Phospholipid	0.43	0.1
Sterol	0.21	0.014
	moles/mole phospholipid	
Sterol	0.98	0.27

[a] Data are from A. G. Ulsamer, P. L. Wright, M. G. Wetzel, and E. D. Korn, *J. Cell Biol.* **51**, 193 (1971).

[b] The phagosome membrane has an identical specific activity for alkaline phosphatase and 5'-nucleotidase but more recent work suggests that both activities are derived from a small percentage of contaminating contractile vacuole membranes (B. Bowers and E. D. Korn, *J. Cell. Biol.*, in press).

TABLE III
PHOSPHOLIPID COMPOSITION[a]

Component	Plasma membrane[b]	Phagosome membrane[b]	Cells[b]
Acidic phospholipids	5	2	2
Diphosphatidyl glycerol	3	1	4
Phosphatidylethanalomine	47	43	33
Phosphatidylserine	27	25	10
Phosphoinositide	0.2	0.6	6
Phosphatidylcholine	19	23	45
Lysophysphatidylcholine	0	5	0.5
Unknown	0	0.7	2

[a] Data are from A. G. Ulsamer, P. L. Wright, M. G. Wetzel, and E. D. Korn, *J. Cell Biol.* **51**, 193 (1971).

[b] Values are expressed as moles percent.

Although the recovery of plasma membrane may be only about 20%, it is reasonable to assume that this is a truly representative fraction. Approximately another 20% of the plasma membrane is found at the 35%/40% and 45%/50% interfaces in the final sucrose density centrifugation, and although these fractions are somewhat less pure than the plasma membranes isolated at the 40%/45% interface, their analyses are very similar. The remainder of the plasma membranes that are not recovered have probably been excluded because they have been fragmented into smaller pieces and are therefore discarded during the low speed centrifugations.

In addition to the well-characterized components the isolated plasma membranes contain a high concentration (31%) of a partially characterized lipophosphonoglycan that consists of 26% neutral sugars, 3.3% aminosugars, 10% aminophosphonic acids, 3.2% phosphate, and 14% long chain normal and branched fatty acids and 2-hydroxy fatty acids.[12,12a,12b,12c,12d] This is clearly seen in analysis of lipid-free residues obtained by extraction of the membranes by chloroform–methanol, 2:1 (Table IV). Some of the carbohydrate is exposed at the outer surface of the membrane in intact amoebae since the cells are agglutinated by concanavalin A (unpublished observations). Recent studies of the macromolecular composition of the plasma membrane show the presence of only one major membrane protein, the frequent association with isolated mem-

[12] E. D. Korn and T. Olivecrona, *Biochem. Biophys. Res. Commun.* **45**, 90 (1971).

[12a] E. D. Korn and P. L. Wright, *J. Biol. Chem.* **248**, 439 (1973).

[12b] E. D. Korn, D. G. Dearborn, H. M. Fales, and E. A. Sokoloski, *J. Biol. Chem.* **248**, 2257 (1973).

[12c] E. D. Korn and D. Dearborn, *J. Biol. Chem.* in press.

[12d] T. D. Pollard and E. D. Korn, *J. Biol. Chem.* **248**, 448 (1973).

TABLE IV

ANALYSIS OF *Amoeba* PLASMA MEMBRANES AND LIPID-EXTRACTED
PLASMA MEMBRANES[a]

Component	Plasma membrane	Lipid-extracted plasma membrane
Membrane protein (%)	100	95
Total phosphorus (μmoles/mg protein)	1.6	1.1
Lipid phosphorus (μmoles/mg protein)	0.65	0
Sterol (μmole/mg protein)	0.65	0
Nonlipid phosphorus (μmoles/mg protein)	—	1.1
Carbohydrate (μmoles/mg protein)	—	1.1

[a] Proteins were analyzed by the method of O. H. Lowry, N. J. Rosebrough, A. L. Fain, and R. J. Ransell, *J. Biol. Chem.* **193**, 265 (1951). Phosphorus was analyzed by the method of B. N. Ames and D. T. Dubin, *J. Biol. Chem.* **235**, 769 (1960). Sterols were analyzed by gas–liquid chromatography according to A. G. Ulsamer, P. L. Wright, M. G. Wetzel, and E. D. Korn, *J. Cell Biol.* **51**, 193 (1971). Carbohydrate was analyzed by the anthrone reaction. All analyses are the mean values of all least six experiments.

branes of cytoplasmic actin filaments, and the above-mentioned lipophosphonoglycan.[12a,12d]

Isolation of the Phagocytic Vacuole (Phagosome) Membrane

Solutions

Tris·HCl buffer, 20 mM, pH 7

10%, 20%, 25%, and 30% sucrose in Tris·HCl buffer

Method. When amoebae are incubated with latex beads of diameter 1 μm or greater, the beads are rapidly ingested and are contained singly within a phagocytic vacuole (phagosome) whose membrane is derived from the plasma membrane.[13] This phagocytic process occurs under a wide variety of conditions, but not under all circumstances,[14] requires O_2, and is inhibited by inhibitors of oxidative metabolism.[14] Although the cells will phagocytose many particulate objects, the use of latex beads is convenient because of their chemical and metabolic inertness and because they provide a simple basis for the isolation of the phagosomes.[15]

Amoebae are grown to a concentration between 0.5 and 1 × 10⁶ cells per milliliter and latex beads (Dow Chemical Company) of approximately 1 μm in diameter are added to a concentration of 1 mg/ml. The culture is allowed to continue rotating for 15–30 minutes (or longer if desired),

[13] E. D. Korn and R. A. Weisman, *J. Cell Biol.* 34, 219 (1967).

[14] R. A. Weisman and E. D. Korn, *Biochemistry* 6, 485 (1967).

[15] M. G. Wetzel and E. D. Korn, *J. Cell Biol.* 43, 90 (1969).

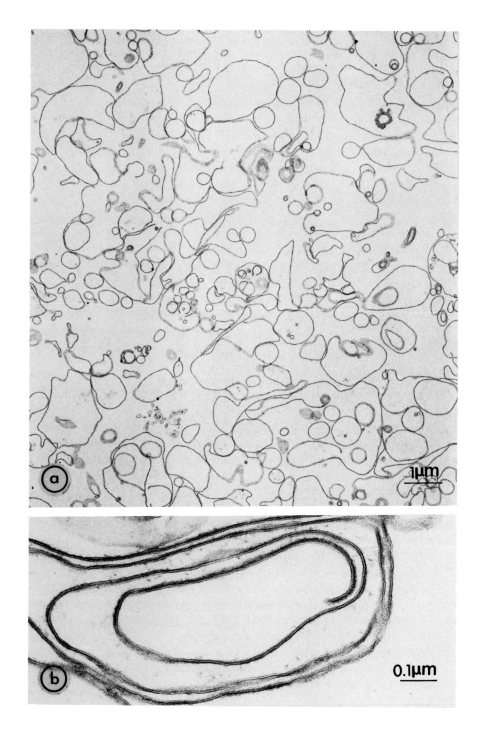

and then the cells are collected by centrifugation at 500 g for 5 minutes. Phagocytosis may be more efficient if the cells are first concentrated to about 3×10^6/ml. After washing the amoebae three times with 20 mM Tris·HCl, pH 7, or other suitable buffer, the cells are homogenized in 30% sucrose in buffer (approximately 5×10^7 cells per milliliter) with 10 cycles of a tight Dounce homogenizer. Approximately 10 ml of the homogenate is placed at the bottom of an appropriate centrifuge tube and overlayered with 9 ml each of 25% sucrose, 20% sucrose, and 10% sucrose. The tubes are then centrifuged at 131,000 g for 90 minutes in an SW 27 rotor. The phagosomes accumulate at the interface between 10% and 20% sucrose from which they are collected after removal of the overlying solution (Table V). If desired the suspension can be diluted with 20 mM Tris buffer and the phagosomes collected as a pellet after centrifuging for 15 minutes at 5000 g. This procedure yields a homogeneous preparation of phagosomes consisting of latex beads tightly surrounded by membrane (Fig. 3a) and also containing hydrolytic enzymes, such as acid phosphatase and acid glucosidases (Table V).

The phagosomes can be disrupted[15] by sonication for two 1-minute intervals at 0°, and the free latex beads removed by centrifuging at 5000 g for 15 minutes. Centrifugation at 100,000 g for 60 minutes produces a homogeneous pellet of membranes (Fig. 3b). These membranes are enzymatically and chemically very similar to plasma membranes insofar as they have been examined (Tables II and III). The soluble acid hydrolases that were contained within the phagosomes are released into the supernatant solution.

TABLE V

ANALYSIS OF PHAGOSOMES ISOLATED FROM AMOEBAE INCUBATED WITH POLYSTYRENE LATEX BEADS

Component	Phagosomes[a]	Control[a]
Polystyrene	80	—
Protein	3.2	0.014
Acid phosphatase	22[b]	0
β-Glucosidase	15[b]	0

[a] Analyses are of the fraction recovered at the interface between 10% and 20% sucrose after centrifuging homogenates of cells which had phagocytosed latex beads and control cells. Values are expressed as percentage of total cell homogenates.
[b] The specific activity of acid phosphatase is about 1.5 units per milligram of protein and the specific activity of β-glucosidase is about 2.4 units per milligram of protein

FIG. 3. Isolated phagocytic vacuoles (phagosomes) containing polystyrene latex beads (a) and the membranes isolated from the sonicated phagosomes (b). a, ×10,000; b, ×100,000.

This procedure for the isolation of phagosomes has also been used for the isolation of phagosomes from L cells.[16] It has also been slightly modified by the use of albumin-coated emulsions of paraffin oil for the isolation of phagosomes from polymorphonuclear leukocytes[17] and alveolar macrophages.[18]

[16] J. W. Heine and C. A. Schnaitman, *J. Cell Biol.* 48, 703 (1971).
[17] T. P. Stossel, T. D. Pollard, R. J. Mason, and M. Vaughan, *J. Clin. Invest.* 50, 1745 (1971).
[18] T. P. Stossel, R. J. Mason, T. D. Pollard, and M. Vaughan, *J. Clin. Invest.* 50, 1745 (1971).

[72] Transport in Isolated Bacterial Membrane Vesicles

By H. R. KABACK

Two general types of transport mechanisms have been described in detail in isolated bacterial membrane vesicles—respiration-coupled systems which are dependent upon the activity of specific dehydrogenases or artificial electron donors for "classic" active transport, and P-enolpyruvate-P-transferase-dependent systems which are dependent upon P-enolpyruvate for "vectorial phosphorylation."[1] A third mechanism involving adenine P-ribosyltransferase-mediated adenine uptake as AMP[2] will not be discussed here.

General Principle

Isolated bacterial membrane vesicles prepared as described in a previous volume of this series[3] (or in *Staphylococcus aureus* and *Salmonella typhimurium,* as described by Short *et al.*[4,5] and Hong and Kaback,[6] respectively) are incubated with a radioactive transport substrate in the presence of a specific energy source. At a given time, the reaction mixtures are diluted to terminate the uptake reaction, and the vesicles are separated from the medium by means of rapid filtration on nitrocellulose filters of a specified pore size. The samples are washed once rapidly in order to lower the background radioactivity adhering to the filters.

[1] H. R. Kaback, *Biochim. Biophys. Acta* 265, 367 (1972).
[2] J. Hochstadt-Ozer and E. R. Stadtman, *J. Biol. Chem.* 246, 5304 (1971).
[3] H. R. Kaback, this series, Vol. 22, p. 99.
[4] S. A. Short, D. C. White, and H. R. Kaback, *J. Biol. Chem.* 247, 298 (1972).
[5] S. A. Short, D. C. White, and H. R. Kaback, *J. Biol. Chem.* 247, 7452 (1972).
[6] J.-s. Hong and H. R. Kaback, *Proc. Nat. Acad. Sci. U.S.* 69, 3336 (1972).

Respiration-Dependent Transport[1,3-26]

Using Specific Dehydrogenase Substrates. Aliquots (25 μl) of membrane vesicles suspended in 0.1 *M* potassium phosphate buffer (pH 6.6) are diluted to a final volume of 50 μl containing (in final concentrations) 50 m*M* potassium phosphate (pH 6.6) and 10 m*M* magnesium sulfate. The samples (in 5-ml test tubes) are incubated at the desired temperature for approximately 30 seconds to 1 minute. At this time, the appropriate dehydrogenase substrate (at 20 m*M* final concentration) and immediately thereafter, a radioactive transport substrate are added, and the incubation is continued for a given period of time. To terminate the reaction, each sample is rapidly diluted with 2.0 ml of a 0.1 *M* solution of lithium chloride at room temperature, immediately filtered through nitrocellulose filters (30 mm diameter, 0.45 μm pore size, obtained from Schleicher and Schuell, Inc., Keene, New Hampshire, or Millipore Filter Corporation, Bedford, Massachusetts), and washed once with an equal volume of 0.1 *M* lithium chloride. The wash volume is added to the reaction vessel and immediately poured over the filter. The dilution, filtration, and washing procedures are conducted in less than 30 seconds; the filters are immediately removed from the suction apparatus, and the bot-

[7] H. R. Kaback and L. S. Milner, *Proc. Nat. Acad. Sci. U.S.* **66**, 1008 (1970).

[8] E. M. Barnes, Jr. and H. R. Kaback, *Proc. Nat. Acad. Sci. U.S.* **66**, 1190 (1970).

[9] E. M. Barnes, Jr. and H. R. Kaback, *J. Biol. Chem.* **246**, 5518 (1971).

[10] H. R. Kaback and E. M. Barnes, Jr., *J. Biol. Chem.* **246**, 5523 (1971).

[11] W. N. Konings, E. M. Barnes, Jr., and H. R. Kaback, *J. Biol. Chem.* **246**, 5857 (1971).

[12] G. K. Kerwar, A. S. Gordon, and H. R. Kaback, *J. Biol. Chem.* **247**, 291 (1972).

[13] W. N. Konings and E. Freese, *J. Biol. Chem.* **247**, 2408 (1972).

[14] G. W. Dietz, *J. Biol. Chem.* **247**, 4561 (1972).

[15] E. M. Barnes, Jr., *Arch. Biochem. Biophys.* **152**, 795 (1972).

[16] H. Hirata, A. Asano, and A. F. Brodie, *Biochem. Biophys. Res. Commun.* **44**, 368 (1971).

[17] G. D. Sprott and R. A. MacLeod, *Biochem. Biophys. Res. Commun.* **47**, 838 (1972).

[18] P. Bhattacharyya, W. Epstein, and S. Silver, *Proc. Nat. Acad. Sci. U.S.* **68**, 488 (1971).

[19] F. J. Lombardi, J. P. Reeves, and H. R. Kaback, *Fed. Proc., Fed. Amer. Soc. Exp. Biol.* **31**, 457 (1972).

[20] C. T. Walsh, R. H. Abeles, and H. R. Kaback, *J. Biol. Chem.* **247**, 7858 (1972).

[21] F. J. Lombardi, J. P. Reeves, and H. R. Kaback, *J. Biol. Chem.* **248**, 3551 (1973).

[22] F. J. Lombardi and H. R. Kaback, *J. Biol. Chem.* **247**, 7844 (1972).

[23] S. Murakawa, K. Izaki, and H. Takahashi, *Agr. Biol. Chem.* **35**, 1992 (1971).

[24] M. K. Rayman, T. C. Y. Lo, and B. D. Sanwal, *J. Biol. Chem.* **247**, 6332 (1972).

[25] W. N. Konings, A. Bisschop, and M. C. C. Daatselaar, *FEBS (Fed. Eur. Biochem. Soc.) Lett.* **24**, 260 (1972).

[26] A. Matin and W. N. Konings, *Eur. J. Biochem.* **34**, 58 (1973).

toms of the filters blotted on tissue paper. The filters are then placed in 30 mm-diameter, 2 mm-deep aluminum or stainless steel planchets, and stainless steel rings which fit into the planchets but do not cover the filtration area of the filters (30 mm outside diameter, 28 mm inside diameter, 2 mm thick) are placed on top of the filters in order to keep them flat during drying and counting. The samples are then placed in an oven at approximately 180° for 1–2 minutes, and counted in a gas flow counter at 10–20% efficiency. Alternatively, the filters may be counted by means of liquid scintillation spectrometry. For multiple samples carried out routinely with transport substrates labeled with sufficiently high-energy emitters (i.e., ^{14}C, ^{32}P, ^{35}S, or ^{86}Rb), the use of a gas flow counter with an automatic sample changer is suggested. This method is faster, more convenient, and less expensive than liquid scintillation for this particular assay. Each set of samples is corrected for a control obtained by diluting the samples before adding the radioactive substrate and omitting the dehydrogenase substrate.

Using Artificial Electron Donors.[1,11,13] Five-milliliter test tubes containing reaction mixtures made up as described above are fitted with perforated rubber stoppers (No. 00) so that the samples can be gassed with oxygen by means of a 20-gauge hypodermic needle inserted through the stopper (Fig. 1). Additions are made to the reaction mixtures with a

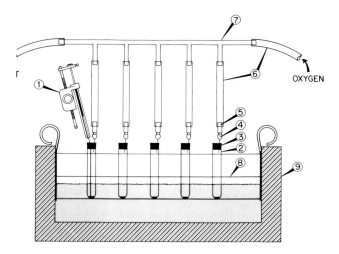

FIG. 1. A simple, inexpensive apparatus for performing assays under various atmospheric conditions. (1) Hamilton microsyringe with dispenser; (2) 12 × 75 mm glass test tube; (3) number 00 perforated rubber stopper; (4) 20-gauge hypodermic needle inserted through the stopper; (5) tip from a 1.0-ml disposable tuberculin syringe; (6) Tygon tubing; (7) glass manifold; (8) test-tube support; (9) water bath.

Hamilton microsyringe (Hamilton Syringe Co., Reno, Nevada) by insertion through the perforation in the stopper. The syringes are made to order with 3-inch, 22-gauge, flat-tipped needles, and are used in conjunction with a dispenser (also made by Hamilton Syringe Co.) that allows multiple microliter additions to be made in rapid succession (Fig. 1). Samples are incubated at the desired temperature, and gassed with pure oxygen for 2–3 minutes. At this time, ascorbate (20 mM, final concentration), phenazine methosulfate (0.1 mM, final concentration), and a radioactive transport substrate are added in sequence. At the desired time, the reactions are terminated and the samples are assayed by the methods described above.

As an alternative to ascorbate-phenazine methosulfate, reduced diphosphopyridine nucleotide (NADH) and pyocyanine perchlorate (K & K Laboratories, Plainview, New York) may be used to drive these transport systems.[27] Since pyocyanine has a lower redox potential than phenazine methosulfate, ascorbate cannot be used as a reductant.

In some bacterial membrane preparations (i.e., those from *Azotobacter vinelandii*[15] and respiratory particles from *Mycobacterium phlei*[16]), N,N,N',N'-tetramethyl-p-phenylenediamine (TMPD) may be used as an electron carrier instead of phenazine methosulfate (Table I). In other preparations (i.e., a marine pseudomonad B-16[17]), TMPD, but apparently not phenazine methosulfate, can be used as an electron carrier. With the exception of duroquinone (K & K Laboratories) (using dithiothreitol as a reductant) which is approximately one-tenth as effective as ascorbate-phenazine methosulfate or NADH-pyocyanine in *E. coli* vesicles, all other electron carriers tested thus far have no effect.

Valinomycin-Induced Rubidium or Potassium Uptake.[18-21] A 1.0-ml aliquot of membrane vesicles (containing 3–5 mg of membrane protein) suspended in 0.1 M potassium phosphate buffer (pH 6.6) is diluted with 10 ml of 0.1 M sodium or choline phosphate buffer (pH 6.6) and incubated at 0° for approximately 5 minutes. The sample is then centrifuged at approximately 25,000 g for 15 minutes at 4°, and the supernatant discarded. The membrane pellet is resuspended in another 10-ml aliquot of sodium or choline phosphate buffer (pH 6.6) and centrifuged as described above. After aspirating the supernatant, the pellet is resuspended to 1.0 ml in sodium or choline phosphate buffer (pH 6.6), and valinomycin (dissolved in dimethyl sulfoxide) is added to give a final concentration of 2 nmoles per mg membrane protein (0.1% dimethyl sulfoxide). It is important to note that the valinomycin must be added to the membrane

[27] H. R. Kaback, unpublished information; this electron donor system was suggested to the author by Dr. David F. Wilson of the Johnson Foundation of the University of Pennsylvania.

suspension rather than vice versa because this hydrophobic compound has a high affinity for glass. Aliquots (25 μl) of this membrane suspension are diluted to a final volume of 50 μl, containing, in final concentrations, 50 mM sodium or choline phosphate buffer (pH 6.6), 10 mM magnesium sulfate, and 1 nmole per mg membrane protein valinomycin. The samples are then assayed for ^{86}Rb or ^{42}K uptake using dehydrogenase substrates or artificial electron donors as described above.

Comments. Although the procedures outlined above are generally applicable to each of the vesicle systems studied thus far, certain aspects of the assay systems are particularly noteworthy.

1. Speed is essential during dilution, filtration, and washing. Once the reaction is terminated by dilution of the transport substrate and electron donor, there is a large diffusion head from within the vesicles which will lead to loss of accumulated substrate. Thus, it is imperative that filtration rates be maximized, and that the filters be removed from the filtration apparatus within a reasonably short period of time (i.e., within 1 minute). In this regard, it should be emphasized that at high vesicle concentrations, filtration rates are diminished due to occlusion of pores in the filters.

2. In most of the transport systems described (Table I), potassium phosphate buffer has been used extensively. However, other buffer systems can be used with little or no loss in activity. With valinomycin-induced rubidium uptake, for instance, sodium or choline are used as cations rather than potassium (so that the specific activity of the isotope is not diluted).[18-21] With the *Escherichia coli* transport systems studied thus far, there does not appear to be a requirement for a specific cation[22]; however, it should be noted that sodium ion (but not lithium) inhibits proline and, to a lesser extent, serine transport, and that TRIS and imidizole buffers are generally inhibitory. With regard to anions, phosphate can be replaced with arsenate and, in some instances, cacodylate.[21]

3. All the transport systems studied thus far in vesicles are sensitive to ionic strength; in *E. coli* membrane vesicles, they exhibit optimal activity at approximately 25 mM potassium phosphate. On the other hand, in *S. aureus* vesicles, where the oxidation of α-glycerol-P is utilized to drive transport,[4,5] the optimal potassium phosphate concentration is 10 mM.[5]

4. Vesicles transport activities measured as a function of pH generally exhibit broad optima over a pH range from 6.0 to 8.0 with a maximum at pH 6.5 to 7.0.[22]

5. The salt solution used for dilution and washing, as well as the temperature at which this procedure is carried out, is important with regard to retention of accumulated solute. Of a number of salts tested, 0.1 M lithium chloride at room temperature has been found to be optimal for dilution, filtration, and washing.

6. The temperature optima for the transport assay per se vary depending upon whether initial rates or steady-state levels of accumulation are studied, and upon the transport system under investigation.[5,10,22] In general, initial rates of transport exhibit optima at 45° to 50°, whereas the optimal temperature for steady-state levels of accumulation are much lower and vary depending upon the transport system. It should also be emphasized that initial rates of uptake at higher temperatures (i.e., above 35°) are linear for only short periods of time (i.e., on the order of seconds), especially when artificial electron donors are used to drive transport, and that the solubility of oxygen in aqueous solutions is inversely related to temperature. These properties of the system frequently produce overshoot phenomena when transport is studied as a function of time at temperatures above 40°.[10] The overshoot phenomena can be alleviated by carrying out the reactions under oxygen as described for the artificial electron donor systems.

7. The choice of a dehydrogenase substrate to drive transport depends upon the organism from which the vesicles are prepared (Table I). It is therefore advisable, when investigating respiration-coupled transport in a totally new system to use one of the artificial electron donor systems initially, and subsequently screen a number of potential dehydrogenase substrates.

8. Apparent Michaelis constants for the amino acid transport systems in isolated bacterial membrane vesicles range from 0.1 μM to 1 μM[1,5,13,22]; for the organic acid transport systems, from 5 μM to 50 μM[24-26]; for the sugar and hexose-P transport systems, from 50 μM to 0.5 mM[1,8,12,14,15]; and the valinomycin-induced rubidium uptake system exhibits apparent K_m's of approximately 0.4 nmoles per mg membrane protein for valinomycin and 0.9 mM for rubidium and potassium.[21] Generally, transport in the vesicles is assayed at concentrations of solute which are at least twice the K_m concentration.

9. Since the respiration-dependent systems utilize oxygen as a terminal electron acceptor, it is important that the reaction mixtures be aerated as well as possible. Thus the use of small reaction volumes (50 μl) in relatively large reaction vessels (5 ml). Although this is especially important with the artificial electron donor system[11,13] (i.e., reduced phenazine methosulfate or pyocyanine react directly with oxygen decreasing the oxygen available to the terminal oxidase), it is also an important consideration when using dehydrogenase substrates to drive transport. Initial rates and steady-state levels of accumulation are linear with membrane protein concentration over a relatively small concentration range. The linearity range of the assay can be increased by carrying out the reactions under oxygen.

TABLE I

RESPIRATION-DEPENDENT TRANSPORT SYSTEMS IN ISOLATED BACTERIAL MEMBRANE VESICLES

Source of vesicles	Best electron donors	Transport systems	Remarks
Escherichia coli	D-Lactate; ASC-PMS; NADH-pyocyanine	β-Galactosides,[a-e] galactose,[d,f] arabinose,[d,g] gluconic acid,[d,g] glucuronic acid,[h] hexose-P,[d,g,i] Pro,[i,k] Gly-Ala,[i,k] Ser-Thr,[i,k] Glu-Asp,[i,k] Phe-Tyr-Trp,[i,k] His,[i,k] Leu-Iso-Val,[i,k] Cys,[i,k] peptides,[g] succinate,[l-o] valinomycin-induced Rb+ or K+ uptake[p,q]	Sugar, hexose-P, peptide, and succinate systems induced; amino acid systems constitutive
	NADH, ASC-PMS	D- and L-lactate[o]	Induced system
	ASC-PMS	Pyruvate[o]	—
Salmonella typhimurium	D-Lactate, ASC-PMS	Various amino acids,[a,c] valinomycin-induced Rb+ uptake[r]	Amino acid systems constitutive
Micrococcus denitrificans	ASC-PMS	Citrate[s]	Induced system
	D-Lactate, formate, ASC-PMS	Gly-Ala, GlN-AsN,[e,t] val-induced Rb+ uptake[u,q]	Constitutive system
Azotobacter vinelandii	Malate (+FAD), ASC-PMS, ASC-TMPD	Glucose[v]	Induced system
Mycobacterium phlei	NADH, ASC-TMPD, ASC-PMS	Pro[w]	Apparently constitutive
Pseudomonas sp.	L-Lactate, ASC-PMS	Succinate[o]	Apparently constitutive
Marine *Pseudomonas* B-16 (ATCC 19855)	NADH, succinate, ASC-PMS NADH, ASC-TMPD	D- and L-lactate[o] Ala, α-aminoisobutyrate[z]	Apparently constitutive
Pseudomonas putida	ASC-PMS	Pro[e]	Apparently constitutive
Proteus mirabilis	ASC-PMS	Pro[e]	Apparently constitutive
Staphylococcus aureus	α-Glycerol-P (vesicles from gluconate-grown cells); L-lactate (vesicles from glucose-grown cells); ASC-P	Ala-Gly,[y,z] Leu-Iso-Val, Ser-Thr,[y,z] Asp-Glu,[y,z] AsN-GlN,[y,z] His,[y,z] Arg,[y,z] Phe-Tyr-Trp,[y,z] Cys,[y,z] Met,[y,z] Pro,[y,z] Val-induced Rb+ uptake[q]	Amino acid systems constitutive
Bacillus subtilis	α-Glycerol-P (vesicles from gly-	Ala-Gly,[aa] Leu-Iso-Val,[aa] Ser-Thr, AsN-GlN,[aa]	Constitutive

				systems
	cerol-grown cells); NADH; ASC-PMS		Asp-Glu,[aa] Lys,[aa] His,[aa] Arg,[aa] Phe-Tyr-Trp,[aa]	Apparently constitutive
	NADH, ASC-PMS		Cys-(Cys)$_2$-Met,[aa] Pro[aa]	—
Bacillus megaterium	ASC-PMS		D- and L-lactate,[o] succinate[o] Pro[e]	Apparently constitutive
Bacillus licheniformis	ASC-PMS		Various amino acids[bb]	Apparently constitutive

[a] J.-s. Hong and H. R. Kaback, *Proc. Nat. Acad. Sci. U.S.* **69**, 3336 (1972).
[b] E. M. Barnes, Jr. and H. R. Kaback, *Proc. Nat. Acad. Sci. U.S.* **66**, 1190 (1970).
[c] E. M. Barnes, Jr. and H. R. Kaback, *J. Biol. Chem.* **246**, 5518 (1971).
[d] H. R. Kaback and E. M. Barnes, Jr., *J. Biol. Chem.* **246**, 5523 (1971).
[e] W. N. Konings, E. M. Barnes, Jr., and H. R. Kaback, *J. Biol. Chem.* **246**, 5857 (1971).
[f] G. K. Kerwar, A. S. Gordon, and H. R. Kaback, *J. Biol. Chem.* **247**, 291 (1972).
[g] H. R. Kaback, *Biochim. Biophys. Acta* **265**, 367 (1972).
[h] G. K. Kerwar and H. R. Kaback, unpublished information.
[i] G. W. Dietz, *J. Biol. Chem.* **247**, 4561 (1972).
[j] H. R. Kaback and L. S. Milner, *Proc. Nat. Acad. Sci. U.S.* **66**, 1008 (1970).
[k] F. J. Lombardi and H. R. Kaback, *J. Biol. Chem.* **247**, 7844 (1972).
[l] S. Murakawa, K. Izaki, and H. Takahashi, *Agr. Biol. Chem.* **35**, 1991 (1971).
[m] M. K. Rayman, T. C. Y. Lo, and B. D. Sanwal, *J. Biol. Chem.* **247**, 6332 (1972).
[n] W. N. Konings, A. Bisschop, and M. C. C. Daatselaar, *FEBS (Fed. Eur. Biochem. Soc.) Lett.* **24**, 260 (1972).
[o] A. Matin and W. N. Konings, *Eur. J. Biochem.* **34**, 58 (1973).
[p] P. Bhattacharyya, W. Epstein, and S. Silver, *Proc. Nat. Acad. Sci. U.S.* **68**, 488 (1971).
[q] F. J. Lombardi, J. P. Reeves, and H. R. Kaback, *J. Biol. Chem.* **248**, 3551 (1973).
[r] J.-s. Hong, and H. R. Kaback, unpublished information.
[s] H. R. Kaback, unpublished information.
[t] A. N. Tucker, D. C. White, and H. R. Kaback, manuscript in preparation.
[u] F. J. Lombardi, J. P. Reeves, and H. R. Kaback, *Fed. Proc., Fed. Amer. Soc. Exp. Biol.* **31**, 457 (1972).
[v] E. M. Barnes, Jr., *Arch. Biochem. Biophys.* **152**, 795 (1972).
[w] H. Hirata, A. Asano, and A. F. Brodie, *Biochem. Biophys. Res. Commun.* **44**, 368 (1971).
[x] G. D. Sprott and R. A. MacLeod, *Biochem. Biophys. Res. Commun.* **47**, 838 (1972).
[y] S. A. Short, D. C. White, and H. R. Kaback, *J. Biol. Chem.* **247**, 298 (1972).
[z] S. A. Short, D. C. White, and H. R. Kaback, *J. Biol. Chem.* **247**, 7452 (1972).
[aa] W. N. Konings and E. Freese, *J. Biol. Chem.* **247**, 2408 (1972).
[bb] W. S. May, Jr. and H. R. Kaback, unpublished information.

P-Enolpyruvate–P-Transferase-Dependent Transport[1,28–31]

Aliquots (25 μl) of membrane vesicles are diluted to a final volume of 50 μl containing, in final concentrations, 50 mM potassium phosphate buffer (pH 6.6), 10 mM magnesium sulfate, 0.3 M lithium chloride, and 0.1 M P-enolpyruvic acid (adjusted to pH 6.5–7 with sodium carbonate). The samples are incubated at a given temperature for 15 minutes. After this period of time, radioactively labeled sugar (glucose, methyl-α-D-gluco-pyranoside, fructose, or mannose) is added, and the incubation is continued for an appropriate period of time. To terminate the reaction, each sample is rapidly diluted with 2.0 ml of a 0.5 M solution of lithium chloride at room temperature, immediately filtered, and washed once with an equal volume of 0.5 M lithium chloride. The filters are removed from the suction apparatus, placed in planchets, dried, and counted as described for the respiration-dependent transport assays.

Comments. The following information may be useful in assaying P-enolpyruvate-P-transferase-dependent transport in isolated membrane vesicles:

1. Although speed is critical once the samples are diluted, this system is not sensitive to oxygenation, and can be assayed under anoxic conditions.

2. These systems exhibit a higher ionic strength optimum than the respiration-dependent transport systems. For this reason, 0.3 M lithium chloride is added to the reaction mixtures. Similarly, 0.5 M lithium chloride is used for dilution, filtration, and washing, as opposed to 0.1 M lithium chloride in the respiration-dependent transport assays.

3. The apparent Michaelis constant for glucose and α-methylglucoside transport in *E. coli* membrane vesicles is approximately 4 μM.[28]

4. Membrane vesicles which have not been adequately washed sometimes exhibit a high endogenous transport activity which can be abolished by the addition of 10 mM sodium fluoride (final concentration) to the reaction mixtures.[28] Addition of sodium fluoride does not impair transport in the presence of P-enolpyruvate.

5. The requirement for high concentrations of P-enolpyruvate, which is a highly charged compound, may result from its poor ability to penetrate the membrane. This permeability barrier can be transiently overcome by subjecting the vesicles to a mild osmotic shock in the presence of P-enolpyruvate as follows[30]: 25 μl of vesicles suspended in 0.1 M potassium phosphate buffer (pH 6.6) are rapidly diluted into 25 μl of 10 mM

[28] H. R. Kaback, *J. Biol. Chem.* **243**, 3711 (1968).
[29] H. R. Kaback, *Annu. Rev. Biochem.* **39**, 561 (1970).
[30] H. R. Kaback, *Curr. Top. Membranes Transport* **1**, 35 (1970).
[31] H. R. Kaback, *in* "The Molecular Basis of Membrane Transport" (D. C. Tosteson, ed.), p. 421. Prentice-Hall, Englewood Cliffs, New Jersey.

P-enolpyruvate; 10 mM magnesium sulfate, and 0.3 M lithium chloride are immediately added; and radioactive substrate is added without a 15-minute preincubation. Under these conditions, the initial rate of α-methylglucoside uptake at 46° is approximately the same as that of vesicles assayed in the usual manner (i.e., with 0.1 M P-enolpyruvate).

Distribution of Respiration-Coupled and P-Enolpyruvate– P-Transferase-Dependent Transport Systems in E. coli Membrane Vesicles

Although the precise mechanism of transport of particular nutrients in various bacterial species has not been studied extensively, a number of transport systems have been studied in detail in vesicles prepared from E. coli. This information is tabulated in Table II. It should be emphasized that these results cannot be generalized.[1] In S. aureus, for instance, although amino acids are transported by respiration-coupled mechanisms (Table I), most sugars are apparently transported by P-transferase-dependent systems, while many obligate aerobes do not exhibit P-enolpyruvate-P-transferase activity.

Membrane Barrier Function[28-32]

Owing to the unique features of the P-enolpyruvate–phosphotransferase system, this system may be utilized to study certain functional aspects of membrane permeability in vesicles from those organisms which possess this system. As opposed to the respiration-dependent active transport systems, sugars transported via the P-transferase system appear in

TABLE II

TRANSPORT SYSTEMS IN *Escherichia coli* MEMBRANE VESICLES

Respiration-coupled systems	PTS-mediated systems
β-Galactosides	Glucose
Galactose	Fructose
Arabinose	Mannose
Glucuronic acid	Mannitol
Gluconic acid	Sorbitol
Hexose-P's	
Amino acids (nine independent systems)	
Peptides	
Succinate	
D-Pyruvate and L-lactate	

[32] E. Shechter, T. Gulik-Kryzwicki, and H. R. Kaback, *Biochim. Biophys. Acta* **274**, 466 (1972).

the intravesicular pool as phosphorylated derivatives. Since the vesicles cannot dephosphorylate sugar phosphates, nor can they transport sugar phosphates unless a specific sugar-P transport system is induced in the parent cells (Table I), the retention of substrates transported via the P-transferase system is a reflection of the passive permeability of the vesicle membrane. Using the assay system described above with a few simple manipulations, the barrier properties of the vesicles can be assayed as an isolated functional entity.[1,24-28]

The experiment presented in Fig. 2 shows the time course of methyl-α-D-glucopyranoside uptake and phosphorylation by E. coli ML 308-225 membrane vesicles at 46° in the presence of P-enolpyruvate. The data are expressed as a percentage of the concentration of α-methylglucoside added to the reaction mixtures, so that the amounts of α-methylglucoside and α-methylglucoside-P recovered from the vesicles can be directly compared with that recovered from the medium. As shown, uptake (α-MGP$_{in}$)

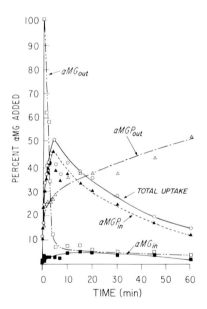

Fig. 2. Time course of methyl-α-D-glucopyranoside uptake and phosphorylation at 46° by Escherichia coli ML 308-225 membrane vesicles. Total uptake (○—○), uptake of α-[14C]methylglucoside by vesicles incubated in the presence of 0.1 M P-enolpyruvate for the times shown; α-MGP$_{in}$ (▲- - - - -▲), α-methylglucoside-P recovered from vesicles; α-MG$_{in}$ (■—■), α-methylglucoside recovered from vesicles; α-MGP$_{out}$ (△— – –△), α-methylglucoside-P recovered from the medium in which the vesicles were incubated; α-MG$_{out}$ (□— · · —□), free α-methylglucoside recovered from the medium in which the vesicles were incubated. The experiment was carried out as described by H. R. Kaback, J. Biol. Chem. 243, 3711.

increases rapidly for about 5 minutes, when it stops abruptly and begins to decrease. By 60 minutes, the vesicles lose over 60% of the α-methylglucoside-P accumulated in 5 minutes.

The loss of α-methylglucoside-P from the vesicles appears to result from increased membrane permeability at 46° since vesicles incubated at lower temperatures are able to retain high intravesicular concentrations for at least 30 minutes.[28] As shown by the curve labeled "α-MGP$_{out}$," the rate of appearance of α-methylglucoside-P in the medium increases rapidly for the first 5 minutes of the incubation, diminishes, and finally achieves an inverse relationship to α-MGP$_{in}$. In other words, α-MGP$_{in}$ has a precursor-product relationship to α-MGP$_{out}$. By 5 minutes, about 95% of the free sugar added to the reaction mixtures has been phosphorylated (α-MGP$_{out}$). Thus, phosphorylation and transport stop at 5 minutes because substrate becomes limiting. Simultaneously, α-methylglucoside-P in the vesicles is redistributed from a higher concentration in the intravesicular pool to a lower concentration in the medium. Despite the large quantity of α-methylglucoside-P that appears in the medium at 60 minutes, the concentration of α-methylglucoside-P is still about 25 times higher inside the vesicles than it is in the medium. If the incubations are continued, the two compartments eventually equilibrate.

Induction and reversal of the leak occur extremely rapidly. When vesicles are incubated with α-methylglucoside at 27° and subsequently transferred to 46°, there is an immediate decrease in the α-methylglucoside-P retained by the vesicles.[1,28] On the other hand, when the vesicles are incubated with α-methylglucoside at 46° for 5 minutes and then shifted to 27°, there is essentially no loss of α-methylglucoside-P from the intravesicular pool.[1,28,30-32] It should be emphasized that the temperature-induced leak described here is not "carrier"-mediated as is the case for the respiration-coupled transport systems.[1] These experimental manipulations allow a functional study of the passive permeability of the vesicles to the efflux of sugar-P in the absence of transport. The vesicles can be preloaded with sugar-P at 46° until all the transport substrate has been phosphorylated. At this point, the concentration of sugar-P in the intravesicular pool is at least 50 times higher than that of the medium. By lowering the temperature, leakage can be abolished, and the effect of various perturbants on the membrane can be studied. Although the time required for a given vesicle preparation to phosphorylate all the sugar in the reaction mixture at 46° varies from one preparation to another and may take as long as 20–30 minutes, effects similar to those shown in Fig. 2 and described above have been demonstrated with vesicles prepared from a number of strains of *E. coli, S. typhimurium,* and *Bacillus subtilis.*

Section VI

General Methodology

[73] Centrifugation

By Eric Reid and Robert Williamson

This article deals with intracellular components, and only scant mention is given to intact cells or viruses. Elsewhere in *Methods in Enzymology* there is a useful article on zonal centrifugation,[1] as well as repeated references to centrifugation as a means of isolating subcellular components. The present article is neither a compendium of such information nor a guide for the absolute beginner. It takes the form of a somewhat dogmatic survey of principles and approaches, giving illustrative examples of separation techniques. No attention is paid to the analytical ultracentrifuge, nor to recently marketed devices that enable the sedimentation of macromolecules to be scanned during centrifugation in a high-speed "preparative" centrifuge.[2]

Types of Approach

"Analytical" and "Preparative" Centrifugation of Cell Components

Although the present article deals only with what are commonly called preparative centrifuges, a distinction can usefully be made between "analytical" and "preparative" conditions of centrifugation in such centrifuges. The distinction is related not to the scale of the procedure, but to the proportion of the product that has to be set aside for characterization by methods that do not allow final recovery. In an analytical run on a mixture of subcellular components, these have to be separated from each other as far as possible, in order that the distribution of a product can be analyzed in relation to the starting material and a "balance sheet" compiled; the conditions are necessarily a compromise.[3] The preparative approach entails the isolation of a particular component for further study, as far as possible free of other components that are of interest only as possible contaminants; usually there has to be some sacrifice of yield for the sake of purity.

For most enzyme preparations, the appropriate subcellular fraction is the starting material of choice; yet a regrettable tradition persists of applying protein fractionation techniques directly to a cell homogenate or to

---

[1] G. B. Cline and R. B. Ryel, this series, Vol. 22, p. 168.

[2] Facilities for *in situ* scanning in a preparative ultracentrifuge are offered by certain manufacturers (Beckman-Spinco, Heraeus-Christ).

[3] D. B. Roodyn (ed.), "Enzyme Cytology." Academic Press, New York, 1967.

a supernatant fraction, possibly hypotonic and containing unwanted proteins inadvertently liberated from organelles such as mitochondria.

With recent advances in centrifuges, rotors, and media, even small and subtle differences between cells, organelles, or molecules can be exploited to give a centrifugal separation. The position to which a particle or molecule sediments in a centrifugal field (an artificially created rotational field greater than 1 g) depends upon its density relative to the medium and its size.[1] If conditions of centrifugation and the density of the medium are controlled, the technique can be used to determine the apparent mass of the particle, a composite of its true mass and its shape. On the other hand, with appropriate conditions based on ascertained properties, different particles can be isolated in useful amount. Centrifuges now exist that can routinely give a force of 500,000 g, enabling macromolecules of molecular weights down to a few thousand to be separated; rotors exist allowing components or organelles present in very low proportions to be concentrated in milligram amounts.

Strategy and Nomenclature

Figure 1 describes in outline form possible centrifugal methods of separating the components of a mixture; it underlies much of what follows. A classic technique developed in the laboratories of W. C. Schneider[5] and de Duve,[3,4] here called *differential pelleting* (the term differential centrifugation being ambiguous)[6] still has an important place and is further considered below. However, it is obviously a crude method, as a slow-moving particle which is already near the bottom of the tube (or the wall, in the case of an angle rotor) at the beginning of a run may be pelleted while a distant fast-moving particle is still in mid-journey. This difficulty can be circumvented by loading a necessarily smaller amount of starting material as a layer over the centrifugation medium, which must be of slightly higher density. This was done in an early method for isolating liver cell nuclei.[5] Such a centrifugation need not in principle entail pelleting; with a suitably short time, components of different sedimentation rate can be stratified at different levels, provided that there is at least a slight graduation in the density of the medium to minimize convective disturbances. Such a separation represents *rate-dependent banding* (Fig. 1).

For banding it may be sufficient, or even advantageous (cf. ref.

[4] C. de Duve, J. Berthet, and H. Beaufay, *Progr. Biophys. Biophys. Chem.* 9, 325 (1959).

[5] W. C. Schneider, G. H. Hogeboom, and M. J. Striebich, *Cancer Res.* 13, 617 (1953).

[6] E. Reid, letter in *Subcell. Biochem.* 1, 217 (1972).

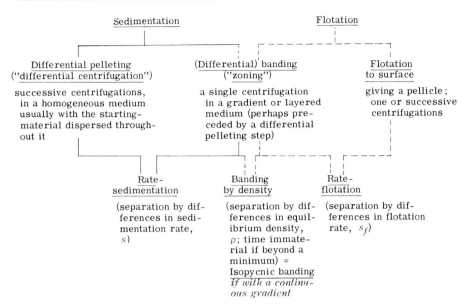

Fig. 1. Centrifugal methods for separating components from a mixture. With some mixtures a hybrid-type separation (S-ρ) may occur, some components being separated on the basis of rate of movement and others on the basis of isopycnic equilibrium. The term "zonal centrifugation" sometimes connotes any banding procedure, but is often used in the special sense of banding with a zonal rotor. The particular unit used for sedimentation rates (strictly speaking, $s_{20,w}$) elsewhere in the chapter is the Svedberg (S), which has the dimensions 10^{-13} second.

28 as cited below in connection with bound ribosomes), to have a series of density "steps" rather than a continuous gradient. In this case a component will ultimately band at a particular interface when the next layer is denser than the particle, giving *density-dependent banding*. With a continuous gradient of suitable steepness, the equilibrium position will be at the density which exactly matches that of the particle, giving *isopycnic banding*. With dense particles such as ribosomes, true isopycnic banding is not possible with a sucrose gradient. Commonly a shallow sucrose gradient is used, and separation is by rate-dependent banding with only a small dependence on density.

In banding runs it is often advantageous to place a dense *cushion* at the bottom of the tube or chamber to band material that would otherwise become pelleted and be difficult to recover quantitatively. Although a theoretical distinction is made between isopycnic and rate sedimentation, most separations in sucrose depend upon a combination of particle density and size.

With lipid-rich material of low density, a *flotation* procedure may be advantageous, or even obligatory, as in the case of serum lipoproteins of the LD or VLD types. For flotation work with low-density lipoproteins it is particularly important to use a rotor giving a high RCF (relative centrifugal force) value, as the gradient materials of low molecular weight commonly used in lipoprotein separations, such as NaBr, diffuse rapidly, and the high force is required to preserve the gradient. In sedimentation equilibrium banding with a salt medium, the same consideration obtains, the shape and steepness of the gradient being dependent on the rotor speed.

It should be stressed that certain elements, notably microsomes, are actually defined in terms of their sedimentation behavior. Others, such as mitochondria, are more rigorously defined, and the term "mitochondrial fraction" is more appropriate for a component isolated by differential pelleting until the absence of contaminants has been demonstrated by morphological studies and/or by assays for marker enzymes or other organelle components. The same is, of course, true for nuclei and lysosomes.

Centrifugal Tools and Media

Conventional Rotors

Most centrifugal procedures have been adapted for both swing-out and angle rotors. Where the fraction to be isolated may give rise to a loose pellet (e.g., nuclei mixed with unbroken cells and cell membranes), swing-out rotors are preferable, as material in angle rotors reorients when the rotor stops and the pellet is disturbed. Swing-out rotors also are often used for high speed, rate-dependent sucrose gradient centrifugation (e.g., polysomes, or RNA). Although isopycnic banding is commonly performed in swing-out rotors using sucrose, NaBr, or CsCl at high speeds, it is also possible to use angle rotors, and in the case of DNA, the resolution obtained is superior.[7] Sucrose gradients are remarkably resistant to disturbance during reorientation or changes in rotor speed, but salt gradients (e.g., CsCl) are easily upset. Therefore they are sometimes stabilized by the addition of sucrose or some other viscous component.

Sucrose gradients are normally formed by a gradient-making device and introduced into the tube prior to running, or in the case of the zonal rotor, pumped directly into the rotor. Salt gradients, on the other hand, are often formed by the centrifugal field during a run, and therefore centrifugation may extend over several days while the gradient is established and

---

[7] W. G. Flamm, M. L. Birnsteil, and P. M. B. Walker, *in* "Subcellular Components: Preparation and Fractionation" (G. D. Birnie, ed.), 2nd ed., p. 279. Butterworth, London, 1972.

components migrate to their equilibrium positions. There are commercially available "gradient engines" for making sucrose gradients of any desired shape, which are particularly useful for zonal runs or for sophisticated equivolumetric or isokinetic gradients. Most laboratories construct their own gradient makers for small swing-out tubes,[8] although diffusion overnight in a two-layer system often gives as linear and reproducible a gradient.

Angle rotors are preferred for preparative work, since greater volumes can be accommodated. When centrifugation is through a gradient, the horizontal path to the wall is essentially the effective centrifugal path length. Once a particle reaches the wall, it may adhere to, or more commonly may migrate to, the foot of the tube. In either case, a certain amount of gradient disturbance is inevitable. Therefore angle rotors are more favored for differential pelleting than for banding.

Whatever the rotor, each laboratory must establish empirically the optimal time and speed (including acceleration/deceleration rates) for any given purpose, and adhere to these rigidly unless it is proved that conditions are not critical. Particularly in the case of differential pelleting, this is essential so that other laboratories can reproduce reported results. The product of the time and gravitational field (g-min) is useful for comparing different centrifuges, although, strictly speaking, the integrated value of acceleration and deceleration should also be incorporated, particularly for short run times. However, even when sedimentation times and forces are equal, different rotors may give different results, and each technique must be verified if a new type of rotor is used.

Zonal Rotors

Zonal rotors are single-chamber rotors of high capacity suitable for banding separations, and can usually be run in conventional centrifuges. Loading and unloading are normally performed through a seal head apparatus,[1] which in high speed models must be removed before acceleration from the idling speed. Loading, particularly of the sample, and unloading are usually performed at a low, idling speed for gradient stability, although newer rotors have been developed that permit static loading (reorientating gradient, or "reograd," rotors, commercially available).

The batch-type zonal rotors fall into three major classes: (1) A-type, 5000 g at edge, contents viewable during the run; used for whole cells, nuclei and (at top speed) mitochondria. (2) HS-type (e.g., Z-XV), 13,000 g, viewability; lysosomes, mitochondria. (3) B-type, up to 165,000 g; mostly used for ribosomes, RNA, DNA, proteins; cannot be

---

[8] *Exempli gratia:* J. Edwards and A. P. Mathias, *Nature (London)* **199**, 603 (1963).

used easily with particles as large as mitochondria; will separate particles down to 5 S; efficacious for microsomes.

There are also continuous flow zonal rotors, which have been extensively used for purifying viruses from cell lysates.

In addition to literature on zonal centrifugation,[1,9-11] any experimenter contemplating purchasing a zonal rotor should consult manufacturers (e.g., Beckman-Spinco, M.S.E., I.E.C., Heraeus-Christ, Sorvall) to determine which of the rotors is most suitable. Although the use of zonal rotors is still regarded as an expert technique, the increased resolution and capacity should lead to its consideration for most centrifugal problems.

Ancillary Equipment

Gradient centrifugation calls for ancillary equipment which can advantageously be "on-line," with short, narrow-bore flow lines, as for the equipment clustered round the zonal rotor in Fig. 2. Programmed gradient makers are available from several manufacturers (e.g., Beckman-Spinco, M.S.E., L.K.B., Buchler/G.D. Searle), and also can be constructed quite easily in a workshop. Any pump of variable flow rate which is not prone to strong pulsation is suitable for pumping solutions into and out of tubes or zonal rotors; in general, peristaltic pumps are used, as the tubing can be easily cleaned and sterilized. Obviously the flow rate of the pump will have to be appropriate for the tube or rotor volume.

If continuous monitoring of some relevant parameter of the solution is feasible, this is very desirable; for instance, analysis of polysome profiles is facilitated when the optical density is continuously monitored. Optical density can be followed using any spectrophotometer with a small-volume flow cell. (Such cells are commercially available, those with variable path length being particularly suitable, but any competent glassblower can make a quartz window flow cell quite easily.) Gradient inversions (points at which a gradient of increasing density flows downward) should be avoided if possible, as turbulence can result. Most spectrophotometers of narrow slit width are suitable for absorbance monitoring. Where the density profile of gradients also has to be determined, this can be done on-line with a flow cell, on the basis of refractometry[12] or of attenuation of radioactivity.[13]

[9] E. Reid (ed.), "Separations with Zonal Rotors." Wolfson Bioanalytical Centre, University of Surrey, Guildford, 1971.
[10] E. Reid (ed.), "Methodological Developments in Biochemistry, Vol. 3: Advances with Zonal Rotors." Longmans, Green, New York, 1973.
[11] G. D. Birnie (ed.), "Subcellular Components: Preparation and Fractionation." Butterworth, London, 1972.
[12] Among commercially available flow refractometers, one claimed to be suitable for zonal rotors is marketed by Winopal (Isernhagen, West Germany).
[13] R. S. Atherton, I. S. Boyce, C. G. Clayton, and A. R. Thomson, p. 71 in ref. 10; see also, J. Cope and H. R. Matthews, p. 77 in ref. 10.

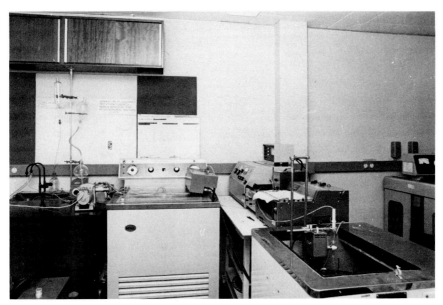

Fig. 2. "On-line" equipment grouped round a zonal rotor (in the centrifuge) in the laboratory of one author (E. R.). Gradient liquid in the round mixing flask with a magnetic stirrer passes through the pump in front of the stirrer and thence into the rotor, normally via a flow refractometer. The latter is shown resting on the top of the centrifuge, ready to deal with the pump-out at the end of the run. The effluent is pumped through the flow cell of a spectrophotometer which is mounted on a trolley together with a 2-pen recorder. Thence the effluent passes to a fraction collector in a cooled cabinet, likewise mobile, the turntable movement being regulated by the effluent volume as sensed in the control device which protrudes above the cabinet. An event marker enables each turntable movement to be recorded by a spike on the same chart as for the refractometer and spectrophotometer traces.

Simple devices for tube piercing and unloading are commercially available and also are simple to build.[7] The contents of tubes can be unloaded from the bottom by dripping or pumping through a narrow tube inserted through the solution or into the bottom. Alternatively the contents can be floated out using a solution more dense than that at the bottom of the tube. Resolution appears almost equivalent, but contamination of tubing, etc., can occur, and the trailing edge tends to be broader than the leading edge owing to drag and turbulence.

Media

Factors affecting the choice of the support medium for centrifugation have been summarized by Cline[1] and other authors. Regard must be paid not only to its intrinsic properties in relation to those of the organelle, but also to the possibility that impurities present at very low concentrations (as in analytical reagent grade chemicals) can be present in media in appreciable amounts because of the high concentrations used in gradients. For instance, in 50% sucrose solution the lead contamination could be as high as 0.5 mg/liter, certainly high enough to inactivate some enzyme preparations. Assay results may have to be corrected for interference due to sucrose itself.[14] Similar considerations obtain for cesium chloride and cesium sulfate. Certain commercially available polymers may have preserving agents present, without mention of this in the specifications (e.g., Ludox, a high viscosity silica gel preparation).

Organic compounds such as sucrose or bovine serum albumin easily become contaminated with bacteria, or may contain enzyme activities deleterious to particular elements. Sucrose in particular often contains ribonuclease, and plant ribonucleases are very resistant to heat inactivation. Autoclaving of sucrose in any case causes variable hydrolysis, with loss of viscosity. A recommended method, albeit not infallible,[15] for inactivating ribonuclease in sucrose (and also eliminating contamination in tubing used for loading and unloading tubes and rotors) is a pretreatment with diethylpyrocarbonate to 0.1%. This compound irreversibly interacts with nucleic acids as well as denaturing proteins, and therefore should be permitted to decompose entirely to carbon dioxide and ethanol before coming into contact with biological preparations or humans. "Proteinase K" has been used in a recent alternative approach.[15]

Gradient Shape

In both isopycnic and rate banding, the shape of the gradient determines the ultimate position of the cellular components and the resolution obtained. Fortunately, comprehensive reviews have recently appeared discussing the factors affecting resolution in isopycnic centrifugation of nucleic acids[6] and in rate sedimentation in zonal rotors.[16] The size and density of the sample obviously affects resolution, as does the rate of sample introduction in the case of zonal rotors. When tubes are used for centrifugation, gradients linear with volume are also linear with distance from the rotor center. This is not true for zonal rotors. A number of different types of convex gradients have been proposed, all of which have a large capacity for sample at the top of the gradient. Once the mixture migrates and separates into a number of zones, less capacity is required and the gradient can become shallower, and thus capable of greater resolution.[10]

Isokinetic gradients (in which the distances migrated by the particle zones are proportional to time and to the sedimentation coefficients of the

[14] R. H. Hinton, M. L. E. Burge, and G. C. Hartman, *Anal. Biochem.* **29**, 248 (1969).
[15] U. Wiegers and H. Hilz, *Biochem. Biophys. Res. Commun.* **44**, 513 (1971).
[16] G. D. Birnie, in ref. 10, p. 17.

particles) are convenient for estimating sedimentation coefficient in tubes and may also give particle zones of constant width.[17] Such gradients give excellent resolution, and can be constructed with simple, commercially available equipment (G. D. Searle). For zonal rotors, where different geometrical considerations apply, "equivolumetric" gradients have been recommended.[18]

Since sucrose gradient centrifugation of cell components often entails a hybrid of rate and isopycnic separation principles, the choice of conditions for complex mixtures invariably must be to some extent a matter of trial and error. Increase in resolution and band broadening both result from the use of shallow gradients. This is most clearly seen in the case of isopycnic separations, where the size of the components and the time of centrifugation also determine resolution. The mean density of the medium should approximate to the density of the particle of major interest, and the centrifugation speed be sufficient to space out the different molecules or organelles. When equilibrium is attained, time no longer matters.

Particular Separations

The following survey of methods is not systematic. Brief consideration is given to nucleic acids, but not to other macromolecules, such as macroglobulins, or to enzyme aggregates and immune complexes, although preparative rotors do have a limited applicability to the separation of such materials.[9,10]

Cells and Nuclei

In common with other starting material, cells can be separated by size or density. Since cell aggregation would alter both of these parameters, and in particular cell size, it is important to ensure that only single cells are present in the separation mixture. For this reason, centrifugation has been particularly useful in the separation of blood cells and cells disaggregated with trypsin.

Red blood cells from anemic chickens contain a mixture of erythrocytes and reticulocytes. The reticulocytes contain larger nuclei which are less pycnotic than those of the erythrocyte. Although the separation of the cell types is only partial, the nuclei from disrupted reticulocytes and erythrocytes can be separated completely by isopycnic centrifugation for 10 hours at 3200 rpm in a 53–63% sucrose gradient containing 1 mM MgCl$_2$, pH 7.2. The reticulocyte nuclei have a banding density of 1.287 g/cm^3 while those from erythrocytes band at 1.294 g/cm^3.[19]

[17] H. Noll and E. Stutz, this series, Vol. 12B, p. 129.
[18] M. S. Pollack and C. A. Price, *Anal. Biochem.* **42**, 38 (1971).
[19] I. R. Johnston and A. P. Mathias, in ref. 11, p. 53.

For isolating nuclei from liver without regard to ploidy or cell origin, an approach due to J. Chaveau is effective. With prolonged centrifugation in dense sucrose (2.2–2.3 M) containing a divalent cation, nuclei are recoverable as a pellet from a liver homogenate which is initially present either throughout the tube or as a top layer; other elements including unbroken cells float up or, particularly if an angle rotor is employed, adhere to the wall of the tube.

In most tissues cells are in a varying state of ploidy, depending upon which stage of the cell cycle they are in at the time of isolation. The amount of nuclear DNA affects the size of the nucleus, and its density, and therefore its behavior on ultracentrifugation. Mathias and his colleagues have been able to separate diploid, tetraploid, octaploid, and hexadecaploid nuclei from adult albino mice of the ANIH strain using a sucrose gradient with an A-XII zonal rotor. Nuclei from stromal and parenchymal cells are also separated in this system, as their shapes differ.

The density of the nuclei may be calculated by extrapolation of the rate of movement in sucrose solutions of differing concentration, as can the proportion of the volume of the nucleus which is impermeable to sucrose.[19] Such nuclear separations permit the analysis of enzymatic activity in the different size classes separated. Johnston and Mathias have discussed at some length the applications of these techniques, and their results for DNA polymerase and RNA polymerase indicate that in nuclei of different ploidy the amount of enzyme per genome remains approximately constant.[19]

Not only is it possible to isolate nuclei centrifugally, but high molecular weight DNA can be isolated directly from cells as well. Sambrook et al.[20] dissolved SV40-infected 3T3 cells in sodium dodecyl sulfate solution and centrifuged on an alkaline sucrose gradient. After centrifugation for 5 hours at 25,000 rpm, the viral DNA banded in the middle of the tube free from cellular contaminants. The particular value of this direct technique is that it permits isolation of completely undegraded viral DNA molecules. It illustrates the particular value of one-step techniques that do not entail preliminary cell lysis by homogenization or hypotonic medium, since such breakage often liberates degradative enzymes which hydrolyze cellular components.

The buoyant density of DNA depends upon its base composition and to a lesser extent on its secondary structure. Analytical analysis of DNA by banding in cesium salts has recently been extensively reviewed by Flamm et al.[7] Similar analysis is also possible in less expensive sodium iodide.[21] Not only the strandedness of DNA affects buoyant density, but also whether

[20] J. Sambrook, H. Westphal, P. R. Srinivasan, and R. Dulbecco, *Proc. Nat. Acad. Sci. U.S.* **60**, 1288 (1968).

[21] R. Anet and D. R. Strayer, *Biochem. Biophys. Res. Commun.* **37**, 52 (1969).

it is in a linear, circular, or supercoiled conformation. Agents such as ethidium bromide, which bind preferentially to linear DNA molecules, increase the separation between them and circular and supercoiled forms. Such techniques have been reviewed in this series.[22]

If DNA is to be used for subsequent molecular hybridization, it is desirable that a particular size class be isolated. This can be accomplished well in a zonal rotor.[23] In the case of DNA, unlike RNA, there are few molecular species of accurately known size, and therefore molecular weights are normally calculated by reference to the standard formulae of Studier.[24]

Ribonucleoprotein Particles and RNA

Ultracentrifugation has always been the method of choice for characterization of these components, usually by rate banding on a sucrose density gradient. The procedures have been reviewed extensively[9–11,17] and are familiar to most workers in the field. Only a few technical points will be discussed below.

If 18 S and 28 S RNA are to be separated in a swing-out rotor, e.g., Spinco SW 39, typically this is run for 4 hours at 39,000 rpm with 15% to 30% (w/w) linear sucrose gradient[25]; optionally there may be a dense cushion at the bottom of the tube. For polysome profile analysis in tubes, the centrifugation time is of the order of 1 hour.

Resolution of RNA and nucleoprotein by rate sedimentation depends upon sample volume and gradient shape to a greater extent than for most centrifugations, since the species being studied are in general of defined size rather than a broad spectrum of sedimentation coefficients (as in the case of mitochondria, or sheared DNA). Therefore a gradient can be constructed quite specifically to give resolution in the region of interest, and other components compressed, pelleted, or floated. It is also not always appreciated that molecules down to a sedimentation value of approximately 10 S can be sedimented at 500,000 g in dilute media overnight, and this can represent a nondestructive way of concentrating molecules, such as minor RNA species.

Isopycnic techniques using CsCl have been applied to nucleoproteins after fixation with formaldehyde.[26] The fixation links the RNA and protein, and the RNA is then not easily detachable. Attempts have been made

[22] W. Szybalski, this series, Vol. 12B, p. 330.
[23] A. Hell, G. D. Birnie, T. K. Slimming, and J. Paul, *Anal. Biochem.* **48**, 369 (1972).
[24] F. W. Studier, *J. Mol. Biol.* **11**, 373 (1965).
[25] E. D. Whittle, D. E. Bushnell, and V. R. Potter, *Biochim. Biophys. Acta* **161**, 41 (1968).
[26] A. S. Spirin, *Eur. J. Biochem.* **10**, 20 (1969).

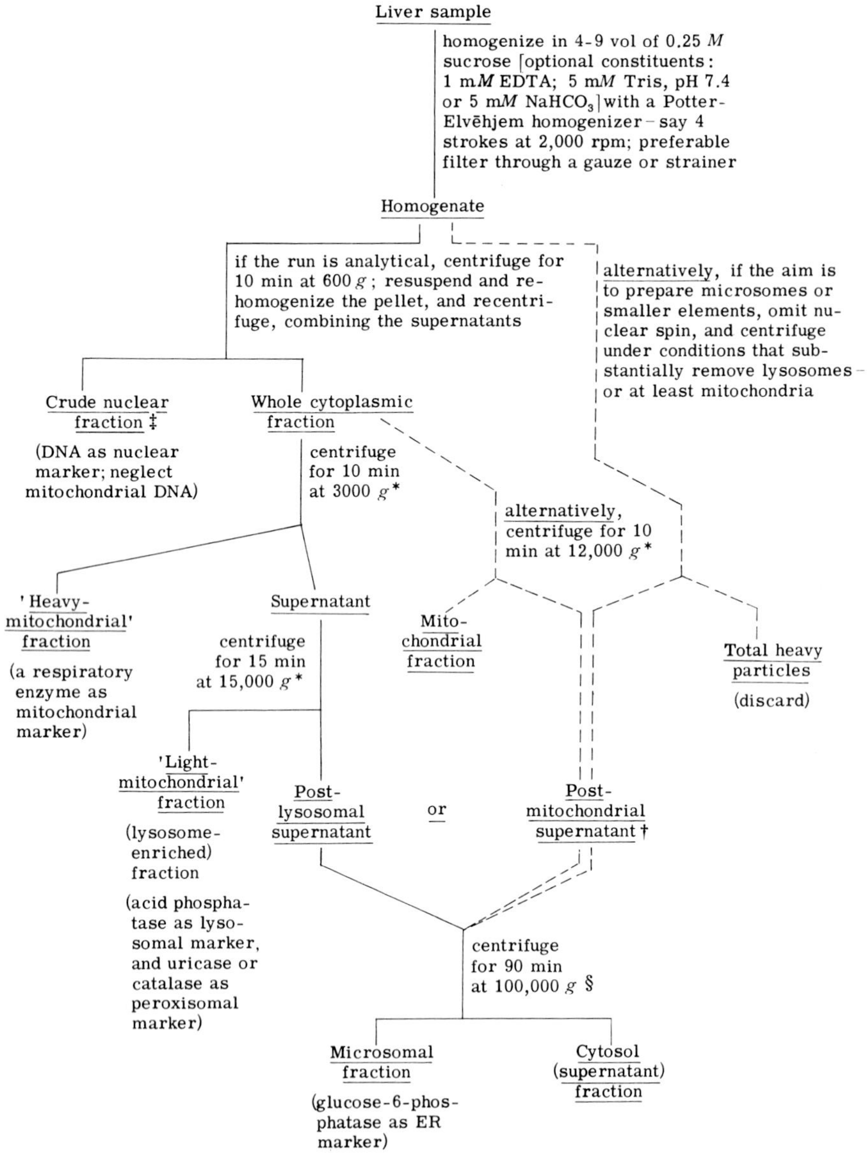

Liver sample

homogenize in 4-9 vol of 0.25 M
sucrose [optional constituents:
1 mM EDTA; 5 mM Tris, pH 7.4
or 5 mM NaHCO$_3$] with a Potter-
Elvēhjem homogenizer – say 4
strokes at 2,000 rpm; preferable
filter through a gauze or strainer

Homogenate

if the run is analytical, centrifuge for
10 min at 600 g; resuspend and rehomogenize the pellet, and recentrifuge, combining the supernatants

alternatively, if the aim is
to prepare microsomes or
smaller elements, omit nuclear spin, and centrifuge
under conditions that substantially remove lysosomes –
or at least mitochondria

Crude nuclear
fraction ‡

(DNA as nuclear
marker; neglect
mitochondrial DNA)

Whole cytoplasmic
fraction

centrifuge
for 10 min
at 3000 g*

alternatively,
centrifuge for 10
min at 12,000 g*

'Heavy-
mitochondrial'
fraction

(a respiratory
enzyme as
mitochondrial
marker)

Supernatant

centrifuge
for 15 min
at 15,000 g*

Mito-
chondrial
fraction

Total heavy
particles

(discard)

'Light-
mitochondrial'
fraction

(lysosome-
enriched)
fraction

(acid phospha-
tase as lyso-
somal marker,
and uricase or
catalase as
peroxisomal
marker)

Post-
lysosomal
supernatant

or

Post-
mitochondrial
supernatant †

centrifuge
for 90 min
at 100,000 g §

Microsomal
fraction

(glucose-6-phos-
phatase as ER
marker)

Cytosol
(supernatant)
fraction

Fig. 3. Differential pelleting ("differential centrifugation") of a liver homogenate, to furnish fractions for an "analytical" study or a fraction to serve as a source of microsomes, polysomes (by deoxycholate treatment at †) or fractions containing smooth endoplasmic reticulum (ER)/rough ER/free ribosomes. In an "analytical" run, pellets marked * should be washed and recentrifuged. Fractions marked ‡ will also contain plasma-membrane fragments (5'-nucleotidase as marker). If polysomes

to separate nucleoproteins of different density (e.g., ribosomes and messenger ribonucleoproteins) using nonionic media such as chloral hydrate or Urografin, and these may provide biologically intact molecules for further experimentation. Metrizamide is a promising alternate medium.

Rate sedimentation, when applied to minor components of an RNA mixture (such as mRNA), is best performed in a zonal rotor, not only for the sake of the considerable increase in capacity, but also because the resolution is greatly improved. RNA components amounting to as little as 0.1% of a mixture can be adequately separated.[27]

Concerning the nature of the starting material for isolation of ribonucleoprotein particles, the centrifugal aspect warrants comment. In anticipation of the following presentation of a "differential pelleting" scheme (Fig. 3), it may be noted that a postmitochondrial (or postlysosomal) fraction is commonly prepared, and treated with deoxycholate, when polysomes together with any monomeric ribosomes have to be isolated from liver. While such an end product will furnish a representative profile for polysomes, it will be biased in favor of polysomes that were originally free as distinct from membrane-bound, since some rough endoplasmic reticulum will have sedimented with mitochondria and lysosomes and been inadvertently discarded. (The integrity of polysomes is jeopardized if lysosomes, which possess latent ribonuclease activity, are still present at the deoxycholate stage.) If, however, the aim is merely to ascertain what proportion of the hepatic ribosome–polysome population is truly free as distinct from membrane bound, a homogenate freed only from nuclei is recommended as the starting material for gradient centrifugation, with no attempt to ascertain the size profile, and no deoxycholate step.[28] In a study[29] of postmitochondrial supernatants as a source of free polysomes (with no deoxycholate treatment), it was found that hold-up of polysomes in the membrane-containing region at the top of the gradient, and hence loss of yield, could be minimized by using fasted rather than fed rats, by centrifuging

[27] R. Williamson, in ref. 10, p. 135.
[28] G. Blobel and V. R. Potter, *J. Mol. Biol.* **26**, 279 (1967).
[29] D. Lowe, E. Reid, and T. Hallinan, *FEBS* (*Fed. Eur. Biochem. Soc.*) *Lett.* **6**, 114 (1970).

or rough ER are being prepared, EDTA should not be added, and a TKM medium is advantageous [D. Lowe, E. Reid, and T. Hallinan, *FEBS* (*Fed. Eur. Biochem. Soc.*) *Lett.* **6**, 114 (1970)] (0.25 M sucrose containing 50 mM Tris, pH 7.5 at 20°, 25 mM KCl in 5 mM MgCl$_2$). The g values represent g_{av}, i.e., values for halfway down the tube, but should be regarded merely as a guide, the exact conditions to depend on the rotor and on the purpose of the experiment. For the centrifugation denoted § a g value of the order of only 25,000 will often be sufficient [E. Reid, *in* "Enzyme Cytology" (D. Roodyn, ed.), Academic Press, New York, 1967].

for as long as 20 hours, and by starting with a dilute supernatant (preferably under 0.25 g of tissue equivalent per milliliter).

Membrane-Bounded Elements

In the following survey the emphasis is on rodent liver (as in a background article by E. Reid[3]). Three features of liver that are relevant to centrifugation of tissue preparations are the presence in variable amount of ferritin granules which tend to contaminate 60 S subribosomal particles, the availability of a useful "marker" (glucose-6-phosphatase) for endoplasmic reticulum fragments, and the presence in fed animals of glycogen granules that sediment at a comparable rate to the latter. When the endoplasmic reticulum fragments are harvested as a microsomal pellet, plasma membrane fragments will also be present. In part, however, the latter appear in the crude "nuclear" pellet which is the first fraction to be obtained in the following scheme (Fig. 3), and which also contains some endoplasmic reticulum fragments.

Differential Pelleting. Typical conditions for centrifuging a liver homogenate are shown in Fig. 3, which is applicable even if the desired separation hinges on subsequent use of a gradient technique. (The latter would give poor resolution if applied directly to a whole homogenate, unless the load were kept low.) The first ('nuclear') pellet still represents a crude mixture of elements, and the best 'lysosome-rich' fraction that is obtainable by differential pelleting may have a lysosome content of less than 10% by weight. Although differential pelleting in a uniform medium represents a pioneer technique[3,5] that will continue to occupy an important place in centrifugal methodology, its effectiveness is inherently limited because of insufficiently sharp differences among different organelles in sedimentation rate.

Properties of Subcellular Components Relevant to Centrifugal Separation. Data for various components are listed in Table I, concerning the differences in sedimentation rate among membrane-bounded elements from liver. Evidently certain separations might be achieved by rate-dependent banding (cf. Fig. 1), although differential pelleting would be predictably inefficient. As an alternative approach, or as a principle applicable to only some of the components in a mixture subjected to an S-ρ hybrid separation,[1] certain separations may be achievable on the basis of differences in equilibrium density.

The values given in Table I are not "constants," being influenced by the history and environment of the tissue material. As a rough generalization, membrane fragments as distinct from intact organelles have a low density because of the high phosphatide content, although this is offset by the high density of the bound ribosomes in the case of rough endoplasmic

TABLE I. FEATURES OF SUBCELLULAR RELEVANT TO THEIR SEPARATION BY BANDING[a]

Component	Value for S	Value for ρ at equilibrium	Remarks
Protein	<50	1.2	Macroglobulins (ca. 19 S) can be isolated with a zonal rotor, on rate basis
RNA	<50	2.0	4 S RNA barely moves even at 100,000 g
Ribosomes (monomers)	80	1.4	Isolated on rate basis—likewise for subribosomal particles (ca. 40 S) and polysomes; "fixed" ribosomes have $\rho = 1.16$ in CsCl
DNA, in CsCl	10^2	1.7 (e.g.)	
Membrane fragments	10^2–10^4		Rate separation is exceptional
Golgi		1.10–1.13	Lysosomal membrane may likewise have $\rho = 1.13$
Smooth ER		1.15	Higher ρ if 5 mM Mg^{2+} present
Plasma membrane		1.17	Lower ρ if from perfused liver
Mitochondrial membranes		Outer, 1.16; Inner, 1.19	
Rough ER		1.22	Unaffected by Mg^{2+}
Lysosomes	10^4	1.20–1.22	Those richest in acid phosphatase have the lower ρ
Peroxisomes		1.23	
Mitochondria	5×10^4	1.18	S not so high as to preclude rate separation
Nuclei	>10^6	1.3	ρ depends on ploidy and on cell origin
Cells	10^8	Up to 1.10	Often isopycnic separation (e.g., in Ficoll)
Synaptosomes		1.15	
Muscle microsomes	5×10^2	1.05–1.2	"Relaxing particles" have ρ near 1.2
Eukaryotic microorganisms			
Mitochondria		1.18–1.22	Peroxisomes similar
Lysosomes		1.13–1.19	Heterogeneous populations
Chloroplasts		<1.15	Hypertonicity undesirable

[a] The values usually refer to liver and to sucrose medium and are not to be regarded as precise or invariable. For documentation other sources may be consulted: for example, D. B. Roodyn (ed.), "Enzyme Cytology," Academic Press, New York, 1967; C. de Duve, J. Berthet, and H. Beaufay, *Progr. Biophys. Biochem. Chem.* **9**, 325 (1959); E. Reid (ed.), "Separations with Zonal Rotors," Wolfson Bioanalytical Centre, University of Surrey, Guildford, 1971; E. Reid (ed.), "Methodological Developments in Biochemistry, Vol. 3: Advances with Zonal Rotors," Longmans, Green, New York, 1973; G. D. Birnie (ed.), "Subcellular Components: Preparation and Fractionation," Butterworth, London, 1972.

reticulum. Evidently reliance can be placed merely on density differences for the separation of smooth from rough microsomes, and even, with care, of mitochondria from plasma membrane fragments. It must be remembered that there may have been some inadvertent damage to delicate organelles, e.g., during homogenization; hence it may be important to examine the products, suitably by marker enzyme assays, not only for the desired elements but also for possible contaminants such as fragments of mitochondrial membranes, taking into account their expected density and sedimentation rate.

Small-scale separations with adequate resolution can often be achieved in tubes, possibly with layers, as distinct from a continuous gradient. Since, however, zonal rotors have marked advantages, the following examples of centrifugal runs have been based on the use of appropriate zonal rotors, each especially or uniquely suitable for the particular separation. The following comments supplement rather than duplicate the legends to the figures, yet barely touch on the important matter of the fate of components of the starting-material other than those evident from the graphs. Nor is consideration given to the usefulness of examining the isolated fractions by cytochemical "staining" for appropriate phosphatases.[30–32]

Use of a B Rotor to Separate Microsomal Membrane Fragments. The run shown in Fig. 4 is presented not as a model separation, but rather as a pointer to difficulties that may be encountered in devising conditions. Fragments of rough endoplasmic reticulum are obtainable in reasonable purity if the medium is free of Mg^{2+} ions, as in the run depicted. The smooth membrane fraction is, however, contaminated not only by ribosomes, but also by plasma membrane fragments as judged by assays for 5′-nucleotidase. (The latter may be present to some extent in endoplasmic reticulum membranes as well as in the plasma membrane,[32] but probably not to a sufficient extent to invalidate its use as a plasma membrane marker.) Approaches to the difficult problem of separating different types of smooth vesicle have been discussed elsewhere.[3,33] Partial separation, albeit rather ineffective, can be achieved either by using a resuspended microsomal pellet as starting material and applying it to the dense end of the gradient in flotation run, or by performing a run of the type shown in Fig. 4, but with 5 mM Mg^{2+} present. In the latter case the smooth endoplasmic reticulum fragments are mainly associated with the rough in the peak near the cushion. This also holds for the most effective solution so far found, where the procedure of Fig. 4 is varied by sonicating a resus-

[30] A. A. El-Aaser, in ref. 9, B-5.
[31] A. A. El-Aaser and S. J. Holt, in ref. 10, p. 85.
[32] C. C. Widnell, *J. Cell Biol.* **52**, 542 (1972).
[33] R. H. Hinton, K. A. Norris, and E. Reid, in ref. 9, S-2.

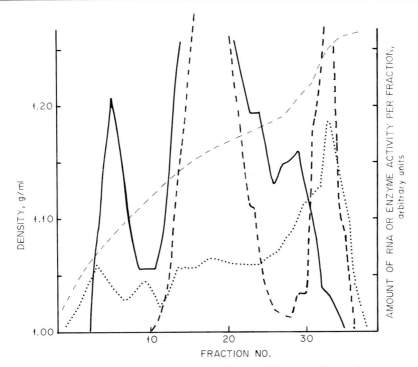

Fig. 4. Zonal rotor run on a liver postlysosomal fraction, illustrating some principles in the separation of membrane fragments (Courtesy of Mr. K. A. Norris). An
amount of the fraction (cf. Fig. 3) equivalent to 8 g of liver was centrifuged for
16 hours at 21,000 rpm in an aluminum B-XV rotor. (With a titanium B-XIV rotor,
the time could have been only 2–4 hours, with benefit to the sharpness of the peaks.)
The operational procedures for the runs shown in Figs. 4–7 were as described elsewhere [G. B. Cline, this series, Vol. 22, p. 168]; the gradient shape was verified
refractometrically during the final pumping out. "Soluble" components, such as tRNA,
have hardly moved from the sample position. The RNA plateau in the central region
is attributable to sedimenting polysomes and smaller ribonucleoprotein particles, including monomeric ribosomes resulting from inadvertent degradation of polysomes in
the Mg^{2+}-free medium. The absence of Mg^{2+}, however, has enabled smooth and rough
endoplasmic reticulum fragments (glucose-6-phosphatase as marker) to be separated
isopycnically; the RNA peak near the cushion is due to the membrane-bound ribosomes. The central peak for 5′-nucleotidase signifies that plasma membrane fragments
are present in the smooth endoplasmic reticulum fraction, although with different
conditions (see text) some separation could have been achieved. — — —, Density;
- - - -, glucose-6-phosphatase; ......, RNA; ————, 5′-nucleotidase.

pended microsomal pellet in the presence of 0.75 mM Pb^{2+}; a 4-hour run at
47,000 rpm in a B-XIV rotor gives a peak at $\rho = 1.14$ containing some of
the 5′-nucleotidase activity (most of the remainder being in the rough endoplasmic reticulum peak) with little glucose-6-phosphatase activity.

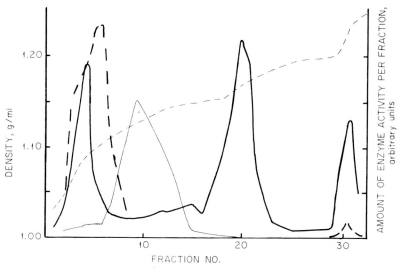

FIG. 5. Zonal rotor run on a crude nuclear fraction from perfused liver, primarily intended to separate plasma membrane fragments isopycnically. (Courtesy of Dr. R. H. Hinton.) An amount of the fraction (cf. Fig. 3) corresponding to 20 g of perfused liver was centrifuged for 1 hour at 3700 rpm in an A-XII rotor. This gave a plasma-membrane peak at tube 21 and a nuclear peak at the cushion (marked by DNA, and associated with endoplasmic reticulum fragments containing glucose-6-phosphatase; assays not shown). Mitochondria, still sedimenting, are almost clear of the sample position, in contrast with lysosomes (acid phosphatase as marker). Erythrocyte contamination of the plasma membrane fragments was minimized by use of a shallow gradient in this region, at the expense of peak sharpness and of freedom from mitochondrial contamination (succinate dehydrogenase as marker). Although in the run depicted the starting material was exposed to hypotonic conditions to reduce the number of intact erythrocytes, such exposure is not necessary. [R. H. Hinton, M. Dobrota, J. T. R. Fitzsimons, and E. Reid, *Eur. J. Biochem.* **12**, 349 (1970). - - - -, Density; — — —, acid phosphatase; ———, succinate dehydrogenase; ———, 5'-nucleotidase.]

It is important to note that fractions described in the literature that supposedly represent smooth endoplasmic reticulum will normally contain plasma membrane fragments also. No method has so far been discovered to furnish smooth and rough endoplasmic reticulum fragments and plasma membrane fragments as three side-by-side products. There is, moreover, the problem of the fate of other types of membrane fragments—not only mitochondrial and lysosomal, but also Golgi. Procedures specially aimed at isolating Golgi fragments are now available, as cited elsewhere[33,34]; one is outlined later in this section.

Use of an A-XII Rotor to Separate Large Fragments of Plasma Membrane. The plasma membrane sheets and large vesicles that pellet in the

[34] R. W. Mahley, R. L. Hamilton, and V. S. Lequire, *J. Lipid Res.* **10**, 433 (1969).

crude nuclear fraction can be separated either in tubes or, as illustrated in Figs. 5 and 6, in a low-speed zonal rotor. The problems involved, and the methodology to overcome them, have been set down elsewhere,[11,33,35] together with a recommended method that obviates the customary exposure of the starting material to a hypotonic environment. It will be noted that

Fig. 6. Appearance of the contents of an A-XII rotor in a run comparable to that shown in Fig. 5, at the start of the pumping out. The band at the periphery contains nuclei and unbroken cells. There is a faint band due to plasma membrane fragments, and a heavy band due to mitochondria. The innermost band represents material, largely microsomal, which has not moved from the initial position of the sample. (Courtesy of Dr. R. H. Hinton, Mr. M. Dobrota, and Mr. C. Aggett; taken on fast film with the aid of the light in the centrifuge bowl.)

[35] R. H. Hinton, M. Dobrota, J. T. R. Fitzsimons, and E. Reid, *Eur. J. Biochem.* **12,** 349 (1970).

mitochondria can be isolated simultaneously, although not lysosomes—the latter having hardly moved from the origin. The plasma membrane product can be further purified by flotation in tubes.

Use of an HS Rotor to Separate Lysosomes. The isolation of lysosomes poses severe problems,[9,11] as is evident from a classical paper concerned with separation of lysosomes and other organelles including peroxisomes by banding in tubes.[36] It is evident from the zonal rotor run shown in Fig. 7 that even with the stratagem of injecting Triton-X-100, which makes the granules containing acid phosphatase sediment faster, the lysosome peak was broad, reflecting heterogeneity in the granule population. At least, however, the lysosomes were substantially free of peroxisomes, and mitochondria were obtained as a separate peak. Material recovered from a lysosomal peak, such as is shown in Fig. 7, may be partly resolved into "subpopulations" by recentrifugation under different conditions.

Separation of Golgi Fragments. The fate of Golgi fragments in zonal rotor runs such as that shown in Fig. 4 is not clear. The special methods recently developed for Golgi fragments are of interest in the present connection only insofar as they illustrate centrifugal methodology. In one method[34] a homogenate in $0.5 M$ sucrose containing 1% dextran, 5 mM $MgCl_2$, and 37.5 mM Tris-malate was subjected to a differential pelleting procedure; the upper one-third of the stratified pellet obtained by centrifugation for 20 minutes at 25,000 rpm in a swing-out rotor was resuspended, and centrifuged on a discontinuous gradient, giving a Golgi fraction at the upper surface of the $1.25 M$ sucrose layer. The particles were harvested by dilution (1:1) with $0.15 M$ NaCl and centrifugation at 25,000 rpm for 40 minutes, and were then suspended in a salt or buffer solution and subjected to freeze-thawing and sonication so as to liberate lipoprotein particles; these were recovered at $\rho = 1.006$ by centrifugation for 26 hours at 100,000 g.

Membrane-Bounded Elements from Sources Other Than Liver. Some of the early literature on the isolation of elements from tissues other than liver does not inspire confidence, not necessarily because of unsuitable centrifugation conditions. A common shortcoming has been inadequate characterization of the product. However, despite inherent difficulties due largely to cellular heterogeneity in the tissue, there are now established methods for various elements, e.g., synaptosomes from brain (cf. Table I) and glomerular fragments from kidney. Progress has also been made with isolating cytoplasmic elements from nucleated microorganisms.

A problem that may be more severe with chloroplasts than with other organelles is their fragility at abnormal tonicities, especially in the case of

---

[36] H. Beaufay, P. Jacques, P. Baudhuin, O. Z. Sellinger, J. Berthet, and C. de Duve, *Biochem. J.* **184**, (1964).

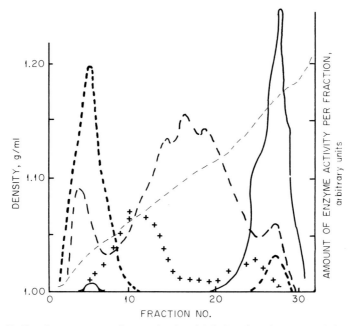

FIG. 7. Zonal rotor on a liver mitochondrial fraction from rats injected with Triton X-100 (2.5 g/kg body weight on each of three successive days) so as to swell the lysosomes and facilitate their separation from other organelles. (Courtesy of Dr. M. L. E. Burge.) An amount of the crude fraction equivalent to 9 g of liver was centrifuged for 45 minutes at 9000 rpm in an HS rotor. Microsomes as identified by glucose-6-phosphatase activity have not moved from the sample position, whereas mitochondria have banded isopycnically near the cushion on the right. Lysosomes are found midway as a broad sedimenting band, with peroxisomes (urate oxidase as marker) at the trailing edge. – – –, Density; — — —, acid phosphatase; - - -, glucose-6-phosphatase; + + +, urate oxidase; ———, succinate dehydrogenase.

an algal source as studied in C. A. Price's laboratory; the problem was overcome by relying on an expensive polymer, Ficoll, to provide an iso-osmotic density gradient.[37] The centrifugation was performed by rate sedimentation in a B-XXXa zonal rotor, run at only 7000 rpm for a time of the order of 15 minutes.

Conditions for various nonhepatic elements will be found elsewhere in this series, and in other literature.[3,9,11] Success has come from intelligent design of conditions for all stages of the procedure (including final characterization) in the light of the properties of the desired elements and of possible contaminants. Published centrifugal conditions have in general been devised on the basis of the principles outlined in the present article.

[37] A. Vasconcelos, M. Pollack, L. R. Mendiola, H.-P. Hoffmann, D. H. Brown, and C. A. Price, *Plant Physiol.* **47**, 217 (1971).

[74] Isolation of Subcellular Organelles of Metabolism on Isopycnic Sucrose Gradients

By N. E. TOLBERT

By buoyant density or isopycnic centrifugation subcellular organelles of a tissue homogenate can be separated on sucrose gradients at densities determined mainly by their protein, lipid, and final bound water composition (see the table). Although the densities for each type of particle will vary somewhat from tissue to tissue, they are remarkably similar throughout the biological world. Only organelles enter the dense sucrose gradient during the centrifugation, and the smaller proteins remain near the top. With less dense gradients for protein separation, the organelles, if present, would be pelleted. Rate separation of organelles by density gradient centrifugation has been used, but will not be considered in this chapter. Equilibrium density is usually reached in 2–4 hours, when using zonal rotors at recommended speeds. If centrifugation is continued too long, particles may rupture and fragments may begin to migrate in the gradient.

Initially sucrose gradients were prepared in 10–55-ml tubes for swinging-bucket rotors, such as the SW 25.2. Anderson[1-3] and others developed zonal rotors, which eliminate edge effects, increase the volume, and adapt to continuous large-scale separations. Procedures for zonal rotors may be scaled down for tube gradients. Differential centrifugation has been used to batch or concentrate organelles, but with the large volume in zonal rotors this may be eliminated to reduce damage and save time. The separation of nuclei, nucleoli, and Golgi bodies are not to be considered, and in fact most of these organelles are removed by an initial low speed differential centrifugation. For nuclei, gradients of higher density are necessary.

Elsewhere in this volume details are presented for zonal rotor and centrifuge design[3] and procedures for running the centrifuge and manipulating of the rotors.[4] In this chapter some of the biological considerations and enzymatic identification of organelles are summarized, and more details are to be found in reviews, particularly by de Duve.[5-7] A large cold chest

[1] N. G. Anderson (ed.), "The Development of Zonal Centrifuges and Ancillary Systems for Tissue Fractionation and Analysis." Nat. Cancer Inst. Monogr. 21 (1966).

[2] N. G. Anderson, Science 154, 103 (1966).

[3] T. D. Tiffany, C. A. Burtis, and N. G. Anderson, this volume [78].

[4] C. A. Price, this volume [52].

[5] F. Leighton, B. Poole, H. Beaufay, P. Baudhuin, J. W. Coffey, S. Fowler, and C. de Duve, J. Cell Biol. 37, 482 (1968).

Marker Enzymes for Particles Separated on Sucrose Gradients

Lysosomes—$d_{10}^0 \cong 1.12$ g $\times$ cm^{-3} for major fraction
$\qquad d_{10}^0 \cong 1.24$–$1.26$ g $\times$ cm^{-3} for minor fraction
  Acid phosphatase, esterases
Microsomes—$d_{10}^0 \cong 1.14$–1.15 g $\times$ cm^{-3}
  Cytochrome c reductase, glucose-6-P dehydrogenase
Broken chloroplasts—$d_{10}^0 \cong 1.14$–1.17 g $\times$ cm^{-3}
  Chlorophyll
Mitochondria—$d_{10}^0 \cong 1.16$–1.18 g $\times$ cm^{-3}
  Cytochrome c oxidase, fumarase, succinic dehydrogenase
Etioplasts, proplastids, whole chloroplasts—$d_{10}^0 \cong 1.22$–1.25 g $\times$ cm^{-3}
  Triose-P isomerase, dihydroxyphenylalanine oxidase, P-glycolate phosphatase
Microbodies—$d_{10}^0 \cong 1.24$–1.26 g $\times$ cm^{-3}
  Catalase, glycolate oxidase, NADH-hydroxypyruvate reductase, malate synthetase, isocitrate lyase
Protein bodies—$d_{10}^0 \cong 1.25$–1.36 g $\times$ cm^{-3}
  Protein peak, protease activity
Starch grains or glycogen—bottom of gradient

adjacent to the zonal facilities is used for storage of rotors, gradient fractions before and after runs, and a fraction collector.

Preparation of Gradients

The density is generally maintained by sucrose, and tables for w/v or percent sucrose, sucrose molarity, and density are available in handbooks.[8] Ficoll gradients have been used, particularly for rate zonal centrifugation,[9] but they are expensive. Such polymers have been added to sucrose gradients to reduce osmotic effects. With sucrose gradients, separation is dependent not only on the particle density, but also upon an osmotic effect, which is less pronounced with Ficoll-sucrose gradients. For instance microbodies in a sucrose gradient lose bound water and have a final density on the gradient higher than mitochondria, but on a Ficoll gradient microbodies are not well separated from other organelles after isopycnic centrifugation. Glycerol gradients have not been used for isopycnic separation, because densities high enough to separate all organelles cannot be obtained.

Sucrose is available in three grades. Concentrated solutions of commercial sucrose are yellowish when prepared and are generally to be avoided. Analytical grade sucrose is inexpensive, but it contains variable

[6] C. de Duve, *J. Cell Biol.* **50**, 200 (1971).
[7] N. E. Tolbert, this series, Vol. 23, p. 665.
[8] A Handbook of Data for Biological and Physical Scientists. Instrumentation Specialties Company, 47000 Superior, Lincoln, Nebraska 68504 (1970).
[9] D. H. Brown, *Biochim. Biophys. Acta* **162**, 152 (1968).

amounts of material absorbing at 260–280 nm. Density-gradient grade sucrose is nearly free of 260–280 nm absorbing material and should be used if possible. A cost-saving operation is to use analytical grade sucrose for the 56–59% sucrose fraction. This large volume at the outer rim of the rotor is displaced by other fractions, or it is used to displace fractions from the core, and particles do not migrate into it. The rest of the gradient ought to be prepared with purest sucrose if nucleic acids and proteins are to be determined by absorbancy, or until it is ascertained that analytical grade sucrose has no effect on the investigation.

Besides sucrose, gradients contain about 20 mM buffers, and the effect of the buffer upon the density is ignored. Because pH electrodes function poorly in dense sucrose solutions, the pH of the gradient fractions is assumed to be that of the buffer. A gradient of sucrose in 20 mM glycylglycine at pH 7.5 is routinely used by us for separation of organelles described in the table. Fractions are prepared the day before use and stored in the cold chest. Sucrose solutions above 35% may be stored in the cold chest for days without spoilage. Large volumes of sucrose solutions may also be stored in sterile culture bottles. Other components often added to gradients are EDTA, various salts, and sulfhydryl compounds. Removal of these factors and sucrose before enzyme assays is not easy. Sephadex G-25 filtration of sucrose gradient fractions is not feasible, for the dense sucrose falls through the usual short columns.

Gradient makers are commercially available for continuous gradients of different profiles[3,4] or simpler designs can be built.[10] Each has some limitations, all are expensive, and their operation at 2° is difficult. A high quality gradient pump with voltage regulator, such as the variable speed Masterflex, is essential. The system must be capable of pumping at a nearly constant rate solutions of rapidly changing densities. Often adequate gradients can be prepared by adding aliquots of different sucrose concentrations directly into the rotor. Although referred to as step gradients, aliquots of 25 ml for the B-30 rotor or 50 ml for the B-29 rotor provides a nearly linear gradient inside the rotor. These fractions are poured into a separatory funnel connected by a short Tygon tubing through the pump to the rotor seal. The fractions may be added nearly as fast as one can measure the aliquots into the funnel, except for the rim pad of dense sucrose. The last 1–2 ml of each aliquot, while still in the funnel, may be mixed with the first few milliliters of the next sucrose concentration in order to blend the interfaces between fractions.

A typical gradient is loaded from the rotor edge by adding some grinding medium which is to be pushed off the gradient when ascertaining that

[10] E. H. McConkey, this series, Vol. 12A, p. 620.

the rotor is full. The number and volume of each sucrose fraction depend on gradient design, on how continuous a gradient is needed, and on rotor capacity and homogenate volume. Generally 10–50 ml fractions may consist of 20%, 25%, 30%, 33%, 36%, 38%, 40%, 41%, 42%, 43%, 45%, 46%, 48%, 49%, 50%, 51%, 52% w/w sucrose. A pad of 56–59% w/w sucrose, equal in volume to the homogenate plus an extra 50 ml, is added last until the core exit line is full. Both core and edge lines are closed with one scissors clamp to exclude air bubbles. The pump is reversed in order to add sample at the core. As the sample is added, part of the sucrose pad is discarded through the exit line. After or simultaneous with gradient preparation, tissue homogenization is done.

Grinding

Most organelles are damaged by the grinding procedure, which therefore must be as brief as possible. A loose-fitting, power-driven Potter-Elvehjem homogenizer with a Teflon pestle is sufficient for liver tissue cut into small pieces and mixed with 2 volumes of medium. The pestle should be pushed down and up only once for recovery of microbodies. Several passes may increase the yield of mitochondria but will break most microbodies as well as many of the mitochondria. A tissue press (Howard Apparatus Co.) or a meat grinder has to be used first for tougher tissues, such as muscle, and the mince is mixed with grinding medium. One pass with a Potter-Elvehjem homogenizer is then possible. Use of a blendor is common, but damage to some particles, particularly microbodies and lysosomes, is severe. Nevertheless, it is necessary to use a blendor or a cutting device (razor blades) for many tissues. The blendor is run only for brief periods (5–10 sec), and total yield of enzyme activity is sacrificed for better particles. Mortar and pestle are also used, but the yield of organelles does not seem to be improved. These problems have been discussed for plant tissue,[7] and for bundle sheath cells, roots, and cotyledons it is necessary to cut them into small pieces by scissors before grinding with a mortar and pestle.

Grinding solutions are similar to the gradient and contain 20–30% sucrose in buffer with additional components such as EDTA, cations, and sulfhydryl agents. A ratio of one weight of tissue to 2 volumes of medium is generally used. For the B-30 and B-29 rotors, samples of 10–200 g of leaf tissue or 5–100 g of liver or kidney are possible. The animal should have been bled or the organs perfused with saline. The homogenate is squeezed through 4–8 layers of cheesecloth and centrifuged for 5–10 minutes at 450–1000 g to remove cell debris. This step also removes most of the nuclear fraction, Golgi bodies, and part of the chloroplasts, but not the smaller organelles. Filtration through Miracloth (Chicopee Mills, Inc.) is an alternative to low speed centrifugation.

Differential Centrifugation

When using the zonal rotors it is not necessary to concentrate a particulate fraction by a differential centrifugation. From 100 to 500 ml of homogenate may be added at the core of the rotors. Besides saving time, avoidance of a differential centrifugation step may provide more intact particles. When organelles are pelleted and resuspended, inactivation, losses, and breakage may occur. Published data to document this concept are not available, but we experience better yields and higher specific activities for some particulate fractions when a prior differential centrifugation is not used. On the other hand, partial fractionation by differential centrifugation can provide a large and cleaner preparation for the sucrose gradient.

Duration of Zonal Run

The total centrifugal force to place a particle at a given gradient location is an integrated product of time and rotor velocity and radius,[3,4] and the whole can be programmed through instrumentation, after which the drive returns to a hold speed of 2000 rpm. For rate-zonal centrifugation careful monitoring is necessary for reproducible results. For isopycnic zonal centrifugation more latitude in the duration of the centrifugation is permissible. The time must be long enough for the particles to reach their sedimentation density, but not so long that particle breakage becomes serious. It is felt that initial rapid acceleration may sheer particles as they cross sucrose interfaces. A continuous gradient and slower acceleration speeds reduce this type of damage. After centrifugation for too long a time, the peaks of activity for marker enzymes of particles are often smeared or shifted. Microbodies and lysosomes may break, and their marker enzymes begin to drift. Mitochondria with time slowly exchange bound water and increase in density during long centrifugation runs. This change in equilibrium density is followed by disruption of the mitochondria.[11] The hydrostatic pressure to which the particles are subjected in the gradient seems to be the main cause of this breakage. For microbodies this phenomenon seems to be even more extensive and serious than with mitochondria. To reduce particle breakage, the duration and speed of the run should be minimal. For the isopycnic separation of the organelles listed in the table, time of centrifugation was 1.5–4 hours at 30,000 rpm with the B-29 and B-30 rotors. Particles from liver homogenates were separated after 1.5–2 hours of centrifugation, from most leaves a run of 2.5–3 hours have been used, and for homogenates of castor bean cotyledons a 4-hour run was best.

[11] R. Wattiaux, S. Wattiaux-DeConinck, and M. F. Ronveaux-Dupol, *Eur. J. Biochem.* **22**, 31 (1971).

Sometimes shorter runs have produced sharper bands of particles, though not at their isopycnic density.

Unloading Gradients

After the gradient development the rotor speed is returned to a hold speed of 2000 rpm. The gradient may be unloaded from the rim by centrifugational force while pushing water into the core or from the core by displacement with dense sucrose at the rim. Fractions of 10–50 ml may be collected manually in a few minutes into graduated tubes in an ice bucket. A refrigerated fraction collector may be more convenient. Samples representing the 56% to 59% sucrose pad at the rim contain no organelles, need not be collected into small fractions, and need not be assayed. Samples from the core contain the soluble components of the original homogenate, and analysis of them is necessary to determine percent distribution of total activity.

The gradient is first characterized by its sucrose concentration and protein profile. Sucrose in each fraction can be measured manually in a short time with a Bausch and Lomb refractometer and converted to density (p) at a given temperature, i.e., d_{10}^0. The sucrose density of the effluent may be monitored continuously by a Waters Milan refractometer. Continuous recording of absorbancy at 280 nm is indicative of particulate bands. The refractive index and 280 nm absorbancy can be collected simultaneously by a two-pen recorder with a fraction indicator. Protein determination for specific activities is done by Lowry's procedure,[12] but samples of 50 μl or less must be used so that the sucrose will not interfere with the assay.[13]

Gradient fractions are generally stored at 0–4°. Depending on the enzymes, activity remains stable for days or weeks in these concentrated sucrose solutions, although the intact organelle may be broken.

Marker Enzymes

Each particle contains numerous enzymes, but the selection of specific ones as markers depend upon organelle specificity and ease of assay, preferably by a spectrophotometric procedure. Fumarase or succinic dehydrogenase for mitochondria meet these requirements. The presence of isoenzymes in different cellular locations excludes such enzymes as malate dehydrogenase, present in both mitochondria and microbodies, or aspartate aminotransferase present in most organelles. Enzyme stability in the sucrose gradient fractions must also be considered. Cytochrome c oxidase

[12] O. H. Lowry, N. J. Rosebrough, A. L. Farr, and R. J. Randall, *J. Biol. Chem.* **193**, 265 (1951).

[13] B. Gerhardt and H. Beevers, *Anal. Biochem.* **24**, 337 (1968).

activity is partially (over 50%) lost in a few days, whereas fumarase and succinic dehydrogenase activity is relatively stable.

It is desirable to use a cytosol marker to evaluate whether these constituents were present in the gradient from mixing or sticking to particles. Choice of marker enzymes for the cytosol are limited as more and more enzymes or isoenzymic forms are found to be in particles, and at present no one cytosol enzyme has been used consistently. Enzymes associated with hexose metabolism have been used, however, for plants the hexose and triose isomerases and dehydrogenases are present in chloroplasts. Lactate dehydrogenase in mammalian tissue is not satisfactory as a cytosol marker, as several percent of this activity is present in fractions containing microbodies and mitochondria.[14]

Characterization of gradients by assays for marker enzymes becomes a time consuming routine with 20-ml fractions from a 1500-ml gradient. We have observed only two solutions to this problem. The number of assays can be reduced by concentrating on only one organelle, so that only one marker enzyme and the relevant part of the gradient are used. On the other hand, a centrifugation center with technical help and automatic analyzers may be established,[5,15] and one well characterized gradient can be used by several projects at a time. Several investigators, each studying one of the organelles cited in the table, as well as the nucleic acid fractions, can combine their effort to provide all with a better characterized gradient in a minimum of time and to permit more exact comparison of changes in activity among organelles.

Purity and Recentrifugation

During the first gradient fraction, particulate bands are hardly ever completely separated from each other. Rather profiles of marker enzymes are plotted to visualize the location and purity of each organelle. In addition contamination between organelles can be estimated by comparing the specific activity (SA) of the marker enzymes in the various fractions.[16] The SA of a marker enzyme will be highest in its organelle. The SA of this enzyme in another fraction is a measure of contamination between the two particles. The ratio between the SA of the marker enzyme in another organelle to its SA in its designate organelle is an expression of contamination. Thus the SA of cytochrome *c* oxidase in the microbody fraction on sucrose gradients of particles from spinach leaves or dog liver was generally 0.01–0.05 that of the SA for this enzyme in the mitochondrial band.

[14] N. E. Tolbert, unpublished.
[15] F. Leighton, B. Poole, F. B. Lazarow, and C. de Duve, *J. Cell Biol.* **41,** 521 (1969).
[16] R. D. Cheetham, D. J. Morré, and W. N. Yunghans, *J. Cell Biol.* **44,** 492 (1970).

Approximation of cross contamination among peak fractions is used to decide whether enzyme activity in a band is from its presence in that organelle or from a contaminating particle. If the ratio of SA is similar to the SA ratio for the marker enzymes, then the activity is due to cross contamination. On the other hand, similar enzyme activity may be in two or more particles. Then the ratio of SA for these enzymes in the other particles will be greater than that for the marker enzyme. As an example, malate dehydrogenase on gradient profiles of particles from spinach leaves or dog liver appeared in both the mitochondrial (cytochrome c oxidase) and microbody (catalase) bands. The ratio of SA for cytochrome c oxidase in the microbody band to its SA in the mitochondria fraction ranged between 0.01 and 0.05. The specific activity of malate dehydrogenase in the microbody fraction from spinach leaves was about 0.9 of that in the mitochondria[17] and in dog liver the ratio was about 0.3.[18]

Analysis of enzyme profiles on the gradient thus serve to establish gradient location, purity, and enzymatic content of the organelles. When an enzyme activity is distributed among 2 or more organelles, the isoenzymic nature of the proteins should be studied. In the case of malic dehydrogenase from spinach leaves, organelle specific isoenzymes are present in the chloroplasts, mitochondria, and peroxisomes as well as in the cytoplasm.[17] On the other hand, in spinach leaves aspartate-α-ketoglutarate aminotransferase activity is also in these same organelles, but the major isoenzyme in each particle has the same electrophoretic and enzymatic properties.[19]

Further purification of the organelles for composition analyses or other more specific studies is necessary after the first sucrose gradient. This cannot be done by differential centrifugation. A second sucrose gradient may be constructed with less change in sucrose concentration so that the particles are separated further apart. Dilution of fractions from the first gradient is necessary but should be minimal to prevent osmotic shock. For some organelles, such as microbodies from leaves, considerable breakage and solubilization of enzymes occur nevertheless.

Large-Scale Preparations

The whole homogenate from 100 g of liver or 200 g of leaves in 200–500 ml can be placed directly over a sucrose gradient in the B-29 or B-30 rotor. In these rotors it is also possible to pump into the core region, a part of the homogenate, centrifuge just long enough to sediment the particles into the gradient, and then push out the supernatant from the core

[17] R. K. Yamazaki and N. E. Tolbert, *Biochim. Biophys. Acta* **178**, 11 (1969).
[18] B. Hsieh and N. E. Tolbert, unpublished.
[19] D. W. Rehfeld and N. E. Tolbert, *J. Biol. Chem.* **247**, 4803 (1972).

by adding 56% sucrose at the edge. By repeating this procedure, particles from large volumes may be put on a gradient, after which the usual gradient development is achieved by centrifugation for several hours at high speed.

The B-24, CF-6, JCF-2, and Z-15 rotors are designed for continuous addition of a suspension, after which the gradient is developed at higher speeds.[3] The homogenate or suspension remains in the rotor core area only long enough for particles to move onto the beginning of the gradient before the supernatant is displaced by more incoming suspension. The large B-24 rotor has been used in this manner mainly for separation of bacteria, virus, and other whole cells. The CF-6 and JCF-2 rotors may be more suitable for biochemical investigation of subcellular organelles.

Mitochondria

The marker enzymes generally are either succinate dehydrogenase,[20] fumarase,[21] or cytochrome c oxidase.[22] These established assays are run in 10-mm cuvettes at 25° or 30°. Cytochrome c oxidase assay is measured at 550 nm with the recorder at 1.0 OD for full scale and the chart speed at 1 inch per minute. In one bottom corner of the cuvette 1–10 μl of enzyme is added to 5 μl of 1% digitonen, mixed by tapping the cuvette, and incubated for 60 seconds. Then 200 μl of 0.1 M phosphate at pH 7.0 is added and 50 μl of reduced cytochrome c (5 mg/ml). The cytochrome c had been reduced with bisulfite until the E_{550}/E_{565} nm was greater than 6 and preferably between 9 and 10. The fumarase assay is run at 240 nm with the recorder settings of 1 OD full scale, and a chart speed of 1 inch per 5 minutes. To the cuvette is added 0.40 ml of 0.1 M TES-buffer at pH 9.5, 0.03 ml of 0.5% Triton X-100, 0.24 ml of 0.1 M $(NH_4)_2SO_4$, and 0.3 ml of enzyme plus water. The reaction is initiated by addition of 0.03 ml of 0.1 M potassium fumarate at pH 7.5. In this assay too much detergent will inhibit fumarase and too little will not break all the mitochondria.

Microbodies

The marker enzymes most used have been catalase and glycolate or α-hydroxy acid oxidase.[7] The assay for glycolate oxidase, as described for plant microbodies,[7] is also applicable to animal microbodies. In addition, NADH:hydroxypyruvate reductase[7] is an active and easy assay for plant microbodies, but it is not particle specific for animal tissue, since lactate dehydrogenase catalyzes the reduction of hydroxypyruvate. Malate syn-

[20] W. D. Bonner, Jr., this series, Vol. 1, p. 722.
[21] E. Racker, *Biochim. Biophys. Acta* 4, 20 (1950).
[22] T. Yonentani, this series, Vol. 10, p. 332.

thetase[23] and isocitrate lyase[24] are marker enzymes for glyoxysomes. Catalase is measured by the initial rate of disappearance of 12.5 mM H_2O_2 as measured by a decrease in absorbancy at 240 nm.[25] The recorder is set at 1.0 OD per full scale and a chart speed of 1 inch per minute. To the cuvette is added 3.0 ml of a solution containing 50 mM phosphate buffer at pH 7.0 to 7.5 and 12.5 mM H_2O_2 so that the initial OD is between 0.6 and 0.8. The reaction is initiated with 1–200 μl of enzyme and a 1-mm change is equivalent to 276 nmoles.

Microsomes or Endoplasmic Reticulum

This rather broad fraction peaks around a density of 1.14 g × cm^{-3}. Glucose-6-phosphatase is a simple marker enzyme[26] which is relative specific, although phosphatase activity in the lysosomes interferes. Antimycin A-insensitive cytochrome *c* reductase is very active in the microsome,[27] but some activity is also present in the outer membranes of other organelles. The microsomal cytochrome *c* reductase can be differentiated from the inner mitochondrial form by gradient profiles of activity with and without antimycin A.[28] This same analysis will reveal a peak of antimycin A-insensitive cytochrome *c* reductase with the microbody fraction and is attributed to its outer membrane.

NADH–cytochrome *c* reductase is measured in 0.3-ml cuvettes at 550 nm.[27] For a final volume of 0.27 ml add 0.1 ml of 0.2 M phosphate or glutamate buffer at pH 7.0, 50 μl of oxidized cytochrome *c* (5 mg/ml), 5 μl of 10 mM KCN (to inhibit any cytochrome *c* oxidase), and 2 μl of antimycin A (2 mg/ml ethanol). After obtaining the NADH-independent rate, the reaction is initiated with 50 μl of NADH (3 mg/ml). Based on an extinction coefficient of 21.1 mM^{-1} cm^{-1} for cytochrome *c*,[29] a change of one OD at 550 nm is equivalent to 12.8 nmoles. If this assay were to be run for mitochondrial activity the antimycin A should be omitted and the pH of the buffer changed to about 9.

Lysosomes

In assays developed by de Duve's group,[5] acid phosphatase is most often used as the marker enzyme. Lysosomal activity peaks in two areas

[23] G. H. Dixon and H. L. Kornberg, this series, Vol. 5, p. 633.

[24] B. A. McFadden, this series, Vol. 13, p. 163.

[25] H. Lück, in "Methods of Enzymatic Analysis" (H. U. Bergmeyer, ed.), p. 886. Academic Press, New York, 1965.

[26] R. C. Nordlie and W. J. Arion, this series, Vol. 9, p. 619.

[27] F. L. Crane, *Plant Physiol.* 32, 619 (1957).

[28] R. Donaldson, N. E. Tolbert, and C. Schnarrenberger, *Arch. Biochim. Biophys.* 152, 199 (1972).

[29] B. F. Van Gelder and E. C. Slater, *Biochim. Biophys. Acta* 58, 593 (1962).

of the gradient. One fraction has nearly the same final density as the microbodies, since both organelles have a single membrane surrounding a stroma of soluble enzymes, and both lose bound water on the sucrose gradient and sediment to a final isopycnic density of about 1.24 g × cm^{-3}. Part of the lysosomes contain engulfed lipids or fats and have, therefore, a much lighter isopycnic density nearer the top of the gradient. The proportion between the two lysomal fractions varies immensely. de Duve's group introduced an *in vivo* treatment whereby nearly all the lysosomes engulf a detergent so that they will float near the top of the gradient.[5] Rats of about 200 g each are injected with 1.5 ml of 10% (w/w) Triton W-1339 (Ruger Chemical Co., Irvington, New Jersey) 3.5 days prior to sacrifice, and they are also starved overnight before sacrifice. In this case 90% or more of the lysosomes will be in the light fraction well removed from the mitochondria and microbodies, but they do contaminate the microsomes. From adult dogs with no prior treatment of any type, the lysosomal activity from liver and kidney is nearly all (>90%) located near the top of the sucrose gradient, and the peak of acid phosphatase in denser sucrose is small and between the microbody and mitochondrial fractions. Thus for dogs the profiles of enzyme activities on the sucrose gradient can distinguish between microbodies, lysosomes, and mitochondria.[14] When starting with liver from young pigs, however, the acid phosphatase activity is nearly equally distributed between the two fractions, and the microbody and lysosomal activity in the denser fraction nearly coincide.

Broken Chloroplasts

This massive visible fraction is quantitated by chlorophyll analyses.[7] Its relatively light density on the gradient reflects the lamallae composition of the broken plastids, whereas the soluble stroma proteins of the chloroplasts have been lost and appear in the supernatant at the top of the gradient.

Whole Chloroplasts, Etioplasts, and Proplastids

Particles of this nature from plant tissue band between the mitochondria and the microbodies. At early stages of development whole plastids band close to the microbodies, but as they develop into chloroplasts with lipid-rich lamella their density shifts to values similar to the mitochondria.[30] At all stages of development whole plastids contain specific stroma enzymes which may be used as marker enzymes to differentiate them from microbodies and mitochondria. We have used triosephosphate isomerase, NADPH glyoxylate reductase, L-dihydroxyphenylalamine oxidase, and

---

[30] C. Schnarrenberger, A. Oeser, and N. E. Tolbert, *Plant Physiol.* **50**, 55 (1972).

P-glycolate phosphatase. Triosephosphate isomerase is coupled to glycerol phosphate dehydrogenase and followed by the oxidation of NADH.[31] The reaction mixture contains in a total volume of 1.0 ml, 0.1 M triethanol-amine pH 7.6, 1 mM KCN, 0.01% Triton X-100, 0.4 unit of glycerol-phosphate dehydrogenase, 250 μM NADH and 1.5 mM DL-glyceralde-hyde-3-P. NADPH glyoxylate reductase is determined by following the oxidation of NADPH at 340 nm with glyoxylate as substrate.[7] L-Dihy-droxyphenylalanine oxidase is measured by the increased adsorbancy at 480 nm from the red-brown reaction product.[32] For a 1.0 ml reaction volume, add 0.5 ml of 0.1 M phosphate and 0.1 M citrate buffer at pH 7.0, and 0.05 ml trypsin (4 mg/ml), 0.20 ml enzyme, and water. The cuvette should be aerated with 100% oxygen for about 20 seconds. The reaction is initiated by adding 0.25 ml of 0.24 M L-dihydroxyphenylalanine. The increase in color due to melanin formation is a relative value that cannot be stated in micromoles. When using the specific P-glycolate phos-phatase assay,[33] other enzymes in this part of a sucrose gradient from plant tissue do not interfere.

Protein Bodies

Plant tissue, particularly cotyledon tissue, contains storage protein bodies which are surrounded by a single membrane. These very dense organelles sediment in a sucrose gradient at high densities of 1.26 to 1.36 g $\times$ cm^{-3}.[34] A wide range in density probably reflects varying ratios of protein to bound water and lipid membrane as the particle is formed or undergoes digestion. If present the protein bodies contaminate the micro-body band resulting in low SA values for the microbody enzymes. Quan-titative enzymatic measurement of protein bodies is not feasible. They contain some activity for an acid proteinase and membrane-bound cyto-chrome c reductase.[34] Unusually high protein content in these dense sucrose fractions is indicative of their presence.

Starch Grains or Glycogen

These particles from plant or animal tissue are the major component of the pellet below the 59% sucrose pad on the gradient. This pellet may also contain nuclei and even whole cells, depending upon the preparation of the homogenate. One would normally use differential centrifugation, perhaps through a layer of very dense sucrose to isolate the glycogen and starch grains. However, their presence in large amounts at the bottom of

[31] G. Beisenberz, this series, Vol. 1, p. 387.
[32] N. H. Horowitz and S.-C. Shen, J. Biol. Chem. 197, 513 (1952).
[33] D. Anderson and N. E. Tolbert, this series, Vol. 9, p. 646.
[34] C. Schnarrenberger, A. Oeser, and N. E. Tolbert, Planta 104, 185 (1972).

sucrose gradients also can provide a relative clean preparation of them. The starch-iodine stain and hydrolysis to glucose has been used to characterize them.[35]

Microbody Membranes

Peroxisomes from rat liver in the sucrose gradient fraction can be broken by dilution with an equal volume of 10 mM pyrophosphate at pH 8.5.[15] After standing overnight, the broken particles when rerun on a sucrose gradient partially fractionated into solubilized stroma catalase at the top of the gradient, core urate oxidase at a density of about 1.23 g × cm^{-3}, and membrane NADH–cytochrome c reductase at a density of 1.173 g × cm^{-3}.[28] Peroxisomes from leaves in sucrose gradient fractions are very easily broken during dilution by osmotic shock, after which differential centrifugation or another sucrose gradient has provided fractions enriched only to a limited extent in core or membrane activity.

[35] D. D. Randall and N. E. Tolbert, *Plant Physiol.* 48, 488 (1971).

[75] The Use of Continuous Preparative Free-Flow
Electrophoresis for Dissociating Cell Fractions
and Isolation of Membranous Components

By K. Hannig and H.-G. Heidrich

In order to obtain an understanding of the specific function of cells and their substructures (the organelles and their membrane systems), research is faced with the problem of separating the particles carrying out individual functions and then of characterizing them. Particularly for studying cooperative interactions in biological processes, it is crucial to isolate the interacting systems in the most homogeneous and still functioning form. Until now, the methods for separating cell particles and membranes have been mainly those in which a separation was based on differences in size and density, i.e., sedimentation and centrifugation methods. However, size and density are not always unequivocal criteria for the homogeneity of such biological particles. Therefore, newly developed separation methods, e.g., the electrophoretic method described in this article, make particular use of characteristics based on functional surface properties. The membrane surface is known to be the site of many important biological activities. Transformation, differentiation, and specialization processes, for example, often go hand in hand with changes in the surface charge, which are then reflected in the electrophoretic behavior. Differences in electrophoretic mobility are also an expression of variation in the chemical composition of biological membranes since their principal constituents exhibit very different electrical charges. The number of possible combinations is large. However, the composition of the membrane surfaces of populations of the same particles is very uniform, as was shown by mobility measurements made using analytical cell electrophoresis.[1]

These facts have inspired us to adapt our preparative electrophoretic separation method, continuous preparative free-flow electrophoresis,[2] to the solution of the more difficult separation problems of particles and to the clarification of unanswered questions concerning the structure and function of higher structures. As this technique utilizes a different principle than the classic separation methods, it may in many cases be used to complement those or to provide new approaches to certain problems in cytology, immunobiology, and general biochemistry. In this article, we will present a survey of the applications and results of "free-flow electrophoresis" in the fields of separation and characterization of cell fractions and membranous components.

Principle and Design of the Apparatus

Figure 1 illustrates the principle of free-flow electrophoresis. An electrolyte solution of suitable pH flows across the lines of force of an electric field. The sample mixture to be separated is injected continuously into the streaming medium at a defined point. Components with differing electrophoretic mobility move along different paths and can be collected continuously at different points at the end of the separation chamber.

Figure 2 shows the latest commercial model of the electrophoresis apparatus (Desaga, Heidelberg). The separation chamber (a) is composed of two parallel glass plates which are 0.5 or 1 mm apart. The height of the chamber is 55 cm and the width 10 cm. A laminar flow of buffer film moves continuously down through the chamber. On both sides of the buffer film there is a membrane above which the film is in electrical contact with the electrode chambers. The mixture or particle suspension to be separated is injected continuously into the separation chamber through openings (b) in the upper part of the chamber by means of a

[1] "Symposium on Cell Electrophoresis" (E. J. Ambrose, ed.), Churchill, London, 1963; "Electrophoresis" (D. C. Shaw, ed.), Academic Press, New York, 1969.
[2] K. Hannig, "Jahrbuch der Max-Planck-Gesellschaft," p. 117, 1968; K. Hannig, "Methods in Microbiology" (J. R. Norris & D. W. Ribbons, ed.), Vol. 5B, p. 513, Academic Press, New York, 1971; K. Hannig, "Techniques of Biochemical and Biophysical Morphology" (D. Glick and R. Rosenbaum, ed.), Vol. 1, p. 191. Wiley, New York.

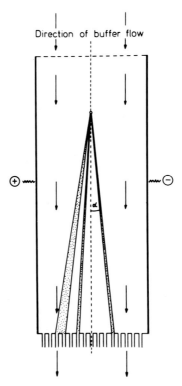

FIG. 1. Principle of continuous free-flow electrophoresis.

peristaltic pump (c). The separated fractions are collected at the lower end of the chamber via a 92-channel peristaltic pump (d) and the same number of this Teflon tubes leading into a fraction collector (e) which can be kept at 4°C. Uniform cooling of the separation chamber is extremely important to avoid disturbances caused by convection currents. For this reason, a particularly effective electronically controlled cooling system is employed to dissipate the Joule heat which forms due to the flow of current in the separation chamber. Field strengths of up to 100 V/cm can be applied and the material to be separated needs usually remain within the electric field for only 2–10 minutes.

Operational Considerations

Generally, in electrophoretic experiments the ionic strength of the separation buffer should be greater than that of the material to be separated. In the electrophoretic separation of particles (cells, cell fractions, and membranous components), the ionic strength of the particles plays

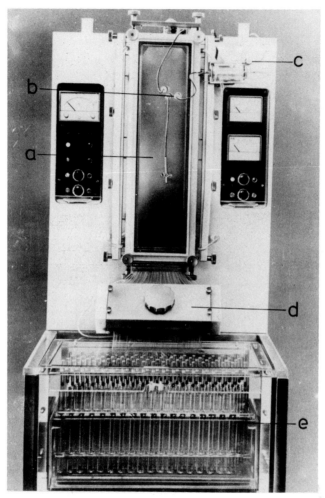

FIG. 2. The latest commercial free-flow electrophoresis apparatus (FF4; Desaga, Heidelberg). a, Separation chamber; b, openings; c, peristaltic pump; d, 92-channel peristaltic pump; e, fraction collector.

only a minor role owing to their relatively small surface charges. Thus, the ionic strength of the separation medium can be kept very low, and this improves the separation. A specific conductivity of the separation buffer between 300 and 1500 μmhos $\times$ cm^{-1} is usually efficient. A higher conductivity of the buffer solution is unfavorable not only because of an absolute and relative slower electrophoretic mobility of the components to be separated.

For soluble mixtures such as peptides, proteins, etc., the standard milieu conditions used for normal electrophoretic separations on carriers can generally be extended to free-flow electrophoresis as well. For particle separation, however, the experimental conditions for electrophoretic separation must be worked out empirically on the basis of pilot experiments. It has been found that these conditions will vary widely from one problem to another, and no generalization should be attempted.

The pH of the separation medium must be kept within the limits physiologically compatible with the particles. In addition, the buffer must constitute an isoosmotic milieu. The correct osmolarity cannot be obtained with buffer salts alone, as the conductivity would be too high. It must be effected by adding chemicals which do not contribute to the conductivity by themselves, i.e., which do not dissociate. Sucrose, glucose, ribose, glycine, etc., are normally used. Sometimes it can be of value to add nutrients or ions that promote the metabolic processes in order to maintain the biological activity of the particles.

The buffer in the electrode vessels is composed of the same buffer salts as that in the separation chamber and in the same relation, but many times (2–10×) more concentrated. The pH of this buffer has to be the same as that of the separation buffer. The electrode buffer does not contain nutrients or other additives.

The separation medium, devised on the basis of all these considerations, should be used as the isolation medium for processing the biological material from the beginning of the experiment, as milieu and pH changes very often damage the sensitive particles. It is self-evident that all the disintegration, homogenization, and centrifugation procedures applied to the material to be separated, have to be performed very carefully, gently, rapidly, and at temperatures below 5°. In all events, special care must be taken not to alter the surface properties of the particles by adsorption of proteins from the already disintegrated cells or by adsorption of complexing ions from the buffer. In order to exclude the first phenomenon, the sample to be separated has to be washed at least three times in the separation medium. Complexing ions, such as EDTA, should be omitted from the isolation and separation medium if possible, as they reduce strikingly the electrophoretic mobility of biological membranes. The frequently observed formation of aggregates during the preparation of the material and during electrophoresis ("flocculation") is probably due to the presence of constituents from damaged nuclei (DNA, RNA, chromatin) and can be avoided almost completely by using very gentle homogenization methods, particularly during the first homogenization steps.

Care should be taken that the density of the suspension injected into the apparatus is never lower than that of the separation buffer. If this is the case, the injected suspension will rise in the separation chamber making a separation impossible. The sample should be suspended in the separation buffer used in the experiment, either by washing or dialyzing prior to injection. The addition of other salts to the sample worsens the separation.

For the purpose of sterile working, the separation chamber can be rinsed with a highly diluted aqueous solution of chlorine dioxide, followed by a rinse with sterile separation medium. Sterile-filters did not give satisfactory results.

A number of milieu compositions and processing methods suitable for the separation of biologically interesting particles is given in the next section and in the literature.

Application

The application of the free-flow electrophoresis technique has been focused on the separation and characterization of intact cells,[3] of cell organelles,[4-8] and of membrane systems of cells and cellular components. Particularly, the studies of cell organelles and their membranes have been carried out in order to obtain some knowledge of a possible relation between these structures (GERL theory[9]). It can be seen from the results illustrated in Fig. 3, that, with exception of the lysosomal (and inner mitochondrial) membranes, all the other membranes of the cellular components (at least from the rat liver cell) are very similar in their electrophoretic behavior. This is particularly emphasized in order to point out the limitations of the free-flow electrophoresis technique. However, for many problems definite solutions could be obtained and, particularly in combination with other separation techniques, very pure organelle and membrane fractions could be prepared, characterized, and then studied with respect to their structure and function. No indication has ever been obtained that this technique injures the morphology or biological activity of the separated biological material.

[3] K. Hannig and W. F. Krüsmann, *Hoppe-Seyler's Z. Physiol. Chem.* **349**, 161 (1968); K. Zeiller, G. Pascher, and K. Hannig, *Hoppe-Seyler's Z. Physiol. Chem.* **351**, 435 (1970); K. Zeiller, K. Hannig, and G. Pascher, *Preparative Biochem.* **2**(1) 21 (1972); K. Zeiller, J. C. F. Schubert, F. Walther, and K. Hannig, *Hoppe-Seyler's Z. Physiol. Chem.* **353**, 95 (1972); K. Zeiller, E. Holzberg, G. Pascher, and K. Hannig, *Hoppe-Seyler's Z. Physiol. Chem.* **353**, 105 (1972).
[4] W. Klofat and K. Hannig, *Hoppe-Seyler's Z. Physiol. Chem.* **348**, 1332 (1967).
[5] A. Schweiger and K. Hannig, *Hoppe-Seyler's Z. Physiol. Chem.* **348**, 1005 (1967).
[6] R. Stahn, K.-P. Maier, and K. Hannig, *J. Cell Biol.* **46**, 576 (1970).
[7] R. Stahn and W. Klofat, in preparation.
[8] H.-G. Heidrich and A. M. Gonzalez, in preparation.
[9] For a recent and complete literature index on this subject, see P. M. Novikoff, A. B. Novikoff, N. Quintana, and J.-J. Hauw, *J. Cell Biol.* **50**, 859 (1971).

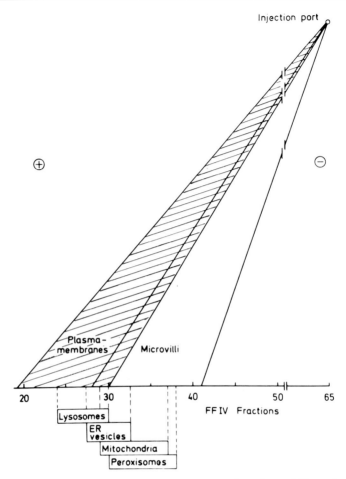

FIG. 3. Schematic representation of the electrophoretic mobility of some cell organelles and membranous components in the FF4 apparatus.

Cell Organelles

Mitochondria. The normally used techniques for the isolation of mitochondria (differential centrifugation and washing steps) are easy to carry out and result in clean mitochondria. However, a further purification, particularly from microsomal vesicles, could be obtained by free-flow electrophoresis. Furthermore, mitochondria from rat liver, rat heart, and rat kidney were compared with respect to their electrophoretic behavior. The thus obtained mitochondria behaved as though they were morphologically and functionally intact. Heart mitochondria possessed a slightly different electrophoretic mobility than liver mitochondria, the latter being less negatively charged. The reason for this different electrical behavior is under investigation.[8] A separation into "small" and "large" mitochondria could not be observed, but the question is being studied whether the orthodox and condensed forms of mitochondria show different electrophoretic mobilities.[8]

BUFFER. Triethanolamine, 10 mM; acetic acid, 10 mM; sucrose, 0.31 M; pH 7.4 (2 N NaOH); 100 V/cm, t = 5–6°, 2.2 ml per fraction per hour.

Lysosomes and Tritosomes. Using differential centrifugation and one free-flow electrophoresis step, lysosomes of rat liver homogenate were prepared in excellent purity (Fig. 4) and morphological integrity.[6] The specific activities of the lysosomal marker enzymes were extremely high. A certain enzymatic heterogeneity within the lysosomes could be demonstrated. Contamination, particularly with mitochondria and vesicles from the endoplasmatic reticulum, was almost zero. For the isolation of these intact lysosomes, no Triton WR-1339 had to be administered to the

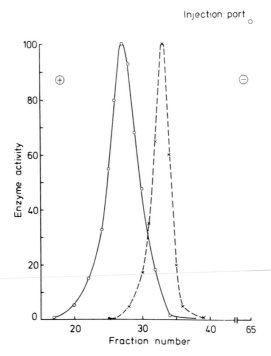

FIG. 4. The distribution of the particle-bound lysosomal (β-glucuronidase, O—O) and mitochondrial (cytochrome c oxidase, ×---×) marker enzymes in a free-flow electrophoresis run of a "light mitochondrial fraction" from a rat liver homogenate [R. Stahn, K.-P. Maier, and K. Hannig, *J. Cell Biol.* **46**, 576 (1970)].

animals as was done in the ingenious isolation method of Baudhuin *et al.*[10] However, when Triton WR-1339 was used, i.e., when tritosomes (lysosomes having incorporated the Triton) were produced by the animals and were enriched by centrifugation techniques, it was shown by free-flow electrophoresis experiments that those particles possessed the same electrophoretic mobility as did lysosomes isolated by free-flow electrophoresis without using Triton WR-1339. It follows that according to these results, Triton WR-1339 does not alter the surface charges of the lysosomal membrane.[7]

BUFFER. Triethanolamine, 10 mM; acetic acid, 10 mM; EDTA, 1 mM; sucrose, 0.33 mM; pH 7.4 (2 N NaOH), 85 V/cm$_{corr}$, $t = 4°$, 2.2 ml per fraction per hour.

Golgi Apparatus. The very few experiments carried out by the authors in order to isolate pure Golgi membranes have been of only very little importance as the electrophoretic behavior of these membranes in the buffer used was so similar to that of mitochondria that an isolation in preparative scale can be achieved much better with the techniques described by Fleischer and Fleischer[11] or Morré *et al.*[11] However, a systematic investigation of the separation medium has not yet been carried out. At least, a final purification of the Golgi membranes, isolated with the techniques mentioned above, via free-flow electrophoresis could be possible and of interest.

Peroxisomes (Microbodies). Microbodies have almost the same electrophoretic mobility as mitochondria although in some experiments an indication for a slightly slower mobility of the peroxisomes was observed. Therefore, an isolation and purification of these particles cannot be achieved by the electrophoresis technique alone. For the preparation of peroxisomes, the procedure of Leighton *et al.*[12] appears to be the better method. However, contaminations of lysosomal and microsomal vesicles can be successfully removed from peroxisomes by free-flow electrophoresis.

Microsomal Vesicles. According to the authors' definition in this article, microsomal vesicles are those which have been derived from the endoplasmatic reticulum (rough and smooth). These vesicles can be found as contamination in almost all the isolated organelles and membranes, and they can be removed from those particles either with difficulty or not at all. The easiest way of removing most of the microsomal

[10] P. Baudhuin, H. Beaufay, and C. de Duve, *J. Cell Biol.* **26**, 219 (1965).

[11] B. Fleischer and S. Fleischer, *Biochim. Biophys. Acta* **219**, 301 (1970); D. J. Morré, R. L. Hamilton, H. H. Mollenhauer, R. W. Mahler, W. P. Cunningham, R. D. Cheetham, and V. S. Lequire, *J. Cell Biol.* **44**, 484 (1970).

[12] F. Leighton, B. Poole, H. Beaufay, P. Baudhuin, J. W. Coffey, S. Fowler, and C. de Duve, *J. Cell Biol.* **37**, 482 (1968).

vesicles from heavier particles (plasma membranes, mitochondria) is by washing. The remaining amount can be separated by free-flow electrophoresis. Among the lighter particles (lysosomes, peroxisomes), only the lysosomes can be easily purified from microsomes by free-flow electrophoresis. This can be also concluded from the flow diagram shown in Fig. 3. As a rule, heavily contaminated particle preparations cannot be freed completely from the microsomal vesicles by free-flow electrophoresis alone, as those vesicles have a tendency to stick to other membrane surfaces and even to influence their electrophoretic mobility. On the other hand, it is easy to obtain relatively pure microsomal vesicles by differential centrifugation alone. In several experiments it was tried to separate such microsomal preparations into rough and smooth particles by free-flow electrophoresis. This could never be achieved. Whether milieu conditions or, in contradiction to theoretical considerations, a very similar surface charge of both types of vesicles is the reason for this behavior, remains to be investigated.

Synaptosomes and Synaptic Vesicles. Recently, Ryan et al.[13] have successfully used the free-flow electrophoresis technique for the isolation of pure synaptosomes and synaptic vesicles from guinea pig brain. Although the fractions obtained cannot be considered to be purer than those particles isolated by centrifugation techniques, the total preparation time is much shorter than that required for the centrifugation method. This is of importance for the isolation of the very sensitive synaptic vesicles. From the published results, synaptic vesicles possess a similar electrophoretic mobility as mitochondria, whereas synaptosomes are deflected more toward the anode and can be obtained in a relatively homogeneous form. In another study by Weiss and Matussek,[14] preliminary results indicated that synaptic vesicles of rat brain can be separated into two different populations representing the γ-aminobutyric acid- and norepinephrine-containing vesicles.

BUFFER. Michaelis-Veronal buffer, pH 7.15 (4.1 × 10⁻⁴ ohm⁻¹ cm⁻¹ conductivity, no detailed recipe given in publication); 100 V/cm, $t = 6°$.

Membranous Components

Plasma Membranes and Junctional Complexes. Several procedures have been described to isolate and purify cell membranes, particularly of rat liver cells. In all of them a crude membrane preparation is isolated in either isoosmotic or hypoosmotic buffer, and this is then purified by density gradient centrifugation. Since a hypoosmotic isolation medium

[13] K. J. Ryan, H. Kalant, and E. Llewellyn Thomas, *J. Cell Biol.* **49**, 235 (1971).
[14] D. Weiss and N. Matussek, in preparation.

produces mitochondrial subparticles which are difficult to remove, most of the applied procedures are quite time consuming. Therefore, it was worthwhile to devise a method for preparing plasma membranes in an isoosmotic buffer by free-flow electrophoresis; these experiments are still under way.[15] Preliminary results have shown that plasma membranes from rat liver (as well as from rat kidney) cells, have relatively high electrophoretic mobilities and can, therefore, be separated from endoplasmatic reticulum vesicles, mitochondria, and microbodies. As their electric properties are similar to those from lysosomes, the latter have to be removed by centrifugation steps prior to electrophoresis. The rat liver plasma membranes thus obtained are high in Na^+-K^+-ATPase and particularly rich in 5'-nucleotidase activity. Morphologically they resemble the "classical" plasma membranes.[16] The junctional complexes can be found indicating that these structures are not damaged or uncoupled by the electrophoretic forces.

BUFFER. Triethanolamine, 10 mM; acetic acid, 10 mM; sucrose, 0.245 M, pH 7.4 (2 N NaOH); 90 V/cm, $t = 5°$, 2 ml per fraction per hour.

Brush Border Fractions. The proximal tubule cell of rat kidney morphologically and functionally represents a polar cell type with the microvilli (brush border) at the luminal, and the basal infoldings at the interstitial side. The localization of several enzymes, particularly that of Na^+-K^+-ATPase, has not been clarified definitely, and therefore, the question whether both parts of the cell are sites of active transport could not be answered unequivocally. Some results of investigators using density gradient centrifugation have given hints that the distribution of, for example, Na^+-K^+-ATPase and alkaline phosphatase is not equal in all areas of the cell envelope. However, Mg^{2+}-ATPase could be found in the same specific activity in all areas of the cell enveloping membrane. Free-flow electrophoresis experiments gave definite answers to these questions.[17] From a rat kidney cortex homogenate, specially prepared for the demands of the free-flow electrophoresis technique, two populations of the proximal tubule cell envelope could be obtained after the free-flow electrophoresis run, and these two fractions could be further purified in a second reelectrophoresis run. One of them consisted of morphologically intact microvilli (Fig. 5a) and was characterized by a high specific activity of alkaline phosphatase, but it was free of Na^+-K^+-ATPase activity (membranes of the luminal part of the proximal tubule cell, brush border). The other

[15] H.-G. Heidrich, R. Illmensee, and K. Hannig, to be published.

[16] E. L. Benedetti and P. Emmelot, *in* "The Membranes" (A. J. Dalton and F. Haguenua, eds.), p. 33. Academic Press, New York, 1968.

[17] H.-G. Heidrich, R. Kinne, E. Kinne-Saffran, and K. Hannig, *J. Cell Biol.* **54**, 232 (1972).

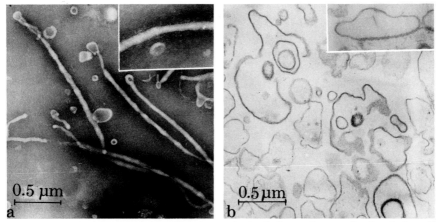

FIG. 5. The two membrane fractions obtained from the proximal tubule cell of rat kidney by free-flow electrophoresis. (a) Plasma membranes of the basal infoldings. Thin section. ×25,350; inset, ×30,500. (b) Microvilli of the apical part. Negative staining. ×25,350; inset, ×59,800. From H.-G. Heidrich, R. Kinne, E. Kinne-Saffran, and K. Hannig, *J. Cell Biol.* **54**, 232 (1972).

fraction (Fig. 5b) represented the membranes of the basal infoldings of the cell (interstitial part of the cell). Morphologically they resemble "classical" plasma membranes[16] but contain only relatively short, tight junctions. Enzymatically, this fraction was characterized by a very high activity of Na^+-K^+-ATPase and no alkaline phosphatase activity. The contamination of both of the fractions with other cell organelles was very low. Brush border fractions from other organs, for example, from intestine, have not yet been investigated with this technique.

BUFFER. Isolation medium: Triethanolamine·HCl, 10 mM; sucrose, 0.25 M, pH 7.6 (1 N NaOH). Separation medium: triethanolamine 10 mM; acetic acid, 10 M; sucrose, 0.33 M, pH 7.4 (2 N NaOH); 90 V/cm, $t = 5°$, 2 ml per fraction per hour.

Mitochondrial Membrane Systems. A couple of years ago some uncertainty was created among investigators of the mitochondrial membrane systems as to whether the localization of some enzymes and metabolic cycles had been determined correctly or not (monoamine oxidase, citric acid cycle, etc.). The crucial question to be solved was, how the outer and the inner mitochondrial membranes could be unequivocally identified as outer and inner membranes without using marker enzymes. Only after such a characterization, should the characteristic enzymes be investigated in the thus defined membrane fractions. Free-flow electrophoresis experiments could solve the problem definitely.[18] Rat liver mitochondria were

[18] H.-G. Heidrich, R. Stahn, and K. Hannig, *J. Cell Biol.* **96**, 137 (1970).

subjected to a very mild swelling-shrinking process omitting any kind of chemical which could possibly alter the membrane surfaces. Then, the suspension containing mitochondrial outer membranes and inner membranes with the matrix was electrophoresed in a free-flow electrophoresis apparatus, and two fractions were obtained. One of them had the same electrophoretic mobility as intact mitochondria. The other fraction showed a completely different electrophoretic behavior and was deflected more toward the anode. Since the electrophoretic properties of particles are determined only by the outer surfaces of these particles, the fraction having the same mobility as mitochondria must be considered as representing the outer enveloping mitochondrial membrane. This was confirmed by electron micrographs (Fig. 6a), and enzymes like monoamine oxidase and rotenone-insensitive NADH–cytochrome c oxidase were found in this fraction. The particles collected closer to the anode consisted of pure inner mitochondrial membranes with their matrix, and their morphology in the electron microscope was characteristic for inner mitochondrial membranes (Fig. 6b). In this fraction the enzymes of the citric acid cycle, of fatty acid oxidation, and of oxidative phosphorylation could be found. Particles with the same electrophoretic behavior and possessing all the other properties of mitochondrial inner membranes could be isolated in the same free-flow electrophoresis fractions when intact rat liver mitochondria were treated with digitonin and then electrophoresed.[19] This is another proof for the reliability of the described method. The results show, in addition, that the digitonin method does not alter the electrical

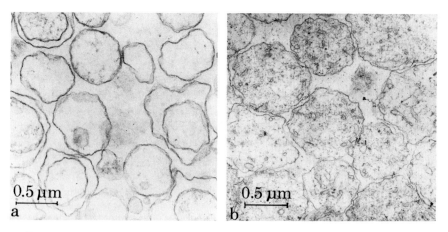

FIG. 6. Membranes from rat liver mitochondria separated with the free-flow electrophoresis technique. (a) Outer membranes. (b) Inner membranes with matrix. a, b, 22,100. From H.-G. Heidrich, R. Stahn, and K. Hannig, *J. Cell Biol.* **46**, 137 (1970).

charge of the membrane surface, and a combination of this technique with the free-flow electrophoresis is a very fast procedure to obtain very pure inner mitochondrial membranes with matrix on a preparative scale.[19]

BUFFER. Triethanolamine, 90 mM; acetic acid, 90 mM; EDTA, 0.9 mM; sucrose, 0.3 M; pH 7.4 (2 N NaOH); 85 V/cm, $t = 5°$, 2 ml per fraction per hour.

Escherichia coli Envelope. The cell envelope of *Escherichia coli* consists of two distinct structures: the rigid, complex, and multilayered cell wall, and the innermost cytoplasmic membrane. By disintegration, both moieties of the envelope are ruptured and disconnected from each other to a large extent. Then they can be separated via density gradient centrifugation. The cytoplasmic membrane fraction thus obtained is not homogeneous, but it no longer contains cell wall particles. The cell wall fraction, however, still possesses succinic dehydrogenase activity (a marker for the cytoplasmic membrane). Consequently, the chemical composition (protein, lipid, etc.) of the cell envelope structures cannot be determined definitely. Using free-flow electrophoresis, three fractions with different electrophoretic mobilities could be isolated from disintegrated *E. coli* cells.[20] One of them represented cell wall particles (free of cytoplasmic membranes), the other contained plasma membrane vesicles and was characterized by a high specific activity of succinic dehydrogenase. The third fraction contained regions of adhesion between the cell wall and the plasma membrane which represent the *E. coli* DNA-binding sites.[20a]

BUFFER. Triethanolamine, 10 mM; acetic acid 10 mM; MgCl$_2$ 0.1 mM; sucrose 0.3 mM, pH 7.4 (2 N NaOH); 115 V/cm, t = 5°, 1.7 ml per fraction per hour.

Inside-out Vesicles

Particularly for the understanding of transport processes, our knowledge of the "other" side of a biological membrane would be of fundamental importance. A precondition for a successful preparation of inside-out vesicles with the free-flow electrophoresis technique is, besides the possibility of turning the biological membrane around, the presence of different electrical charges on both sides of the membrane. For theoretical reasons, this can be expected for the majority of the membranes carrying out active transport. Even though the results which demonstrate the usefulness of the free flow electrophoresis technique for solving such problems are still fragmentary, the possibilities can already be anticipated and the

---

[19] H.-G. Heidrich, to be published.
[20] C. Schnaitman and H.-G. Heidrich, to be published.
[20a] W. L. Olsen and H.-G. Heidrich, *J. Mol. Biol.*, submitted for publication.

following should be considered as an outlook for further applications of the free-flow electrophoresis.

Submitochondrial Particles. It has been shown that mitochondria, when treated by sonication or osmotic shock, disintegrate into small vesicles which possess a conspicuous pattern of 80 Å subunits. These vesicles were demonstrated to be derived from the inner mitochondrial membrane, and efforts have been made in order to show that the side carrying the particles actually is the inner side of the inner mitochondrial membrane.[21] The free-flow electrophoresis technique could contribute some valuable answers to this question.[22] Pure mitochondrial inner membranes (with matrix) obtained by free-flow electrophoresis were sonicated and subjected to free-flow electrophoresis. After the run, a second type of membranous material could be isolated which had a much lower electrophoretic mobility than the intact inner membranes. The longer the sonication was carried out the more this material was formed. Electron micrographs showed definitely that this second membrane fraction consisted of very homogeneous vesicles carrying the above-mentioned subunits on their outer surfaces. The specific activity of the enzyme succinic dehydrogenase (marker for the inner mitochondrial membrane) was increased in comparison with the starting material whereas the activity of the enzyme glutamate dehydrogenase (marker for the matrix) was almost completely lost. Consequently, the material of the second peak consisted of pure inside-out inner mitochondrial membrane vesicles showing a different surface charge than the outer side of the inner membrane. Contamination with outer mitochondrial membranes, with inner outside-out membranes, or with other cell material was completely removed by free-flow electrophoresis, and this vesicle fraction is suitable for further investigation.

BUFFER. Triethanolamine, 10 mM; acetic acid, 10 mM; sucrose, 0.3 M, pH 7.4 (2 N NaOH); 95 V/cm, $t = 5°$, 1.75 ml per hour.

Erythrocytes. From several investigations it can be presumed that the erythrocyte membrane can form vesicles with the original inner surface at the outside. As intact erythrocytes carry almost all of their sialic acid on the outside of their plasma membranes (it can easily be cleaved off by enzymes, and the remainder no longer contains much sialic acid), both sides of the membrane should have different electrical surface properties making electrophoretic separation from each other possible. Experiments carried out recently,[23] gave promising results. From erythrocyte ghost

[21] E. Racker, C. Burstein, A. Loyter, and R. O. Christiansen, *in* "Electron Transport and Energy Conservation" (J. M. Tager, S. Papa, E. Quagliariello, and E. C. Slater, eds.), p. 235. Adriatica Editrice, Bari, 1970.

[22] H.-G. Heidrich, *FEBS* (*Fed. Eur. Biochem. Soc.*) *Lett.* **17**, 253 (1971).

[23] H.-G. Heidrich and G. Leutner, *Europ. J. Biochem.*, in press.

preparations, specially prepared for these experiments, two fractions of membranous material were obtained. One of them had the same electrophoretic mobility as intact erythrocytes; the other was electrophoretically different. The particles in both fractions contained the same total amount of sialic acid, but the sialic acid from the fraction which had the same electrophoretic behavior as the erythrocytes could be cleaved off easily by enzymes. This indicated that the sialic acid was located on the accessible outer surface of the vesicles. The sialic acid from the second type of membrane vesicles could not easily be removed from the membrane. It was either masked, hidden, or enclosed in the vesicles. Further experiments are necessary to clarify and elucidate the mechanism of the formation and nature of these vesicles.

BUFFER. Triethanolamine, 10 mM; acetic acid, 10 mM; $MgSO_4$ 0.1 mM; 85 V/cm, $t = 6$–$7°$, 2.7 ml per fraction per hour.

[76] Countercurrent Distribution of Cells and Cell Organelles

By PER-ÅKE ALBERTSSON

Cell particles can be separated by partition between two immiscible liquid phases composed of different aqueous polymer solutions.[1] By this technique particles are separated according to their surface properties and it therefore complements centrifugation methods where size and density are the determining factors.

Phase partition has been applied to a number of cell particles such as blood cells,[2-6] algae,[7,8] bacteria,[2] chloroplasts,[9-11] mitochondria,[12] cell

[1] P.-Å. Albertsson, "Partition of Cell Particles and Macromolecules," 2nd ed. Almqvist & Wiksell, Stockholm; Wiley, New York, 1971.

[2] P.-Å. Albertsson and G. D. Baird, *Exp. Cell Res.* **28**, 296 (1962).

[3] H. Walter and F. W. Selby, *Biochim. Biophys. Acta* **112**, 146 (1966).

[4] H. Walter, F. W. Selby, and R. Garza, *Biochim. Biophys. Acta* **136**, 148 (1967).

[5] H. Walter, E. J. Krob, R. Garza, and G. S. Ascher, *Exp. Cell Res.* **55**, 57 (1969).

[6] H. Walter, E. J. Krob, and G. S. Ascher, *Exp. Cell Res.* **55**, 279 (1969).

[7] H. Walter, G. Eriksson, Ö. Taube, and P.-Å. Albertsson, *Exp. Cell Res.* **64**, 486 (1971).

[8] H. Walter, G. Eriksson, Ö. Taube, and P.-Å. Albertsson, *Exp. Cell Res.* **77**, 361 (1973).

[9] B. Karlstam and P.-Å. Albertsson, *FEBS (Fed. Eur. Biochem.) Lett.* **5**, 360 (1969).

[10] B. Karlstam and P.-Å. Albertsson, *Biochim. Biophys. Acta* **255**, 539 (1972).

[11] C. Larsson, C. Collin, and P.-Å. Albertsson, *Biochim. Biophys. Acta* **245**, 425 (1971).

[12] I. Ericson, unpublished.

membranes,[13-13b] and also enzymes and nucleic acids.[1] For a recent general review, see the reference cited in footnote 1. This paper will deal with the application of phase partition on chloroplasts, mitochondria, and microbial cells. Fractionation of mammalian blood cells by the same technique is described in the article by Walter.[14]

Principle of Fractionation

Unlike soluble materials which partition between the bulk phases, suspended particles often distribute between one of the phases and the interface. Thus, if a material distributes between the upper phase and the interface the percent in upper phase will be characteristic for the material in a given phase system. If two types of particles distribute differently enough, they may be almost completely separated by one partition step. If they are closer to each other in their behavior, a multistage procedure such as countercurrent distribution has to be resorted to. The technique of countercurrent distribution is described elsewhere; the reader is referred to the book by Hecker[15] and the articles by Craig and Craig[16] and von Tavel and Signer.[17] By countercurrent distribution, mixtures can be completely resolved which are only partly separated by one partition step. Each substance travels along a train of tubes with a characteristic speed and gives rise to a peak in the countercurrent distribution diagram. Thus, different substances are separated from each other in a similar fashion as by chromatography.

Several factors determine the distribution of particles in a phase system. The most prominent factors are particle surface properties, such as surface charge, type and concentration of polymers, and the ionic composition of the phase system. These factors are treated in detail in the reference cited in footnote 1. To adjust the distribution of particles for maximal separation one can, therefore, vary either the polymer composition or the ionic composition. It has been demonstrated that inorganic

---

[13] D. M. Brunette and J. E. Till, *J. Membrane Biol.* **5**, 215 (1971).

[13a] L. Lesko, M. Donlon, G. V. Marinetti, and J. D. Hare, *Biochim. Biophys. Acta* **311**, 173 (1973).

[13b] T. Russell and I. Pastan, *J. Biol. Chem.* **248**, 5835 (1973).

[14] H. Walter, this series, Vol. 32, [63] in press.

[15] E. Hecker, "Verteilungsverfahren im Laboratorium," Monographien zu *Angewandte Chemie* und *Chemie-Ingenieur-Technik,* No. 67, Verlag Chemie, GMBH, Weinheim/Bergstr., Germany, 1955.

[16] L. C. Craig and D. Craig, *in* "Technique of Organic Chemistry" (A. Weissberger, ed.), 2nd ed., Vol. III, Part 1. Wiley (Interscience), New York, 1956.

[17] P. von Tavel and R. Signer, *Advan. Protein Chem.* **11**, 237 (1956).

salts differ significantly from each other in their partition.[18] This can be interpreted as resulting from different affinities of different inorganic ions for the two polymer phases. If so, an electrical potential difference between the phases is created. Its sign and relative value have been calculated.[1] The electrical potential difference is very small, of the order of 1–2 mV, yet it has a very strong influence on the partition of charged macromolecules or particles since these carry a large number of charges.[1]

It can be deduced from partition studies on inorganic ions that salts like Na+, or Li+, phosphate, or sulfate create a larger electrical potential difference between the phases than NaCl. In the former salts, therefore, one would expect surface charge differences between particles to be more prominent in determining their partition behavior. On the other hand, in NaCl, when the electrical potential difference is smaller, other factors such as hydrophobicity might be of more importance.

It should be pointed out that the electrical charges that are important in partition are not necessarily the same as those that are effective in electrophoresis. The electrophoretic mobility depends on the zeta potential of the surface layer of the particles whereas partition depends on the potential difference between the two bulk phases and the particle charges which are exposed to the phases.

Choice of Phase System

Most countercurrent experiments have employed the dextran–polyethylene glycol system since this has a relatively short settling time.

The following stock solutions are prepared.

Dextran; 20% (w/w). Dextran T 500 (Pharmacia, Uppsala, Sweden) is dissolved directly in water. It is desirable to heat the solution up to boiling to minimize subsequent microbiological contamination. The moisture content of the dextran powder can be supposed to be 10%. The exact concentration of dextran in the solution is determined polarimetrically as follows: 10 g of the solution is diluted to 25 ml; the optical rotation is determined with a tube 20 cm long, and the concentration is calculated using an $[\alpha]_D^{25}$ value of $+199°$.

PEG 6000 or PEG 4000; 40% (w/w)

PEG (Carbowax 6000 or 4000 from Union Carbide, New York) does not contain moisture and may be dissolved directly in water.

Salt and buffer concentrations. Stock solutions of these are made in concentrations 10 times higher than that of the final concentration in the phase systems.

[18] G. Johansson, *Biochim. Biophys. Acta* **222**, 381 (1970).

Example of a Scheme for Investigating the Partition Behavior of Cell Organelles in the Dextran–Polyethylene Glycol System[a]

Dextran 500, 20% w/w (g)	PEG 6000, 40% w/w (g)	PEG 4000, 40% w/w (g)	Sodium phosphate, 0.4 M, pH 7.8	0.4 M NaCl (ml)	Sucrose (g)	H_2O (ml)
(1) 1.0	0.35	—	0.1	—	0.4	2.15
(2) 1.0	0.4	—	0.1	—	0.4	2.1
(3) 1.4	0.44	—	0.1	—	0.4	1.66
(4) 1.2	—	0.6	0.1	—	0.4	1.7
(5) 1.24	—	0.62	0.1	—	0.4	1.64
(6) 1.28	—	0.64	0.1	—	0.4	1.58
(7) 1.32	—	0.66	0.1	—	0.4	1.52
(8) 1.36	—	0.68	0.1	—	0.4	1.46
(9) 1.4	—	0.7	0.1	—	0.4	1.4
(10) 1.0	0.4	—	0.1	0.2	0.4	1.9
(11) 1.0	0.4	—	0.1	0.5	0.4	1.6
(12) 1.0	0.4	—	0.1	1.0	0.4	1.1

[a] In tubes 1–3 and 4–9, the effect of polymer concentration in the PEG 6000 and PEG 4000 systems, respectively, is studied. In tubes 2 and 10–12 the effect of NaCl is studied. The cell particles are included in the phase systems by replacing a given volume of H_2O with a particle suspension in an appropriate medium, for example, phosphate buffer plus sucrose. Alternatively a pellet may be suspended in the top phase.

In order to select a phase system with a particle partition suitable for countercurrent distribution, a series of single tube experiments are set up (see the table). Preferably a sodium phosphate buffer in the pH range 7–8 at a final concentration of 5–10 mM is chosen and the polymer concentration is varied. The molecular weight of the polymers can also be varied. In another series of experiments the polymer composition is kept constant, and the ionic composition is varied.

Figure 1 shows the result of such an experiment with chloroplasts where the ionic composition is kept constant and the polymer composition is varied. Generally particles are more adsorbed at the interface the higher the polymer composition, as shown for chloroplasts in Fig. 1. The result of another experiment with chloroplasts where the salt composition is varied is shown in Fig. 2. Generally, for negatively charged particles, such as cells and cell organelles, the following ions facilitate partition into the upper phase Li^+, Na^+, HPO_4^{2-}, SO_4^{2-}, while K^+, Cl^-, Br^-, have the opposite effect. If PEG 6000 is replaced by PEG 4000 partition into the upper phase is favored while lower molecular weight of dextran favors partition into the lower phase.

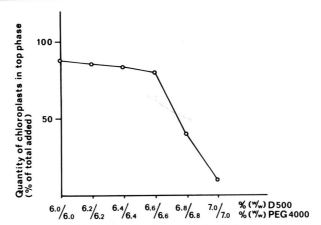

FIG. 1. Partition of chloroplasts in dextran–polyethylene glycol two-phase system at constant ionic composition (5 mM sodium phosphate, pH 7.8) but different polymer concentrations. ◯—◯, NaCl; ×—×, KCl. From B. Karlstam and P.-Å. Albertsson, *Biochim. Biophys. Acta* **255**, 539 (1972).

From experiments like those shown in Figs. 1 and 2 one can select a composition which yields a suitable distribution of material between the upper phase and the interface. Usually a system where 30–70% of the particles are in the upper phase is then selected for countercurrent distribution.

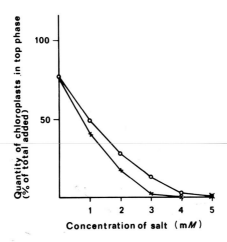

FIG. 2. Effect of NaCl and KCl on partition of chloroplasts in 6.4% dextran–6.5% PEG 4000 in 5 mM sodium phosphate, pH 7.8. From B. Karlstam and P.-Å. Albertsson, *Biochim. Biophys. Acta* **255**, 539 (1972).

Chloroplasts

A chloroplast preparation obtained by differential centrifugation is very heterogeneous, consisting of chloroplasts with very different appearance in the phase contrast and electron microscope. By countercurrent distribution, such a preparation has been separated into three distinct classes differing with respect both to morphology and to enzyme content.[9-11] Here follows a procedure which has been used for chloroplasts from *Spinacia oleracea*.[10,11]

Preparation of Chloroplasts. Twenty grams of leaves are cut into pieces with scissors and then blended in a chilled knife blendor (Turmix) with 125 ml of a solution at pH 7.8 containing 0.4 M sucrose and 50 mM potassium phosphate buffer, pH 7.8. The duration of blendor treatment can be varied, but a treatment of 3–10 times 2 seconds gives a good yield of chloroplasts. The blended material is filtered through four layers of perlon net (Monodur 31, Vereinigte Seidenweberein AG, Krefeld, West Germany). The filtrate is then centrifuged for 10 minutes at 400 g, and the pellet is washed once in the same medium as that used for blending, before its contents are resuspended in about 0.5 ml of this medium. All operations are carried out at 2°.

Countercurrent Distribution. Phase system A: 63 g of 20% (w/w) Dextran 500; 31.5 g of 40% (w/w) PEG 4000; 50 ml of 30% (w/v) sucrose; 5 ml of 0.2 M potassium phosphate buffer pH 7.8 (ratio 1 KH_2PO_4 to 10 K_2HPO_4); H_2O up to a total weight of 201.2 g. The mixture is shaken in a cold room (2°) and allowed to separate in a funnel. The greater parts of the two phases are then collected (interfacial material discarded) and stored separately.

Chloroplast phase system, A_{Chl}, for loading the countercurrent distribution apparatus: 1.26 g of 20% (w/w) Dextran 500; 0.63 g of 40% (w/w) PEG 4000; 0.8 ml of 30% (w/w) sucrose. Mix, and add 0.5 ml of chloroplast suspension (see above) and 0.66 ml of the bottom phase from A. Except for the presence of chloroplasts and the 0.66 ml of bottom phase added this system is almost identical to phase system A. The 0.66 ml addition is made to make the volume ratio top:bottom equal to 1.

An automatic thin-layer countercurrent distribution apparatus with 120 chambers[19,20] is used. The bottom phase chamber has a capacity of 0.7 ml. Since the chloroplasts partition between the upper phase and the interface, the method of liquid-interface countercurrent distribution[2] is employed. The chambers are numbered 0–119. The tubes numbered 10–119 are first filled with 0.6 ml each of top and bottom phase of system A.

[19] P.-Å. Albertsson, *Anal. Biochem.* **11**, 121 (1965).
[20] P.-Å. Albertsson, *Science Tools*, **17**(3), 56 (1970).

Then each of tubes numbered 0–9 are filled with 1.2 ml of chloroplast phase system A_{Chl}. The shaking time is 30 seconds, and the settling time is 8 minutes; temperature is 2°. After 120 transfers, the chambers are emptied and the contents are diluted 3-fold with buffered 0.4 M sucrose to break the phase. The absorbance of the diluted fractions at 550 and 680 nm is measured. The absorbance at 680 nm gives the overall concentration of chloroplasts while the ratio between the absorbances at 550 and 680 nm give the percentage of intact chloroplasts.[21]

The diagram in Fig. 3 shows three well separated peaks indicating three classes of chloroplasts. These have been characterized with respect to

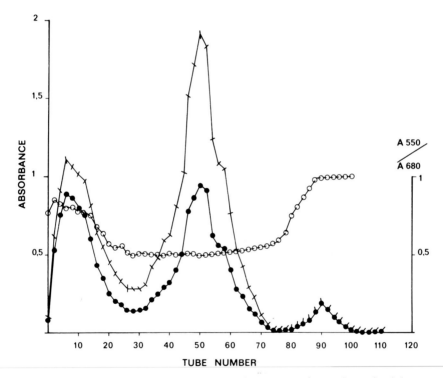

FIG. 3. Countercurrent distribution of chloroplasts. The peak to the left represents class I chloroplasts; the peak in the middle, class II chloroplasts; and the peak to the right, intact chloroplasts surrounded by an extra thin cytoplasmlike layer with an additional membrane. ●—●, A_{550}; —, A_{680}; ○—○, A_{550} : A_{680}. From B. Karlstam and P.-Å. Albertsson, FEBS (Fed. Eur. Biochem. Soc.) Lett. 5, 360 (1969); C. Larsson, C. Collin, and P.-Å. Albertsson, Biochim. Biophys. Acta 245, 425 (1971).

[21] B. Karlstam and P.-Å. Albertsson, Biochim. Biophys. Acta 216, 220 (1970).

morphology and enzymatic content. The peak I chloroplasts are intact (class I) with a high ribulosediphosphate carboxylase activity, high CO_2-fixation, and low malate dehydrogenase activity; the second peak contains broken chloroplasts (class II) with little or none of these activities. The third peak consists of intact chloroplasts surrounded by an extra layer of cytoplasmlike material; these chloroplast particles have high CO_2 fixation high ribulose diphosphate carboxylase activity and, unlike peak I chloroplasts, a high malate dehydrogenase activity.

Microorganisms

Bacteria, algae, and yeast cells have been partitioned in the dextran–polyethylene glycol system.[1,2,7] Different strains of bacteria display different partition behavior. Even closely related strains may be separated. Countercurrent distribution of a culture of a single strain of *Escherichia coli* often shows at least two peaks indicating a separation of different classes of cells.[2] In the case of *Chlorella pyrenoidosa*, it has been demon-

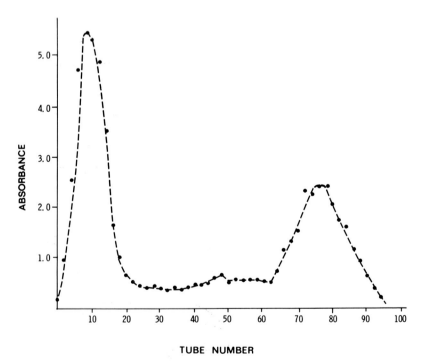

FIG. 4. Separation of young (left) from old (right) cells of *Chlorella pyrenoidosa* by countercurrent distribution. From H. Walter, G. Eriksson, Ö. Taube, and P.-Å. Albertsson, *Exp. Cell Res.* **64**, 486 (1971).

strated that cells in different stages of the growth cycle behave differently, and these may therefore be separated. This was shown with synchronized cultures of *Chlorella*.[7] When an exponentially grown culture of *Chlorella* was subjected to countercurrent distribution the diagram shown in Fig. 4 was obtained. It shows two major peaks representing young and old cells which therefore must have different surface properties. That distinct peaks are obtained, indicates that there are different classes of cells and that the surface properties of the cells change in steps rather than continuously when the cells grow. A detailed study of the countercurrent distribution of synchronized *Chlorella* cultures at different stages has recently been published.[8]

General Comments

Since particles are separated according to their surface properties by countercurrent distribution this technique complements centrifugation. A combination of centrifugation and countercurrent distribution can therefore yield considerable information on the composition of a population of cell particles. An illustrative example is given by the experiment shown in Fig. 3, where a chloroplast preparation obtained by centrifugation is separated into three distinct classes by countercurrent distribution due to differences in the properties of the particle surfaces. Particles having the same size and density may therefore be separated by countercurrent distribution provided the surface properties differ.

Although the choice of a suitable phase system involves some preliminary work of trial and error, an experiment like that outlined in the table is a useful guide. In fact the dextran–polyethylene glycol system seems to be a generally applicable system which has been applied on such widely different substances as proteins, nucleic acids, virus, mitochondria, chloroplasts, bacteria, algae, and blood cells. The phase partition method seems to be gentle since most of the biological activities tested withstand treatment by the phases. The activities include various enzyme activities,[1] oxidative and photosynthetic phosphorylation, CO_2 fixation of chloroplasts and viability of cells. In many cases the polymers seem to exert a protective effect on cell organelles. Intact chloroplasts, for example, retain their structure longer in the presence than in the absence of the dextran–polyethylene glycol polymers. The main limiting factor in the application of countercurrent distribution on cell organelles is the long time needed for an experiment, 1 hour to 20 hours depending on the number of transfers and settling time. Once a particle population has been characterized by countercurrent distribution, one can modify the phase system, however, to make it more efficient so that only a small number of transfers are needed. The time needed for a separation can thereby be reduced to less than an hour.

[77] Destabilization of Membranes with Chaotropic Ions

By Y. HATEFI and W. G. HANSTEIN

The purpose of this article is to discuss in general the utility of chaotropic ions (Fig. 1) for the resolution of membranes and enzyme complexes and the isolation of membrane-bound proteins. The fundamental considerations underlying the mechanism of action of chaotropes are as follows.

Hydrophobic interactions make a major contribution to the stability of most membranes, multimeric proteins, and the native conformation of biological macromolecules. Other stabilizing forces are hydrogen bonds, Coulombic effects, disulfide bonds, and London dispersion forces. According to Kauzmann,[1] apolar groups form hydrophobic bonds mainly as a result of their thermodynamically unfavorable interaction with water, rather than as a consequence of attraction for each other. This repulsion of the apolar groups from the aqueous phase is related mainly to water structure. The transfer of an apolar molecule from a lipophilic surrounding to water is associated with a unitary entropy decrease of 10–20 eu/mole, and is endergonic by 2–6 kcal/mole.[1] It might be expected, therefore, that disordering of water would lower these thermodynamic barriers, facilitate the transfer of apolar groups to the aqueous phase, weaken hydrophobic interactions, and lead to the destabilization of membranes and biological macromolecules.

Several lines of evidence have indicated that chaotropic ions decrease water structure,[2,3] and other studies have shown that the relative water structure-breaking potency of the chaotropic ions listed in Fig. 1 is well correlated with their ability to destabilize membranes and enzyme complexes and increase the water solubility of various nonelectrolytes.[2,4,5] The list of chaotropic ions shown in Fig. 1 is by no means exhaustive. In general, ions with low charge density (e.g., large radius, single charge) appear to be chaotropic, whereas ions with high charge density appear to have the opposite effect (see below). Among a series of related ions (e.g., halides

---

[1] W. Kauzmann, *Advan. Protein Chem.* **14**, 1 (1959).

[2] Y. Hatefi and W. G. Hanstein, *Proc. Nat. Acad. Sci. U.S.* **62**, 1129 (1969).

[3] These include positive entropies of aqueous ions, proton NMR shifts to higher field strengths, facilitated water self-diffusion, negative viscosity B coefficients, diminution of surface tension rise, and increased surface potential when chaotropic ions are introduced into water.

[4] F. A. Long and W. F. McDevit, *Chem. Rev.* **51**, 119 (1952).

[5] Even in the case of simple hydrocarbons, chaotropes have a smaller salting-out effect than most other salts, and $HClO_4$ salts in benzene.[4]

$CBr_3COO^- > CCl_3COO^- > SCN^-$, guanidinium $> I^-$, $ClO_4^- > CHCl_2COO^- >$
NO_3^-, $Br^- > CF_3COO^- > CH_2ClCOO^-$

FIG. 1. Chaotropic ions in decreasing order of potency as found in most cases.

or haloacetates), relative chaotropic potency is also inversely related to their charge densities (Fig. 1).

Application

Variations in composition and stability of biological materials preclude the formulation of a general procedure for the application of chaotropic agents. Furthermore, chaotropes can be used for a large variety of purposes in addition to resolution of membranes and isolation of membrane-bound enzymes. They can be used for increasing the water solubility of many nonelectrolytes,[2,4] controlled perturbation of the native structure of proteins and nucleic acids,[6,7] depolymerization of protein aggregates and multimeric enzymes,[8] resolution of antigen–antibody complexes,[9] solubilization of cell surface antigens,[10] study of the kinetics and the thermodynamics of resolution processes involving protein ± lipid aggregates, such as membranes and multicomponent enzymes,[11-13] investigating the effect of substrates, inhibitors, allosteric effectors, and pH on the structural (tertiary and quaternary) stability of membranes and multimeric enzymes[11,14-16] as well as the conformational (secondary and tertiary) stability of monomeric

[6] P. H. von Hippel and K.-Y. Wong, *Science* **145**, 577 (1964); F. J. Castellino and R. Barker, *Biochemistry* **7**, 4135 (1968).

[7] K. Hamaguchi and E. P. Geiduschek, *J. Amer. Chem. Soc.* **84**, 1329 (1962).

[8] A. Holtzer, T.-Y. Wang, and M. E. Noelken, *Biochim. Biophys. Acta* **42**, 453 (1960); J. Wolff, *J. Biol. Chem.* **237**, 230 (1962); B. Nagy and W. P. Jencks, *J. Amer. Chem. Soc.* **87**, 2470 (1965).

[9] W. B. Dandliker, R. Alonso, V. A. deSaussure, F. Kierszenbaum, S. A. Levison, and H. C. Shapiro, *Biochemistry* **6**, 1460 (1967).

[10] H. Marquardt and C. B. Wilson, *Fed. Proc., Fed. Amer. Soc. Exp. Biol.* **31**, 765, Abstract (1972); H. Marquardt, C. B. Wilson, and F. J. Dixon, *Biochemistry* **12**, 3260 (1973).

[11] K. A. Davis and Y. Hatefi, *Biochemistry* **8**, 3355 (1969).

[12] K. A. Davis and Y. Hatefi, *Arch. Biochem. Biophys.* **149**, 505 (1972).

[13] W. G. Hanstein, K. A. Davis, and Y. Hatefi, *Arch. Biochem. Biophys.* **147**, 534 (1971).

[14] J. S. Rieske, H. Baum, C. D. Stoner, and S. H. Lipton, *J. Biol. Chem.* **242**, 4854 (1967).

[15] Y. Hatefi and W. G. Hanstein, *Arch. Biochem. Biophys.* **138**, 73 (1970).

[16] W. G. Hanstein, K. A. Davis, M. A. Ghalambor, and Y. Hatefi, *Biochemistry* **10**, 2517 (1971).

proteins.[17,18] In addition, most chaotropes are good permeant ions and can be used for studying nonspecific transmembrane ion movement, collapsing transmembrane gradients, and for inducing membrane lipid autoxidation.[15,19,20] The following sections illustrate some of the examples enumerated above.

Membrane Resolution and Protein Extraction

Nearly 70% of the proteins of freshly prepared ETP from bovine heart mitochondria suspended in 0.25 M sucrose can be rendered soluble[21] in the presence of 2 M NaClO$_4$.[22] While such a massive protein extraction demonstrates the potency of chaotropic agents for resolution of membranes, it is often the specific isolation of a membrane-bound protein or resolution of an enzyme complex that is more desirable. Many examples of the use of chaotropes for selective protein extraction or controlled resolution are found in the literature, both prior and subsequent to our systematic application of these ions and formulation of their mechanism of action. However, the following examples regarding the resolution of complex II (succinate–ubiquinone reductase) with respect to succinate dehydrogenase (SD),[23,24] the extraction of SD from *Rhodospirillum rubrum* chromatophores,[25] and the resolution of bovine SD into its subunits[16,24] should suffice in pointing out the usefulness of chaotropic salts for selective resolution and enzyme purification. The example of SD was selected in part because chaotropes made it possible for the first time to isolate and completely purify SD without damaging this labile enzyme.

Figure 2 shows the chaotrope-induced resolution of complex II with respect to SD. The ordinate is a measure of succinate-Q or succinate–PMS reductase activity. The rapid loss of the former activity upon addition of chaotropes indicates the resolution of complex II into water-soluble SD

---

[17] R. Cammack, K. K. Rao, and D. O. Hall, *Biochem. Biophys. Res. Commun.* **44**, 8 (1971).

[18] J. A. Gordon, *Biochemistry* **11**, 1862 (1972).

[19] R. S. Cockrell, *Biochem. Biophys. Res. Commun.* **46**, 1991 (1971).

[20] W. G. Hanstein and Y. Hatefi, *Arch. Biochem. Biophys.* **138**, 87 (1970).

[21] That is, not sedimentable by 2 hours centrifugation at 105,000 g.

[22] As might be expected, some of the proteins so solubilized become once again water-insoluble after removal of NaClO$_4$ by dialysis.

[23] Abbreviations: SD, succinate dehydrogenase; PMS, phenazine methosulfate; Q, ubiquinone (coenzyme Q); SDS, sodium dodecyl sulfate; TBA, tribromoacetate; TCA, trichloroacetate; TFA, trifluoroacetate; DCA, dichloroacetate; MCA, monochloroacetate; Gu-HCl, guanidine hydrochloride; AcO⁻, acetate; ETP, electron transport particles.

[24] K. A. Davis and Y. Hatefi, *Biochemistry* **10**, 2509 (1971).

[25] Y. Hatefi, K. A. Davis, H. Baltscheffsky, M. Baltscheffsky, and B. C. Johansson, *Arch. Biochem. Biophys.* **152**, 618 (1972).

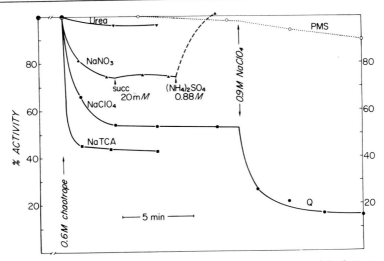

FIG. 2. Resolution of complex II with respect to succinate dehydrogenase by various chaotropes. Complex II was suspended in 50 mM Tris·HCl, pH 8.0. After addition of 0.6 M chaotrope the concentration of complex II was 8 mg/ml. After addition of the salts, samples were taken at the intervals indicated and assayed for succinate-Q and succinate—phenazine methosulfate (PMS) reductase activities. Solid lines, succinate–ubiquinone (Q) reductase activity; dotted line, succinate-PMS reductase activity; dashed line, reconstitution of sucinnate–Q reductase activity upon addition of ammonium sulfate. Resolution temperature, 0°; assay temperature, 38°. The complex II preparations used in the experiments of this and Figs. 12 ,16, and 17 had specific activities between 40 and 45 μmoles of Q$_2$ reduced by succinate per minute per milligram of protein. From K. A. Davis and Y. Hatefi, *Arch. Biochem. Biophys.* **149**, 505 (1972).

and a particulate fraction containing cytochrome b_{557},[26] lipids, and other proteins. As shown by the dotted line, however, the PMS reductase activity of the system remains unchanged, because this activity is the same in the membrane-bound and the soluble forms of SD. The resolution of complex II with respect to SD is an equilibrium process (see below). The yield of solubilized SD is a function of the concentrations of both complex II and the added chaotrope (Fig. 3A and 3B). As seen in Fig. 2, both the rate and the extent of complex II resolution are also functions of the potency of the added chaotrope (TCA$^-$ > ClO$_4^-$ > NO$_3^-$). The selectivity of the resolution of complex II with respect to SD is depicted on SDS-acrylamide gels in Fig. 4. Tube 1 is complex II, tube 2 is the first extraction of SD (2

[26] One of three *b*-type cytochromes found in beef-heart mitochondria.[27] The designation 557 refers to the position of the major α peak in nanometers at 77°K.

[27] K. A. Davis, Y. Hatefi, K. L. Poff, and W. L. Butler, *Biochem. Biophys. Res. Commun.* **46**, 1984 (1972).

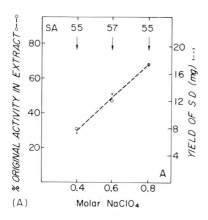
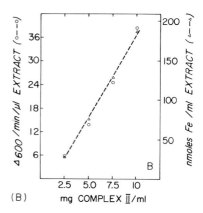

Fig. 3. (A) Correlation between the concentration of NaClO₄ and the amount of succinate dehydrogenase (SD) extracted from complex II at 12 mg of protein per milliliter. Left-hand ordinate: percent of complex II phenazine methosulfate (PMS) reductase activity found in the 49,000-rpm supernatant after extraction with NaClO₄. Right-hand ordinate: yield of succinate dehydrogenase after ammonium sulfate fractionation. SA: specific activity of succinate dehydrogenase isolated at the NaClO₄ concentrations indicated. (B) Correlation between concentration of complex II and the amount of succinate dehydrogenase extracted from it with 0.4 M NaClO₄. Left-hand ordinate: PMS reductase activity of the extract. Right-hand ordinate: Amount of iron in the extract, which is proportional to the amount of succinate dehydrogenase rendered soluble. From K. A. Davis and Y. Hatefi, *Biochemistry* **10**, 2509 (1971).

subunits) by 0.4 M NaClO₄, tube 3 is the second extraction of SD by 0.75 M NaClO₄, and tube 4 is the remainder of complex II after the two extractions. It is seen that only the two subunits of SD are extracted from complex II, and all the other protein components of complex II are essentially undiminished in going from tube 1 to tube 4.

A similar procedure was applied subsequently for the isolation of SD from *R. rubrum* chromatophores.[25] This example demonstrates the use of chaotropes for selective extraction of a single enzyme from whole membranes. The chromatophore preparation was washed once with 0.35 M NaClO₄ in 0.1 M glycylglycine, pH 7.4. At this level of chaotrope, a considerable amount of protein became soluble, but SD remained almost completely membrane bound. The chromatophores, so depleted of easily extractable proteins, were then reextracted with 0.85 M NaClO₄. At this point SD became soluble, and was collected by ammonium sulfate precipitation (35–50% saturation). This simple procedure resulted in an 80-fold purification and a 40% yield of SD based on the assayable activity of chromatophores.

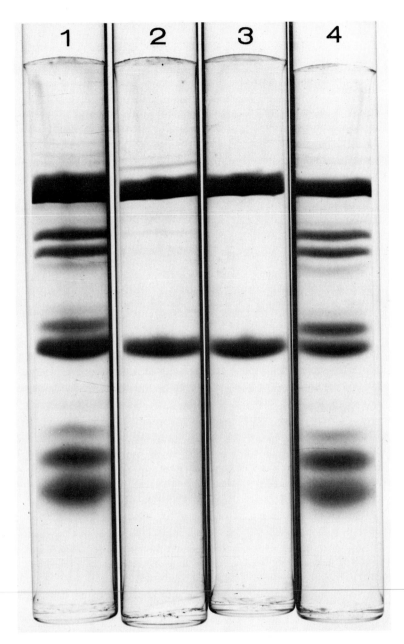

Fig. 4. Profile of the purification procedure of succinate dehydrogenase (SD) depicted on polyacrylamide gels. 1, complex II; 2, first extraction of SD with 0.4 M $NaClO_4$; 3, second extraction of SD with 0.75 M $NaClO_4$; 4, particulate fraction of complex II remaining after twice removal of SD. All protein samples were treated according to the method of K. Weber and M. Osborn, *J. Biol. Chem.* **244**, 4406 (1969). From K. A. Davis and Y. Hatefi, *Biochemistry* **10**, 2509 (1971).

The third example concerns the resolution of bovine SD into its sub-units.[16,24] High concentrations (6–8 M) of guanidine and urea, or potent detergents such as SDS, are commonly used for the resolution of multimeric enzymes into their subunits. However, these procedures usually result in completely denatured products and the loss of loosely bound cofactors, such as iron and acid-labile sulfide found in SD. In contrast to these drastic methods, the resolution of SD in the presence of chaotropes allowed the separation of the two subunits of the enzyme, and their identification as an iron-sulfur flavoprotein and an iron-sulfur protein. The spectra of SD and its two subunits, and their molecular weights and composition are shown respectively in Fig. 5 and Table I. It might be of interest to note that the resolution of SD by chaotropes required repeated freeze-thawing, respec-tively, in liquid nitrogen and at room temperature (Figs. 6 and 7). Neither freeze-thawing nor chaotrope treatment alone was effective (Fig. 7). Facili-tation of the resolution of SD by cold treatment is reminiscent of the cold lability of F_1 (mitochondrial ATPase) and the effect of chaotropic ions on its cold-induced depolymerization.[28] It is also in agreement with the con-

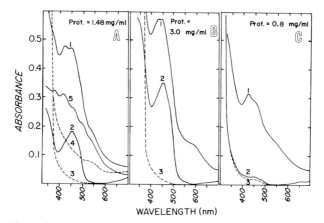

FIG. 5. Absorption spectra of succinate dehydrogenase (A) and its iron-sulfur flavoprotein (B) and iron-sulfur protein (C) subunits. (A) Trace 1, oxidized SD; trace 2, flavin contribution to spectrum of oxidized SD; trace 5, iron-sulfur con-tribution to spectrum of oxidized SD; trace 4, SD treated with dithionite; trace 3, SD treated with sodium mersalyl and dithionite. (B) Trace 1, oxidized iron-sulfur flavo-protein; trace 2, flavin contribution to trace 1; trace 3, iron-sulfur flavoprotein after treatment with sodium mersalyl and dithionite. (C) Trace 1, oxidized iron-sulfur protein; trace 2, after treatment of 1 with mersalyl; trace 3, after treatment of 2 with dithionite. For details see K. A. Davis and Y. Hatefi, *Biochemistry* **10**, 2509 (1971).

[28] H. S. Penefsky and R. C. Warner, *J. Biol. Chem.* **240**, 4694 (1965).

TABLE I

Molecular Properties of Succinate Dehydrogenase from Bovine Heart

Molecule	Molecular weight	Flavin	Iron	Labile sulfide
Succinate dehydrogenase	$97,000 \pm 4\%$	10.3^a	70–80	70–80
Iron-sulfur flavoprotein	$70,000 \pm 7\%$	12–13	45–55	45–55
Iron-sulfur protein	$27,000 \pm 6\%$	<0.5	95–110	90–100

[a] Average of 12 preparations. Values are expressed as nanomoles (ng atoms) per milligram of protein.

clusions of Kauzmann[1] and Scheraga et al.[29] regarding the inverse relationship between temperature and the strength of hydrophobic bonds.

Kinetics and Thermodynamics of Membrane Resolution

Kinetic and thermodynamic characteristics of the chaotrope-induced resolution of membranes and multimeric systems can be studied with considerable accuracy. These measurements not only are interesting per se, but can be used to obtain important information regarding the effect of

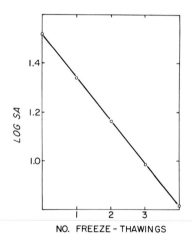

NO. FREEZE – THAWINGS

Fig. 6. Effect of freeze-thawing on the chaotrope-induced resolution of succinate dehydrogenase. The enzyme in 50 mM sodium phosphate (pH 8.0), 20 mM succinate, and 5 mM dithiothreitol (protein 5.3 mg/ml) was treated with 0.3 M sodium trichloroacetate, frozen in liquid nitrogen, thawed at room temperature, and assayed for reduction of phenazine methosulfate as indicated. From W. G. Hanstein, K. A. Davis, M. A. Ghalambor, and Y. Hatefi, *Biochemistry* **10**, 2517 (1971).

[29] H. A. Scheraga, G. Nemethy, and I. Z. Steinberg, *J. Biol. Chem.* **237**, 2506 (1962).

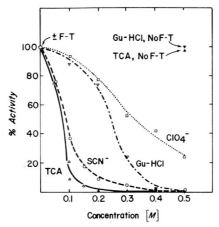

FIG. 7. Effect of chaotropic agents and freeze-thawing on the resolution of succinate dehydrogenase. The enzyme at a protein concentration of 1.48 mg/ml of 50 mM sodium phosphate (pH 7.0) and 20 mM succinate, was treated with various chaotropic agents as shown in the figure, and frozen in liquid nitrogen, and thawed at room temperature three times. Since in the presence of higher concentrations of chaotropes the iron-sulfur flavoprotein subunit precipitated after freeze-thawing, all samples were centrifuged for 1 minute at 35,000 rpm, and only the clear supernatant was assayed for PMS reductase activity. F-T, freeze-thawing. From W. G. Hanstein, K. A. Davis, M. A. Ghalambor, and Y. Hatefi, *Biochemistry* **10**, 2517 (1971).

substrates, allosteric effectors, inhibitors, and pH on the structure and stability of multicomponent systems.

Figure 8 shows first-order plots of the resolution of complex I (DPNH-Q reductase) with respect to solubilization of DPNH dehydrogenase at 0.4, 0.45, and 0.5 M $NaClO_4$, and Fig. 9 shows plots of the first-order rate constants so obtained versus the concentration of various chaotropes. Similar results (Fig. 10) were also obtained when chaotrope-induced lipid autoxidation in complex I was used as an index of its destabilization. These figures show (a) the potency order of various chaotropes for the resolution of complex I (compare with Fig. 1), (b) the accuracy with which the rate of complex I resolution can be controlled over a very wide range, and (c) the cooperative nature of complex I resolution, which is in agreement with the destabilization characteristics of biological macrostructures.[30]

[30] The four plateau regions in Fig. 10 at approximately 20, 40, 70, and 100 nmoles O_2/min/mg are also interesting, since it has been shown elsewhere that complex I contains 4 iron-sulfur center[31] and that destabilized iron-sulfur proteins are potent catalysts of membrane lipid autoxidation.[20]

[31] N. R. Orme-Johnson, W. H. Orme-Johnson, R. E. Hansen, H. Beinert, and Y. Hatefi, *Biochem. Biophys. Res. Commun.* **44**, 446 (1971).

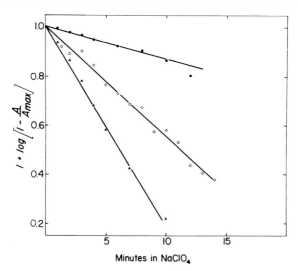

FIG. 8. First-order plot of the kinetics of complex I resolution at various concentrations of NaClO₄. Rate constants (k) were calculated from $k = -2.303 \times$ slope. A is specific activity at time t, A_{max} is maximum dehydrogenase activity releasable from complex I, and k is in $min^{-1} \times mg\ protein^{-1}$. ●—●, 0.4 M, $k = 0.030$; ○—○, 0.45 M, $k = 0.104$; ▲—▲, 0.5 M, $k = 0.184$. From K. A. Davis and Y. Hatefi, *Biochemistry* **8**, 3355 (1969).

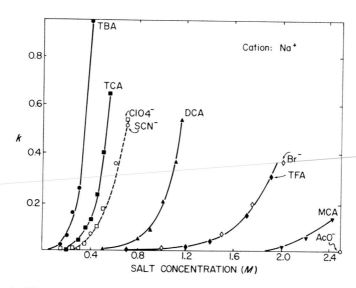

FIG. 9. Effect of various chaotropes on the first-order rate constants (k) of the resolution of complex I. The dimension of k is the same as in Fig. 8.

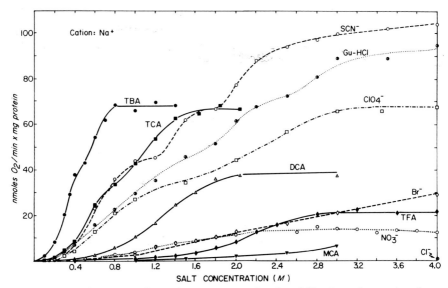

FIG. 10. Effect of various chaotropes on the destabilization of complex I as measured by lipid autoxidation. TBA, tribomoacetate; TCA, trichloroacetate; TFA, trifluoroacetate; DCA, dichloroacetate; MCA, monochloroacetate; Gu-HCl, guanidine hydrochloride. From W. G. Hanstein, K. A. Davis, and Y. Hatefi, *Arch. Biochem. Biophys.* **147**, 534 (1971).

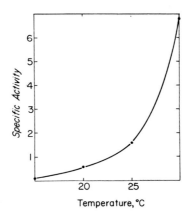

FIG. 11. Kinetics of the resolution of complex I as a function of the temperature of the incubation medium. The release of menadione reductase activity was measured 2.5 minutes after incubation of complex I with 0.47 M NaClO$_4$ at the temperatures indicated. From K. A. Davis and Y. Hatefi, *Biochemistry* **8**, 3355 (1969).

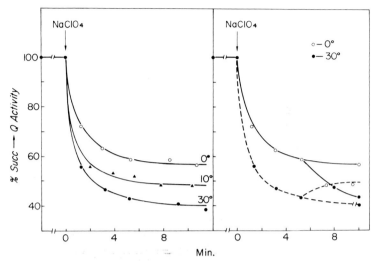

Fig. 12. Effect of temperature on the resolution of complex II. Complex was suspended in a solution containing 30 mM Tricine, 100 mM boric acid, 20 mM succinate, and 5 mM dithiothreitol. pH of the buffer was adjusted to 8.1 with NaOH. Complex II was resolved by addition of 1.0 M NaClO$_4$ at the temperatures indicated. After addition of NaClO$_4$ the concentration of complex II was 8 mg/ml. Right-hand panel shows increased resolution and reconstitution as the temperature was changed, respectively, from 0° to 30° and from 30° to 0°. Assays were performed at 38°. From K. A. Davis and Y. Hatefi, *Arch. Biochem. Biophys.* **149**, 505 (1972).

The temperature dependencies of NaClO$_4$-induced complex I and complex II resolution are shown in Figs. 11 and 12. It is seen that the resolution of complex I is highly temperature dependent and can be completely inhibited, under the conditions used, at temperatures below 15°, whereas complex II resolution is only slightly affected by decreasing the tempera-

TABLE II
Activation and Thermodynamic Parameters for the Resolutions
of Complexes I and II

Complex	Parameter	Value
I[a]	$\Delta H^{\ddagger}$	+37 kcal/mole
	$\Delta S^{\ddagger}$	+53 eu
II[b]	ΔH	+4 kcal/mole
	ΔS_u	−15 eu[c]

[a] Conditions were the same as in Fig. 11.
[b] Conditions were the same as in Fig. 12. For other details, see K. A. Davis and Y. Hatefi, *Arch. Biochem. Biophys.* **149**, 505 (1972).
[c] Under the assay conditions ΔS was approximately +13 eu.

ture from 30° to 0°. The activation and thermodynamic parameters calculated, respectively, for the resolutions of complex I and complex II are
shown in Table II. The large activation enthalpy of the resolution of complex I agrees with the high stability of this complex in aqueous media, and
the small ΔH value associated with the resolution of complex II adds further support to our finding that the binding of SD to particles is mainly
hydrophobic and cannot involve too many ionic bonds.[12]

Substrates and Inhibitors. ETP, and complexes I, II, and III are structurally stabler in the presence of substrates, and Rieske *et al.*[14] have shown
that antimycin A inhibits the chaotrope-induced b-c_1 cleavage of complex
III. Representative data concerning the effect of reductants on complexes
I and III, and the effect of antimycin A on complex III, are shown in
Figs. 13, 14, and 15, respectively. With regard to Fig. 15, it is important
to note that antimycin A completely inhibits both the enzymatic activity
and the chaotrope-induced b-c_1 cleavage of complex III when it is added
in amounts stoichiometric with cytochrome c_1 (the arrow in Fig. 15 indicates the c_1 content of complex III). As pointed out by Slater,[32] these
and other findings indicate that antimycin A is an allosteric inhibitor of
complex III.

pH. Regarding the effect of pH, it was mentioned earlier that Coulombic forces can make significant contributions to the stability (or destabiliza-

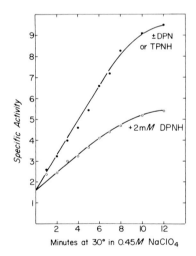

FIG. 13. Inhibition of $NaClO_4$ induced resolution of complex I by DPNH. Experimental conditions were the same as in Fig. 11. From K. A. Davis and Y. Hatefi,
Biochemistry **8**, 3355 (1969).

[32] J. Bryla, Z. Kaniuga, and E. C. Slater, *Biochim. Biophys. Acta* **189**, 317 (1969).

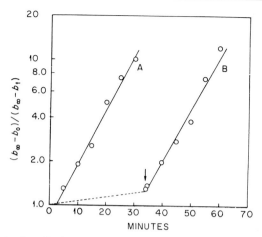

FIG. 14. Effect of preliminary exposure of reduced complex III to guanidine on the subsequent cleavage at pH 7.0 of oxidized complex III by guanidine. The reaction mixture was prepared by combining of 2.0 ml of a solution of complex III (21 mg of protein per milliliter in sucrose·Tris); 0.1 ml of 1.0 M phosphate buffer, pH 7.0; 0.1 ml of 30% sodium taurocholate (to prevent nonspecific precipitation); 0.15 ml of water; 0.25 ml of 6.0 M guanidine·HCl; 0.2 ml of 0.1 M dithiothreitol; and 0.2 ml of 0.5 M potassium ferricyanide. After the reaction was started by the addition of guanidine·HCl, 0.3-ml samples were removed at desired intervals and were treated with 0.1 μmole of antimycin. Each sample (0.3 ml) was treated with 0.02 ml of 2 M Tris·HCl (pH 8), 0.025 ml of water, and 0.055 ml of 6.0 M guanidine·HCl, and then was incubated for 5 minutes at 20°. The cleaved cytochrome b was precipitated by dilution of each sample with 0.5 ml of water and estimated. In Reaction A, both dithiothreitol and potassium ferricyanide were added prior to addition of guanidine·HCl. In Reaction B, complex III was kept in the reduced state after addition of guanidine·HCl; after 34 minutes the complex was oxidized by addition of ferricyanide (arrow). The reaction temperature was 0°. The ordinate is a plot of cleaved cytochrome b on a log scale. b_0, b_t, and b_∞ are the amounts of cytochrome b cleaved, respectively, at times zero, t, and infinity. From J. S. Rieske, H. Baum, C. D. Stoner, and S. H. Lipton, *J. Biol. Chem.* **242**, 4854 (1967).

tion) of the secondary, tertiary, and quaternary structure of biological materials.[33] Thus, changes of pH can be used in appropriate systems to accentuate or inhibit their chaotrope-induced destabilization. For example, the resolution of SD into its subunits by the application of chaotropes (e.g., 0.4 M Na-TCA) plus freeze-thawing occurs easily at the pH range 6.5–7.3, but becomes considerably inhibited at pH values above 8.3. Since potent

[33] In this regard, it might be added that metal ions can also contribute to the structural stability of membranes. In fact, chelators have been used to extract membrane-bound proteins.[34]

[34] V. T. Marchesi and E. Steers, *Science* **159**, 203 (1968); G. L. Nicolson, V. T. Marchesi, and S. J. Singer, *J. Cell Biol.* **51**, 265 (1971).

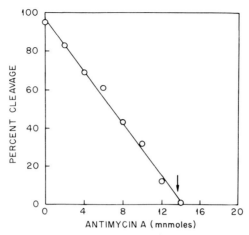

Fig. 15. Effects of preliminary treatment of complex III with graded amounts of antimycin on the cleavage of complex III by guanidine. Samples of complex III (0.2 ml 20 mg of protein per milliliter in sucrose·Tris) were treated with 2.0 μl of ethanolic solutions of antimycin (1–10 μmoles/ml). To these samples were added 0.04 ml of Tris·HCl (2 M, pH 8), 0.06 ml of water, and 0.07 ml of 6 M guanidine·HCl, in the order listed. The mixtures were incubated at 20° for 10 minutes, then diluted with 1.0 ml of water; the precipitated cytochrome b was estimated. The arrow points to the estimated content (in nanomoles) of cytochrome c_1 in each sample of complex III. From J. S. Rieske, H. Baum, C. D. Stoner, and S. H. Lipton, *J. Biol. Chem.* **242,** 4854 (1967).

chaotropes are neutral salts of strong acids, their ionization would not be affected by pH studies appropriate for biological systems.[35] Therefore, any pH effect on the chaotrope-induced destabilization of a system might be considered as a reflection of the Coulombic effects involved in the structural stability of the system itself. In the example given above, pH 8.3 coincides with the pK of the cysteine sulfhydryl group. However, in a protein another group (e.g., α-amino or imidazole) might also have a pK near 8.

Antichaotropes and D_2O

The foregoing sections described the effect of decreased water structure, brought about by chaotropic ions, on the destabilization of membranes, enzyme complexes, and macromolecules. In this section, it will be shown that increased water structure can result in increased stability of biological macrostructures. Thus, water structure-forming ions, such as SO_4^{-2}, HPO_4^{-2}, and F^-, have a strong salting-out effect on nonelectrolytes, and tend to strengthen hydrophobic interactions in proteins and multicomponent systems. In contrast to chaotropes, the water structure-forming ions have high

[35] This comment also applies to the guanidinium ion whose pK is 13.65.

charge densities, very small (and often negative) entropies of aqueous ions,[2] positive viscosity B coefficients,[36] and retard water self-diffusion.[2] These *anti*-chaotropic ions can be used, in fact, to reverse the effect of water structure-breaking ions and reconstitute an enzyme complex which has been resolved by chaotropes. Figures 16 and 17A show the resolution of complex II by chaotropes followed by its reconstitution upon addition of antichaotropes, and Fig. 17B shows that the complex can be reconstituted also by removing the chaotrope by the addition of KCl and precipitation of KClO$_4$. It should be pointed out that (a) the levels of antichaotropes used in the above experiments are far below precipitating concentrations, and (b) the relative potencies of antichaotropes for reconstitution are not correlated with their relative ionic strengths (Fig. 17).

Another way of increasing the stability of membranes and enzyme complexes in aqueous media is by using D$_2$O as solvent. Liquid D$_2$O is considered to be more structured than H$_2$O.[37] Therefore, it might be expected that D$_2$O would have a strengthening and stabilizing effect on the hydrophobic interactions of biological macrostructures. As seen in Table III, the

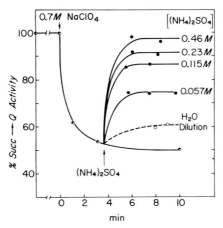

FIG. 16. Resolution of complex II by NaClO$_4$, and reconstitution by increasing concentrations of ammonium sulfate as indicated. Complex II was suspended in a solution containing 50 mM Tris·HCl, pH 8.0, 20 mM succinate, and 5 mM dithiothreitol. After the addition of NaClO$_4$, the concentration of complex II was 8 mg/ml. Other conditions were the same as in Fig. 2. From K. A. Davis and Y. Hatefi, *Arch. Biochem. Biophys.* **149**, 505 (1972).

[36] W. P. Jencks, "Catalysis in Chemistry and Enzymology." McGraw-Hill, New York, 1969; R. A. Robinson and R. H. Stokes, "Electrolyte Solutions." Butterworth, London, 1968.
[37] E. M. Arnett and D. R. McKelvey, in "Solute-Solvent Interaction" (J. F. Coetzee and C. D. Ritchie, eds.). Dekker, New York, 1969.

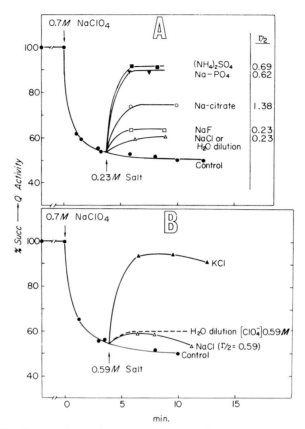

Fig. 17. Resolution of complex II by $NaClO_4$ and reconstitution by (A) various antichaotropic salts, and (B) removal of ClO_4^- with K^+. Conditions were the same as in Fig. 16. At the concentrations used here and in Fig. 16, the antichaotropic salts had no effect on the original activity of unresolved complex II. From K. A. Davis and Y. Hatefi, *Arch. Biochem. Biophys.* **149**, 505 (1972).

$NaClO_4$-induced resolution of complex II with respect to succinate dehydrogenase is extensively inhibited (50% and 40% in the presence of 0.4 and 0.75 M $NaClO_4$ respectively) when the complex is suspended in D_2O instead of H_2O. Similar results are shown at 30° and 20° in Fig. 18 for the effect of D_2O on the resolution kinetics of complex I at different $NaClO_4$ concentrations. Thus at 30° and 0.7 M $NaClO_4$, the resolution rate of complex I in D_2O is only 38% of that in H_2O, and a visual inspection of Fig. 18 suggests that, at a given chaotrope concentration, the temperature of the D_2O medium would have to be raised by approximately 4–6° in order to achieve the same rate of resolution as in H_2O. It should be pointed

TABLE III
EFFECT OF D₂O ON THE RESOLUTION OF COMPLEX II BY NaClO₄

Medium	NaClO₄ (M)	Succinate → PMS[a] activity of extract (μmoles/min × μl)	Yield of SD[a] (mg)	Succinate → PMS activity of SD (μmoles/min × mg)
H₂O	0.4	0.142	4.47	70
D₂O	0.4	0.075 (53%)	2.31 (52%)	65
H₂O	0.75	0.270	5.29	76
D₂O	0.75	—	3.92 (62%)[b]	76

[a] PMS, phenazine methosulfate; SD, succinate dehydrogenase.
[b] Based on the amount of SD left in each complex II pellet after the first extraction.

out that the stability differences observed between H₂O and D₂O media are actually considerably minimized when measured in the presence of effective concentrations of chaotropes. This is true also for the stability differences discussed above with respect to substrates and other additives.

Weak Chaotropes

In most cases the relative effectiveness of chaotropes follows the order of potency given in Fig. 1. However, there are instances in which a weak

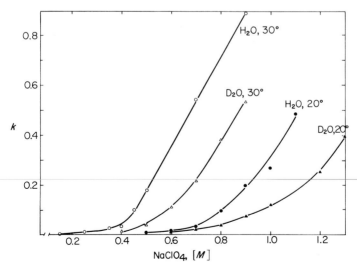

FIG. 18. Effect of D₂O on the first-order rate constants (k) of the resolution of complex I with NaClO₄ at 20° and 30°. The dimension of k is the same as in Fig. 8. From W. G. Hanstein, K. A. Davis, and Y. Hatefi, unpublished.

chaotrope, such as Br^- or NO_3^- (occasionally even Cl^-), is more effective than a potent chaotrope for membrane resolution and enzyme extraction. For example, it has been found by Marquardt et al.[10] that KBr is more effective than Na-TCA in solubilizing human glomerular basement membranes (GBM), which contain antigens concerned with the autoimmune disease generally referred to as antiglomerular basement membrane nephritis. Thus, 3.5 M KBr could be used at 36° to solubilize approximately 25% of GBM proteins, whereas at the optimal concentration of Na-TCA only 10% GBM protein was solubilized.[38]

We have not investigated the reasons for the greater effectiveness of weak chaotropes in certain instances. However, one of the contributing factors might be as follows. If both electrostatic and hydrophobic forces contribute significantly to the association of a protein with a membrane, then it is conceivable that chaotropes with greater charge density might be more effective than ClO_4^-, TCA^-, and TBA^- for destabilizing this association. As pointed out by Kauzmann,[1] the unitary free energy change for the interaction of two ions is approximately

$$\Delta F_u = -\frac{e_1 e_2}{D}\left(\frac{1}{r_1} + \frac{1}{r_2}\right) \tag{1}$$

where e_1 and e_2 are the charges of the ions, r_1 and r_2 are their radii, and D is the dielectric constant of the solvent. Assuming r_1 and r_2 are independent of temperature, then in water at 25°,

$$-T\Delta S_u = 1.37\Delta F_u \tag{2}$$

whereas

$$\Delta H = -0.37\Delta F_u \tag{3}$$

Equations (2) and (3) and the data presented earlier indicate, therefore, that the stabilities of both electrostatic and hydrophobic interactions are related mainly to entropy changes. However, there is a major difference in the effect of salts on electrostatic and on hydrophobic bonds. In the case of chaotropic salts with a common cation, this difference would arise mainly from the charge densities of the anions.[39] It was pointed out earlier that there is an inverse relationship between chaotropic potencies and the charge densities of related chaotropes. Thus, $I^- > Br^- > Cl^-$, and $TBA^- > TCA^- > TFA^-$. However, Eqs. (1) and (3) indicate that the effect of

---

[38] That the difference in GBM solubilization by KBr and Na-TCA is not caused by the different cations has been ascertained.

[39] This is because the dielectric constant of water is decreased roughly to the same extent by various binary salts at least up to about 2 M concentration.[40]

[40] G. Kortüm, "Treatise on Electrochemistry." Elsevier, Amsterdam, 1965.

these ions for destabilization of electronstatic interactions would be in the opposite direction. Therefore, where both hydrophobic and electrostatic association forces are involved, the destabilizing effect of a weak chaotrope exerted on both forces might be greater in the balance than that of a strong chaotrope. In these considerations, it should be kept in mind that for secondary-tertiary-quaternary structures the thermodynamics of association-dissociation equilibria might be so poised that a small change in the attraction-repulsion forces can bring about a significant change in these equilibria. This is clearly evident from the cooperative nature of dissociation or denaturation effects with respect to the concentration of various perturbants.

Conclusions

The utility of chaotropic salts for selective and controlled resolution of membranes, enzyme complexes, and multimeric proteins has been discussed and examples have been given. It has also been shown that chaotropes can be used for studying the stability changes of the above systems as might be affected by substrates, allosteric effectors, inhibitors, and pH. That other perturbants of biological macrostructures, such as ultrasonic irradiation, detergents, organic solvents, etc., are generally less effective for the purposes enumerated above is obvious and does not require elaboration here. However, the limited effectiveness of certain perturbants is often of considerable advantage. Examples are the use of digitonin for isolation of intact mitochondrial inner membranes,[41] and the use of bile salts for isolation of the four complexes of the mitochondrial respiratory chain.[42]

It is important to point out that by definition chaotropes are potent denaturants. However, most proteins appear to withstand up to about 2 M concentration of strong chaotropes. Another hazard which attends the use of chaotropes and other perturbants is membrane lipid autoxidation, especially at ambient temperatures.[15,20] This problem pertains mainly to mitochondria, chloroplasts, chromatophores, and microsomes, because cytochromes and destabilized iron-sulfur proteins are potent catalysts of lipid autoxidation.[15,20] Once initiated, membrane lipid autoxidation has a vast destructive potential. The proteins and nucleic acids are also modified by the reactive products (e.g., malonaldehyde) of lipid autoxidation.[43,44]

[41] C. C. Cooper and A. L. Lehninger, *J. Biol. Chem.* **219**, 489 (1956); T. M. Devlin and A. L. Lehninger, *J. Biol. Chem.* **233**, 1586 (1958); W. B. Elliot and D. W. Haas, this series, Vol. 10, p. 179.
[42] Y. Hatefi, *Comp. Biochem.* **14**, 199 (1966).
[43] K. S. Chio and A. L. Tappel, *Biochemistry* **8**, 2821, 2827 (1969).
[44] B. R. Brooks and O. L. Klamerth, *Eur. J. Biochem.* **5**, 178 (1968).

Therefore, in the resolution of membranes, anaerobic conditions or the use of antioxidants are strongly recommended.

While chaotropes are useful for perturbation and resolution of macrostructures, often the contrary problem also concerns the investigator. An isolated enzyme might be unstable in solution; it might denature or depolymerize into inactive subunits. In these instances, addition of antichaotropic salts and/or D_2O can be beneficial.

[78] Fast Analyzers for Biochemical Analysis[1]

By T. O. TIFFANY, C. A. BURTIS, and N. G. ANDERSON

The combination of high-resolution separation methods which produce large numbers of fractions for analysis, the requirement for precise kinetic assays, the lability of many biological materials, the requirement in many situations for minute quantities of sample per assay, the use of more complex analytical procedures, and the trend toward the demand for more reliable and definitive results through the use of statistical data reduction routines point toward the need for new fast computer-interfaced microanalytical systems which provide more experimentalist interaction with the experiment in progress and which minimize the tedium involved in the actual analysis. One such analytical system is the Fast Analyzer, developed at Oak Ridge National Laboratory.[1a-8] It has application in a variety of ways to many areas of biochemical analysis. This chapter is devoted to the description of the Fast Analyzer and to the discussion of the use of computer-interfaced Fast Analyzers in biochemical analysis.

Fast Analyzer Instrumentation

The Fast Analyzer system can be thought of as consisting of the following: (1) an analytical module, (2) a data-gathering and -processing

[1] Research sponsored jointly by NIGMS and the U.S. Atomic Energy Commission under contract with Union Carbide Nuclear Corporation.
[1a] N. G. Anderson, Anal. Biochem. 28, 545 (1969).
[2] N. G. Anderson, Anal. Biochem. 32, 59 (1969).
[3] N. G. Anderson, Anal. Biochem. 31, 272 (1969).
[4] N. G. Anderson, Science 166, 317 (1969).
[5] N. G. Anderson, Clin. Chem. Acta 25, 321 (1969).
[6] N. G. Anderson, Amer. J. Clin. Pathol. 53, 778 (1970).
[7] D. N. Mashburn, R. H. Stevens, D. D. Willis, L H. Elrod, and N. G. Anderson, Anal. Biochem. 35, 98 (1970).
[8] J. M. Jansen, Jr., Clin. Chem. 16, 515 (1970).

system, and (3) an automated sample-reagent pipetting and dispensing system.

Analytical Module

The analytical module consists of an analytical rotor and removable transfer disk, an optical system (spectrophotometric, photometric, or fluorometric), some means for temperature regulation and control, and provision for synchronization of the output signal with data-taking devices.

Multiple-Cuvette Rotor

The rotor is constructed by compressing an annulus containing several slots (i.e., one slot per cuvette) between a top window annulus and a bottom window disk. Pyrex glass, quartz, or ultraviolet-transmitting acrylic plastic have been used to fabricate the window components. Rotors

FIG. 1. Disassembled 42-cuvette Fast Analyzer rotor. (A) Lower rotor housing; (B) lower and upper glass windows; (C) Teflon cuvette spacer; (D) upper rotor housing; (E) 42-place Teflon transfer disk; (F) rotor sealing cover. [From C. A. Burtis, W. F. Johnson, J. E. Attrill, C. D. Scott, N. Cho, and N. G. Anderson, *Clin. Chem.* **17**, 686 (1971).]

are available with 15, 16, 30, and 42 cuvettes. A disassembled 42-cuvette rotor is shown in Fig. 1.

One of the unique features of the rotor design is that each cuvette contains an individual siphon. As shown in Fig. 2, the outer edge of each cuvette (the bottom of an ordinary cuvette) is sloped, and a small siphon is connected to the point of greatest radius. These siphons (one per cuvette) are bifunctional in that they are used to mix as well as to dump the contents of the individual cuvettes. By application of either suction at the center of the rotor or air pressure at the edge of the rotor, a burst of air or other gas may be drawn through each cuvette by way of the siphons. The turbulence induced by these air bubbles streaming through the cuvettes ensures thorough and complete mixing of the reaction mixtures. After mixing, the bubbles are easily removed from the optical paths of the cuvettes by rapid acceleration of the rotor. The siphons are arranged so that under normal conditions they will not drain the contents of the cuvettes.

Once the analyses are completed, the reaction mixtures are emptied

FIG. 2. Teflon annulus used to define the cuvettes. Note the individual siphons.

from the cuvettes through the siphons by application of suction at the rotor edge or air pressure at the rotor center. The siphons may drain into a stationary collecting ring around the rotor edge, in which case the center rotor chamber is closed, and a rotating seal is provided to allow air or vacuum to be applied to the rotor center. An alternative approach is to lead the siphon lines back to the center underneath the rotor and attach the air and vacuum lines through a rotating seal at that point.

A removable component of the rotor is the transfer disk, which is used to introduce samples and reagents into the analytical system. A disk contains three concentric rings of discrete cavities; for each cuvette in the rotor, there are three corresponding radial cavities in the disk. Small aliquots of samples (1–50 μl) and reagent (300–500 μl) are discretely loaded into the innermost and central cavities, respectively. The third cavity may be used to contain a second reagent. Its primary function, however, is as a mixing and guide chamber through which the solutions are directed to their respective cuvettes upon rotation. A transfer disk used in a 42-cuvette Fast Analyzer is shown in Fig. 3.

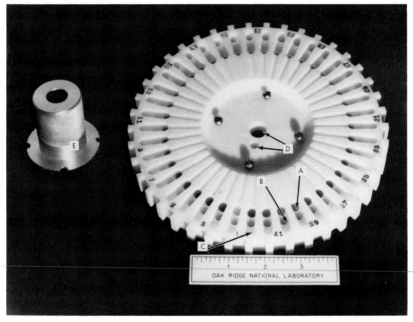

FIG. 3. Transfer disk used in a 42-cuvette Fast Analyzer. (A) Reagent cavity, here filled with 400 μl of biuret solution; (B) sample cavity; (C) mixing and transfer cavity; (D) centering and indexing holes; (E) positioning tool used for carrying, placing, and removing transfer disk from rotor (engages four screws shown in middle of transfer disk). [From C. A. Burtis, W. F. Johnson, J. E. Attrill, C. D. Scott, N. Cho, and N. G. Anderson, *Clin. Chem.* **17**, 686 (1971).]

Optical Systems

After transfer of solutions into their respective cuvettes, the rotational motion of the rotor is used to move the cuvettes rapidly and continuously through a single fixed optical system, as shown in Fig. 4. The normal operating speed of the rotor is in the range of 500–1500 rpm, with a provision for acceleration up to 5000 rpm. This means that the time required for one complete revolution is of the order of 50–100 msec; during this time 15–42 cuvettes can be monitored, depending on the analytical rotor in use. Furthermore, it is possible to obtain several optical readings over the approximately 4–5 msec that each cuvette spends passing by the stationary detection system during each revolution of the analytical rotor.[9] Two optical systems have been developed for use on the Fast Analyzer: photometric and fluorometric.

Photometric. By rapidly passing the cuvettes through a single photometric system, the Fast Analyzer combines the simplicity of a single-beam photometric system with the continuous referencing advantages of a

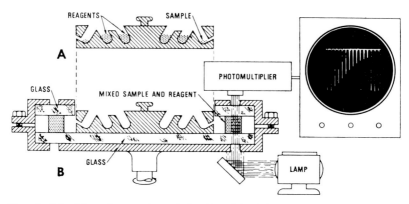

FIG. 4. Principles of operation of the GeMSAEC Fast Analyzer. Measured volumes of reagents and samples are placed in depressions in the fluorocarbon transfer disk shown in (A). The depressions are arranged so that reagents and samples are unmixed at rest, but can move radially without mixing with adjacent sets of reagents and samples. The transfer disk is placed in a cuvette rotor, as shown in (B), and rotated. Centrifugal force moves all liquids out into the rotor cuvettes which rotate past a stationary light beam as shown in (B). The signal obtained from the photomultiplier is displayed on an oscilloscope continuously. Fifteen separate reactions are therefore started at the same time and may be continuously followed. Gross errors in pipetting or reagent or standard preparation may be observed, in the case of fast reactions, after a few seconds. [From C. A. Burtis, W. F. Johnson, J. C. Mailen, and J. E. Attrill, *Clin. Chem.* 18, 433 (1972).]

[9] R. L. Coleman, Dissertation, University of Tennessee, 1971.

double-beam system. This is best illustrated by observing the electrical signal obtained directly from the photomultiplier tube of the optical system. As shown in Fig. 5, the oscilloscope simultaneously displays a series of peaks which are proportional to the light transmission of the solutions contained in each of the individual cuvettes. (For illustrative purposes, the signals generated from a 17-cuvette rotor are shown.) In this type of display the upper base line represents 0% transmittance (infinite absorbance), while the full-scale signal obtained from the first cuvette containing a "blank" sample represents 100% transmittance, which becomes the reference. In addition, the signal obtained when the solid portion of the rotor (i.e., between any two cuvettes) is in the optical light path is proportional to the dark current and can be used to correct the cuvette transmission signals for dark current fluctuations. Thus every 50 to 100 msec, the following signals are obtained: (1) dark current, (2) signal from reference (100% T) cuvette, and (3) sample signals from

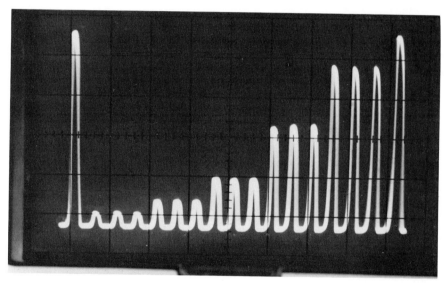

FIG. 5. Photomultiplier signals obtained at 405 nm from a Fast Analyzer as displayed on an oscilloscope. The first cuvette is filled with water, and the remaining cuvettes are filled with solutions containing various quantities of p-nitrophenol in 0.1 M Tris buffer. In this scope display the upper base line represents 0% transmittance (infinite absorbance), while the full-scale signal obtained from the solution in cuvette 1 represents 100% transmittance (zero absorbance). The signals from the other cuvettes are compared with the 100% transmittance signals of cuvette 1 to obtain their relative transmittance. These values are subsequently converted to absorbance and used for calculations.

all other cuvettes. The relative transmittance of each of the reaction mixtures can then be obtained by:

$$\% \, T = \frac{(\text{sample signal} - \text{dark current})}{(\text{reference signal} - \text{dark current})}$$

These values may then be converted to absorbance values, statistically evaluated, and use in various quantitative calculations. Furthermore, the data obtained from successive revolutions of the rotor may be accumulated and processed. This technique of signal averaging further improves the signal-to-noise performance of the analyzer. The net effect of combining continuous referencing with signal averaging is that the analyzer is in essence a multicuvette, high-performance spectrophotometer or photometer which is unaffected by minor signal drift in the system. Consequently, Fast Analyzers are capable of performing precise kinetic enzyme assays,[10,11] since they allow accurate and precise monitoring of small absorbance changes occurring in several simultaneous reactions. This matter will be considered in much more detail later in this chapter.

Monochromatic light for a particular reaction may be obtained either with a monochromator or with interference filters which may be mounted within a filter wheel and controlled either manually or by a computer. By fabricating the cuvette windows from quartz and using the proper light source, reactions may be monitored in the 200–700 nm region of the energy spectrum. A flexible monochromator system consisting of high-intensity grating monochromator (Bausch and Lomb 33-86-07; 200–700 nm),[12] variable slits, and a choice of three lamp sources (tungsten quartz iodine, xenon, and deuterium) has been adapted to the ORNL Fast Analyzer.[10] High-intensity narrow band pass interference filters which have provided good optical performance even at 340 nm have been obtained from Spectro-Film[13] and from the Ditric Company.[14]

Fluorometric Measurements. A basic objective in the development of a fluorometric Fast Analyzer is measurement of solute concentration over a broad concentration range from 10^{-12} to 10^{-5} M. The problem of spanning this concentration range or a greater one is 2-fold and requires (1) sufficient sensitivity of the optical system at the lower range of concentrations to provide sensitivity above the background, and (2) minimum inner filter effect in the higher concentration ranges. The inner

[10] T. O. Tiffany, J. M. Jansen, Jr., C. A. Burtis, J. B. Overton, and C. D. Scott, *Clin. Chem.* **18**, 829 (1972).
[11] C. A. Burtis, W. F. Johnson, J. E. Attrill, C. D. Scott, N. Cho, and N. G. Anderson, *Clin. Chem.* **17**, 686 (1971).
[12] Bausch and Lomb Inc., Rochester, New York.
[13] Spectro-Film, Inc., Winchester, Massachusetts.
[14] Ditric Co., Marlboro, Massachusetts.

filter effect as it is related to right-angle and frontal or surface fluorescence is therefore considered. In an excellent discussion of inner filter effects, Brand and Witholt[15] have presented the requirements for optimization of fluorescence measurements to extend the measurable concentration range. Their consideration was based on the relationship between excitation light intensity and the position in the cuvette from which fluorescence emission is detected. If I_0 is the intensity in quanta available at the cuvette sample interface, the intensity at any given distance y in the cuvette from the cuvette surface can be determined from Beer's law. From this relationship can be obtained,

$$N_a = I_0 e^{-\epsilon c y}(1 - \epsilon^{-\epsilon c \Delta y}) \tag{1}$$

where N_a represents the number of photons absorbed in a volume element at a depth y_1 from the cuvette sample interface. The instrumental interpretation of Eq. (1) is for a right-angle detection system in which Δy is equivalent to the emission slit width and y_1 represents the position of the slit relative to the excited cuvette surface. The measured relative fluorescence intensity is proportional to the number of photons absorbed, and from Eq. (1) it is apparent that the relative intensity can never be exactly linear with respect to concentration. However, by controlling instrumental parameters the measurable concentration range can be extended. When y_1 is set equal to zero and $ec\Delta y$ is very small, an idealized expression can be obtained,

$$N_a = I_0 ec\Delta y \tag{2}$$

which is linear with respect to concentration. Expression (2) can be approached instrumentally by (1) observing the fluorescence emission at the surface of the cuvette (i.e., by setting $y_1 = 0$) and (2) holding $ec\Delta y$ very small. The latter requirement is satisfied when the concentration is very low, or by reducing Δy when working at higher solute concentrations. In this manner the measurable linear concentration range can be extended.

The above considerations concerning instrumental parameters for fluorescence concentration measurements were incorporated into an existing Fast Analyzer; the resulting instrument is shown in Fig. 6. Excitation is achieved with a 150-W xenon-mercury source coupled with a small monochromator (Schoeffel Instrument Corp.),[16] while a quartz optical fiber bundle[17] 10 cm in length and 4 mm in diameter provides the optical path between the monochromator and the cuvette surface. Excitation is

[15] L. Brand and W. Witholt, this series, Vol. 11 [87].
[16] Schoeffel Instrument Co., West Wood, New Jersey.
[17] Schott Glass Co., Duryea, Pennsylvania.

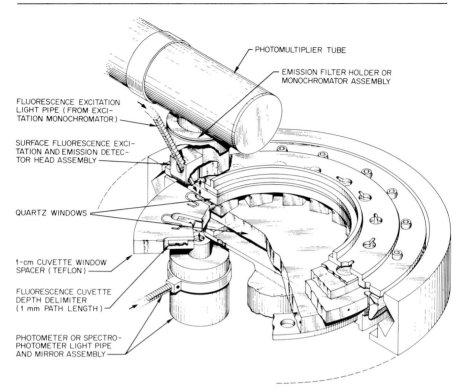

PHOTOMULTIPLIER TUBE

EMISSION FILTER HOLDER OR
MONOCHROMATOR ASSEMBLY

FLUORESCENCE EXCITATION
LIGHT PIPE (FROM EXCI-
TATION MONOCHROMATOR)

SURFACE FLUORESCENCE EXCI-
TATION AND EMISSION DETEC-
TOR HEAD ASSEMBLY

QUARTZ WINDOWS

1-cm CUVETTE WINDOW
SPACER (TEFLON)

FLUORESCENCE CUVETTE
DEPTH DELIMITER
(1 mm PATH LENGTH)

PHOTOMETER OR SPECTRO-
PHOTOMETER LIGHT PIPE
AND MIRROR ASSEMBLY

FIG. 6. Isometric drawing of the Fluorometric Fast Analyzer. [From T. O. Tiffany, C. A. Burtis, J. C. Mailen, and L. H. Thacker, *Anal. Chem.* 45, 1716 (1973).]

at an angle of incidence of 30°, and the emitted signal is detected normal to the rotating cuvette surface. Barrier filters (Baird Atomic, Inc.),[18] cut-on, cut-off filters (Ditric Co.),[14] or sharp-cut filters are used in place of a monochromator for isolation of the emission signal to obtain greater sensitivity. This optical configuration permits one either to use the analyzer for fluorescence measurements or to shift the optical fiber to a mirror assembly which directs the signal through the rotating cuvettes for absorbance measurements after removal of the emission filters. After amplification, the emission signal of the fluorescence analyzer is monitored on-line by a computer which will be described in a following section; this permits the same type of data reduction that is available for spectrophotometry.

Two concerns arise from the use of a multiple-cuvette fluorometer. One is unique to the instrument and the other is common to most instru-

---

[18] Baird-Atomic, Inc., Cambridge, Massachusetts.

ments used for fluorescence measurements. The first concern is the excitation and emission geometry with respect to obtaining a constant identical signal for the same solution from cuvette to cuvette. Slight variations in the excited volume element from cuvette to cuvette can produce a variation in the emitted signal which, if large, can reduce the effectiveness of this approach. The measured uncorrected emission intensity for the 15-cuvette analyzer is about 1%. Through the use of the computer, all cuvettes can be normalized with respect to the first cuvette, and a correction factor for emission intensity variation can be applied to each cuvette. This permits calibration of the fluorescence analyzer with the same reference fluorescence solution in all cuvettes. These calibration factors, when automatically applied in the computer estimation of relative intensity, reduce the variation from cuvette to cuvette to about 0.3%. The second concern is related to the calculation of solute concentration or enzyme activity directly from the emitted intensity of the unknown sample or enzyme assay, without using calibration curves obtained from the use of expensive substrates or products. The multicuvette Fast Fluorometric Analyzer enables one to first obtain the ratio of relative intensity of a standard reference solution prepared under defined conditions to that of the desired solute of enzyme substrate product or coenzyme under defined conditions of buffer, pH, and temperature. This is expressed as the ratio of relative intensity per mole of reference standard to the relative intensity per micromole of solute. Two reference cuvettes can then be used in each run—a buffer (nonfluorescent) blank and a reference standard, respectively. The difference in relative intensity of these ($I_{reference} - I_{blank}$) times the ratio provides an immediate value for the relative intensity of the solute (substrate product, coenzyme, etc.) in terms of relative intensity per micromole of substrate. From this value enzyme activity or substrate concentration can be determined directly using the relative intensity in much the same manner as in spectrophotometry using molar absorptivity. (This will be discussed in more detail in a later section on applications.) Thus, the use of the on-board reference standard compensates for intensity variations, and the use of the ratio permits rapid calculation of enzyme activity in terms of international units per liter directly from the stored relative intensity data.

Temperature Control

Several different methods for accurately and precisely measuring and controlling the temperature of cuvette rotors and transfer disks have been developed by the Oak Ridge National Laboratory and by the manufacturers of commercial versions. Most include some type of temperature sensing probe, usually a calibrated thermistor in the vicinity of the rotor or within the rotor proper (one manufacturer places a thermistor directly in one of the cuvettes). The temperature sensed by the probe is displayed; it is also sensed by a controlling circuit to maintain it at a preset level. Various methods are used to control the temperature of the rotor, including control of the rotor environment temperature by means of either an air or liquid bath or electrical heating of the rotor. By combining monitoring and control circuits, the rotor temperature can be maintained to within ± 0.1–$0.5°$ of the preset temperature.[11,19] In general, this preset temperature ranges from $25°$ to $37°$, with provisions added to allow one to change it as needed. There is, however, considerable variation in the time required to change from one temperature to another in different instruments.

Synchronization

As previously discussed, the basic raw data generated by a Fast Analyzer consists of a series of sequential voltage signals originating from the photomultiplier tube of the analyzer's optical system. In order that these signals may be acquired and processed by the computer in the proper sequence, a series of synchronization signals are required.

In the ORNL-designed Fast Analyzers, synchronization signals are generated by a radially slotted disk attached to the lower end of the rotor shaft, which then rotates between two stationary photodetectors. One of the slots is elongated and produces a separate signal once every revolution of the rotor. This serves to signal the computer when the dark current reading should be taken and when a new set of cuvette readings is about to be initiated. It also synchronizes the rotor speed and photomultiplier output with the oscilloscope display. The second set of signals (i.e., one per each cuvette in the rotor) are synchronized with the cuvette transmission signals and allow for digitization at a precise rotor position (usually two-thirds of the way through a transmission peak plateau) or at the maximum transmission peak as determined by a peak-following circuit.[7,20]

Data System

Fast Analyzers were designed to yield output data in a form and at a rate suitable for direct input into a small computer.[1,4,6] Consequently,

[19] Instrument names and manufacturers of Fast Analyzers are as follows: "Rotochem," American Instrument Co., Silver Spring, Maryland; "GeMSAEC," Electro-Nucleonics, Inc., Fairfield, New Jersey; "CentrifiChem," Union Carbide Corporation, Tarrytown Technical Center, Tarrytown, New York.

[20] N. G. Anderson and W. W. Harris, "Real Time Data Reduction with Fast Analyzers," *Proc. 8th Int. Conf. Med. Biol. Eng. and 22nd Ann. Conf. Eng. Med. Biol.*, 1970.

these analyzers have been successfully interfaced with small digital computers,[8,20,21] and such systems are available commercially.[19]

Description of System

Depending on the application, many options are available in assembling the data system. Consequently, the commercial versions of Fast Analyzers utilize data systems that range from a dedicated, hard-wired digital module to that of a versatile computer system. For research purposes an analyzer-computer interface, a dedicated small computer, and one or more peripheral display devices are essential.

One of the four data systems that have been developed at the Oak Ridge National Laboratory for use with Fast Analyzers[4,7,11,20] is shown in Fig. 7. This system is built around a PDP-8/I computer[22] having 8K of 12-bit core memory and 64K of 13-bit (12 bits + parity) disk memory. The computer is equipped with a photoelectric paper tape reader, Teletype, eight-channel analog signal multiplexer, analog-to-digital converter, 60-Hz real-time clock, 12 output relays and control, 12 relay contact sensors, storage display scope, and oscilloscope. The computer also includes appropriate buffering hardware plus special analog and digital circuits to interface the computer to the GeMSAEC Fast Analyzer. This data system has been demonstrated to be quite powerful and versatile[10,11]; consequently, it is used routinely as well as in a research environment.

The most useful peripheral device is a teletypewriter (TTY), since it allows for both written data input and output through its keyboard and also through its associated paper tape unit. If one requires a more rapid data output, the TTY can be replaced with a high-speed printer. In addition to the written output provided by either a TTY or a high-speed printer, other types of data display are often desirable. An analog display scope that provides a "real-time" display (Fig. 7) of the light transmission of each cuvette is very useful for semiquantitative and diagnostic purposes. With it an operator is immediately able to view what is occurring in each cuvette, providing an instant check of technique and chemistry. An additional peripheral device that has been shown to be quite useful under both routine and research conditions is the storage scope. This type of computer-controlled display provides a convenient means of presenting data in an absorbance-vs.-time format. In rate analysis is provides a "real-time" check on the progression of a reaction; it may be photographed to provide a permanent record of the course of several reactions occurring simultaneously (Fig. 8). In essence it replaces a strip chart recorder, except that

[21] M. T. Kelley and J. M. Jansen, Jr., *Clin. Chem.* **17**, 701 (1971).
[22] Digital Electronics Corporation, Maynard, Massachusetts.

Fig. 7. Computer system developed for use with GeMSAEC Fast Analyzer.

several reactions are displayed simultaneously instead of one as with a strip chart recorder. An X-Y recorder can also be used as an inexpensive means of providing a permanent record of absorbance vs. time of stored data, but it has the disadvantage that plots must be made sequentially for each cuvette.

Programs and Concepts

It is well known that the most efficient and economical use of computer memory as well as minimum computer manipulation and computation time can be obtained by writing the programs directly in assembler language. Therefore, two commercial manufacturers of Fast Analyzers have taken this approach and found 4K of computer memory adequate

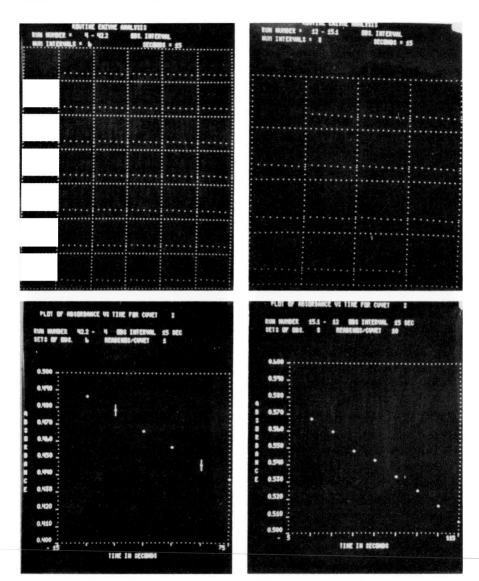

Fig. 8. The use of a digital scope to display data generated by a Fast Analyzer. In the upper two figures the plots of all the individual cuvettes are displayed simultaneously. With the aid of an additional FOCAL program, the operator is able to display the individual rate plot of the desired cuvette. As an example, in the lower figures cuvette 3, contained in each of the upper figures, is displayed individually. Left, 42-place; right, 15-place. [From C. A. Burtis, W. F. Johnson, J. E. Attrill, C. D. Scott, N. Cho, and N. G. Anderson, *Clin. Chem.* **17**, 686 (1971).]

for routine operations. However, assembler-language programming is inflexible except for the specialist. The use of the conversational interpretive language FOCAL[23] readily provides a high degree of flexibility to the data system, but requires a larger memory.

Several basic subroutines exist that allow the experimentalist to set up new programs as they are needed. These include: (1) clock and data-taking functions, (2) a subroutine for converting transmission data to absorbance or relative intensity values, (3) linear least-squares routines for obtaining slope and intercepts, (4) second-degree polynomial least-squares data routine for rate determination at any given time (t), and (5) a variety of subroutines for interpolation, calculation of enzyme activity, substrate concentration, etc. The formation of a new program often involves merely the use of these basic subroutines with minor alterations to subroutines falling under group (5). Some examples of the use of these in programming for different analytical approaches will be presented later in this chapter.

In developing the Fast Analyzer data system shown in Fig. 7, we have utilized both machine and FOCAL language programming to obtain a system that combines the efficiency of machine language with the flexibility and versatility of the FOCAL language. In this system a machine language program controls and directs the basic data acquisition and storage functions of the system. To retrieve, manipulate, and compute these data into a meaningful form, various operational programs have been developed and written in FOCAL. Operation of the system is very simple, and all that is required to bring any of the operational programs into the core of the computer for an analysis is a two-letter command and the number of the program that is desired. Only a few milliseconds are required to transfer a program from the disk into core. This provides a powerful tool for programming and the utilization of larger more complex analytical or statistical data schemes. Several smaller programs can be linked by the library GET command for reduction of acquired data. The statistical enzyme package described in the latter portion of the chapter is an example of such programming.

The two disk units on this system provide generous storage space for 43 FOCAL programs of 2300_8 characters each, with ample space for calibration data for several analyzers and for a complete core image. The program repertoire includes calibration programs with output either on a TTY or a cathode ray tube (CRT). Other programs are available for routine end-point analysis of substrate, enzyme activity assays, statistical evaluation of enzyme kinetic parameters, and linear regression analy-

[23] FOCAL (FOrmula CALculator) is a conversational language useful in mathematical and statistical problems; it is a registered trade name of the Digital Electronics Corporation, Maynard, Massachusetts.

sis; there are also several other programs for the development of analytical experimental procedures.

To operate a Fast Analyzer, the first program required is a calibration program. With this, absorbance of each cuvette is measured from 2 to 100 times for each data point, and the mean values and relative standard deviations (CV) of these readings are printed by the Teletype or displayed on the CRT in both mV and absorbance units. The resulting averaged transmission signal for each cuvette is normalized with respect to cuvette one, and this value is stored for each cuvette as a correction factor. The calibration program thus corrects for small intensity differences that occur from one cuvette to another. After the analyzer (spectrophotometric or fluorometric) is calibrated, any one of several analytical programs can be rapidly loaded off the disk or tape for use.

Sample-Reagent Loader

One of the basic procedures in the operation of a Fast Analyzer is the loading of discrete aliquots of samples and reagent into their respective cavities of a transfer disk. A variety of manual and automatic devices have been utilized or developed to perform this function.[24-26] The one developed at the Oak Ridge National Laboratory (Fig. 9) automatically loads a 15-place transfer disk in 3.25 minutes with a precision of from 0.25 to 0.92%, depending on the sample volume dispensed.[24] Extrapolating to an hour basis, the 3.25-minute loading time of this device corresponds to 18.5 disks per hour or a loading rate of 259 samples per hour (18.5 × 14, since one position of the disk is reserved for referencing purposes). Therefore, since a total of 4–6 minutes is required to analyze and process a single transfer disk, the 3.25 minutes required to automatically load a transfer disk is not the rate-limiting factor in determining the sample analysis rate of the analyzer. It should be noted that all three manufacturers of Fast Analyzers have an automated loading system as a component of their systems.[19]

The Fast Analyzer: Application to Biochemical Analysis and Enzymology

Multiple Parallel Analysis

The use of centrifugal force to rapidly add and mix each sample and reagent simultaneously into the appropriate cuvettes establishes a unique

[24] C. A. Burtis, W. F. Johnson, J. C. Mailen, and J. E. Attrill, *Clin. Chem.* **18**, 433 (1972).

[25] A. G. Skinner, R. Holder, B. E. Northam, and T. P. Whitehead, *Scand. J. Clin. Lab. Invest.* **29**, 16.5 (1972).

[26] C. D. Scott and J. C. Mailen, *Clin. Chem.* **18**, 749 (1972).

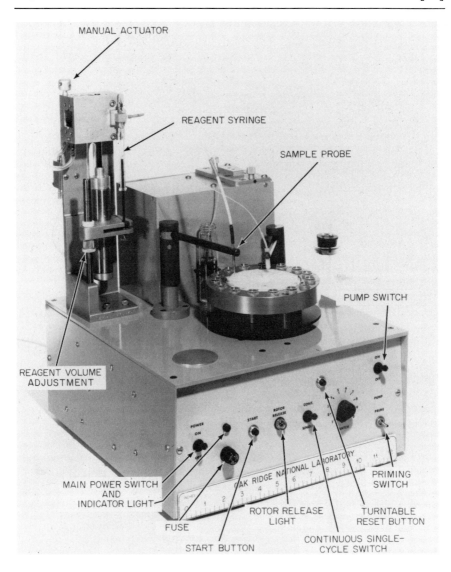

FIG. 9. Front view of automated sample-reagent loader developed to automatically load samples and reagent into a 15-place transfer disk. [From C. A. Burtis, W. F. Johnson, J. C. Mailen, and J. E. Attrill, *Clin. Chem.* **18**, 433 (1972).]

initial reaction starting time (t_0), with all reactions proceeding essentially in parallel under similar conditions of time, temperature, and reaction composition. This concept differentiates the Fast Analyzer from the sequential discrete and sequential segmented flowing-stream analyzers. Furthermore, parallel analysis gives to it a unique capability in performing kinetic assays and many multiple forms of biochemical analyses in a single, well-defined time dimension which not only increases significantly the rate of sample analysis, but also introduces a simple sophistication in approach to analysis which will be discussed at some length in the next few sections. Burtis et al.[11] have discussed the rate of analysis of the Fast Analyzer as a function of cuvette number; they have demonstrated that a rate of analysis of 108 samples per hour can be achieved for a 15-cuvette analyzer, and that this can be increased to over 300 samples per hour when a 42-cuvette analyzer is used and when automated sample and reagent pipetting is used to load the transfer disks.

General Bioanalytical Applications

Colorimetric and Spectrophotometric Analysis

The Fast Analyzer, as a multicuvette spectrophotometer, can perform most of the functions of a spectrophotometer with good precision and accuracy, at a much higher sample analysis rate, and on a microchemical scale requiring only 10–50 μl of sample per assay. This is an important consideration when performing destructive type colorimetric procedures where the sample cannot be recovered. Among the types of photometric analysis that have been adapted for use on the Fast Analyzer are colorimetric end-point[9,27] and spectrophotometric enzymatic end-point analysis,[10] kinetic spectrophotometric enzyme assay,[11,28-30] kinetic substrate analysis[9,10,31] (enzyme catalyzed and noncatalyzed), spectrophotometric multiple wavelength scanning and multiwavelength analysis, and enzyme kinetic parameter evaluation using linear regression analysis for the statistical evaluation of these parameters through the use of the Fast Analyzer's computer for both rapid on-line data taking and immediate off-line statistical evaluation. All these forms of analysis as applied to the analyzer will be discussed in this section except the latter, which will be discussed in a section devoted specifically to enzyme kinetic parameter evaluation.

Enzyme Activity Assays

Important to the consideration of enzyme activity assays on the Fast Analyzer are: (1) instrumental parameters and how they affect the determination of enzyme activity with respect to accuracy and precision; (2)

[27] D. W. Hatcher and N. G. Anderson, Amer. J. Clin. Pathol. 52, 645 (1969).
[28] T. O. Tiffany, G. F. Johnson, and M. E. Chilcote, Clin. Chem. 17, 715 (1971).
[29] D. L. Fabiny-Byrd and G. Ertingshausen, Clin. Chem. 18, 841 (1972).
[30] B. E. Statland and A. L. Louderback, Clin. Chem. 18, 845 (1972).
[31] D. L. Fabiny and G. Ertingshausen, Clin. Chem. 17, 696 (1971).

some approaches that have been made to the kinetic evaluation of enzyme activity through the determination of initial slope or rate and determination of zero-order reaction rate; and (3) the types of enzyme activity assays that have been adapted to the Fast Analyzer.

Instrumental Parameters. Most of the ultraviolet and visible spectrophotometric enzyme activity assays are optimized for zero-order kinetics, i.e., zero-order with respect to substrate ($S \gg K_m$) where the concentration of substrate is much greater than the Michaelis-Menten constant of the enzyme for that particular substrate. Under these conditions the zero-order reaction rate is maintained over a considerable absorbance change or change in substrate concentration, and the reaction rate is considered to be proportional to the activity of the particular enzyme being determined. Determination of this zero-order rate frequently requires that a small absorbance change with time be measured at low and moderately high absorbances. For example, optimized procedures for enzyme activity assays that utilize the conversion of NADH to NAD, either as a primary substrate of the enzyme itself or as a substrate in a coupled assay system, require an initial NADH substrate concentration in the range of 0.17–0.20 mmoles/liter, and the additive absorbance of the sample and the assay solution can be above 1.5 absorbance units. It is apparent that in performing routine enzyme activity assays where several different types of assays are involved, a rather large range of initial absorbance values will be encountered; it therefore becomes imperative to consider both the photometric error and the optical linearity of the instrument over the range of absorbances that are expected to be encountered in a given set of assays and to determine their contribution to the overall accuracy and precision of the assay.

Most instrumentalists are familiar with the derivation of the error in photometric analysis derived from the fundamental Bouguer-Beer law of spectrophotometry.[32,33] If the spectrophotometric analysis obeys Beer's law, the error in photometric analysis will be at a minimum at 0.4343 absorbance unit; the percent relative error in concentration measurements (coefficient of variation of the analysis) will remain near this minimum in the range of 0.2–0.8 absorbance units. In the evaluation of the relative error of measurement of absorbance as a function of absorbance or transmittance, a photometric error of 1% is generally assumed in order to make this parametric evaluation. However, it should be noted that although the minimum in percent relative error of the analysis will be at an absorbance of 0.4343, the magnitude of the percent relative error

[32] G. H. Ayres, *Anal. Chem.* **21**, 652 (1949).
[33] H. K. Hughes, *Appl. Opt.* **2**, 937 (1963).

in absorbance measurements is a function of photometric error; if the photometric error is very small with respect to the total error of the analysis, the useful absorbance range is extended beyond 0.2–0.8 absorbance units.

The photometric error for the Fast Analyzer can be considered to be a function of the electrooptical noise[34] of the analyzer and the resolution of the analog-to-digital (A/D) converter which interfaces the analyzer to the computer. If one considers the latter, the maximum resolution for a typical A/D converter used in Fast Analyzers (12 bit with 4096 words full scale) depends on whether the signal to the A/D converter is a transmission signal (logarithmic) or an absorbance signal (linear). For a full-scale signal across the A/D converter, the resolution for 0–2.0 absorbance units for a linear signal is 0.0005 absorbance unit per word $(2.0/4096 \simeq 0.0005)$; this is the maximum resolution obtainable for any reading from 0 to 2.0 absorbance units. However, for a transmission or logarithmic signal, the relationship becomes more complex. This resolution as a function of absorbance from 0 to 2.0 absorbance units is shown in Fig. 10. These values are for the uncertainty of photometric measurement due to resolution of the A/D converter, and represent the

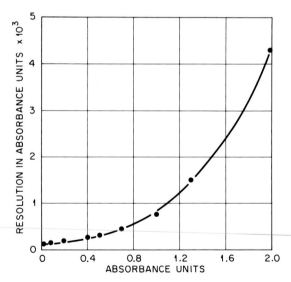

FIG. 10. The resolution of the analog-to-digital converter (A/D) as a function of absorbance units for a transmission signal. (From T. O. Tiffany, unpublished data.)

[34] E. Maclin, *Clin. Chem.* **17**, 707 (1971).

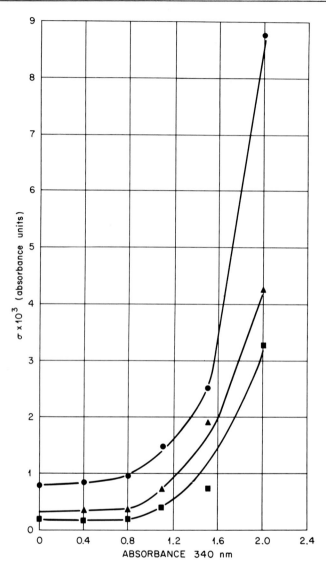

FIG. 11. Standard deviation of absorbance measurements vs. absorbance units as a function of digital averaging. ●, $N = 1$; ▲, $N = 4$; ■, $N = 16$. [From T. O. Tiffany, C. A. Burtis, J. M. Jansen, J. B. Overton, and C. D. Scott, *Clin. Chem.* **18,** 829 (1972).]

maximum obtainable precision if instrumental optical noise from the analyzer is much smaller than the A/D converter resolution. However, random variations do exist in the photometric signal of the analyzer; the standard deviation of absorbance as a function of absorbance is shown in Fig. 11. Also shown in Fig. 11 is the effect of digital averaging. Such averaging is performed readily by the computer, as described in the first part of this chapter, and should reduce the uncertainty in absorbance measurement due to instrumental noise by a factor of $\sigma/\sqrt{N}$, where σ is the uncertainty in absorbance measurement for $N = 1$. This relationship is seen to hold, and as N increases, the uncertainty of measuring absorbances decreases, as expected. At large values of N the standard deviation in absorbance approaches the calculated uncertainty in absorbance due to the resolution of the A/D converter. Confirming data are shown in Table I. The data in Fig. 11 were obtained at 340 nm with the ORNL Fast Analyzer, and the absorbance measurements were made using NADH solutions ranging in concentration from 0.032 to 0.32 mmoles/liter and in absorbance from 0.2 to 2.0 absorbance units. The linearity of absorbance, as has been discussed, is better than 1% over this range.

If the standard deviations involved in measuring two different absorbances are known, as in our case, then the standard deviation of the measurement of the difference between these two absorbance values can be calculated from a consideration of propagation of error.[28] This is shown in the following expression:

$$S_{\Delta A}^2 = S_{A1}^2 + S_{A2}^2 \tag{3}$$

where S_{A1} and S_{A2} are the standard deviations in the separately measured values, and $S_{\Delta A}$ is the standard deviation in the measurement of the difference between these values. The percent relative error or the coefficient of variation due to noise that would be found upon replicate determinations

TABLE I

THE RESOLUTION OF THE ANALOG-TO-DIGITAL CONVERTER AS A FUNCTION OF TRANSMISSION SIGNAL

Percent transmittance	Absorbance	Resolution (absorbance units)
100	0.000	
95	0.022	0.00011
90	0.046	0.00012
80	0.097	0.00013
70	0.155	0.00015
60	0.222	0.00017
50	0.301	0.00020
40	0.398	0.00024
30	0.523	0.00031
20	0.699	0.00044
10	1.000	0.00075
5	1.301	0.00151
1	2.000	0.00437

of A can be expressed as $\sqrt{S_{\Delta A}^2} \times 100/\Delta A$. In the case where a small absorbance change is measured, a useful approximation is $S_{A1} = S_{A2}$. The percent relative error due to photometric error (instrumental noise and interface resolution) can then be expressed as $S_{\Delta A} \times 100 \times \sqrt{2}/\Delta A$. For example, the uncertainty in measurement of absorbance in the range of 1.3–1.6 absorbance units from Table I is about 1.5–2.0 $\times$ 10^{-3} absorbance units, and the uncertainty in the measurement of the change in absorbance will be less than 3.0 $\times$ 10^{-3} absorbance unit in this absorbance range.

Enzyme activity is determined spectrophotometrically from the zero-order rate using an expression similar to the following:

$$\text{activity (IU/liter)} = \Delta A \, \frac{(1)}{a} \, \frac{(V_t)}{V_e} \, (1000) \tag{4}$$

where ΔA is the absorbance change between two fixed times or represents the linear regression slope of absorbance vs. time, V_t is the total volume of reaction, V_e is the volume of enzyme added, and a is the molar absorptivity of substrate or product monitored during the assay. If pipetting errors and errors due to temperature fluctuation can be ignored, for the sake of this discussion, then the uncertainty in enzyme activity will be due to the uncertainty in measuring absorbance at each of two fixed times or due to uncertainty in the determination of a slope using least-squares when N readings are used to determine the activity. For the case of a two-point fixed-time kinetic assay, if the enzyme reaction is allowed to proceed beyond an absorbance change in most cases of 0.200 absorbance unit, then the photometric error will be below 2%, and such factors as pipetting and temperature variations become the limiting error factor. If several points are used to determine the absorbance change by a least-squares method, then the photometric error contribution should be even less over the same 0.200 absorbance change.

To illustrate this discussion experimentally, a premix experiment[28] can be performed in which the enzyme and the assay solution are mixed before analysis and dispensed into each of the sectors of the Fast Analyzer rotor; the analyzer is then accelerated such that all solutions will be distributed identically in all cuvettes to be monitored. Thus the reaction proceeds uniformly in all cuvettes at a uniform temperature without pipetting errors. Data from such a premix experiment are shown in Table II for the assay of glutamic–oxaloacetate transaminase where the starting reaction absorbance was about 1.0 absorbance unit. The data illustrate that the instrumental noise or photometric error is in the range of 1.5 $\times$ 10^{-3} absorbance units, which is close to the expected error in this absorbance range.

From this discussion it should be apparent that if the conditions of

TABLE II
Percent Relative Error as a Function of Total Absorbance Change
for an Enzyme Assay with a High Starting Absorbance, A_0

Reaction time (min)	Total absorbance change	Coefficient of variation ($N = 13$)
1	0.017	19.13
2	0.036	7.13
3	0.056	4.03
4	0.073	3.21
5	0.089	1.99
6	0.103	1.63
7	0.116	1.42
8	0.129	1.33
9	0.141	1.06
10	0.153	1.06

enzyme activity assays are properly set up with regard to pipetting, temperature control, and extent of absorbance change, the precision in determining enzyme activity even for low-activity samples should approach 1% or better. This discussion can be extended to the consideration of the precision with which one can obtain initial velocities for enzyme kinetics experiments. An example of the precision obtainable for several enzyme activity assays using the Fast Analyzer is shown in Table III.

Enzyme Activity Assay Techniques. The Fast Analyzer can be used both for sampling type enzyme activity assays and for continuous monitoring of enzyme activity. The flexibility and sophistication of the analyzer is utilized more fully, however, in the latter type of enzyme activity assay. When coupled to a small computer, the Fast Analyzer can be utilized in

TABLE III
Analytical Precision of the Fast Analyzer in Performing an
Enzyme Assay

Enzyme	Average within-run variation[a]		Run-to-run variation[b]	
	ΔA/minute	Percent R.S.D.	ΔA/minute	Percent R.S.D.
Alkaline phosphatase[c]	0.0353	0.75	0.0353	1.51

[a] Sixteen replicate aliquots were assayed during a single run, and the resultant data were statistically analyzed.
[b] One hundred twelve replicate aliquots were assayed over a 7-run period, and the resultant data were statistically analyzed.
[c] Reaction conditions: wavelength = 400 nm; sample volume = 0.010 ml; total volume = 0.130 ml; total run time = 200 seconds.

conjunction with different software packages to obtain initial velocity measurements, zero-order reaction rate measurements over some defined time period Δt, and zero-order reaction rate at a preset absorbance change ΔA over a varying time period Δt. When performing an enzyme activity assay that involves a coupled assay where a lag phase might be present due to low sample enzyme activity, a program can be used that establishes when the lag phase is over and then determines the zero-order reaction rate. The time required to initiate an enzyme activity assay, or any reaction on the Fast Analyzer, is of the order of 3.2 to 13 seconds, depending on the type of mixing involved before initial readings are made. If care is taken to have the sample, reactants, rotor, and analyzer equilibrated before initiating the analysis, initial monitoring times of 5–10 seconds can be obtained from the start of the analysis. While this is relatively slow in terms of rapid reaction kinetics and stop-flow systems, it represents a very rapid entry into the monitoring of steady-state reactions, particularly when one considers that some 14 to 41 reactions can be monitored simultaneously under these conditions.

A variety of algorithms have been written for monitoring enzyme activity. These have basically been designed around a least-squares approach for determining the slope of a best-fit line through the absorbance at several different times. Statland and Louderback[30] have looked into a variety of different data-fitting routines in an attempt to find the best means of obtaining initial velocity data over a broad range of enzyme activity. A program used routinely at Oak Ridge establishes a preset absorbance change which is determined from the error consideration previously discussed; it uses this ΔA change as a criterion for the length of time the computer will monitor the reaction before calculating the enzyme activity. This procedure works well when using substrate saturated or near-saturated enzyme assay methods. It was set up to eliminate two technical problems in enzyme assays: (1) to avoid loss of accuracy resulting from high-activity samples since only a very small absorbance change is required ($\simeq 0.200$ absorbance unit) before activity is calculated, and (2) to enable the slower reactions to be monitored over a long enough period of time to assure reasonable precision of the resulting activity. Figure 12 demonstrates a reaction proceeding to substrate depletion. The cursors on the rate curves marked fixed and variable demonstrate one technical problem associated with the enzyme activity assays. As discussed in the first sections of this chapter, sufficient programming flexibility exists with FOCAL to allow the researcher to take a variety of approaches to the measurement and estimation of enzyme activity according to the needs and requirements of the particular assay.

Spectrophotometric Enzyme Activity Assays. The number of adapted

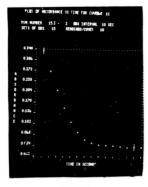 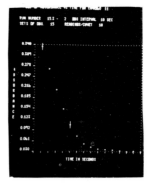

FIXED   VARIABLE

EFFECT OF INTERVAL ON RATE ANALYSIS

FIG. 12. Effect of interval on enzyme activity assays. (From C. A. Burtis, un-published data.)

or potentially adaptable enzyme activity assays obviously will include those assays that can be monitored in various ways spectrophotometrically. Brief general discussions of types of enzyme activity assays which are adapted to spectrophotometric instruments have been made by Dixon and Webb[35] and by Guilbault.[36] Most commercial Fast Analyzers have an optical capability in the range of 340–700 nm, since commercial demand at present is for enzyme activity assays either at 340 nm for dehydrogenase assays or at 405 nm for alkaline phosphatase and acid phosphatase assays. Monochromators and power supplies are available for use in the range from 200 to 400 nm,[12,16] and these have been adapted and used on the ORNL Fast Analyzer Systems.[10] A partial list of some of the enzyme assays adapted is given in Table IV.

Substrate Analysis

A general discussion of the application of the Fast Analyzer to spectrophotometric biological analysis would not be complete without some consideration of substrate analysis. The approaches to substrate analysis can be conceived of as falling under two broad categories: (1) destructive colorimetric type analysis applied either directly to the sample or preferably after some initial extraction has been performed to increase the specificity of the method, and (2) enzymatic substrate analysis. Both approaches can be adapted rather easily to the analyzer, but in certain

[35] M. Dixon and E. C. Webb, "Enzymes," 2nd ed. Academic Press, New York, 1964.
[36] G. C. Guilbault, "Enzymatic Methods of Analysis." Pergamon, Oxford, 1970.

TABLE IV
Enzyme Assays Adapted to the Fast Analyzer

I. Oxidoreductases
   Alcohol dehydrogenase
   Glycerol dehydrogenase
   Lactate dehydrogenase
   Isocitrate dehydrogenase
   Glucose-6-phosphate dehydrogenase
   Glutamate dehydrogenase
II. Hydrolases
   Cholinesterase
   Alkaline phosphatase
   Acid phosphatase
   Lysozyme
   Trypsin
   Glutamyltranspeptidase
III. Transferases
   Aminotransferases
      (a) Aspartate aminotransferase
      (b) Alanine aminotransferase
   Hexokinase
   Pyruvate kinase
   Creatinine kinase
IV. Lyases
   Aldolase

cases severe heating conditions must be used to develop color prior to analysis. The use and adaptation of general colorimetric procedures to the Fast Analyzer is obvious and analogous to the preparation required to adapt any such method to a spectrophotometer. The advantages gained are the improvement of precision of analysis through elimination of tedious repetitive manual steps; the capability of analyzing several samples in a single run; the requirement of semimicro reaction volumes, greatly decreasing sample volumes needed per assay; and the advantage of running several reactions under the same conditions. The adaptation of enzymatic substrate analysis to the Fast Analyzer, however, while similar to the adaptation of such procedures to a spectrophotometer, enables the analyst to take a certain sophistication of analytical approach and yet utilize the analyzer in a routine practical manner.

The use of enzyme assays for substrate analysis of single components in complex biological matrices has been of interest to biological analysis for some time due to the selectivity and relative sensitivity of these assays. Scopes[37] has emphasized the point that the introduction of enzymatic

[37] R. K. Scopes, *Anal. Biochem.* 49, 73 (1972).

methods for most compounds of biological interest has greatly advanced our knowledge of metabolic processes and has made the determination of metabolites a simple routine task. Many of the enzymatic analyses involve the use of nicotinamide adenine nucleotide reduction or oxidation, either as an excess substrate in the primary enzymatic reaction or in a coupled enzymatic assay system, to quantitate the substrate. Thus spectrophotometry has become a very common method of biochemical analysis. Compilation of spectrophotometric methods of enzymatic substrate analysis and discussions concerning the use of these assays have been made first and in great detail by Bergmeyer[38] and more recently by Guilbault.[36] The use of enzymatic substrate analysis in biochemical research has been extensive, with some examples of its use by Lowry et al.,[39,40] Newbold and Scopes,[41] Guillory and Fisher,[42] and Scopes.[43] One primary drawback to using enzymatic substrate anlyses has been the relatively high cost per assay, the low sample analysis rate, and in some cases the lack of sensitivity for substrates present in extremely low concentrations. The capability of the Fast Analyzer for multiple chemical analyses on small volumes of sample and reagent (e.g., 500 μl total reaction volume on the larger analyzers and 125 μl on the miniature analyzer) allows enzymatic substrate analyses to be done routinely at a reduced cost per analysis. With the addition of the fluorometric optical system, sensitivity can be increased as discussed by Scopes.[37] The adaptation of enzymatic substrate analysis to the Fast Analyzer has been discussed in some detail by one of us[10]; it was pointed out that the approach to substrate analysis could be made using either conventional end-point analysis or kinetic methods.

End-Point Enzymatic Substrate Analysis. The most common type of enzymatic substrate analysis is the end-point or equilibrium method, in which the reaction is allowed to go to completion. In this type of spectrophotometric substrate assay, the initial absorbance (A_0) and the final absorbance (A_f) of a coenzyme, substrate, or product are obtained, and the initial concentration of the unknown substrate is related directly to the absorbance change ($A_f - A_0$) either by using a standard curve or by multiplying ΔA by a substrate factor.

[38] H. U. Bergmeyer, ed., "Methods of Enzymatic Analysis." Academic Press, New York, 1970.
[39] O. H. Lowry, J. V. Passonneau, F. X. Hasselberger, and D. W. Schulz, *J. Biol. Chem.* 239, 18 (1964).
[40] O. H. Lowry and J. V. Passonneau, *J. Biol. Chem.* 239, 31 (1964).
[41] R. P. Newbold and R. K. Scopes, *Biochem. J.* 105, 127 (1967).
[42] R. J. Guillory and R. R. Fisher, *Biochem. J.* 129, 471 (1972).
[43] R. K. Scopes, *Anal. Biochem.* 49, 88 (1972).

One problem associated with conventional enzymatic end-point determinations is the requirement to analyze a separate sample without reagent to determine the initial absorbance. A significant feature of a computer-interfaced Fast Analyzer is that the initial absorbance readings after initiation of the reaction can be obtained in the first few seconds of the reaction and extrapolated back to zero time in order to obtain A_0, thus eliminating measurement of a separate transfer disk containing sample without reagent.

After values have been obtained for A_0 and A_f, the substrate concentration is determined by multiplying ΔA by a substrate factor (F) that is calculated by solving Beer's law for absorbance in terms of micromole/liter concentration units as shown in the following discussion:

$$F \text{ (micromole/liter}^{-1} A^{-1}) = \frac{(V_t)(1000)}{(V_s)(a)(b)}, \tag{5}$$

where V_t = total reaction volume, V_s = sample volume (actual volume, not diluted volume), a = molar absorptivity of the component being monitored spectrophotometrically, and b = path length in centimeters. The proper use of such a substrate factor demands that the spectrophotometric system be properly calibrated and that volumetric devices used for measuring the sample and reagent volumes also be calibrated.

Enzymatic Substrate Analysis Program in FOCAL. The enzymatic substrate programs have a unique element—the use of a clock-reset subroutine that allows a variable rate of data acquisition. This function allows several absorbance readings to be obtained in the early stages of a reaction; then, with the clock reset, considerable time can pass before the final data point is taken. This conserves core data storage by obtaining the maximum amount of useful information during the initial reaction, ignoring the intermediate data, and obtaining a final accurate absorbance reading. Figure 13 shows the reaction progress curve of a glucose determination with hexokinase, illustrating how the initial data could be used, by extrapolation to $t = 0$, to obtain an A_0. The clock reset routine can also be used to obtain a series of reaction rates at various time intervals throughout the reaction.

Kinetic Enzymatic Substrate Analysis. It is of interest to approach the analysis of substrate concentration by the use of rate measurements instead of by the equilibrium procedures that are commonly employed. Some advantages in the rate analysis approach are: a marked decrease in overall analysis time, the elimination of the sample blanks, and the cancellation of competing reaction rates due to the presence of enzyme impurities present in some enzyme substrate analysis reagents. Due to the complexity of enzymatic reactions, particularly coupled enzymatic reactions,[37] kinetic substrate analysis has not been used extensively. How-

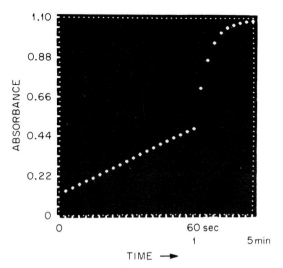

FIG. 13. Variable clock-reset function used to vary the rate of data acquisition (glucose hexokinase reaction). [From T. O. Tiffany, J. M. Jansen, C. A. Burtis, J. B. Overton, and C. D. Scott, *Clin. Chem.* **18**, 829 (1972).]

ever, enzymatic substrate rate measurement analyses have been discussed by Ingle and Crouch,[44] and to some extent by Guilbault.[36] Discussions of enzymatic kinetic substrate analysis focus on the Michaelis-Menten expression:

$$V_0 = -\frac{dA}{dt} = \frac{dP}{dt} = \frac{V_{max}S_0}{K_m + S_0} \tag{6}$$

for the case when $K_m \gg S_0$, $K_m + S_0 \simeq K_m$, and the initial velocity of the reaction is first order with respect to initial substrate concentration, S_0, or

$$V_0 = \frac{dS}{dt} = \frac{dP}{dt} = \frac{V_{max}S_0}{K_m} = K'S_0 \tag{7}$$

where $K' = V_{max}/K_m$.

It should be emphasized that Eq. (7) is valid only under conditions where K_m is approximately 100 times S_0. For enzymatic reactions where $S_0 = 0.01\ K_m$, Eq. (7) can be integrated and rearranged to give the following equation in terms of S_0:

$$S_0 = \frac{\Delta S}{e^{-K't_2} - e^{-K't_1}}. \tag{8}$$

A basic property of a first-order or pseudo-first order reaction is evident in Eq. (8). If t_1 and t_2 are accurately defined and all other variables such

[44] J. D. Ingle and S. R. Crouch, *Anal. Chem.* **43**, 697 (1971).

as reaction pH, temperature, etc., are held constant, then the initial substrate concentration, S_0, is linearly related to the change in substrate concentration during that time period. Since the absorbance change, ΔA, is proportional to the change in substrate concentration, it can be used to calculate S_0. A Fast Analyzer is ideally suited for this approach, since t_1 and t_2 are accurately defined for all determinations in one rotor due to the unique parallel analyses of these samples under the same constant conditions. All that is required of the analyst to obtain substrate concentration is to include a calibration standard in the analytical run with the remaining unknown samples. The absorbance change of the known substrate is related by the computer to the concentration of the substrate, and a factor is obtained which, when multiplied by the absorbance change for each of the remaining cuvettes, results in the calculation of the initial substrate concentrations of the unknown samples.

Integration of the Michaelis-Menten single substrate–enzyme expression yields a mixed first-order–zero-order expression, and the validity of the linear rate method and fixed-time method based on Eq. (8) is limited by the substrate: K_m ratio. These limits for either the rate method or the fixed-time method have not been well defined. Guilbault[36] has suggested that the region in which reaction rate and substrate concentration are linearly related, and in which an analytical determination of substrate concentration is achieved, is for initial substrate concentrations below 0.2 K_m. However, it appears that for initial rate methods the limitation is closer to the 0.01 K_m value, while for fixed-time methods the substrate: K_m ratio can be apparently somewhat higher, in the range of 0.01–0.15 if some initial delay time is allowed before measuring t_1.[10]

Single Enzyme Substrate Analysis; Fixed Time. An example of the use of Eq. (8) for the analysis of substrate for a single enzyme assay is the analysis of uric acid at 292 nm.[45,46] The end-point analysis of uric acid with porcine liver uricase requires 20 minutes of analysis time; with bacterial uricase the analysis time is 10 minutes.[10] However, use of fixed-time enzymatic substrate analysis allows one to permit the reaction to proceed long enough to ensure adequate resolution as well as a sufficiently large absorbance change to minimize the instrumental uncertainty in determining ΔA and yet shorten the analysis time to 2 or 3 minutes.

The introduction of bacterial uricase[47] as an analytical reagent has made it possible to analyze for uric acid at 292 nm by using linear fixed-time analysis. The K_m for this enzyme is 1.0×10^{-4} mole/liter as compared

[45] K. M. Kalchar, *J. Biol. Chem.* **167**, 429 (1947).
[46] E. Praetorius and H. Poulsen, *Scand. J. Clin. Lab. Invest.* **5**, 273 (1953).
[47] J. L. Mahler, *Anal. Biochem.* **38**, 65 (1970).

with 1.7×10^{-5} mole/liter for porcine liver uricase[48]; this increases 6-fold the concentration of initial substrate that can be used and still maintain linear kinetics. Figure 14 shows the absorbance change vs. uric acid concentration at various fixed-time intervals with t_1 equal to 4 seconds. The reaction concentration of uric acid at the highest concentration employed was 1.4×10^{-5} mole/liter, which is a molar substrate concentration:K_m ratio of 0.14.

Coupled Assay Substrate Analysis. The use of kinetic enzyme substrate

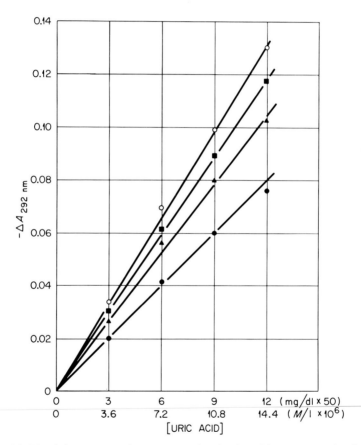

FIG. 14. Fixed-time enzymatic rate analysis of uric acid. t_1, 4 seconds; $\triangle t =$ ●, 90 seconds; ▲, 150 seconds; ■, 210 seconds; ○, 330 seconds. [From T. O. Tiffany, J. M. Jansen, C. A. Burtis, J. B. Overton, and C. D. Scott, *Clin. Chem.* **18**, 829 (1972).]

[48] H. R. Mahler, *in* "The Enzymes" (P. D. Boyer, H. Lardy, and K. Myrbäck, eds.), Vol. 8. Academic Press, New York, 1963.

analysis becomes more complex in the case of coupled enzymatic substrate assays. They can be treated by assuming that all substrates are in excess except the initial unknown substrate, which is held at an initial concentration $S_0 \ll K_m$ as previously discussed. Under these conditions the reactions can be treated as general-series first-order reactions in which a set of m substrates with concentrations $S_1, S_2, \ldots S_m$ react by a first-order or pseudo-first-order process to form each of the other m substrates. A treatise of general-series first-order equations is presented by Frost and Pearson[49] and will not be discussed in detail here.

The use of enzymatic rate analysis and fixed-time analysis for the determination of substrate concentration using coupled-assay systems has been examined using the Fast Analyzer.[10] It was shown that for the analysis of urea by the coupled urease/glutamate dehydrogenase assay an expression similar to Eq. (8) could be derived relating the change in formation of product in the second reaction (in this case NAD) to the initial substrate concentration of the first reaction (urea); therefore a fixed-time rate analysis approach can be used for the coupled assay. Furthermore, a rate expression relating the rate of formation of product to the initial substrate concentration was derived which was different from the rate expression for single substrate reaction assays in that the rate overall time (t) of the reaction, not just the initial rate of the reaction, could be used for the determination of initial substrate concentration. This means the reaction rate of series first-order reactions can be monitored for a considerable time period, similar to zero-order enzyme activity assays, resulting in greater precision of analysis. The use of the simple, irreversible, series first-order approach appears to be valid as long as (1) all reactions in the series are first-order or pseudo-first-order, (2) the coupled enzymes are in sufficient concentration to rapidly remove the forming products, and (3) the products formed in any of the reactions that are themselves not involved directly in the following consecutive reactions do not activate or inhibit any of the reactions of the sequence. A discussion of these latter two points is presented by Scopes.[37]

Figures 15 and 16 demonstrate the use of the two approaches (series enzymatic rate analysis and fixed-time analysis) for relating either the rate of formation of NAD, or the absorbance change in conversion of NADH to NAD at a given fixed-time interval, to the initial substrate concentration of urea. The precision of analysis of urea using fixed-time procedures is of the order of 1% and falls to 2 or 3% when using the rate approach. The end-point analysis of urea requires approximately 15 minutes to perform, but the fixed-time assay gives similar accurate and precise results in an assay time of 90 seconds.[10]

[49] A. A. Frost and R. C. Pearson, "Kinetics and Mechanism." Wiley, New York, 1963.

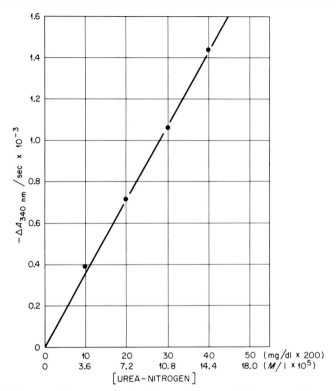

FIG. 15. Enzymatic rate analysis of urea using coupled enzymatic reactions.

End-Point vs. Kinetic Analyses. In summary, both enzymatic end-point and kinetic substrate analyses can be adapted readily to the Fast Analyzer. There are many examples of the use of enzymatic end-point analysis of substrate in biochemical research, and there is a growing interest in adapting kinetic methods to the Fast Analyzer as these approaches become better defined through their use. One important factor now present is the ability to perform rapidly off-line and immediately after analysis sophisticated computation and data-fitting routines with the Fast Analyzer computer. Now one does not demand necessarily that the absorbance change ΔA be linear with respect to initial substrate concentration if he knows what the nonlinear function is or if he chooses a sufficient range of substrate standards in the same run with his unknown. An additional factor is the introduction of the fluorometric detection system which enables the reduction of substrate concentration to meet the requirement that $S_0 \ll K_m$ and also allows the use of much less of costly substrates and cofactors in the analysis, further reducing the cost of analysis. The advantage of using kinetic assays over the end-point method is mainly practical, in that they can mean the

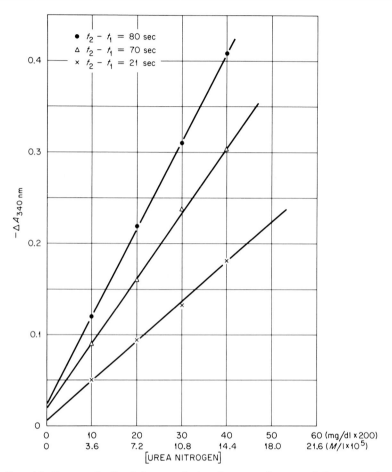

FIG. 16. Enzymatic fixed-time analysis of urea using coupled enzymatic reactions. $t_2 - t_1 = \bullet$, 80 seconds; $\triangle$, 70 seconds; $\times$, 21 seconds. [From T. O. Tiffany, J. M. Jansen, C. A. Burtis, J. B. Overton, and C. D. Scott, *Clin. Chem.* **18**, 829 (1972).]

reduction of analysis time from 15–20 minutes to 1–2 minutes. There are few reasons why one should expect better precision and accuracy from an end-point procedure than from a kinetic fixed-time analysis under controlled conditions—proper standardization and instrumental noise considerations to ensure a sufficient fixed-time absorbance change, as discussed in the section concerning enzyme activity assays.

Multiple Wavelength Scan of Absorbance

To conclude the discussion of general colorimetric and spectrophotometric biochemical analyses and some of their applications to the Fast

Analyzer, we wish to briefly discuss the use of the analyzer as a multi-cuvette recording spectrophotometer.

An attractive element of the analyzer is that prior to obtaining cuvette transmission readings during any given revolution of the rotor, a dark current reading and the reference cuvette reading are taken, then the remaining 2 through 15 (or 42) transmission signal readings are taken; the absorbances of these cuvettes are obtained based on the dark current and reference signal, as described in the first section of this chapter. This provides a constantly updated absorbance with respect to reference intensity and dark current. This enables sequencing the monochromator setting to the rotor pulse and provides a readily available means of obtaining the absorbance of n cuvettes as a function of wavelength for several different wavelength settings. A total of 48 wavelength settings can be made for a 15-cuvette analyzer, therefore one can obtain approximately 2-nm increments over a 100-nm span, and of course better resolution at much narrower spans. In addition, at each wavelength setting several readings of each cuvette are possible for number-averaging purposes to minimize instrumental noise contributions. One additional feature of the Fast Analyzer has to do with the relationship of light source intensity and detector sensitivity as a function of wavelength: Since the intensity of cuvette 1 is used as a reference, it is possible to maintain this intensity constant over a given wavelength increment by automatically adjusting the photomultiplier voltage, thus providing the means of producing a computer-controlled automatically scanning multicuvette recording spectrophotometer.

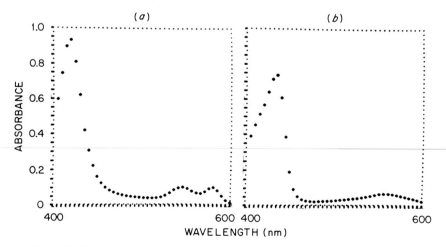

Fig. 17. Simultaneous multiple cuvette spectrophotometric scans of human (a) oxyhemoglobin (in 0.08 M phosphate buffer, pH 7.0) and (b) deoxyhemoglobin (as in "a" with the addition of sodium dithionite) obtained using the Fast Analyzer. (From T. O. Tiffany, unpublished data.)

Figure 17 is an example of the use of the analyzer as a scanning spectrophotometer; two simultaneous scans of oxyhemoglobin and methemoglobin are shown which were scanned from 395 to 600 nm in 5-nm increments. The data were plotted on the CRT scope for convenience of observation. The instrument has been valuable in studying the effects of pH on hemoglobin and bilirubin separately and in the same reaction mixture. An additional study one can perform is to change the nature of gaseous atmosphere in contact with the cuvette solution and monitor gas effects on a variety of reactions; in the case of hemoglobin one can measure these effects as they change the spectral response. The Fast Analyzer applied as a scanning multicuvette spectrophotometer could be extremely useful for obtaining difference spectra, particularly in conjunction with the computer, where extreme care could be exercised in eliminating artifactual contributions due to experimental or procedural errors. Another use of such a system would be to perform complicated multiwavelength multicomponent analyses, where not only could the appropriate data be obtained, but the simultaneous solutions of the multiple equations involved could then be performed off-line through use of the analyzer computer.

Fluorescence Assays in Biochemical Analysis

Fluorescence assays in biochemical analysis have been surveyed in depth by Udenfriend.[50] The subject of fluorescence enzyme activity assays has been reviewed by Roth,[51] and the subject of fluorometric substrate analysis (enzymatic substrate analysis) has been discussed by Scopes.[37] It is obvious from the magnitude of these reviews that the role and importance of fluorescence in biochemical analysis is increasing due to the increased sensitivity of fluorescence assays and the development of a large number of sensitive and specific assay procedures.

The discussion of the application of the Fast Analyzer to fluorescence assays will be limited to the potentially significant features of the analyzer for fluorescence assays and some applications under consideration.

Fluorescence Enzyme Activity Assays

The major problem of using fluorescence for enzyme activity assays is the difficulty of obtaining the results in international units which are accurate and can be reproduced from laboratory to laboratory. The use of molar absorptivity in spectrophotometry, when available for the assay, has provided a way of reporting enzyme activity in international units that in

[50] S. Udenfriend, "Fluorescence Assay in Biology and Medicine." Academic Press, New York, 1969.
[51] M. Roth, "Methods of Biochemical Analysis," Vol. 17, p. 189. Wiley (Interscience), New York, 1969.

theory, at least, should be reproducible from laboratory to laboratory through the use of properly calibrated spectrophotometers. However, in practice this apparently is seldom achieved. The same approach is difficult to achieve with fluorometers due to the varied intensity response from instrument to instrument, therefore activity must be calculated from calibration curves using the monitored cofactor, substrate, or product under the conditions of the assay when these substances are available in pure enough form. An alternative procedure is to obtain the ratio of the intensity of the given substrate, etc., in relative intensity per micromole of substrate, etc., under the conditions of the assay and relate that to the intensity of a fluorescence standard. The ratio of the two factors can then be used along with the continued use of the fluorescence standard during the time of assay to calculate enzyme activity in international units.

The multicuvette fluorometric Fast Analyzer enables the analyst to rapidly obtain the ratio of the relative intensity per micromole of the monitored component of the reaction to the relative intensity per micromole of a fluorescence standard (e.g., quinine sulfate in $0.1 \, N \, H_2SO_4$) by running standard curves of each under controlled conditions in a single run. The fluorescence standard is then included in each analytical run, and the enzyme activity is quickly calculated in international units from the following expression:

$$\text{enzyme activity} = \Delta I / \min \, (R)^{-1} \, (I_{\text{std}})^{-1} (V_t / V_s) \qquad (9)$$

where R is the ratio in I substrate per micromole to I standard per micromole, I_{std} is the relative intensity of the fluorescence standard, V_t is total reaction volume, and V_s is the sample volume. We have used such procedures to establish fluorescence ratios for 4-methylumbelliferone in basic and acidic buffers to quinine sulfate and NADH in various buffers to quinine sulfate in order to perform assays on alkaline phosphatase and acid phosphatase using 4-methylumbelliferyl phosphate as a substrate,[52] and for several enzyme assays involving either dehydrogenase as the primary enzyme of interest or in coupled-assay systems. Fluorescence assays can be checked in several cases against similar spectrophotometric assay conditions with the exception of higher enzyme concentration, and the results can be shown to correlate very nicely.

Fluorescence Enzymatic Substrate Analyses

It has been previously pointed out that one limitation of using kinetic enzyme substrate analyses is the requirement that $S_0 \ll K_m$ for the pseudo-first-order substrate relationship to hold. This is a problem with a number

[52] H. N. Fernely and P. G. Walker, *Biochem. J.* **97**, 95 (1969).

of spectrophotometric enzymatic substrate analyses where K_m is of the order of 10^{-4} to 10^{-5} mole/liter; reducing the range of the substrate concentration in the assay to these levels as the upper limit of the initial substrate concentration in the reaction mixture results in a loss of sensitivity and precision for the assay. The use of fluorescence substrate assays minimizes this problem owing to increased sensitivity of the assay[37] and thus enhances the attractiveness of using kinetic fixed-time or rate enzymatic substrate analyses.

Specific Applications to Enzymology

Dixon and Webb[35] have described the areas of investigation that must be carried out if one is to regard any given enzyme study as thorough. These areas of investigation must include: studies of protein properties, structure, enzyme properties, the active center, thermodynamics, kinetics, and biological properties. While it is understood that such investigations must involve a variety of experimental approaches and different types of instrumentation, a multicuvette spectrophotometer/spectrofluorometer such as the Fast Analyzer, which provides both data acquisition and data reduction, could be used to eliminate much of the error-prone tedium and to add sophistication of analytical approach previously available only with off-line large computer systems. To digress, it might be emphasized that with the use of larger computer systems much analysis time is spent preparing the data and getting it fed into the computer via Teletype terminal, magnetic tape, or key-punched cards; this can become a limiting factor in the rate of analysis and in the time it takes to obtain kinetic enzyme parameters.

Determination of Multiparameters

After an isolated enzyme has been purified to a given stage, a variety of determinations are required to characterize it. These are generally long and tedious experiments in which the enzyme is probably labile and denatures with time. Therefore experiments involving the determination of enzyme activity as a function of pH, of ionic strength of a given buffer, of different metal ions at different concentrations, of various substrate analogs, of chelators, of substrate concentration, and of various inhibitors and activators can be each relegated to one of a series of single analytical runs where the variations of the particular parameter of interest are made in that particular determination. The pH optima of an enzyme, as an example, can be determined on a 15-cuvette Fast Analyzer in a single run at 14 different pH levels or 7 levels of pH in duplicate, and the results can be plotted automatically on the analyzer's CRT scope and photographed from this display for permanent record and publication. Performing experiments in parallel in such a manner not only greatly reduces time of analysis, but it can also reduce manual manipulation and data-handling errors and, equally important, make it possible to minimize artifactual effects on the experiment due to the time-dependent denaturation of isolated and partially purified enzymes. Having isolated the enzyme and determined several of its properties such as active site, free sulfhydryl groups, disulfide bonds, the analyzer can be used to look at the kinetics of the enzyme in more detail.

The Fast Analyzer and Enzymatic Catalysis

The investigator's control of an experiment while it is in progress is an important element in obtaining more information efficiently from a given set of experiments. A desirable goal in the design and conception of experimental instrumentation is to be able to detect and explore subtle changes and differences in the experiment upon evaluation of the data minutes after the analysis is completed and while reaction conditions such as a labile enzyme and substrates are essentially unchanged. The use of the Fast Analyzer in the study of enzymatic catalysis allows the enzymologist more time and opportunity to interact with the experiment in progress. While other instrumentation has been designed for total automation of enzymatic catalyzed reactions,[53,54] the Fast Analyzer offers the advantages of (1) the determination of initial velocity data on several reactions proceeding essentially in parallel, (2) the use of more than one substrate and/or inhibitor during a single enzyme run, and (3) a choice of statistical data-fitting routines to perform the data reduction. How these are practically applied will be the subject of the remainder of this chapter.

Automated Determination of Enzyme Kinetic Parameters

The Fast Analyzer has been programmed for the study of enzymatic catalysis. The program consists of the following subroutines: (1) gathering of initial experimental information and setting up of the reaction time, number of data points to be taken, and delay time interval, (2) actual acquisition of reaction velocity data in terms of initial velocities and printout of these data with the respective substrate concentration, (3) plotting of reciprocal plots (s/v_0 vs. $[s]$, or $1/v_0$ vs. $1/[s]$) for rapid visual inspection of the data, and (4) the statistical evaluation of the enzyme data as suggested by Wilkinson,[55] Johansen and Lumry,[56] and Cleland.[57,58]

[53] A. A. Eggert, G. P. Hicks, and J. E. Davis, *Anal. Chem.* 43, 736 (1971).
[54] G. P. Hicks, A. A. Eggert, and E. C. Toren, *Anal. Chem.* 42, 729 (1970).
[55] G. N. Wilkinson, *Biochem. J.* 80, 324 (1961).
[56] G. Johansen and R. Lumry, *C. R. Trav. Lab. Carlsberg* 32, 185 (1961).
[57] W. W. Cleland, *Advan. Enzymol.* 29, 1 (1967).
[58] W. W. Cleland, *in* "The Enzymes" (P. D. Boyer, ed.), 3rd ed., Vol. II, p. 1. Academic Press, New York, 1970.

Initialization of the Experiment. Substrates are chosen, dilutions are made, and proper dilution of the enzyme is chosen. A practical discussion of setting up enzymatic catalysis experiments to attain more meaningful data is given by Cleland.[57] The program is called into the computer, and the initial information such as buffer, pH, velocity, enzyme, and a factor for converting rate of absorbance change to change in molar concentration per minute is entered into the initial portion of the program. The enzymologist then must make a decision as to whether he wants to run two or more sets of the same substrate dilutions on different enzymes, run duplicate or triplicate substrate dilutions with the same enzyme, run two or more sets of different substrates on the same enzyme, or run one substrate and one or more sets of the same substrate dilutions plus different concentrations of inhibitors in a single run. For example, with a 15-place analyzer two sets of different substrate dilutions (7 per set) could be run for a two-substrate enzyme, or with the 42-place analyzer one set of 7 enzyme dilutions and at

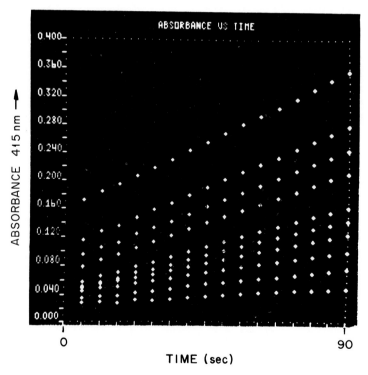

Fig. 18. Simultaneous alkaline phosphatase (placental) reactions at various dilutions of *p*-nitrophenylphosphate over a 90-second time period. (From T. O. Tiffany, unpublished data.)

least five different sets of dilutions with varying inhibitor concentrations could be determined in a single run. After entering in the experiment protocol and substrate concentrations as a function of cuvette number, information to set up a delay time, time of reaction, and number of data points to be taken, the experiment is ready to begin. After placing the loaded transfer disk containing enzyme and substrate dilutions into the analyzer, the next phase of the experiment is begun.

Acquisition of Initial Velocity Data. Upon acceleration of the rotor, all reactions proceed in parallel, with the computer monitoring all reactions as they progress in time. Various data-fitting routines can be used to obtain initial velocity data; we have chosen to use a second-degree polynomial fit of several data points over the first 30–60 seconds of the enzyme reaction. The velocities are obtained and plotted on the scope for inspection by the analyst. An example is shown in Fig. 18 for placental alkaline phosphatase. The substrate concentration and initial velocity data are printed out simultaneously (Table V) as the plots are made on the scope. When this is finished, the data-reduction phase of the experiment begins.

Reciprocal Plots. Reciprocal plots like that shown in Fig. 19 are then made for quick check of the data. Statistical evaluation of the data is then begun.

Statistical Evaluation. The small computer is able to perform nonlinear regression iterative data-fitting routines to simple and complex enzyme systems. This offers the direct advantage of allowing the data to be evaluated immediately after the experimental data are obtained in a sophisticated form, and if different directions in the experiment are called for, these can

TABLE V

EVALUATION OF PLACENTAL ALKALINE PHOSPHATASE VELOCITY AGAINST
INCREASING CONCENTRATION OF p-NITROPHENYL PHOSPHATE

Cuvette No.	Substrate concentration (mmoles)	Millimole ml^{-1} min^{-1}
2	0.110	0.0123
3	0.220	0.0244
4	0.330	0.0311
5	0.550	0.0404
6	0.770	0.0550
7	1.100	0.0617
8	2.200	0.0773
9	4.667	0.0838
10	5.500	0.0938
11	11.000	0.1041

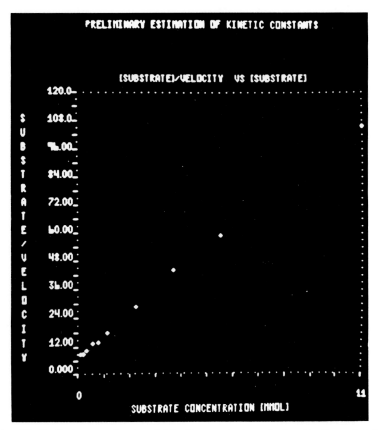

Fig. 19. Preliminary estimate of kinetic constants and visualization of data for placental alkaline phosphatase using a [s/v] vs. [s] plot. (From T. O. Tiffany, unpublished data.)

be quickly made while everything is at hand. Table VI shows the data from a placental alkaline phosphatase experiment in its evaluated form. This is a simple fitting routine patterned after that of Wilkinson.[55] More complex data routines are possible on the small computer. We have developed a nonlinear regression iterative best-fit routine involving matrix solution for the evaluation of parameters for two enzymes acting upon the same substrate (see Cleland[57]).

The total analysis time from the initiation of the reaction to final printout of data, picture taking, and visual data inspection is 10–15 minutes. It is obvious that several runs can be made in a single afternoon and the results neatly printed out with an estimate of acceptability of the enzyme parameters.

TABLE VI

EVALUATION OF K_m AND V_{max} FOR PLACENTAL ALKALINE PHOSPHATASE[a]

A. Calculation of [substrate]/velocity

Substrate concentration	[Substrate]/velocity
0.110	8.944
0.220	9.028
0.330	10.619
0.550	13.609
0.770	13.989
1.100	17.830
2.200	28.466
3.667	43.787
5.550	58.652
11.000	105.685

B. Statistical evaluation of the data
1. Provisional K_m = 0.832 mmole
   Provisional V_{max} = 0.109 mmole ml^{-1} min^{-1}
2. Final K_m = 0.839 mmole
   Final V_{max} = 0.108 mmole ml^{-1} min^{-1}
   Standard error for K_m = 0.057 mmole
   Standard error for V_{max} = 0.002 mmole^{-1} min^{-1}
   Iteration number = 3.

[a] Reaction in 0.5 M 2-amino-2-methyl-1-propanol, pH 10.2, 30.0°.

Summary

We have described the Fast Analyzer system and have indicated some of its uses for biochemical analysis. It is a centrifugal, multicuvette, computerized spectrophotometer/fluorometer. Its uniqueness is in the synergistic interaction of centrifugal field, multicuvette optical system, and computer to provide an automated fast microanalyzer useful in many areas of analysis including biochemical research.

Acknowledgments

The development of a sophisticated device such as a GeMSAEC Fast Analyzer requires the cooperation, advice, and assistance of many individuals representing a broad spectrum of technical specialties. At ORNL we are fortunate to have available to us such a staff of highly competent and trained personnel. We gratefully acknowledge the assistance of C. D. Scott, J. E. Attrill, E. L. Candler, N. Cho, E. L. Collins, L. H. Elrod, A. S. Garrett, W. W. Harris, D. W. Hatcher, J. M. Jansen, W. F. Johnson, M. T. Kelley, T. A. Lincoln, J. C. Mailen, D. N. Mashburn, W. A. Walker, and D. D. Willis. We greatly appreciate the technical advice and constructive criticism provided by Dr. G. F. Johnson of Johns Hopkins and Dr. R. S. Melville of the National Institute of General Medical Sciences, Bethesda, Maryland.

Author Index

Numbers in parentheses are reference numbers and indicate that an author's work is referred to, although his name is not cited in the text.

Steinberg, I. Z., 777(29)
Steiner, D. F., 377(7, 9, 10, 11), 378(9, 15, 16)
Stenlid, G., 531(16)
Stephenson, M., 653(3)
Stern, H., 246(1), 250(1), 252(1), 272(23)
Stern, M. D., 442(35)
Stevens, B. J., 288(17)
Stevens, J. G., 445(5)
Stevens, R. H., 790(7), 800(7), 801(7)
Still, C. C., 545(13)
Stirpe, F., 254(10), 270(14)
Stocking, C. R., 546(20, 21)
Stoeckenius, W., 20(45), 23(45), 293(11), 667(5, 6), 668, 670(12), 672(12), 679(7)
Stokes, D. M., 539, 592(27)
Stoner, C. D., 184(8), 387(12), 591(17), 592(17), 597(17), 598(17), 771(14), 782(14)
Storey, B. T., 501(2), 536(43)
Stossel, T. P., 341(5), 343(5), 698(17, 18)
Straus, J. H., 67(13), 172(1), 173(1), 174(1), 175(1, 6), 178(1), 180(1)
Straus, W., 330(2, 3)
Strayer, D. R., 722(21)
Strehler, B. L., 106(12), 108(12), 114(13), 425(3)
Striebich, M. J., 253(8), 714(5), 726(5)
Studier, F. W., 723
Stumpf, P. K., 547(24), 565(10), 655(12)
Stutz, E., 721(17), 723(17)
SubbaRow, Y., 92(9), 246(20)
Suit, S. C., 633(4)
Sulkowski, E., 629(18)
Summers, D. F., 162(15)
Suss, R., 227(9), 228(9)
Sutherland, E. W., 103(1, 2), 111(1), 145(8), 146(8)
Sutton, D. D., 612
Suzuki, K., 438(31), 444(41, 42)
Swanson, M. A., 20(42)
Swift, H., 369(6), 370(6)
Szabo, A., 508(21), 510(21)
Szabo, A. S., 566(24)
Szabo, E., 460(18), 473(18), 475(18)

Szego, C. M., 461(23)
Szybalski, W., 723(22)

T

Tadano, H., 368(4), 369(4), 370(4)
Täljedal, I.-B., 378(13)
Takabatake, Y., 406(7)
Takahashi, H., 699(23)
Takayama, K., 184, 387
Takeby, I., 583(14)
Takenaka, Y., 60
Takeuchi, M., 148(15)
Tamaoki, B., 237(34, 35)
Tamburrini, O., 149(18)
Tan, L. Y., 102(20)
Tannhauser, S. J., 410
Tappel, A. L., 330(6), 522(17), 523(17), 789(43)
Tarjan, E. M., 600(43)
Tartakoff, A. M., 23
Tashiro, Y., 211(22), 217(3)
Tata, J. R., 8(19), 254(16), 255(19, 20, 22), 256(23), 258(19, 20), 261, 262(16)
Taube, Ö., 761(7, 8), 768(7), 769(7, 8)
Tauro, P., 508(19), 510(19)
Tautvydas, K. J., 508(17), 533, 542, 562(6)
Tavaststjerna, M. G., 444(43)
Tavlitzki, J., 627(12)
Taylor, D., 446(16, 19)
Taylor, I. F., 612(11)
Taylor, R. E., 686(3)
Tenenbaum, I. L., 555(5)
Terayama, H., 148(15)
Terebus-Kekish, O., 148(16)
Terry, R. D., 436(5), 438(5), 444(5)
Tesar, J. T., 416(13), 418(13)
Tester, C. F., 585, 586(10)
Tetas, M., 25(53), 201(49)
Thatcher, F. S., 633(1)
Theimer, R. R., 565(3), 567(3)
Thimann, K. V., 536(40)
Thinès-Semploux, D., 202(5), 324
Thomas, E. L., 460(17)
Thompson, J. E., 689(10)
Thompson, J. S., 661(31)
Thomson, A. R., 718(13)
Thomson, R. H., 528(4)

Subject Index

in sarcoplasmic reticulum, 241
  assay, 245
in synaptosome plasma membrane,
  450, 452
Auto Densiflow apparatus, 652
Axenic growth medium, for amoeba,
  composition of, 687
Azotobacter agilis, cell-wall isolation of,
  655
Azotobacter vinelandii
  membrane isolation from, 701
  transport systems in, 704
Azurophile granules, isolation of, 345–
  353

B

Bacillus sp., cell-wall isolation from, 661
Bacillus cereus, cell-wall isolation of, 655,
  656
Bacillus circulans, cell wall hydrolyzing
  enzyme from, 610
Bacillus licheniformis, transport systems
  in, 705
Bacillus megaterium
  cell-wall isolation of, 654, 663
  transport systems in, 704
Bacillus subtilis, transport systems in,
  704, 709
Bacteria
  countercurrent distribution of, 761, 768
  gram-negative, cytoplasmic membrane
    isolation from, 642–653
  as plant fractionation contaminant,
    549, 553
  in sucrose gradients, 720
  transport in membrane vesicles of,
    698–709
Bacterioruberin, in *Halobacterium* red
  membrane, 677
Bacteroids, isolation of, 518
Ball mills, for cell rupture, 660
Bananas, cell organelle studies on, 532,
  541
Barley
  aleurone grain from, 576
  protoplast isolation from, 580
Batten's disease, *see* Neuronal ceroid
  lipofuscinosis
Beads, for column sieving fractionation,
  549–551

Bean
  aleurone grain from, 576
  chloroplast isolation from, 602
  enzyme studies on, 522, 523
  organelles of, 491, 493, 533, 542
Beaufay's rotor, isopycnic equilibration
  in, 349–351, 357, 367
Beckman-Spinco centrifuge, liver homogenization in, 196–197
Beer's law, 818
Bentonite, in plant-cell fractionation,
  504, 505
Benzopyrene, as mixed-function oxidase
  substrate, 233
BHK21/C_{13} cells, plasma membrane isolation from, 161
Bicarbonate reagent, for plasma mem-
  brane isolation, 77
Bicine buffer, 20
Bilirubin, as mixed-function oxidase sub-
  strate, 233
Bile salts, in membrane isolation, 789
Biuret method, for plant protein deter-
  mination, 542–544
Block staining, for electron microscopy
  of liver cells, 22–23
Blood, for erythrocyte ghost isolation,
  171
Blood banks, as platelet source, 151
Blood cells, countercurrent distribution
  of, 761
Blue-green algae, gas-vesicle isolation
  from, 678–686
Borate, as antioxidant, 539–540
Bovine serum albumin (BSA), 720
  as liver sample preservative, 19
  in mitochondria isolation procedure,
    306, 310, 531
  in plant-cell extraction, 524, 525, 536–
    537, 592, 599, 603
Brain
  lysosome isolation from, 457–477
  nuclei isolation from cells of, 452–457
Branton-de Silva extraction apparatus,
  541, 570
Braun shaker, 658
  properties of, 662
Brendler homogenizing apparatus, 43
Brij 35, for microsome isolation, 216, 224
Browning, of plant tissue, 532

Protein kinase, for cyclic AMP assay, 106

Protein kinase inhibitor, for cyclic AMP assay, 105

"Proteinase K," for ribonuclease inactivation, 720

Proteinoplasts, *see* Proteoplasts

Proteolytic enzymes, plasma membrane and, 90

Proteoplasts, in plant cells, 495

Proteus mirabilis, transport systems in, 704

Protoplasm, in plant cells, protection of, 529

Protoplasts. (*See also* Spheroplasts)
　isolation of, 578–583
　nuclei extraction from, 563–564

Protuberances, of plant cells, as fractionation contaminant, 545

Pseudomonas B-16, transport systems in, 704

Pulegone reductase, extraction of, 537–538, 539

Purple membrane, of *Halobacterium,* isolation and properties of, 670–672

PVP, as plant-cell protectant, 533, 536, 537, 592, 599, 603

Pyocyanine perchlorate, as electron donor, 701

Pyrogallol, in polyphenol oxidase assay, 534

Pyrophosphatases, in plasma membranes, 88, 90

Pyrosulfite, *see* Metabisulfite

Pyruvate kinase, assay by Fast Analyzer, 816

Q

Quick-freeze procedure, for liver, 5

Quick-thaw procedure, for liver, 5, 19

Quinones
　in plant-tissue extraction, 528–544
　protein reactions, 529–530, 537

R

Rabbit, liver, microsome isolation from, 199

Radish, organelles of, 491

Rat liver
　homogenates prepared from, 7–8

plasma membrane from, 75–102
preservation of, 3–6

Rate-dependent banding, in centrifugation, definition of, 714–715

Rate-zonal centrifugation, of plant cells, 545, 549

Raytheon sonic oscillators, properties of, 662

RBA virus, cells transformed by plasma membrane isolation from, 167, 168

RC-2 centrifuge, 96

Red cells, *see* Erythrocytes

Red membrane, of *Halobacterium,* isolation and properties of, 670–672

Refrigerator, liquid nitrogen type, 4

"Release action," by platelets, 149

Renograffin, as gradient material, 509

Resin acids, as plant extraction interferents, 531

Resins, in plant-tissue fractionation, 535–536

Reticulocytes, differential centrifugation of, 721

Retinyl ester hydrolase, in microvillous membrane, 130

Rhodospirillum rubrum, enzyme extraction from, 772

Ribi Press
　for bacterial cell rupture, 659
　for plant-cell rupture, 502
　properties of, 663

Ribonucleases
　in plasma membranes, 88
　protective agents for, 504, 505
　in sucrose, 720

Ribonucleoprotein, centrifugal separation of, 723–726

Ribosomes
　centrifugal separation of, 717, 727
　of plant cells, 489, 491, 495, 583
　　core particles and, 516
　　isolation, 501, 504, 509, 510, 514, 533, 534, 585–586

Ribulose diphosphate carboxylase, 768

Rice, aleurone grain from, 576

Rice polish, protein bodies in, 518

RNA
　centrifugal separation of, 723–726
　in chromatin, 278, 279
　density gradient studies on, 510

A 4
B 5
C 6
D 7
E 8
F 9
G 0
H 1
I 2
J 3